4

5(a)

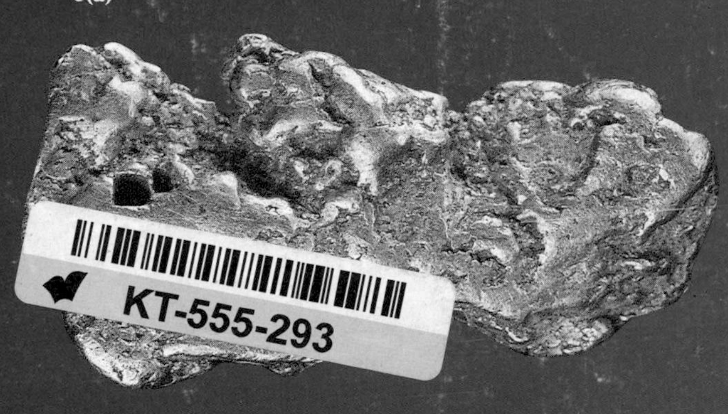

5(b)

5(c)

5(d)

5(e)

Colour 1 Atomic spectra of (a) hydrogen, (b) sodium

Colour 2 Separation of the pigments in an extract from annual meadow grass (*Poa annua*) by thin-layer chromatography, using a mixture of propanone and petroleum ether as the solvent: the top band (yellow) is carotene, and the second and third green bands are chlorophylls

Colour 3 Poppies and cornflowers: the different colours of the flowers are due to the different acidities of their saps, the same pigment being present in both

Colour 4 These gems are all corundum, Al_2O_3, their different colours being due to different impurity atoms

Colour 5 Minerals from which elements are extracted: (a) native gold, (b) zinc blende (yellow), galena (grey), pyrites ('fools' gold) and quartz (white), in association, (c) fluorite, Iceland spar and pyrites, (d) haematite, (e) galena with calcite on siderite

~~Jessica Pemlyn-Jones~~

~~VII HB~~

Katie Woolf VII Pd.

CHEMISTRY

Principles and Applications

CHEMISTRY

Principles and Applications

P. W. Atkins M. J. Clugston
M. J. Frazer R. A. Y. Jones

London and New York

Longman Group UK Limited,
Longman House, Burnt Mill, Harlow,
Essex CM20 2JE, England
and Associated Companies throughout the world.

Published in the United States by Longman Inc,
The Longman Building, 95 Church Street, White Plains, New York 10601

First published 1988
ISBN 0 582 35590 7

Set in 10/11pt Times New Roman

Produced by Longman Group (FE) Ltd
Printed in Hong Kong

British Library Cataloguing in Publication Data

Chemistry: principles and applications.
1. Chemistry
I. Atkins, P. W.
540 QD33

ISBN 0-582-35590-7

Library of Congress Cataloging-in-Publication Data

Chemistry: principles and applications.
Includes index.
1. Chemistry. I. Atkins, P. W. (Peter William), 1940-
QD31.2.C4314 1988 540 87-3919

ISBN 0-582-35590-7 (pbk.)

Contents

PART THREE ORGANIC CHEMISTRY 391

Preface

A message to the student

We wrote this book to help you learn what many regard as the central science. When you have mastered chemistry, you stand at the meeting point of physics on the one hand and biology on the other. You will know how and why matter behaves as it does, and how one substance can be changed into another. Chemists have become almost magicians with matter, for they can transform one substance into another (but not, some regret, lead into gold!). But it is not magic that they practise, it is science. Chemists have discovered what changes are possible, and are learning to build ever more complex substances. Even as we were writing this book, one of the most exciting developments for years was announced: scientists discovered how to make a class of material that conducts electricity without resistance at temperatures not far below room temperature, instead of the extremely low temperatures that were needed for this superconductivity before. The scientists who made the materials were chemists, and the materials will change your life. You may be one of the people who find ways of using these new and extraordinary substances. You may even be lucky enough to make a discovery of similar importance.

Even if you decide that your career will not be in chemistry itself, your understanding of the world around you will remain incomplete unless you know something of the subject. Its explanations depend on the concepts of physics, and its applications extend over the whole spectrum of human activities, from the production of new materials destined for use in engineering and electronics, to the unravelling of the biochemical processes that underlie that most extraordinary collection of chemical reactions, which we term 'life'.

You should not think that all this knowledge and power – power to create new substances, that is, as well as power to understand the behaviour of existing ones – will come without a great deal of effort. Even we authors think that chemistry is a subtle and difficult subject. We thought it was difficult when we were at the same stage as you are now. Although we found it challenging, we had a glimmering of insight, and could see that there was a great opportunity open to us if we stuck with it, an opportunity to see into the secrets of the world and to comprehend the behaviour of substances, be they the substances that we tread on as we walk across a landscape or a carpet, or the substances that industry manufactures, such as plastics and drugs. There is no such thing as a dull subject: there are only subjects that you don't know enough about!

We wrote this book to help you master the enthralling subject we call chemistry. You have a lot of hard work ahead of you, but the rewards at the end are extraordinary. Even if you don't go on with the subject, once you have mastered it you will have a grasp of why things behave as they do. Your understanding of the everyday will be enhanced. You will have a knowledge that in some cases even a decade or so ago did not exist.

Good luck! The future is yours.

A message to the teacher

We have sought to write a book that students will enjoy reading, and which will give them all they need to know about modern chemistry for A-level examinations or their equivalents in other countries. We have tried everywhere to avoid old-fashioned explanations and outdated theories, and have sought to produce a book that describes chemistry as it is in the late twentieth century.

The starting point for the text was *Principles of Physical Chemistry*, which two of us (Atkins and Clugston) wrote a few years ago and which aimed to present what was needed for A-level courses in that branch of chemistry. The writing team has been expanded, and all four of us have worked at revising the physical chemistry section in the light of comments we obtained from the many people who used that book. Even more important, however, has been our expansion of the text to include the other two branches of chemistry: inorganic and organic. We have all worked on all three sections, and have tried to produce a text which is not only authoritative and well informed, but uniform in style and level and in its deployment of explanations.

We believe that the most important device for making a textbook attractive and helpful for the reader is an interesting style coupled with a clear discussion. We have aimed to attract and hold the reader's interest in this way. To enhance the clarity of the text, we have paid special attention to the illustrations. Some of these are in colour, for chemistry is a colourful subject and many are first attracted to it by the range of colours that can be produced. The others include clear line diagrams, half-tones of modern equipment and applications, and marginal structures.

Each chapter has a variety of devices to help the reader thread his or her way through what everyone has to admit is an intricate subject. We set the stage in each chapter with a few introductory sentences, which explain the general thrust of what is to follow. The text of each chapter is interspersed with Examples, so that a particular point can be explained in more detail, a concept illustrated, or an application described. We also use Boxes to explain the background to the development of a concept, or to one of its modern applications in industry, medicine or some other aspect of everyday life. Each chapter concludes with what we intend to be a handy revision aid: a succinct Summary of the points that have been made. A real grasp of the subject comes only by doing numerous problems, and each chapter has a collection. Many of these are original; the others are taken from a wide variety of examining boards. At the end of the book there is another revision aid. Here we have added a few pages that give a synopsis of the central points of the three parts of the book: they include the most important definitions and formulae from physical chemistry, a summary of inorganic chemistry, and some reaction schemes from organic chemistry.

We believe that we have produced an authoritative and enjoyable text. We have tried very hard to keep our feet on the ground and to produce one that teachers and students alike will be able to use with confidence and pleasure. We span a wide range of interests and expertise, but were very conscious of the need to make sure that our ideas were workable in the classroom and matched the requirements of examinations, both as they are at present and as they are evolving. We could not do this alone, but were very fortunate to be able to rely on the wisdom, advice, and help of a board of advisors who were not only deeply committed to the success of the project but were highly experienced in both teaching and examining. They included Dr. T. P. Borrows, Science Adviser for the London Borough of Waltham Forest and formerly Head of Science, Pimlico School, Dr. C. L. Mason of Strode's College, Egham, Dr. D. Nicholls of the University of Liverpool, and N. Rowbotham of Loughborough Grammar School. All four of us would like to say how greatly we valued their contribution, which started at the inception of the project and continued throughout. You, the reader, will not be able to distinguish their input from ours, but we wish to emphasize that theirs was considerable and has made us confident that the book will be not only innovative but usable.

P.W.A.
M.J.C.
M.J.F.
R.A.Y.J.

Acknowledgements

We are grateful to the following examination boards for permission to reproduce questions from past papers:

The Associated Examining Board; the Cambridge Tutorial Representatives; Joint Matriculation Board; Oxford and Cambridge Schools Examination Board; Oxford Colleges Admissions Office; Southern Universities Joint Board; University of Cambridge Local Examinations Syndicate; University of London School Examinations Board; University of Oxford Delegacy of Local Examinations; Welsh Joint Education Committee.

We are also grateful to the following for permission to reproduce photographs; Aerofilms, 7.25; Professor B. J. Alder, Lawrence Radiation Laboratory, University of California, 6.1; Professor J. F. Allen, Department of Physics, University of St. Andrews, 24.6; Allyn and Bacon Inc, *Chemistry*, Ronald J. Gillespie, David A. Humphreys, N. Colin Baird, Edward A. Robinson, 4.13, 21.6, colour 6 and 7; Heather Angel, colour 3; Aviation Photographs International/J. Flack, 25.6; Bancroft Library, University of California, 2.2; Francis T. Bacon, 13.18; Barnaby's Picture Library, 34.6 (a) (Ken Lambert) and (b) (Bill Meadows); Professor N. Bartlett, University of California, 24.2; British Alcan Aluminium, 8.3; British Coal, 33.5; British Museum (Natural History) Geological Museum, 18.10, 20.8, colour 4 and colour 5 (a), (c) and (e); British Petroleum, 27.3; British Steel, 25.11 (Appleby Frodingham) and colour 9; Camera Press, 3.13 and 16.3 (Norman Sklarewitz); J. Allan Cash, 17.8 and 33.8; Chemical Design, pages 9 and 233; Professor A. V. Crewe, Enrico Fermi Institute, University of Chicago, 2.1; Vivien Fifield Picture Library, 18.2 and 34.2; General Motors Research Laboratories, Warren, Michigan, 5.14 and 8.4; Geoscience Features, 2.18, 5.13 (b), 22.3 (a), colour 5 (b) and (d); Glaxo Holdings, 35.4; Hewlett-Packard, 17.9; John Hillelson Agency/Sygma/Regis Bossu, 12.8; Her Majesty's Stationery Office, 20.4 (b); J. Holloway, Department of Chemistry, University of Leicester, 23.5; IBM Zurich Research Laboratory, 1.1 (b); ICI Chemicals and Polymers Group, 14.18, 21.10 and 21.11; Ind Coope Burton Brewery, 29.6; Inmos/Jerry Cottignies, 20.5; International Tin Research Institute, 6.9 and 7.3; JEOL (UK) 3.12; Suzi Kennett, 19.8; Kodak, 25.7; Laboratory of the Government Chemist, 10.11; Andrew Lambert, 1, 1.6 (a), 10.6, 11.13, 12.1, 12.3, 13.13 and 33.4; Massachusetts Audubon Society, 21.7; Nuffield-Chelsea Curriculum Trust, *Nuffield Advanced Science Revised Book of Data*, colour 1; Open University, 17.7; Pilkington Glass Museum, 23.7; Pilkington Group Services, 6.3; Popperfoto, 19.4; Ann Ronan Picture Library, 28.7; Royal Institution/Sir George Porter, 14.3; Royal Observatory, Edinburgh, 4.10; Science Museum, 1.6 (b); Science Photo Library, 1.20, 5.13 (a) (Dr. Jeremy Burgess), 5.15, 9.7 (Don Fawcett), 13.1, 16.4, 21.4 (Burgess), page 391 (Laboratory of Molecular Biology, University of Cambridge), 30.3 (Burgess), 32.4 (Dr. Arthur Lesk), 34.3 (Martin Dohrn), colour 2 (Sinclair Stammers) and colour 8 (NASA); Scott Polar Research Institute, Cambridge, 5.10; Shirley Institute, Manchester, 32.5; Frank Spooner Pictures/Gamma, 11.8; St. Bartholomew's Hospital, 18.8; Telegraph Colour Library/Space Frontiers, 3.1; Texaco, 27.5; Thermit Welding, 19.6; J. M. Thomas/Royal Institution, 1.1 (a); United Distillers Group, 29.1; United States Department of the Interior, Bureau of Reclamation, 7.24; C. James Webb, 13.9, 31.6 and 32.3; Professor M. H. F. Wilkins, Biophysics Department, King's College, London, 2.3 (h); reproduced by gracious permission of Her Majesty the Queen, Royal Archives, Windsor Castle, 20.4 (a).

Figures 2.3 (a)–(g), 26.3, 26.4, 27.1 (b) and 31.2 by Peter Atkins; 26.1, 26.5 by Richard Jones; 10.7, 22.3 (b), 25.9, 26.2, 27.1 (a), 27.4, 31.1, 31.3 (courtesy European Fragrances) by Longman Photographic Unit.

The cover photograph of a macrophotograph of sapphire or corundum is by John Walsh/Science Photo Library.

The mineral corundum or aluminium oxide occurs as well developed hexagonal crystals. It is a mineral valued both as an abrasive and as a gemstone. Transparent corundum when tinted blue is known as sapphire.

INTRODUCTION

The mole

Chemistry, like other physical sciences, depends on measurements. Chemists express their measurements in the same way that physicists do, using metres for length, kilograms for mass, and seconds for time. Chemists also deal with large numbers of atoms, molecules and ions, and find it very convenient to introduce a special unit. This typically chemical unit is called the *mole* (symbol: mol), a name derived from the Latin for 'heavy mass'. The mole is used throughout chemistry, and this introductory chapter explains what it means and how it is used.

The **mole** is the chemist's version of the shopkeeper's dozen. A shopkeeper often finds it convenient to sell objects (more formally: **entities**) by the dozen, using the exact relation:

1 doz corresponds to 12 entities

A chemist deals with myriads of atoms and molecules, with typical samples having more atoms in them than there are stars in the Galaxy. The chemist's shorthand unit is the mole:

1 mol corresponds to 6.022×10^{23} entities

That is, 1 mol copper atoms corresponds to 6.022×10^{23} Cu atoms, and 1 mol water molecules corresponds to 6.022×10^{23} H_2O molecules.

The number 6.022×10^{23} may seem an odd choice, but it is based on a careful definition:

1 mol of entities is the amount of substance that contains the same number of entities as there are atoms in exactly 12 g of ^{12}C

(^{12}C is a specific isotope of carbon – see Section 1.1.) That is, because the number of atoms of carbon in a 12 g sample can be counted experimentally (in principle at least, although in practice indirectly) a mole is a unit just like a metre and a kilogram. The best experimental determination is currently that 1 mol corresponds to 6.022×10^{23} entities. We will always state specifically what those entities are: it is useless saying '1 mol of hydrogen' because that could mean 6.022×10^{23} atoms or molecules of hydrogen, or even, more fancifully, 6.022×10^{23} truckloads of cylinders of hydrogen. **Always specify the entities precisely** and state '1 mol H' if atoms are meant, or '1 mol H_2' if molecules are meant.

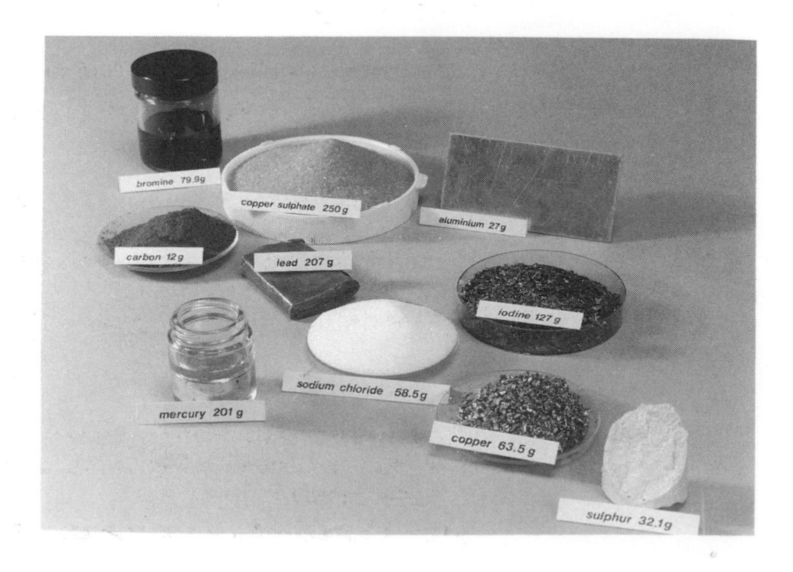

Figure 1 A mole of some common substances: each of these piles contains (approximately) 6.022×10^{23} entities

Amount of substance

Just as the kilogram is the unit for the measurement of a physical quantity (mass), the mole is the unit for the measurement of another physical quantity, the **amount of substance** (symbol: n). Thus, the amount of substance in a sample of copper might be reported as 2.0 mol of copper atoms (written $n(\mathrm{Cu}) = 2.0\,\mathrm{mol}$), just as the mass of something might be 2.0 kg (written $m = 2.0\,\mathrm{kg}$). Throughout this book 'amount' will always mean amount of substance, and will be reported in moles.

Avogadro's constant

A very useful fundamental constant in chemistry is **Avogadro's constant** (symbol: L), which is defined as follows:

$$L = 6.022 \times 10^{23}\,\mathrm{mol}^{-1}$$

This constant is used to convert an amount expressed in moles to an actual number, N, of entities and to express a number as an amount:

$$N = n \times L \qquad (1a)$$

$$n = N/L \qquad (1b)$$

EXAMPLE

Calculate the amount of hydrogen chloride molecules in a sample that contains 1.2×10^{24} HCl molecules.

METHOD Use equation 1b with $L = 6.022 \times 10^{23}\,\mathrm{mol}^{-1}$.

ANSWER The amount of hydrogen chloride corresponding to 1.2×10^{24} HCl molecules is

$$n(\mathrm{HCl}) = \frac{1.2 \times 10^{24}}{6.022 \times 10^{23}\,\mathrm{mol}^{-1}} = 2.0\,\mathrm{mol}$$

COMMENT Notice how the use of moles greatly simplifies the numbers involved, just as a shopkeeper often finds that book-keeping is simplified by expressing items in dozens.

Moreover, just as multiplying the cost of a single egg by the constant $12\,\mathrm{doz}^{-1}$ converts it into the cost per dozen eggs, so multiplying the mass of a single entity (typically an atom or molecule) by L converts it into the mass per mole of entities (the mass per mole of atoms or molecules). The mass per mole of entities is called the **molar mass**. Similarly, the **molar volume** is the volume occupied per mole of entities (per mole of H_2 molecules, for instance).

EXAMPLE

Calculate the molar mass of argon atoms, given that the mass of a single argon atom is $6.634 \times 10^{-26}\,\mathrm{kg}$.

METHOD Multiply the mass of a single atom by Avogadro's constant, which will give the mass per mole of argon atoms. The symbol for molar mass is M; that of a single atom is m, so that $M = mL$.

ANSWER Since $m = 6.634 \times 10^{-26}\,\mathrm{kg}$,

$$M = 6.634 \times 10^{-26}\,\mathrm{kg} \times 6.022 \times 10^{23}\,\mathrm{mol}^{-1}$$
$$= 3.995 \times 10^{-2}\,\mathrm{kg\,mol}^{-1} = 39.95\,\mathrm{g\,mol}^{-1}$$

COMMENT Each mole of argon atoms (that is, each 6.022×10^{23} Ar atoms) has a mass of 40 g. It is conventional to express molar masses in grams per mole ($\mathrm{g\,mol}^{-1}$).

Molar mass

The molar mass (symbol: M) of a compound has the units $g\,mol^{-1}$ and the same numerical value as the sum of the relative atomic masses of the elements present in it. The latter are all listed on the Periodic Table on page 544, so that it is easy to calculate the molar mass of any compound. For example, the molar mass of methanol, CH_3OH, is

$$\begin{aligned} M(CH_3OH) &= M(C) + 3M(H) + M(O) + M(H) \\ &= M(C) + 4M(H) + M(O) \\ &= (12.01 + 4 \times 1.008 + 16.00)\,g\,mol^{-1} \\ &= 32.04\,g\,mol^{-1} \end{aligned}$$

That is, 1 mol of CH_3OH molecules has a mass of 32.04 g.

Although it is not possible to speak of 'molecules of sodium chloride' (or of other ionic compounds) the **formula unit**, NaCl, is a definite entity. Its molar mass is also the sum of the molar masses of its elements:

$$\begin{aligned} M(NaCl) &= M(Na) + M(Cl) \\ &= (22.99 + 35.45)\,g\,mol^{-1} \\ &= 58.44\,g\,mol^{-1} \end{aligned}$$

That is, 1 mol of NaCl formula units has a mass of 58.44 g.

The concept of molar mass makes it very easy to determine the amount of substance from the mass of a sample. Since the molar mass of methanol is $32.04\,g\,mol^{-1}$, the amount of substance in a 50 g sample is

$$n(CH_3OH) = \frac{50\,g}{32.04\,g\,mol^{-1}} = 1.6\,mol$$

In general, if the molar mass of a substance is M, then the amount of substance in a sample of mass m is

$$n = \frac{m}{M} \qquad (2)$$

EXAMPLE

Calculate the amount of substance in 73 g of hydrogen chloride.

METHOD Calculate the molar mass of hydrogen chloride, HCl, from the information on page 544, then substitute into the equation above.

ANSWER $M(HCl) = M(H) + M(Cl)$
$= (1.008 + 35.45)\,g\,mol^{-1} = 36.46\,g\,mol^{-1}$

Since the mass is $m = 73$ g,

$$n = \frac{73\,g}{36.46\,g\,mol^{-1}} = 2.0\,mol$$

COMMENT The molar masses of atoms are discussed more fully in Chapter 1.

Moles and reactions

A chemist finds it very helpful to use moles when interpreting a chemical equation such as

$$H_2 + Cl_2 \rightarrow 2HCl$$

The **molecular interpretation** of this equation is

1 hydrogen molecule reacts with 1 chlorine molecule to produce 2 hydrogen chloride molecules.

It is therefore also true that in a shopkeeper's terms

> 1 doz hydrogen molecules reacts with 1 doz chlorine molecules to produce 2 doz hydrogen chloride molecules,

and that in a chemist's terms

> 1 mol hydrogen molecules reacts with 1 mol chlorine molecules to produce 2 mol hydrogen chloride molecules.

EXAMPLE

What amount of substance of NH_3 is produced by 1 mol H_2 in the synthesis of ammonia?

METHOD Write the chemical equation, and decide from it the relation between amount of NH_3 and amount of H_2; then express that relation in terms of 1 mol H_2.

ANSWER The equation for the synthesis of ammonia is

$$N_2(g) + 3H_2(g) \rightarrow 2NH_3(g)$$

Therefore, since 3 mol H_2 produces 2 mol NH_3, it follows that 1 mol H_2 produces $\frac{2}{3}$ mol NH_3.

COMMENT Once again, dealing in moles greatly simplifies (and makes clearer) the relations between the amounts of substances involved in chemical reactions.

Chemical reactions and reacting masses

Using the concepts of moles and molar masses, a chemical equation can be used to calculate the mass of one reactant that reacts with a specified mass of another. This is an extremely important bridge between the relatively abstract ideas of atoms and molecules and the down-to-earth measurements of masses in grams. In the following paragraphs we explain how to calculate either (a) the mass of a product that can be obtained from a specified mass of a reactant, or (b) the mass of a reactant needed to make a specified mass of a product. In each case:

1. Convert the given information into an amount of substance.
2. Identify the relations between amounts from the chemical equation.
3. Use the relations to convert the amount of substance of the known reactant (or product) into the amount of substance of the product (or reactant) of interest.
4. Convert the amount of substance of the product (or reactant) of interest into a mass (in Chapter 4 it is shown how to make the conversion into other quantities, such as the volume or pressure of a gas).

Suppose that we require the mass of hydrogen chloride that can be prepared from 100 g of chlorine gas:

$$H_2(g) + Cl_2(g) \rightarrow 2HCl(g)$$

1. Convert from 100 g of chlorine to the amount of substance of Cl_2 molecules. Since the amount of substance of Cl_2 is related to the molar mass of chlorine by

$$n = \frac{\text{mass of chlorine}}{\text{molar mass of } Cl_2}$$

and the latter is $70.9\,\text{g mol}^{-1}$, the conversion is

$$n = \frac{100\,\text{g}}{70.9\,\text{g mol}^{-1}} = 1.41\,\text{mol}$$

2 The chemical equation shows that 1 mol Cl_2 produces 2 mol HCl.
3 It follows that 1.41 mol Cl_2 produces 2 × 1.41 mol HCl, or 2.82 mol HCl.
4 Since the molar mass of HCl is 36.46 g mol^{-1}, it follows that the mass of 2.82 mol HCl is

$$2.82\,\text{mol} \times 36.46\,\text{g mol}^{-1} = 103\,\text{g}$$

That is, when 100 g of chlorine are consumed, 103 g of hydrogen chloride are produced.

EXAMPLE

Calculate the mass of sodium chloride formed when 10 g of sodium hydroxide reacts with excess hydrochloric acid.

METHOD The mass of product is governed by the mass of sodium hydroxide. Therefore, work through the procedure set out above for the reaction

$$NaOH(aq) + HCl(aq) \rightarrow NaCl(aq) + H_2O(l)$$

Calculate the molar masses of the formula units NaOH and NaCl from the information on page 544.

$$\begin{aligned} M(\text{NaOH}) &= M(\text{Na}) + M(\text{O}) + M(\text{H}) \\ &= (22.99 + 16.00 + 1.008)\,\text{g mol}^{-1} \\ &= 40.00\,\text{g mol}^{-1} \\ M(\text{NaCl}) &= M(\text{Na}) + M(\text{Cl}) \\ &= (22.99 + 35.45)\,\text{g mol}^{-1} \\ &= 58.44\,\text{g mol}^{-1} \end{aligned}$$

ANSWER
1 The amount of substance of NaOH in 10 g is

$$n(\text{NaOH}) = \frac{10\,\text{g}}{40.00\,\text{g mol}^{-1}} = 0.25\,\text{mol}$$

2 The chemical equation shows that 1 mol NaOH produces 1 mol NaCl.
3 It follows that 0.25 mol NaOH produces 0.25 mol NaCl.
4 The mass corresponding to 0.25 mol NaCl is

$$0.25\,\text{mol} \times 58.44\,\text{g mol}^{-1} = 15\,\text{g}$$

COMMENT The method of dealing with volumes of gaseous products (as, for example, in the reaction $2Na(s) + 2H_2O(l) \rightarrow 2NaOH(aq) + H_2(g)$) depends on taking the step of converting amounts of substance to volumes: this conversion is described in Section 4.1.

Solutions

Chemists often transfer substances from one container to another by preparing solutions. The **solvent** is typically water; the **solute** is the substance dissolved. If the **concentration** of the solution – the amount of substance of the solute per unit volume of the solution – is known, then a known amount of solute can be transferred by measuring out the appropriate volume of solution (for instance, from a burette or a pipette).

The concentration is normally reported in mol dm^{-3}. If 1.00 mol NaOH (40.0 g of sodium hydroxide) is added to a flask and water added until the *total* volume is 1.00 dm^3, the concentration would be 1.00 mol dm^{-3}. Then, if 0.100 dm^3 of that solution were measured out from a burette, we would know that the amount of substance of NaOH transferred is

$$n(\text{NaOH}) = 0.100\,\text{dm}^3 \times 1.00\,\text{mol dm}^{-3} = 0.100\,\text{mol}$$

Although concentrations are commonly reported in moles per dm^3, volumes of solutions are commonly reported in cm^3. These two units of volume can easily be interconverted: $1\,dm^3 = 1000\,cm^3$. For example, $1600\,cm^3 = 1.6\,dm^3$. Hence, cars with $1600\,cm^3$ (1600 cc) engines are often labelled '1.6'.

EXAMPLE

Calculate the volume of $1.00\,mol\,dm^{-3}$ aqueous sodium hydroxide that is neutralized by $200\,cm^3$ of $2.00\,mol\,dm^{-3}$ aqueous hydrochloric acid, and the mass of sodium chloride produced.

METHOD Follow the same technique as for calculations involving masses. The amount of substance is calculated by multiplying the volume in dm^3 (which is found by dividing the volume in cm^3 by 1000) by the concentration in $mol\,dm^{-3}$, i.e. $n = V \times c$.

ANSWER

1 The amount of substance of HCl is

$$n(\text{HCl}) = 0.200\,\text{dm}^3 \times 2.00\,\text{mol dm}^{-3} = 0.400\,\text{mol}$$

2 The chemical equation is

$$\text{NaOH(aq)} + \text{HCl(aq)} \rightarrow \text{NaCl(aq)} + \text{H}_2\text{O(l)}$$

and shows that 1 mol HCl neutralizes 1 mol NaOH to produce 1 mol NaCl.

3 It follows that 0.400 mol HCl neutralizes 0.400 mol NaOH to produce 0.400 mol NaCl.

4 The volume of $1.00\,mol\,dm^{-3}$ NaOH is found from the equation above, in the form $V = n/c$:

$$V(\text{NaOH}) = \frac{0.400\,\text{mol}}{1.00\,\text{mol dm}^{-3}} = 0.400\,\text{dm}^3,\ \text{or } 400\,\text{cm}^3$$

The mass that corresponds to 0.400 mol NaCl is

$$m(\text{NaCl}) = 0.400\,\text{mol} \times (22.99 + 35.45)\,\text{g mol}^{-1} = 23.4\,\text{g}$$

COMMENT Acids and alkalis are commonly measured out by volume rather than by mass, as solutions are easier to transfer than gases or solids.

EXAMPLE

Calculate the mass of iron(II) hydroxide that can be precipitated by $10\,cm^3$ of $2.0\,mol\,dm^{-3}$ aqueous sodium hydroxide.

METHOD The amount of substance is calculated by multiplying the volume in dm^3 (which is found by dividing the volume in cm^3 by 1000) by the concentration in $mol\,dm^{-3}$, i.e. $n = V \times c$.

ANSWER

1 The amount of substance of NaOH is

$$n(\text{NaOH}) = 0.010\,\text{dm}^3 \times 2.0\,\text{mol}\,\text{dm}^{-3} = 0.020\,\text{mol}$$

2 The chemical equation is

$$\text{FeCl}_2(\text{aq}) + 2\text{NaOH}(\text{aq}) \rightarrow \text{Fe(OH)}_2(\text{s}) + 2\text{NaCl}(\text{aq})$$

and shows that 2 mol NaOH precipitates 1 mol $Fe(OH)_2$.

3 It follows that 0.020 mol NaOH precipitates 0.010 mol $Fe(OH)_2$.

4 The mass that corresponds to 0.010 mol $Fe(OH)_2$ is

$$m(\text{Fe(OH)}_2) = 0.010\,\text{mol} \times (55.85 + 2 \times 16.00 + 2 \times 1.008)\,\text{g}\,\text{mol}^{-1} = 0.90\,\text{g}$$

COMMENT The green colour characteristic of iron(II) hydroxide provides a useful qualitative test for distinguishing iron(II) salts from iron(III) salts, as described in Section 25.6.

The figure on the opposite page shows a computer graphic of a zeolite with benzene in the central channel.

PART ONE

PHYSICAL

CHEMISTRY

1

Atomic structure

In this chapter we meet the evidence for the modern view of an atom as a central massive nucleus surrounded by a cloud of electrons. We see how the electrons are arranged, and meet the description of atomic structure in terms of orbitals. The basis of the periodic classification of the elements is explained in terms of these ideas. We see how atomic spectroscopy is used to identify atoms and explore their structures, and how mass spectrometry is used to determine their masses. Finally we examine the nucleus more closely, and see how it gives rise to the three kinds of radioactivity.

The dark spots in the photograph in Figure 1.1(a) are images of atoms. The idea that matter is built up from these fundamental building blocks is now almost second nature to us, but it is only recently that such direct evidence for their existence has become available. In the late eighteenth and early nineteenth centuries the Manchester teacher John Dalton assembled indirect evidence in favour of the existence of atoms (see Box 1.1). Now, in the late twentieth century, we can actually 'see' them.

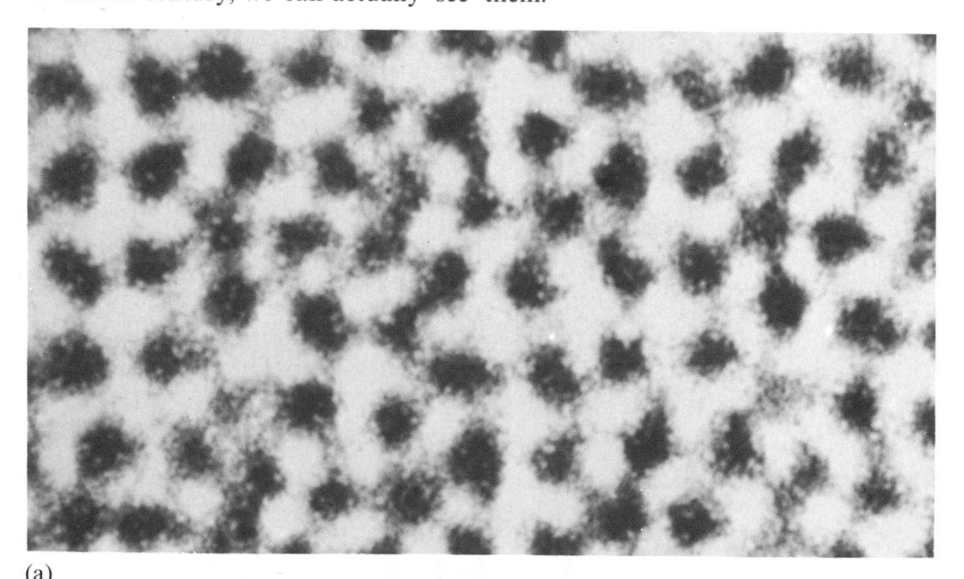

(a)

(b)

Figure 1.1 Modern techniques provide direct evidence for the existence of atoms: (a) electron microscope image of atoms, (b) arrangement of individual atoms on the surface of silicon, as 'seen' by the scanning tunnelling microscope

Dalton thought that atoms were indivisible (which is what *atomos* means in Greek). This view could not be held once electrons were discovered: if electrons could be knocked out of atoms, atoms must have an internal structure. Modern chemistry began when it was understood how atoms are built up, because understanding atoms is the key to understanding chemistry.

BOX 1.1

The origins of the atomic hypothesis

The concept of 'atom' was originated by the Greeks, especially Democritus and Leucippus, and Lucretius's poem 'On the nature of the universe', which was written in about 56 B.C., uses images that bear a striking resemblance to our modern view. Nevertheless, while it is true that the Greeks put forward the idea that matter is composed of atoms, it was only a speculation. The scientific, as distinct from the philosophical, evidence for the existence of atoms was accumulated about 200 years ago. There were three principal contributions. The **law of conservation of mass** (expressed by Lavoisier in 1774 before he fell to the guillotine in 1794) states that matter is neither created nor destroyed in a chemical reaction. The **law of constant composition** (expressed by Proust in 1799) states that all pure samples of the same compound contain the same elements combined in the same proportion by mass. The **law of multiple proportions** (Dalton, 1803) states that when the elements A and B combine together to form more than one compound, the various masses of A that combine with a fixed mass of B are in a simple ratio. (Dalton, who worked in Manchester, came to chemistry through his interest in meteorology.)

The first of these laws suggests that chemical reactions involve the reorganization of things, not their destruction or creation. The second suggests that, since mass characterizes the nature of a compound, the things that are reorganized have characteristic masses. The third suggests that the things are indivisible, so that a compound consists of 1, 2, 3, ... and not some intermediate, fractional number of them. Dalton accounted for all these observations by proposing the existence of atoms, the indivisible fundamental units of elements.

1.1 The nuclear atom

The first sub-atomic particle to be discovered was the **electron** (see Box 1.2). It is negatively charged, and its charge and mass have been measured by observing its deflection in electric and magnetic fields.

When a hydrogen atom loses its single electron a **proton** is left. This is the second fundamental sub-atomic particle of central importance to chemistry. The proton is positively charged (exactly balancing the electron's charge, so that the atom itself is electrically neutral). It is 1836 times heavier than the electron, so that protons, together with the neutrons we shall meet shortly, account for most of the mass of an atom.

BOX 1.2

The discovery of the electron

The discovery of the electron depended on the technological advance of being able to produce a high vacuum. When that had been achieved in the middle of the nineteenth century it was not long before **cathode rays** were discovered by passing an electric discharge through a low-pressure gas. By 1895 it was known that cathode rays were negatively charged, but their nature was unknown. The Germans argued that they were waves of some kind; the French and the British thought they were streams of particles. In 1897 J. J. Thomson scored a resounding victory for the latter view by showing that the rays were deflected by electric and magnetic fields, and that the deflection led to a value of the mass-to-charge ratio. Since it had been shown (by Lenard) that the rays were absorbed by all materials, Thomson put forward the view that these negatively charged particles, or **electrons**, were universal constituents of matter. The charge of the electron was later measured (using Millikan's experiment and the photoelectric effect), and hence the mass itself could be determined. The victorious British–French view survived until 1927, when it was shown that electrons could be diffracted, and hence were waves after all! The crucial experiments were performed by Davisson and Germer and by G. P. Thomson, the son of J. J. Indeed, J. J. Thomson won the Nobel prize for proving that the electron is a particle and his son, G. P. Thomson, received it for proving that the electron is a wave. The resolution of the paradox was achieved with quantum mechanics, which shows how to regard objects, including electrons, as both particles and waves: this is the **wave–particle duality** of matter.

The Geiger–Marsden experiment

In the early years of this century there was controversy about the arrangement of protons and electrons in atoms. The model that originally attracted attention was the 'plum pudding', in which the electrons were pictured as lumps in a jelly-like background of positive charge. This was discredited by the results of a classic experiment carried out in 1909 by Geiger and Marsden working under the guidance of Ernest Rutherford (see Figure 1.2(a)).

The experiment involved shooting projectiles at atoms and measuring their deflection. As projectiles Geiger and Marsden used **α-particles** (alpha-particles, the nuclei of helium atoms – see Section 1.4). As a target they used a very thin gold foil about 1000 atoms thick. The deflection of the particles by the atoms in the metal foil was detected by observing the **scintillations**, bright flashes of light, emitted where they struck a fluorescent screen. Most of the α-particles were deflected through angles of less than 1°. Some – about 1 in 20 000 – were deflected through 90° or more (see Figure 1.2(b)). These big deflections eliminated the plum pudding from science. Rutherford expressed his surprise at the result with the remark that 'it was almost as incredible as if you fired a 15-inch shell at a piece of tissue paper and it came back and hit you'.

(a)

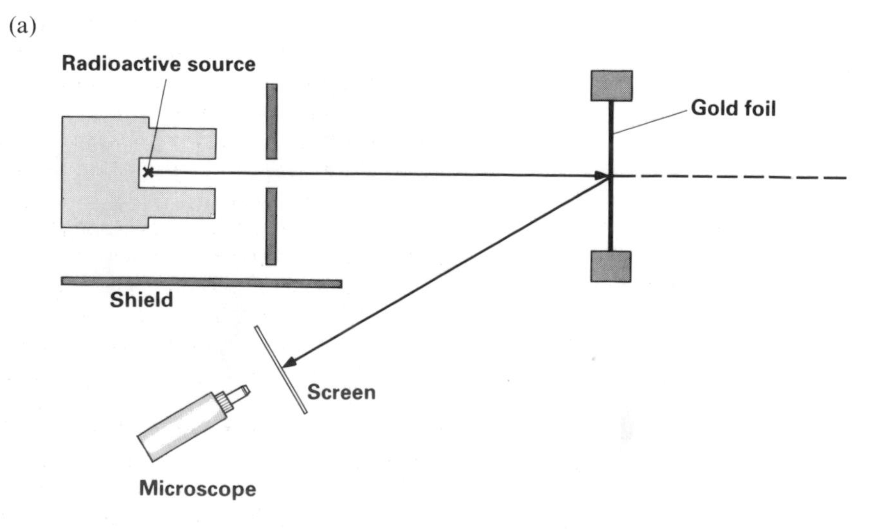

Figure 1.2(a) The apparatus used in the Geiger–Marsden experiment

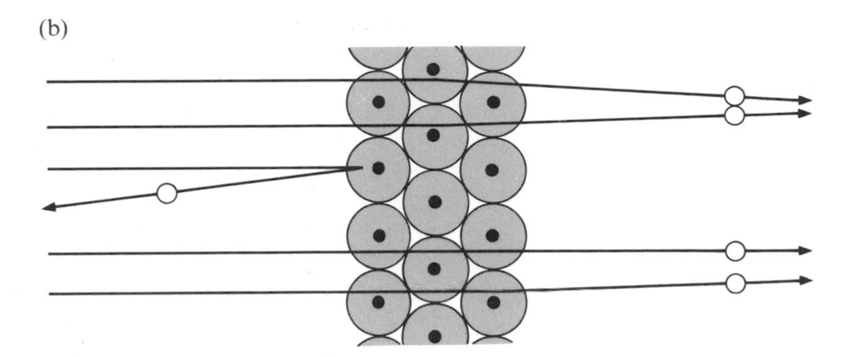

Figure 1.2(b) The interpretation of the results of the Geiger–Marsden experiment

Rutherford went on to argue (in 1911) that (a) an atom is mostly empty space (to account for the undeflected passage of most particles) but (b) most of the mass, and all the positive charge, is in a minute central **nucleus** (to account for the large deflections observed when the α-particle scored an occasional direct hit and experienced a strong electrostatic repulsion). The **nuclear atom**, a positive nucleus surrounded by electrons – the former accounting for the atom's mass and the latter for its size – had entered science.

Atomic numbers and atomic masses

The **atomic number** (Z) was originally a label indicating the place of the element in the Periodic Table. It is now defined as *the number of protons in the nucleus* of an atom of the element (and so it is also sometimes called the **proton number**). The determination of the value of Z was made possible by Moseley's work on X-rays in 1913. It was known that when elements are bombarded with fast electrons they emit X-rays, and that there are two intense components, the ***K* and *L* lines**. Moseley measured their frequencies (ν, nu) and found that when the square root of the frequency (measured in hertz, Hz) is plotted against atomic number, a straight line is obtained (see Figure 1.3). This meant that Z could be found simply by measuring an element's characteristic X-ray frequency and reading off the appropriate value of Z from Moseley's graph.

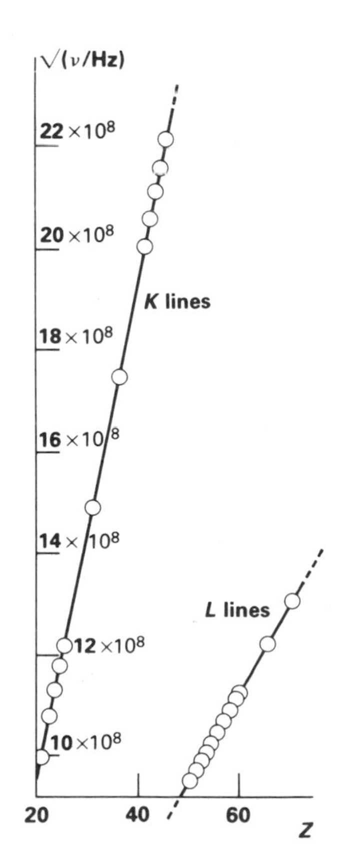

Figure 1.3 Some of Moseley's X-ray results

Hydrogen has $Z = 1$ and so it has one proton (and one electron around it). Carbon has $Z = 6$, and its nucleus contains six protons (with six electrons around it). Until 1940 the element with the largest known atomic number was uranium with $Z = 92$, but since then new elements have been made both as a result of nuclear explosions and through the use of particle accelerators, and all the elements up to $Z = 109$ are now known. The symbol for an atom of an element E is ${}_Z\mathrm{E}$. Therefore a hydrogen atom is written ${}_1\mathrm{H}$ and a carbon atom is written ${}_6\mathrm{C}$.

Evidence that the proton is not the only constituent of a nucleus came when the masses of **ionized atoms** (atoms that have lost one or more electrons) were measured using the technique of deflection by electric and magnetic fields. In 1913 Thomson found that neon ($Z = 10$) is a mixture of atoms of different masses. These various alternative versions of an element are called **isotopes** (from the Greek for 'equal place', since they all correspond to the same place in the Periodic Table).

The explanation for the existence of isotopes became clear when the **neutron** was discovered (by Chadwick in 1932). The neutron is very similar to the proton, having almost the same mass, but it is electrically neutral. Protons and neutrons are collectively known as **nucleons**.

Thomson's observations are explained if a sample of neon contains two types of atom. Both have $Z = 10$ (and hence 10 protons) and are denoted ${}_{10}\mathrm{Ne}$. One type of atom has 10 neutrons in the nucleus and the other has 12. Since there are 10 protons there are also 10 electrons in each case, and so the chemical properties of the atoms are almost the same. The total numbers of nucleons in the two isotopes are 20 and 22, respectively. These are the **mass numbers** (or **nucleon numbers**) of the atoms. The mass number of an element is written A, and an atom of an element of a specified mass number is called a

nuclide and denoted ${}^{A}_{Z}E$. The isotopes of neon are therefore the two nuclides ${}^{20}_{10}Ne$ and ${}^{22}_{10}Ne$.

The isotopes of an element have different masses because they have different numbers of neutrons. Because a neutron has approximately the same mass as a proton, the mass of the nuclide ${}^{20}_{10}Ne$ is about 20 times greater than that of ${}^{1}_{1}H$, while the mass of ${}^{22}_{10}Ne$ is about 22 times greater.

Masses and relative masses

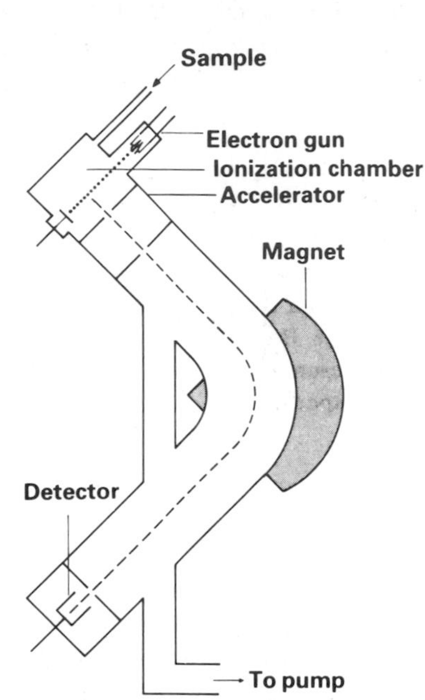

Figure 1.4 The layout of a mass spectrometer

Atomic masses are measured with a **mass spectrometer**. The principle is illustrated in Figure 1.4. The sample is introduced in the form of a low-pressure vapour (that involves vaporizing a sample if it is liquid or solid). In the ionization chamber it is exposed to an electron beam, which ionizes the atoms by colliding with them and knocking out one or more electrons. The positive ions so formed are then accelerated through a large potential difference, and the speed they acquire depends on their mass (and charge). This beam of fast-moving ions then enters the region of the magnet, where it is deflected, the deflection being greatest for ions of lowest mass. As the strength of the magnetic field is varied, ions of different mass are focused in turn on to the ion detector. The output from the apparatus, the **mass spectrum**, is a series of peaks corresponding to the masses present in the sample, the peak heights giving their relative abundances.

The mass spectrometer actually responds to the mass-to-charge ratio of the ions, m/z; hence ions of mass m and unit charge contribute to the same peak as those of mass $2m$ and twice the charge. At low ionization beam energies only singly charged ions are produced and so this is not a serious complication.

The mass spectrum of a typical sample of neon is shown in Figure 1.5. The two peaks corresponding to the isotopes ${}^{20}Ne$ and ${}^{22}Ne$ are plainly visible, and the abundances 91% and 9% can be deduced from the peak heights. (Modern spectrometers also show a weak ${}^{21}Ne$ peak; the abundances are actually 90.92%, 0.257% and 8.82% for ${}^{20}Ne$, ${}^{21}Ne$ and ${}^{22}Ne$, respectively.) Measurements of the positions of the peaks show that the masses of the principal nuclides are 3.32×10^{-26} kg and 3.65×10^{-26} kg, respectively. The same technique can be applied to other elements, and the masses of all the known nuclides have been measured.

Instead of working with the actual masses of atoms, it is much more convenient to set up and use a relative scale. The **relative atomic mass,** A_r, is the mass of an atom of a specified nuclide relative to the mass of a single atom of ${}^{12}C$ taken as 12 exactly. Thus, the relative atomic mass of the nuclide ${}^{20}Ne$ is obtained from

$$\frac{A_r({}^{20}\mathrm{Ne})}{A_r({}^{12}\mathrm{C})} = \frac{3.32 \times 10^{-26}\,\mathrm{kg}}{1.99 \times 10^{-26}\,\mathrm{kg}} = 1.67$$

Since $A_r({}^{12}C) = 12$ exactly, it follows that $A_r({}^{20}Ne) = 1.67 \times 12 = 20.0$. A typical sample of an element is a mixture of isotopes, and so its relative atomic mass is a weighted average over all the nuclides present. Values for natural samples of the elements are given in the Periodic Table printed on page 544. (The **relative molecular mass**, or the **relative formula mass**, M_r, is the sum of the relative atomic masses of the component atoms.)

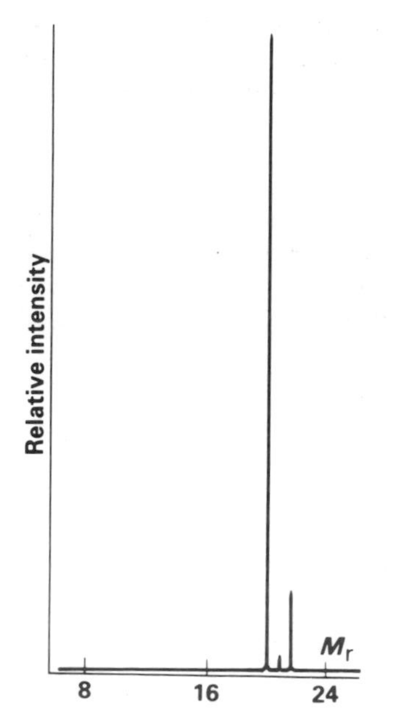

Figure 1.5 The mass spectrum of neon

EXAMPLE

Predict the form of the mass spectrum of Br_2.

METHOD Assume that only singly charged ions are present. There will be Br_2^+ ions arising from the ionization of Br_2 molecules, and Br^+ ions from the fragmentation of Br_2^+ ions. The naturally occurring bromine isotopes ${}^{79}Br$ and ${}^{81}Br$ are present in almost equal abundance. Lines can be expected for each isotopic entity.

ANSWER Entities and mass numbers present are as follows:

Entities:					
$^{79}Br^+$	$^{81}Br^+$	$(^{79}Br^{79}Br)^+$	$(^{79}Br^{81}Br)^+$	$(^{81}Br^{79}Br)^+$	$(^{81}Br^{81}Br)^+$
Mass numbers:					
79	81	158	160	160	162

The heights of the lines due to ions of mass numbers 79 and 81 will be in the ratio 1:1. The other four entities will also give line intensities in the ratio 1:1:1:1, because of the equal abundances of the two isotopes, but the lines from $(^{79}Br^{81}Br)^+$ and $(^{81}Br^{79}Br)^+$, both with mass number 160, coincide to give a single line of double intensity. Therefore the $Br_2{}^+$ ions give rise to three lines at mass numbers 158, 160 and 162 in the intensity ratio 1:2:1.

COMMENT The positions of the lines can be used to measure the masses of the bromine nuclides. There is no simple way of predicting the relative intensities of different types of ion, and so we cannot predict the intensities of the Br^+ peaks relative to the $Br_2{}^+$ peaks.

Now that the relative atomic masses of the elements are known, the role of the mass spectrometer has changed. It is now used to identify the nuclides present in a sample from their masses – for example, in partially spent nuclear fuel.

1.2 The Periodic Table

Chemists of the nineteenth century had no mass spectrometers, yet they obtained moderately good values of relative atomic masses (or *atomic weights* as they were then called) by measuring the masses of elements that combined together. Several people (among them Döbereiner, Newlands and Meyer) noticed regularities in the properties of elements when they were listed in order of their relative atomic masses. The credit for the first useful form of the Periodic Table goes, however, to Mendeleyev, who set out the elements in a table and had the confidence to leave gaps for elements which were then unknown (see Table 1.1) but which seemed to be necessary (see Box 1.3).

A modern Periodic Table is constructed using the atomic numbers of the elements, not their relative atomic masses. The fact that relative atomic mass generally (but not always) increases with atomic number is the coincidence responsible for Mendeleyev's success, but the explanation of periodicity lies, as we shall see below, in the atomic numbers, for these determine how many electrons an atom possesses.

Table 1.1 Mendeleyev's predictions for eka-silicon (germanium)

Property	eka-Silicon (E)	Germanium (Ge)
A_r	72	72.6
density	$5.5\,g\,cm^{-3}$	$5.35\,g\,cm^{-3}$
melting point	high	937 °C
appearance	dark grey	grey-white
oxide	EO_2; white solid, amphoteric, density $4.7\,g\,cm^{-3}$	GeO_2; white solid, amphoteric, density $4.23\,g\,cm^{-3}$
chloride	ECl_4; boiling below 100°C, density $1.9\,g\,cm^{-3}$	$GeCl_4$; boils at 84°C, density $1.84\,g\,cm^{-3}$

BOX 1.3

The development of theories about atomic structure

Numerous people had noticed trends and regularities in the properties of the elements, but the credit for organizing them into an arrangement that was powerful enough to be used to make predictions goes to Dmitri Ivanovich Mendeleyev. Mendeleyev was born in Siberia in 1834 and arranged his table on a single day, 17 February 1869. (According to Russian Orthodox law he was a bigamist, because he had divorced his wife and had married an art student. Tsar Alexander II rejected criticism with the remark 'Yes, Mendeleyev has two wives, but I have only one Mendeleyev'.)

After the discovery that atoms contained electrons, the explanation of chemical periodicity was seen to lie in the way the electrons are arranged. The most important contributions came from Ernest Rutherford. He was a student of J. J. Thomson's, having obtained a scholarship to Cambridge from New Zealand (where he had been born in 1871) only because the winner of the competition had dropped out in order to marry. Rutherford arrived in Cambridge when there was high interest in radioactivity, and the study of radioactive rays led to their use as probes of atomic structure. (The direct descendants of this technique are today's particle accelerators, which use fast particles to explore the structure of the fundamental particles: the faster the particle, the deeper into the structure of matter it can probe.) Rutherford also achieved the first transmutation of an element (in 1919), turning nitrogen into oxygen by bombarding it with α-particles. He died in 1937 and lies beside Newton in Westminster Abbey. Rutherford's nuclear atom led Niels Bohr to propose his model of the hydrogen atom. Bohr, a Dane, was working with Rutherford in Manchester at the time. Note the continuity of ideas (and the importance of being in the right place at the right time).

Electrons in atoms

The key to the explanation of chemical periodicity and of all the chemical properties of the elements is the arrangement of electrons around the nucleus. Once Rutherford had established the nuclear atom it was natural for people to speculate that the electrons were arranged like planets around the central nucleus. This speculation was unacceptable, however, because electrons are electrically charged. When anything charged is accelerated (and an orbiting electron, like anything in circular motion, is accelerating all the while) it generates electromagnetic radiation. Therefore a planetary electron would radiate, lose its energy and spiral into the central nucleus. No electron could survive in orbit for more than a fraction of a second. Consequently, the nuclear atom cannot be a planetary atom.

The first attempt at a modern description of the atom was made by Bohr (1913). In effect he dealt with the problem of electrons collapsing into the nucleus merely by saying that they couldn't. He asserted that the electron in a hydrogen atom could exist in one of a number of **stationary states,** or **orbits**, and that there was an orbit of lowest energy. An electron in that orbit could not lose energy, and so its spiralling collapse into the nucleus was prevented. Bohr was able to employ the newly emerging **quantum theory** to find an expression for the energies of the orbits, and they turned out to agree almost exactly with the values obtained from a study of the atomic spectrum of hydrogen (see below). Nevertheless, his theory did not really explain the structure of atoms, and is now of little more than historical interest.

The current theory of atomic structure is based on **quantum mechanics**, and in particular on solutions of the Schrödinger equation (see Box 1.4). The Schrödinger equation is as central to quantum mechanics as Newton's equations are to classical mechanics. The main change of viewpoint, as far as we are concerned, is that whereas Newton's equations can be solved to predict the precise paths of particles, the solutions of the Schrödinger equation give only *probabilities* of finding particles at various places. So, instead of being able to predict exactly where a particle will be at some instant, we can predict only the chance of it being there. This probability is expressed by a mathematical function which in atoms is called an **atomic orbital** (a name selected to convey a less precise impression than 'orbit').

The energy of an electron in an orbital depends on the electrostatic attraction between the negatively charged electron and the positively charged nucleus, and different orbitals (which correspond to the electron having different spatial distributions around the nucleus) correspond to different energies.

BOX 1.4

Particles, waves and the Schrödinger equation

The lack of precision about the location of a particle is attributed by quantum theory to its **wave nature**, its possession of the characteristics of a wave. As a result, the particle's location is blurred, just as a wave cannot be thought of as occupying a single, well-defined location.

One of the first predictions of quantum theory was that a particle travelling with momentum p (the product of its mass and its velocity) should be regarded as being a wave of wavelength λ given by the **de Broglie relation**,

$$\lambda = h/p$$

where h is **Planck's constant**, a fundamental constant with the value 6.626×10^{-34} J s. For a tennis ball during a typical volley, the wavelength is so small (about 10^{-34} m) that the imprecise location of the ball provides no excuse for missing it. For an electron in an atom, however, the wavelength is of the same order of magnitude as the diameter of the atom, and its wave nature must be taken into account. That electrons do have a wave nature was confirmed when it was shown that they can be diffracted like light (see Figure 1.6), and electron diffraction is now an important technique for investigating the structures of molecules.

The location of a particle in quantum theory is described by its **wavefunction** (ψ, psi). Max Born suggested that the square of the wavefunction, ψ^2, at any point should be interpreted as the probability of finding the electron there: where ψ^2 is large (in illustrations, where the shading is dense), the probability of finding the electron is large.

The wavefunction, which is called an atomic orbital when it refers to an electron in an atom, is obtained by solving the **Schrödinger equation**. Erwin Schrödinger proposed his equation in 1926 after he had been shown de Broglie's doctoral thesis and had responded by saying 'That's rubbish'. He was advised to look at it again, realized that it made sense after all, and so set out on the track of his equation. The Schrödinger equation is a differential equation of the form

$$-\frac{h^2}{8\pi^2 m}\frac{d^2\psi}{dx^2} + V\psi = E\psi$$

where h is Planck's constant, V is the potential energy and E is the total energy. When the potential energy has a simple form the equation can be solved exactly – for the hydrogen atom, for instance. In more complicated systems, such as in many-electron atoms and molecules, there is no hope of finding exact solutions, but computers are used to obtain highly precise numerical solutions for the energies and wavefunctions. Modern theoretical chemistry is largely the exploration of the solutions of the Schrödinger equation for complicated molecules. Even solutions for parts of the DNA molecule can now be obtained.

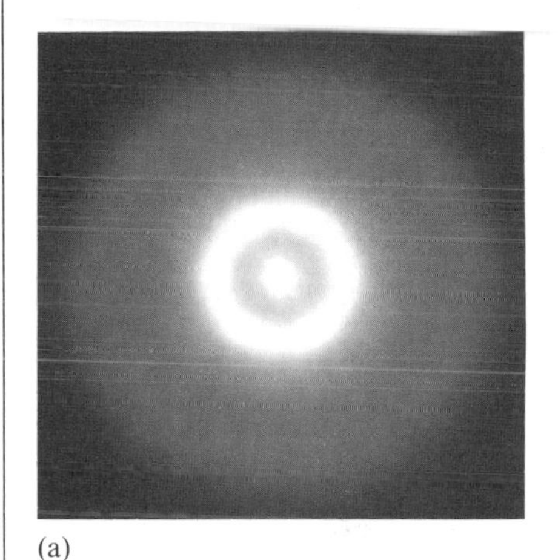

(a)

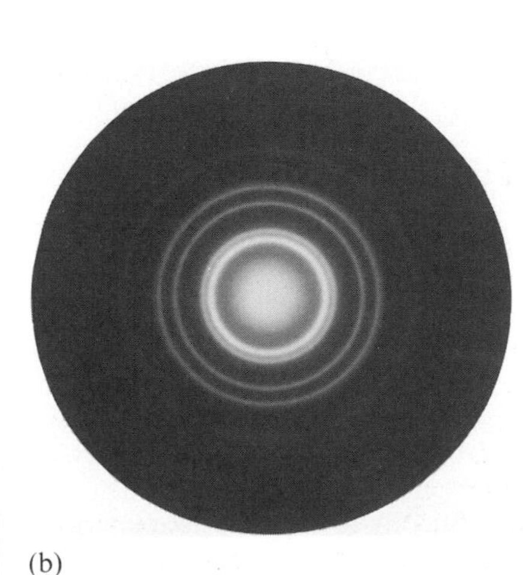

(b)

Figure 1.6 Similar diffraction patterns can be obtained with (a) electrons and with (b) light

The structure of the hydrogen atom

Orbitals give the probability of finding an electron at any point. In illustrations the probability can be represented by the density of shading. This is the technique used in Figure 1.7, which shows some of the orbitals of the hydrogen atom and how they are labelled. When the location of an electron is described by one of these orbitals we say that the electron *occupies* or *is in* that orbital. We often refer to the 'energy of an orbital' as a convenient shorthand for the energy of an electron that is occupying that orbital.

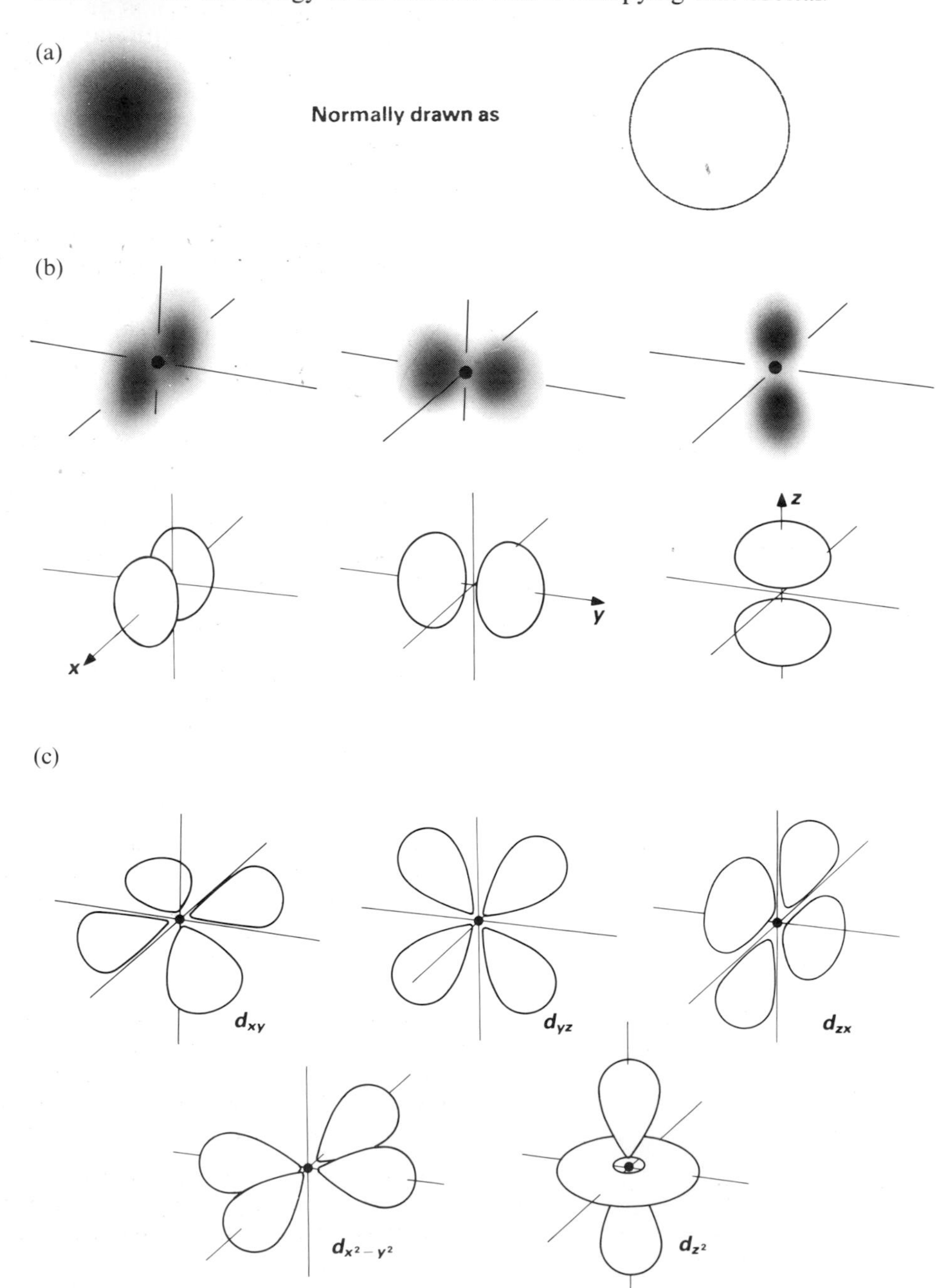

Figure 1.7 (a) An s-orbital, (b) the three p-orbitals, (c) the five d-orbitals

When an electron occupies the **1s-orbital** it has its lowest possible energy and the atom is in its **ground state**. Free hydrogen atoms (such as those responsible for the intergalactic radio waves used in radioastronomy) are almost always found in this state. Figure 1.7(a) shows that a **1s-electron** (the electron in a 1s-orbital) is spread spherically symmetrically around the nucleus. The electron should not be thought of as actually circulating round the nucleus, however, but as clustering in its vicinity.

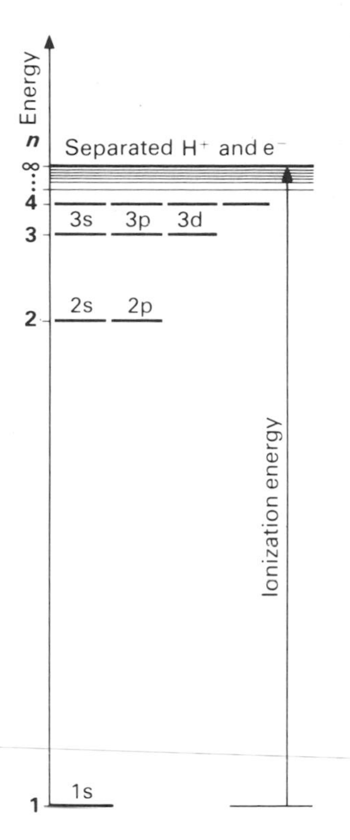

Figure 1.8 The energies of the orbitals of the hydrogen atom

When an electron occupies other orbitals it has a higher energy than when it is in the 1s-orbital, as shown in Figure 1.8. The next lowest orbitals form a group of four, called the second **shell** of the atom. They fall into two **subshells**; the spherical one is the 2s-orbital and the three others are the **2p-orbitals** (see Figure 1.7(b)). The lobes of the p-orbitals lie along the x, y and z axes, and so they are called $2p_x$, $2p_y$ and $2p_z$, respectively. Higher still in energy is the third shell of the atom; this is a group of nine orbitals (note the progression 1, 4, 9, ...), all having the same energy. One is spherical, and is called 3s. Three have two lobes each, like the 2p-orbitals, and are called 3p. Five have four lobes each and are called the **3d-orbitals** (see Figure 1.7(c)). The s-, p- and d-orbitals are all we need for most of chemistry but there are others, which have more complicated shapes.

The structures of many-electron atoms

The atomic orbitals of hydrogen are used to describe the structures of more complicated atoms, and we continue to speak of 1s-, 2s-, 2p-,... orbitals. These orbitals are broadly the same as hydrogen's but differ in detail. For instance, a nucleus with a greater charge draws the inner electrons closer, so that their orbitals spread over a smaller volume. The innermost electron in uranium, for example, is almost 100 times closer to the nucleus on average than in hydrogen, and its orbital is correspondingly more compact. The outer electrons do not experience the full electric charge of the nucleus, however, because it is partly hidden behind (or **shielded** by) the inner electrons.

There is one important difference between hydrogen and many-electron atoms (in this context, 'many' means 'more than one'): the energies of electrons in 2s- and 2p-orbitals are no longer equal, nor are those of the 3s-, 3p-, 3d-orbitals (see Figure 1.9).

Atomic spectroscopy (discussed in Section 1.3) shows that the electrons of a many-electron atom do not all occupy its 1s-orbital, even though it is the orbital of lowest energy. A remarkable feature of nature is that *no more than two electrons can occupy the same orbital.* This is the **Pauli exclusion principle**, which can be traced to a subtle connection between relativity and quantum theory. In fact, the rule the electrons obey is more interesting than this statement of the principle implies. It is known (from spectroscopy) that an electron possesses an angular momentum called its **spin**, which for elementary purposes can be pictured as a spinning motion. Every electron spins at exactly the same rate, but its orientation can take only one of two different directions, which are denoted ↑ 'up' and ↓ 'down' (anticlockwise and clockwise spin respectively, as seen from above). The Pauli principle goes on to state that *if two electrons occupy the same orbital, they must have opposite spins,* denoted by two opposed arrows (↑↓).

The **building-up principle** (which is also called the **aufbau principle**, from the German for 'building-up') converts this description into a set of rules for predicting the electronic structure of any atom. By 'structure' is meant its **electron configuration**, the list of the orbitals occupied by its electrons. The principle is expressed in terms of the following rules:

1. If the atom has atomic number Z, then Z electrons have to be accommodated (in the case of the neutral atom; the principle can also be applied to the atom's positive and negative ions by subtracting or adding electrons).
2. Add the electrons one at a time to the lowest energy orbitals available. The order of orbital energies is 1s, 2s, 2p, 3s, 3p, 4s, 3d, 4p.
3. No more than two electrons may occupy any given orbital.
4. When several orbitals of the same energy are available, electrons enter different orbitals (to keep as far apart as possible). Hence nitrogen has the configuration $1s^2 2s^2 2p_x^{\ 1} 2p_y^{\ 1} 2p_z^{\ 1}$.
5. **Hund's rule** states that *electrons in singly occupied orbitals have parallel spins* (↑↑).

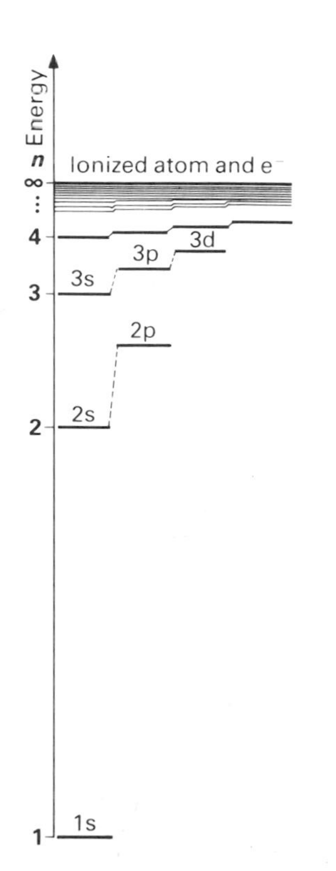

Figure 1.9 The energies of the orbitals in a typical many-electron atom

In the case of helium ($Z = 2$) two electrons have to be accommodated. The first enters the 1s-orbital, and the second joins it. Hence the configuration of helium in its ground state is $1s^2$. (Figure 1.10 illustrates the electron configurations of the first ten elements in the Periodic Table.) In lithium ($Z = 3$) the first two electrons enter the 1s-orbital, but the third is excluded from it and enters the next lowest orbital, which is the 2s-orbital of the second shell; hence the lowest energy configuration of lithium is $1s^2 2s^1$. Beryllium ($Z = 4$) has another electron, and so its configuration is $1s^2 2s^2$. Both lithium and beryllium are therefore members of the **s-block** of the Periodic Table, the elements in which an s-subshell is being filled.

Boron ($Z = 5$) has one more electron, which enters the next lowest orbital, one of the three 2p-orbitals. Its configuration is therefore $1s^2 2s^2 2p^1$, which makes it a member of the **p-block**, the elements in which a p-subshell is being filled. There are three 2p-orbitals, and so up to six electrons can enter them.

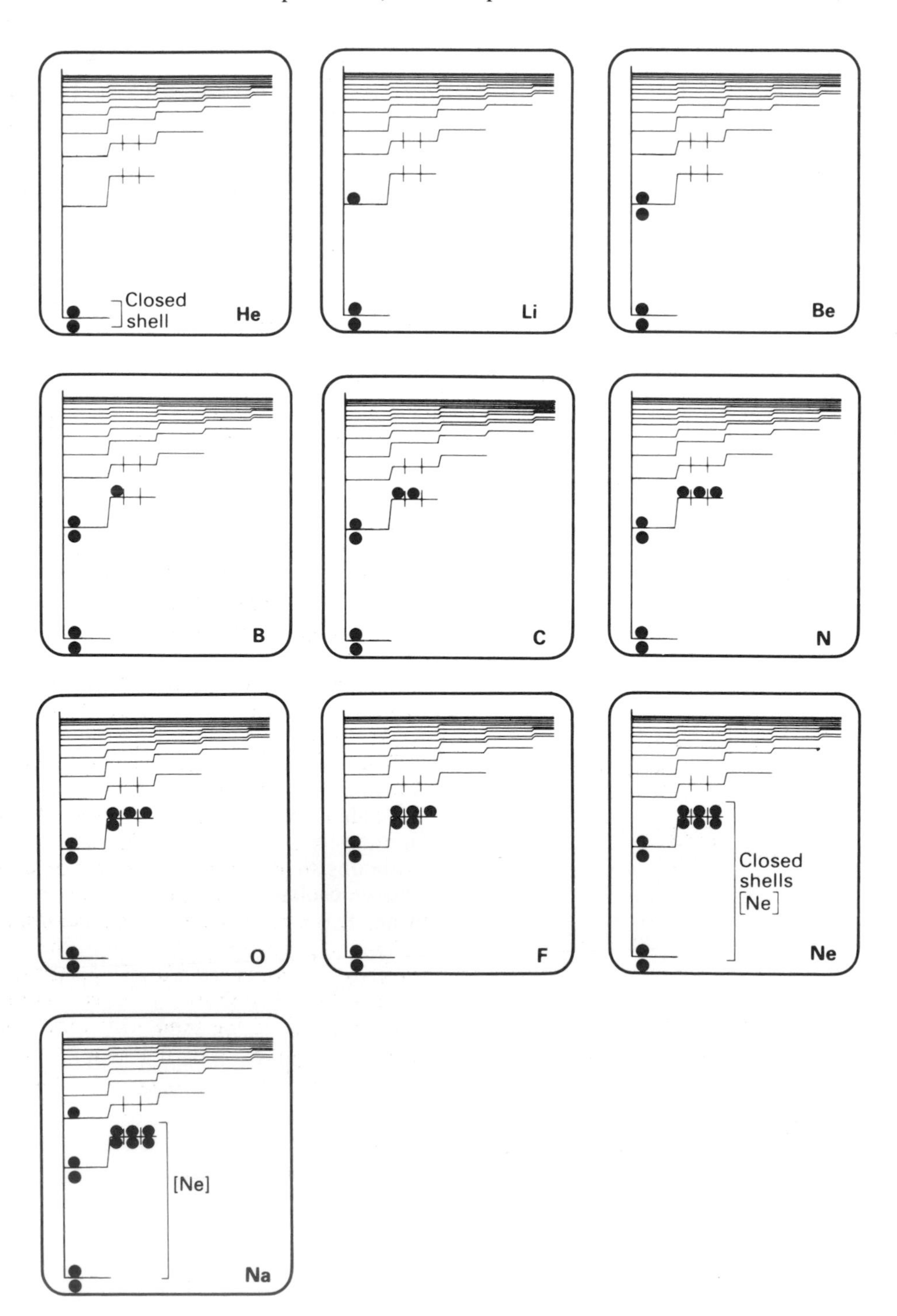

Figure 1.10 The building-up principle and the configurations of atoms

They are therefore gradually filled as we go from boron, carbon, nitrogen, oxygen and fluorine, to neon, all of which belong to the p-block. Neon has $Z = 10$, and its configuration is $1s^22s^22p^6$. This configuration, in which all the orbitals of the outer s- and p-subshells are full, is called a **closed-shell configuration**. The next element, sodium, has one more electron, which has to go into the next lowest orbital, 3s. Its configuration is therefore $1s^22s^22p^63s^1$, and it belongs to **Period 3** (see Figure 1.11).

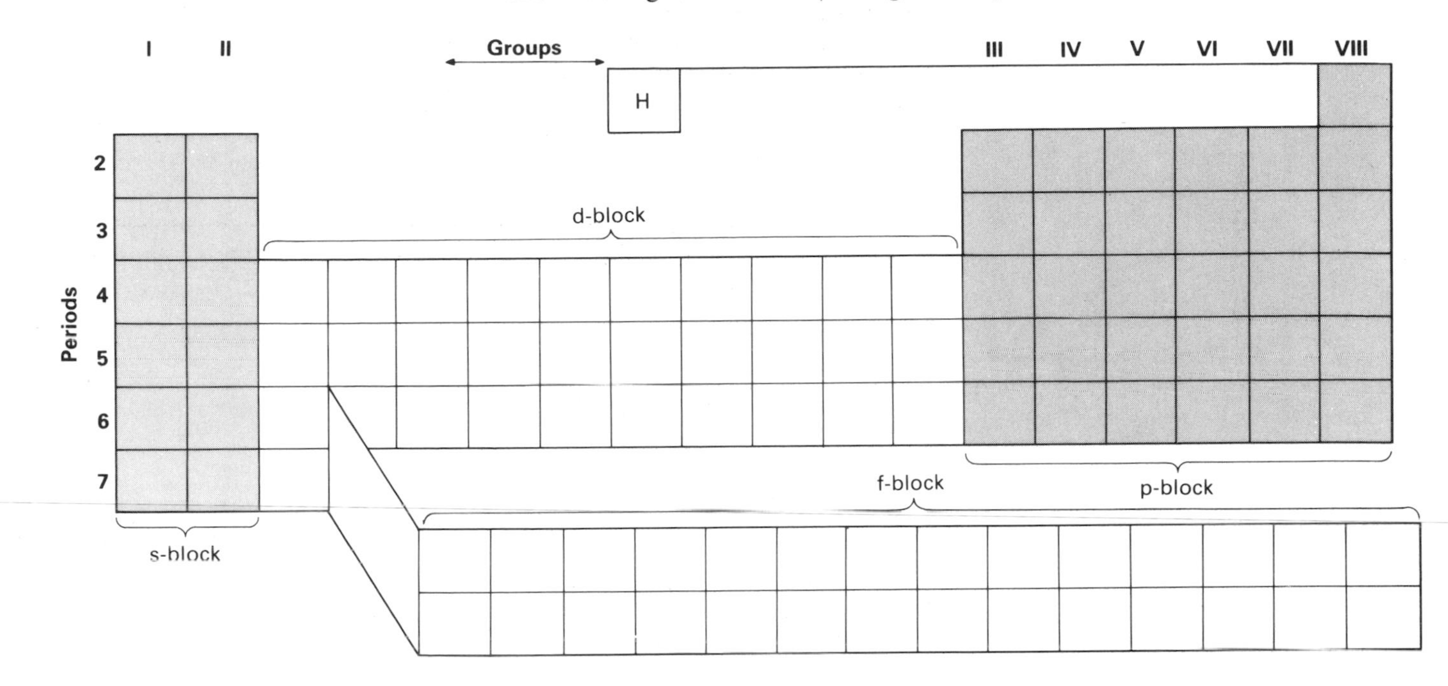

Figure 1.11 Blocks, Groups and Periods in the Periodic Table

We have here the basis of the explanation of the periodicity of the elements. Lithium ($1s^22s^1$) has a single s-electron outside an inner, complete shell ($1s^2$); sodium ($1s^22s^22p^63s^1$) also has a single s-electron outside a closed shell. The same pattern is repeated between beryllium and magnesium, between boron and aluminium, and so on up to $Z = 18$, argon, which is a closed-shell species, like neon.

The building-up principle also neatly accounts for the occurrence of the elements belonging to the **d-block**. Potassium ($Z = 19$) has the configuration $1s^22s^22p^63s^23p^64s^1$, or $[Ar]4s^1$ for short, where [Ar] denotes its **core**, the closed-shell argon configuration. Calcium has the configuration $[Ar]4s^2$ (see Figure 1.12). At this point the 3d-orbitals are next in line to be filled. There are five of them, which hold the next ten electrons. Ten is exactly the number of the d-block elements of Period 4. At zinc ($Z = 30$) the 3d-orbitals are full (we say the 3d-subshell is complete), and the next electron enters a 4p-orbital. The 4p-orbitals accommodate the electrons for the elements up to krypton, which has a closed-shell configuration, $[Ar]\ 3d^{10}4s^24p^6$. Krypton resembles argon, and completes this row of the Periodic Table. All the subsequent Periods, including the **f-block elements**, the lanthanoids and the actinoids (where seven f-orbitals are being occupied with up to fourteen electrons), are accounted for in the same way. Since electronic structure is the basis of an element's chemistry, the periodic repetition of configuration accounts for the periodicity of the chemistry of the elements.

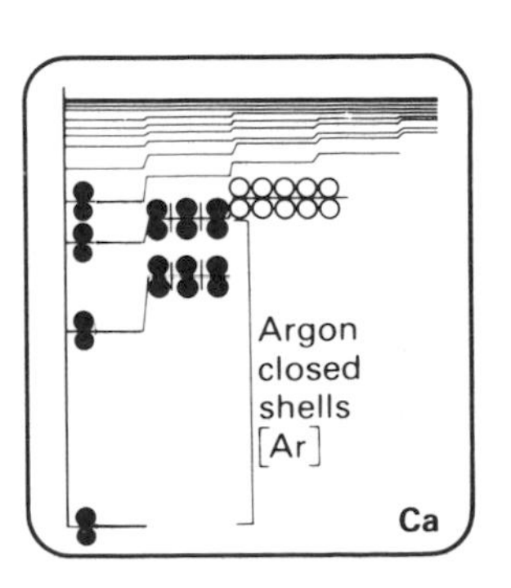

Figure 1.12 The configuration of the calcium atom

EXAMPLE

State the electron configurations of the silicon and chlorine atoms.

METHOD Decide how many electrons are involved in each case by noting the atomic numbers (see the copy of the Periodic Table on page 544). Take the preceding closed-shell configuration and write it [Ne] ([Ne] = $1s^22s^22p^6$); that accounts for ten electrons. Feed in the remaining electrons, permitting up to two to occupy 3s, and up to six to occupy the 3p-subshell.

ANSWER Z(Si) = 14, leaving four electrons to be accommodated outside the closed shell. Two enter 3s, leaving two to enter 3p. Hence silicon has the configuration $[Ne]3s^23p_x{}^13p_y{}^1$. Z(Cl) = 17, leaving seven electrons to be accommodated outside the closed shell. Two enter and fill 3s, leaving five to enter 3p. The configuration of chlorine is therefore $[Ne]3s^23p_x{}^23p_y{}^23p_z{}^1$.

COMMENT Since the partly filled orbitals of both elements are p-orbitals, both silicon and chlorine are p-block elements.

One important property that determines an element's chemical behaviour is the ease with which its outermost or **valence** electrons can be removed. The **first ionization energy** is the minimum energy required to remove an electron from a neutral atom in the gas phase (denoted g). That is, it is the energy required for the process

$$E(g) \rightarrow E^+(g) + e^-(g)$$

It is normally expressed as an energy per unit amount (for example, per mole) of atoms. The **second ionization energy** is the energy required to remove a second electron. It is always larger than the first because more energy is needed to remove an electron from a positively charged ion than from a neutral atom (see Table 1.2).

Figure 1.13 shows how the first ionization energy varies for the first few elements, and Figure 1.14 shows the variation throughout the Periodic Table.

Table 1.2 First and second ionization energies and electron-gain energies of atoms/kJ mol^{-1}

	Ionization energy First	Ionization energy Second	Electron-gain energy
H	1312		−74
Li	520	7300	−56
Na	496	4562	−71 ←
C	1086	2353	−121
N	1402	2856	+26
O	1314	3388	−141
O$^-$	141	1314	+844
S	1000	2258	−200 ←
F	1681	3375	−333
Cl	1251	2296	−349 ←
Br	1140	2084	−328

A negative electron-gain energy signifies that energy is released when the electron attaches. The older term *electron affinity* had the opposite sign, so that a positive electron affinity implied a release of energy.

Mg, Al, Si, P,

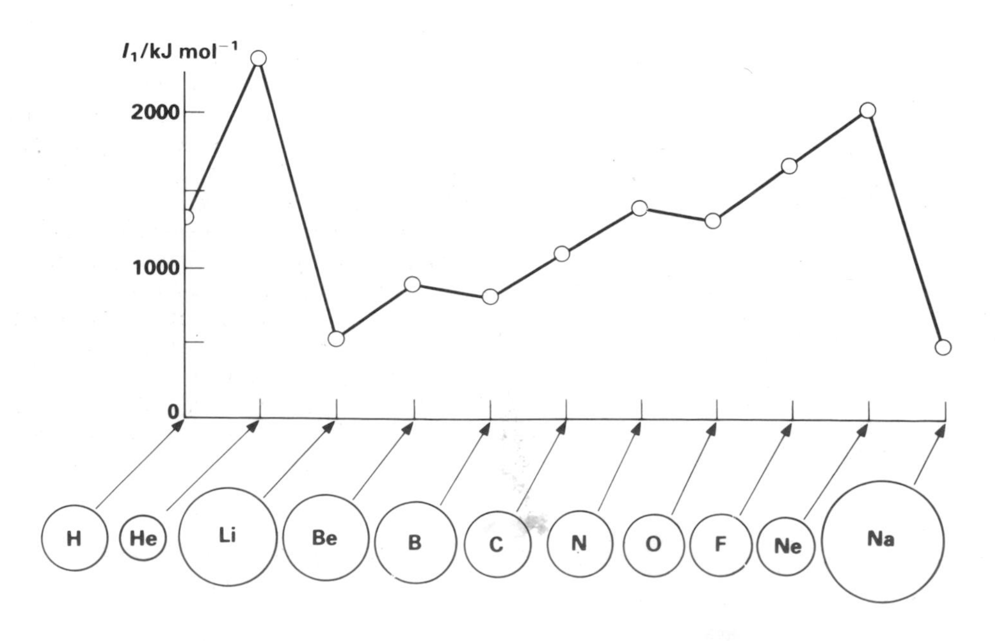

Figure 1.13 First ionization energy and atomic size

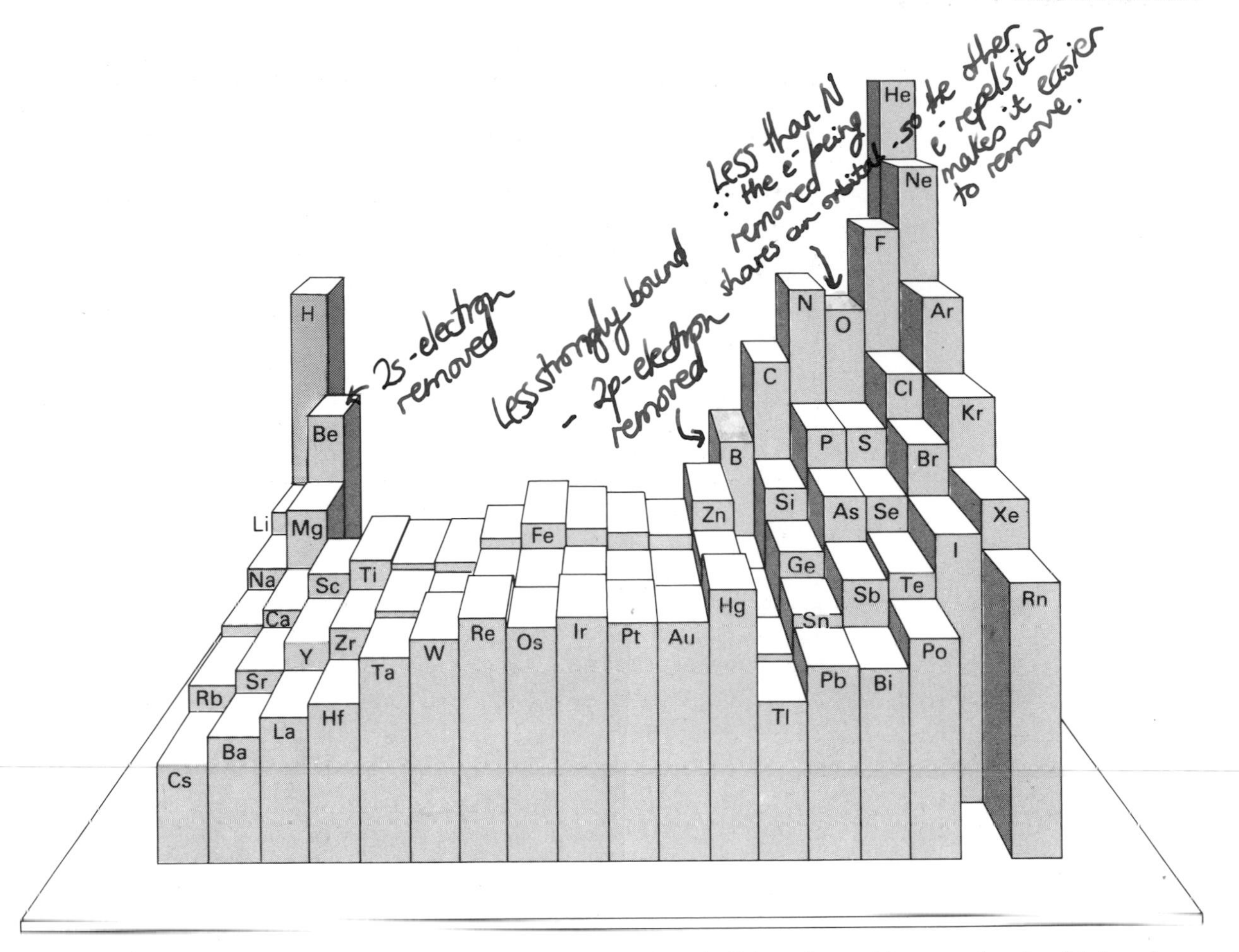

Figure 1.14 The first ionization energies of the elements. Values decrease down Groups, and generally increase across Periods; those of the d-block elements are similar

There is a pronounced periodicity, the noble gases having high values, showing that the electron is difficult to remove, while the alkali metals have relatively low values, so that much less energy is needed to remove the single outermost electron (nevertheless, 500 kJ mol^{-1} is a lot of energy per mole, and so the outer electrons do not simply 'drop off'). Why this is so can be seen from Figure 1.13, which shows the atomic sizes of the first few elements; the atoms shrink across a Period (for example, from lithium to neon) because the increasing nuclear charge draws in the electrons which, as a result, are increasingly difficult to remove (see Section 15.4). On going from neon to sodium the additional electron occupies a shell on the outside of the atom and is quite well shielded from the nucleus by the inner electrons. The ionization energy therefore falls back to a smaller value. The ionization energy of boron is slightly less than that of beryllium; the reason is that in beryllium a 2s-electron is being removed but in boron a 2p-electron (see Figure 1.10 above), and the latter is less strongly bound than the former. The ionization energy of oxygen is less than that of nitrogen because in oxygen the electron that is being removed shares an orbital with another (again, compare Figure 1.10), which repels it and makes it easier to remove.

One method for measuring the ionization energy of gases uses a **thyratron**, which is a valve containing two electrodes in the gas of interest, such as xenon. The current through the circuit is monitored as the potential difference is increased. When the potential difference is great enough, it strips electrons from the atoms, and the thyratron suddenly becomes highly conducting. Spectroscopy is also used to measure ionization energies, as we describe later.

The **electron-gain energy** of an atom is the change in energy when an electron is attached to an atom to form a negative ion in the gas phase, that is, in the process

$$E(g) + e^-(g) \rightarrow E^-(g)$$

The electron-gain energy is negative when energy is released. Halogen atoms have highly negative electron-gain energies (see Table 1.2), because the added electron enters a gap in a shell and interacts strongly with the nucleus. The noble gas atoms have low electron-gain energies because the added electron must enter a new shell a long way from the nucleus and outside the shielding core electrons. The concept of electron-gain energy is discussed in more detail in Section 9.4.

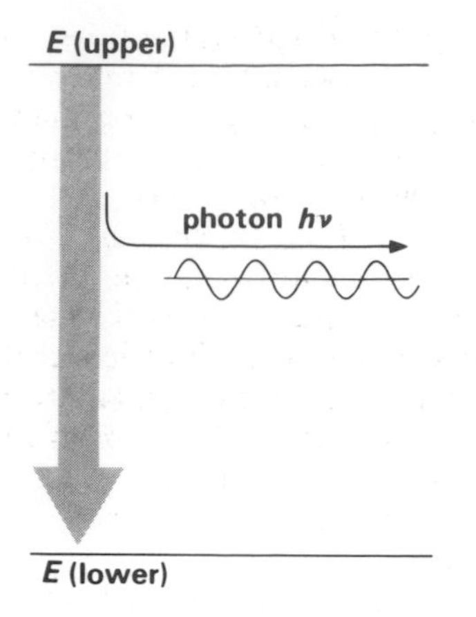

Figure 1.15 The emission of radiation

1.3 Atomic spectroscopy

The key to understanding the chemistry of the elements is the electron configurations and energies of their atoms: that is what the Periodic Table summarizes so neatly. **Atomic spectroscopy** is used to determine electron configurations experimentally.

The essential feature of atomic spectroscopy is that, when an electron changes from one orbital to another of lower energy, the discarded energy is carried away as radiation (see Figure 1.15). A ray of light of frequency ν (sometimes the symbol f is used) is a stream of **photons**, each one of energy $h\nu$, where h is Planck's constant (see Box 1.4). An intense ray of light is a dense stream of photons, whereas a weak ray contains relatively few. Photons of blue light (which lies at the high-frequency end of the visible region of the electromagnetic spectrum) each carry more energy than do photons of red light (which lies at the low-frequency end). When an atom loses energy as radiation it does so by generating a single photon of light, and so it is easy to see that the frequency emitted by the atom when it makes a transition from a state of energy E(upper) to one of energy E(lower) is given by the **Bohr frequency condition**

$$h\nu = E(\text{upper}) - E(\text{lower}) \qquad (1.3.1)$$

Sometimes it is more convenient to discuss radiation in terms of its wavelength λ (lambda). Wavelength and frequency are related through

$$c = \nu\lambda$$

where c is the speed of light ($c = 2.998 \times 10^8\ \text{m s}^{-1}$).

In practice, atoms are generated in their electronically **excited** states by passing a high-voltage electric discharge through a gaseous sample or by burning a sample in a flame. For instance, the spectrum of atomic hydrogen can be generated in an electric discharge through low-pressure molecular hydrogen. The discharge, in effect an electric storm of electrons and ions, rips the molecule apart, and the atoms so formed are initially in a variety of high-energy, excited states. As their electrons make transitions to states of lower energy, they generate photons of various frequencies (see Figure 1.16) and hence give rise to a purple glow. The radiation emitted by the sample is in fact a collection of different colours, and if it is passed through a prism or a diffraction grating, as shown in Figure 1.17, the individual frequencies may be resolved and recorded photographically or electronically. That is how the spectra shown in Colour 1 were obtained.

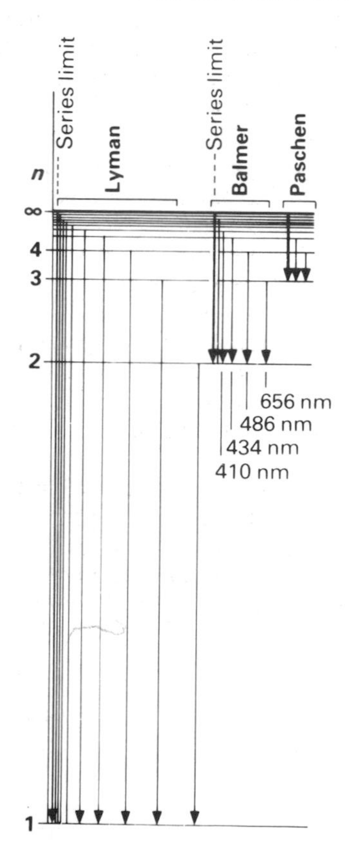

Figure 1.16 The transitions responsible for the spectrum of atomic hydrogen

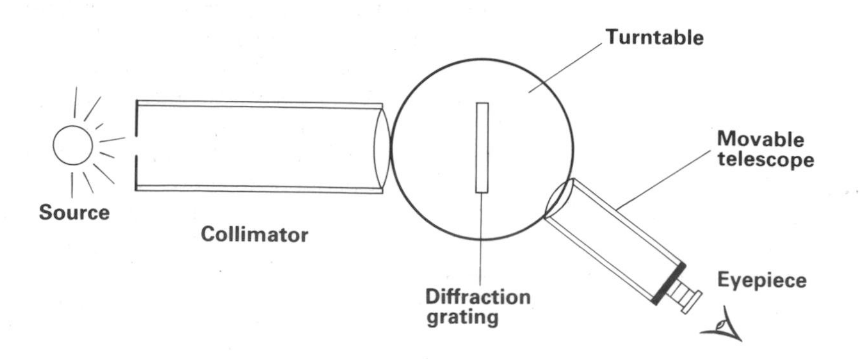

Figure 1.17 A simple atomic emission spectrometer: the collimator produces a parallel beam of light

Atomic spectra can also be observed in absorption. If a photon of light of frequency ν hits an atom, it may cause the atom to make a transition from its ground state to an excited state an energy $h\nu$ higher. If the light's frequency ν, and therefore the photon's energy $h\nu$, does not correspond to an excitation energy of the atom, the photon is not absorbed, but if it does match it is absorbed. If the sample is illuminated with a spread of frequencies, therefore, (with white light, for example), there is a reduction of intensity of the beam at every frequency that corresponds to an absorption. If after passing through the sample the beam of light is passed through a prism or a diffraction grating and the spectrum photographed, there will be dark lines, corresponding to the frequencies of the absorbed photons, against an otherwise uniform spectrum of colours. The bright yellow lines in the emission spectrum of sodium atoms, which account for the illumination from sodium street lighting, therefore appear as dark lines in its absorption spectrum.

Since atoms give rise to characteristic spectra, both absorption and emission spectroscopy are applied in analysis. In one type of spectrometer, a hollow cathode is made of the element of interest, and is heated to a high temperature so that it emits its characteristic spectrum. The sample being analysed for the presence of the element is burnt in a cool flame under carefully controlled conditions, and its absorption of the incident light from the hot cathode is monitored electronically. A modern commercial **atomic absorption spectrometer** can analyse as many as twenty samples for up to twenty elements automatically, and can detect concentrations as low as one part in a billion (1 in 10^9). Emission spectra are used in the **flame test**, where excitation is caused by heating in a bunsen flame, and is used to test for the presence of Group I and Group II metals (for more details see Sections 17.6 and 18.6). Helium was identified on the Sun from its emission spectrum thirty years before it was found on Earth. Both emission and absorption spectroscopy are very useful for the detection of trace elements in water and food and in the analysis of industrial materials, including steels. One of the first forensic applications of atomic spectroscopy was its use to identify the bullets fired in the St Valentine's Day massacre in Chicago in 1929.

An atom's spectrum can be analysed in terms of its allowed energy levels and can be used to measure its ionization energy. A particularly important case is hydrogen, where late in the nineteenth century Balmer, a Swiss schoolmaster with a passion for numbers, noticed that the spectral lines then known (of wavelengths 656.3 nm, 486.1 nm, 434.0 nm and 410.2 nm) fitted the formula

$$1/\lambda = R_{\mathrm{H}}\{(\tfrac{1}{4}) - (1/n^2)\}, \qquad n = 3, 4, 5, 6 \qquad (1.3.2)$$

R_{H} being a constant. When detectors became available that were sensitive to the infrared and ultraviolet regions of the spectrum (see Section 3.2), other wavelengths were discovered. All were found to fit a more general form of this expression known as the **Rydberg–Ritz formula**:

$$1/\lambda = R_{\mathrm{H}}\{(1/n_1{}^2) - (1/n_2{}^2)\} \qquad (1.3.3)$$

R_{H} is the **Rydberg constant** for hydrogen (its value is $1.097 \times 10^7\,\mathrm{m}^{-1}$). Different values of n_1, which is a **quantum number** (specifically, the **principal quantum number**), correspond to different series of lines in the spectrum. For instance, the Lyman series ($n_1 = 1$ and $n_2 = 2, 3, 4, \ldots$) has wavelengths in the ultraviolet region, the Balmer series ($n_1 = 2$ and $n_2 = 3, 4, 5, \ldots$) has wavelengths in the visible range, and the Paschen, Brackett, Pfund and Humphreys series ($n_1 = 3, 4, 5$ and 6 respectively, with n_2 starting with the value $n_1 + 1$ in each case) all lie in the infrared.

The Rydberg–Ritz formula can be deduced from the Schrödinger equation, and the value of the Rydberg constant can be predicted very precisely. The agreement between theory and experiment for hydrogen and for many-electron atoms shows beyond reasonable doubt that our theories of atomic spectra and structure are now substantially correct.

A feature of the Balmer series that is apparent from Figure 1.16 is that the wavelengths of the Balmer series come closer together or **converge.** The same is true of all the series. The line of shortest wavelength of each series is called

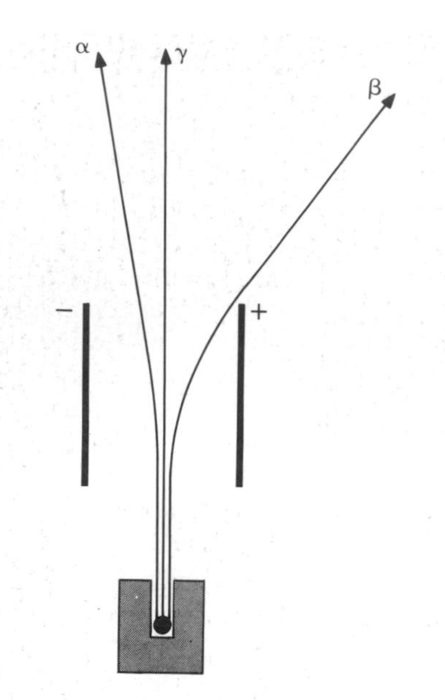

Figure 1.18 The effect of an electrostatic field on α-, β- and γ-radiation

the **series limit** (or **convergence limit**). In an absorption experiment using wavelengths shorter than the series limit, each photon carries so much energy that when it is absorbed it ionizes the atom. This is the basis of the spectroscopic method of measuring ionization energies. On comparing Figure 1.16 with Figure 1.8 we see that the Lyman series limit corresponds to the ionization energy of the atom. Therefore, to find the ionization energy, we note the wavelength of the Lyman series limit, where the spectrum merges into a continuum, and convert it to an energy.

EXAMPLE

The Lyman series limit is at wavelength 91.2 nm. Calculate the ionization energy of the hydrogen atom.

METHOD The energy needed to ionize the atom is equal to the excitation energy that corresponds, via the Bohr relation, to the frequency of the series limit. Hence, combine $E = h\nu$ with the relation between frequency and wavelength, $\nu = c/\lambda$:

$$E = hc/\lambda$$

ANSWER Substitute the data:

$$E = \frac{6.626 \times 10^{-34}\,\text{J s} \times 2.998 \times 10^{8}\,\text{m s}^{-1}}{91.2 \times 10^{-9}\,\text{m}}$$
$$= 2.18 \times 10^{-18}\,\text{J}$$

COMMENT The ionization energy per mole of hydrogen atoms is obtained by multiplying the energy for one atom by Avogadro's constant (see the *Introduction* to this book). In this case:

$$I = 2.18 \times 10^{-18}\,\text{J} \times 6.022 \times 10^{23}\,\text{mol}^{-1} = 1310\,\text{kJ mol}^{-1}$$

1.4 The nucleus in chemistry

Most of chemistry takes place without any change occurring in the structure of the nucleus; its role is to govern the tightness with which the electrons are bound. Nevertheless nuclei *can* change their energy states. Sometimes the nucleons within nuclei change their energy and, like atoms, emit the excess as radiation. Sometimes nuclei eject charged particles, and therefore change their chemical identity. These changes, and the corresponding emissions, are called **radioactivity**.

Radioactivity: α-, β- and γ-rays

The early investigators of radioactivity found that radioactive emissions could be classified into three groups according to their response to electric and magnetic fields; these are shown diagrammatically in Figure 1.18.

β-rays (or **β-particles**) were easily identified because they behaved like electrons, which were already known. β-particles are fast electrons ejected from within the nucleus as a result of the decay of a neutron into a proton. Since they carry unit negative charge, the nucleus is left with one additional unit of positive charge. Therefore, when an element emits a β-particle it changes from atomic number Z to atomic number $Z + 1$ (see Figure 1.19). For instance, the nuclide $^{14}_{6}C$ is β-active (it is a **radionuclide**), and each atom that loses an electron from its nucleus becomes an atom of $_{7}N$. Note that, because the electron has such a minute mass, even relative to a proton, the mass number does not change on β-emission; $^{14}_{6}C$ therefore decays into $^{14}_{7}N$, the common isotope of nitrogen. In general an element $^{A}_{Z}E$ **transmutes** into $^{A}_{Z+1}E'$ on β-emission.

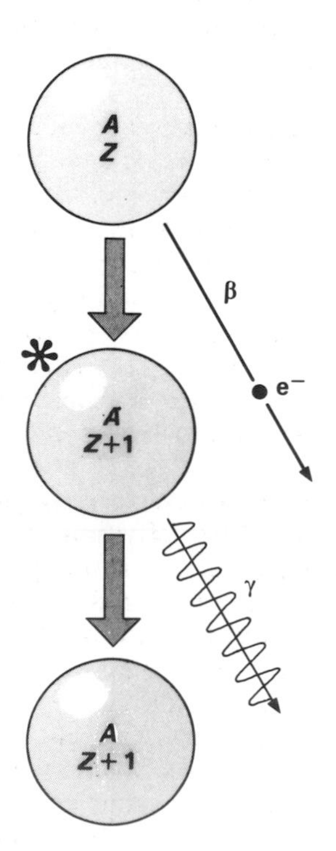

Figure 1.19 β-decay (accompanied by γ-emission); the asterisk indicates an excited nucleus

γ-rays are undeflected by either electric or magnetic fields: they are **electromagnetic waves**, and a part of the electromagnetic spectrum. They are of

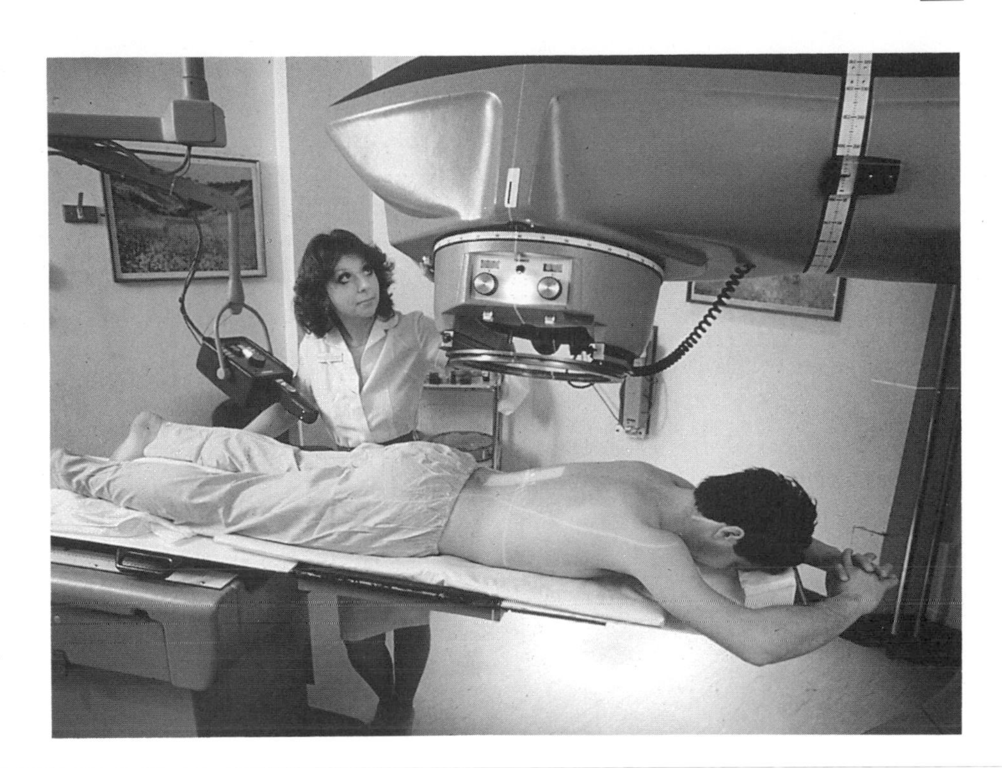

Figure 1.20 Radiation treatment: the γ-rays from a cobalt-60 source can be used to kill cancerous cells, which are more sensitive to radiation than healthy cells are

much higher frequency than visible light, and hence of much shorter wavelength. Their wavelengths are shorter even than the wavelengths of X-rays (shorter than 100 pm), and since they are very energetic, they penetrate deep inside the body and damage living tissue by causing ionization. γ-rays destroy rapidly metabolizing malignant cells more effectively than healthy cells, and so the radiation treatment of cancer patients, in which they are exposed to rays from cobalt-60 (^{60}Co), can be helpful. In industry γ-radiation is used to measure the thickness of steel plates: the thicker the plate, the greater the absorption. A similar relation, using β-rays, is used for quality control during the manufacture of paper, and to check that bottles and fire extinguishers are filled.

A photon of γ-radiation is generated when a nucleus changes its energy state and the excess energy is carried away as radiation. Since the nucleons are bound together by the **strong force** (as distinct from the much weaker electrostatic force between charged particles that accounts for the electronic structure of atoms), the energies involved, and therefore the frequencies, are much greater than in atomic spectra. γ-rays usually accompany β-rays (and α-rays) because the latter's emission leads to the formation of a new element in an energetically excited state (compare Figure 1.19). Since a γ-ray is the result of an energetic state of a nucleus losing energy, and involves no change of nuclear charge, its generation does not result in the formation of a new element.

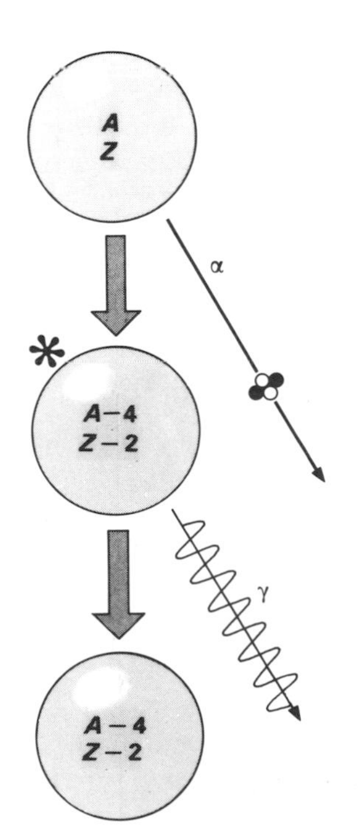

Figure 1.21 α-decay (accompanied by γ-emission)

α-rays (or **α-particles**) were the most difficult to identify because nothing similar was known when they were first discovered. Their deflection by an electric field showed that they were positively charged. The crucial experiment was inspired, once again, by Rutherford. It showed that when radon decayed by α-emission, helium was formed in the vicinity. Since the mass-to-charge ratio of α-particles had been measured in a deflection experiment, the conclusion could be reached at once that an α-particle is $^{4}_{2}He^{2+}$, or the bare helium nucleus, a tight cluster of two protons and two neutrons. This combination of nucleons is very stable, like a closed-shell electronic structure.

An α-particle emitted by a nucleus carries away two units of positive charge and so the atomic number of the element falls from Z to $Z-2$ (see Figure 1.21). Since an α-particle consists of four nucleons it carries away four mass units and the mass number of the element also changes from A to $A-4$. An element $^{A}_{Z}E$ therefore transmutes into $^{A-4}_{Z-2}E'$ on α-emission.

EXAMPLE

Predict the nuclides produced on (a) β-decay of $^{24}_{11}Na$ and (b) α-decay of $^{226}_{88}Ra$.

METHOD Use the rules stated above: on β-decay $^{A}_{Z}E$ changes to $_{Z+1}^{\ A}E'$ and on α-decay $^{A}_{Z}E$ changes to $^{A-4}_{Z-2}E'$. Identify the element from the information in the Table on page 544.

ANSWER (a) $A = 24$, $Z = 11$; therefore the product is $^{24}_{12}E'$. The element with $Z = 12$ is magnesium. Hence the decay is $^{24}_{11}Na \rightarrow {}^{24}_{12}Mg + \beta$.
(b) $A = 226$, $Z = 88$; therefore the product is $^{222}_{86}E'$. $Z = 86$ identifies the product as radon. Therefore the decay is $^{226}_{88}Ra \rightarrow {}^{222}_{86}Rn + \alpha$.

COMMENT The radium decay reaction was used by Rutherford and Royds to generate the radon in their identification of the α-particle.

When an element decays radioactively, it may transmute to another radioactive element which can itself decay. Such a sequence of transmutations of the elements is illustrated by the **uranium series** shown in Figure 1.22. Other natural radioactive elements of mass numbers similar to uranium's also decay to some isotope of lead, $_{82}Pb$, and lead's unusual stability, together with some theories of nuclear structure, has led people to speculate that the element with $Z = 114$ would also be stable, if only it could be prepared. Elements are synthesized in the interiors of stars and during their explosions, but on Earth we have to use less dramatic methods, and new nuclides are prepared by collisions in particle accelerators. The **nuclear reaction** that takes place on impact is written in the form

target nuclide (incoming particle, outgoing particle) product nuclide.

The 'equation' is balanced so that the sums of the atomic numbers and the mass numbers are the same before and after the collision; electrons, protons and neutrons are written $_{-1}^{\ 0}e$, $^{1}_{1}p$ and $^{1}_{0}n$ respectively.

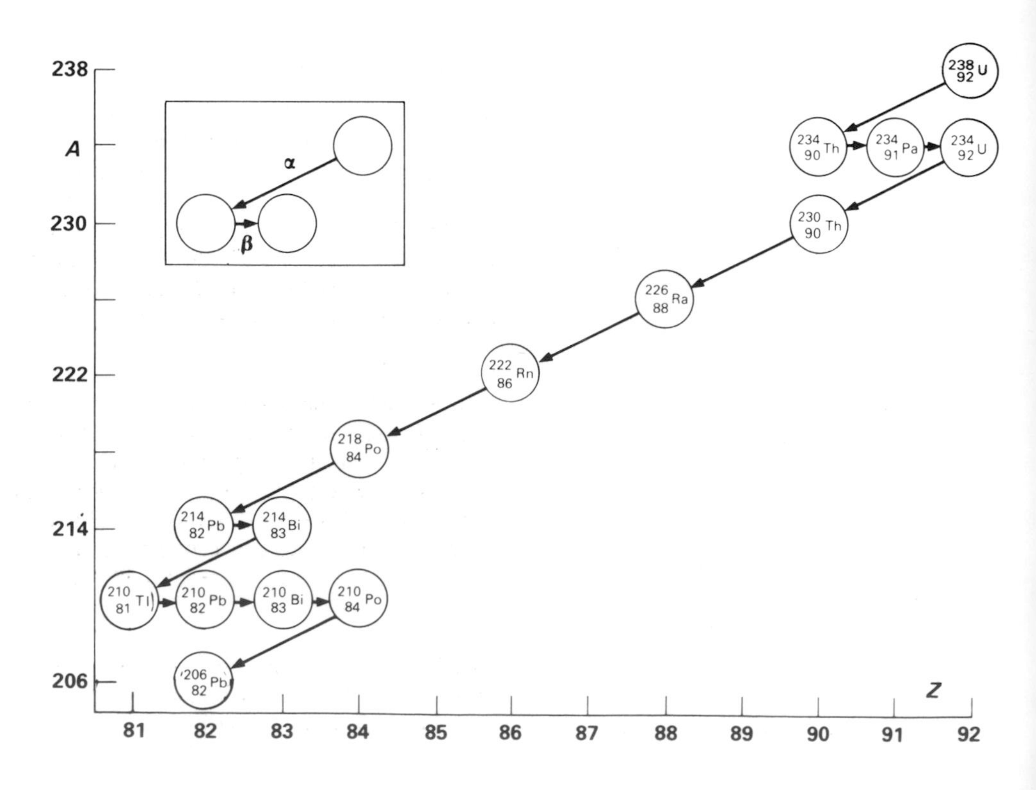

Figure 1.22 A radioactive series

EXAMPLE

What nuclides are produced in the following nuclear reactions:
(a) $^{14}_{7}N(^{1}_{0}n, ^{1}_{1}p)$; (b) $^{9}_{4}Be(^{4}_{2}\alpha, ^{1}_{0}n)$; (c) $^{7}_{3}Li(^{1}_{1}p, ^{4}_{2}\alpha)$?

METHOD The sum of the mass numbers in the products must equal their sum in the two colliding particles; the same is true of the atomic numbers. Therefore balance the reaction by finding $^{A}_{Z}E$ for which this is true, and identify the nuclide from the Table on page 544.

ANSWER (a) $^{14}_{7}N(^{1}_{0}n, ^{1}_{1}p)^{A}_{Z}E$ requires $A = 14$, $Z = 6$; the product is $^{14}_{6}C$.
(b) $^{9}_{4}Be(^{4}_{2}\alpha, ^{1}_{0}n)^{A}_{Z}E$ requires $A = 12$, $Z = 6$; the product is $^{12}_{6}C$.
(c) $^{7}_{3}Li(^{1}_{1}p, ^{4}_{2}\alpha)^{A}_{Z}E$ requires $A = 4$, $Z = 2$; the product is $^{4}_{2}He$.

COMMENT Reaction (b) was used by Chadwick (in 1932) to identify the neutron. Reaction (c) was the first transmutation achieved using artificially accelerated ions (in Cockcroft and Walton's original 'atom-smasher' of 1932). Heavy elements are formed using heavy projectiles fired at heavy targets. Unnilhexium (element 106), for example, is produced in the reaction $^{249}_{98}Cf(^{18}_{8}O, 4^{1}_{0}n)^{263}_{106}Unh$.

The energy released in some nuclear reactions is used in the generation of nuclear power. In **nuclear fission**, collision of a neutron with a fissionable nucleus, such as ^{235}U or ^{239}Pu, results in the fracture of the heavy nucleus with the liberation of nuclear binding energy (and a consequent overall loss of mass described by $E = mc^2$) and more neutrons. These neutrons in turn cause fission in other nuclei, and a **chain reaction** sets in. If many of the neutrons are absorbed by non-fissionable material (as in a reactor) the chain reaction is controlled. If they are not absorbed the chain reaction rapidly escalates into an explosion. In **nuclear fusion** energy is released when two small nuclei (such as deuterium, ^2H, and tritium, ^3H) fuse into a single nucleus, as in the interiors of stars and soon, it is hoped, under controlled conditions on Earth.

Radioactive half-lives

The time it takes for half the radioactive atoms in any sample to decay is called the radioactive **half-life** of the radionuclide and is denoted $t_{\frac{1}{2}}$. Half-lives can vary from less than 1 s (e.g., 3.04×10^{-7} s for $^{212}_{84}Po$) to more than a thousand million million years (e.g., 6×10^{15} years for $^{50}_{23}V$) (some are listed in Table 1.3). The rate of radioactive decay is independent of the conditions, such as the temperature and the type of chemical bonding.

The **law of radioactive decay** is that *the rate at which a radioactive sample decays is proportional to the number of radioactive atoms present:*

$$\text{rate} = \lambda N \qquad (1.4.1)$$

λ (lambda) is the **decay constant** and is related to the half-life by $\lambda = (\ln 2)/t_{\frac{1}{2}}$. Why this should be the law will become clear once we have dealt with reaction rates (in Chapter 14): the essential feature is that nuclei decay *at random*. The importance of the law is that it can be used to predict the proportion of atoms of some initial sample that have not yet decayed. This follows from the solution of equation 1.4.1 for the number of atoms present at time t given that N_0 were present initially:

$$N = N_0 e^{-\lambda t} \qquad (1.4.2)$$

The form of this *exponential decay* is shown in Figure 1.23(a). Note that N/N_0 falls to half its initial value in one half-life, to one quarter in two half-lives, and so on (Figure 1.23(b)).

Table 1.3 Radioactive half-lives

Nuclide	$t_{\frac{1}{2}}$
^{3}H	12.3 yr
^{14}C	5730 yr
^{24}Na	15.0 h
^{40}K	1.3×10^9 yr
^{60}Co	5.25 yr
^{90}Sr	27 yr
^{131}I	8.06 day
^{253}No	10 min

(a)

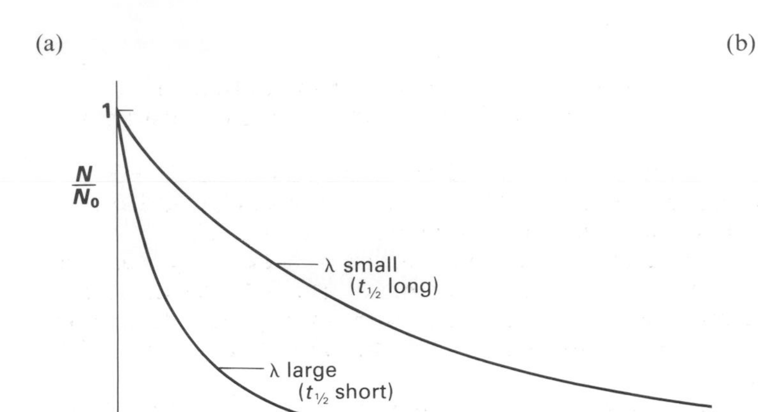

(b)

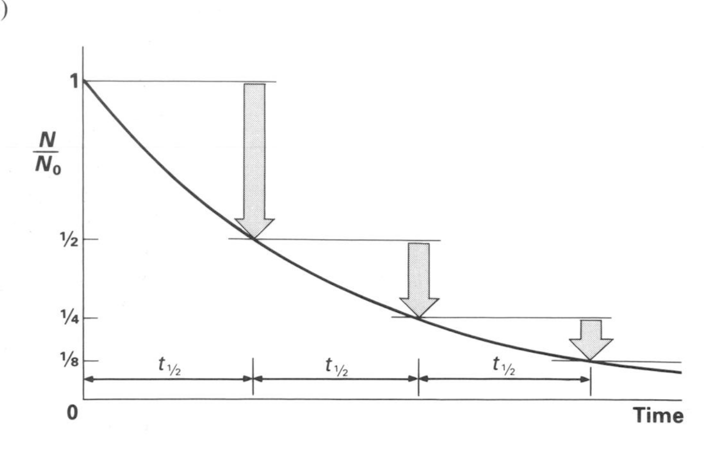

Figure 1.23 (a) Radioactive decay has (b) a constant half-life

EXAMPLE

The nuclide ^{60}Co is a significant product of nuclear explosions. Its half-life is 5.25 years. A nuclear device was tested in 1988; what fraction of the initial amount of nuclide produced will remain 20 years later?

METHOD Use equation 1.4.2 to find the fraction N/N_0 remaining at $t = 20\,\text{yr}$. For λ use $(\ln 2)/t_{\frac{1}{2}}$.

ANSWER $\lambda = (\ln 2)/(5.25\,\text{yr}) = 0.132\,\text{yr}^{-1}$. Then from equation 1.4.2,

$$N/N_0 = e^{-(0.132\,\text{yr}^{-1})\times(20\,\text{yr})} = e^{-2.64} = 0.0714, \text{ or } 7.14\%$$

COMMENT This result gives some idea of the great lengths of time it takes for some dangerous radionuclides to disappear. The fraction of ^{60}Co does not fall below 0.01 per cent until 70 years after the explosion.

An important application of the law of radioactive decay is in the dating of archaeological remains. The nuclide ^{14}C is continuously generated in the atmosphere as a result of the bombardment of nitrogen by cosmic rays:

$$^{14}_{7}N(^{1}_{0}n, ^{1}_{1}p)^{14}_{6}C$$

Therefore, to a good approximation its abundance is constant, and the fraction in atmospheric CO_2 is also constant (about 1 molecule in 10 000). Since living things continuously ingest carbon compounds as complex molecules or as carbon dioxide (in photosynthesis) they contain a constant fraction of ^{14}C. When they die they stop ingesting, and so the ^{14}C content is no longer replenished but decays according to the radioactive decay law with a half-life of 5730 years. Therefore, if the proportion of ^{14}C to ^{12}C is determined (using a mass spectrometer – see Section 1.1) after the object's death, the time since death can be determined. The error in the determination arises partly from the fact that the proportion of ^{14}C in the atmosphere is not quite constant, because it depends on both the solar sunspot activity and the Earth's magnetic field. Hence it is necessary to calibrate the method against a more direct dating method, such as counting tree rings. Other nuclides (such as ^{40}K, for which $t_{\frac{1}{2}} = 1.3 \times 10^9\,\text{yr}$) are also used for dating rocks.

There are numerous applications of radioactive isotopes in chemistry (and also in medicine). In particular radioactive **labelling** lets chemists follow the course of an element through a series of reactions. This is especially important in biochemistry, where metabolic pathways can be unravelled by following ^{14}C from its ingestion to its excretion.

Summary

☐ The atom consists of a central nucleus surrounded by electrons. This nuclear atom was suggested by the Geiger–Marsden α-particle scattering experiment.

☐ The identity of an element is determined by its atomic number, Z, the number of protons in its nucleus. Different isotopes of an element have different numbers of neutrons but the same number of protons.

☐ The mass number of an atom is the total number of nucleons (protons and neutrons) in the nucleus. A nuclide is an atom of specified mass number.

☐ The relative atomic mass, A_r, of an element is the average mass of an atom of the element in a typical sample relative to the mass of one atom of ^{12}C taken as 12 exactly.

☐ Mass numbers, actual masses and abundances of nuclides are determined with a mass spectrometer.

☐ In a mass spectrometer, a vaporized sample is ionized, and the ions are accelerated and then deflected; the mass spectrum is a series of peaks corresponding to the mass-to-charge ratios of the ions present.

☐ Atomic orbitals give the distributions of electrons in atoms.

☐ s-orbitals are spherically symmetrical, p-orbitals are two-lobed and d-orbitals are four-lobed.

☐ The electronic structure of atoms is expressed in terms of the configuration predicted using the building-up principle. This is based on the Pauli exclusion principle, that no more than two electrons may occupy an orbital. When in the same orbital, their spins are paired.

☐ Chemical periodicity arises from the periodicity of the electron configurations of the elements, and is explained by the building-up principle.

☐ The first ionization energy is the energy required per unit amount of atoms for the process $E(g) \rightarrow E^+(g) + e^-(g)$.

☐ First ionization energies are periodic, and are related to the electrostatic control of the nucleus over its outermost, valence, electrons.

☐ The configurations $1s^2$ and $1s^2 2s^2 2p^6$ are examples of closed-shell configurations. Closed-shell species are stable mainly because they are so compact and all the electrons are close to the nucleus.

☐ Atomic spectroscopy is used to identify elements, to investigate the electron configurations and energy levels of atoms, and to measure ionization energies. The analysis is based on the Bohr frequency condition.

☐ Light of frequency ν and wavelength $\lambda = c/\nu$ may be regarded as a stream of photons, each one of energy $h\nu$, where h is Planck's constant.

☐ The spectrum of atomic hydrogen is summarized by the Rydberg–Ritz formula (equation 1.3.3).

☐ Radioactive emissions from nuclei fall into three groups. α-rays are bare helium nuclei, ${}^4_2He^{2+}$. β-rays are electrons ejected from the nucleus. γ-rays are short-wavelength electromagnetic radiation.

☐ When an α-particle is emitted a nuclide ${}^A_ZE$ transmutes to ${}^{A-4}_{Z-2}E'$; when a β-particle is emitted a nuclide changes from ${}^A_ZE$ to ${}_{Z+1}^{A}E'$; when a γ-ray is emitted a nucleus changes its energy but not its identity.

☐ The radioactive half-life is the time it takes for the number of radioactive atoms of a radionuclide in the sample to decrease by half. It is independent of the quantity of radionuclide. The law of radioactive decay implies an exponential decrease in the abundance of a radioactive species.

PROBLEMS

The solution of problems in chemistry often calls for the bringing together of information from various parts of the subject. Many of the chemistry problems in this book require knowledge of material from earlier chapters as well as from the chapters in which they appear, and some require knowledge of material from later chapters too.

1 Write an essay on the development of the modern theory of atomic structure.

2 State the numbers of protons, neutrons and electrons in each of the following atoms: (a) ^{16}O, (b) ^{18}O, (c) ^{27}Al, (d) ^{35}Cl, (e) ^{40}Ca, (f) ^{235}U.

3 Write an essay on isotopes. Include an account of their discovery, their significance, the measurement of their relative atomic masses and their uses.

4 On the basis that a sample of sulphur had the following composition, calculate its relative atomic mass: ^{32}S, $A_r = 31.972$, 95.0%; ^{33}S, $A_r = 32.972$, 0.8%; ^{34}S, $A_r = 33.969$, 4.2%.

5 Describe the appearance of the mass spectrum of argon.

6 Give the electron configurations of the ground states of the following atoms: (a) O, (b) Si, (c) K, (d) Fe.

7 A 100 W light bulb gives out about 2×10^{18} photons of yellow (550 nm) light each second. What is the energy carried by each photon? What fraction of the total energy emitted by the lamp do they carry? (Use $1\,W = 1\,J\,s^{-1}$.)

8 Explain why the minimum energy to remove an electron from the ground state of a hydrogen atom can be calculated from the Rydberg–Ritz formula by setting $n_1 = 1$ and $n_2 = \infty$. Calculate the ionization energy of hydrogen.

9 The longest-wavelength radiation that can be used to ionize a sodium atom in its ground state is 241 nm. What is the ionization energy of sodium?

10 What nuclides are formed by the following radioactive decays: (a) β-decay of ^{131}I, (b) α-decay of ^{251}Cf, (c) β-decay of ^{90}Sr?

11 A ^{60}Co source was installed in a hospital radiation unit in 1985. Given that the half-life of the nuclide is 5.25 years, calculate the fraction of the source that will remain in (a) 1995, (b) 2085.

12 (a) The relative atomic mass of the element palladium is 106.40. State concisely what this statement means.
(b) Using mass spectrometry, the element gallium has been found to consist of 60.4 per cent of an isotope of atomic mass 68.93 and 39.6 per cent of an isotope of atomic mass 70.92. Calculate, to three significant figures, the relative atomic mass of gallium. *(Oxford)*

13 (a) List the three main fundamental particles which are constituents of atoms, and give their relative masses and charges.
(b) Similarly, name and *differentiate* between the radiations emitted by naturally occurring radioactive elements.
(c) Complete the following equations using your Periodic Table to identify the elements X, Y, Z, Q and R. Add atomic and mass numbers where these are missing:

(i) $^{24}_{11}Na \rightarrow X + {}^{0}_{-1}e$ (ii) $^{14}_{7}N + {}^{1}_{0}n \rightarrow {}^{14}Y + {}_{1}Z$
(iii) $Si \rightarrow {}^{27}_{13}Q + {}^{0}_{+1}e$ (iv) $R + {}^{4}_{2}He \rightarrow {}^{13}_{7}N + {}^{1}_{0}n$

(d) Refer to (c)(i) and (c)(iii) above. For each of these two processes, briefly describe *one* chemical test which could be used to confirm that a change of chemical element has occurred. *(SUJB; continued as Question 10.21)*

14 (a) Explain the meaning of the term *first ionization energy* of an atom.
(b) The first, second, third and fourth ionization energies of the elements X, Y, Z are given below:

	Ionization energies/kJ mol⁻¹			
	First	*Second*	*Third*	*Fourth*
X	738	1450	7730	10550
Y	800	2427	3658	25024
Z	495	4563	6912	9540

Using this information, state, giving your reasons, which element is most likely
(i) to form an ionic univalent chloride;
(ii) to form a covalent chloride;
(iii) to have +2 as its common oxidation state. *(Oxford)*

15 (a) Explain what is meant by each of the following terms: (i) electron, (ii) proton, (iii) neutron, (iv) isotopes.
(b) The atomic number provides three pieces of information about an element. What are they?
(c) The radioactive atom $^{224}_{88}Ra$ decays by α-emission with a half-life of 3.64 days.
(i) What is meant by 'half-life' of 3.64 days?
(ii) Referring to the product of the decay, what will be its mass number and its atomic number?
(iii) Radium is in Group II of the Periodic Table. In what Group will the decay product be?
(d) Explain briefly the principles underlying (i) the use of radioactive isotopes as 'tracers', (ii) the dating of dead organic matter using radiocarbon, $^{14}_{6}C$. *(London)*

16 Discuss the evidence for the existence of isotopes. The nuclear reaction

$$^{6}Li + {}^{2}H \rightarrow 2\,{}^{4}He$$

is proposed as a possible future energy source. Given the isotopic masses ^{6}Li (6.015 06), ^{2}H (2.014 07) and ^{4}He (4.002 63) calculate the energy released by consumption of 1 g ^{2}H. What daily consumption of ^{2}H corresponds to a power of 1 MW?

Outline a scheme for the production and concentration of ^{2}H, explaining the relevant physical and chemical principles. *(Oxford Entrance)*

17 Discuss the atomic spectrum of hydrogen and its relation to our understanding of the electronic structure of atoms.

Suggest explanations for the following observations:
(a) The atomic spectrum of hydrogen contains lines in the radiofrequency region of the electromagnetic spectrum.
(b) A line in the spectrum of atomic hydrogen on a distant object in the universe occurs at a wavelength of 300 nm though it is known to occur in the laboratory at 121.6 nm. *(Oxford Entrance)*

2

Chemical bonding

In this chapter we see why atoms form bonds. We see that the formation of ionic and covalent bonds can be explained in terms of the electrostatic interactions between electrons and nuclei. There are simple rules for predicting the shapes of molecules and for discussing the extent to which a bond is ionic or covalent. The modern description of bonding and molecular structure in terms of molecular orbitals is introduced. Molecular orbital theory provides a straightforward explanation of the exceptional stability of aromatic compounds, and is readily extended to account for the structures of metals.

The world around us takes on its richness because atoms cluster together into compounds. The view that molecules are definite arrangements of atoms grew during the nineteenth century as a result of careful observations and measurements. Modern techniques have brought about a revolution in the kind of evidence. Originally molecules were suggested to account for observations on the masses of elements that combined together (see Box 2.1) but now we can 'see ' them. An example is shown in Figure 2.1: the coiled strand of atoms in the molecule is just visible.

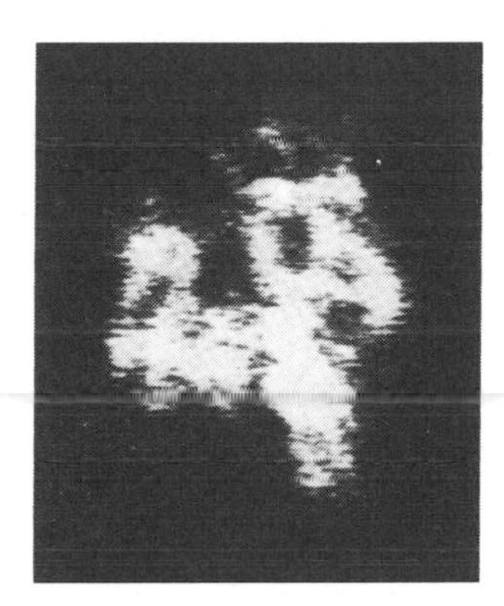

Figure 2.1 Electron microscope image of haemoglobin

BOX 2.1

The development of the concept of the molecule

The name 'molecule' (which is derived from the Latin for 'little mass') was introduced by Amedeo Avogadro. The line symbol for the chemical bond, as in H—Cl, was first used in 1858 by Archibald Couper, who also introduced what we now call a 'structural formula'. The first reference to a 'molecular structure' seems to have been made in 1861 by the Russian chemist Alexander Butlerov. August Kekulé proposed his structure for benzene in 1865, and so by then the view that molecules had not only a definite composition but also atoms in a definite spatial arrangement was well established.

Once the structures of atoms were understood, attempts were made to explain molecular structure and bond formation in similar terms. A major contribution was made by Gilbert Newton Lewis (see Figure 2.2) when he identified the importance of the electron pair. Much credit for the rationalization of bonding theory in terms of quantum mechanics is due to Linus Pauling, who won two Nobel prizes (one for chemistry, the other for peace). He exerted considerable influence on chemical thought through his book *The nature of the chemical bond.*

Modern studies of the chemical bond depend on careful experimental determinations of structures using spectroscopy and X-ray diffraction, and on detailed calculations using computers. It is now possible to compute the details of the electron distributions in complex molecules and to predict their bond angles and bond lengths. The major unsolved problem is the prediction of reactions, but even that is beginning to become possible. Despite all these great computational achievements, however, the problems of interpreting, understanding and using the results remain as challenging as ever.

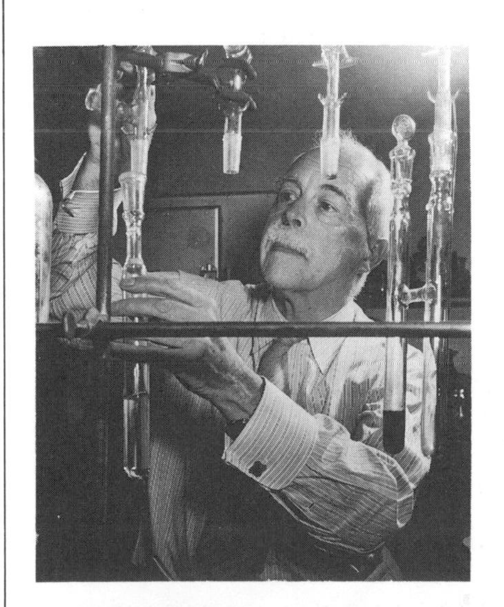

Figure 2.2 Gilbert Newton Lewis, who proposed the idea of covalent bonding and showed the importance of the electron pair in bonding (Lewis structures are named after him). He also introduced an important definition of acids and bases, and was the first person to isolate heavy water

An impression of the range of molecular structures and their relative sizes is given by the photographs of molecular models in Figure 2.3. The smallest molecule shown is that of hydrogen (H_2). Since it consists of two identical atoms it is called a **homonuclear diatomic molecule** (*homo* comes from the Greek for 'same'). The oxygen molecule (O_2) is also a homonuclear diatomic molecule, but it is larger than the H_2 molecule because it has more electrons.

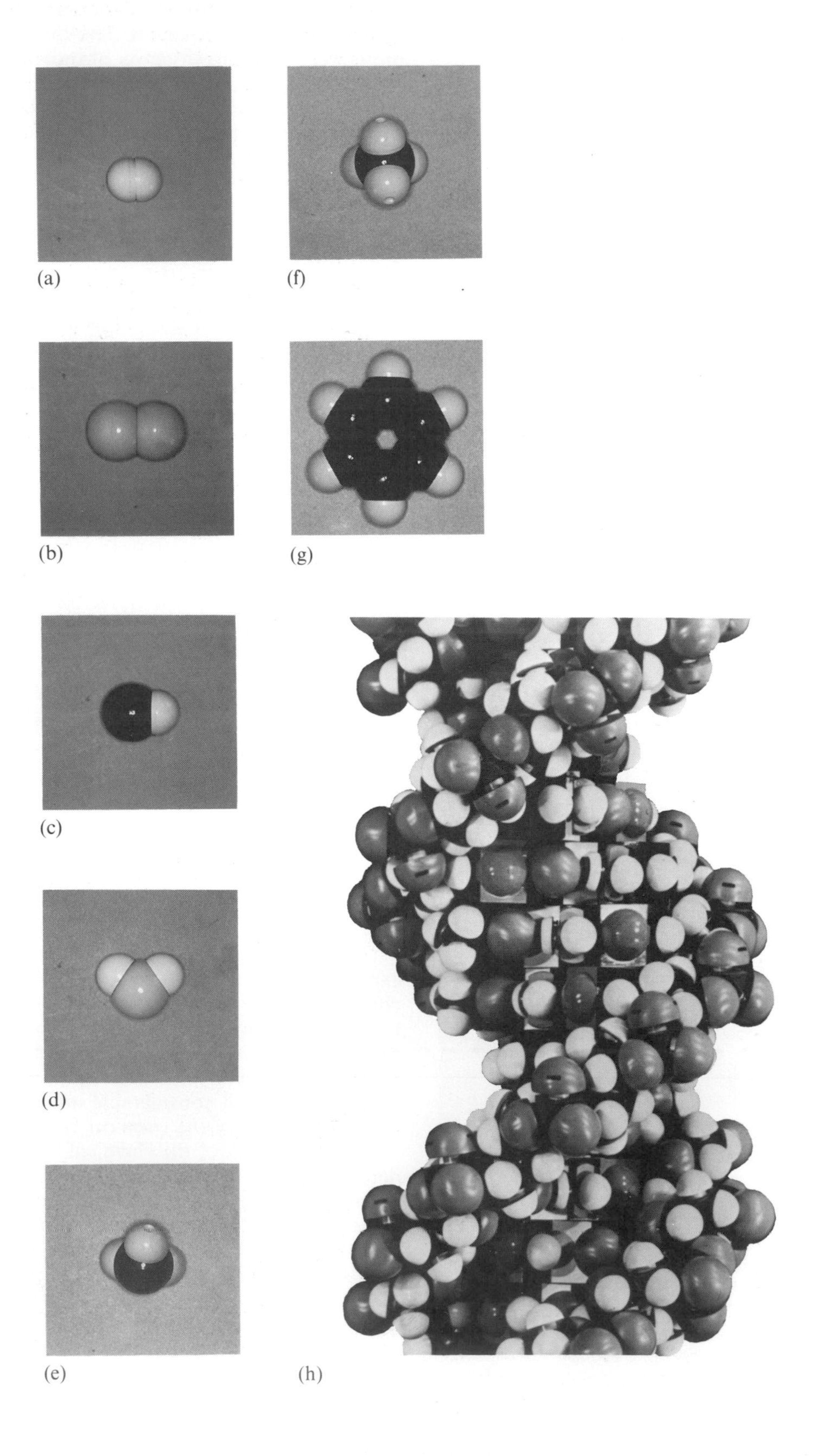

Figure 2.3 Molecular models of (a) H_2, (b) O_2, (c) HCl, (d) H_2O, (e) NH_3, (f) CH_4, (g) C_6H_6 and (h) DNA

The hydrogen chloride molecule (HCl) is a **heteronuclear diatomic molecule** (*hetero* comes from the Greek for 'different'), and the difference in the sizes of the two atoms should be noted. Water (H_2O), ammonia (NH_3) and methane (CH_4) are all examples of **polyatomic molecules**. The hexagonal benzene molecule (C_6H_6) is important to chemistry because it is the parent of the family of **aromatic compounds**, discussed in Chapters 34 and 35. Almost the extreme example of molecular complexity is shown by the structure of the genetic material DNA, and the illustration shows only a fragment of this huge molecule: human DNA, fully stretched out, would be about 2 m long. The fact that hydrogen is a gas is related to the simplicity of its molecule and the very few electrons it contains; the complexity of DNA is related to its function, for it must convey detailed structural information from one generation to the next.

The energies of atoms and their compounds are controlled largely by the way their electrons are arranged, and a general principle in **valence theory**, the theory of bond formation, is that *a bond will form if this will result in a lower-energy arrangement of electrons and nuclei.* This electron rearrangement can occur in two ways. An atom may *transfer* one or more of its electrons to another atom. Alternatively, two atoms may *share* electrons.

When electron transfer is complete, the atoms A and B become the ions A^+ and B^- (or ions such as A^{2+}, B^{2-}, if more than one electron is transferred). The ions then stick together as a result of the electrostatic Coulombic attraction between opposite charges. The resulting link is called an **ionic bond**, and the resulting substance is an **ionic compound**.

When electrons are shared the link is called a **covalent bond**, and the substance is a **molecular compound** (or a **covalent compound**). In some substances the covalent bonding is so extensive that a crystal is a single huge molecule: this special case of an **extended covalent structure** is described more fully in Section 5.1.

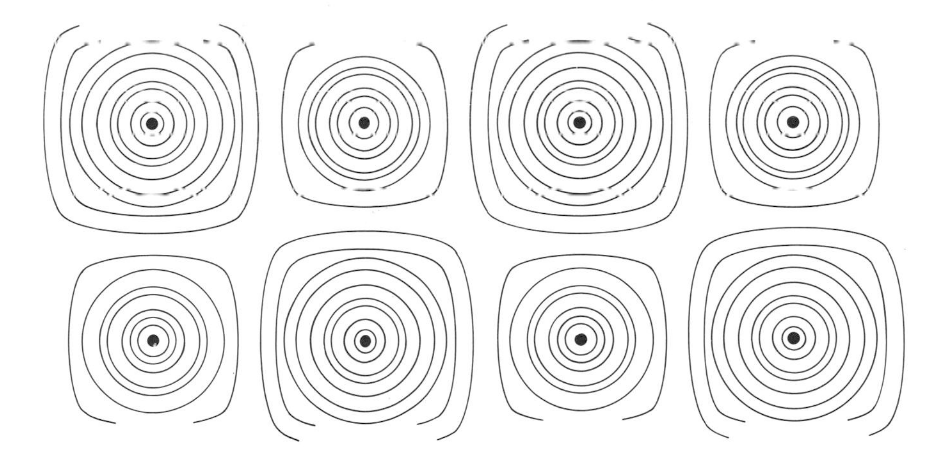

Figure 2.4 Electron density plot for sodium chloride: the Cl^- ions are larger than the Na^+ ions

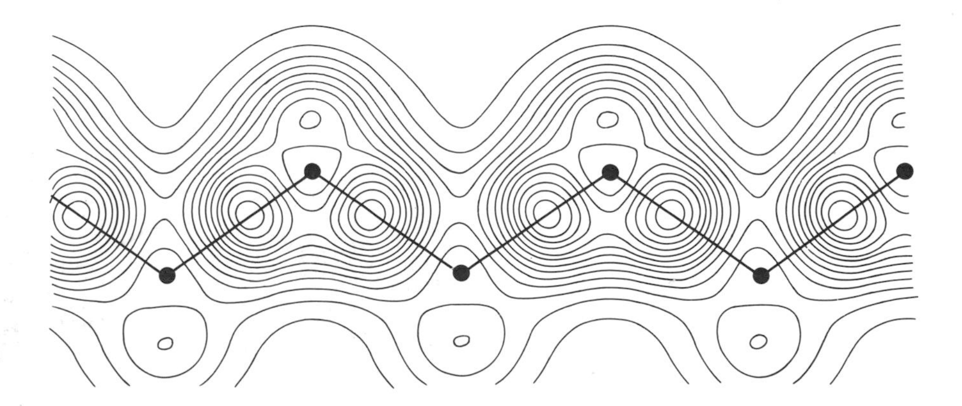

Figure 2.5 Electron density plot for gallium arsenide: the maximum electron density is a little closer to the arsenic atoms than to the gallium atoms (the dots indicate the location of the atomic nuclei, and the lines are the conventional representation of the covalent bonds)

Electron density plots show the difference between ionic and covalent compounds. Figure 2.4 shows that in sodium chloride the electron density between the ions is very small. On the other hand, Figure 2.5 shows that in gallium arsenide, GaAs – a predominantly covalent compound used in some semiconductor devices, including the lasers in compact disc players – the electron density between the atoms is significant. Detailed examination of the plots shows, however, that in sodium chloride there is *some* electron density between the ions, and that in gallium arsenide the electrons are not shared quite equally. That is, ionic and covalent bonding are two extreme models, and in all ionic compounds there is some electron sharing, and in all covalent compounds built from different elements there is some electron transfer. In elementary valence theory one or the other type of bonding is emphasized; in advanced valence theory the two approaches blend into a single description.

2.1 The ionic bond

Two gas-phase atoms A and B form a gas-phase **ion pair** held together by an ionic bond if the unit A^+B^- (or $A^{2+}B^{2-}$ etc) has a lower energy than the separated atoms. Ionic compounds are crystals composed of millions of ions, and all their interactions must be taken into account when assessing the energy changes. That is a somewhat complex problem, and is left until Sections 5.1 and 9.4; for the present we consider the energy changes accompanying the formation of a single ion pair.

The energy contributions to ionic bond formation

For convenience of discussion, ionic bond formation may be analysed into three steps, illustrated diagrammatically in Figure 2.6, and only if the *overall* energy change is favourable will the bond form. The first step is the removal of an electron from A to form A^+. This requires the investment of the ionization energy of A. The second step is the addition of the electron to B to form B^-. This results in an energy change equal to the electron-gain energy of B (both ionization energy and electron-gain energy are discussed in Section 1.2). When the electron-gain energy is negative (as for halogen atoms), the energy of B^- is lower than that of the separated B atom and the electron. When the electron-gain energy is positive (as for the addition of an electron to O^- to form O^{2-}), the energy change is positive and energy must be supplied to form the negative ion. The third step takes into account the attractive electrostatic interaction between the two oppositely charged ions formed in the first two steps.

The three steps suggest that the overall energy will be lowered if A has a low ionization energy (so that it is energetically easy to remove an electron), if B has a large negative electron-gain energy (so that energy is released when the negative ion is formed) or at least not too strongly positive a value, and if the ions are so close together that their mutual attraction overcomes the net energy needed to make the ions. Similar considerations apply in the discussion of ionic crystals, the only difference lying in the large numbers of ions that are interacting with each other.

Atoms with low ionization energies are found in Groups I and II: their **valence electrons** (the electrons in their outermost shells) can be removed with relatively little energy investment, and so they can be expected to take part in ionic bonding. Once these electrons have been lost, however, further ionization demands so much energy that it cannot be recovered from the increased attraction that the ion would have with negatively charged ions, so core electrons do not take part in bonding (of any kind, either ionic or covalent).

Atoms with negative (or small positive) electron-gain energies are those of Group VII (the halogens) and Group VI, because the added electron enters a gap in the valence shell where it is strongly attracted by the opposite charge of the nucleus. A closed-shell negative ion, such as a halide ion, has a strongly positive electron-gain energy because the additional electron must enter a new shell, where it can interact only weakly with the distant, shielded nucleus.

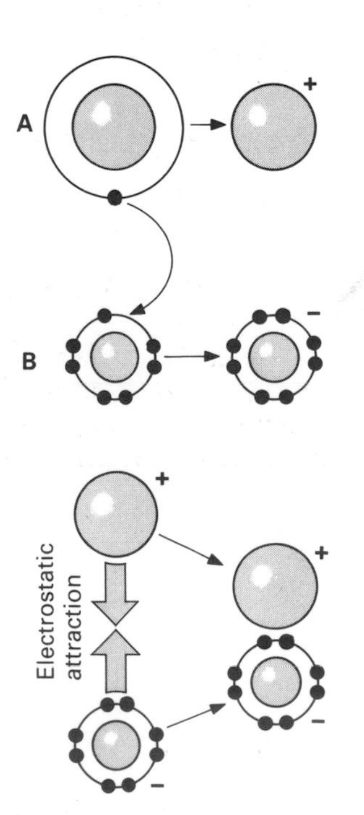

Figure 2.6 The formation of an ionic bond

The octet rule

The preceding discussion suggests that ionic bonds are likely to form between the atoms of elements on the left and elements on the right of the Periodic Table, and that electron transfer will continue until the atoms have lost or gained enough electrons to reach closed-shell configurations. This is normally summarized by the rule that

atoms have a tendency to attain a noble gas configuration.

This is the origin of the **octet rule**, formulated by G. N. Lewis in 1916, and expressed by him in the form

atoms have a tendency to acquire a stable octet of electrons.

In modern atomic theory, the 'stable octet' is the s^2p^6 valence shell configuration typical of the noble gas atoms. The word 'tendency' must not be misinterpreted: the energy of a gas-phase O^{2-} ion is much greater than that of a gas-phase O atom, and an O^{2-} ion exists only because the *overall* energy change, including the interactions in the solid compound, is favourable.

The octet rule should not be taken to mean that Group I and II atoms 'want' to lose electrons, or that Group VI and VII atoms 'want' to gain them. It means that the electrostatic interaction between ions is strong enough to overcome the combined demands of the ionization and electron-gain energies so long as only the valence electrons are transferred between the atoms. Going beyond closed shells – removing core electrons or adding electrons to closed shells – demands so much energy that it cannot normally be recovered from the stronger electrostatic interactions that would result, and so *overall* there is no release of energy.

Lewis structures

Lewis structures, which are sometimes called **electron dot diagrams**, summarize ionic bond formation simply, give a method of deducing the formula of an ionic compound and (as explained in Section 2.2 below) are particularly useful for discussing bonding in molecular compounds. To write the Lewis structure of an ionic substance, the valence electrons of each atom are shown as dots, and the gaps in the valence shell of one atom are completed by transferring valence electrons from the other. Sometimes more than one atom of each kind is needed in order to accommodate all the electrons without breaking the octet rule. This is illustrated in the following Example.

EXAMPLE

Write Lewis structures for the following compounds: (a) KBr, (b) K_2O, (c) $CaCl_2$, and (d) MgO.

METHOD All the substances are compounds of metals from Groups I or II with non-metals from Groups VI and VII, and so they are all ionic. Write the Lewis structures by transferring enough electrons to complete octets or to leave closed shells. Group I atoms can lose one electron and Group II atoms can lose two; Group VI atoms can gain two electrons and Group VII atoms can gain one. Show only the valence electrons in each case.

ANSWER

(a) $\mathrm{K}\cdot \quad \cdot\ddot{\underset{\cdot\cdot}{\mathrm{Br}}}: \quad \rightarrow \quad \mathrm{K}^{+}\,:\ddot{\underset{\cdot\cdot}{\mathrm{Br}}}:^{-}$

(b) $2\,\mathrm{K}\cdot \quad \cdot\ddot{\underset{\cdot\cdot}{\mathrm{O}}}\cdot \quad \rightarrow \quad 2\,\mathrm{K}^{+}\,:\ddot{\underset{\cdot\cdot}{\mathrm{O}}}:^{2-}$

(c) $\cdot\mathrm{Ca}\cdot \quad 2\,\cdot\ddot{\underset{\cdot\cdot}{\mathrm{Cl}}}: \quad \rightarrow \quad \mathrm{Ca}^{2+}\;2\,:\ddot{\underset{\cdot\cdot}{\mathrm{Cl}}}:^{-}$

(d) $\cdot\mathrm{Mg}\cdot \quad \cdot\ddot{\underset{\cdot\cdot}{\mathrm{O}}}\cdot \quad \rightarrow \quad \mathrm{Mg}^{2+}\,:\ddot{\underset{\cdot\cdot}{\mathrm{O}}}:^{2-}$

COMMENT Sometimes the electrons are distinguished according to their source using dots and crosses. This is misleading, because once the compound has formed it is impossible to say which electron came from which atom.

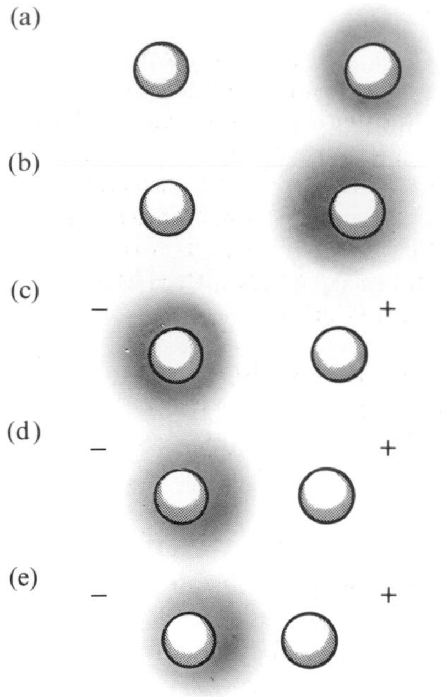

Although Lewis structures are primarily an accounting method, they do capture the principal features of ionic bond formation, the transfer of electrons and the electrostatic attraction between ions. They appear to treat only isolated ion pair formation, but in fact they summarize the essentials of ionic crystal formation where large numbers of ions, formed by octet completion, interact electrostatically.

That electron transfer does take place is confirmed experimentally by X-ray diffraction of ionic solids (see Section 5.2), and supported theoretically by solving the Schrödinger equation for a pair of atoms as they approach each other, as in Figure 2.7. This type of computation also emphasizes the point made earlier that no bond is purely ionic. Initially, at (a), the two atoms have their usual numbers of electrons and are neutral (and note that the Li atom is bigger than the F atom). As the atoms approach and reach (c) the three individual energy contributions discussed above favour electron transfer: the Li valence electron hops on to the F atom, leaving the small Li^+ ion nearby. As the Li^+ ion moves closer to the F^- ion it begins to distort the latter's electron distribution, and at (e) the Li^+ ion is beginning to recover a significant share in its lost electron. At this stage the ionic bond is starting to take on some of the character of a covalent bond.

Figure 2.7 Electron density changes during the formation of an ionic bond

2.2 The covalent bond

Lewis proposed that a covalent bond is formed when two atoms share a pair of electrons:

H· ·H forms H:H

H· ·C̤̈l: forms H:C̤̈l:

EXAMPLE

Write the Lewis structures for the following compounds: (a) H_2O, (b) CO_2 and (c) HCl.

METHOD All three compounds are covalent; therefore write Lewis structures with enough shared electron pairs to complete the octets of the atoms. It is helpful to begin by writing the atoms in the locations they have in the molecules (which in these simple cases can normally be anticipated).

ANSWER (a) H· ·Ö̤· ·H → H:Ö̤:H, corresponding to H—O—H

(b) :Ö̤ C̈ Ö̤: → :Ö̤:C:Ö̤:, corresponding to O=C=O

(c) H· ·C̤̈l: → H:C̤̈l:, corresponding to H—Cl

COMMENT Lewis structures do not in general show the spatial arrangement of the atoms, and H_2O, for example, is angular, not linear. Nor do they show the **polarization** of bonds, that is, the extent to which the bonding pair of electrons is unequally shared.

The simplest covalent molecule is the hydrogen molecule, H_2, which has only two electrons, one supplied by each atom, and the bond consists of an electron pair lying between the atoms. This simple picture is supported by detailed solution of the Schrödinger equation, for the computed electron density in H_2, shown in Figure 2.8, is high between the nuclei as well as close to them. That the shared electron pair lies mainly *between* the nuclei helps to explain the stability of the bond: the pair lie where they can attract both nuclei strongly, and hold them together as a kind of 'electrostatic glue'. This is the common feature of all covalent bonds, for the shared electrons lie between the nuclei, attract them both and bind the atoms together.

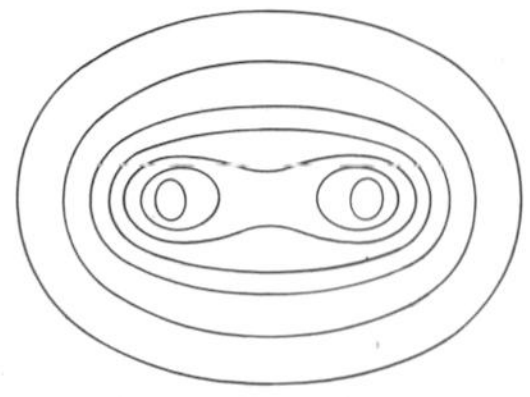

Figure 2.8 Electron density distribution in the H_2 molecule

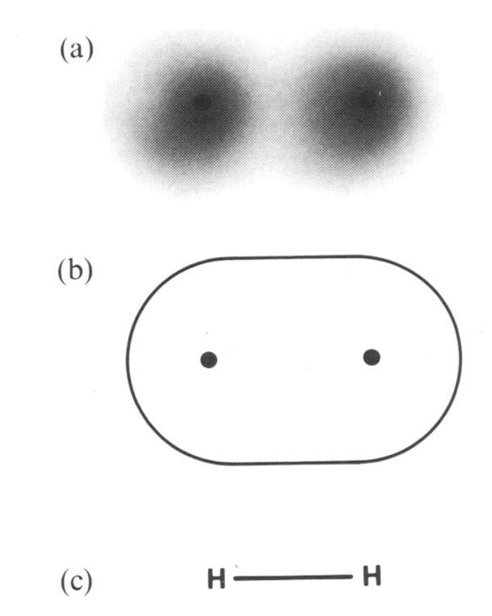

Figure 2.9 Three ways of drawing a covalent bond

The representation of covalent bonds

There are several ways of drawing a covalent bond between two atoms. The most detailed description is obtained by showing the computed electron density, often as a density of shading – the method used in Figure 2.9(a). This method is too detailed for most applications, and so only the general shape of the density distribution is indicated, as in Figure 2.9(b), where the sausage shape (which does not have a sharp boundary) suggests the region where the electron pair is most likely to be found. Even this representation is too detailed for most purposes, and so it is common merely to indicate which atoms are bonded by drawing a line between them (Figure 2.9(c)). In this simplest representation a hydrogen molecule is denoted H—H, the line denoting the shared pair of electrons. Always bear in mind, however, that a line between two atoms is only shorthand, and the *explanation* of the formation of the bond must be found in the energy changes that occur on rearrangement of the electron distributions of the original atoms.

Molecular potential energy curves

Covalent molecules are characterized by their **bond lengths** and (for polyatomic molecules) their **bond angles**. The bond length is the separation of the nuclei in the lowest energy state of the molecule, and the bond angle is the angle between the lines joining an atom to its two neighbours.

The bond length has a characteristic value because at very small separations the attraction between the electron pair and the nuclei is outweighed by the repulsion between the two nuclei. Moreover, the electron pair would eventually be squeezed out of the space between the nuclei, and its bonding effect would be reduced. This suggests that the energy of the molecule should depend on the distance between the nuclei as shown in Figure 2.10, which is called a **molecular potential energy curve**. This shows that as the atoms approach there is first a lowering of energy as the electron pair is shared and interacts favourably with both nuclei, but when the nuclei become very close the energy of the molecule rises. The minimum of the curve corresponds to the equilibrium bond length of the molecule. Modern computer techniques can calculate molecular potential energy curves very accurately, and can find the bond lengths corresponding to their minima. These are usually in good agreement with the bond lengths measured experimentally using the techniques described in Chapter 3.

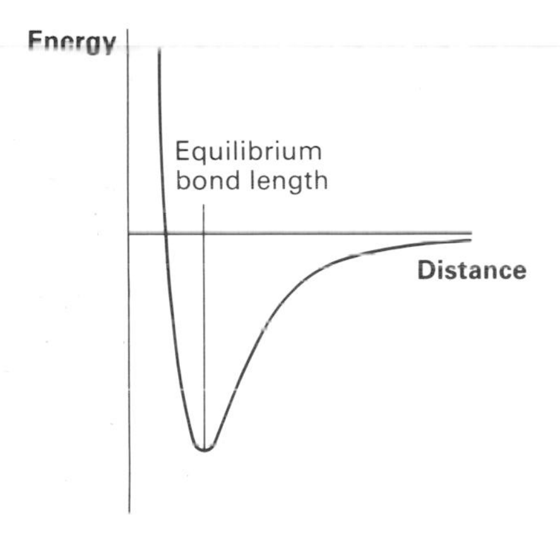

Figure 2.10 A molecular potential energy curve

It is very difficult to predict bond lengths from simple arguments, but surprisingly simple to predict the approximate bond angles of some molecules. This is described in the following section.

2.3 The shapes of polyatomic molecules

The shapes of polyatomic molecules are largely a result of the electrostatic repulsions between electron pairs in the valence shell. The theory of shapes built up by concentrating on the effects of these repulsions is called **valence shell electron pair repulsion theory**, or **VSEPR theory** for short. The theory was introduced by Sidgwick and Powell and refined by Gillespie and Nyholm, and is sometimes named after one or other of the partnerships.

VSEPR theory

The first stage in VSEPR theory is to use the molecule's Lewis structure to identify the total number of valence electron pairs that surround a given atom. (The following rules apply when a given pair of atoms share *one* electron pair: when atoms share more than one electron pair the rules are slightly modified, as explained in Section 2.4.)

Consider the example of methane (CH_4, **1**), where four electron pairs surround the central carbon atom. The electron pairs repel each other, and so move to positions where they are furthest apart; in the case of four electron

H
H:C:H
H

1

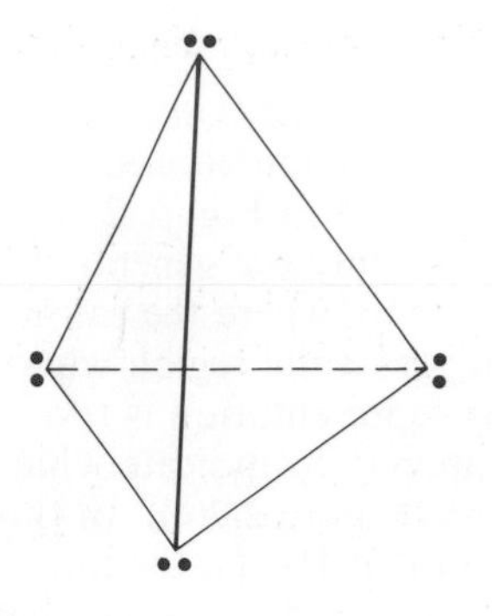

Figure 2.11 Four electron pairs repel each other into a tetrahedral shape

pairs, they move to the corners of a regular tetrahedron (see Figure 2.11). The arrangement adopted by the electron pairs is called the **basic shape** of the molecule. The basic shapes for all numbers of pairs from two to six are shown in Figure 2.12: each corresponds to an arrangement in which the electrostatic repulsion between the electron pairs is a minimum.

Once the basic shape has been decided, the **actual shape** of the molecule can be predicted.

When all the electron pairs are **bonding pairs**, and there is an identical atom attached to each one, then the actual shape is the same as the basic shape. For example, the structure of CH_4 is predicted (and found) to be tetrahedral (see Figure 2.13). When one or more of the electron pairs are **lone pairs** (unshared electron pairs belonging to only one atom), it is important to take into account their greater repulsive effect, which arises because lone pairs are more diffuse than bonding pairs are, and hence repel other electron pairs from a

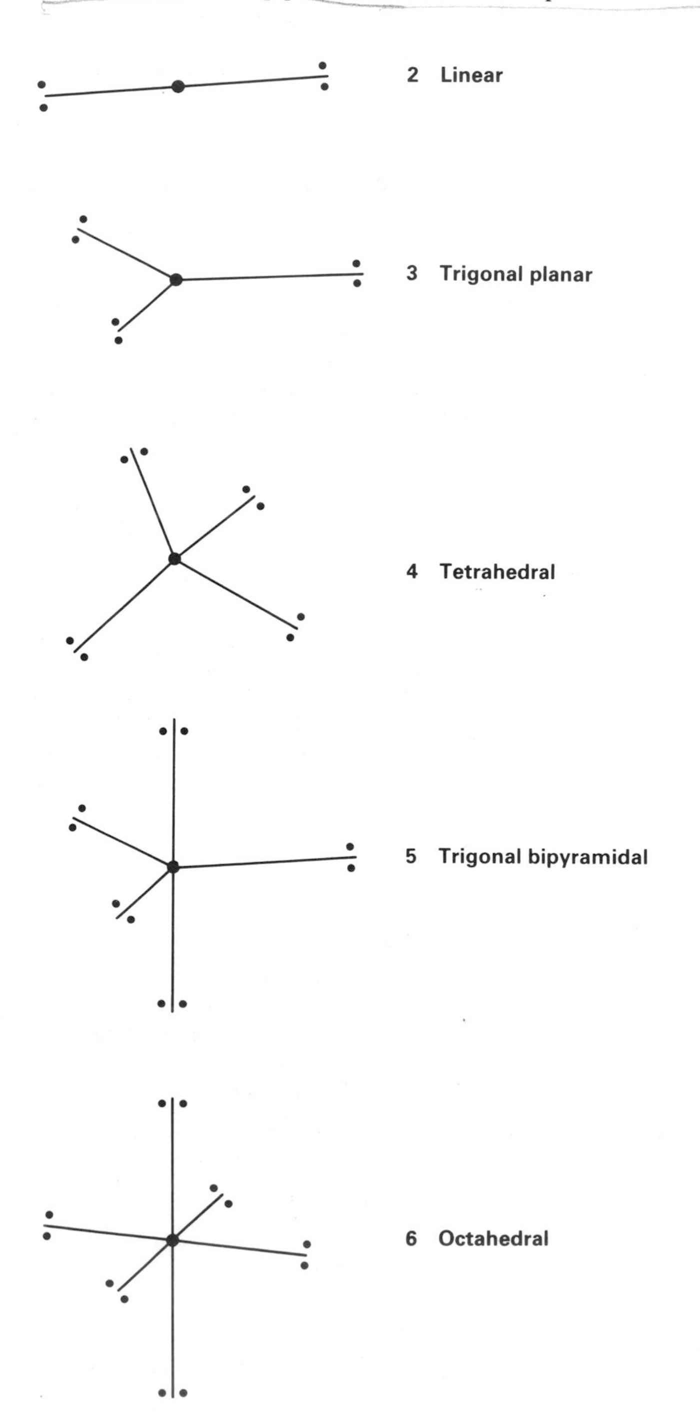

Figure 2.12 The energetically most favourable ways of arranging pairs of electrons

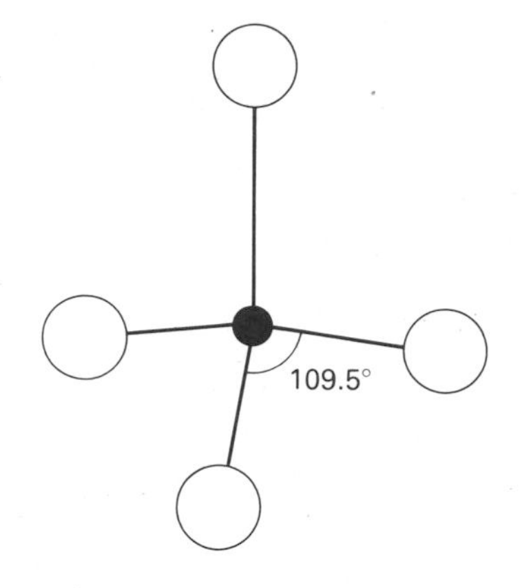

Figure 2.13 The structure of methane

bigger region of space. VSEPR theory allows for this by supposing that the order of repelling power is

(BP, BP) < (BP, LP) < (LP, LP)

where LP stands for lone pair and BP for bonding pair. As a result, bonding pairs tend to move as far apart as possible from lone pairs even though that might result in them moving slightly closer to other bonding pairs. The number of lone pairs is found as follows:

number of lone pairs = (number of electrons in valence shell − number of electrons in bonding pairs)/2

The shapes of NH_3 and H_2O

The VSEPR prediction for NH_3 can now be made. First, the number of electron pairs is counted from the Lewis structure:

H:N̤̈:H   4 electron pairs (1LP + 3BP)
  H

The basic shape is tetrahedral because there are four electron pairs (see Figure 2.12). One of the pairs is a lone pair, and the three bonding pairs move away from it, as shown in Figure 2.14. This brings the three N—H bonds closer to each other, like an umbrella closing, so that the angle between them is expected to be less than that in a regular tetrahedron (109.5°). The molecule is therefore predicted to be pyramidal; this is confirmed experimentally, and the angle is found to be 107.3°.

When a proton attaches to the lone pair of ammonia to form the ammonium ion, NH_4^+, all four electron pairs become bonding pairs, and the ion is predicted (and found) to be a regular tetrahedron. Note that CH_4 and NH_4^+ have the same number of electrons (ten) and hence are **isoelectronic**: it is often helpful to note that isoelectronic molecules and ions have closely similar and sometimes identical shapes.

Another important application of VSEPR theory is to the H_2O molecule. The Lewis structure is

 H
:Ö̤:H   4 electron pairs (2LP + 2BP)

The basic shape is therefore tetrahedral. The two lone pairs repel each other strongly, and move apart; the two bonding pairs escape from the lone pair repulsion by moving together slightly (see Figure 2.15). This predicts that the H—O—H angle will be slightly less than 109.5°. The experimental value of 104.5° is in agreement with this prediction, and the molecule is non-linear (or *angular*).

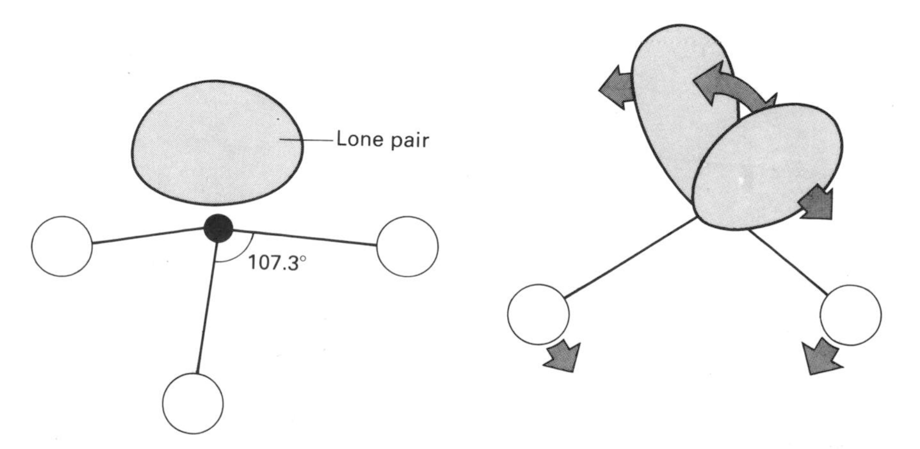

Figure 2.14 The structure of ammonia

Figure 2.15 The effect of lone pair repulsions in H_2O

EXAMPLE

Predict the shapes of the following molecules: (a) CCl_4, (b) BF_3, (c) SF_6 and (d) PH_3.

METHOD Consider only the valence electrons on the central atom. Write a Lewis structure for each molecule, count the number of electron pairs, and deduce the basic shape from Figure 2.12. Allow for the stronger (LP, LP) and (LP, BP) repulsions by allowing lone pairs to be furthest from each other and allowing bonding pairs to move away from lone pairs.

ANSWER (a) A Lewis structure for CCl_4 is

```
   Cl
   ..
Cl:C:Cl
   ..
   Cl
```

so that the central carbon has four bonding pairs. Hence, from Figure 2.12, its basic shape is tetrahedral. All four pairs are bonding pairs, so that the molecule is predicted to be a regular tetrahedron.

(b) A Lewis structure for BF_3 is

```
F:B:F
  ..
  F
```

so that the central boron atom has three electron pairs and its basic shape is therefore trigonal planar. All three pairs are bonding pairs, so that the molecule is predicted to be trigonal planar.

(c) A Lewis structure for SF_6 is

```
 F..F
F:S:F
 F''F
```

so that the central sulphur atom has six electron pairs and its basic shape is therefore octahedral. All six pairs are bonding pairs, so that the molecule is predicted to be a regular octahedron.

(d) A Lewis structure for PH_3 is

```
  ..
H:P:H
  ..
  H
```

so that the central phosphorus atom has four electron pairs and its basic shape is therefore tetrahedral. One of the pairs is a lone pair, and so the P—H bonds will move away from it, resulting in a pyramidal molecule.

COMMENT VSEPR theory is suitable for making *qualitative* predictions of bond angles, but for *quantitative* (numerical) predictions it is necessary to solve complicated equations on a computer. Very accurate predictions can now be made in this way.

2.4 Multiple bonding

In some cases more than one electron pair must be shared between a pair of atoms in order to complete their octets. A single shared pair of electrons is responsible for a **single bond** between the atoms it holds together. Two shared pairs give rise to a **double bond**, and three shared pairs to a **triple bond**. For example carbon dioxide (CO_2) has the following Lewis structure:

```
:O::C::O: , written O=C=O
```

ethene (C_2H_4) is

```
H     H             H       H
 .   .               \     /
  C::C   , written    C=C
 .   .               /     \
H     H             H       H
```

and ethyne (C_2H_2) is

H:C⋮C:H, written H—C≡C—H

The minimum energy required to break a bond is called the **dissociation energy** of the bond; the greater the dissociation energy, the stronger the bond. Double bonds are weaker than the sum of two single bonds because the two electron pairs cannot both be in the optimum bonding position, lying halfway between the two nuclei. Similarly, triple bonds are weaker than the sum of three single bonds: ethyne (old name: acetylene) reacts very vigorously with oxygen, and the oxyacetylene flame is the hottest flame readily achievable.

The effect of multiple bonding on shape

When multiple bonds are present the VSEPR rules are extended as follows:

1 *A multiple bond is treated as though it were a single electron pair.* The carbon atoms in both carbon dioxide and ethyne are treated as having *two* electron pairs, so that the basic shapes of the molecules are linear. Since neither molecule has any lone pairs of electrons on carbon, the actual shapes of the molecules are also predicted (and found) to be linear. A double bond is actually slightly more repulsive than a single bond, so that in methanal (**2**) the H—C—H angle is slightly less than 120°; the H—C—H angle in ethene is 117°.

2 H—C(=O)—H, 118°

2 *A double bond is rigid to twisting*, that is, a double bond is **torsionally rigid.** Each carbon atom in ethene is treated as having *three* pairs of electrons: the arrangement around each atom is trigonal planar (compare Figure 2.12), and they are linked together as shown in Figure 2.16(a). However, by the second rule, the two CH_2 groups are predicted to be held in the arrangement shown in Figure 2.16(b) (a perspective view of the molecule). This also agrees with experiment.

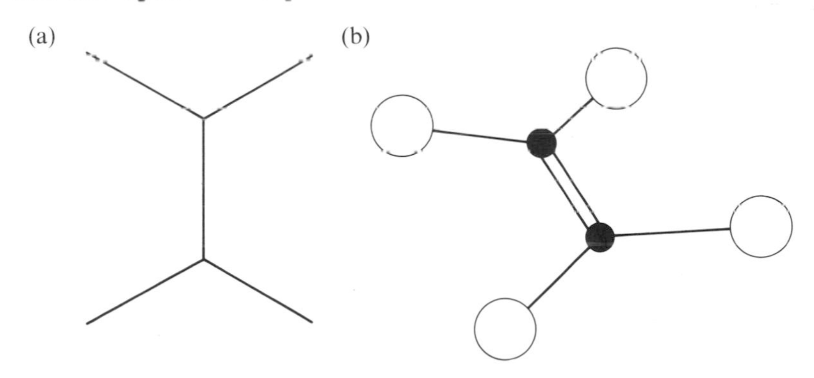

Figure 2.16 Ethene: (a) the σ-framework, (b) the double bond

2.5 Covalent bonds: the modern view

The modern description of bonds in molecules is in terms of **molecular orbitals**, the analogues of the atomic orbitals used for describing atoms, discussed in Section 1.2. Just as an atomic orbital gives the probability of finding an electron at each point in an atom, a molecular orbital (MO) gives the distribution throughout a molecule. Molecular orbitals are calculated by solving the Schrödinger equation for the molecule (using a computer), but it is not difficult to see pictorially how they arise and how they can be described.

σ-bonds

An MO can be thought of as being formed by the interference between atomic orbitals, just like the interference that occurs between any waves, such as ripples on a pond (see Figures 2.17 and 2.18): in places where their peaks coincide with each other the waves add to give a wave of greater amplitude (this is **constructive interference**), and in others, where their peaks coincide with their troughs, the amplitudes cancel (this is **destructive interference**).

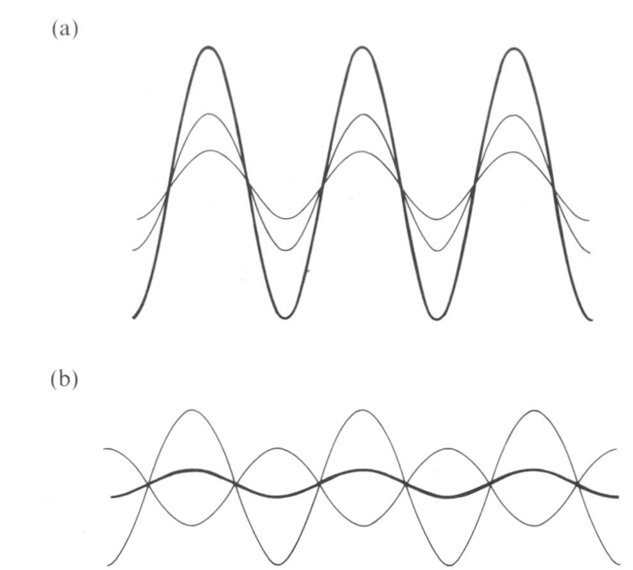

Figure 2.17 (a) Constructive and (b) destructive interference of waves

Figure 2.18 Interference of water waves on a pond: the peaks of the waves are higher where the ripples coincide

Consider two hydrogen atoms approaching each other. Their 1s-orbitals **overlap** when the nuclei are close enough, and if they interfere constructively (see Figure 2.19) the wave amplitude is increased, that is, the chance of finding the electron between the nuclei is increased. This interference gives rise to a

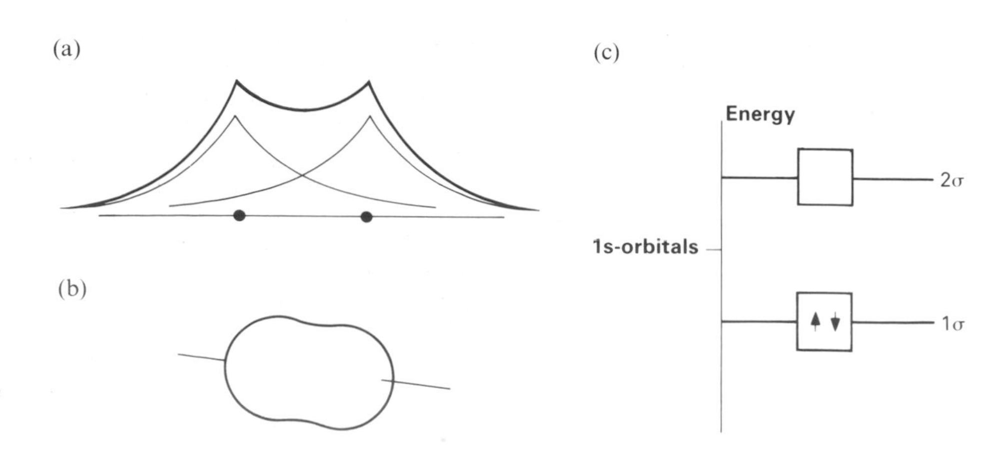

Figure 2.19 (a) Constructive overlap of atomic orbitals leads to (b) a molecular orbital (c) lying lower in energy than the atomic orbitals

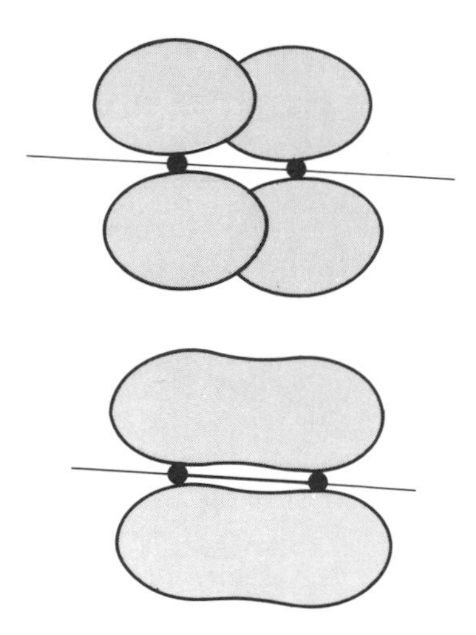

Figure 2.20 Two p-orbitals overlap to give a π-orbital

bonding molecular orbital which is cylindrically symmetrical around the internuclear axis, and which is called a **σ-orbital** (sigma-orbital), the Greek letter σ echoing the 's' in the name of an atomic s-orbital. Two electrons can occupy a σ-orbital (the maximum allowed by the Pauli principle for any orbital), and when they do the atoms are held together by a **σ-bond**. As the energy of the molecule is lower when two electrons occupy a σ-orbital than when they are on separate atoms, energy is released on bond formation.

π-bonds

When two atoms are close together, their p-orbitals may overlap broadside on (see Figure 2.20) and give rise to two regions of increased electron density above and below the internuclear axis. The resulting orbital is called a **π-orbital** (pi-orbital), the Greek letter π echoing the 'p' of atomic p-orbitals (because the π-orbital looks like a p-orbital when viewed along the axis). If two electrons occupy this orbital, then the two atoms are drawn together more strongly than by the electrons in the σ-bond alone, and the atoms are joined by a double bond. Note that a double bond consists of a σ-bond plus a π-bond – two electrons in a σ-orbital, and two in a π-orbital.

There are three points that should be noted about a π-orbital and its contribution to a double bond.

First, although the π-orbital has *two* regions of enhanced electron density, one above and one below the plane containing the σ-orbital, it is a *single* orbital, and can hold up to two electrons.

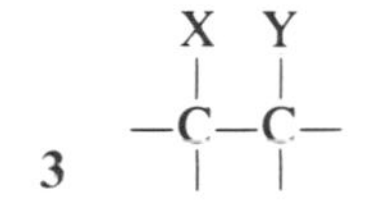

Second, because the regions of enhanced electron density are not directly between the nuclei, a π-bond is not as strong as a σ-bond, which means that a double bond is less strong than two single σ-bonds. This characteristic has important consequences in organic chemistry, because it means that C═C bonds are reactive (see Section 33.3) and when attacked by a reagent X—Y are likely to form compounds such as **3**, where all the bonds are single.

Third, a double bond is resistant to twisting because as one end of the molecule is rotated relative to the other, the broadside overlap of the two p-orbitals of the π-orbital is decreased, the bond is weakened and the energy of the molecule rises.

The formation of a π-bond by broadside overlap also accounts for the importance of multiple bonding between Period 2 elements (such as C═C, C═O and N═O) compared with later Periods. In order for a π-orbital to form, the two atoms must come close enough together for their p-orbitals to overlap significantly. This is possible for the small atoms of Period 2, but the atoms of later Periods are simply too fat.

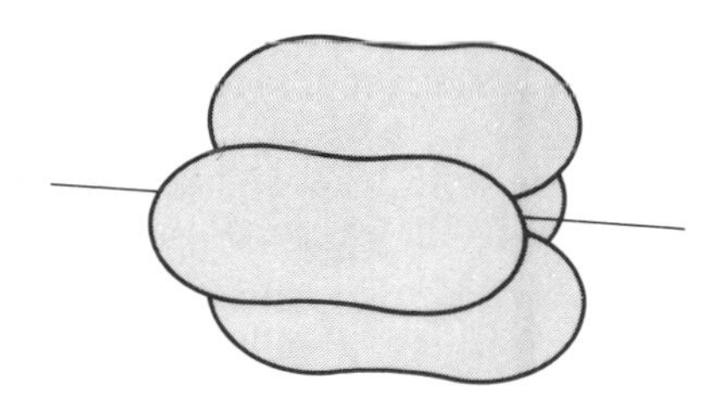

Figure 2.21 Two π-orbitals

A triple bond is like a double bond, but is composed of a σ-bond and two π-bonds formed by the broadside overlap of two pairs of p-orbitals (see Figure 2.21). The σ-orbital contains two electrons, and each of the π-orbitals contains two, so a triple bond consists of three shared pairs of electrons, as in the simple Lewis picture.

Delocalized bonding

The MO description of benzene is a very natural extension of the MO description of ethene. The essential feature is shown in Figure 2.22(a): a **σ-framework** of σ-bonds joining each carbon to its two neighbours and to one hydrogen atom. This arrangement leaves one 2p-orbital unused on each carbon atom, lying perpendicular to the hexagonal ring of atoms (see Figure 2.22(b)). Each 2p-orbital overlaps *both* its two neighbours broadside on, and forms π-orbitals that extend round the ring (Figure 2.22(c)); since the π-orbitals are not confined between two atoms, they are called **delocalized orbitals**.

The overlap of the six carbon 2p-orbitals thus leads to six delocalized π-orbitals, and the orbital of lowest energy binds the six atoms together very strongly. There are six electrons to accommodate, one provided by each carbon atom (the other three valence electrons of each carbon are used for two C—C σ-bonds and one C—H σ-bond). Two electrons enter the lowest-

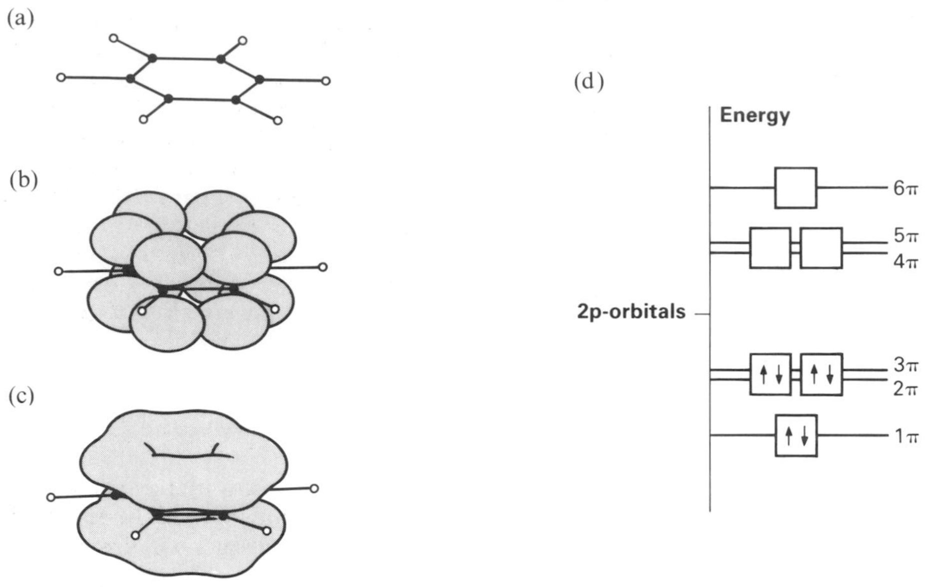

Figure 2.22 (a) The σ-framework of benzene leaves (b) six p-orbitals which overlap (c) to form π-orbitals, (d) three of which are lower in energy than the atomic orbitals

energy delocalized π-orbital and the remaining four enter the next two orbitals (Figure 2.22(d)). That is, *all six electrons occupy delocalized bonding orbitals*, a feature that is the major contribution to benzene's stability.

Although the electronic structure of benzene is normally represented as in Figure 2.22(c), this picture should be interpreted with care. First, the bands of π-electron density represent six electrons in *three* molecular orbitals that spread round the ring. Second, each π-orbital corresponds to a region of enhanced electron density both above and below the ring.

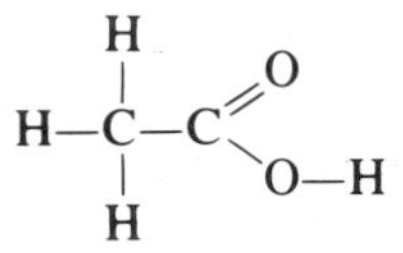

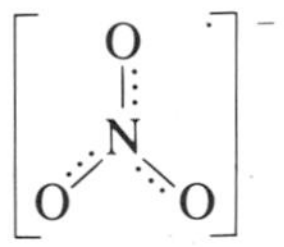

Delocalized orbitals occur in many substances, and are responsible for the stabilization of many molecules and ions. They are, for example, the reason why ethanoic acid (**4**) is a much stronger acid than ethanol (**5**): the ethanoate ion (**6**) formed when ethanoic acid loses a proton has a delocalized orbital spreading over both oxygen atoms, and the negative charge of the ion is spread equally over these two atoms. On the other hand, in the ion left when ethanol loses a proton (**7**) the charge is *localized* on one oxygen atom. The energy of the ion **6** is lower than that of the ion **7**, and so ethanoic acid has a greater tendency to lose its proton and hence to behave as an acid (see Section 12.2).

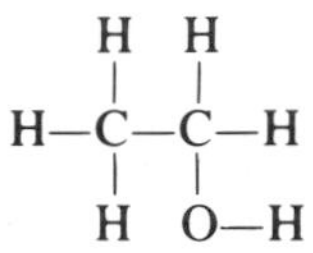

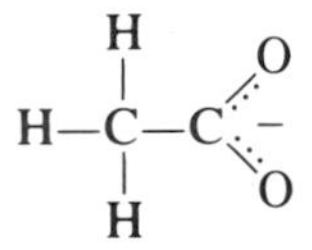

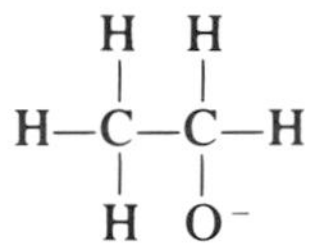

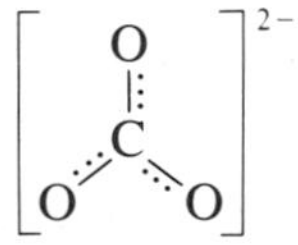

Delocalization also occurs in inorganic compounds, and has a similar stabilizing effect. An example is the stability it brings to the isoelectronic carbonate ion (CO_3^{2-}, **8**) and nitrate ion (NO_3^-, **9**). The electronic structure of both these anions involves delocalized π-orbitals constructed from a p-orbital on the central atom and p-orbitals on the surrounding oxygen atoms (see Figure 2.23).

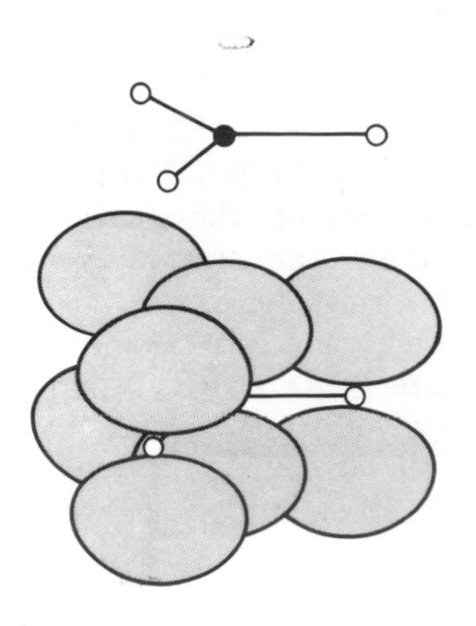

Figure 2.23 Overlap of p-orbitals in carbonate or nitrate ions

Extreme delocalization: metallic bonding

Whereas in individual molecules and ions delocalization extends over only a few atoms, in metals delocalization extends over as many atoms as there are in the sample.

Bonding in Group I metals, such as sodium, can be pictured as follows (the same principles apply to other metals, but there are sometimes complications stemming from the presence of p- and d- as well as s-orbitals). The s-orbital on one atom overlaps with those on all its immediate neighbours (see Figure 2.24), and those in turn overlap with each of theirs. This continues right up to the boundaries of the sample, and the molecular orbitals so formed extend throughout the solid. From N atomic orbitals N molecular orbitals can be formed, and their energies lie so close together that instead of an energy level diagram like benzene's (shown in Figure 2.22(d)) with well-separated energies, the MO energy levels form a virtually continuous band. This gives rise to the alternative name **band theory** for the MO theory of metals.

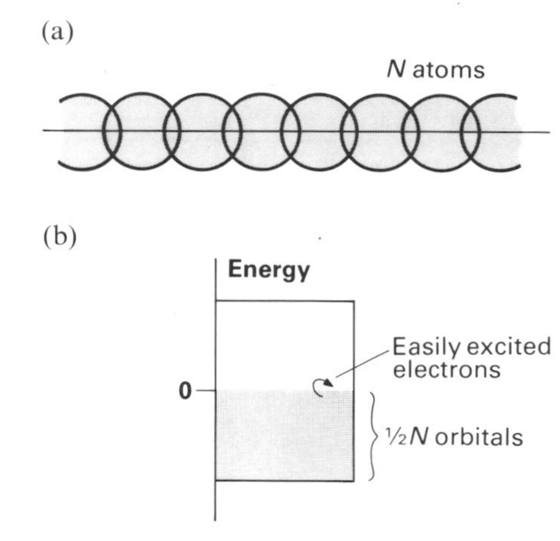

Figure 2.24 Overlap of (a) N atomic orbitals gives (b) N molecular orbitals, half of which are occupied

Although the energies lie in a virtually continuous band, the band is in fact composed of individual orbitals, and each one can accommodate up to two electrons. When each of the N atoms supplies one electron (and in the process becomes a positive ion), the building-up principle results in the lower $\frac{1}{2}N$ orbitals being fully occupied and the ones above empty (see Figure 2.24(b)).

The ease with which electrons can move through a metal (and hence carry an electric current) is easily explained on this model. The uppermost filled orbitals have almost the same energy as the unoccupied orbitals immediately above them, and the electrons at the top of the band can be moved into the empty orbitals when a voltage is applied across the sample. That is, the uppermost electrons are **mobile**. This mobility also accounts for the ductility, malleability and reflectivity of metals (see also Section 5.1).

2.6 Other features of bonding

The outstanding characteristic of covalent bonding is that since there are specific locations for the atoms at which they share the electrons most effectively, the bonds are **directional**. As a result, a molecule has a definite shape, with definite bond lengths and bond angles. Furthermore, once all the electron pairs have been shared there is no possibility of forming more covalent bonds, so that covalent compounds are generally well-defined, discrete (but not necessarily small) groups of atoms.

In contrast, ionic bonds do not have definite directional characteristics because the electrostatic interaction is the same in all directions. Moreover, a single ion may attract more than one ion of opposite charge and extensive ionic structures may form (see Section 5.1).

Coordinate bonds

A **coordinate bond** (which is also called a **coordinate–covalent bond** or a **dative covalent bond**) is formed when one atom contributes both electrons to a covalent bond. An example is the bond between ammonia (NH_3) and boron trifluoride (BF_3) leading to the formation of the **addition compound** $H_3N{\rightarrow}BF_3$:

```
  H      F              H F
H:N:     B:F   ──→    H:N:B:F
  H      F              H F
```

There is no fundamental difference between a coordinate bond and any other covalent bond, and the reason for its stability is the same: the *only* difference is the origin of the two electrons.

Coordinate bonds are responsible for the formation of **complex ions**, such as $[Fe(H_2O)_6]^{2+}$, in which each H_2O molecule uses one of its lone pairs to form a coordinate bond to the central metal ion (**10**). The water molecules are called the **ligands** of the metal ion. Compounds containing coordinate bonds are discussed in Section 25.7.

10 Fe^{2+} (:) O(H)(H)

Polar covalent bonds

Covalent bonds may be **polar** or **non-polar**. In a non-polar bond, there is no net redistribution of charge when electrons are shared, and the atoms remain electrically neutral. The bonds in all homonuclear diatomic molecules are non-polar, and each C—C bond in benzene is non-polar.

A bond is polar if its formation leaves partial charges on the two atoms. The hydrogen chloride molecule, for example, has a polar covalent bond because the shared electron pair is closer to the chlorine atom, resulting in a molecule that may be represented as in **11**. A similar charge redistribution occurs in the water molecule (**12**). The six C—H bonds in benzene are all slightly polar.

11 $\overset{\delta+}{H}$—$\overset{\delta-}{Cl}$

12 $\overset{2\delta-}{O}$ with $\overset{\delta+}{H}$ and $\overset{\delta+}{H}$

Electronegativity

The **electronegativity** (symbol: χ, chi) of an element is a measure of its ability to attract electron pairs in a covalent compound: the higher its value, the greater the element's attracting power. Some electronegativity values are given in Table 2.1, and their variation across the Periodic Table is illustrated in Figure 2.25.

Table 2.1 Pauling electronegativities, χ

H 2.20						
Li 0.98	Be 1.57	B 2.04	C 2.55	N 3.04	O 3.44	F 3.98
Na 0.93	Mg 1.31	Al 1.61	Si 1.90	P 2.19	S 2.58	Cl 3.16
K 0.82	Ca 1.00					Br 2.96
Rb 0.82	Sr 0.95					I 2.66

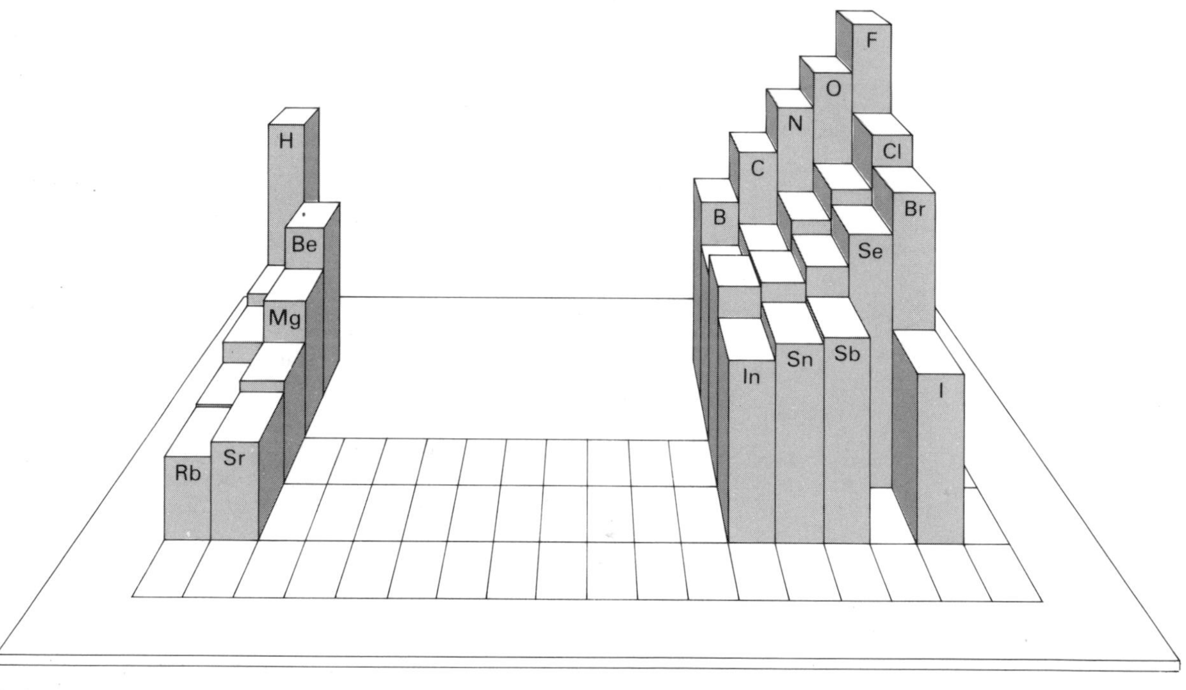

Figure 2.25 The variation of electronegativities through the Periodic Table

An atom with a high ionization energy (which makes its electrons difficult to remove) is unlikely to lose electrons to another atom in a molecule; an atom with a strongly negative electron-gain energy (signifying that energy is released when electrons attach to it) is likely to attract electrons from another atom. Therefore, elements with high ionization energies and negative electron-gain energies have high electronegativities. The **Mulliken scale** of electronegativity recognizes this connection by defining the electronegativity of an element as proportional to the difference between its first ionization energy and its electron-gain energy. The **Pauling scale**, which is used in Table 2.1, is defined in a more complex way but results in similar values.

The variation of ionization energy and electron-gain energy through the Periodic Table was described in Section 1.2, and accounts for the variation illustrated in Figure 2.25, with high electronegativities at the top right of the Table and low values at the lower left.

The electronegativity of an element can be used to predict the degree of polarity of the bonds it forms. Electrons tend to accumulate near the atom

13 $\overset{\delta-}{C}-\overset{\delta+}{H}$

14 $\overset{\delta-}{O}-\overset{\delta+}{H}$

with the greater electronegativity, so that such atoms tend to acquire a partial negative charge. For example, since $\chi(C) = 2.55$ and $\chi(H) = 2.20$, a C—H bond is very slightly polar with the partial negative charge on the carbon atom (**13**). The electronegativity difference is greater in an O—H bond, and so it is more strongly polar (**14**).

If the electronegativity difference between two atoms is very large (as in potassium fluoride), one atom acquires the bonding electrons almost completely, and the bonding is better described as ionic. This emphasizes, yet again, that pure covalent and pure ionic bonding are only convenient starting points for the description of a compound, and that in real compounds the bonding is intermediate: in some cases (HCl, SO_2) the electronegativity difference is small enough for the bonding to be closer to covalent, while in others (NaCl, MgO) the electronegativity difference is so large that the ionic model is more appropriate.

Another application of electronegativities, the calculation of oxidation numbers, is described in Section 15.2.

Dipole moments and polar molecules

A **polar molecule** is one with a net **dipole moment** (symbol: μ, mu). A dipole moment $\mu = qR$ arises when electric charges $+q$ and $-q$ are separated by a distance R (see Figure 2.26). For example, an isolated gas-phase ion pair Na^+Cl^- with charges $+e$ and $-e$ separated by 251 pm (the distance deduced from spectroscopy) corresponds to a dipole moment given by

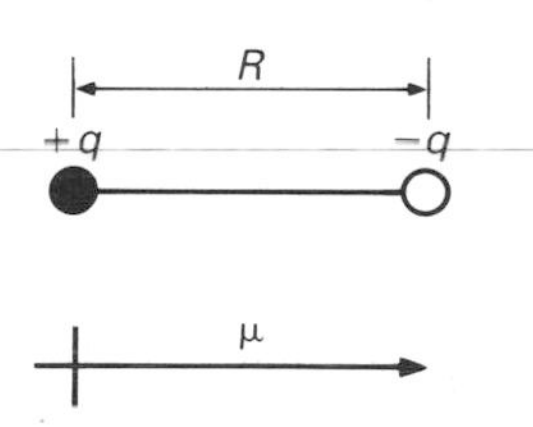

Figure 2.26 The structure of an electric dipole of magnitude $\mu = qR$

$$\mu = 1.602 \times 10^{-19}\,\text{C} \times 251 \times 10^{-12}\,\text{m} = 4.02 \times 10^{-29}\,\text{C m}$$

Dipole moments are normally expressed in debyes (symbol: D), a unit named after Peter Debye, a pioneer in the study of dipole moments:

$$1\,\text{D} = 3.336 \times 10^{-30}\,\text{C m}$$

In the present example, the dipole moment is 12.1 D. This is a huge value compared with those of most molecules (around 1 D) because Na^+Cl^- is an ion pair with full charges on each ion, whereas in HCl, for example, the charges are only partial.

The dipole moment of any homonuclear diatomic molecule is zero because the partial charges on the atoms are zero. The partial charges on a heteronuclear diatomic molecule are proportional to the electronegativity difference of the two atoms. A *very approximate* relation is

$$\mu = [\chi(A) - \chi(B)]\,\text{D}$$

EXAMPLE

Estimate the dipole moments of hydrogen chloride and hydrogen bromide.

METHOD Refer to the electronegativities in Table 2.1, and substitute into the equation above. The negative end of the dipole lies at the more electronegative atom.

ANSWER From Table 2.1 the dipole moments are

$$\mu(HCl) = (3.16 - 2.20)\,\text{D} = 0.96\,\text{D}$$
$$\mu(HBr) = (2.96 - 2.20)\,\text{D} = 0.76\,\text{D}$$

In each case the halogen is the more electronegative atom, and corresponds to the negative end of the dipole.

COMMENT The experimental values are $\mu(HCl) = 1.08$ D and $\mu(HBr) =$ 0.82 D. The greater the electronegativity difference in a diatomic molecule, the greater the dipole moment.

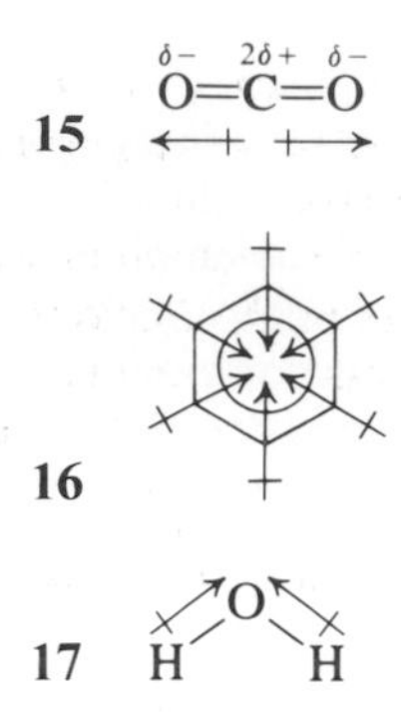

It is important to distinguish between the polarity of a *bond* and the polarity of a *molecule*. A molecule cannot be polar unless it has at least one polar bond, but a molecule with polar bonds need not be polar. This is because the dipole moments of each bond may be arranged so symmetrically that they cancel. The two C=O bonds in carbon dioxide are polar, but they point in opposite directions (**15**) and *overall* the molecule is non-polar. The six C—H bonds in benzene are polar, but opposite pairs cancel (**16**) and *overall* the molecule is non-polar and has zero dipole moment. The four C—H bonds in methane are polar, but their symmetrical, tetrahedral arrangement leads to their cancellation, and the CH_4 molecule is non-polar. The water molecule is non-linear, and the polarities of the two O—H bonds do not cancel (**17**); H_2O is a polar molecule (with $\mu = 1.84\,\mathrm{D}$).

A **polar solvent** (one composed of polar molecules, such as water) can interact electrostatically with ions and compensate, at least partially, for the loss of the ion–ion interactions in the solid when it dissolves (see Section 8.1). A **non-polar solvent** (one made up of non-polar molecules, such as benzene) is not able to do so. Thus water can dissolve many ionic solids, but benzene cannot.

One test of whether or not a liquid is polar is to run a stream of the liquid vertically past a charged rod: if the molecules are polar, the stream is deflected.

Summary

- ☐ A homonuclear diatomic molecule consists of two identical atoms, a heteronuclear diatomic molecule consists of two different atoms, and a polyatomic molecule is one that consists of more than two atoms.
- ☐ Bond formation occurs if it leads to a lower-energy arrangement of electrons and nuclei.
- ☐ On a simple view, bonds are either ionic (when there is electron transfer) or covalent (when pairs of electrons are shared); in reality, almost all bonds are intermediate between these extremes.
- ☐ An ionic bond is likely to form when one atom has a low ionization energy and the other a strongly negative electron-gain energy.
- ☐ The tendency of atoms to form bonds ends once they have reached a noble gas closed-shell electron configuration (a 'complete octet').
- ☐ The strength of a covalent bond (its dissociation energy) is due to the accumulation of electrons between the two nuclei.
- ☐ In valence shell electron pair repulsion theory (VSEPR theory), the basic shape is the one in which all electron pairs have minimum mutual repulsion (see Figure 2.12).
- ☐ The actual shape of a molecule in VSEPR theory is predicted using the principle that repulsions are in the order (BP, BP) < (BP, LP) < (LP, LP); multiple bonds are treated as effectively single bonding pairs.
- ☐ A molecular orbital describes the distribution of the electrons in the molecule; it can be regarded as being formed by the overlap and constructive interference of atomic orbitals.
- ☐ One molecular orbital can hold up to two electrons, with paired spins.
- ☐ A σ-orbital is a cylindrical molecular orbital; a π-orbital is an orbital with two lobes, one on either side of the internuclear axis.
- ☐ A single bond is one shared pair of electrons, a double bond is two shared pairs, and a triple bond is three shared pairs.
- ☐ In MO terms, a single bond consists of two electrons in a σ-orbital (which is called a σ-bond), a double bond consists of a σ-bond and a π-bond, and a triple bond of a σ-bond and two π-bonds.
- ☐ A π-bond is torsionally rigid, but not as strong as a σ-bond.
- ☐ A double bond is weaker than two single bonds, and a triple bond is weaker than three single bonds.
- ☐ Electron delocalization lowers the energy of a molecule, and contributes to the stability of the benzene molecule.

- ☐ Ionic bonding is non-directional and leads to the formation of extensive aggregates.
- ☐ Covalent bonding normally leads to well-defined, individual molecules with specific shapes.
- ☐ A polar bond is a covalent bond in which the electron pair is shared unequally; a molecule is polar if it has polar bonds in an arrangement that does not lead to a cancellation of their dipole moments.
- ☐ The electronegativity of an element is a measure of the electron-attracting power of an atom in a molecule; atoms near fluorine in the Periodic Table have high electronegativities and form the negative end of dipoles.

PROBLEMS

1 Write an essay on the modern theory of chemical bonding. Include an account of ionic and covalent bonding, the use and justification of Lewis structures, and the role of electron-pair repulsions.

2 List the factors that favour the formation of ionic compounds.

3 Account for the tendency of atoms to lose or gain electrons until they have reached a closed-shell configuration.

4 Account for the bonding in (a) H_2, (b) N_2, (c) O_2, (d) Cl_2 in terms of Lewis structures and electron distributions.

5 Use Lewis structures to account for the bonding in the following compounds: (a) KF, (b) CaO, (c) H_2S, (d) PCl_3.

6 On the basis of electron-pair repulsion (VSEPR) theory, predict the shapes of the following molecules: (a) H_2S, (b) SO_3, (c) PCl_3, (d) PCl_5, (e) ClF_3.

7 Explain the nature of the bonding in (a) C_2H_4, (b) C_6H_6, (c) NO_3^-, (d) CH_3COOH, (e) $CH_3CO_2^-$.

8 The polymer PVC, polyvinyl chloride, is manufactured from the monomer $H_2C{=}CHCl$, chloroethene ('vinyl chloride'). Describe the bonding in the monomer, and predict its shape.

9 Ethyne ($HC{\equiv}CH$, 'acetylene') burns with a very hot flame and is used in oxyacetylene welding. Comment on the nature of the bonds in the molecule and account for the large quantity of energy evolved when it is burnt in oxygen.

10 Explain why the electronegativities of atoms affect the extent to which bonds are polar. State which member of the following pairs of molecules has the larger dipole moment: (a) HF, HCl; (b) CO_2, CS_2; (c) NH_3, PH_3.

11 State which of the following molecules have dipole moments: (a) CO_2, (b) NO_2, (c) HBr, (d) PCl_3, (e) BF_3, (f) C_6H_6, (g) $C_6H_5CH_3$. Suggest directions for the dipoles.

12 The balance between the abundances of metal ions is essential to the proper functioning of the nervous system, and their transfer through cell membranes depends on their sizes. Account for the fact that the order of sizes is: $Mg^{2+} < Ca^{2+} < Na^+ < K^+$. (We develop this point in Chapter 8.)

13 (a) The atomic radii of sodium and magnesium are 0.157 nm and 0.136 nm, respectively. Why is the magnesium atom smaller than the sodium atom?
(b) The *ionic* radius of sodium is 0.095 nm. The *atomic* radius of neon is 0.160 nm. Comment on the difference between these figures. (*Welsh part question*)

14 Describe the shape of each of the following molecules and account for each shape by considering the electronic repulsions within each molecule:

$SiCl_4$ PCl_3 PF_5 H_2S (*JMB part question*)

15 (a) What is understood by the term *electronegativity* as applied to an element?
(b) Say how the electronegativities of the elements change
(i) as Group I (alkali metals) is descended;
(ii) as Group VII (halogens) is descended;
(iii) as Period 2 (lithium to fluorine) is crossed.
Comment on your answers to (i) and (ii) in the light of the electronic configuration of these elements.
(c) Why do the transitional elements, of any one series, have similar electronegativity values?
(d) In what way is the *type* of chemical bond formed between two atoms related to their electronegativities? Having made a general statement, illustrate your answer by referring to (i) chlorine, (ii) magnesium chloride, (iii) hydrogen chloride. Draw a simple bond diagram in each case.
(e) (i) Using bond diagrams for water and hydrogen chloride, show how hydrogen chloride gas becomes ionized when dissolved in water.

(ii) $ICl + H_2O \rightarrow IOH + HCl$
$ICl + H_2O \rightarrow ClOH + HI$

Which of the two equations above is more likely to represent the hydrolysis of iodine(I) chloride? Give reasons for your answer. (*SUJB*)

16 (a) The bond lengths (in nanometres) and also the dipole moments (in debye units) of the gaseous hydrogen halides are shown below.

	HF	HCl	HBr	HI
Bond length	0.092	0.127	0.141	0.161
Dipole moment	1.91	1.05	0.80	0.42

(i) What is meant by 'dipole moment'?
(ii) Comment briefly on the reasons for the decrease in dipole moment from HF to HI.

(b) (i) State whether the following molecules would have a dipole moment or not, giving reasons.

CH_3Cl CCl_4 BCl_3 NH_3

(ii) CO_2 has no dipole moment, while SO_2 has quite a large one. What difference in structure does this suggest?

(c) Many of the differences between water and hydrogen sulphide are ascribed to the fact that the former has hydrogen bonds but the latter has not.
(i) Explain what is meant by 'hydrogen bonds'.
(ii) List three differences between water and hydrogen sulphide which can be ascribed to hydrogen bonding in the former. (*London*)

17 Discuss how the properties of *four* of the following substances reflect their structure and bonding:

anhydrous aluminium chloride, benzene, caesium chloride, copper(II) sulphate pentahydrate crystals, ice, quartz, poly(propene) (polypropylene), silk. (*Nuffield*)

18 (a) Give the number of bonding pairs and the number of lone pairs of electrons in CH_4, NH_3, H_2O and AlH_4^-.
(b) For the hydrides methane, ammonia and water, state how the bond angle varies, giving a reason for this variation.
(c) Describe the type of bonding and structure present in each of the following solids in their usual state at room temperature: (i) iodine, (ii) sulphur, (iii) graphite, (iv) diamond. (*AEB 1985*)

19 Show how knowledge of the spatial distribution of electrons in atoms may be used to explain the bonding in the following molecules:

N_2, O_2, NO, BF_3, SF_4, CO_2. (*Oxford Entrance*)

3 The determination of molecular structure

In this chapter we see how mass spectrometry is used to identify molecules, to measure their molar masses and to determine their molecular formulae. We see how spectroscopy can be used to measure bond lengths and bond angles, and to identify molecules. We also see how nuclear magnetic resonance and X-ray diffraction, two of the most useful techniques of modern chemistry, are used to determine molecular structures.

Modern techniques can reveal the arrangement of atoms in molecules as complicated as the ones that control biological functions, such as proteins, enzymes and nucleic acids. Molecular biology could not have developed without the spectroscopic and diffraction methods we describe here. The development of new materials, such as semiconductors, synthetic polymers and fibres, and the improvement of constructional materials like cement and concrete, would be excessively slow if these techniques were not available, and a very hit-or-miss affair. They are also among the most important (and sensitive) techniques available for chemical analysis. They can be used both to study complex material in accessible places, such as proteins, and to study simple material in inaccessible places, such as the small molecules present in interstellar space and planetary atmospheres (see Figure 3.1, for example).

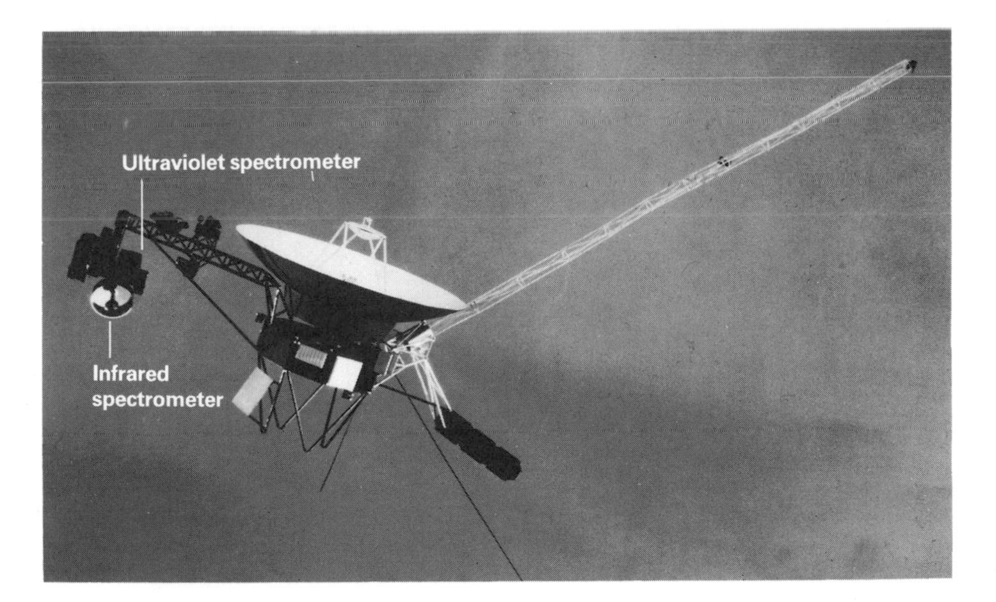

Figure 3.1 The instruments on board the *Voyager* spacecraft are used to investigate the atmospheres of distant planets

3.1 Mass spectrometry

The principles behind the mass spectrometer were described in Section 1.1, where we saw how it is used to measure atomic masses. The spectrometer is sensitive to the mass-to-charge ratio, m/z, of species, and depends on the way that electric and magnetic fields affect the paths of positive ions. In molecular mass spectrometry, as in the atomic version, the vaporized sample is ionized (usually with an electron beam). Vaporizing molecules with high relative molecular masses is very difficult, but they can be dislodged from a solid

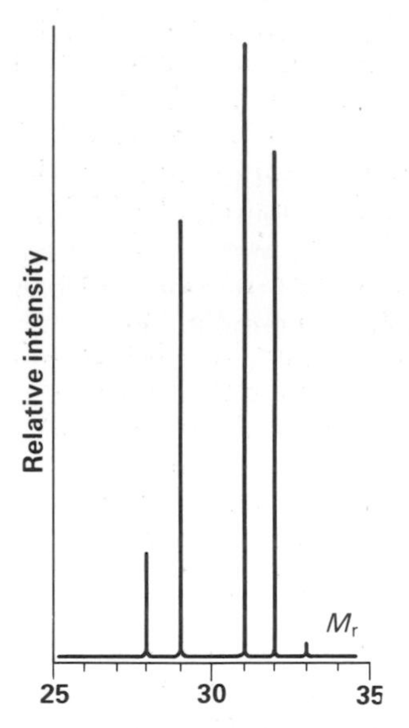

Figure 3.2 The mass spectrum of methanol

sample by a stream of rapidly moving atoms in the **fast atom bombardment** (FAB) technique. The ions formed in the ionization chamber are normally so highly excited that they break up into fragments. Consequently, a plot of signal strength against magnetic field, the **mass spectrum**, is a series of peaks corresponding to the values of m/z in the ionized sample. The technique consists therefore of ionizing the molecule, identifying the fragment ions, and inferring the structure of the original molecule by accounting for all the fragments.

There are simplifying features in the technique as well as complications. For example, the ionizing electron beam is usually not energetic enough to remove more than one electron from each sample molecule, and so we rarely need to worry about the presence of multiply charged ions such as A^{2+}, A^{3+} and so on. The mass spectrum of A therefore consists of peaks corresponding to the m/z ratio for the **parent ion** or **molecular ion** A^+ together with whatever singly charged species it falls apart into (sometimes the parent ion A^+ is so unstable that only its fragments are detected).

As an example of the type of information obtained, consider the case of methanol, CH_3OH. Its mass spectrum is shown in Figure 3.2. Since $M_r = 32$ we might expect a single peak at an m/z ratio corresponding to $M_r = 32$ (colloquially this is referred to as 'a peak at $M_r = 32$') due to the singly ionized molecule $(CH_3OH)^+$. The parent ion is present, but there are also other peaks at $M_r = 31$, 29 and 28. These are consistent with the presence of the fragments CH_3O^+, CHO^+ and CO^+, respectively.

1

H H H H
| | | |
H—C—C—C—C—OH
| | | |
H H H H

Molecules of similar constitution may give rise to quite different mass spectra. This is illustrated by the spectra of two substances both having molecular formula $C_4H_{10}O$: butan-1-ol (**1**) and 2-methylpropan-2-ol (**2**); the mass spectrum of **1** is shown in Figure 3.3(a) and is quite different from the spectrum of **2** shown in Figure 3.3(b). The explanation is that the two molecules break up in different ways when smashed by an energetic electron beam. This is helpful because spectra can be interpreted by referring to a library of patterns kept in a computer, and each substance can be identified by taking its **fragmentation fingerprint**.

2

H OH H
| | |
H—C—C—C—H
| | |
H | H
CH$_3$

Mass spectrometry is used to determine the **molecular formula** of a sample. At first sight the technique might appear to be completely useless except for very small molecules. Take, for instance, a molecule found to have $M_r = 194$.

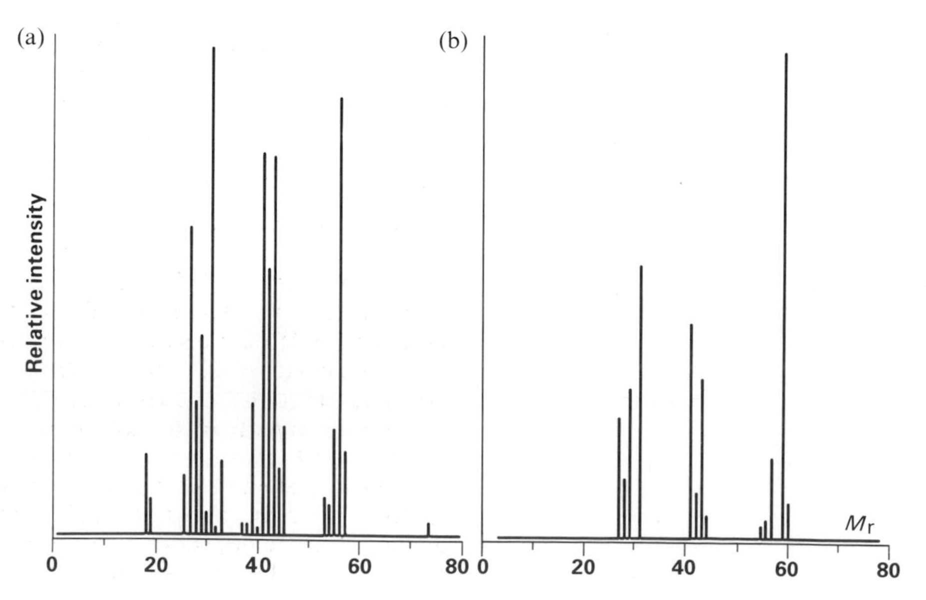

Figure 3.3 The mass spectra of two isomers: (a) butan-l-ol, **1**, and (b) 2-methylpropan-2-ol, **2**

Its molecular formula could be $C_{10}H_{18}N_4$, or $C_8H_6N_2O_4$, or $C_{10}H_{10}O_4$, or any one of about another thirty-five combinations of C, H, N and O (not to mention dozens of other possibilities if the presence of other elements is suspected). However, mass spectrometers can now be built with such high resolution that they can distinguish very small differences of mass. Modern spectrometers can resolve m/z ratios to better than 1 part in 100 000. This overcomes the identification problem because the M_r values for the molecules mentioned above differ when they are calculated more precisely. Thus $M_r(C_{10}H_{18}N_4) = 194.1531$, $M_r(C_8H_6N_2O_4) = 194.0328$, and $M_r(C_{10}H_{10}O_4) = 194.0579$, and so a resolution of 1 part in 10 000 is sufficient to distinguish them. The identification can then be confirmed by seeing whether a peak is accompanied by another arising from the presence of molecules containing a naturally occurring isotope. Thus, if there is a $C_8H_6N_2O_4$ peak with the normal ^{12}C nuclide, there also ought to be a peak with one of the ^{12}C atoms replaced by ^{13}C (which occurs naturally in 1.1 per cent abundance). Figure 3.2 shows a low-intensity peak at m/z corresponding to $M_r = 33$, due to $(^{13}CH_3OH)^+$.

Apart from its use in analysis, mass spectrometry has many other applications. For instance, it is used for quality control in steelmaking. Also, small mass spectrometers tuned to $M_r = 4$ are extremely sensitive leak detectors in vacuum systems. The outside of the apparatus is sprayed with helium (mass number 4), and if any atoms are sucked in they give rise to a signal in a small spectrometer attached to the system. This method can detect leaks of down to $10^{-12}\,cm^3\,s^{-1}$ of gas at 1 atm pressure (at that rate of leakage, a bicycle tyre would stay inflated for about ten million years!). Mass spectrometers are also used to follow the course of chemical reactions, because some of the reaction mixture can be drawn off continuously and analysed both for the components present and, from the intensities of the peaks, the time variation of their concentrations. Mass spectrometers are frequent passengers on planetary missions. The Viking spacecraft carried a mass spectrometer to Mars, and sent back a detailed analysis of the composition of its atmosphere (which is mostly carbon dioxide, as is Venus's).

3.2 Molecular spectroscopy

The principles of molecular spectroscopy, in which spectra are almost always observed in absorption, are the same as those of atomic spectroscopy. The basic equation is the Bohr frequency condition for the frequency ν of the light absorbed when a molecule changes its energy from E(lower) to E(upper) (or vice versa if the photon is emitted):

$$h\nu = E(\text{upper}) - E(\text{lower}) \qquad (3.2.1)$$

The origin of this expression was described in Section 1.3: when a molecule undergoes a transition to a state of lower energy, the excess energy is carried away as a photon of light. In an absorption step the photon must bring the necessary energy if the molecule is to make a transition to a higher energy state. One difference between the spectra of atoms and molecules is that whereas changes in the energies of atoms arise only from changes in the distributions of their electrons, energy is also required to make molecules rotate more rapidly or to vibrate more vigorously, and is discarded when rotations and vibrations decelerate. In other words, the energies of molecules can be changed in three ways: by shifts of their electrons, by changes in their vibrations and by changes in their speeds of rotation. A result is that the spectra of molecules show broad *bands*, rather than the sharp lines characteristic of atomic spectra.

EXAMPLE

Calculate the change in energy that occurs when a molecule absorbs a photon of blue (450 nm) light.

METHOD Use equation 3.2.1 with $h = 6.626 \times 10^{-34}$ J s and relate wavelength to frequency through $\nu = c/\lambda$, where c is the speed of light, $c = 2.998 \times 10^{8}$ m s^{-1}.

ANSWER The frequency of the light is

$$\nu = \frac{2.998 \times 10^{8}\,\mathrm{m\,s^{-1}}}{450 \times 10^{-9}\,\mathrm{m}} = 6.66 \times 10^{14}\,\mathrm{s^{-1}}\,(6.66 \times 10^{14}\,\mathrm{Hz})$$

The energy change is

$$E(\text{upper}) - E(\text{lower}) = 6.626 \times 10^{-34}\,\mathrm{J\,s} \times 6.66 \times 10^{14}\,\mathrm{s^{-1}}$$
$$= 4.41 \times 10^{-19}\,\mathrm{J}$$

COMMENT 4.41×10^{-19} J may not seem like very much energy, but when expressed in terms of the energy per mole it corresponds to $6.022 \times 10^{23}\,\mathrm{mol^{-1}} \times 4.41 \times 10^{-19}\,\mathrm{J} = 266\,\mathrm{kJ\,mol^{-1}}$, a very significant quantity.

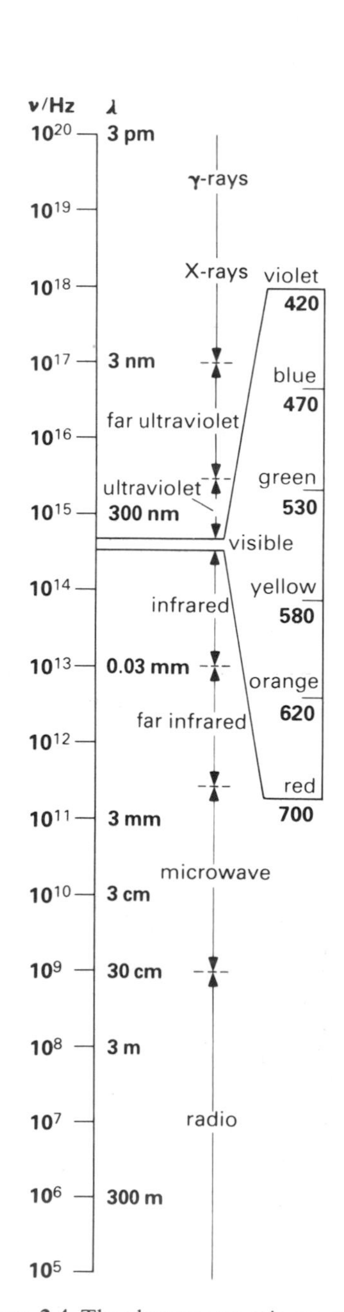

Figure 3.4 The electromagnetic spectrum: $E = h\nu$

The energy changes due to molecular rotation and vibration are much smaller than those due to shifts of electrons. As a result, the absorption frequencies arising from vibrational and rotational changes lie at much lower frequencies (longer wavelengths). The electromagnetic spectrum, together with the names given to its various regions, is set out in Figure 3.4. We shall see that whereas changes in electron distribution (**electronic transitions**) are responsible for absorptions in the high-energy **visible and ultraviolet regions** of the spectrum, vibrational changes are responsible for absorption in the **infrared region** and rotational changes for absorption in the **microwave region**.

The layout of a typical spectrometer for observing spectra in the visible, ultraviolet or infrared regions (microwave techniques tend to be more specialized) is illustrated in Figure 3.5. Radiation from an appropriate source is passed through a prism or a diffraction grating, and the selected monochromatic (single-frequency) component is then divided into two; one beam is passed through a sample made up in a solvent or a gel and the other is passed through the pure solvent or gel. The difference in the intensities is measured, and plotted as a function of the frequency or wavelength to produce the absorption spectrum. A typical spectrum (of chlorophyll) is shown in Figure 3.6.

Microwave spectra

When microwaves (electromagnetic radiation in the wavelength range between 1 mm and 30 cm) pass through samples they stimulate molecular rotation. A gas has to be used for quantitative work because then molecules can rotate freely. In a liquid they jump round and are jostled by their neighbours and the sample simply gets hot (as in a microwave oven). In a solid the molecules are normally frozen into fixed orientations.

Molecular rotational energies are **quantized**; that is, they can take only various discrete values. These allowed energies depend on bond angles and bond lengths, and both can be measured very accurately. Some of their values are listed in Table 3.1.

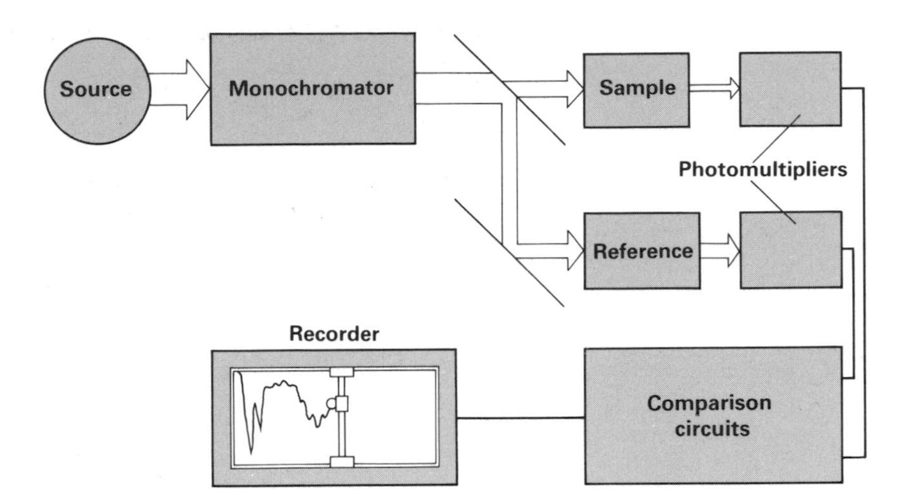

Figure 3.5 The layout of a typical double-beam spectrometer

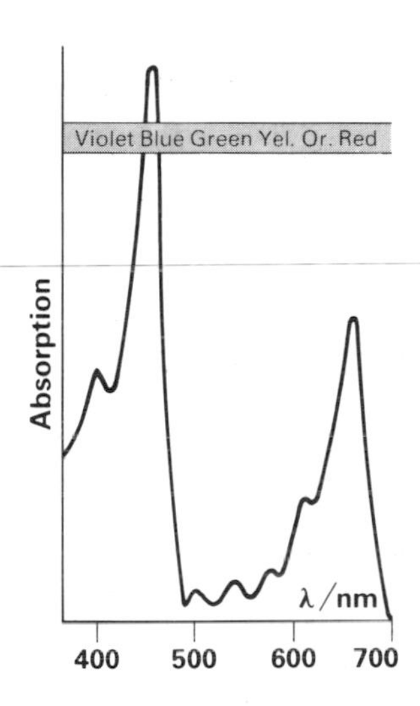

Figure 3.6 The absorption spectrum of chlorophyll in the visible region

Table 3.1 Some molecular dimensions

$O\overset{121}{-}O$	$N\overset{110}{-}N$	$C\overset{113}{-}O$	$H\overset{92}{-}F$
H_2O: O–H 95.8; H–O–H 104.5°	NH_3: N–H 101; H–N–H 107°	$O\overset{116}{-}C-O$	$H\overset{128}{-}Cl$
H_2S: S–H 133; H–S–H 92°	$[NH_4]^+$: N–H 103; H–N–H 109.5°	CH_4: C–H 109; H–C–H 109.5°	$H\overset{141}{-}Br$
SO_2: S–O 143; O–S–O 119°	NO_2: N–O 120; O–N–O 134°	C_6H_6: C–H 109; C–C 139; 120°; 120°	$H\overset{161}{-}I$
SO_3: S–O 143; O–S–O 120°	N_2O_4: N–N 175	C_2H_6: C–C 154; C–H 110	

Note: The nature of bonds (single, double, etc.) has not been specified. Bond lengths are in picometres (1 pm = 10^{-12} m)

Infrared spectra

Molecules vibrate as well as rotate, and this motion also is quantized. Electromagnetic radiation can force a molecule into more energetic vibration (and in the process be absorbed) only if its frequency has the appropriate value. Most vibrational excitations require infrared radiation (radiation with a frequency in the range 10^{12} Hz to 4×10^{14} Hz) and vibrational spectroscopy is also called **infrared spectroscopy**. When you get warm in front of a radiant heater, including the Sun, you are behaving like a vibrational spectroscopy sample, with the molecules in your body capturing the infrared radiation and vibrating more energetically.

In infrared spectroscopy, it is usual to express absorptions in terms of their **wavenumber** ($\tilde{\nu}$), the reciprocal of the wavelength. A wavenumber may be visualized as the number of waves of radiation per unit length, so that a wavenumber of $100\,\text{cm}^{-1}$ means that the radiation has 100 wavelengths per centimetre. Since wavelength and frequency are related by

$$c = \nu\lambda$$

it follows that wavenumber and frequency are related by

$$\tilde{\nu} = 1/\lambda = \nu/c$$

Thus a frequency of 6.0×10^{13} Hz corresponds to the wavenumber

$$\tilde{\nu} = \frac{6.0 \times 10^{13}\,\text{Hz}}{3.00 \times 10^{10}\,\text{cm s}^{-1}} = 2000\,\text{cm}^{-1}$$

(because $1\,\text{Hz} = 1\,\text{s}^{-1}$). Wavenumbers are usually expressed in reciprocal centimetres (cm^{-1}) to make the numbers manageable. When interpreting wavenumbers, remember that they are directly proportional to frequency.

Atoms vibrate relative to their neighbours like particles attached to springs. Just as the natural frequency of a particle on a spring depends on its mass and the stiffness (technically, the **force constant**) of the spring, so the natural frequencies of atoms depend on their masses and the stiffnesses of the bonds connecting them to the rest of the molecule. Hydrogen atoms are very light, and so vibrate at high frequencies. For example, the C—H stretching and contracting motion called a **C—H stretch** occurs near 9.0×10^{13} Hz ($3000\,\text{cm}^{-1}$) whereas the C—Cl stretch occurs at about 2.1×10^{13} Hz ($700\,\text{cm}^{-1}$).

A double bond is stiffer than a single bond between the same two atoms, and as the restoring force is greater it can be expected to vibrate (and hence to absorb) at a higher wavenumber. This is generally found in practice; whereas C—C absorbs at about $1000\,\text{cm}^{-1}$, C═C absorbs near $1650\,\text{cm}^{-1}$ and C≡C at about $2200\,\text{cm}^{-1}$. A bond is usually less rigid to bending than to stretching, so that bending vibrations generally absorb at lower wavenumbers than those of stretching vibrations. For example, whereas the C—H stretch absorbs at about $3000\,\text{cm}^{-1}$, the C—H bending motion (the **C—H bend**, when the hydrogen atom wags from side to side) absorbs at $1400\,\text{cm}^{-1}$.

The atoms in the benzene ring undergo a complex set of motions called **vibrational modes**, in which the bonds stretch and bend with a frequency near 4.5×10^{13} Hz ($1500\,\text{cm}^{-1}$), and detecting an absorption at that frequency is a very good indication of the presence of an aromatic ring. Figure 3.7 shows some of the vibrational modes of groups of atoms in molecules. Between about $1000\,\text{cm}^{-1}$ and $1400\,\text{cm}^{-1}$ a molecule's vibrational spectrum is often very complex. Although it cannot easily be analysed into individual contributions, the pattern is characteristic of the molecule as a whole. This region is called the **fingerprint region**, since recognizing the pattern (using a computer) identifies the molecule.

A final point is that the O—H stretch in molecules that can form hydrogen bonds (discussed in Section 5.1) often absorbs at a lower wavenumber than expected (at about $3200\,\text{cm}^{-1}$ in place of $3600\,\text{cm}^{-1}$) because the hydrogen-bonded group (the X in O—H---X) increases the effective mass of the hydrogen atom and slightly weakens the O—H bond.

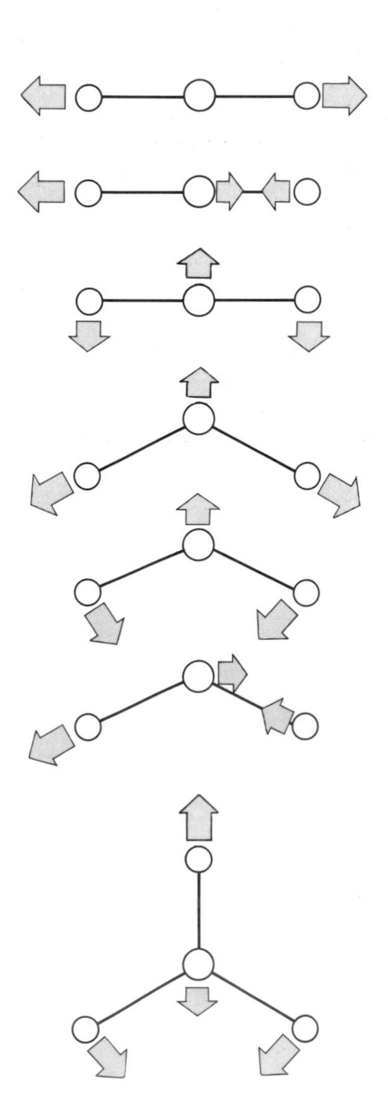

Figure 3.7 Some typical vibrations

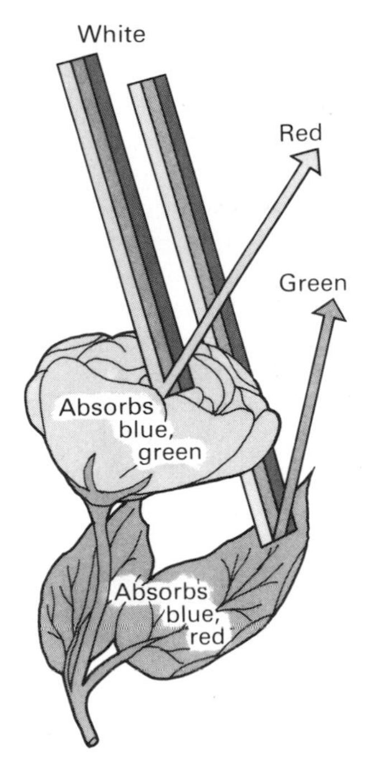

Figure 3.8 Objects appear coloured as a result of selective absorption

Ultraviolet and visible spectra

Photons of visible light carry a lot of energy (about 200 kJ mol^{-1} for yellow light and 260 kJ mol^{-1} for blue), and ultraviolet photons carry even more. When photons of this energy are absorbed by molecules, they cause electrons to move from one region to another. For instance, the spectrum of chlorophyll in Figure 3.6 shows that the molecule can absorb in the red and in the blue; as a consequence, the photons corresponding to these two colours are extracted from sunlight, and the reflected light is green (see Figure 3.8). In general, the colour of a substance in white light is the **complementary colour** of the light absorbed. Complementary colours are best displayed as diametrically opposite pairs on a **colour wheel**, as shown in Figure 3.9: that aqueous copper(II) sulphate is blue implies that it absorbs orange light, the colour opposite blue across the wheel, and a substance that appears orange, such as potassium dichromate(VI), absorbs blue light (see also Box 3.1).

Absorption in the visible and ultraviolet regions is often a result of the presence of particular groups of atoms in the molecule. These groups are called **chromophores** (from the Greek for 'colour bringer'), and ultraviolet and visible absorption spectra can be used to identify their presence in a compound. For instance, the carbonyl group, $>C{=}O$, has an absorption maximum at $\lambda = 185$ nm, and the carbon–carbon double bond $>C{=}C<$ has one at $\lambda = 170$ nm. Azo dyes (see Box 35.3) depend on the presence of the —N=N— chromophore for the colours: this typically absorbs blue light, and the dyes are often orange-red. Complex ions containing d-block metals often have characteristic colours, as we shall see in Section 25.7.

Information from electronic absorption spectra is used to understand the chemical reactions brought about by light. The initial step in these **photochemical reactions** is the absorption of a photon by a molecule, which is followed by the reaction of the molecule as a result of its high energy. Photochemical reactions are essential to all aspects of life (even to organisms like people and fungi that do not photosynthesize). Not only do they capture the radiant energy of the Sun, the first step in the sequence of events that makes life possible, but they are of increasing importance as an energy source. Photochemical reactions also play many roles in everyday affairs, such as the fading of dyes and photography. The absorption in the ultraviolet region by 4-aminobenzoic acid (PABA, NH_2—C_6H_4—COOH) is the basis of some sun-tan creams; detergents that wash 'whiter-than-white' increase the optical brightness of clothes by using additives that absorb in the ultraviolet and re-emit in the visible region.

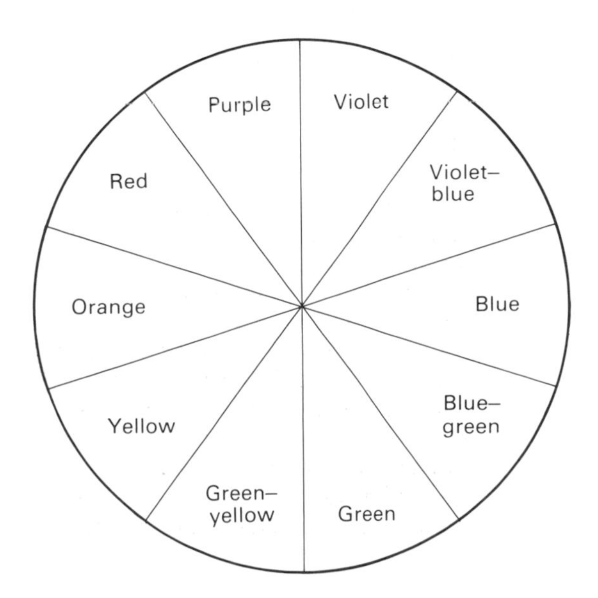

Figure 3.9 A colour wheel

BOX 3.1

Colour and vision

The everyday world is full of colour, and in many cases the source can be traced to electronic transitions. Exceptions include colours arising from light **scattering** from (bouncing off) particles, which accounts for the blue of the sky, and from **interference** effects between light waves reflected by layers of material, which accounts for the colours of butterfly wings and of thin films of oil, and for the lustre of hair.

The green of the chlorophyll of leaves and the red of the haemoglobin of blood are due to the electronic excitation that occurs when light strikes structures similar to the **porphyrin** ring (**3**). In the free porphyrin the presence of the two central hydrogen atoms distorts the ring so that bonds are twisted out of the plane. As a result, the delocalization cannot extend through all four nitrogen atoms, only ultraviolet light – not visible light – is absorbed, and the substance is colourless. However, when a central metal ion is present (magnesium in chlorophyll and iron in haemoglobin) and the two protons are absent, all four nitrogen atoms lie in the same plane (**4**). This opens up a different path for the delocalization, and absorption occurs in the visible region of the spectrum.

That we see at all is due to the excitation of electrons. The retinas of our eyes contain **rhodopsin** (sometimes called **visual purple**), which consists of a protein called **opsin** in combination with the molecule **retinal** (**5**). The retinal acts as an antenna for a passing photon, and if one strikes the retina the retinal at that point is excited, as an electron in its delocalized double-bond system makes a transition to a higher energy state. The removal of that electron from the chain of double bonds makes the molecule flexible, and it rotates into the shape shown in **6**. This motion triggers an impulse in the optic nerve, which is interpreted by the brain as a visual signal.

3

4

5

6

3.3 Nuclear magnetic resonance

Nuclear magnetic resonance (n.m.r.) is one of the most widely used techniques, especially in organic chemistry. Each of the three words in its name conveys one important feature and we deal with them in turn.

The technique is sensitive to some of the *nuclei* present in a molecule, but only to nuclei that possess the property of spin (the same property possessed by electrons, discussed in Section 1.2). Not all nuclei possess spin: protons (hydrogen atom nuclei) do spin, and so do ^{13}C nuclei, but the nuclei of ^{12}C and ^{16}O do not. We shall concentrate on **proton magnetic resonance**, but n.m.r. techniques also tune into other nuclei, especially ^{13}C and ^{31}P.

The main characteristic of proton spin is that, like electron spin, it can have only two orientations, called **spin-up** (↑) and **spin-down** (↓) (see Figure 3.10).

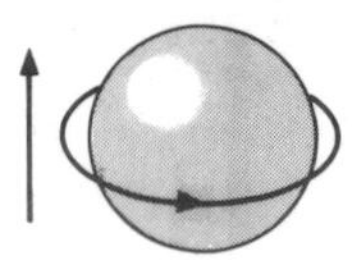
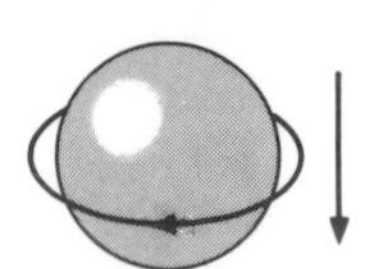

Figure 3.10 Spin-up and spin-down

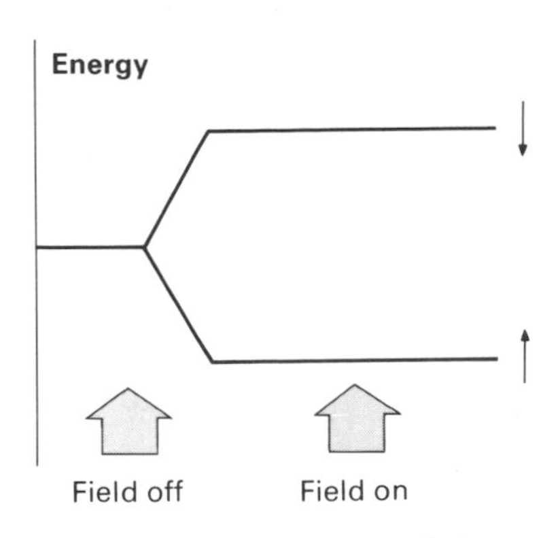

Figure 3.11 The spin states split in energy in a magnetic field

Since nuclei are electrically charged, a spinning nucleus behaves like a tiny bar magnet that can point in either of two directions.

In the absence of any externally applied *magnetic* field, the two orientations have exactly the same energy. When a magnetic field is applied, one orientation corresponds to a lower energy than the other (see Figure 3.11). As a result, in the presence of a magnetic field, energy is required to change the nucleus from a spin-up state to a spin-down state.

The term *resonance* denotes the strong coupling that occurs between two oscillating systems having the same frequency. An example is the coupling between a distant radio transmitter and a receiver: although the signal is very weak, the receiver responds strongly when its circuits are tuned to the transmitter's frequency.

An n.m.r. spectrometer, shown in Figure 3.12, consists of a powerful magnet (a superconducting magnet producing about 7 tesla or more is often used) and a radiofrequency transmitter and receiver linked through the sample cavity.

Figure 3.12 An n.m.r. spectrometer: the large vessel on the right contains a superconducting magnet

It is now possible to record n.m.r. spectra of living bodies, and to analyse organs non-destructively (see Figure 3.13). N.m.r. scanning is beginning to revolutionize the diagnosis of disease, and the tingling sensation in metal dental fillings caused by the high magnetic field is a small price to pay for avoiding the damaging effects of X-radiation.

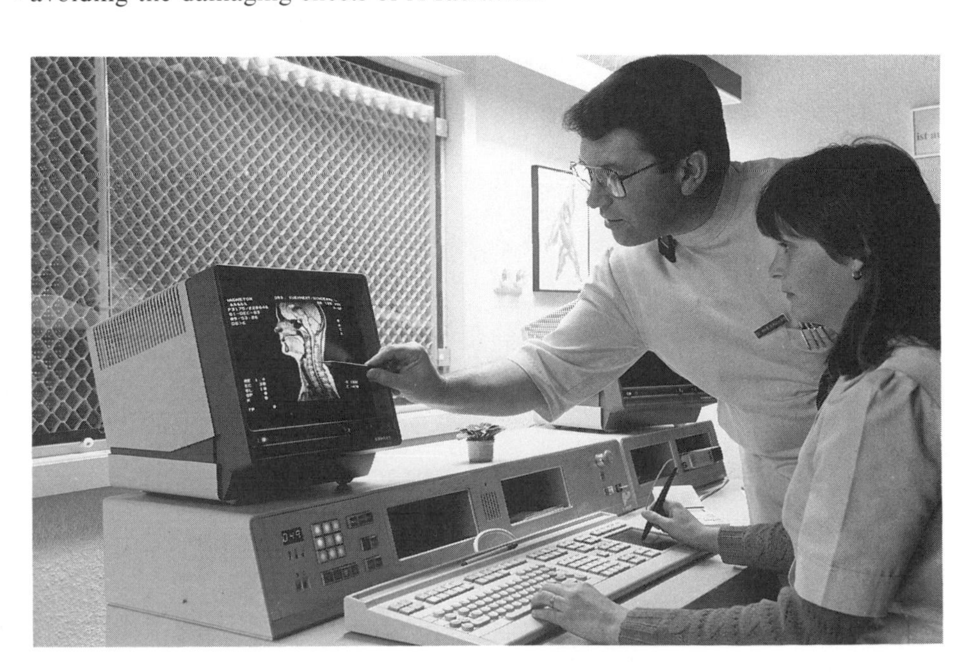

Figure 3.13 The data obtained from an n.m.r. scanner can be analysed to produce extremely detailed images

The transmitter produces low-energy photons of frequency about 300 MHz, corresponding to wavelengths of around 1 m. Either the frequency is kept at a constant value and the magnetic field is varied (in a **field-sweep spectrometer**) or the field is constant and the frequency is swept (in a **frequency-sweep spectrometer**). As the magnetic field increases, the energies of the spin-up and spin-down states move further apart. When the energy of the photons supplied by the transmitter exactly matches the energy difference between the two spin states the nuclei and the photons resonate and the photons are absorbed. This causes a sudden drop in received power, but as the applied field continues to increase resonance is lost and the power recovers. The chart recorder registers the absorption, producing a record like that in Figure 3.14, and then returns to its base-line.

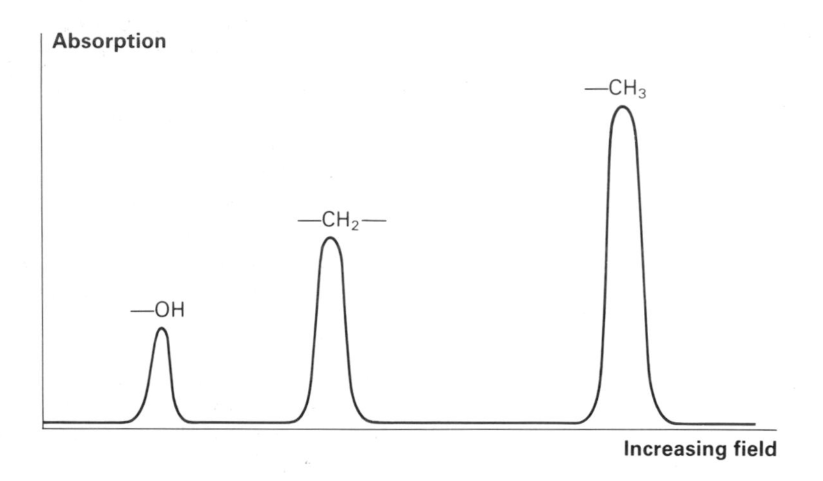

Figure 3.14 A low-resolution n.m.r. spectrum of ethanol

Almost every molecule of interest to the organic chemist contains more than one hydrogen atom. For reasons connected with the details of the molecule's electronic structure, their nuclei come into resonance at different values of the applied field, and are said to show different **chemical shifts**. The area under each peak is proportional to the number of protons having that chemical shift. A low-resolution n.m.r. spectrum of ethanol, like that in Figure 3.14, shows three peaks: they are due to the —OH, the —CH_2— and the —CH_3 protons in the ethanol molecules, and have areas in the ratio 1:2:3.

At higher resolutions, which are today obtained routinely in all commercial spectrometers, a more detailed structure, the **fine structure**, of the spectrum appears (see Figure 3.15). The fine structure, which is due to the interaction of the nuclear spins with each other, gives additional information about the way the protons are grouped together, and for more complicated molecules it gives invaluable information about their structures.

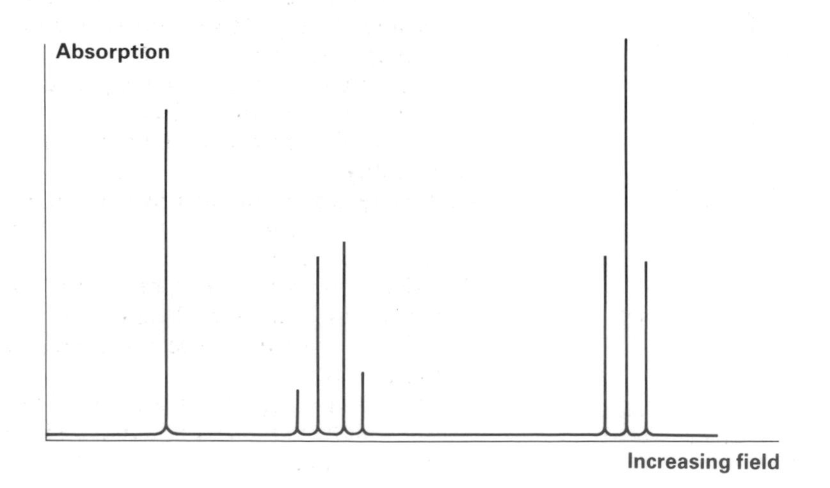

Figure 3.15 A high-resolution n.m.r. spectrum of ethanol

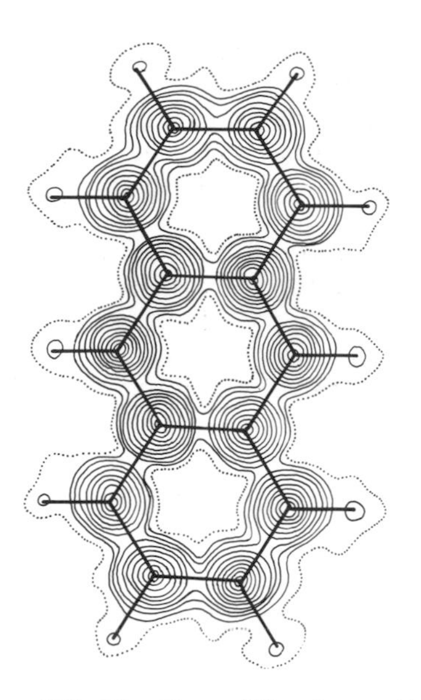

Figure 3.16 The X-ray diffraction pattern of anthracene

3.4 X-ray diffraction

X-ray diffraction is used to obtain the kind of diagram shown in Figure 3.16. The lines on the diagram are contours of equal electron density, and so not only can bond angles and lengths be measured but detailed information about the variation of the electron density throughout the molecule can be obtained. The technique has found its most exciting applications in molecular biology. X-ray diffraction is the principal source of our understanding of the structure of proteins, enzymes and nucleic acids. **X-ray crystallography** is the application of the technique to the study of crystal structure (see Section 5.2).

Diffraction patterns arise from the **interference** between waves (see Figure 2.17). If the crests of two or more waves coincide, then there is an increase in amplitude, but if a crest coincides with a trough there is a cancellation. X-rays, with wavelengths of about 100 pm (0.1 nm), are used to study molecules because only wavelengths as short as this can show up structural details.

The interference can be used to explore the structure of a molecule. Figure 3.17 shows what happens to a beam of X-rays when it is directed on to a molecule held in a rigid orientation in a crystal. The radiation is scattered from the molecule in many different directions. The illustration shows two of the scattered waves. For a detector positioned at O_1 the crests are coincident, and so it registers a strong X-ray intensity. When it moves to O_2 the scattered waves travel different distances to reach it, and the crests of one coincide with the troughs of the other. As a result the detector records zero X-ray intensity. By measuring the intensity of radiation at all orientations around the molecule, taking into account the scattering from every region and not just the two shown, the observer records a very complicated intensity distribution. When it is analysed (always on a computer because the calculations are so complicated) the output is the kind of diagram shown in Figure 3.16.

Electrons accelerated to high speeds behave like waves with wavelengths of about 10 pm (the wave–particle duality was mentioned in Box 1.2), and so can be used in diffraction experiments – see Figure 1.6(a). Electron diffraction is used to study molecules in the gas phase, because the beam does not penetrate solids. Neutrons from reactors also have suitable wavelengths and are used in **neutron diffraction** studies of solids. Their main advantage over X-rays is that they show up the locations of the hydrogen atoms more readily. The structure of ice was determined in this way. The most important technique remains X-ray diffraction, however, and nothing else has done more to make the invisible open to inspection and measurement.

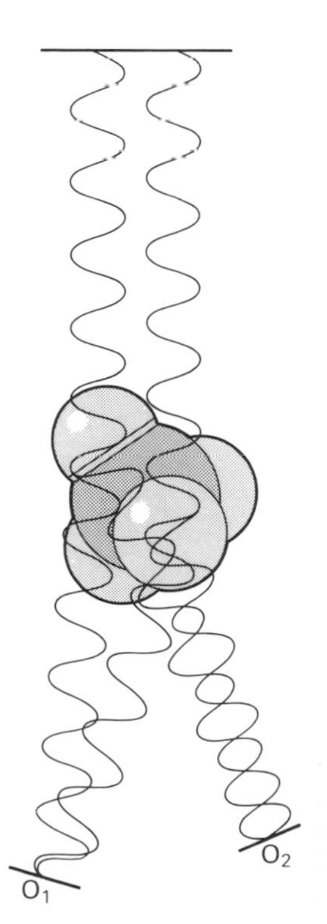

Figure 3.17 Scattering from different parts of a molecule, and the interference at two points

Summary

- Mass spectrometry is used to determine the masses of molecules, their molecular formulae (at high resolution) and their constitution (from their fragmentation patterns).
- The mass spectrometer is sensitive to the mass-to-charge ratio, m/z, of positive ions produced either directly by electron bombardment or by subsequent fragmentation.
- Molecular spectra can be explained in terms of the Bohr frequency condition (equation 3.2.1).
- Electronic, rotational and vibrational energy levels of molecules are quantized.
- Absorption in the microwave region of the electromagnetic spectrum can be accounted for in terms of the stimulation of rotational excitation of gas-phase molecules.
- Microwave spectra can be interpreted in terms of the geometry of molecules, their bond lengths and bond angles.
- Infrared absorption spectra are due to the stimulation of molecular vibrations.
- Infrared spectra are used for the identification of molecules in gases, liquids and solids (mainly in liquids), and for finding the strengths and stiffnesses of their bonds.

- Ultraviolet and visible absorption spectra are due to the stimulation of electronic transitions by the incident radiation.
- Groups that absorb at characteristic frequencies are called chromophores.
- Electronic spectroscopy is used to obtain information on the strengths of bonds, to identify species present and as essential information for the study of photochemical reactions.
- The nuclear magnetic resonance (n.m.r.) spectrum is used to identify organic molecules and to determine their structures, from the chemical shifts and fine structure.
- X-ray diffraction is based on interference between waves scattered by different parts of a molecule. It gives details of electron density distributions.
- Electron diffraction and neutron diffraction depend on the wave characteristics of fast electrons and neutrons, and are used to obtain structural information.

PROBLEMS

1 Write an essay outlining the principles of spectroscopy. Cover the basic process, account for absorption and emission spectroscopy, explain how different regions of the spectrum arise from excitation of different transitions, and mention some applications.

2 Explain the usefulness of the fragmentation pattern in mass spectrometry.

3 Suggest the form of the mass spectrum of (a) methane, (b) water.

4 What extra peaks would be expected on the basis that the methane used in the last question contained 1 per cent of ^{13}C?

5 How might deuteriation of methanol (the substitution of one proton by one deuteron, $^{2}_{1}H^{+}$) affect its mass spectrum?

6 Predict the form of the mass spectrum from gaseous molecular chlorine of natural isotopic abundance.

7 State which spectroscopic technique would be suitable for the investigation of each of the following problems:
(a) the presence of trace amounts of arsenic in the blood stream;
(b) the composition of the atmosphere of Venus;
(c) the nature of exhaust gases from a car engine;
(d) the vibrational frequency of the O—H bond in ethanol;
(e) the bond length of HCl;
(f) the purity of ethanol.

8 Explain the terms *chromophore, vibrational mode, stretching vibration, breathing mode.*

9 Suggest what will happen to the frequency of vibration of C—H on deuteriation.

10 Describe any *one* spectrometric method for the determination of molecular structure (possible methods include infrared spectroscopy or mass spectrometry but *not* diffraction methods). For the method of your choice you should deal with the practical procedure *and* with the interpretation of the results. (*Nuffield*)

4 The properties of gases

In this chapter we look at gases, the simplest state of matter. We see how experiments led to the idea of the 'perfect gas' and to the perfect gas equation and how the basic ideas are readily extended to mixtures. The model of a gas as a collection of chaotically moving particles provides an explanation of the perfect gas equation. It can also be used to relate the speeds of the particles to the temperature. The perfect gas equation is only approximately applicable to real gases under normal conditions, and so we also examine how and why real gases behave differently.

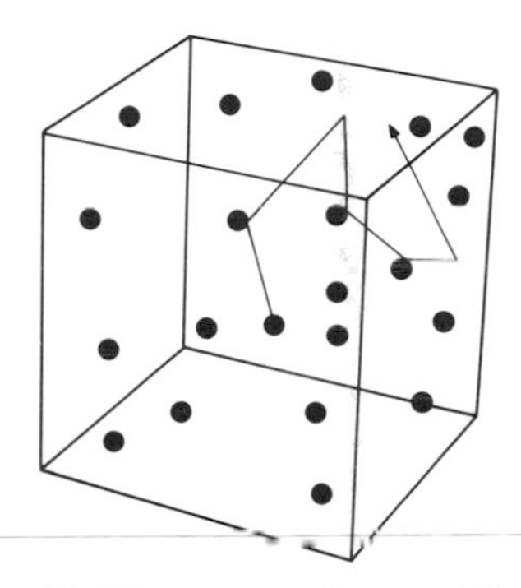

Figure 4.1 The nature of a gas (all particles move)

The name **gas** comes from the word chaos, which neatly summarizes the principal feature of this the simplest state of matter. We picture a gas as a swarm of particles moving randomly and chaotically, constantly colliding with each other and with the walls of the containing vessel (see Figure 4.1). The pressure exerted by the gas is a result of the never-ending battering of the particles on the walls. The tendency of any gas to fill the available volume is a result of the freedom that the particles have to move everywhere open to them. The easy compressibility of gases is due to the presence of so much empty space between their particles (see Figure 4.2).

The chaotic structure of gases does not mean that they cannot be discussed scientifically; in fact some of the earliest experiments in the field that has since become physical chemistry were done on gases, and led to the formulation of Boyle's law (in 1662) and Charles's law (in 1787) (see Box 4.1).

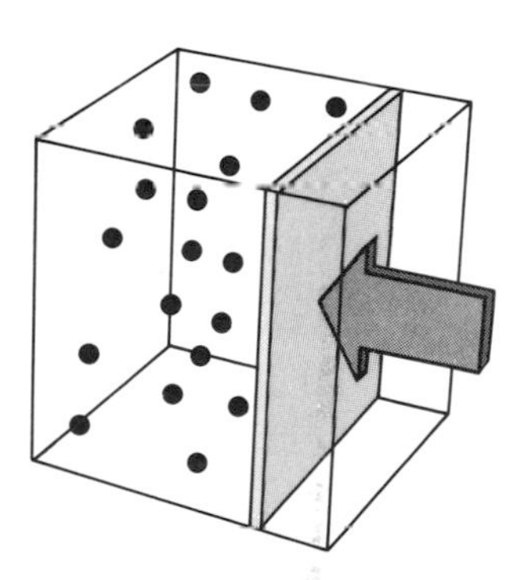

Figure 4.2 Gases are easy to compress

BOX 4.1

The development of the understanding of gases

Robert Boyle was born in Ireland in 1627 and went to Oxford in 1654. There he constructed a pump with the help of Robert Hooke, and set about his studies of the properties of gases. His avowed motivation for science was religious, in the sense that he wanted to understand the working of the creation. In the course of his work he attacked Aristotle's views on the constitution of the world in terms of earth, air, fire and water, and argued in favour of a corpuscular nature of matter. He proposed his law in 1662; it was proposed independently in France by Edmé Mariotte (in 1676) where it bears his name along with Boyle's. France was the source of further understanding of gases, for Jacques-Alexandre-César Charles was a Frenchman with a consuming interest in ballooning (the early hydrogen balloons were often referred to as *charlières*), which stimulated him to study the effect of temperature on gases. He did not publish his findings (perhaps because he was more interested in breaking world altitude records, which he did, on 1 December 1783, by a factor of 25!), and they were re-discovered by Joseph Gay-Lussac, another Frenchman, who did. The study of real gases and their liquefaction was put on secure foundations by the Irishman Thomas Andrews, and the first largely successful attempt to account for their properties was made by the Dutchman Johannes van der Waals in 1873. His was the first attempt to establish an equation of state for a real gas, but his successors have suggested another hundred or so different ways of describing them. Modern work concentrates on two aspects of gases. One is the description of their imperfections in terms of the **virial equation of state** (which is normally associated with another Dutch physical chemist, Heike Kamerlingh Onnes, who proposed it in 1901) and the other concerns the dynamical properties of flowing gases, such as their viscosities.

4.1 The gas laws

There are three characteristic physical properties of a gas: its response to pressure, its response to temperature, and the dependence of its volume on the amount of substance in the sample. Many gases also have characteristic smells: odour is related to the ability of molecules to attach to nerve endings in the membranes of the nose, which depends on the structures of the molecules (for example, their shape and the presence or absence of lone pairs of electrons). In this chapter we concentrate on the physical properties.

The pressure dependence

The physical property we meet throughout this chapter is the **pressure.** Pressure is defined as *force per unit area*, and so with force measured in newtons (symbol: N, $1\,N = 1\,kg\,m\,s^{-2}$), pressure is reported in newtons per square metre, $N\,m^{-2}$. However, there are three more convenient versions of this unit. One is the **pascal** (symbol: Pa), which is another name for $N\,m^{-2}$ (so that $1\,Pa = 1\,N\,m^{-2}$): it will be helpful to remember that the pressure of the atmosphere at sea level is about 100 kPa (1×10^5 Pa). A second is the **bar**, which is defined as 100 kPa exactly (1 bar = 10^5 Pa): weather forecast charts often quote pressures in millibars (mbar), and so atmospheric pressure is close to 1000 mbar. The third unit is called the **atmosphere** (symbol: atm), and is defined rather awkwardly as 101.325 kPa. This odd value is the pressure needed to push a column of mercury in a barometer to a height of 760 mm, which is 'standard atmospheric pressure'. We will use all three units as appropriate, but the modern tendency is to move from the atmosphere to the bar for precise work and to retain the atmosphere only for more casual usage. Pressures are measured by attaching a manometer to the apparatus: this is a vertical column of liquid (often mercury), like a barometer, and its height is proportional to the pressure.

Robert Boyle measured the volume of a sample of gas subjected to different pressures. He found that, at a fixed temperature, the volume is *inversely proportional* to the applied pressure as shown in Figure 4.3:

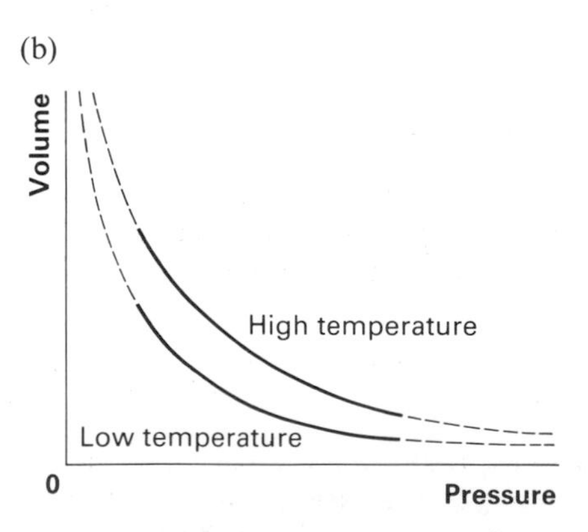

Figure 4.3 Boyle's law: (a) V plotted against $1/p$, (b) V against p

Boyle's law: At constant temperature and for a fixed mass of gas, $V \propto 1/p$. (4.1.1)

An equivalent statement is that, where the temperature and mass of the sample are fixed,

Boyle's law: $pV = constant$ (4.1.2)

We now know that this law is only an approximation. Nevertheless there are two points of great significance which stop us dismissing it as a 300-year-old historical irrelevance.

In the first place Boyle's law is very useful, even though it is an approximation. It gives a simple way of predicting the volume of a gas when it is subjected to pressure, and the deviations from predictions are important only when the gas is dense or cold. The second point has a wider significance. Careful measurements have shown that, as the density of a sample of any gas is reduced, Boyle's law is obeyed increasingly accurately. The law is a statement about an ideal fluid called a **perfect gas** (or an **ideal gas**), which is obeyed exactly when the density of the gas is zero.

EXAMPLE

What is the volume remaining in one cylinder of a four-cylinder $1300\,cm^3$ petrol engine when the pressure increases by a factor of 9.0 at the top of the stroke?

METHOD This is an application of Boyle's law, stated in equation 4.1.2, in the form $(pV)_{\text{top of stroke}} = (pV)_{\text{bottom of stroke}}$. The volume at the bottom of the stroke is one-quarter of the total specified volume of the four-cylinder engine.

ANSWER Volume of one cylinder $V_{\text{bottom}} = \frac{1}{4} \times 1300\,\text{cm}^3 = 325\,\text{cm}^3$. Ratio of pressures $p_{\text{top}}/p_{\text{bottom}} = 9.0$. Application of Boyle's law then gives

$$V_{\text{top}} = \frac{(pV)_{\text{bottom}}}{p_{\text{top}}} = \frac{325\,\text{cm}^3}{9.0} = 36\,\text{cm}^3$$

COMMENT In a working engine the compression stroke also heats the gas. We shall see how to take this into account next. It is important to realize that Boyle's law applies specifically to a compression or an expansion at constant temperature. The **compression ratio**, the ratio of the volume at the bottom of the stroke to that at the top, is 9.0:1.

The temperature dependence

The experiments of Gay-Lussac and Charles led them to the conclusion that a gas responds to temperature in a very simple way:

Charles's law: The volume of a gas depends linearly on the temperature, the pressure and mass being fixed.

More recent observations have shown that **real gases** (that is, actual gases) deviate from this law too. Nevertheless, under normal conditions the deviations are small, and disappear in the extreme case of a gas at zero density.

It is found experimentally that the volume of a sample of any gas extrapolates to zero at $-273\,°\text{C}$ (see Figure 4.4). (Once again this is strictly true only when the density is so low that the gas is behaving perfectly.) This suggests the existence of a natural **absolute zero of temperature**, and the natural temperature scale, or **Kelvin scale**, takes $T = 0$ as the temperature at which a perfect gas occupies zero volume. Temperatures on the **Celsius (centigrade) scale** are converted to the Kelvin scale simply by adding 273. Thus $0\,°\text{C}$ corresponds to 273 K, and $25\,°\text{C}$ to 298 K. Human blood temperature, $37\,°\text{C}$, is 310 K.

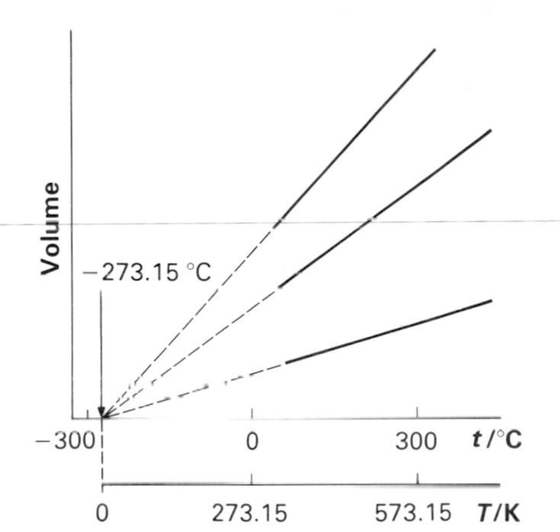

Figure 4.4 Charles's law (different samples correspond to different masses of gas)

Using the Kelvin scale, Charles's law can be expressed very simply as

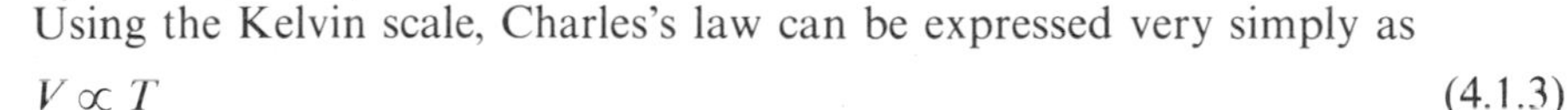

$$V \propto T \tag{4.1.3}$$

where the pressure and mass of the sample are fixed.

The dependence on amount of substance

The third piece of information is due to the Italian chemist Amedeo Avogadro. A 1 mol sample of $H_2(g)$ is found to occupy the same volume as a 1 mol sample of $O_2(g)$ (which at $20\,°\text{C}$ and 1 atm pressure is about $24\,\text{dm}^3$). This observation generalizes to:

Avogadro's law: Gases containing equal amounts of substance occupy the same volume at the same temperature and pressure.

In other words, the volume occupied per unit amount of substance, the **molar volume**, is the same for all gases at the same temperature and pressure. If the amount of substance of gas is n, and its molar volume is V_m, then the volume of the sample is

$$V = nV_m \tag{4.1.4}$$

EXAMPLE

What volume is occupied by 2.70 g of hydrogen gas, given that the molar volume of a perfect gas is $24\,\text{dm}^3\,\text{mol}^{-1}$ under the same conditions of pressure and temperature?

METHOD Assume that hydrogen behaves perfectly and use Avogadro's law in the form $V = nV_m$, where n is the amount of substance of the gas and V_m its molar volume. Find n on the basis that the molar mass of H_2 is $2.016\,\text{g mol}^{-1}$ (i.e., $2 \times 1.008\,\text{g mol}^{-1}$, see page 4).

ANSWER The amount of substance of H_2 in a sample of mass 2.70 g is

$$n = \frac{2.70\,\text{g}}{2.016\,\text{g}\,\text{mol}^{-1}} = 1.34\,\text{mol}$$

Since $V_m = 24\,\text{dm}^3\,\text{mol}^{-1}$, it follows that

$$V = 1.34\,\text{mol} \times 24\,\text{dm}^3\,\text{mol}^{-1} = 32\,\text{dm}^3$$

COMMENT Using Avogadro's constant (see the introductory chapter, 'The mole') we can also state that 2.70 g of hydrogen gas contains $1.34\,\text{mol} \times 6.022 \times 10^{23}\,\text{mol}^{-1} = 8.07 \times 10^{23}$ H_2 molecules. Since the sample occupies $32\,\text{dm}^3$, an individual molecule has about $(32 \times 10^{-3}\,\text{m}^3)/(8.07 \times 10^{23}) = 4.0 \times 10^{-26}\,\text{m}^3$ of space to itself. This corresponds to a cube of side 3.4×10^{-9} m, or 3.4 nm. Since the diameter of the molecule is about 0.1 nm we can begin to visualize the diffuseness of the gas: its nearest neighbour is about 30 molecular diameters away under the conditions of this Example, which are similar to normal atmospheric conditions.

Avogadro's law is an idealization, and is known not to be strictly true. Nevertheless it is a very helpful broad summary of the behaviour of gases under normal conditions, and it is exactly true when the density of the gas is vanishingly small. In other words, it is another property of a perfect gas.

The perfect gas equation

The experimental observations described above have identified three aspects of the behaviour of a perfect gas: under the appropriate conditions they are $pV = \text{constant}$, $V \propto T$, and $V \propto n$. All three can be combined into the single, simple expression $pV \propto nT$. On writing the coefficient of proportionality as R, this becomes the **perfect gas equation** (or **ideal gas law**):

$$pV = nRT \tag{4.1.5}$$

R, which is the same for every gas, is the **molar gas constant** (more usually called the **gas constant**). It can be determined once and for all by evaluating $R = pV/nT$ for some gas. For example, when 0.20 g of hydrogen (corresponding to 0.10 mol H_2) is confined in a $1.0\,\text{dm}^3$ vessel at 0 °C (273 K) it is found to exert a pressure of 227 kPa. Therefore

$$R \approx \frac{2.27 \times 10^5\,\text{N}\,\text{m}^{-2} \times 1.0 \times 10^{-3}\,\text{m}^3}{0.10\,\text{mol} \times 273\,\text{K}}$$

$$= 8.3\,\text{N}\,\text{m}\,\text{K}^{-1}\,\text{mol}^{-1} = 8.3\,\text{J}\,\text{K}^{-1}\,\text{mol}^{-1}$$

As progressively smaller amounts of gas are confined in the vessel at the same temperature, the value of pV/nT approaches a constant. When the value of pV/nT is extrapolated to $n = 0$, it is found that

$$R = 8.314\,\text{J}\,\text{K}^{-1}\,\text{mol}^{-1}$$

The same value of R is found whatever gas is used.

The perfect gas equation makes it easy to predict the state of a gas under different conditions. For instance, it is sometimes convenient to express the volume of a gas in terms of its volume under a standard set of conditions, namely 298 K and 1 bar (or 100 kPa). If the molar volume of a gas is V_m at a temperature T and a pressure p, the conversion to standard conditions is

$$V_m^{\ominus} = \frac{298\,\text{K}}{T} \times \frac{p}{100\,\text{kPa}} \times V_m \tag{4.1.6}$$

The perfect gas equation is also used in the measurement of the molar masses of volatile liquids. First, the equation is rearranged into $1/n = RT/pV$. Then the amount of substance is found from the mass of the sample m and its molar mass M, using $n = m/M$. The volume of a known mass of gas is measured at a known temperature and pressure, and the molar mass is obtained by combining these two equations into

$$M = mRT/pV$$

In the modern version of the **Victor Meyer method** for the determination of molar mass, a known mass of the volatile liquid is injected into a gas syringe (see Figure 4.5), the temperature and pressure are measured, and the volume occupied by the vapour is noted. This technique is now used mainly for analysing the compositions of mixtures of gases because there are better methods (such as mass spectrometry) for measuring the molar masses themselves.

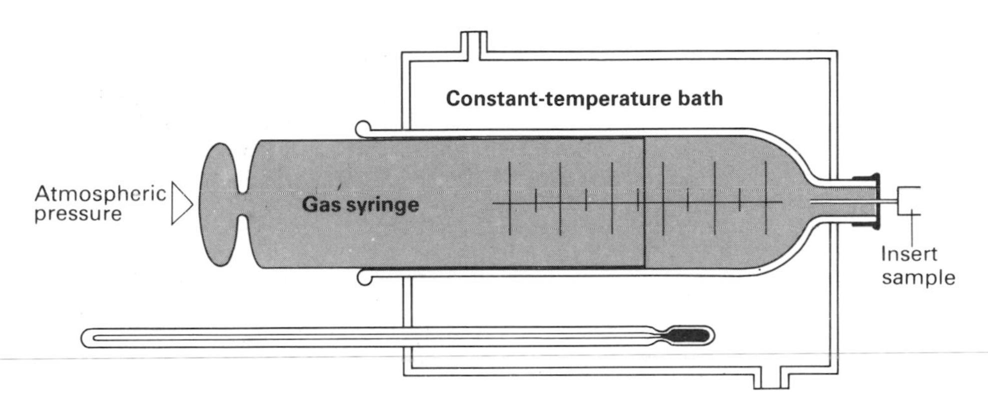

Figure 4.5 A gas syringe

EXAMPLE

An ammonia plant produces 1000 tonnes per day by the Haber–Bosch process. The plant operates at 200 atm and 525 °C. What volume of ammonia at that pressure is produced each day? What volume does that correspond to at 1.0 atm pressure and 25 °C?

METHOD Assume that the gas behaves perfectly and apply equation 4.1.5 in the form $V = nRT/p$, with $R = 8.206 \times 10^{-2}\,\text{dm}^3\,\text{atm}\,\text{K}^{-1}\,\text{mol}^{-1}$. Convert the temperature to the Kelvin scale and calculate n on the basis that the molar mass of ammonia is $17.0\,\text{g mol}^{-1}$ (see page 4). For the second part use the gas law in the form $p_1V_1/T_1 = p_2V_2/T_2$ for a given amount of gas.

ANSWER The temperature of the plant is $T = (273 + 525)\,\text{K} = 798\,\text{K}$. The amount of substance of NH_3 corresponding to 1000 tonnes ($10^3 \times 10^3\,\text{kg} = 10^6\,\text{kg}$) is

$$n = \frac{10^6\,\text{kg}}{17.0\,\text{g mol}^{-1}} = \frac{10^9\,\text{g}}{17.0\,\text{g mol}^{-1}} = 5.88 \times 10^7\,\text{mol}$$

The perfect gas law (equation 4.1.5) then gives

$$V = nRT/p$$
$$= \frac{5.88 \times 10^7\,\text{mol} \times 8.206 \times 10^{-2}\,\text{dm}^3\,\text{atm}\,\text{K}^{-1}\,\text{mol}^{-1} \times 798\,\text{K}}{200\,\text{atm}}$$
$$= 1.93 \times 10^7\,\text{dm}^3$$

The volume corresponding to the same amount of substance but at 1.0 atm pressure and 25 °C (298 K) is

$$V_2 = (T_2/T_1) \times (p_1/p_2) \times V_1$$
$$= \frac{298\,\text{K}}{798\,\text{K}} \times \frac{200\,\text{atm}}{1.0\,\text{atm}} \times 1.93 \times 10^7\,\text{dm}^3 = 1.4 \times 10^9\,\text{dm}^3$$

COMMENT These huge volumes give some idea of the large scale of industrial processes: $10^9\,dm^3$ corresponds to a cube of side 100 m. In real life the ammonia has to be treated as a real gas, and at the pressures dealt with in this Example the deviations from perfect behaviour are significant. Nevertheless, the numbers we have arrived at give an idea of the magnitudes involved.

Mixtures of gases

In chemistry we often have to deal with mixtures of gases (the atmosphere itself is an example). A law introduced by Dalton as a result of his experimental studies extends the perfect gas laws to mixtures (recall his interest in meteorology mentioned in Box 1.1).

The problem can be expressed as follows. Suppose an amount of substance of gas n_A (0.1 mol N_2, for example) is introduced into a container of volume V. Then the perfect gas equation predicts that the pressure it exerts is $p_A = n_A RT/V$. If another gas B had been introduced instead it would have exerted a pressure $p_B = n_B RT/V$, where n_B is its amount of substance (such as 0.2 mol O_2). But suppose the second gas had been introduced into the container that already had the first gas inside. What pressure would each gas exert, and what would the total pressure be?

Dalton's observations led him to conclude that, provided no chemical reaction occurred, neither gas was affected by the presence of the other. Therefore, even though gas A is present, gas B exerts the pressure p_B given by the expression above. This observation is expressed as follows:

Dalton's law: Each gas in a mixture exerts the same pressure as when it alone occupies the container at the same temperature.

The pressures p_A and p_B exerted by the individual gases are called their **partial pressures.** The total pressure, the pressure observed if there is a pressure gauge attached to the vessel, is the sum of these partial pressures:

$$p = p_A + p_B \tag{4.1.7}$$

The gas laws provide a simple way of calculating the partial pressure of each component, given the composition of the mixture and the total pressure. First we introduce the **mole fractions** x_A and x_B. These are defined as

$$x_A = n_A/n \quad \text{and} \quad x_B = n_B/n \tag{4.1.8}$$

with $n = n_A + n_B$ and express the amount of substance of each type of particle in the mixture as a fraction of the total amount of substance. Then since $p_A = n_A RT/V$, $p_B = n_B RT/V$, and $RT/V = p/n$, it follows that

$$p_A = x_A p \quad \text{and} \quad p_B = x_B p \tag{4.1.9}$$

(see Figure 4.6). This is a very useful (and simple) way of arriving at partial pressures when the composition and total (measured) pressure of a mixture are known.

Dalton's law is another idealization that applies exactly to a perfect gas, but only approximately to real gases. It captures the essential features of the behaviour of gases when their density is so low that their particles are completely independent.

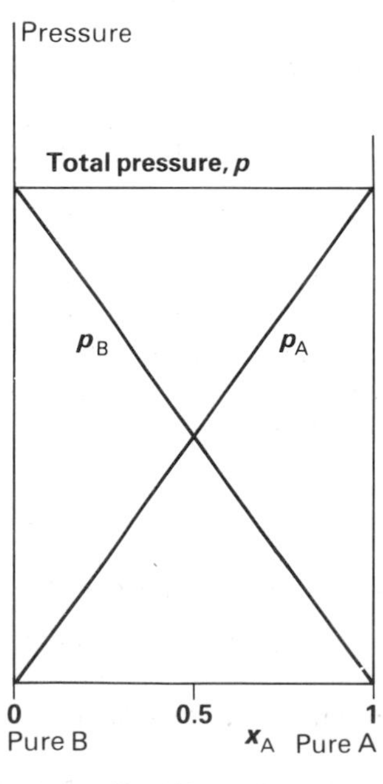

Figure 4.6 Dalton's law of partial pressures

EXAMPLE

In simple applications it is often sufficient to regard the atmosphere as a mixture of 76.8% by mass of nitrogen and 23.2% by mass of oxygen. Calculate the partial pressure of each gas when the total pressure is 100 kPa (1 bar).

METHOD Use equation 4.1.9 in the form $p(N_2) = x(N_2)p$ and $p(O_2) = x(O_2)p$, where x is the mole fraction of the component. Find the mole fractions from equation 4.1.8, but first obtain the amounts of substance of each component. Since the composition is expressed as percentages, the simplest procedure is to consider a sample of 100 g, which then contains 76.8 g nitrogen and 23.2 g oxygen. Then use $M(N_2) = 28.02\,\text{g mol}^{-1}$ and $M(O_2) = 32.00\,\text{g mol}^{-1}$ (see the Periodic Table on page 544.

ANSWER The amounts of substance present in 100 g of air are

$$n(N_2) = \frac{76.8\,\text{g}}{28.02\,\text{g mol}^{-1}} = 2.74\,\text{mol}$$

$$n(O_2) = \frac{23.2\,\text{g}}{32.00\,\text{g mol}^{-1}} = 0.725\,\text{mol}$$

The mole fractions of the components are therefore

$$x(N_2) = \frac{2.74}{2.74 + 0.725} = 0.791 \quad \text{and} \quad x(O_2) = \frac{0.725}{2.74 + 0.725} = 0.209$$

The partial pressures are therefore given by equation 4.1.9 as

$$p(N_2) = 0.791 \times 100\,\text{kPa} = 79.1\,\text{kPa}$$

$$p(O_2) = 0.209 \times 100\,\text{kPa} = 20.9\,\text{kPa}$$

COMMENT A more complete description of the composition of dry air at sea level is as follows (in mass per cent): N_2(75.52), O_2(23.15), Ar(1.28), CO_2(0.046), corresponding to the partial pressures (in bar) N_2(0.782), O_2(0.209), Ar(0.009), CO_2(0.0003). There are also traces of other gases, especially in towns. The partial pressure of water varies, depending on the humidity.

4.2 Accounting for the perfect gas equation

The pressure of a gas – the average force it exerts per unit area – is due to the collisions its particles make with the walls of the container. Their constant battering on the walls exerts a force (the same resisting force felt when testing an inflated bicycle tyre), and there are so many collisions that it appears to us as a steady pressure. In this section we see how this picture is used to account for the perfect gas equation.

The kinetic theory

The **kinetic theory of gases** is based on the following model of a perfect gas:

1. The gas is pictured as a collection of identical particles of mass m in continuous random motion.
2. The particles are regarded as having zero volume (are considered 'pointlike').
3. They move without interacting with each other except for the collisions that maintain their random motion.
4. All collisions between particles and with the walls of the container are **elastic**, which means that the particles' total translational kinetic energy after a collision is the same as it was before (similar to collisions between snooker balls but unlike those between cars).

Using this model it is possible to derive the following expression for the pressure of the gas:

$$pV = \tfrac{1}{3}Nm\overline{c^2} \qquad (4.2.1)$$

where V is the volume of the container, N is the number of particles present and $\overline{c^2}$ is their **mean square speed**. (The mean square speed is the average value of the squares of the speeds of the particles.) This result is derived in

Appendix 4.1. The number of particles can be expressed as an amount of substance n through $N = nL$, L being Avogadro's constant. The product Lm that then occurs in the equation is the molar mass M of the particles, and the equation becomes

$$pV = \tfrac{1}{3}nM\overline{c^2} \qquad (4.2.2)$$

The pressure–volume relation of a perfect gas is correctly reproduced by this expression. The mean square speed of the particles is independent of the pressure and the volume (they go on hurtling around with the same speeds when the gas is compressed, so long as the temperature is the same) and so the expression has the form pV = constant, in accord with Boyle's law.

The mean square speed rises when the temperature is increased and the particles move more vigorously. The precise dependence can be deduced by comparing equation 4.2.2 with $pV = nRT$. It follows that

$$\tfrac{1}{3}nM\overline{c^2} = nRT \quad \text{or} \quad \overline{c^2} = 3RT/M$$

That is, the **root mean square speed** (r.m.s. speed) c_{rms} is

$$c_{rms} = \sqrt{(\overline{c^2})} = \sqrt{(3RT/M)} \qquad (4.2.3)$$

There are two immediate conclusions from this result. The first is that *the r.m.s. speed of particles is proportional to the square root of the temperature.* This means that on average the molecules in air move about 5 per cent faster on a hot day (30 °C) than on a cold day (0 °C), and that a particle on the surface of the Sun (6000 °C) moves about $4\frac{1}{2}$ times faster on average than the same particle in the atmosphere of the Earth. The second conclusion is that *the r.m.s. speed is inversely proportional to* $\sqrt{M}$. On average, heavy molecules move more slowly than light molecules do (see Figure 4.7). For instance, since $M = 44\,\text{g mol}^{-1}$ for carbon dioxide and $18\,\text{g mol}^{-1}$ for water, on average the speed of a CO_2 molecule is about two-thirds that of an H_2O molecule in the atmosphere. The values of c_{rms} are calculated simply by substituting the values of the molar mass and the temperature into equation 4.2.3. For carbon dioxide at 25 °C:

$$c_{rms} = \sqrt{\left(\frac{3 \times 8.31\,\text{J K}^{-1}\,\text{mol}^{-1} \times 298\,\text{K}}{44.0 \times 10^{-3}\,\text{kg mol}^{-1}}\right)} = 411\,\text{m s}^{-1}$$

or about 920 m.p.h.

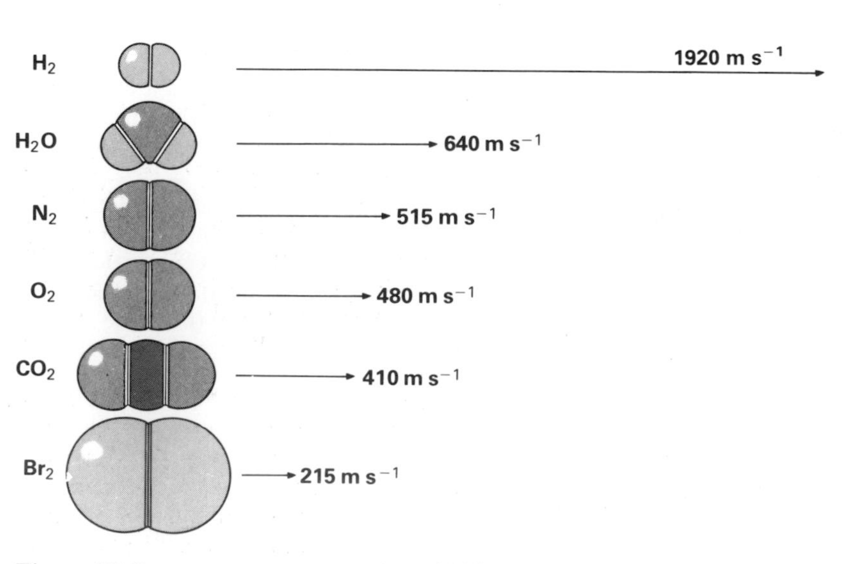

Figure 4.7 Root mean square speeds at 25 °C

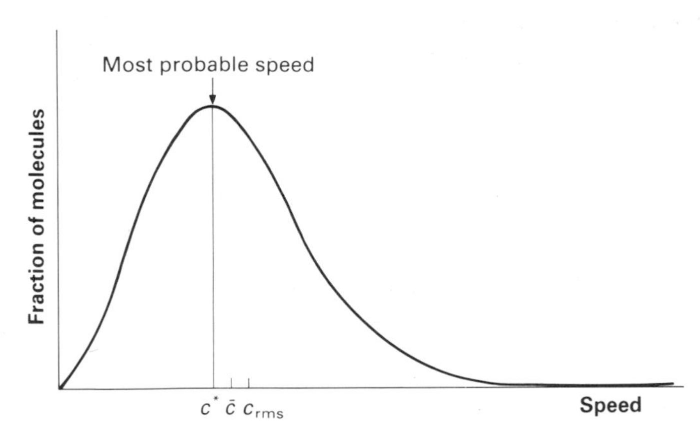

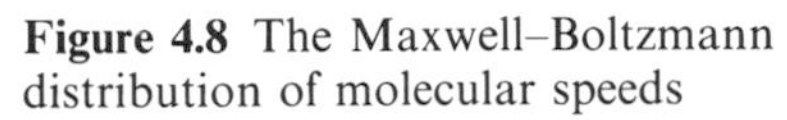

Figure 4.8 The Maxwell–Boltzmann distribution of molecular speeds

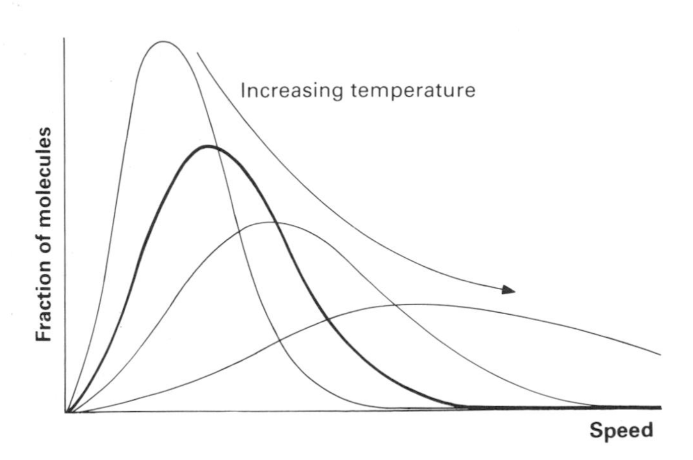

Figure 4.9 The spread of the Maxwell–Boltzmann distribution increases with increasing temperature

The Maxwell–Boltzmann distribution

So far we have discussed the r.m.s. speeds of particles, but in a gas the particles have a range of speeds, and it is often necessary to have more detailed information. The spread of speeds is given by the **Maxwell–Boltzmann distribution**. Its predictions are sketched in Figures 4.8 and 4.9. The point to note from Figure 4.8 is that *a high proportion of molecules have speeds significantly different from the r.m.s. speed.* That many molecules are moving very much faster than the r.m.s. speed has an important effect on the rates at which chemical reactions take place, a topic discussed in Section 14.4. The point to note from Figure 4.9 is that *the higher the temperature, the greater the spread of speeds* (as well as the greater the r.m.s. speed).

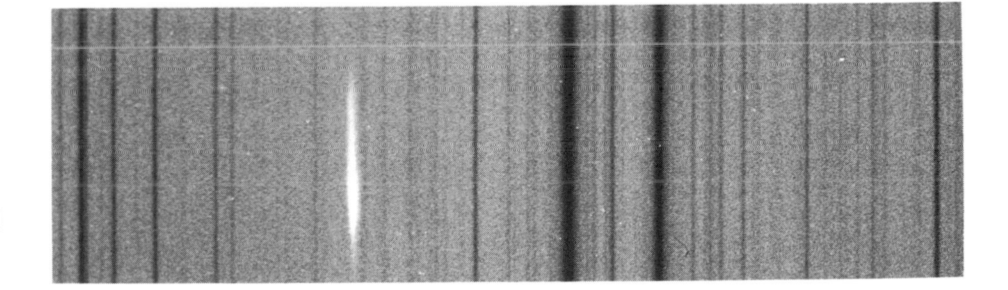

Figure 4.10 A partial spectrum of the Sun: the two prominent dark lines are due to absorption by sodium atoms, and the bright line is due to emission from helium

The escape velocity from Earth – the velocity a particle must reach if it is to escape from the gravitational attraction of the Earth – is about 11 km s^{-1} (25 000 m.p.h.). Although the r.m.s. speeds of atmospheric gas atoms and molecules do not reach this value, the Maxwell–Boltzmann distribution shows that a proportion of them may do so in the hot regions of the upper atmosphere where the Sun's radiation causes reactions that heat the low-pressure gases to several hundred degrees Celsius. This accounts for the absence of the lightest gases, hydrogen and helium, from the atmospheres of the terrestrial planets (helium was detected – by spectroscopy, see Figure 4.10 – on the Sun before it was identified on Earth) and for the almost total absence of an atmosphere on the Moon, where the escape velocity is only 2.4 km s^{-1}.

The spread of speeds in a gas can be measured experimentally by squirting a stream of gas (a **beam**) through a grooved cylinder, as in the Miller–Kusch experiment (see Figure 4.11). The intensity of the beam that completes the journey is proportional to the number of particles that have just the right speed to pass along the groove as the cylinder rotates. It is found that the observed distribution matches the Maxwell–Boltzmann distribution very closely.

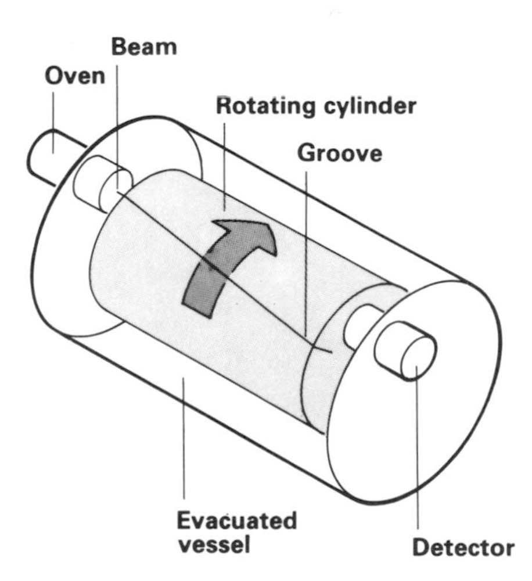

Figure 4.11 The Miller–Kusch experiment

Diffusion and effusion

The process of **diffusion** is the spreading of particles from a region of high concentration to one of low concentration, possibly through a porous barrier (see Figure 4.12(a)). (The explanation for diffusion is that as the particles spread out, there is an increase in the system's entropy – see Chapter 11.) Diffusion occurs in liquids as well as in gases (the spreading of sugar through an unstirred cup of tea takes place by diffusion). **Effusion** is the escape of a gas through a small hole (see Figure 4.12(b)) such as a puncture in a tyre or a spacecraft.

The experimental observation of the rate of diffusion of gases through a porous plug led the Scottish scientist Thomas Graham to formulate the following law in about 1830:

Graham's law: At constant temperature and pressure, the rate of diffusion (or effusion) of a gas is inversely proportional to the square root of its molar mass.

It follows that the molar mass of a gas can be measured by comparing its rate of diffusion with that of another, so long as they have the same pressure and temperature (see Figure 4.13). If the time for a given volume of gas A to diffuse is t_A, and the time for the same volume of gas B to diffuse is t_B (at the same temperature and pressure), then since the time required for a change is inversely proportional to the rate of the process,

$$\frac{t_A}{t_B} = \frac{\text{rate}_B}{\text{rate}_A} = \sqrt{\left(\frac{M_A}{M_B}\right)} \qquad (4.2.4)$$

Hence, if the molar mass of B is known, that of A can be determined.

(a)

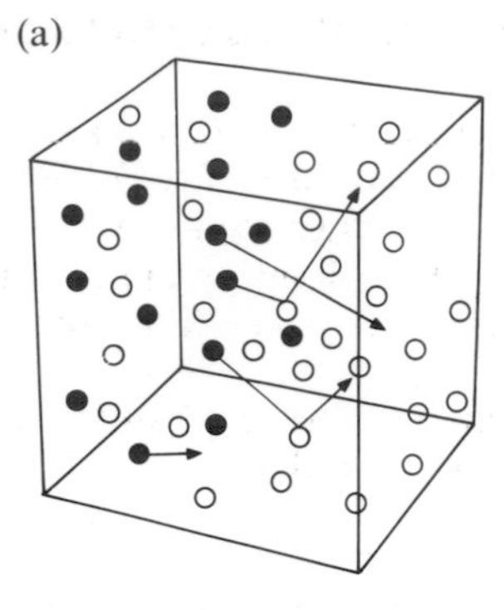

(b)

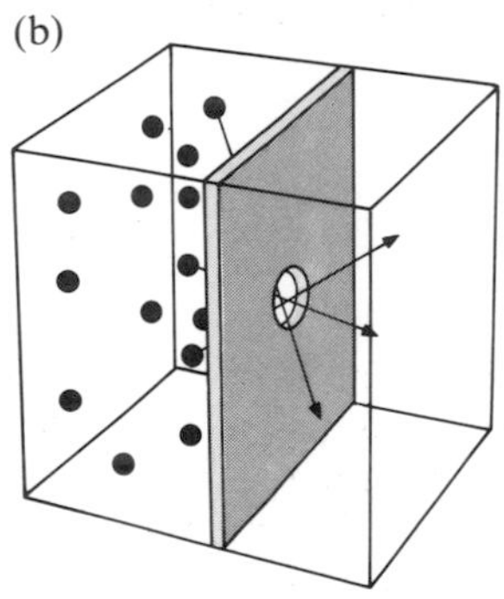

Figure 4.12 (a) Diffusion and (b) effusion

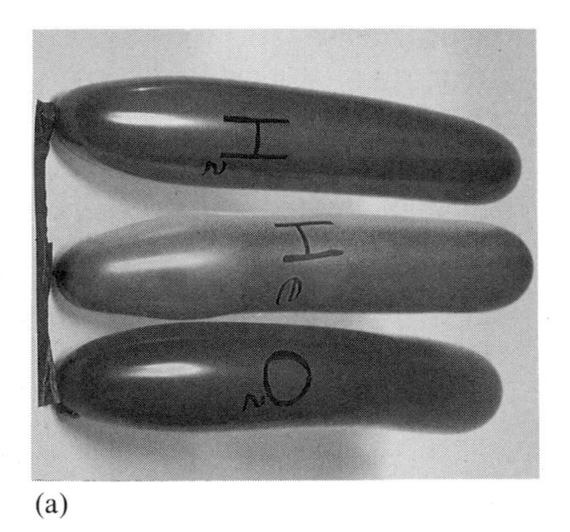

(a)

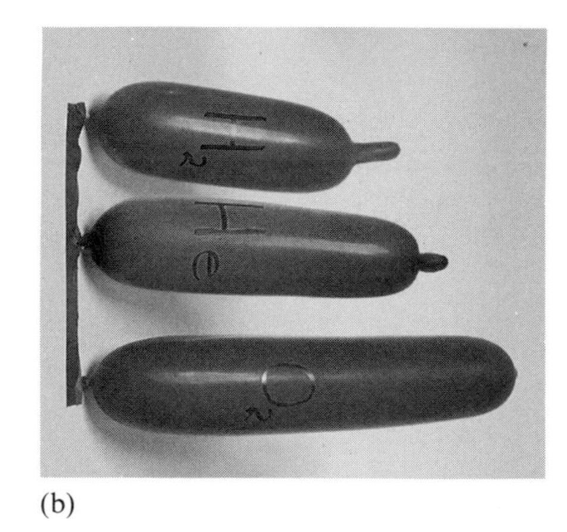

(b)

Figure 4.13 An experiment to demonstrate differential rates of effusion: (a) shows the start of the experiment, with the balloons containing equal volumes of the three gases, and (b) shows the same balloons a few hours later. The rates of effusion are $H_2 > He > O_2$

The different rates of diffusion are easily demonstrated with a long tube having at one end a plug soaked in concentrated hydrochloric acid and at the other a plug soaked in concentrated aqueous ammonia (see Figure 4.14). A white deposit of ammonium chloride forms after some time, and is found further from the ammonia plug because NH_3 molecules are lighter than HCl molecules, and diffuse faster.

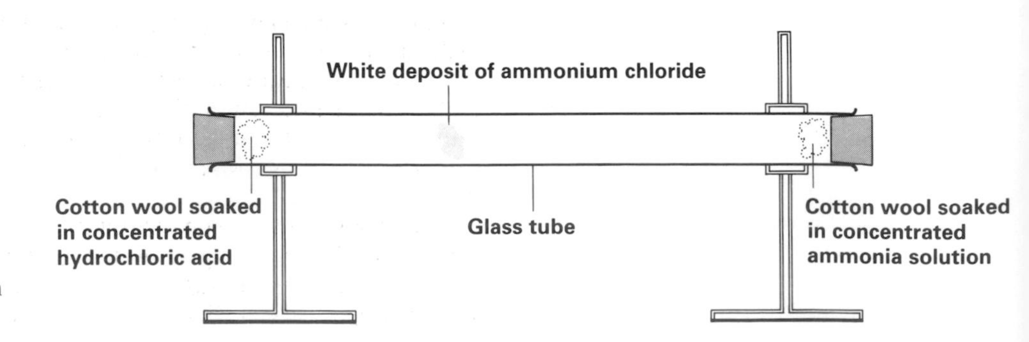

Figure 4.14 Differential rates of diffusion of ammonia and hydrogen chloride

EXAMPLE

Oxygen was confined inside a vessel in the wall of which there was a small hole. It was found that $50\,cm^3$ of the gas escaped in 20 s. When bromine vapour was confined to the same vessel under the same conditions of temperature and pressure, the same volume escaped in 45 s. What are the molar mass and the relative molecular mass of bromine?

METHOD Assume both gases behave perfectly and use Graham's law in the form of equation 4.2.4. Take $M_r(O_2) = 32.0$; note that $M = M_r\,g\,mol^{-1}$.

ANSWER From equation 4.2.4, with A identified as Br_2 and B as O_2,

$$\sqrt{\left(\frac{M(Br_2)}{32.0\,g\,mol^{-1}}\right)} = \frac{45\,s}{20\,s} = 2.25$$

Therefore $M(Br_2) = (2.25)^2 \times 32.0\,g\,mol^{-1} = 162\,g\,mol^{-1}$. It follows from $M = M_r\,g\,mol^{-1}$ that M_r of Br_2 is 162.

COMMENT The true M_r value is 159.8. Instead of dealing with a vessel with a small hole in the wall, an easier experiment is to confine the gas in a porous pot; the same equations apply because the porosity is like a very large number of holes through which effusion can occur. Note that only a small amount of gas must be allowed to escape so that the gas pressure remains almost constant throughout the experiment.

The kinetic theory of gases easily accounts for Graham's law. The rate of diffusion is proportional to the frequency with which molecules strike the area of the pores. That frequency is proportional to their r.m.s. speed, which is proportional to $1/\sqrt{M}$. Therefore the rate of diffusion is proportional to $1/\sqrt{M}$, in accord with Graham's law.

Molecular diffusion through a porous barrier is used in one method of isotope enrichment, especially for uranium. The working material is the volatile, highly corrosive UF_6. The enrichment depends on $^{235}UF_6$ ($M = 349\,g\,mol^{-1}$) diffusing slightly more rapidly than $^{238}UF_6$ ($M = 352\,g\,mol^{-1}$). The rates of diffusion are in the ratio $\sqrt{(352/349)} = 1.004$, so that in order to achieve commercially acceptable separation the vapour has to be allowed to diffuse many times through porous barriers. Calculations like this show why uranium separation plants are so huge (and difficult to conceal).

4.3 Real gases

For a perfect gas $pV_m/RT = 1$. Measured values of pV_m/RT are plotted in Figure 4.15. Note that the pressures are very large: the imperfections are almost negligible at room temperature and pressure. In some cases (generally for pressures below about 350 atm) pV_m/RT is less than 1, while at higher pressures it is greater than 1. What can be the reasons?

Molecules interact with each other (see Figure 4.16). That is an obvious conclusion from everyday experience: gases can be liquefied, their molecules sticking together on account of the attractive forces between them. (We investigate the nature of these forces in the next chapter.) Furthermore, molecules are not points; they occupy a definite volume. Therefore statements **2** and **3** defining the kinetic theory are only approximately true and are significantly in error when the density of the gas is so high that its molecules are close together.

The attractions between molecules have the effect of reducing the pressure of the gas. The product pV_m is therefore expected to be *less* than the perfect gas value when the attractive forces are important. This is the case at pressures up to about 350 atm. At even higher pressures, when the gas is

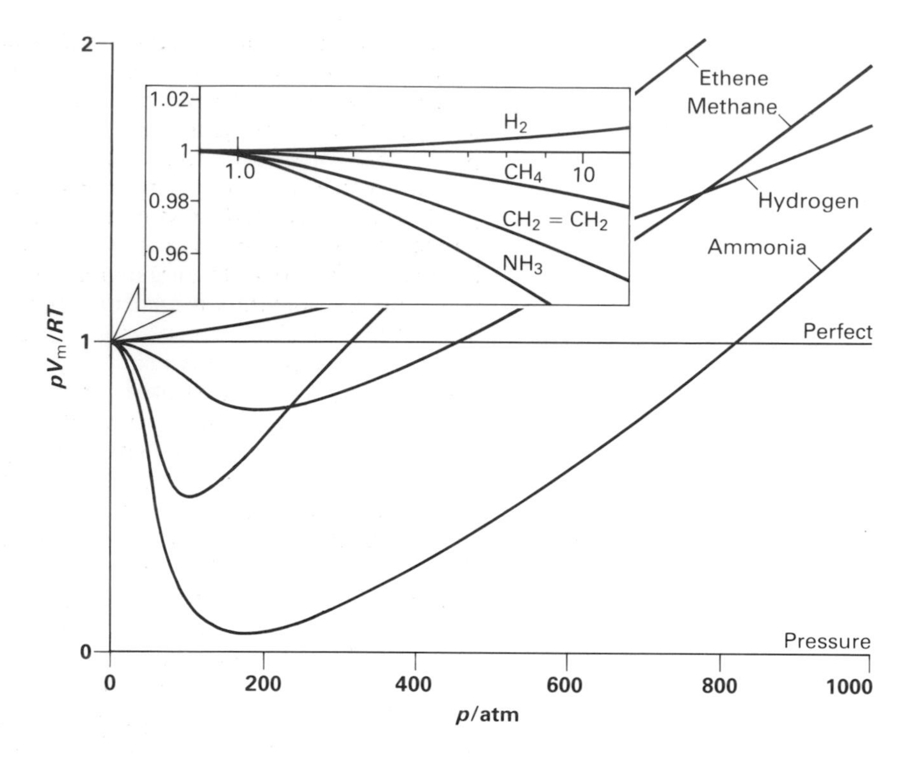

Figure 4.15 The dependence of pV_m/RT on the pressure for real gases

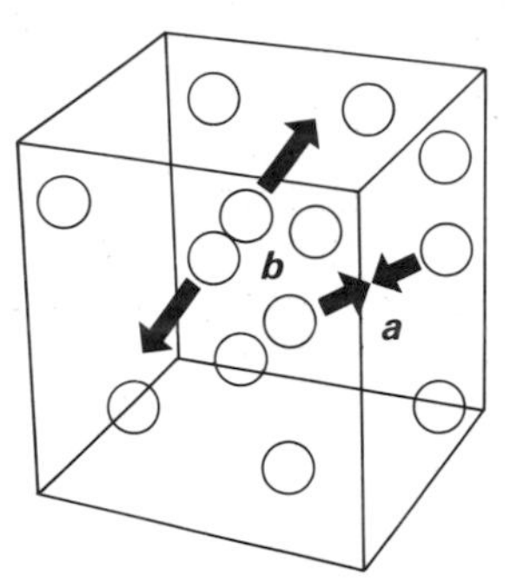

Figure 4.16 (a) The attractive and (b) the repulsive terms in the van der Waals equation

denser, the non-zero volume of the molecules becomes more important and tends to drive them apart. As a result, the product pV_m becomes greater than its perfect gas value. This is why pV_m/RT changes from being less than 1 at high pressures to being greater than 1 at very high pressures. Figure 4.17 shows that as the temperature is increased, the imperfections become less significant. The effect of the intermolecular forces is reduced because of the greater kinetic energy of the molecules.

Many people have tried to modify the perfect gas equation in order to obtain a better description of real gases. The first and most famous attempt was made by van der Waals (1873) who proposed the equation

$$(p + a/V_m^2)(V_m - b) = RT \qquad (4.3.1)$$

in place of $pV_m = RT$. The constant a takes account of the attractive forces

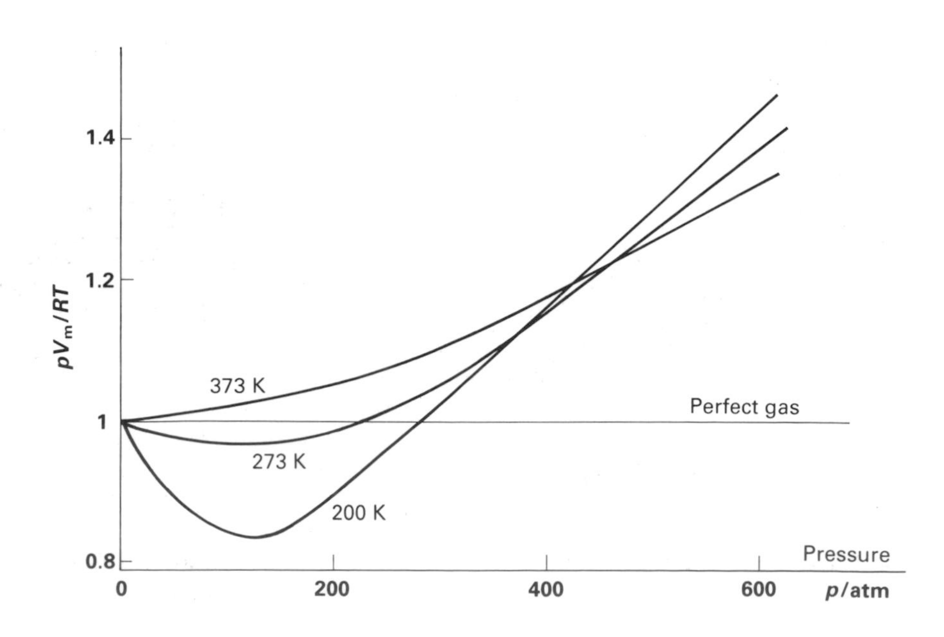

Figure 4.17 The effect of temperature on the pV_m/RT plot for nitrogen

between particles. The constant b takes account of the repulsive forces by supposing that, as the particles themselves occupy a volume, the free volume in the container is reduced from V_m to $V_m - b$. Values of a and b have been determined for many gases, and the van der Waals equation can be used to predict the properties of real gases in a handy but still approximate way.

Some idea of the success of the van der Waals equation is obtained by plotting the calculated pressure against the molar volume at a series of temperatures, in order to compare its predictions with experiment. Each curve is called an **isotherm** (from the Greek for 'equal temperature'). The experimental isotherms for carbon dioxide are shown in Figure 4.18, and those calculated from the van der Waals equation in Figure 4.19. There is reasonable agreement between them, apart from the strange behaviour for isotherms at temperatures lower than the **critical temperature**, T_c.

The significance of the experimental isotherms can be illustrated by following the curve labelled ABCD in Figure 4.18. Initially, close to A, the volume decreases as the pressure is increased (roughly in accord with the perfect gas equation, but in better agreement with the van der Waals equation). At B there is a sudden reduction of volume without the exertion of extra pressure: the gas has condensed to a liquid. (The van der Waals equation does not pretend to be a good description of a liquid, and so it is best to ignore the oscillatory behaviour of the calculated curves in Figure 4.19.) Any further reduction in volume requires great pressure, which accounts for the very steep rise of the isotherm from C to D.

The behaviour of the gas depends on the temperature. At a higher temperature than that for the ABCD isotherm the gas condenses to a liquid at a higher pressure (B′ on the next isotherm) and the volume of the liquid produced, C′, is greater than before. These changes continue at higher temperatures, until on the **critical isotherm**, the one at T_c, the density of the liquid produced, C″, is exactly the same as that of the gas from which it condenses, B″. In other words, at the critical temperature the change from gas to liquid is continuous, and we cannot at any stage call one region of the sample a gas and another a liquid. At the critical temperature there is no surface dividing the two states of matter, and above the critical temperature the gas does not condense.

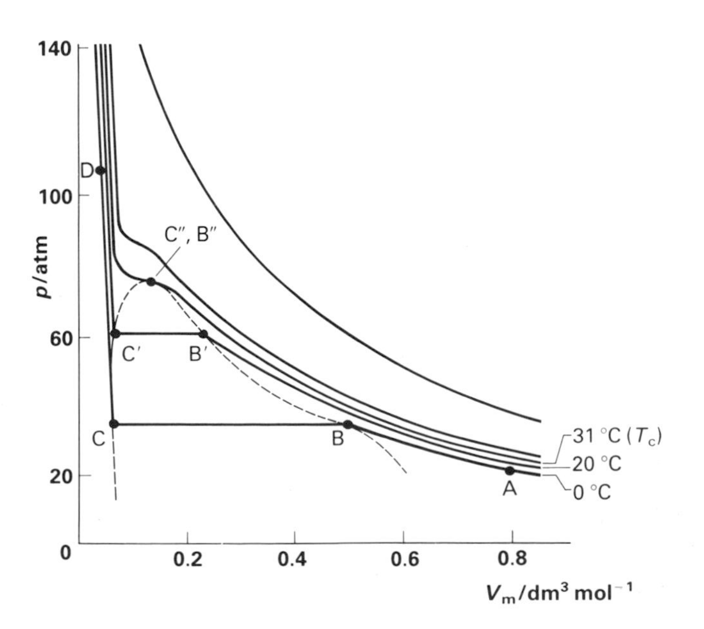

Figure 4.18 Experimental isotherms for carbon dioxide

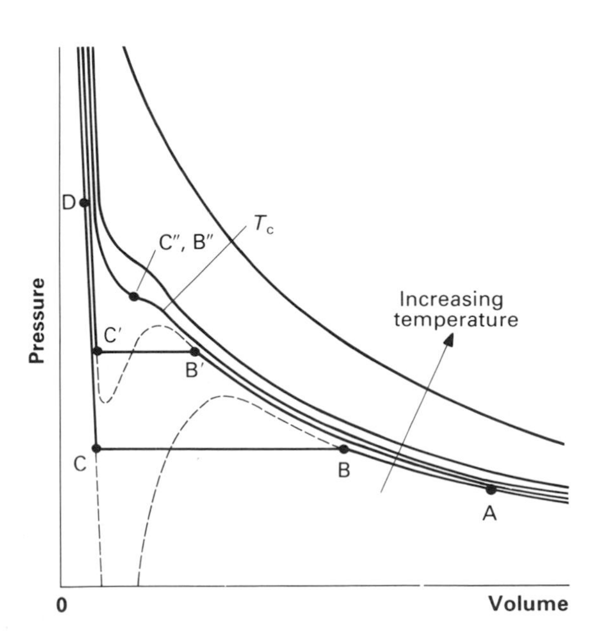

Figure 4.19 Calculated isotherms for carbon dioxide

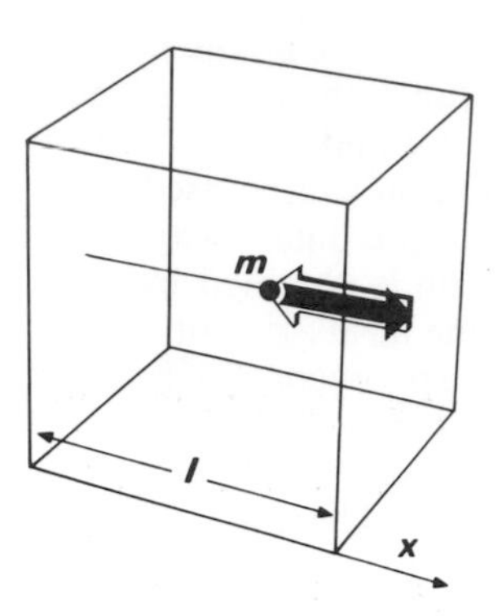

Figure 4.20 The model for calculating pressure; the arrows represent momenta before and after a collision with the wall

Appendix 4.1: The calculation of the pressure

Pressure is *force per unit area.* In order to calculate the pressure exerted by the particles inside a container of volume V, we calculate the average force they exert per unit area of the walls. For simplicity, take the container to be a cube of side l (and so $V = l^3$). If the total force exerted on one of the walls is f, the pressure on the wall is force/area, or f/l^2.

The force is calculated using Newton's laws of motion. According to his second law, *the rate of change of momentum of a particle is equal to the force* acting on it. Consider, therefore, a single particle of mass m travelling towards the wall perpendicular to the x-axis (see Figure 4.20). If its velocity along x is v_x, its momentum along x is mv_x. Immediately after the collision it has reversed its direction, and is travelling backwards with the same speed as before (the collision is elastic). Its momentum along x is now $-mv_x$. As a result of the collision with the wall there has been a change of momentum of magnitude $2mv_x$.

In order to calculate the rate of change of momentum we need to know how often the collisions occur. The molecule revisits the wall after it bounces off the one opposite (we deal below with the possibility of a collision during its journey), and the total time for the round trip is $2l/v_x$. The frequency of collisions it makes with the wall is therefore 1/(time between collisions) or $v_x/2l$. The rate of change of momentum of the particle is therefore the change of momentum on one collision multiplied by the frequency of the collisions:

$$\textit{Rate of change of momentum} = (2mv_x) \times (v_x/2l) = mv_x{}^2/l$$

There are N particles present in the container. They have a wide range of different velocities along the x-direction. Therefore the total rate of change of momentum of all the particles colliding with the wall is the sum of all the individual contributions:

$$\textit{Total rate of change of momentum} = (m/l)\{v_x{}^2(1) + v_x{}^2(2) + \ldots + v_x{}^2(N)\}$$

where $v_x(1)$ is the velocity of particle 1 in the x-direction, and so on. The *mean square velocity* along x is the average value of $v_x{}^2$ for the entire sample:

$$\overline{v_x{}^2} = (1/N)\{v_x{}^2(1) + v_x{}^2(2) + \ldots + v_x{}^2(N)\}$$

(The bar indicates an average value.) Therefore

$$\textit{Total rate of change of momentum} = (Nm/l)\overline{v_x{}^2}$$

This average value is valid even when particles collide in flight between the walls because the collisions are assumed to be elastic.

The total rate of change of momentum is equal to the total force acting on the particles. But according to Newton's third law ('action = reaction') this must also be the magnitude of the force experienced by the wall where the collisions are taking place. Therefore

$$\textit{Average force exerted on wall} = (Nm/l)\overline{v_x{}^2}$$

$$\textit{Average pressure exerted on wall} = \textit{force} \div \textit{area} = (Nm/l)\overline{v_x{}^2} \div l^2 = (Nm/l^3)\overline{v_x{}^2}$$

This is the force exerted on the wall perpendicular to the x-direction. The same force is exerted on the walls perpendicular to the y- and z-directions, and so the *average pressure on any wall* is

$$p = (Nm/l^3)\overline{v_x{}^2} = (Nm/l^3)\overline{v_y{}^2} = (Nm/l^3)\overline{v_z{}^2}$$

We now introduce the *mean square speed* of the particles as

$$\overline{c^2} = \overline{v_x{}^2} + \overline{v_y{}^2} + \overline{v_z{}^2}$$

According to the preceding expression all three contributions are the same, and so $\overline{c^2} = 3\overline{v_x{}^2}$. Consequently

$$p = \tfrac{1}{3}(Nm/l^3)\overline{c^2}$$

Since $l^3 = V$, the volume of the container,

$$pV = \tfrac{1}{3}Nm\overline{c^2}$$

which is the equation used in the text.

Summary

- ☐ Boyle's law states that the product pV is a constant for a fixed temperature and mass of gas.
- ☐ Charles's law states that V is proportional to T for a fixed pressure and mass of gas, with T on the Kelvin scale.
- ☐ Avogadro's law states that gases containing equal amounts of substance occupy the same volume at the same temperature and pressure.
- ☐ The zero of the Kelvin scale is the temperature at which a perfect gas occupies zero volume; the Kelvin and Celsius scales are related by $T/\mathrm{K} = 273.15 + t/°\mathrm{C}$
- ☐ The perfect gas equation is $pV = nRT$; real gases obey this equation increasingly closely as their density is reduced.
- ☐ Dalton's law of partial pressures states that a gas in a mixture exerts the same pressure (its partial pressure) as when it alone occupies the same container at the same temperature.
- ☐ The mole fraction of a component in a mixture is defined as $x_A = n_A/n$, where n is the total amount of substance; the partial pressure is related to the total pressure by $p_A = x_A p$.
- ☐ The kinetic theory of gases is based on a model in which the only contribution to the energy is the kinetic energy of the chaotically moving particles.
- ☐ The root mean square speed of molecules in a gas is proportional to the square root of the temperature and inversely proportional to the square root of the molar mass.
- ☐ The Maxwell–Boltzmann distribution gives the proportion of particles in a gas having a given speed; the higher the temperature, the greater is the spread of speeds.
- ☐ Graham's law states that at constant temperature and pressure the rate of diffusion of a gas is inversely proportional to the square root of its molar mass.
- ☐ The value of pV_m/RT for real gases is generally less than unity at high pressures (up to about 350 atm) on account of the attractions between the molecules; at very high pressures it is greater than unity on account of their repulsive interactions.
- ☐ The van der Waals equation is an approximate description of real gases in which the constant a represents the effects of attractions and the constant b those of repulsions.
- ☐ An isotherm shows the dependence of the pressure of a gas on its volume at a constant temperature.
- ☐ Above the critical temperature a gas cannot be condensed to a liquid by the application of pressure.

PROBLEMS

1 Outline the experiments that led to the formulation of the perfect gas equation.

2 Explain non-mathematically why (a) a gas fills any container it occupies, (b) a gas exerts a pressure on the walls, (c) the pressure of a given amount of gas increases when its temperature is increased in a container of constant volume.

3 A sample of gas supports a column of mercury 720 mm high in an open-ended manometer; what is the pressure of the gas? Express the answer in atm and kPa. (Density of mercury, $13.6\,g\,cm^{-3}$; acceleration due to gravity, $9.81\,m\,s^{-2}$; 1 atm corresponds to 101.3 kPa.)

4 The cathode-ray tube of a television set has a volume of about $5\,dm^3$. The pressure inside it is typically 0.10 Pa at 25 °C. Estimate the number of molecules of air it contains.

5 When the set in Question 4 is operating the temperature inside the tube rises to 80 °C. What is the pressure?

6 A typical plant used for the production of ammonia by the Haber–Bosch process stores the ammonia at 720 K and 3×10^7 Pa before transferring it to a cooling plant for storage as liquid. On the assumption that the gas behaves perfectly under these conditions, estimate the amount of substance of NH_3 that the $100\,m^3$ vessel stores. What mass of ammonia does that represent?

7 A balloon of volume $1000\,m^3$ was filled with helium at 27 °C to a pressure of 1.01 atm. Upon ascending the temperature fell to −10 °C while the external pressure fell to 0.740 atm. What mass of helium must be released in order to keep the volume constant?

8 A vessel of volume $5\,dm^3$ was filled with (a) 2 mol H_2, (b) then a further 1 mol Cl_2, and then (c) the mixture was sparked, leading to the formation of 2 mol HCl(g). What are the total pressure and the partial pressures of the components at each stage, the volumes being measured at 25 °C throughout?

9 Calculate the root mean square speeds of (a) Cl_2, (b) CH_4 molecules at 298 K and 1 bar. What is the influence on the speeds of an increase of pressure at constant temperature? What is the ratio of the speed of $^{37}Cl_2$ to that of $^{35}Cl_2$?

10 Observation of the spectra of ^{57}Fe atoms on the surface of the Sun shows (from the Doppler shift) that they have a root mean square speed of about $1.6\,km\,s^{-1}$. What is the temperature of the Sun's surface?

11 A small volume of gas of unknown molar mass was allowed to effuse from a container during a measured interval, which was compared with the time that it took the same volume of hydrogen chloride to effuse under the same conditions. While hydrogen chloride took 90 s to effuse, the other gas required only 60 s. What is its molar mass?

12 List the assumptions on which the kinetic theory of gases is based. Which assumptions are false? In what ways do real gases differ from the ideal behaviour predicted from these false assumptions?

13 The molar volume of carbon dioxide was found to be $1.32\,dm^3\,mol^{-1}$ at 48 °C and 18.4 atm pressure. Compare this observation with the pressure calculated on the basis of (a) the perfect gas equation, (b) the van der Waals equation (the values of the constants for carbon dioxide are $a = 3.59\,dm^6\,atm\,mol^{-2}$ and $b = 4.27 \times 10^{-2}\,dm^3\,mol^{-1}$).

14 (a) List the main assumptions of the kinetic theory for ideal [perfect] gases.
(b) Give the origins of two of the main differences between the behaviour of real and ideal gases.
(c) State the dependence of the root mean square velocity of a gas on (i) the relative molecular mass, (ii) the temperature.
(d) Calculate the root mean square velocity of dinitrogen (N_2) molecules at 7 °C. *(Welsh part question)*

15 Discuss briefly, but critically, the basic postulates underlying the kinetic theory of gases and explain (a) gas pressure, (b) thermal expansion at constant pressure, (c) the use of $\overline{c^2}$, the mean square velocity, in the equation

$$pV = \tfrac{1}{3}mN\overline{c^2}$$

where N molecules of ideal [perfect] gas each of mass m, occupy a volume V at a pressure p.

Calculate:
(i) the kinetic energy (in joules) of the molecules in one mole of ideal gas at 47 °C.
(ii) the *root* mean square velocity of hydrogen iodide molecules, HI, at 47 °C in the gaseous phase (in this calculation the molar mass of hydrogen iodide must be expressed in the appropriate SI unit, i.e., *kilograms per mole*).
(iii) the *ratio* of the *root* mean square velocities of oxygen, O_2, and hydrogen iodide at 47 °C;
(iv) the time expected to be taken for a given volume of hydrogen iodide at 47 °C to effuse (or diffuse) through a pin-hole if the same volume of oxygen under the same conditions takes 60 s.
(v) Compare this with that predicted by Graham's law of diffusion, which should be stated. *(SUJB)*

16 In the kinetic theory of gases, what basic assumptions are made in the derivation of the expression $pV = \frac{1}{3}Nm\overline{c^2}$ for an ideal [perfect] gas?

Use this expression (a) to derive Graham's law of diffusion, (b) to calculate the root mean square speed ($\sqrt{\overline{c^2}}$) for argon at s.t.p. [273 K, 1 atm], and (c) to calculate the kinetic energy for one mole of an ideal gas in terms of the gas constant, R, and the temperature, T.

Sketch the curve showing the distribution of molecular speeds in a gas and mark on the curve the position of the *average speed*, the *most probable speed* and the *root mean square speed.*

Describe how the distribution of molecular speeds changes when the temperature is increased.

Explain how the gas equation $pV = nRT$ has been modified to take account of the behaviour of *real* gases. *(Oxford and Cambridge)*

17 State the ideal (perfect) gas laws. Show how they must be modified to explain the behaviour of real gases.

Two glass bulbs of capacity $0.2\,dm^3$ and $0.1\,dm^3$ joined only by a narrow capillary contain air at 288 K and 1 atmosphere pressure. Find the pressure of the air when the larger bulb is heated by steam to 373 K whilst the smaller remains at 288 K. State any simplifying assumptions you may make. *(Oxford Entrance)*

5 Solids

In this chapter we look at the structures of solids. We examine the forces that hold atoms, ions and molecules together, and look at ways of discovering their arrangement inside solids by X-ray diffraction. Then we investigate crystals, and see something of the wide range of structures that are responsible for most of what is permanent in the world around us.

Matter often comes in lumps. In contrast to gases, where the particles are dispersed almost completely chaotically and are in continuous motion, the particles inside solids are almost stationary and are usually located in ordered arrays.

The importance of solids in everyday life hardly needs to be stressed. Much of modern technology is concerned with the modification of the structures of solids, either in their external form, as in the manufacture of cars out of ingots of steel, or in their internal structures in order to achieve special properties. Examples include the development of steels with improved corrosion resistance and greater strength, the development of new semiconductors and the manufacture and application of synthetic polymers.

5.1 The forces of aggregation

There are forces that cause particles to stick together even when the valencies of their atoms have been fully satisfied. The clearest evidence for this is the existence of **condensed states** (i.e., liquid and solid states) of the noble gases. At low enough temperatures these forces can overcome the chaotic thermal motion of the atoms and cause them to aggregate into liquids and solids (see Figure 5.1). There are several forces that can bind particles together as solids, and we examine them in turn.

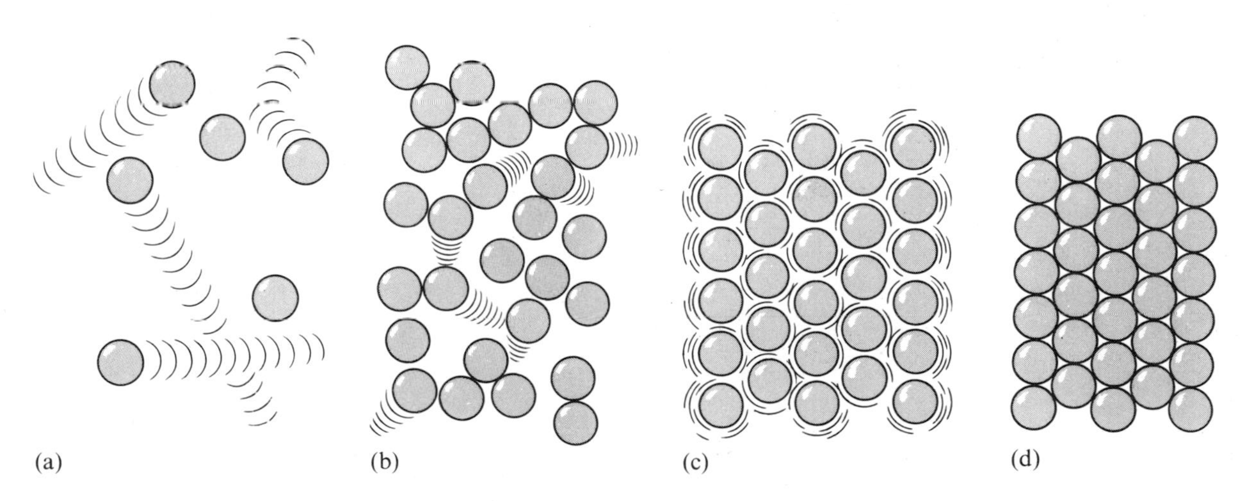

Figure 5.1 (a) Lowering the temperature of a sample of a gas leads to condensation to (b) a liquid, which freezes to (c) a solid, which would only be totally free of vibrational excitation at (d) absolute zero

Covalent structures

Some lumps of matter are single giant molecules, in which the atoms are linked together into extensive covalently bonded structures. An example is **diamond**, where each carbon atom forms a bond with its four neighbours at the points of a regular tetrahedron, as shown in Figure 5.2. This structure is repeated throughout the crystal, so that a 1 carat diamond (of mass 200 mg) can be thought of as a single molecule containing about 1.0×10^{22} atoms. Since it is difficult to break the strong C—C bonds, and because the structure

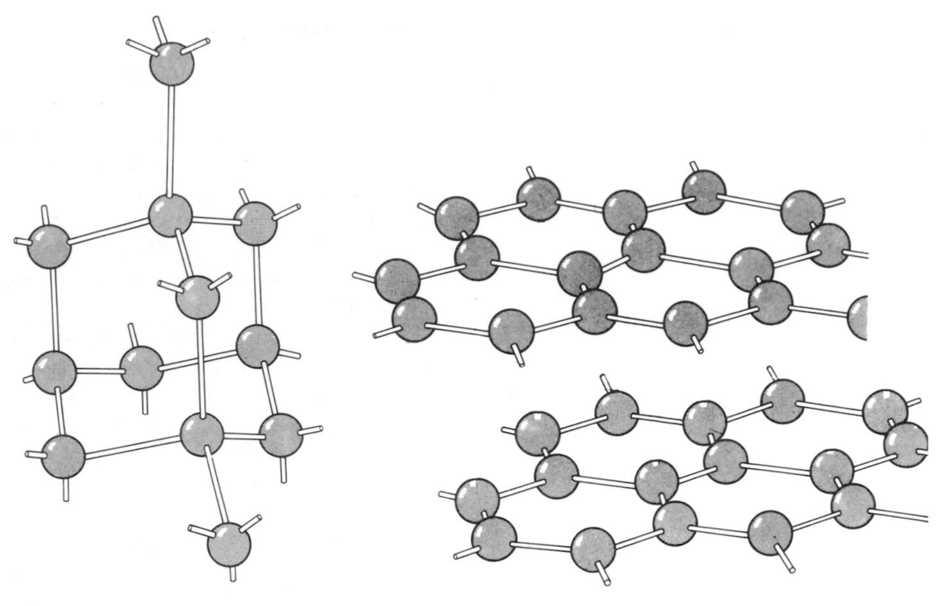

Figure 5.2 The structure of diamond **Figure 5.3** The structure of graphite

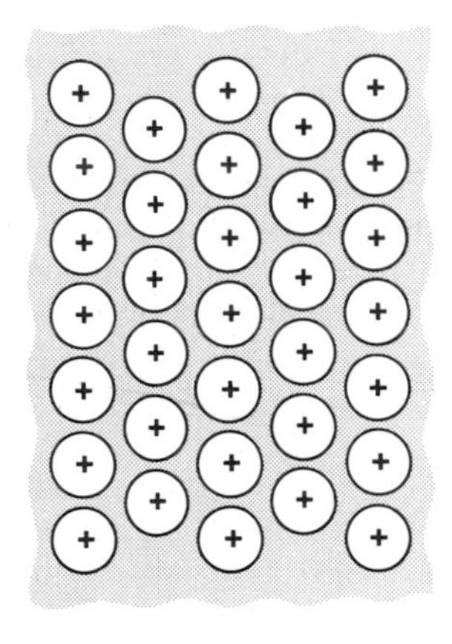

Figure 5.4 A Group I metal

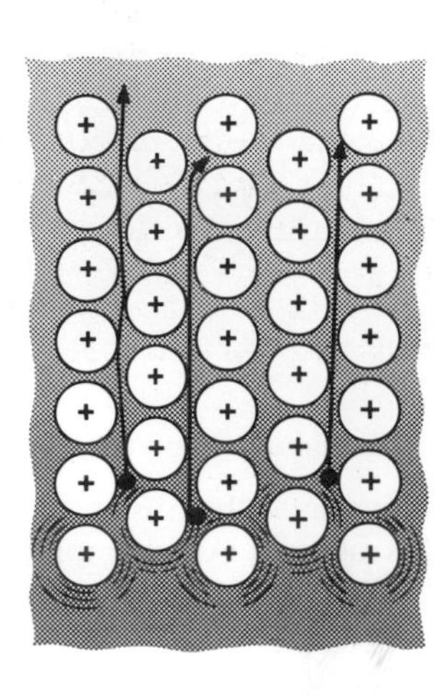

Figure 5.5 The conduction of heat by electrons

is like the steel framework of a large building, the diamond crystal is hard and rigid: the rigidity is typical of other covalently bonded solids (such as quartz, described in Section 20.3). The **graphite** structure is quite different. As Figure 5.3 shows, it consists of planar sheets of carbon atoms arranged in hexagons (like chicken wire). The layers are only weakly bound together; they slide over each other easily, which accounts for graphite's slipperiness.

Metals

A metal is an orderly collection of positive ions surrounded by and held together by a sea of delocalized electrons (see Figure 5.4 and Section 2.5). The diagram shows that the bonding is not directional, as in a covalent solid like diamond, so that the sea of electrons allows the groups of ions to be pushed into a new arrangement. The difference between the directional nature of a covalent bond and the non-directional nature of a metallic bond is the reason why it is far easier to bend a piece of metal than to bend a diamond or a crystal of quartz. Shifting groups of ions through the electron sea can be brought about in a number of ways. These include hammering, rolling, extruding, stretching and bending. The properties of **ductility** (the ability to be drawn out into wires) and **malleability** (the ability to be formed into thin sheets by hammering) are typical of metals.

The presence of the sea of mobile electrons also accounts for the other typical characteristics of metals, including their high electrical and thermal conductivities and their ability to reflect light. High **electrical conductivity** arises from the ability of the electrons to flow through the ion framework when a potential difference is applied across the ends of the metal (ends that may be miles apart if the metal is a part of a national grid). High **thermal conductivity** arises in a similar way. The electrons pick up kinetic energy from the vibrations of the ions where the metal is hot, pass rapidly through the ion framework, collide with a distant ion and cause it to vibrate more vigorously (see Figure 5.5). In other words, the mobile electrons are efficient transporters of energy. The **reflectivity** of metals is another aspect of the mobility of the electrons: an incident light wave forces the electrons near the surface of a metal to oscillate and the incident light is reflected back out instead of penetrating into the bulk. Your reflection in a mirror is the oscillation of the metal's mobile electron sea.

Ionic structures

Common salt is not a gas that has to be bubbled through food: the reason lies in the electrostatic **Coulomb forces** arising from the charges on the sodium and chloride ions. These forces act in all directions, and so it is energetically favourable for other ions to cluster near the first pair, and then for others to cluster near them. The usual form of sodium chloride is therefore an extensive aggregate of positive and negative ions held together by their mutual electrostatic attraction. This is the common feature of all ionic species, including the other alkali metal halides, calcium nitrate and magnesium sulphate.

An **ionic crystal** is a three-dimensional orderly arrangement of ions. Its stability and rigidity are due to the electrostatic interactions between the ions, and only when the crystal has been heated to a high temperature (801 °C in the case of sodium chloride) do the ions shake around so violently that the electrostatic forces are overcome and the crystal melts into a liquid. In Section 9.4 we look at ways of measuring the strengths of binding in ionic crystals.

Van der Waals forces

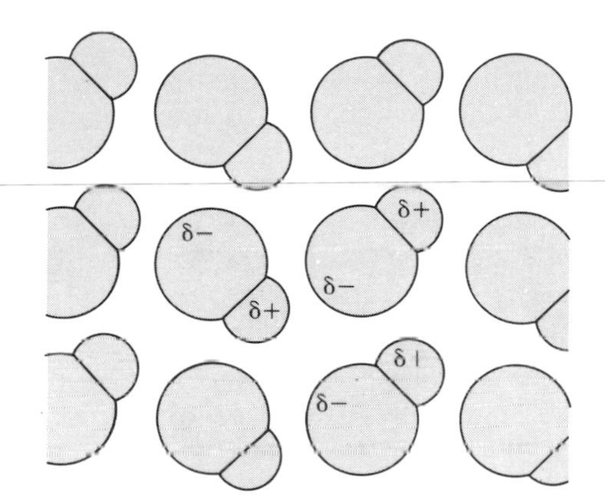

Figure 5.6 Dipole–dipole forces and the structure of solid hydrogen chloride

We now consider solids formed from discrete, non-ionic particles. When the particles are molecules these are called **molecular solids**. How, for instance, do molecules of methane or of benzene stick together? Also, what forces are responsible for the condensation of the noble gases?

The **van der Waals forces** between discrete covalent molecules are due to interactions between dipoles and are the forces responsible for the gas imperfections discussed in Section 4.3. There are two principal types, the **dipole–dipole forces** and the **dispersion forces** (or **London forces**).

Dipole–dipole forces can occur only when both molecules are polar (have permanent electric dipoles – see Section 2.6). From Figure 5.6 it can be seen that when the molecules are in the orientations shown, there is a favourable electrostatic interaction between neighbouring opposite partial charges. As a result they stick together.

Dispersion forces occur not only between polar molecules but also between non-polar molecules, and their effect normally dominates any dipole–dipole forces that are also present. They even occur between noble gas atoms. Their origin is found in the fact that the electron distribution of a molecule (or an atom) should not be thought of as being static and frozen, but as continuously flickering from one arrangement to another. Two nearby molecules interact through their *instantaneous* dipoles, and when the temperature is low enough the attraction is strong enough to account for the condensation of the noble gases, the hydrocarbons, iodine (see Figure 5.7) and so on.

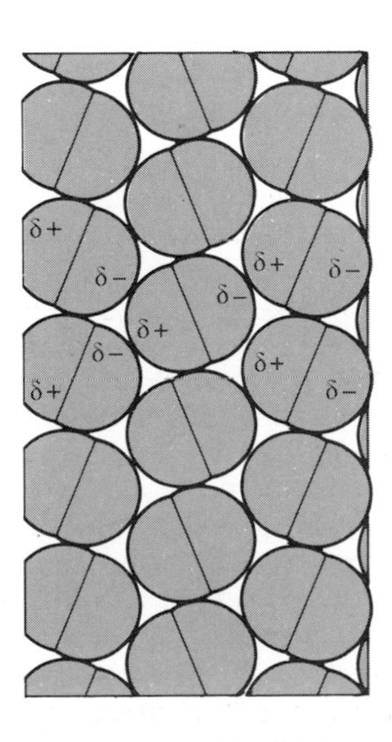

Figure 5.7 Dispersion forces and the structure of solid iodine (the dipoles are instantaneous, *not* permanent)

Van der Waals interactions are much weaker than ionic forces, and so molecular solids are often soft. Margarine, for instance, is a collection of molecules sticking together through van der Waals interactions, and a knife blade can readily reorganize the arrangement. The boiling and melting points of substances held together by van der Waals forces are also low (in comparison with ionic crystals) because little thermal agitation is needed to overcome the interactions. Van der Waals interactions are largest between large molecules, because dispersion forces are largest when a molecule's electron clouds are extensive. The nuclei then have only a weak grip on the outer electrons, and so the instantaneous dipole moment can flicker between large values. A molecule with easily distorted electron clouds is said to be highly **polarizable**.

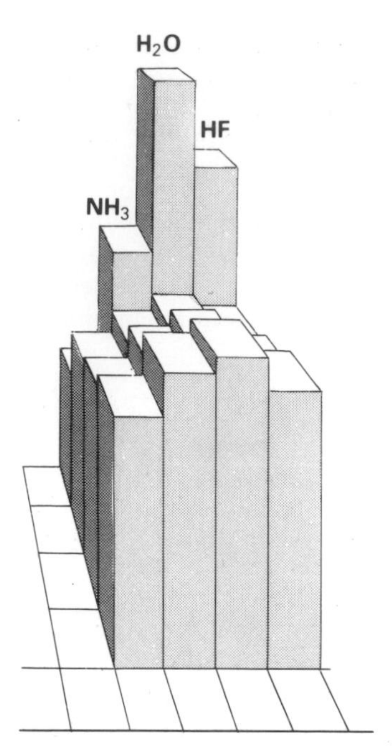

Figure 5.8 The boiling points of some hydrides

EXAMPLE

The boiling points of three of the hydrogen halides under 1 atm pressure are HCl: 188 K; HBr: 206 K; HI: 238 K. Account for the trend in values.

METHOD Note that all three compounds are discrete, covalent molecules in both the liquid and the vapour. The intermolecular forces that must be overcome are dipole–dipole forces and dispersion forces, ignoring hydrogen bonding (see below). Think about the permanent dipole moments on the basis of the electronegativities of the halogen atoms involved (see Table 2.1).

ANSWER The electronegativities of the halogen atoms increase along the series I < Br < Cl, and therefore the magnitudes of the permanent dipole moments of the molecules increase along the series HI < HBr < HCl. On the basis of dipole–dipole interactions we would expect the boiling points to increase in the same order. This is contrary to observation. Dispersion forces are largest between large, many-electron molecules. The number of electrons increases along the series HCl < HBr < HI, and so on the basis of dispersion forces we would expect the boiling points to increase in the same order; this is in agreement with the data. We therefore conclude that the dominant contribution to the intermolecular forces in these compounds is from the dispersion forces.

COMMENT This Example reinforces the point made in the text that dipole–dipole forces rarely exert a dominating influence on boiling points.

Intermediate in structure between molecular solids and covalent solids are many of the polymeric materials that are such a characteristic feature of modern life. As an example, consider polythene (more formally, poly(ethene)). This consists of long chains of covalently bonded —CH_2CH_2— units. The chains themselves stick together as a result of their van der Waals interactions. Some plastics also contain additives and covalent cross-links between the chains; in rubber, sulphur–sulphur cross-links are introduced during vulcanization (see Box 33.3).

Hydrogen bonds

A **hydrogen bond** is a special kind of bond in which the link is due to a proton lying between two very electronegative atoms, such as fluorine, oxygen or nitrogen. We have seen (in Section 2.6) that the O—H bond is polar in the sense $O^{\delta-}$—$H^{\delta+}$: the electron-withdrawing effect of oxygen is so great that the proton is partially exposed and can attach itself electrostatically to some other region of high electron density, such as the lone pairs of other oxygen or nitrogen atoms. The hydrogen bond is often denoted $X^{\delta-}$—$H^{\delta+}$---$Y^{\delta-}$, where X and Y are strongly electronegative atoms.

Hydrogen bonding is of great importance for the structure of the condensed states of water. Its role in the liquid can be seen by examining the boiling points of the hydrides of some elements, shown in Figure 5.8: liquid water has an anomalously high value (so too have ammonia and hydrogen fluoride). The explanation is that in these compounds the hydrogen bonds bind the molecules into a liquid instead of leaving them free as in a gas.

The hydrogen bonds of water remain when it solidifies. They then hold the molecules into an open crystal structure, shown diagrammatically in Figure 5.9. The openness of this structure is also important for the existence of life. This is because the hydrogen bonds not only help to hold the water molecules rigidly as a solid, but also hold the molecules apart, and ice is less dense than water. As a result, in winter the tops of ponds and oceans freeze first and help to insulate the lower regions, and aquatic life is not inhibited (see Figure 5.10).

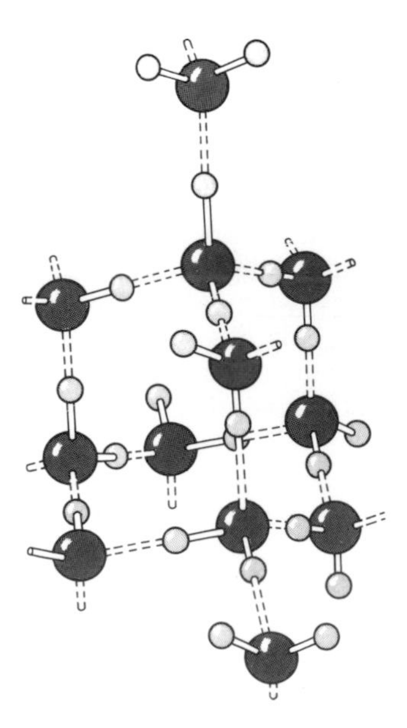

Figure 5.9 The structure of ice

Figure 5.10 The water on the surface of the sea freezes first, forming 'pancake ice'

Hydrogen bonding often survives even in the gas phase. For example, ethanoic acid vapour contains a high proportion of dimers (see Figure 5.11), and this increases its apparent molar mass (see Section 31.2). In liquid hydrogen fluoride the hydrogen bonding is so strong that the molecules stick together into

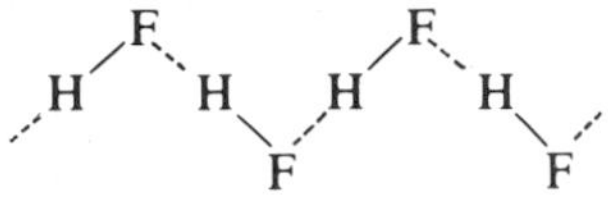

clusters. The structures and physical properties of several important solids are determined largely by hydrogen bonds (see Section 30.5). One example is sugar, which is built from molecules (sucrose) bristling with oxygen and hydrogen atoms. Another is the cellulose of wood, which consists of long chains of glucose units and gets its strength from the hydrogen bonds between the chains. Hydrogen bonds also play a role in proteins (see Section 32.5), in which they organize the long flexible polypeptide chains into definite arrangements; in biology the shape of a molecule plays an important role in its function. The structure of DNA, the molecule central to heredity, is also largely controlled by hydrogen bonding. In this molecule, four nitrogenous bases (adenine, A; cytosine, C; guanine, G; thymine, T) fit neatly together in A–T and G–C pairs (see Figure 5.12) to build up a double-stranded molecule. The specificity of this **base-pairing** is due to the excellent match of each pair by mutual hydrogen bonding and, as a result, each strand can be used as a template for building the other when the molecule replicates.

A summary of the wide variety of solid forms, and some of their typical properties, is given in Table 5.1.

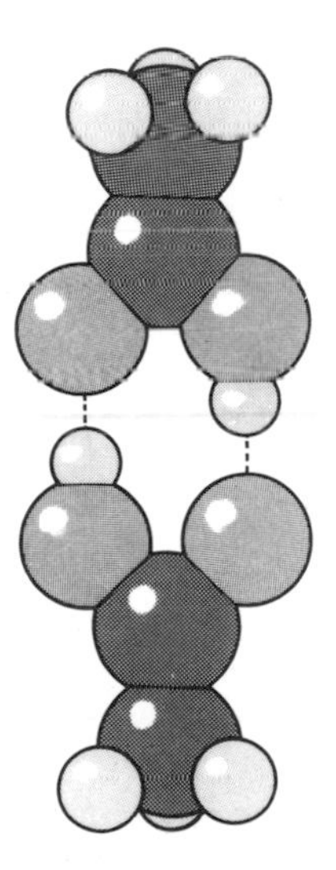

Figure 5.11 The hydrogen-bonded dimer of ethanoic acid

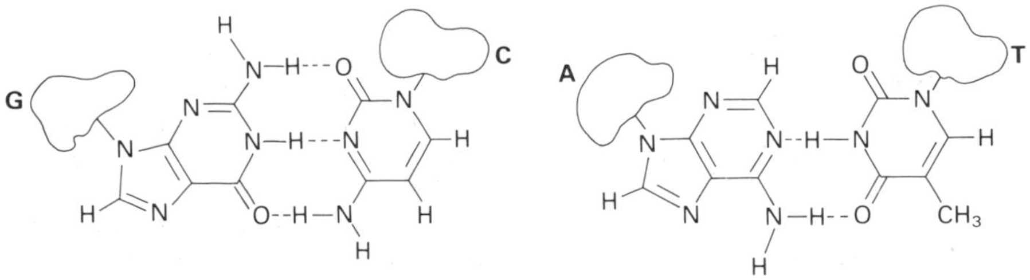

Figure 5.12 Base-pairing in DNA

Table 5.1 Bonding and the nature of solids

	Giant molecules	Metals	Ionic crystals	Simple molecular solids	Polymers
Constituents	atoms	Positive ions and electrons	ions	molecules	molecules
Bonding	covalent between atoms	metallic	coulombic	covalent within molecules, van der Waals (and H-bonding) between molecules	covalent within molecules, van der Waals between molecules (and entanglement)
Examples	SiO_2, C, SiC	Cu, Mg, Au	NaCl, CaO, Na_2SO_4	sucrose, I_2, S_8	poly(ethene), PVC, rubber
Melting point	v. high	high	high	low	moderate
Boiling point	v. high	high	high	low	often decomposed
Physical nature	hard	ductile, malleable	hard but brittle	soft	plastic
Usual solubility:					
(a) water	insoluble	insoluble	soluble	insoluble	insoluble
(b) CCl_4	insoluble	insoluble	insoluble	often soluble	insoluble
Electrical conduction	insulators (except graphite)	conductors	conductors when molten or in solution	insulators	insulators

Solubility and conduction are discussed in more detail in Chapter 8.

(a)

(b)

Figure 5.13 Particles of crystalline materials, such as sodium chloride (a), often have regular shapes, while those of non-crystalline materials, such as obsidian (b), do not

5.2 Investigating crystals

The information on the structure of the solids that we have been describing has come from experimental observations. The most helpful technique has been a version of X-ray diffraction (Section 3.4). X-ray diffraction was first used to discover the way that particles are stacked together to form simple crystals, and in this application is known as X-ray crystallography (see Box 5.1).

Figure 5.14 Etch pits on the surface of a cadmium sulphide crystal, indicating the orderly nature of its structure

BOX 5.1

The development of X-ray crystallography

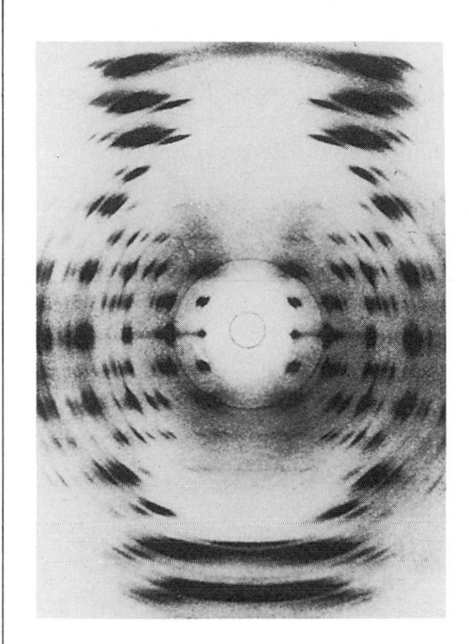

Figure 5.15 An X-ray photograph of DNA: the pattern indicates that the molecule is helical

The origin of X-ray crystallography was a remark made by Max von Laue concerning the then recently discovered X-rays. He put forward the view (in 1912) that, since radiation is diffracted by objects of a size similar to its wavelength, X-rays (which were thought to have wavelengths of the order of 100 pm) ought to undergo diffraction by the parallel planes of ions in a crystal, which are spaced by approximately that distance. The remark was taken up both by von Laue himself and his collaborators Friedrich and Knipping, and also by the Braggs, William and his son Lawrence, who later were to share the Nobel prize for their joint work. Initially the technique was used to obtain the dimensions of crystals and to establish the arrangement of ions in simple compounds. Immense progress came with the introduction of electronic computers, which are able to handle the vast amount of data that comes from the analysis of single crystals of substances diffracting in all directions, and to deal with the complicated and extensive calculations that have to be done in order to extract the detailed information lying within the data. The gathering of data is now extensively automated, and extremely complex molecules can be investigated. Among the landmarks of the applications of the technique are the determination of the structures of penicillin and vitamin B_{12} (by Dorothy Hodgkin in 1949 and 1956, respectively), of lysozyme (by D. C. Phillips in 1965), of myoglobin and haemoglobin (by J. C. Kendrew and M. F. Perutz in the late 1950s), and of the most famous biological molecule of all, DNA (by M. H. F. Wilkins, F. H. C. Crick and J. D. Watson in 1953). Elucidating all these structures took many years of work, even with the use of computers, yet the enormous advances that have taken place in molecular biology in the last decade are a direct consequence of the information they provide.

X-ray crystallography

The basic idea behind the application of X-rays to the determination of the structures of crystals is the interference that occurs when the rays are diffracted – that is, scattered, though they are often (loosely) said to be 'reflected' – by the layers of atoms or ions forming the crystal. This is illustrated in Figure 5.16.

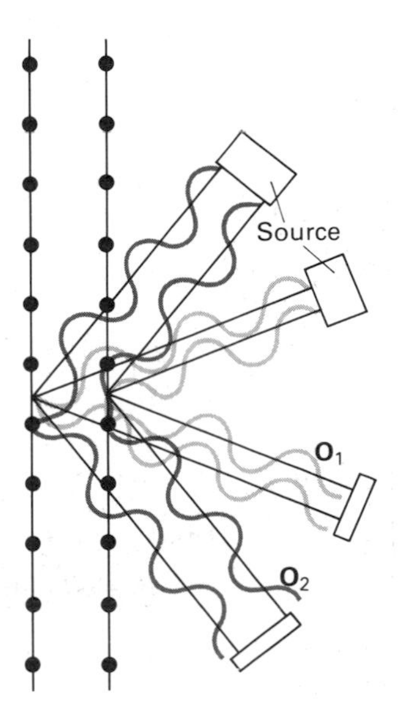

Figure 5.16 X-ray diffraction by a crystal

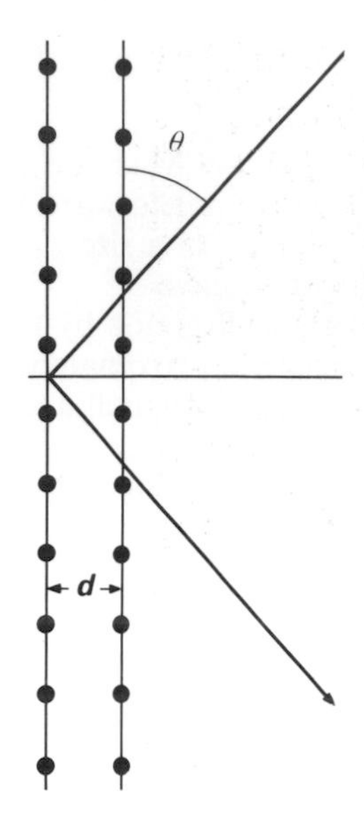

Figure 5.17 The geometry for the Bragg equation (d, layer spacing; θ, glancing angle)

The rows of dots represent the centres of the particles of the crystal, and the observer at O_1 sees waves that have been diffracted by all the layers (only two are shown in the illustration). Although the two rays have travelled different distances, their crests or troughs coincide at O_1, and so the observer detects X-rays there. An observer at O_2, on the other hand, receives waves that are exactly out of step ('out of phase'); the crests cancel the troughs, and the intensity is zero. If the spacing between the planes is different from that shown, the detector has to be moved to a different place in order to give a signal. The location of the detector depends on the spacing between the layers, which can therefore be measured.

The basic equation of X-ray diffraction is the **Bragg equation** (see Figure 5.17), which relates the angle θ, at which diffraction occurs, to the spacing, d, between the layers:

Bragg equation: $\lambda = 2d \sin\theta$ (5.2.1)

λ is the wavelength of the X-rays, and is typically about 100 pm.

EXAMPLE

When a sodium chloride crystal is investigated with X-rays of wavelength 154 pm (generated by electron bombardment of a copper target) a strong diffraction intensity is observed at $\theta = 15.9°$. What is the separation between neighbouring layers of ions?

METHOD Use the Bragg equation (equation 5.2.1), to find the layer separation, d.

ANSWER Equation 5.2.1 rearranges to

$$d = \frac{\lambda}{2\sin\theta} = \frac{154 \times 10^{-12}\,\text{m}}{2\sin 15.9°} = 2.8 \times 10^{-10}\,\text{m}$$

This distance corresponds to 280 pm since 1 pm = 10^{-12} m.

COMMENT It is possible to draw many different types of planes through the ions in a crystal, and so in a practical analysis of a diffraction pattern we have to decide which planes we are investigating before their separations can be interpreted in terms of the crystal structure. In modern work this kind of analysis is virtually fully automated and carried out on a computer. Sometimes the Bragg equation is written

$$n\lambda = 2d \sin\theta$$

where n is a positive integer (1, 2, 3 ... and so on).

In practical applications of X-ray diffraction the intensities are recorded all over a sphere surrounding the crystal, and the crystal itself is rotated. The pattern of intensities then gives very detailed information about the spacing of all the layers. This is the principal source of modern knowledge about the interiors of solids.

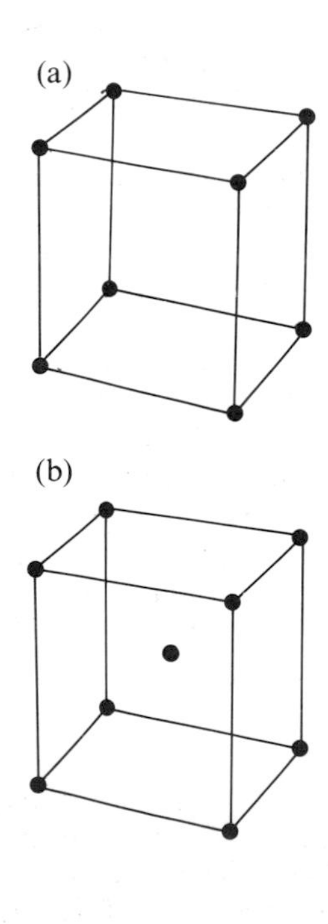

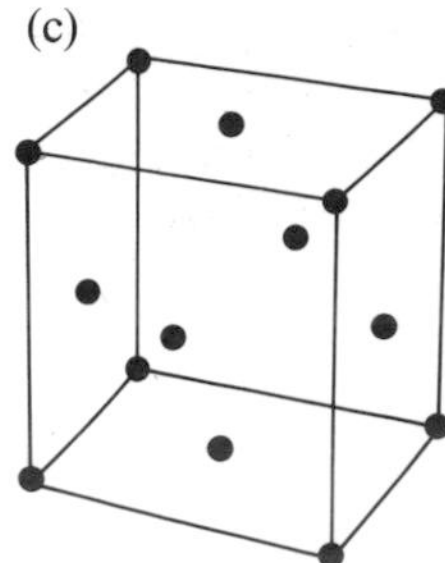

Figure 5.18 Unit cells: (a) primitive cubic, (b) body-centred cubic, and (c) face-centred cubic

The structures of metallic crystals

The description of crystals is simplified by deciding on the size and shape of the **unit cell**. Unit cells are the fundamental blocks which can be stacked together to construct the entire crystal. It turns out that there are only fourteen possible shapes (but they can come in any size). Three of the simplest are illustrated in Figure 5.18.

In the simplest crystal structures, all the particles (ions, for example) are the same. This is the case with metals. Every positive ion can be regarded as a small sphere, and the crystal structure is the result of stacking them together in large numbers in a regular pattern.

Figure 5.19 Hexagonal close packing

Figure 5.20 Cubic close packing

How this works in practice can be seen by considering Figure 5.19, which shows several layers of closely packed spheres. The three-dimensional structure is built up by placing layers one on top of each other. The spheres of the second layer lie in the dips of the bottom layer. The spheres of the third layer lie directly over the spheres of the bottom layer. This gives a closely packed structure known as **hexagonal close packing** (h.c.p.). If the bottom layer is denoted A and the next B, then, since the third layer repeats the first, h.c.p. can be expressed as ABABAB.... This is the structure shown by magnesium and zinc, among others. (Solid helium also has an h.c.p. structure.)

Alternatively, the third layer may be placed over positions occupied by neither A nor B, as shown in Figure 5.20. The structure is denoted ABCABC... and is called **cubic close packing** (c.c.p.) or **face-centred cubic** (f.c.c.). Among the metals having this structure are aluminium, copper, silver and gold. (Solid neon and argon also have c.c.p. structures.)

Careful inspection of the two close-packed structures shows that each sphere has 12 nearest neighbours (6 in its own layer, 3 above and 3 below). This number is its **coordination number**. Some metals have less densely packed structures. For instance, the ions of sodium, potassium and iron (in one of its forms) form **body-centred cubic** (b.c.c.) structures, as shown in Figure 5.18(b). They have a coordination number of 8. Only one metal, polonium, is known to have the **primitive cubic structure**, a cube with an ion at each corner (see Figure 5.18(a)). Its coordination number is 6, and so the structure is relatively loosely packed.

The structures of ionic crystals

Whereas metals have structures built from identical particles, ionic crystals have at least two kinds, which have opposite charges. As a result, their structures are more complicated. Since opposite charges attract each other there is a tendency for ions of one charge to group round ions of the other charge. A further complication is that different ions have different sizes, and so the structure has to take into account the geometrical problem of stacking together spheres of different sizes. The problem is rather like finding the best way of packing oranges and grapefruit together in the same box.

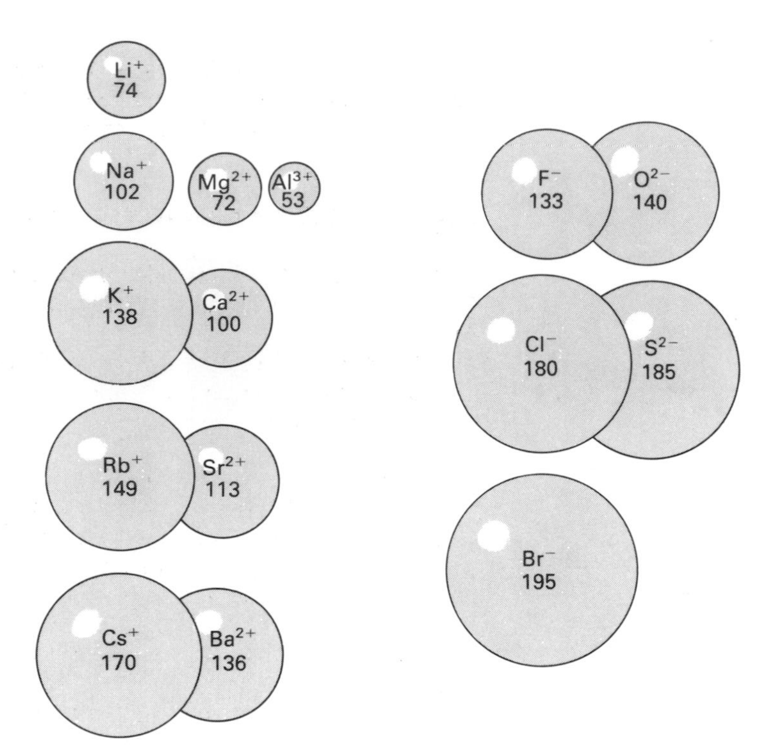

Figure 5.21 Ionic radii (in picometres, 10^{-12} m)

The relative sizes of some ions are illustrated in Figure 5.21. Positive ions are smaller than their parent atoms because the same nuclear charge attracts fewer electrons. Negative ions tend to be larger than their parent atoms because more electrons are under the control of the same nuclear charge.

The problem is to find ways of stacking ions of different size together so as to arrive at a structure that is both electrically neutral overall and has the lowest possible energy, and the solutions to the problem can be illustrated by the structures of sodium chloride and caesium chloride. In the case of sodium chloride the ratio of the radii of the ions, the **radius ratio** (the ratio [radius(smaller)]/[radius(bigger)]) is 0.57, the radius of Na^+ being only about half that of Cl^-. In contrast the radius of Cs^+, an ion lower in Group I, is similar to that of Cl^- and the radius ratio is 0.94. X-ray crystallography shows that, although the ions are of the same charge type, the crystals have markedly different structures.

Figure 5.22 shows the structure adopted by sodium chloride (**rock-salt**). Concentrate first on the arrangement of the chloride ions. They are arranged at the corners of a cube and at the centres of each face. This arrangement is face-centred cubic (see Figure 5.18(c)). The Cl^- array is open and the ions are not touching (that would be energetically very unfavourable). The Na^+ ions are also in an open f.c.c. arrangement and fit comfortably into the holes

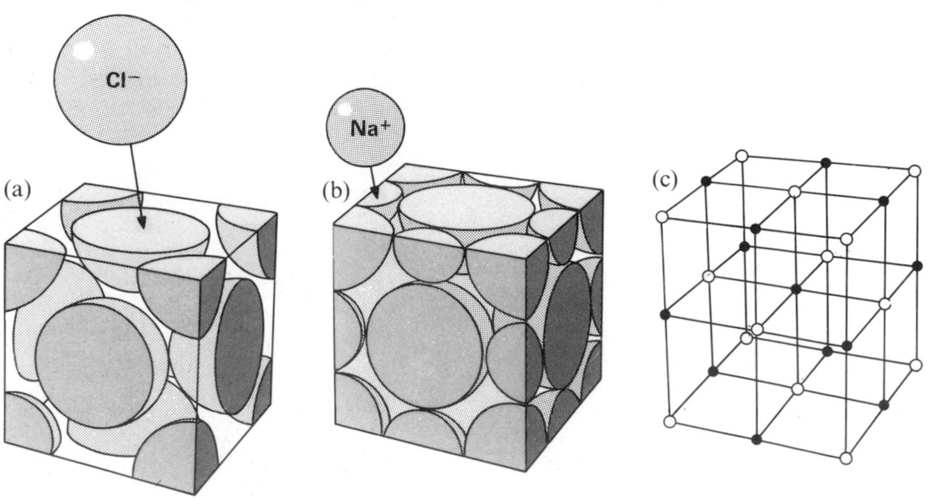

Figure 5.22 The sodium chloride structure: (a) the f.c.c. Cl^- array, (b) the Na^+ ions enter the holes, (c) the resulting two interpenetrating f.c.c. arrays

between the Cl^- ions. Thus the structure consists of *two interpenetrating f.c.c. arrays*. In the complete structure each Na^+ ion is surrounded by six nearest neighbour Cl^- ions and vice versa, and so rock-salt has **6:6-coordination**.

EXAMPLE

The density of sodium chloride is $2.163\,\mathrm{g\,cm^{-3}}$ and the size of its unit cell is 564.1 pm. Deduce a value for Avogadro's constant.

METHOD Calculate the density of the crystal from the molar mass and the volume of the unit cell, and then compare that density with the measured value.

ANSWER Refer to Figure 5.22. Each unit cell contains 4 Na^+ ions and 4 Cl^- ions. This is arrived at by noting that the 8 Cl^- ions at the corners of the cell each project only one octant (one-eighth of a sphere) into the cell, and so contribute one Cl^- overall, while the 6 Cl^- on the faces each project one hemisphere into it, so contributing 3 Cl^- overall, and hence there are 4 Cl^- in total. For electrical neutrality there must also be 4 Na^+ ions. As the side of the unit cell is 564.1 pm the volume occupied by these 4 NaCl units is $(564.1 \times 10^{-12}\,\mathrm{m})^3 = 1.795 \times 10^{-28}\,\mathrm{m^3}$. Therefore the volume occupied by unit amount (i.e., 1 mol NaCl) is $\frac{1}{4} \times L \times (1.795 \times 10^{-28}\,\mathrm{m^3})$. The mass per unit amount of NaCl is the molar mass, $58.44\,\mathrm{g\,mol^{-1}}$. Therefore the density is the ratio

$$\frac{\text{molar mass}}{\text{molar volume}} = \frac{58.44\,\mathrm{g\,mol^{-1}}}{\frac{1}{4} \times L \times 1.795 \times 10^{-28}\,\mathrm{m^3}}$$

Since we also know that the density is $2.163\,\mathrm{g\,cm^{-3}}$ (or $2.163 \times 10^6\,\mathrm{g\,m^{-3}}$) the two expressions can be equated and solved for L:

$$L = \frac{58.44\,\mathrm{g\,mol^{-1}}}{\frac{1}{4} \times 2.163 \times 10^6\,\mathrm{g\,m^{-3}} \times 1.795 \times 10^{-28}\,\mathrm{m^3}}$$
$$= 6.021 \times 10^{23}\,\mathrm{mol^{-1}}$$

COMMENT The accepted value of Avogadro's constant is $6.022 \times 10^{23}\,\mathrm{mol^{-1}}$ (see the introductory chapter, 'The mole').

Now consider the caesium chloride structure. X-ray diffraction shows that it is different from the sodium chloride structure, the explanation being that the Cs^+ ions are large enough to have eight Cl^- ions around them. Figure 5.23 shows how the structure is assembled. The Cl^- ions are at the corners of a primitive cubic array (see Figure 5.18(a)). The Cs^+ ions also form a primitive cubic array. There is a big central empty region in each Cl^- array and the Cs^+ ion fits into it. Therefore the structure consists of *two interpenetrating primitive cubic arrays*. Each Cs^+ ion is surrounded by eight Cl^- ions, and vice versa, and so the structure has **8:8-coordination**.

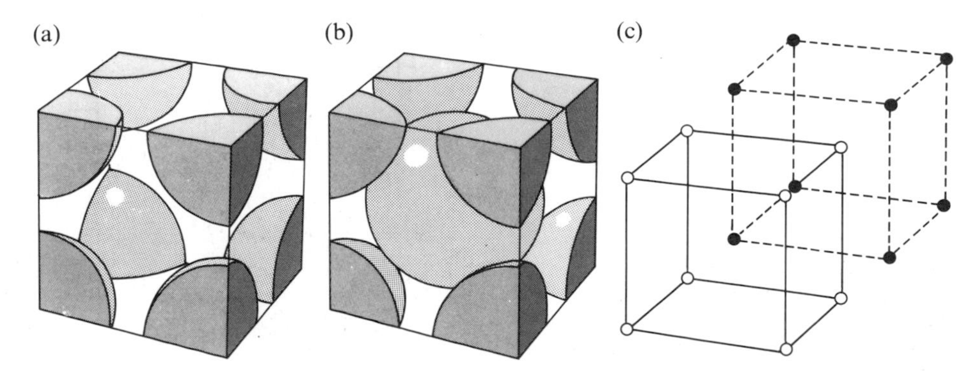

Figure 5.23 The caesium chloride structure: (a) the primitive cubic Cl^- array, (b) a Cs^+ ion enters the hole, (c) the resulting two interpenetrating primitive cubic arrays

The radius ratio gives an important clue to the likely nature of the crystal structure of ionic materials. It follows from geometrical considerations that a coordination number of 8 is expected when the radius ratio exceeds $\sqrt{3} - 1 = 0.732$; for a ratio below 0.732 and above $\sqrt{2} - 1 = 0.414$ a coordination number of 6 is expected, and a coordination number of 4 is likely for a ratio of less than 0.414.

EXAMPLE

Predict the likely crystal structure of potassium bromide.

METHOD Make the prediction on the basis of the radius ratio $r(\text{smaller})/r(\text{bigger})$. If the ratio is less than 0.732 the likely structure is the rock-salt structure; if it exceeds 0.732 the likely structure is the caesium chloride structure. Ionic radii are given in Figure 5.21.

ANSWER The radius ratio is

$$\frac{r(\text{smaller})}{r(\text{bigger})} = \frac{r(\text{K}^+)}{r(\text{Br}^-)} = \frac{138\text{ pm}}{195\text{ pm}} = 0.708$$

Since the radius ratio is less than 0.732 we predict that potassium bromide should have a rock-salt structure.

COMMENT X-ray diffraction studies confirm this prediction. Sometimes, though, the radius-ratio prediction fails. In the case of potassium fluoride the radius ratio is 0.964 yet it has the rock-salt structure.

Although we have described the c.c.p. and h.c.p. structures as close-packed, in fact only 74 per cent of the total space of a crystal is occupied by the spheres, the gaps between them accounting for the rest. These gaps come in two distinct types. In one type, there are six surrounding spheres, giving **octahedral holes**, and in the other there are only four, giving **tetrahedral holes** (see Figure 5.24). There are twice as many tetrahedral as octahedral holes.

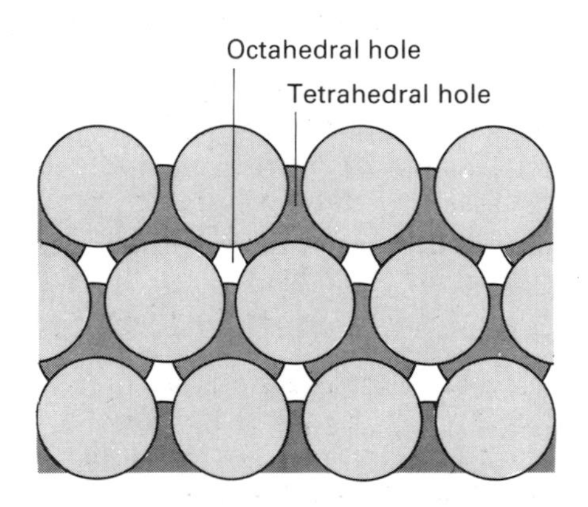

Figure 5.24 Octahedral and tetrahedral holes in a close-packed structure

The rock-salt structure can be reinterpreted in terms of holes: sodium ions occupy the octahedral holes in a c.c.p. array of chloride ions. The advantage of this description is that it reveals rock-salt's relation to some other important structures. In the **fluorite** (CaF_2) structure, shown in Figure 5.25, the F^- ions occupy all the tetrahedral holes in a c.c.p. array of Ca^{2+} ions. In the **zinc blende** (ZnS) structure (Figure 5.26) only half the tetrahedral holes are occupied by the small Zn^{2+} ions in a c.c.p. array of S^{2-} ions. Zinc sulphide can also have the **wurtzite** structure, in which half the tetrahedral holes are again occupied by Zn^{2+} ions, but the S^{2-} ions are in an h.c.p. array. (Note that although we refer to ions, in each case, and especially in zinc sulphide, there may be substantial covalent character in the bonding.)

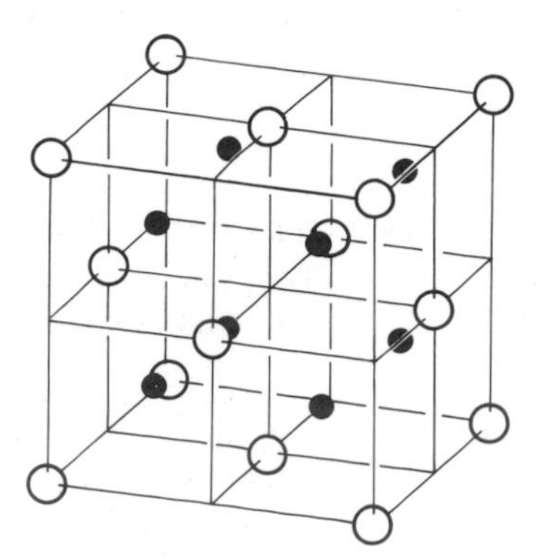

Figure 5.25 The fluorite (CaF_2) structure

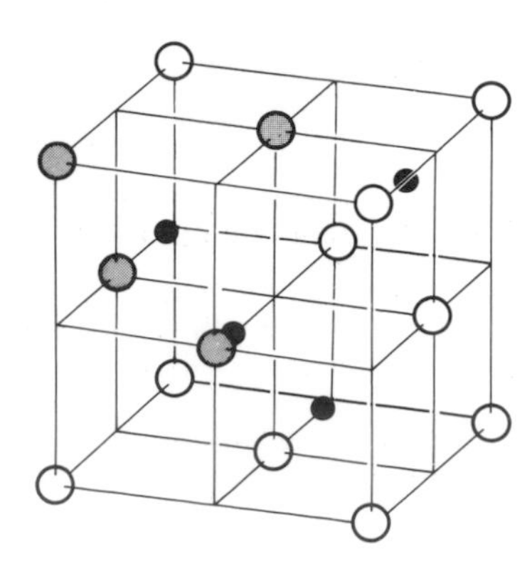

Figure 5.26 The zinc blende (ZnS) structure

Summary

- ☐ Covalent structures are essentially single giant molecules. The structures are hard and rigid.
- ☐ A metal is an orderly array of positive ions in a sea of delocalized electrons. The bonds are non-directional, and so metals are malleable and ductile. The free electrons account for the high thermal and electrical conductivities of metals and for their reflectivity.
- ☐ Ionic structures are held together by the electrostatic attraction between oppositely charged ions.
- ☐ A crystal is the three-dimensional arrangement of ions (or atoms) in space.
- ☐ Van der Waals forces are the forces that bind covalent molecules into a condensed state: they are classified as dipole–dipole forces and dispersion forces.
- ☐ The dipole–dipole forces are the forces between the permanent electric dipoles of polar molecules.
- ☐ The dispersion forces, or London forces, arise from the interaction of instantaneous dipoles in the charge distributions of molecules. They occur between all types of molecules (and atoms).
- ☐ Solids bound together by van der Waals forces tend to be soft and to have low melting and boiling points.
- ☐ A hydrogen bond is the link $X^{\delta-}$—$H^{\delta+}$---$Y^{\delta-}$ where X and Y are strongly electronegative atoms. Hydrogen bonds are important in the structure of ice and in the shapes adopted by biological molecules such as proteins.
- ☐ The principal method for the determination of the structures of crystals is X-ray crystallography, and the basic equation is the Bragg equation (equation 5.2.1).
- ☐ The unit cell of a crystal is the basic structure which, if stacked together, reproduces the structure of the crystal. There are only 14 types of unit cell.
- ☐ The structures of metals can be explained in terms of the geometrical problem of packing together identical spheres. Two close-packed structures are possible.
- ☐ One close-packed structure is hexagonal close packing (h.c.p.). This is an ABABAB... collection of close-packed layers; the particles have a coordination number (the number of nearest neighbours) of 12.
- ☐ The other close-packed structure is cubic close packing (c.c.p.). This is an ABCABC... collection of close-packed layers; the particles have a coordination number of 12.
- ☐ The particles in a body-centred cubic (b.c.c.) structure have a coordination number of 8 and are less closely packed.
- ☐ The structures adopted by ionic crystals depend on the radius ratios r(smaller)/r(bigger) of the ions. The alkali metal halides have the sodium chloride or rock-salt structure (two interpenetrating f.c.c. arrays) when the radius ratio is less than 0.732 or the caesium chloride structure (two interpenetrating primitive cubic arrays).
- ☐ The rock-salt structure has 6:6-coordination; the caesium chloride structure has 8:8-coordination.
- ☐ Close-packed structures have both octahedral holes and tetrahedral holes.

PROBLEMS

1 List the properties characteristic of (a) ionic materials, (b) materials formed from individual covalent molecules, (c) giant covalent structures, (d) metals.

2 Why does copper expand when it is heated? Why does ice contract when it melts?

3 Why are metals (a) good conductors of electricity, (b) good conductors of heat, (c) highly reflective? Explain why some metals are coloured (e.g., yellow gold and pink copper).

4 Describe the structures of (a) diamond and (b) graphite, and list and explain their different physical properties.

5 Describe the structure of the hydrogen bond. State when it is likely to occur, and show how its presence affects the physical properties of (a) water, (b) ethanoic acid, (c) sugar, (d) wood. Why is it easier to chop and saw wood along the grain rather than across?

6 Classify into types of solid the following materials: (a) table salt, (b) wax, (c) hair, (d) wood, (e) detergents, (f) soap, (g) bricks.

7 Explain how the close-packing of identical spheres can lead to either a cubic close-packed or a hexagonal close-packed structure. Give two examples of each type of structure.

8 Explain why the following statements are wrong: (a) the rock-salt (sodium chloride) structure is face-centred cubic; (b) the metal polonium, which has a primitive cubic unit cell with the atoms arranged at the corners of a cube, has a close-packed structure.

9 State the different types of van der Waals interactions that may occur between covalent molecules. Which is usually the more important?

10 Explain the term *radius ratio* and its significance. Calculate the ratio for the alkali metal bromides on the basis of the information in Figure 5.21 and predict the form of the crystal structure in each case.

11 Describe the forces between the constituent species in solids considering the relative magnitude of each type of force by reference to the table of enthalpies of vaporization ($kJ\,mol^{-1}$) given below.

Solid	He	Xe	HCl	Cl_2	CO_2	H_2O	Na	NaCl
$\Delta_{vap}H^{\ominus}$	0.10	15.0	21.1	26.8	26.0	50.0	107.8	222

$\Delta_f H^{\ominus}_{(298K)}(NaCl(s)) = -411\,kJmol^{-1}$

$\Delta H^{\ominus}(\frac{1}{2}Cl_2(g) + e = Cl^-(g)) = -240\,kJ\,mol^{-1}$

First ionization energy of Na is $494\,kJ\,mol^{-1}$.
The heterolytic bond energy of Na—Cl(g) giving $Na^+(g)$ and $Cl^-(g)$ has been estimated to be $556\,kJ\,mol^{-1}$. (*Welsh S*)

12 'The particles in a substance (which may be atoms, molecules or ions) may be held together by ionic, covalent or hydrogen bonds, or by van der Waals' forces.'
(a) Illustrate this statement by reference to carbon, carbon dioxide, water, sodium chloride and a noble gas. Electron configurations should be given wherever possible.
(b) Relate the physical properties of the substances named in (a) to the type of bonding present.
(c) Suggest a reason why 2-nitrophenol,

NO_2
OH

is more volatile than its isomer, 4-nitrophenol,

NO_2
OH

(*London*)

13 Discuss the types of bonding which occur in the following solid compounds: (a) $CuSO_4.5H_2O$, (b) C_6H_5COOH (do not discuss the bonding within the benzene ring), (c) P_4O_{10}.
(*Oxford and Cambridge S part question*)

14 Describe the use of X-rays to determine the structure of solids. Illustrate and discuss the different types of X-ray diffraction pattern that can be obtained and mention any limitations of the method.

X-ray studies of lithium chloride using X-rays of wavelength 0.0585 nm produced a strong diffraction at an angle of diffraction, θ, of 6.3° and another diffraction at 8.8°. There were also two related weaker diffractions at 5.4° and 10.9°. Use the Bragg diffraction equation to calculate the separation of the crystal planes in lithium chloride.

Draw suitable diagrams to show how the three sets of crystal planes are related to the unit cell of lithium chloride [rock-salt structure]. (*Nuffield S*)

15 'It is probably no exaggeration to say that in living processes the hydrogen bond is as important as the carbon–carbon bond.' Justify this statement by making detailed reference to a suitable range of examples. (*Nuffield S*)

16 (a) Draw clear diagrams to show the shape of a molecule of each of the following compounds: (i) CH_4, (ii) NH_3, (iii) H_2O, (iv) HF. Explain the shape of the NH_3 molecule as you have shown it.
(b) The graph shows the variation in boiling point of the hydrides in Groups IV to VII of the Periodic Table.

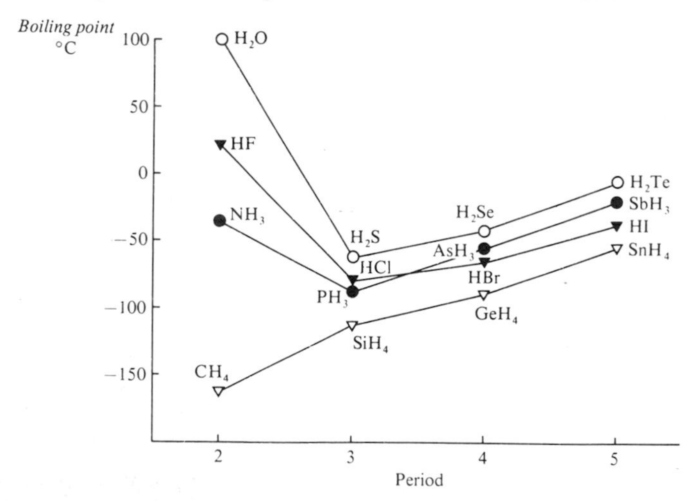

(i) Suggest a reason for the regular increase in boiling point from CH_4 to SnH_4.
(ii) What is responsible for the anomalously high boiling points of NH_3, HF and H_2O?
(iii) Carefully explain, with the aid of a diagram, the nature of the forces *between* molecules of water.
(*AEB 1985*)

17 Give an account of the various bonds formed between ions, atoms and molecules in the solid state.
(*Cambridge Entrance part question*)

6 Changes of state

In this chapter we look at the structure of the most puzzling state of matter, the liquid state. We see how a liquid forms when a solid melts, and go on to see what happens when the liquid evaporates and forms a gas. Phase diagrams provide a convenient way of discussing the conditions for the existence of different phases, and we see what information some of them contain.

Matter often exists as pools instead of lumps. Under the right conditions solids melt to liquids, and gases condense. In this chapter we look at ways of describing these changes of state, which are called **phase transitions**.

The liquid state is the least understood of the states of matter. This is because it is intermediate between a state with a regular structure, such as a crystal, and a state that is almost completely without structure, a gas. Nevertheless, progress has been made, and we shall look briefly at the modern description of liquids. Most of the discussion in this chapter, however, concerns how phase transitions depend on conditions. For instance, it is well known that the boiling point of water is lower at high altitudes than at sea level, and we shall show how to depict this dependence by drawing a **phase diagram**. Melting points also depend on pressure (which is one reason why skaters move so easily over ice), and we shall explore the reasons.

6.1 The structures of liquids

We can arrive at a model of the nature of liquids by thinking about what happens when solids melt and liquids evaporate.

Solids melting

When a solid is heated there is at first very little noticeable change. A block of copper at 100 °C looks much the same as one at 25 °C. Measurements, however, show that the block has expanded slightly. The reason for the expansion is easy to find: the particles (ions in this case) are vibrating more vigorously and so are taking up more space. Computer simulations of the motion show that although the vibrations swing the particles backwards and forwards they do not move away from their original locations (see Figure 6.1(a)).

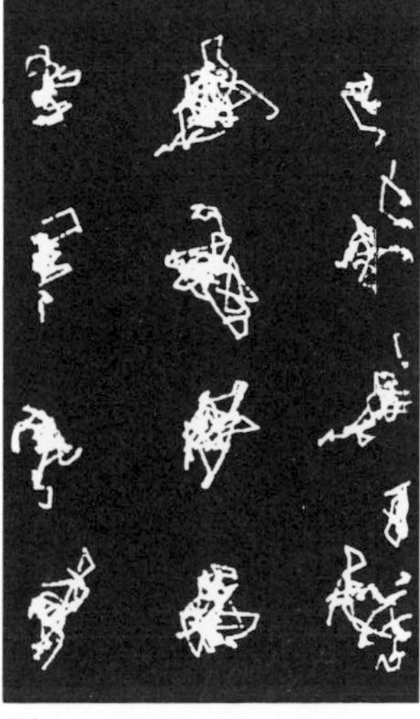

(a)

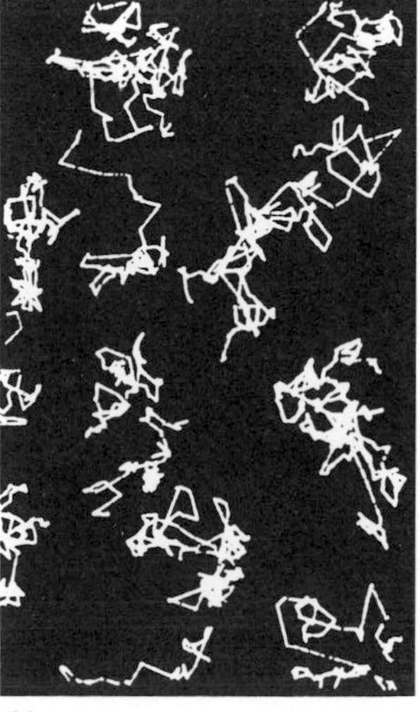

(b)

Figure 6.1 A computer simulation of a solid (a) just below and (b) just above its melting point

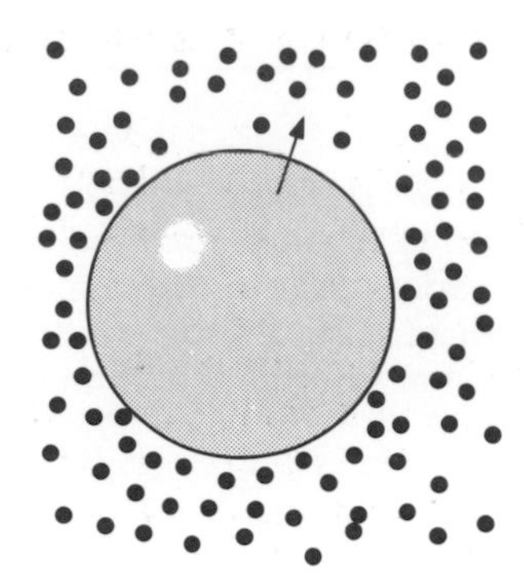

Figure 6.2 The molecular explanation of Brownian motion

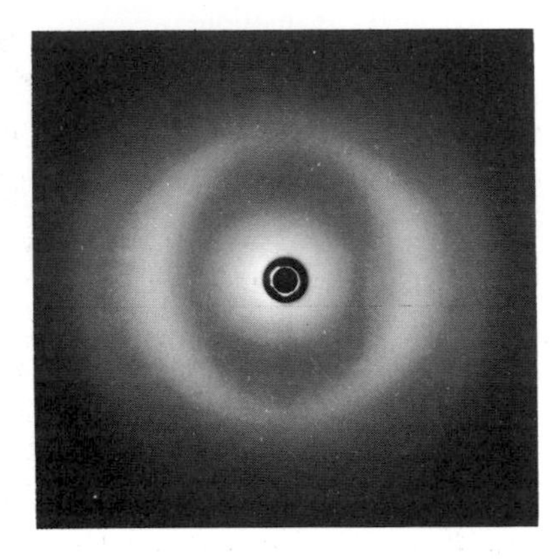

Figure 6.3 X-ray diffraction by a liquid (distilled water), indicating its minimal structure

As the temperature is raised the vibrations become more vigorous. The shape of the solid remains much the same because the motion swings the particles around their original locations, but its dimensions increase. Finally, there comes a point at which the vibrations are so vigorous that the particles break away from their original locations and the solid melts into a liquid. The computer simulation of the motions of the particles just above the melting point shows that they are now free to move over large distances (see Figure 6.1(b)). Because its structure is so loose, a liquid adopts the shape of its container and flows easily. Liquids are much less compressible than gases because their particles are already close together and, unlike gases, they contain little unfilled space.

Experimental evidence that molecules in liquids are in a state of constant motion was first put forward by the botanist Robert Brown (in 1827). He observed the motion of small grains suspended in water and noticed through a microscope that they constantly jittered around. This **Brownian motion** fits in very well with the modern ideas on structure, because every grain is subjected to the battering of large numbers of water molecules, but the battering is not uniform. For a short time the many impacts on one side may exceed those on the other, so that the grain is driven in that direction. Then another side receives most impacts, and so the grain is driven in another direction – see Figure 6.2.

Modern techniques (particularly special developments of spectroscopy) have displaced this type of crude observation, and we now have detailed information about molecular motion in liquids. These observations show that molecules in liquids move in a manner intermediate between the nearly free motion of molecules in gases and the almost total lack of motion in crystals. When molecules move in liquids they do so in short steps, each step being through a fraction of the radius of the molecule (through about 10 pm). The steps are in random directions and occur every 10^{-13} s or so. Figure 6.1(b) should be interpreted with that in mind.

Liquids evaporating

Molecules of a liquid can escape into the gas phase and become a **vapour**, and the number of molecules in the vapour rises until the vapour and the liquid reach a dynamic equilibrium, that is, until molecules continually interchange between the vapour and the liquid in such a manner that the rate of evaporation is equal to the rate of condensation. The pressure exerted by the vapour in equilibrium with a liquid in a closed container is called the **vapour pressure** of the liquid at the temperature of the sample (see Figure 6.4).

The vapour pressure of water at 25 °C is only 3.2 kPa (0.03 atm) because only a small proportion of molecules can escape. When the temperature is 100 °C the vapour pressure rises to 101 kPa (1 atm). Under normal conditions this pressure drives back the atmosphere and the water evaporates freely. That is what happens when a kettle boils. The **normal boiling point** is the temperature at which the vapour pressure of a liquid reaches 1 atm. Under different conditions (up mountains and down mines, for example) the vapour pressure matches the local atmospheric pressure at different temperatures, and so the boiling points are different. The vapour pressures of liquids more volatile than water reach 1 atm at temperatures below 100 °C; the vapour pressure of liquid helium, for example, reaches that value when the temperature rises to only 4.2 K.

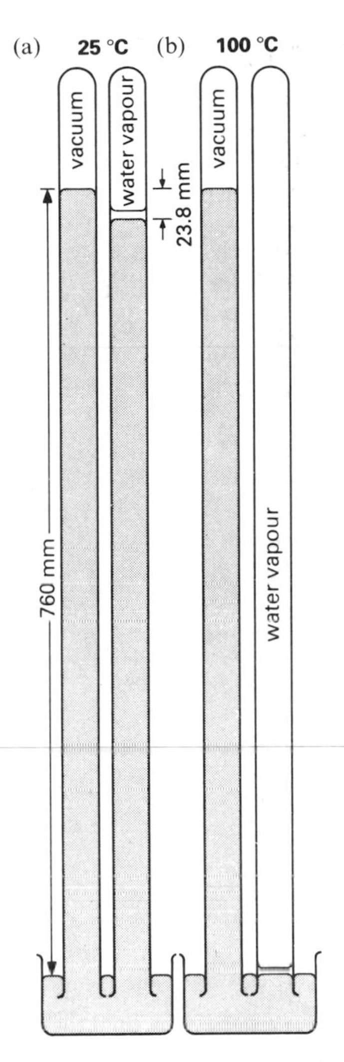

Figure 6.4 The vapour pressure of water at two different temperatures

EXAMPLE

The vapour pressure of mercury at 25 °C is 0.24 Pa. What is the pressure in the 'vacuum' at the top of a mercury barometer tube? A few drops of mercury are spilt on the floor of a 30 m^3 room. What mass of mercury will be found in the atmosphere within the room?

METHOD Mercury evaporates until it exerts a partial pressure of 0.24 Pa. For the second part of the problem assume that the vapour acts as a perfect gas, and calculate the amount of substance in the atmosphere of the room using the perfect gas equation in the form $n = pV/RT$. Convert to a mass on the basis that $A_r(\mathrm{Hg}) = 200.6$.

ANSWER The pressure in the 'vacuum' is the vapour pressure of the mercury, or 0.24 Pa. Inside a room of volume 30 m^3 a partial pressure of mercury of 0.24 Pa corresponds to an amount of substance of Hg given by

$$n = \frac{pV}{RT} = \frac{0.24\,\mathrm{N\,m^{-2}} \times 30\,\mathrm{m^3}}{8.314\,\mathrm{J\,K^{-1}\,mol^{-1}} \times 298\,\mathrm{K}} = 2.9 \times 10^{-3}\,\mathrm{mol}$$

(We have used 1 N = 1 J m^{-1}.) This amount of substance corresponds to a mass

$$m = n \times (A_r\,\mathrm{g\,mol^{-1}}) = 2.9 \times 10^{-3}\,\mathrm{mol} \times 200.6\,\mathrm{g\,mol^{-1}} = 0.58\,\mathrm{g}$$

COMMENT 0.58 g of mercury vapour is dangerous: always clear up any spilt mercury. The vapour pressure of mercury is low, but not nearly low enough for experiments involving a very high vacuum. Other liquids have even lower vapour pressures. Tricresyl phosphate, for instance, has a vapour pressure of only 4.1 × 10^{-5} Pa at 25 °C. That corresponds to a column of mercury only two atoms high! In modern high-vacuum work liquids are not used at all.

The behaviour of a liquid heated in a *sealed* container is quite different. The vapour pressure increases as the temperature is raised, but free boiling cannot occur because the vapour is trapped. As the temperature is increased and the vapour pressure rises, the density of the vapour increases. There comes a stage when the density of the vapour is the same as that of the liquid. At this stage, the sample becomes uniform and no one region can be identified as a liquid or as a gas. The temperature at which this occurs is the **critical temperature** (see Section 4.3) of the substance, and the corresponding vapour pressure is the **critical pressure**. The critical temperature and pressure define the **critical point** of the substance.

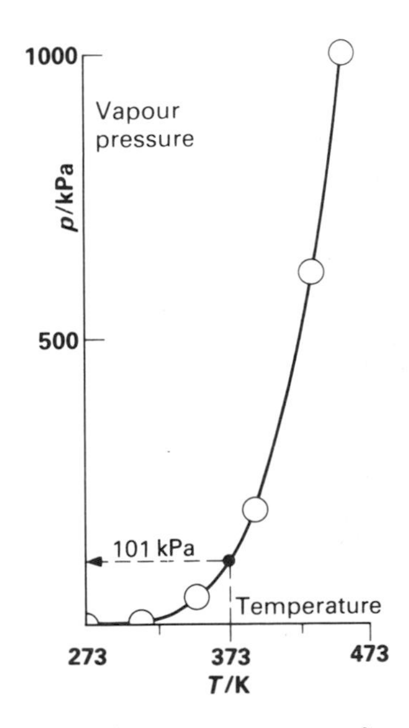

Figure 6.5 The vapour pressure of water

6.2 Plotting the changes: phase diagrams

A simple way of measuring the vapour pressure of a liquid is to float some on to the surface of the mercury in a barometer tube, as shown in Figure 6.4(a). The liquid evaporates into the vacuum until it reaches equilibrium with its own vapour. The resulting vapour pressure depresses the column of mercury, and the change of level gives its magnitude. At 25 °C, for example, the change of level caused by water is only 23.8 mm. At 100 °C the change is 760 mm, and the mercury inside the tube is level with the mercury outside – see Figure 6.4(b). The vapour pressures at a series of temperatures can be measured and the points plotted on a graph, giving a curve like that in Figure 6.5. This is the start of the construction of a **phase diagram**.

The phase diagram of water

The complete phase diagram for water near room temperature and pressure is shown schematically in Figure 6.6. The lines on it mark the conditions under which different pairs of phases are in equilibrium. The line separating liquid and vapour is the **vapour pressure curve** (it is Figure 6.5 drawn on a different scale). It ends at the critical point of water (647 K, 218 atm). The line separating the solid and vapour regions is the **sublimation pressure curve** of ice, and marks the conditions under which ice and water vapour exist in equilibrium. (**Sublimation** is the direct change from solid to vapour.)

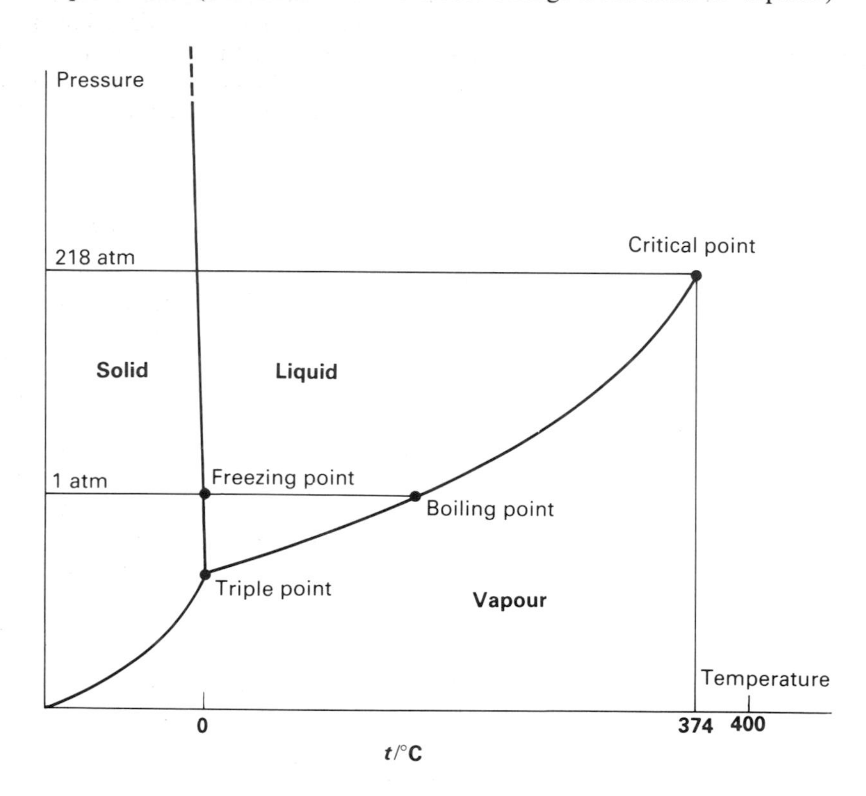

Figure 6.6 The phase diagram for water

The almost vertical line in the diagram marks the conditions under which ice and liquid water are in equilibrium. It is the **melting point curve** of ice. The slope of the line shows that the melting point falls as the pressure is increased. When the pressure is 7×10^6 Pa (approximately the pressure exerted by a 70 kg ice skater with 1 cm^2 contact area) the melting point is about -1 °C. This is because the high pressure tends to crush the hydrogen bonds that hold the water molecules in the comparatively open structure of ice, and so the denser liquid phase is formed.

There is one set of conditions where ice, liquid water and water vapour can exist together in equilibrium. This is the **triple point** shown on Figure 6.6. It occurs at 611 Pa (0.006 atm) and 273.16 K. There is no way of altering these conditions while having all three phases present in equilibrium. Because the point is unique it is used as a second fixed point in the definition of the Kelvin temperature scale (see Section 4.1).

The phase properties of water make an appearance in everyday affairs in several ways. For instance, frost can form from water vapour in the atmosphere in two ways. In one the first step is the formation of dew, when the temperature of damp air falls and the vapour condenses as a liquid. As the temperature falls still further the dew freezes, leading to a layer of frost. When the vapour pressure of water is less than 611 Pa (the pressure of the triple point), lowering the temperature leads directly from the vapour to the solid region of the phase diagram. This gives rise to hoarfrost, ice that forms directly without an intermediate liquid state appearing.

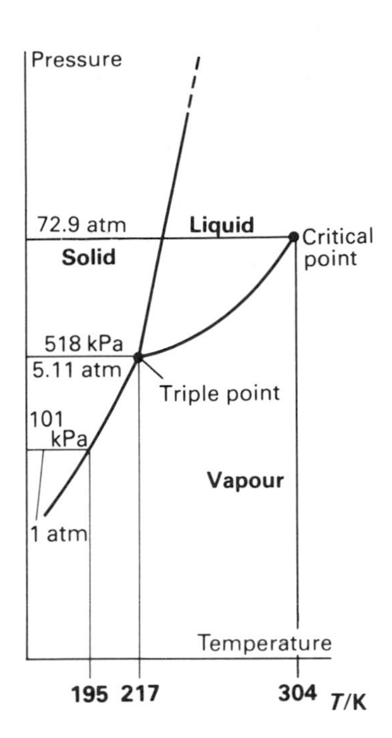

Figure 6.7 The phase diagram for carbon dioxide

The phase diagram of carbon dioxide

The experimentally determined phase diagram for carbon dioxide is shown schematically in Figure 6.7. Two features make it significantly different from that of water. In the first place it shows that solid carbon dioxide sublimes if left in the open air (hence the name 'dry ice'). The liquid phase does not exist below 518 kPa (5.11 atm), and so at least that pressure must be exerted if the liquid is required. The vapour pressure of the solid is greater than 1 atm at all temperatures above 195 K, and so it simply vaporizes if open to the atmosphere. The second point is that increasing the pressure *increases* the melting point of carbon dioxide. This is because the solid is denser than the liquid since the small rod-like molecules pack together more closely. The effect of pressure is to favour the formation of the solid, and so it remains solid until higher temperatures. This behaviour is typical of most materials; water is unusual (yet again).

The phase diagram of sulphur

The existence of more than one form of an element in the same physical state is called **allotropy**: the differences can arise from bonding, as in O_2 and O_3, or from crystal form, as in rhombic and monoclinic sulphur (see Section 22.2). The term allotropy is applied only to elements and should be distinguished from **polymorphism**, the possession of different crystal forms by any substance. Sulphur is both allotropic and polymorphic; oxygen gas is allotropic; zinc sulphide (see Section 5.2) is polymorphic.

The phase diagram for sulphur in Figure 6.8 shows what conditions are needed in order to ensure that any given phase is stable and the conditions under which pairs of phases are in equilibrium. For example, at A, monoclinic sulphur is the stable form. When the temperature is raised to B, monoclinic sulphur and liquid sulphur are in equilibrium: this is the melting point of monoclinic sulphur. (Note that sulphur has three different triple points.)

Other elements showing allotropy include tin, phosphorus and carbon. The term **monotropy** is used to signify that only one allotrope is the stable form at all temperatures under 1 atm pressure (as with phosphorus). **Enantiotropy** signifies that different forms are stable at various temperatures and 1 atm pressure (as with sulphur). The enantiotropy of tin gives rise to 'tin plague' or 'tin pest' (see Figure 6.9, and Section 20.2): the transition occurs at 13 °C but is rapid only at temperatures well below 0 °C.

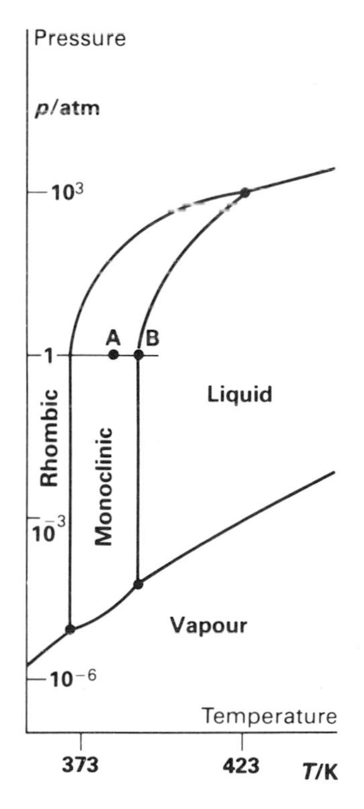

Figure 6.8 The phase diagram for sulphur

Figure 6.9 'Tin plague': this sample of pure tin, which was originally solid, was stored at −20°C and bent. This initiated the phase transition to grey tin, and the sample crumbled

(a)

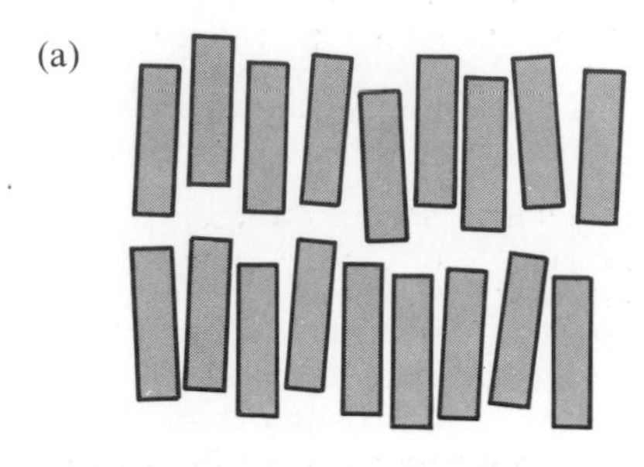

(b)

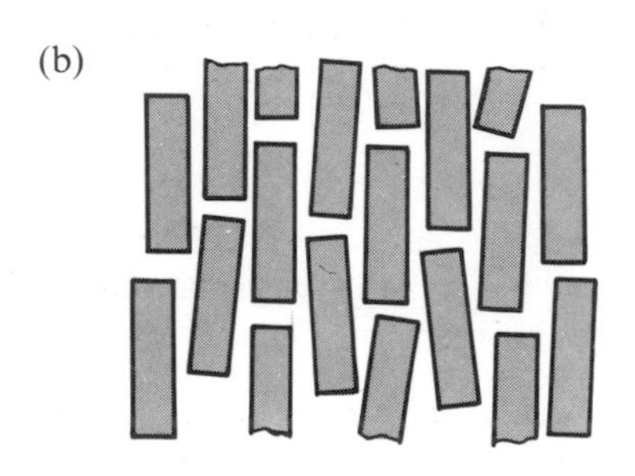

Figure 6.10 (a) Smectic and (b) nematic liquid crystals

6.3 Other phases

There are three types of phase that do not readily fit into the scheme of solid, liquid and gas. One unusual phase is a **plasma**. This is an ionized gas formed when high temperatures strip electrons from atoms. The importance of plasmas lies in their role throughout high-temperature physics, such as in stellar atmospheres and rocket exhausts, and in nuclear fusion research. Another unusual phase is a **glass**. Glasses are supercooled liquids, that is, liquids that have been cooled below their freezing points but have not crystallized. Although apparently solid, structural studies show that glasses resemble liquids because their component particles are in a random jumble; they are *amorphous*, or shapeless. Glass does in fact flow, but only very slowly.

The third unusual phase is a **liquid crystal**. Although liquid crystals flow like ordinary liquids their molecules congregate in swarms, showing some structure (see Figure 6.10). They are therefore intermediate in structure between true crystals and ordinary liquids.

There are several types of liquid crystal phase, and they differ in the way the molecules stack together. In a **smectic phase** (from the Greek for 'soapy') the rod-like molecules form aligned regions like those shown in Figure 6.10(a). In a **nematic phase** (from the Greek for 'thread'), there is less order but the molecules still lie approximately parallel to each other (Figure 6.10(b)). In a **cholesteric phase** (so called because derivatives of cholesterol take this form) there are layers of smectic-like phases, but each layer is twisted relative to its neighbour. The presence of the layers gives rise to colours in the light scattered from them, and since the spacing of the layers depends on the temperature, so too do the colours: the effect is used to make liquid crystal thermometers. The optical properties of nematic liquid crystals are put to use in calculator and watch displays: the display depends on the orientating effect of very small electric fields and the resulting change in the optical properties of the liquid crystal. The great advantage of a liquid crystal display (LCD) is that it consumes very little power. One point about the second industrial revolution – the microelectronic revolution now in progress – is that, unlike the first, it is making extremely small demands on power and is leading to the control rather than the generation of pollution.

Summary

- ☐ Phase transitions are the changes of state of materials, and include melting and freezing, condensation and vaporization (evaporation) and changes of one solid phase into another.
- ☐ The temperature of a phase transition depends on the pressure.
- ☐ In solids, particles vibrate at their original locations; at the melting points they are free to wander away.
- ☐ In liquids molecules move in rapid small steps: this appears as the Brownian motion of suspensions of larger particles.
- ☐ The vapour pressure of a liquid (or a solid) at a specified temperature is the pressure at which the liquid (or solid) is in equilibrium with its vapour.
- ☐ The normal boiling point of a liquid is the temperature at which its vapour pressure is 1 atm.
- ☐ A phase diagram shows the regions of temperature and pressure under which a phase is stable, and the lines show the conditions of temperature and pressure at which two phases exist in equilibrium.
- ☐ The triple point is the condition of pressure and temperature at which three phases simultaneously exist in equilibrium.
- ☐ Sublimation is the vaporization of a solid without an intervening liquid phase.
- ☐ When the pressure is increased the melting point of ice is lowered. This is related to the contraction that occurs when ice melts.
- ☐ When the pressure is increased on solid carbon dioxide its melting point rises. This is related to the expansion that occurs when the solid melts. Most substances behave like carbon dioxide in this respect.

- ☐ Different forms of an element in the same physical state are called allotropes. Monotropy signifies that only one allotrope is stable under 1 atm pressure at any temperature; enantiotropy signifies that different forms can be stable at various temperatures.
- ☐ A plasma is an ionized gas.
- ☐ A glass is a supercooled liquid. It is amorphous and not crystalline.
- ☐ Liquid crystals flow like ordinary liquids but have molecules arranged with a degree of order and are classified as smectic, nematic or cholesteric.

PROBLEMS

1 Explain the terms *phase transition, vapour pressure, phase diagram, normal boiling point, melting point.*

2 Explain the terms *critical temperature* and *triple point.*

3 Why do the melting lines in the water and carbon dioxide phase diagrams slope in opposite directions? Which is more common?

4 Explain the terms *allotropy, monotropy* and *enantiotropy.* Name four elements that show allotropy.

5 Account for the following properties of water: (a) ponds freeze from the surface down, (b) ice-skaters glide easily over ice, (c) a weighted wire passes through a block of ice (this is the phenomenon of *regelation*).

6 Account (a) for the formation of (i) dew, (ii) hoarfrost and (iii) frost, and (b) for the non existence of liquid carbon dioxide under atmospheric conditions.

7 The vapour pressure of water at 18 °C is 2.06 kPa. What depression of the column of mercury in a barometer would be observed if water is introduced and reaches equilibrium with its vapour?

8 What is the mass of water in the atmosphere of a 30 m^3 room when the relative humidity is 50% and the temperature 25 °C (when the vapour pressure is 3.17 kPa)?

9 The vapour pressure of carvone, which is responsible for the odour of spearmint, is about 15 Pa at 25 °C. How many molecules would be present in the atmosphere of a 30 m^3 room at equilibrium with a piece of chewing gum?

10 (a) Sketch the pressure–temperature phase diagram for water. Label your diagram to show the significance of the lines and areas in it.
(b) Comment on the following statements.
(i) It is very unusual for the melting point of a pure substance to fall as pressure is applied.
(ii) Water boils at a much higher temperature than hydrogen sulphide. *(Oxford part question)*

7 The properties of mixtures

In this chapter we deal with the properties of simple non-reacting mixtures. The behaviour of mixtures can be summarized neatly in terms of phase diagrams, and we see how to construct and interpret them. These diagrams can be used to discuss the techniques that are widely used for separating mixtures, both in the laboratory and in industry. We see too how the boiling points, freezing points and osmotic pressures of solutions depend on the amount of solute present, and how they can be used to determine molar masses.

In this chapter we investigate the properties of mixtures, and look at the ways in which they differ from the pure materials discussed so far. The easiest way of depicting their properties is in terms of phase diagrams constructed experimentally. As well as summarizing experiments we should also be able to *predict* what happens when one substance is dissolved in another. For instance, we should be able to predict what will happen to the freezing and boiling points of a solvent when a solute is added. In the process of describing these effects we shall see that the measurement of the properties of a solution can be used to determine the molar mass of the solute, especially solutes consisting of the large molecules typical of polymers and biological materials.

7.1 Phase diagrams of mixtures

We shall normally express the compositions in terms of **mole fractions**, x (see Section 4.1). The mole fractions of A and B in a two-component mixture are defined as

$$x_A = \frac{\text{amount of substance of A}}{\text{total amount of substance}} = \frac{n_A}{n_A + n_B} \qquad x_B = \frac{n_B}{n_A + n_B}$$

n_A and n_B are normally expressed as so many moles of A or of B; x_A is a dimensionless number between 0 and 1. In a mixture of two components, a **binary mixture**, $x_A + x_B = 1$.

EXAMPLE

Calculate the mole fractions of H_2O and CH_3CH_2OH in a mixture containing equal masses of water and ethanol.

METHOD Convert the masses to amounts of substance by making use of $n = m/M$, where m is the mass of the entity and M its molar mass. Then convert to mole fractions using the definition above. Relative atomic masses are shown in the Periodic Table on page 544.

ANSWER $M_r(H_2O) = 2(1.008) + 16.00 = 18.02$; therefore $M = 18.02\,\text{g mol}^{-1}$. $M_r(C_2H_5OH) = 2(12.01) + 6(1.008) + 16.00 = 46.07$; therefore $M = 46.07\,\text{g mol}^{-1}$. For simplicity take the mass of each component in the sample as 100 g; then the amounts of substance present are

$$\left.\begin{aligned} n(H_2O) &= \frac{100\,\text{g}}{18.02\,\text{g mol}^{-1}} = 5.55\,\text{mol} \\ n(C_2H_5OH) &= \frac{100\,\text{g}}{46.07\,\text{g mol}^{-1}} = 2.17\,\text{mol} \end{aligned}\right\} \text{total } n = 7.72\,\text{mol}$$

The mole fractions are therefore

$$x(\mathrm{H_2O}) = n(\mathrm{H_2O})/n = \frac{5.55\,\mathrm{mol}}{7.72\,\mathrm{mol}} = 0.719$$

$$x(\mathrm{C_2H_5OH}) = n(\mathrm{C_2H_5OH})/n = \frac{2.17\,\mathrm{mol}}{7.72\,\mathrm{mol}} = 0.281$$

COMMENT We chose 100 g as the mass of each component of the sample for numerical convenience: the mole fractions are independent of the overall mass of the sample and depend only upon the *relative* abundances of the components.

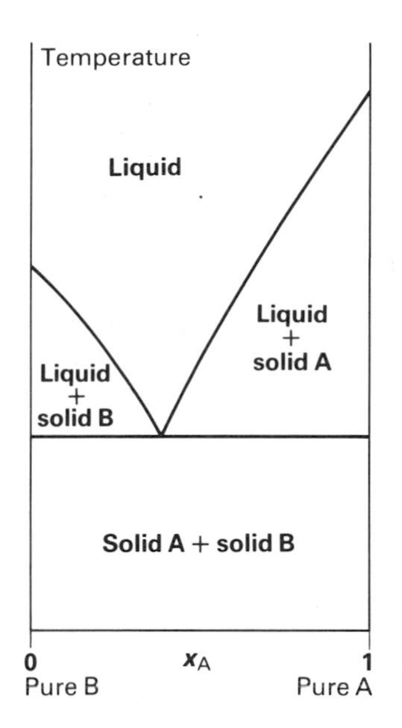

Figure 7.1 A typical solid/liquid two-component phase diagram (at constant pressure)

Phase diagrams are determined experimentally by noting the temperatures at which phase changes occur. For instance, we might note the boiling point of a liquid of some known composition. Then repeating the observation for samples of different compositions lets us plot the **liquid/vapour equilibrium line** on the phase diagram. Similarly, we might note the temperature at which a solid precipitates out of solutions of various compositions and hence find the **liquid/solid equilibrium line**.

Liquid/solid phase diagrams

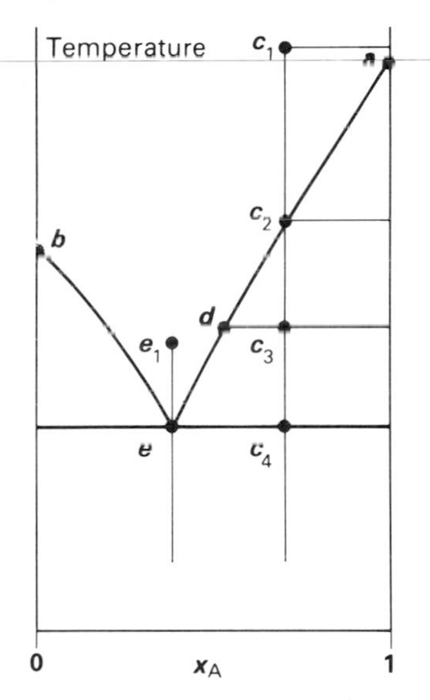

Figure 7.2 Changes represented by the phase diagram

Figure 7.1 shows a typical phase diagram for a system consisting of two components. Examples of this type of behaviour are potassium chloride/silver chloride and tin/lead. The interpretation of the diagram can be understood by thinking about how it is constructed.

Consider the point marked *a* in Figure 7.2. This marks the melting point of pure A at the pressure of the experiment; point *b* marks the melting point of pure B. Now consider c_1. Under these conditions of temperature and composition the sample is entirely liquid. If the sample is allowed to cool its state is represented by the points lying on the vertical line below c_1. Point c_2 marks the temperature below which the sample is no longer entirely liquid; pure solid A begins to precipitate, leaving the liquid richer in B. Put another way, the composition at c_2 denotes the **solubility** of A in liquid B at the temperature corresponding to c_2: at this composition the solution is **saturated** with A.

When the temperature corresponds to c_3 the sample consists of some solid A plus liquid of composition *d*. On cooling the mixture still further, more solid A precipitates and the composition of the remaining solution becomes even richer in B. At the temperature corresponding to c_4 the last of the liquid (now of composition *e*) freezes. The solid then consists of regions of the two immiscible solids A and B, and these remain however low the temperature.

The composition corresponding to *e* is special. This can be appreciated by thinking about what happens when a sample is prepared with exactly that composition. At a high enough temperature, corresponding to e_1, the mixture is entirely liquid. As it is cooled it remains liquid, and no change is observed until the temperature corresponds to *e*, when the entire solution solidifies into solid A plus solid B; at no earlier stage does any A or B precipitate. *All* the solution solidifies at a *definite* temperature, just like a pure material. Furthermore, this temperature is lower than the freezing point of either pure component. A mixture (it is *not* a compound because its composition depends on the pressure – see Figure 7.3) showing this behaviour is called a **eutectic** (from the Greek for 'easily melted'). The eutectic can be distinguished from pure A or pure B by adding a little of either pure component: the melting point of the eutectic *rises* whereas that of a pure component either remains unchanged (for A added to A) or goes down (for A added to B).

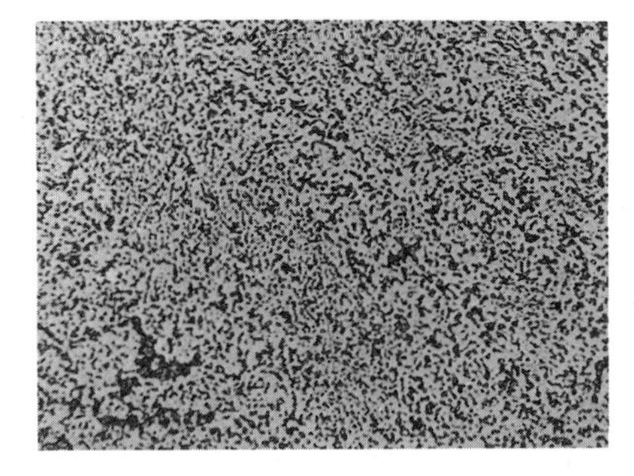

Figure 7.3 Eutectics can be seen to be a mixture of two different crystalline forms when examined under a microscope; here, globules of lead-rich solid solution (dark) are distributed in a tin-rich matrix

Eutectics have a number of commercial applications. For instance, tin and lead form a eutectic that melts at the conveniently low temperature of 183 °C; it is used as electrical solder. The eutectic mixture of common salt and water freezes at −21 °C, significantly below the freezing point of pure water; when salt is spread on wet roads, ice does not form until the temperature falls to

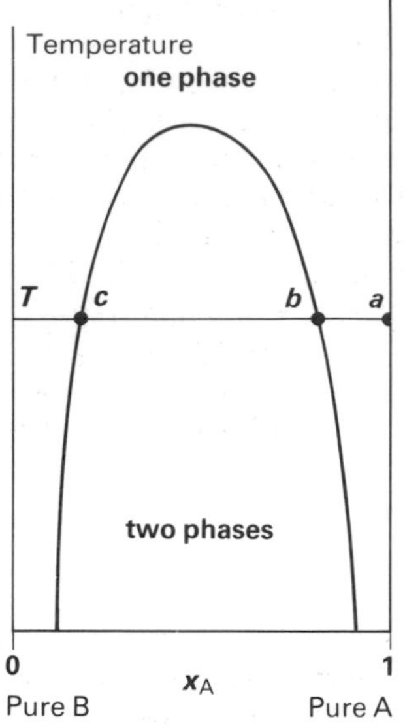

Figure 7.4 A phase diagram for two partially miscible liquids (at constant pressure)

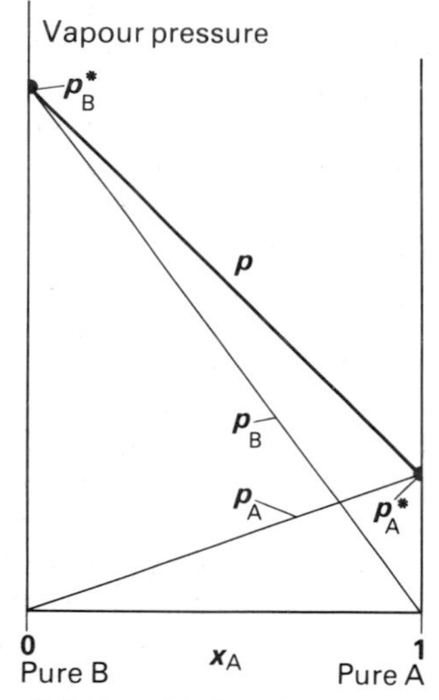

Figure 7.5 Raoult's law (at constant temperature)

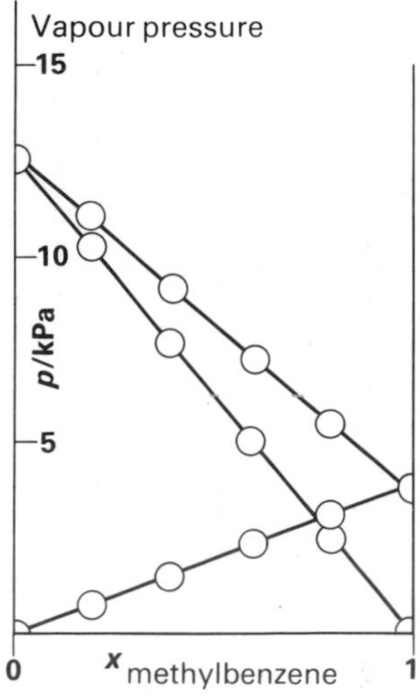

Figure 7.6 An almost ideal benzene/methylbenzene mixture

well below 0 °C (although the spreading does not give the eutectic composition, except by chance).

Liquid/liquid phase diagrams

Figure 7.4 shows a typical phase diagram for a system composed of two **partially miscible liquids** (liquids that do not mix in all proportions at all temperatures). Examples include phenylamine/hexane and phenol/water above room temperature. The interpretation of the diagram follows from the way it is constructed experimentally.

The diagram summarizes the behaviour of the phenol/water system (or a similar system) when it is at some constant pressure (such as 1 atm). Consider the line at the temperature *T*. Point *a* corresponds to pure A (water). As substance B (phenol) is added at constant temperature, the composition moves towards *b*. At *b* itself we notice a change. Instead of there being only a single liquid there are now two. The point *b* marks the composition at which the sample splits into two phases: one liquid phase is a saturated solution of B in A, the other is a saturated solution of A in B. In principle (if we wait long enough) the less dense phase floats on the other, but in practice the sample usually becomes cloudy. The new layer is not pure B; B can dissolve A to some extent, and so the new phase is a saturated solution of A in B, with a composition denoted by *c*.

As still more B is added the new layer increases, but its composition, B saturated with A, remains constant. The *proportion* of the original layer decreases, but its composition, A saturated with B, remains constant.

When enough B has been added to bring the *overall* composition to *c*, it can dissolve all the A present. Therefore the two-phase liquid becomes a single phase again, and remains a single-phase liquid however much more B is added. Repeating the experiment at a series of temperatures then lets us draw the entire curve shown in Figure 7.4.

Liquid/vapour phase diagrams

The French chemist François Raoult spent much of his life measuring vapour pressures of liquid mixtures, and summarized his results (about 1887) as

Raoult's law: The partial vapour pressure of a component in a mixture is equal to the vapour pressure of the pure component multiplied by its mole fraction in the liquid.

If, at the temperature of interest, the vapour pressure of the pure component A is p_A^*, and A is present in a mixture with a mole fraction x_A in the liquid phase, then it follows that the partial vapour pressure of A is

$$p_A = x_A p_A^* \qquad (7.1.1)$$

The total vapour pressure of a mixture of two liquids is the sum of their partial vapour pressures: $p = p_A + p_B$ (Dalton's law, discussed in Section 4.1). Therefore, if a mixture consists of A and B with mole fractions x_A and x_B, the total vapour pressure is

$$\begin{aligned} p &= x_A p_A^* + x_B p_B^* \\ &= x_A p_A^* + (1 - x_A) p_B^* \\ &= p_B^* + (p_A^* - p_B^*) x_A \end{aligned} \qquad (7.1.2)$$

This shows that the vapour pressure of the mixture is p_B^* when A is absent ($x_A = 0$), and changes *linearly* as x_A increases (see Figure 7.5). When B is absent ($x_A = 1$) the vapour pressure is p_A^*. Liquids that show this behaviour are called **ideal solutions**.

In practice, ideal solutions are closely similar pairs of liquids because their van der Waals interactions are similar. For example, linear dependence of the vapour pressure on the composition is illustrated for benzene/methylbenzene in Figure 7.6.

The normal boiling point of a liquid mixture is the temperature at which its total vapour pressure is 1 atm. Mixtures have vapour pressures that depend on their composition, and so the boiling point of a mixture also depends on its

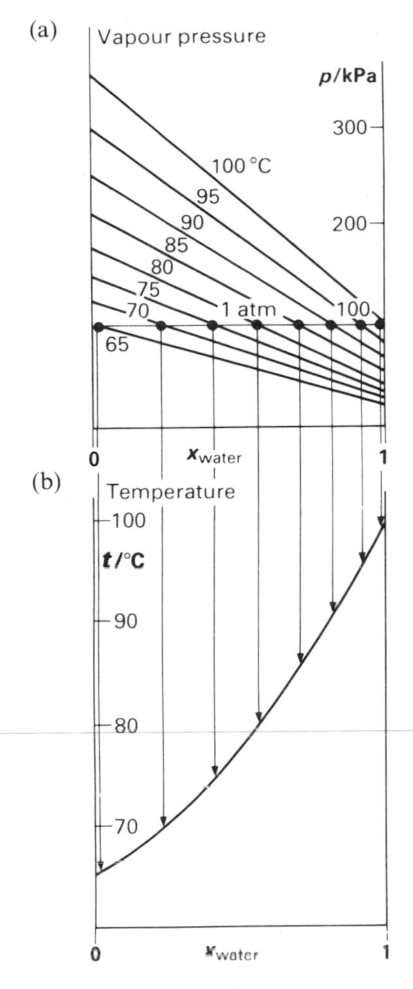

Figure 7.7 Converting a vapour pressure/composition diagram to a boiling point/composition diagram (for water/methanol)

composition. The dependence can be worked out by considering how diagrams like Figure 7.5 change with temperature.

Figure 7.7(a) shows how the total vapour pressure of an ideal solution changes as the temperature is increased. The points on the vertical axes show the rising vapour pressures of pure water and pure methanol. Raoult's law has been used to draw the lines connecting them. The diagram can be converted into a boiling point/composition plot by noting where the lines cut the horizontal line at 1 atm. This leads to the *curve* shown in Figure 7.7(b). All we have to do to find the boiling point of a mixture of given composition is to read off the corresponding temperature. This type of diagram is a **liquid/vapour phase diagram**, because at all temperatures below the line the stable state of the sample is as a liquid, and at all temperatures above the line the stable state is as a vapour. The practical importance of diagrams like this is that they let us discuss distillation in a systematic way.

Distillation

The vapour in equilibrium with a liquid mixture is richer in the more volatile component, and we need a second point on the phase diagram to denote its composition. Figure 7.8 shows how this can be done. Point a indicates the boiling point of a mixture of composition $x_A(a)$. Point a' indicates the composition of the vapour in equilibrium with the liquid at that temperature. It is determined by withdrawing a little of the vapour when the sample is boiling steadily, and analysing it. The line joining the two points, one denoting the composition of the boiling liquid and the other the composition of its vapour, is called a **tie-line**. The entire vapour composition curve may be constructed by repeating the procedure with liquids of different compositions.

In distillation we expect the more volatile component to evaporate preferentially. In terms of the phase diagram, we start with the mixture with the initial composition $x_A(a)$ and heat it to boiling. The vapour has the composition $x_A(a')$ and is richer in the more volatile component B. If that vapour is removed and condensed, it gives a condensate, also of composition $x_A(a')$, that is richer in B than is the original mixture (see Figure 7.9). (If the less volatile component is a non-volatile solid, only the solvent is distilled.)

In order to separate two volatile liquids completely it is necessary to do a series of distillations. The condensate from the first distillation, of composition $x_A(a')$, is itself distilled. From Figure 7.9, we see that it boils to give a vapour of composition $x_A(a'')$, which is even richer in the more volatile component. This condensate may be distilled again, and so on. After several such steps a liquid rich in B is obtained. In practice this process of **fractional distillation** is carried out in a single piece of apparatus, which may be of either laboratory

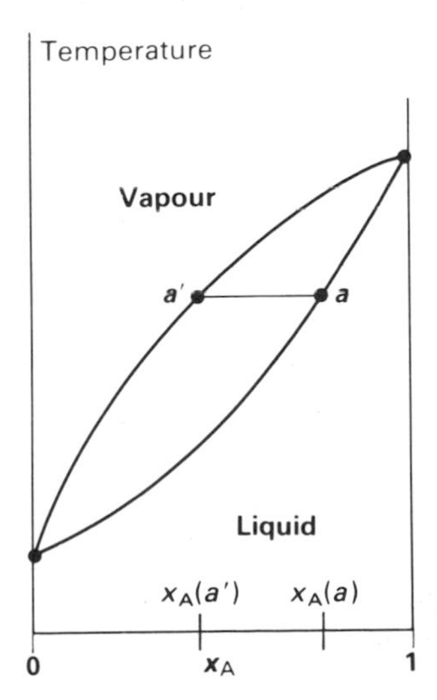

Figure 7.8 A liquid/vapour temperature/composition phase diagram (at constant pressure)

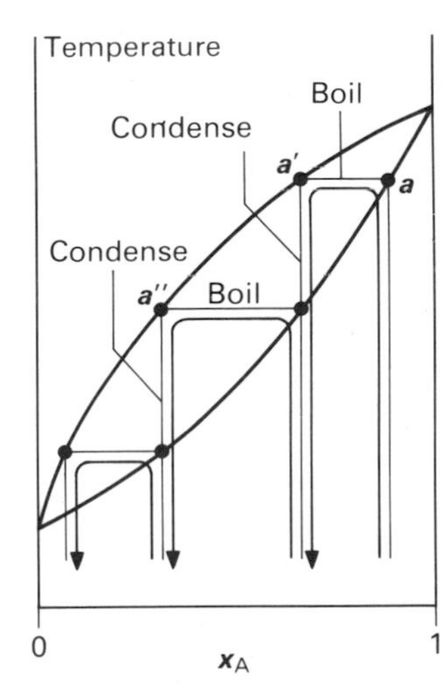

Figure 7.9 Fractional distillation

or industrial dimensions (see Figures 7.10 and 7.11 respectively). The sequence of boiling and condensation steps takes place on the plates or the packing of the columns where the rising vapour meets the falling liquid and reaches equilibrium. This equilibration is facilitated on the industrial scale by the **bubble caps**, which force the ascending hot vapour to pass through the descending liquid. Fractional distillation is essential for the separation of crude oil into its components (see Section 27.4) and to isolate nitrogen, oxygen and the noble gases from liquid air.

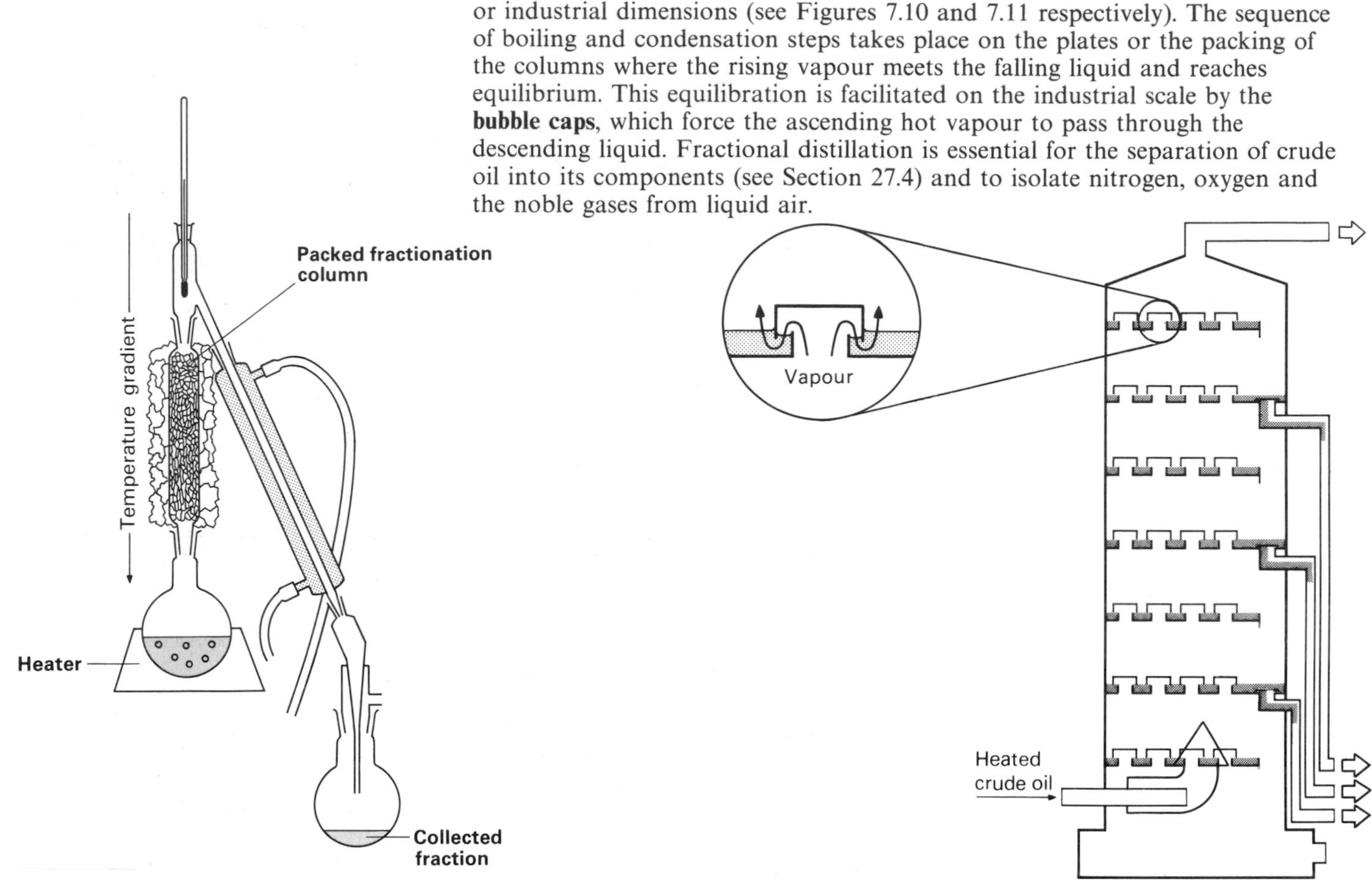

Figure 7.10 A laboratory-scale fractional distillation apparatus

Figure 7.11 Industrial-scale fractional distillation

Deviations from ideality

Some liquid mixtures have vapour pressures greater than Raoult's law predicts (see Figure 7.12); these **non-ideal solutions** are said to show **positive deviations** from the law. The molecules in these mixtures have energetically unfavourable interactions with each other, and hence a greater tendency to vaporize than in an ideal solution. A higher vapour pressure corresponds to a lower normal boiling point, and so mixtures showing positive deviations from Raoult's law have lower boiling points than expected. It is interesting to demonstrate this effect by pouring ethanal into pentane (with some porous pot to provide sites for bubble formation): the mixture spontaneously boils. As in this example, sometimes the deviations are so great that the boiling point curve passes through a minimum, and the mixture boils at a temperature lower than the boiling point of either pure component. Mixtures showing this behaviour are said to form a **minimum boiling azeotrope**. Ethanol and water form an azeotrope (or **constant boiling mixture**) of this kind (see Figure 7.13); the minimum boiling point is 78.2 °C, and occurs when the composition corresponds to 95.6 per cent ethanol (by mass).

The reason for the name **azeotrope** (which comes from the Greek for 'not changed on boiling'), and its importance for distillation, can be understood by thinking about distilling a mixture of composition $x_A(a)$ (see Figure 7.14). This mixture boils at the temperature corresponding to a, and the vapour, point a', is richer in ethanol. Condensation of the vapour and redistillation further increases the ethanol concentration. At b, however, the compositions of the vapour and liquid are almost the same (b' is close to b). At z they are exactly the same: a liquid of composition $x_A(z)$ boils to give a vapour of exactly the same composition. Distillation cannot concentrate the mixture beyond this **azeotropic composition**. Anhydrous ethanol therefore has to be prepared by chemical drying, such as by standing over calcium oxide.

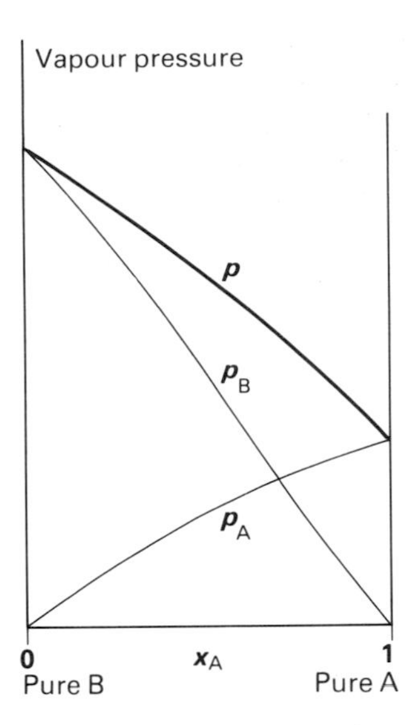

Figure 7.12 Positive deviations from Raoult's law (at constant temperature)

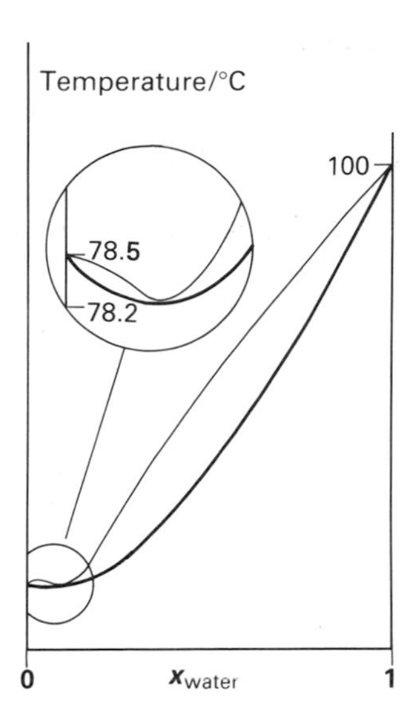

Figure 7.13 The formation of a minimum boiling azeotrope (ethanol/water) (at constant pressure)

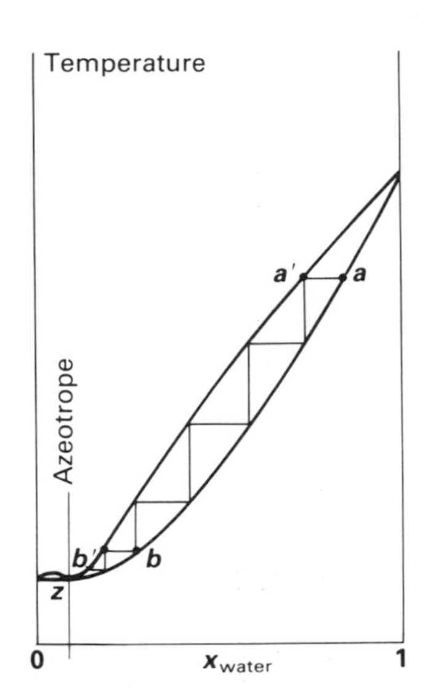

Figure 7.14 Fractional distillation of a mixture with a minimum boiling azeotrope

A few pairs of liquids form mixtures having vapour pressures lower than Raoult's law predicts (see Figure 7.15). These non-ideal solutions show **negative deviations** on account of the favourable interactions between the components. Lower vapour pressure implies a higher normal boiling point, and so such mixtures boil at higher temperatures than expected. When the deviations are very large the boiling point may be higher than that of either pure component, and a **maximum boiling azeotrope** is obtained, as shown in Figure 7.16. The example illustrated here is that of a nitric acid–water mixture: 68 per cent nitric acid (by mass) boils unchanged at 121 °C. (That the interactions are favourable is indicated by the exothermicity of the mixing of nitric acid and water.) Distillation of a mixture of nitric acid and water with $x_{water} = 0.5$ gives pure nitric acid at first (see Figure 7.17). The remaining liquid is therefore richer in water, and as distillation continues the nitric acid concentration decreases until the azeotropic composition is reached. The mixture then boils unchanged.

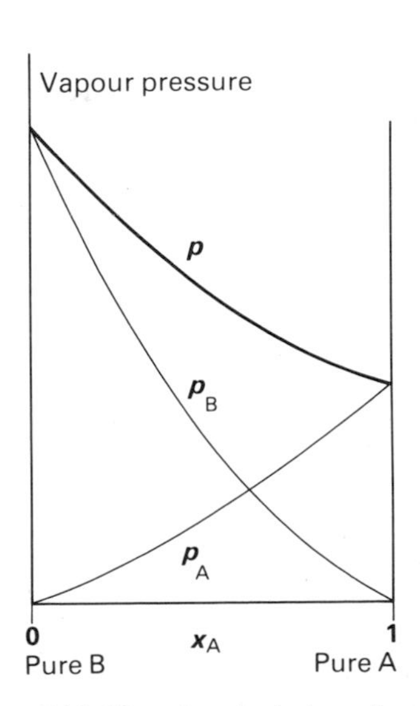

Figure 7.15 Negative deviations from Raoult's law (at constant temperature)

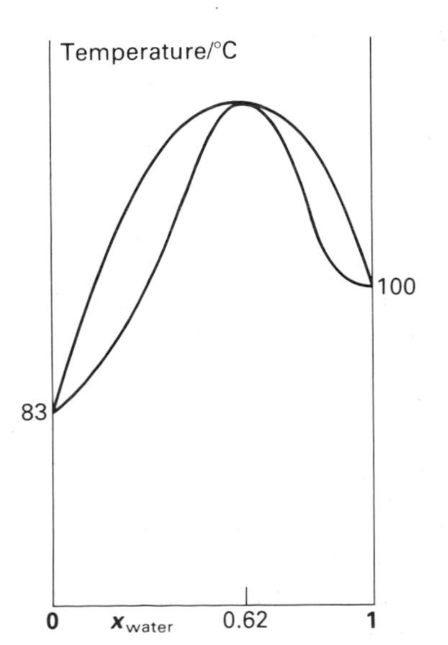

Figure 7.16 The formation of a maximum boiling azeotrope (nitric acid/water) (at constant pressure)

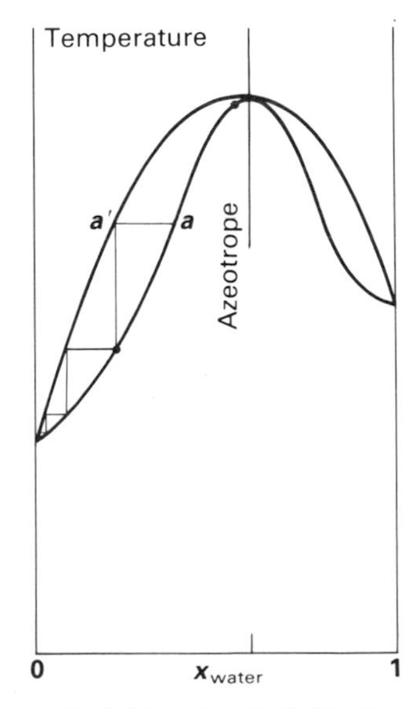

Figure 7.17 Fractional distillation of a mixture with a maximum boiling azeotrope

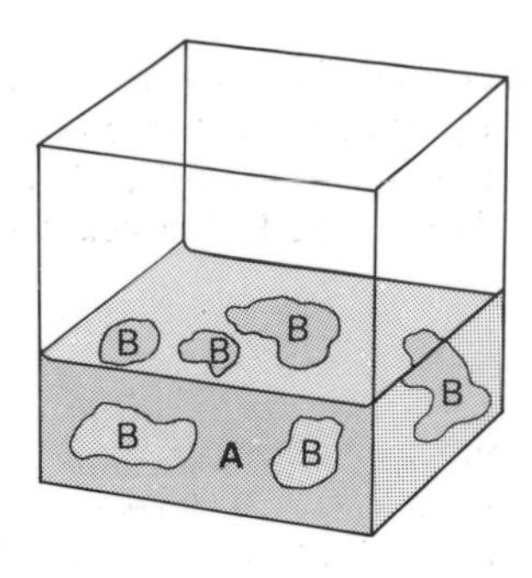

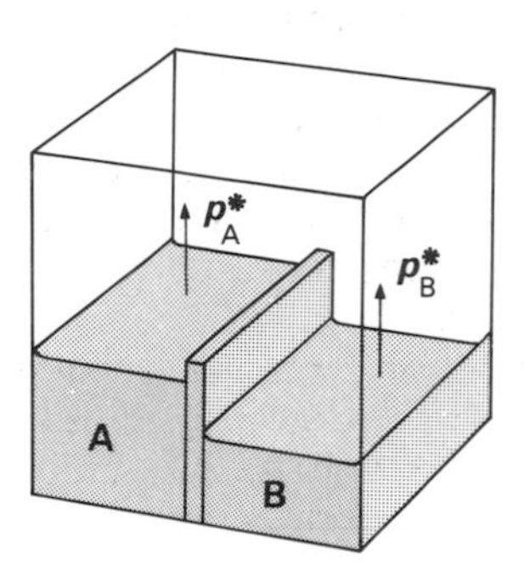

Figure 7.18 The vapour pressure of immiscible liquids

Immiscible liquids

When two liquids are immiscible, like oil and water, the total vapour pressure of an agitated sample is the sum of their individual vapour pressures. Figure 7.18 shows that a sample of two immiscible liquids is equivalent to two separated liquids inside the same container. Liquid A exerts its vapour pressure (p^*_A) and liquid B exerts its vapour pressure (p^*_B); the total vapour pressure is the sum of the two vapour pressures, $p^*_A + p^*_B$, and it is the same whether the liquids are separated or whether one is a collection of blobs floating around in the other, so long as both have access to the vapour.

This behaviour of immiscible liquids is the basis of **steam distillation.** Since a sample of immiscible liquids exerts the sum of their vapour pressures, it boils at a lower temperature than either component alone. Therefore a sample can be distilled (by bubbling steam through it) at a lower temperature, and organic molecules, such as phenylamine, can be distilled under conditions that do not cause them to decompose. Boiling cabbages results in the steam distillation of some of their more disagreeable constituents.

Steam distillation was used in the past to obtain molar masses; but it is very inaccurate, messy and wholly unnecessary now that mass spectrometry, osmometry (described below) and other highly developed techniques are available.

7.2 Colligative properties

Phase diagrams are constructed from experimental observations. When the diagrams are inspected it is found that there are several regularities. For instance, it is found that the presence of a solute always lowers the freezing point of a solvent (which is why we use antifreeze) and, if the solute is non-volatile, always raises the solvent's boiling point.

Raoult's law is particularly simple when one of the components, say A, is not volatile (or, at least, much less volatile than is the other component at the temperature of interest; everything is volatile at high enough temperatures). Since A is not volatile, its vapour pressure is effectively zero at the temperature of interest, and so we may set $p^*_A = 0$ in equation 7.1.2. Then the vapour pressure of a solution of a *non-volatile* solute in a volatile solvent (B) is

$$p = x_B p^*_B \quad \text{or} \quad \frac{p^*_B - p}{p^*_B} = 1 - x_B = x_A \tag{7.2.1}$$

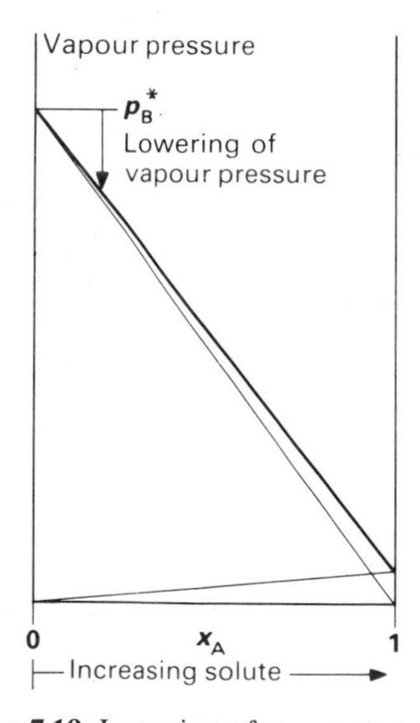

Figure 7.19 Lowering of vapour pressure by an (almost) non-volatile solute (A)

This equation shows that, as x_A is positive, there is a *lowering of vapour pressure*, from p^*_B to p, proportional to the mole fraction of the non-volatile solute. This is illustrated in Figure 7.19. Since a lower vapour pressure implies a higher boiling point, we can also conclude that *the presence of a non-volatile solute raises the boiling point of a solvent* (see Figure 7.20). Notice that the

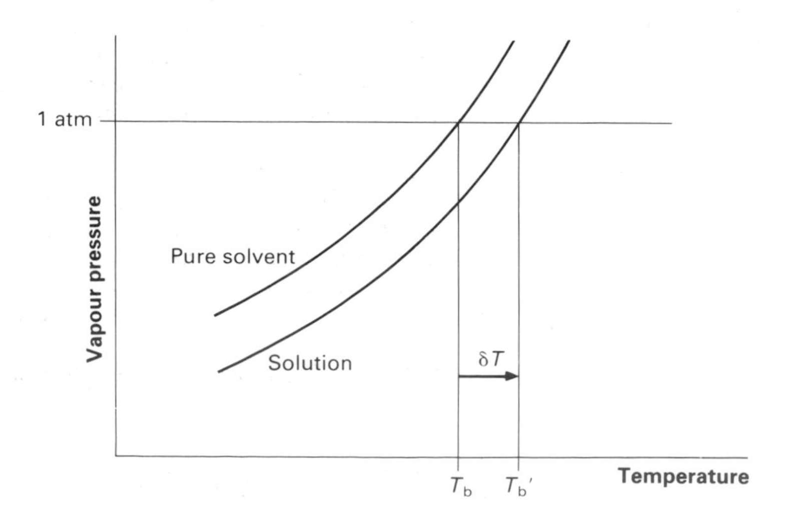

Figure 7.20 The lowering of vapour pressure by addition of a non-volatile solute elevates the boiling point

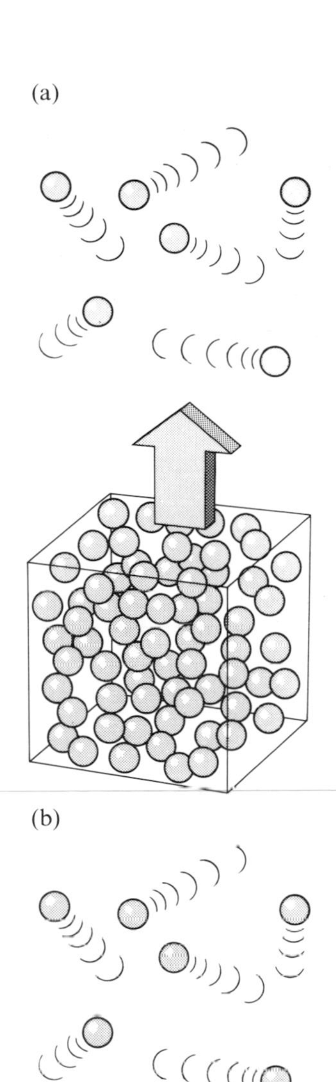

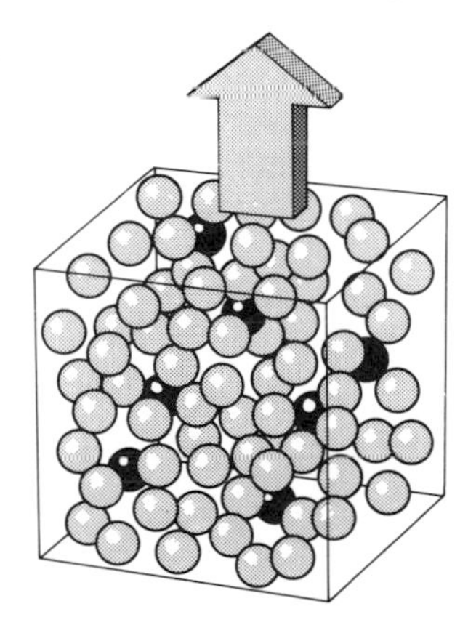

Figure 7.21 Vaporization of (a) pure solvent, (b) solution

vapour pressure lowering depends on the mole fraction of the solute, x_A, but is independent of its chemical composition. Such properties are called **colligative properties**. Thus, for example, if we add N molecules of cholesterol to benzene we get the same decrease in vapour pressure as when we add N molecules of anthracene.

The elevation of boiling point

The reason why a pure solvent gives rise to a vapour pressure lies in the tendency of its molecules to escape and to disperse as a gas, as illustrated in Figure 7.21(a). A gas is a chaotic collection of particles (see Section 4.2). When a non-volatile solute is added to a solvent the solution becomes more chaotic than was the original solvent (Figure 7.21(b)). Since the solution is already chaotic as a result of the presence of the solute, there is a lower tendency for the solvent molecules to escape as a gas. That is, the vapour pressure is lower. (The role of chaos in determining the direction of natural change, such as vaporization, is explained in Chapter 11. In the language developed there, the presence of the solute increases the entropy of the solution.)

The measurement of the elevation of boiling point is a way of determining the molar mass of the solute. The procedure is called **ebullioscopy**. When some simplifications are made, principally that the solution is dilute, it turns out that the elevation of boiling point, δT, is proportional to the molality–see page 518–of the solute A, m_A, the relation being

$$\textit{Elevation of boiling point: } \delta T = K_b m_A \qquad (7.2.2)$$

K_b is called the **ebullioscopic constant** (or **boiling point constant**), and depends on the solvent; some values are listed in Table 7.1. The method of using this expression is described in the following Example.

Table 7.1 Cryoscopic and ebullioscopic constants

	$t_f/°C$	$\dfrac{K_f}{\text{K kg mol}^{-1}}$	$t_b/°C$	$\dfrac{K_b}{\text{K kg mol}^{-1}}$
Benzene	5.5	5.12	80.1	2.53
Camphor	179.5	40.0	208.3	6.0
Ethanoic acid	16.6	3.90	118.1	3.07
Water	0.0	1.86	100.0	0.51

EXAMPLE

The normal boiling point of benzene is 80.1 °C and its ebullioscopic constant is 2.53 K kg mol^{-1}. When 0.668 g of camphor is dissolved in 20 g of benzene the boiling point of the solvent is raised by 0.612 K (*differences* of boiling points can be measured more precisely than the boiling points themselves). What is the molar mass of camphor?

METHOD Equation 7.2.2 is used to find the molality of the solute: 0.668 g in 20 g of solvent corresponds to 33.4 g in 1 kg. A mass of 33.4 g corresponds to an amount of substance (33.4 g/M) = (33.4/M_r) mol, and the molality of the solute is therefore (33.4/M_r) mol kg^{-1}. On equating the two expressions for the molality the only unknown is M_r; hence this may be found. Finally use $M = M_r$ g mol^{-1}.

ANSWER The molality of camphor, m_A, is obtained from equation 7.2.2 as

$$m_A = \delta T/K_b = \frac{0.612\ \text{K}}{2.53\ \text{K kg mol}^{-1}} = 0.242\ \text{mol kg}^{-1}$$

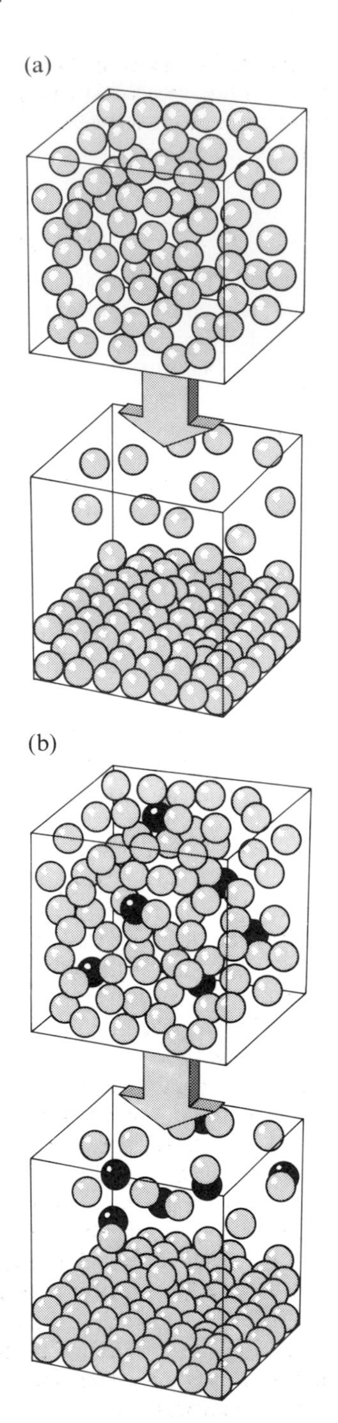

Figure 7.22 Freezing of (a) pure solvent, (b) solution

The molality is also given as $(33.4/M_r)\,\text{mol kg}^{-1}$. Equating these two expressions leads to

$$M_r = \frac{33.4}{0.242} = 138$$

Therefore, $M = 138\,\text{g mol}^{-1}$.

COMMENT The formula for camphor is $C_{10}H_{16}O$; therefore its calculated molar mass is $152.24\,\text{g mol}^{-1}$, and so the result is inaccurate. This technique is rarely used for molar mass determinations because elevations of boiling point are very small and some materials are temperature-sensitive.

The depression of freezing point

We have seen that the presence of a solute makes a solution more chaotic (that is, it increases its entropy). Melting is similar to evaporation in that the particles of a melting solid acquire a less regular arrangement. Freezing is the converse, for the chaos of the liquid gives way to the orderliness of the solid. Figure 7.22 illustrates what happens when there is a solute present that does not dissolve in the solid solvent. Note that the solute need not be non-volatile; indeed, the liquid ethane-1,2-diol (ethylene glycol – see Section 29.5) is the common antifreeze used to protect engines in winter.

In the pure solvent an ordered structure comes about at a fixed temperature, the normal freezing point (see Figure 7.22(a)). When a solute is present there is more randomness than in the original pure solvent, and so the temperature has to be lowered further in order to bring about the transition to a more orderly state. In other words, *the presence of a solute lowers the freezing point of the solvent.* This is illustrated in Figure 7.22(b).

The depression of freezing point depends on the mole fraction of solute present and its measurement can be used to determine the latter's molar mass. The technique is called **cryoscopy**. After some calculation and simplification on the basis that the solution is dilute, it turns out that the depression of freezing point of a solution of molality m_A is

Depression of freezing point: $\delta T = K_f m_A$ (7.2.3)

K_f is called the **cryoscopic constant** (or **freezing point constant**), and depends on the solvent; some values are included in Table 7.1. The value of K_f for camphor is large, and so it is frequently used in cryoscopy (when it is called **Rast's method**). Freezing-point depressions are generally greater than boiling-point elevations, and so cryoscopic values of molar masses are more reliable than are ebullioscopic values. The Example that follows illustrates how equation 7.2.3 is used.

EXAMPLE

The freezing-point depression produced by dissolving 2.16 g of sulphur in 100 g of a solvent was 0.58 K. When 3.12 g of iodine were dissolved in 100 g of the same solvent the depression was 0.85 K. On the grounds that it is known that iodine is present as I_2 in this solvent, deduce the molecular form of the sulphur.

METHOD The freezing-point depressions are given by equation 7.2.3 in each case, with the same value of K_f. Therefore the ratio of the two depressions is equal to the ratio of the molalities of the two solutes. Since the mass of solvent is the same in each case, the ratio of the molalities is equal to the ratio of the amounts of substance. The amounts of substance are given by (mass)/M in each case. Therefore

$$\frac{\delta T(\text{sulphur})}{\delta T(\text{iodine})} = \frac{n(\text{sulphur})}{n(\text{iodine})} = \frac{\text{mass(sulphur)}}{M(\text{sulphur})} \times \frac{M(\text{iodine})}{\text{mass(iodine)}}$$

Since the only unknown in this expression is the molar mass of the sulphur, it can be determined from the data. The relative atomic mass is known (32.06) and so the number of atoms per molecule can be determined. The relative atomic mass of iodine is 126.9 and so $M(\text{iodine}) = M(I_2) = 253.8\,\text{g mol}^{-1}$.

ANSWER Substitution of the data into the expression just derived gives

$$\frac{\delta T(\text{sulphur})}{\delta T(\text{iodine})} = \frac{0.58\,\text{K}}{0.85\,\text{K}} = 0.68$$

$$\frac{\text{mass(sulphur)}}{M(\text{sulphur})} \times \frac{M(\text{iodine})}{\text{mass(iodine)}} = \frac{2.16\,\text{g}}{M} \times \frac{253.8\,\text{g mol}^{-1}}{3.12\,\text{g}} = (176\,\text{g mol}^{-1})/M$$

Therefore $M = (176\,\text{g mol}^{-1})/0.68 = 260\,\text{g mol}^{-1}$. This molar mass corresponds to a molecular formula S_8, since $M(S_8) = 256.5\,\text{g mol}^{-1}$.

COMMENT Sulphur often exists as S_8, with the atoms forming a puckered ring resembling a crown (see Section 22.2).

Osmosis

A third important colligative property is very useful for the measurement of molar mass and is also of great biological importance. This is the property of **osmosis** (from the Greek for 'push'). Osmosis is the tendency of solvent molecules to pass through a membrane from a more dilute to a more concentrated solution. The membrane is special in the sense that it is semi-permeable, that is, it allows solvent but not solute to pass through. In some cases the **semi-permeable membrane** or **selectively permeable membrane** may allow water to pass but not dissolved ions; in others it may allow a solvent composed of small molecules to pass but not large polymer or biological molecules. Cellulose acetate (see Section 30.5) is often used as a membrane. Cell membranes are naturally occurring semi-permeable membranes, which is why osmosis is important in biology.

When a solvent or a dilute solution is brought into contact with a concentrated solution through the semi-permeable membrane in the apparatus shown in Figure 7.23, the solvent flows through. As it does so, the height of the concentrated solution increases, and this gives rise to an opposing hydrostatic pressure. As the flow of solvent continues there comes a point when the extra pressure on the right of the membrane is so great that the two sides of the membrane are in equilibrium, and the net flow is zero. The excess pressure at equilibrium is called the **osmotic pressure**, and is denoted Π (Greek capital pi). The same process contributes to the uptake of water by the root systems of plants, as the concentration of salts is higher in root cells than in the soil water. An egg out of its shell swells when immersed in water, but shrinks if immersed in strong salt solution (the shell membrane is semi-permeable). 'Chemical gardens' are produced by immersing crystals of various metal salts in water-glass (sodium silicate) solutions. As the crystals slowly dissolve, the metal silicates formed by precipitation act as semi-permeable membranes – water passes through and into them, causing them to swell and burst, producing weird shapes.

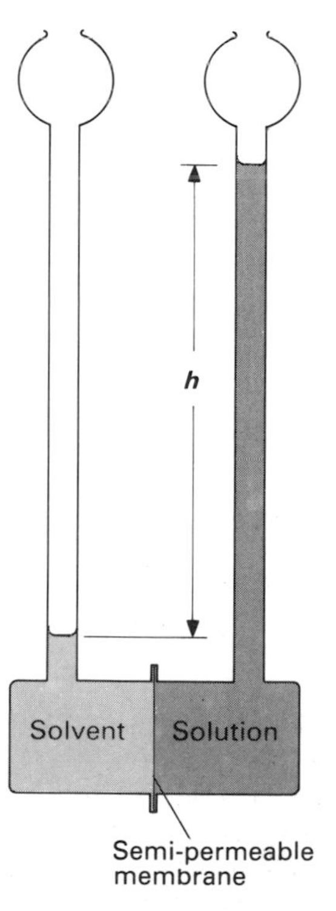

Figure 7.23 The development of osmotic pressure ($\Pi = \rho g h$: ρ = density, g = gravitational acceleration)

Osmosis is a colligative property because the osmotic pressure of a solution depends on the number of solute particles present and is independent of their nature. When a solution containing an amount of substance n of solute in a volume V is in contact with the pure solvent at a temperature T, the osmotic pressure is given by the **van't Hoff equation**

$$\Pi V = nRT \qquad (7.2.4)$$

where R is the molar gas constant. Note the striking similarity of the van't Hoff equation to the perfect gas equation ($pV = nRT$) even though it applies to an apparently quite different system.

Osmosis is very useful for the determination of molar mass. Measurements of osmotic pressure can be made at room temperature, so that fragile, temperature-sensitive biological molecules can be studied. It is also a very sensitive technique, and so can be applied to the determination of the molar masses of proteins, polymers and other large molecules. Thus a 1 mol dm^{-3} solution of sucrose boils at about 0.5 K above the boiling point of water, which is just measurable, and freezes at 1.9 K below its freezing point, which is significant but not dramatic, but exerts an osmotic pressure sufficient to sustain a column of water 250 m high!

EXAMPLE

1.10 g of a protein was dissolved in 100 cm^3 of water at 20 °C. The osmotic pressure of the solution was found to be 395 Pa. What is the molar mass of the protein?

METHOD Use equation 7.2.4 to find the amount of substance of the protein in the solution, then relate that to the molar mass through $n = \text{mass}/M$. Note that 1 Pa = 1 N m^{-2} = 1(J m^{-1}) m^{-2} = 1 J m^{-3}. Insert the data $T = 293$ K, $V = 100 \times 10^{-6}$ m^3, $\Pi = 395$ Pa into equation 7.2.4.

ANSWER

$$n = \frac{\Pi V}{RT} = \frac{395\,\text{J m}^{-3} \times 1.00 \times 10^{-4}\,\text{m}^3}{8.314\,\text{J K}^{-1}\,\text{mol}^{-1} \times 293\,\text{K}} = 1.62 \times 10^{-5}\,\text{mol}$$

The mass of the sample is 1.10 g; therefore the molar mass of the protein is

$$M = \frac{\text{mass}}{n} = \frac{1.10\,\text{g}}{1.62 \times 10^{-5}\,\text{mol}} = 6.8 \times 10^4\,\text{g mol}^{-1}$$

Therefore the relative molecular mass is 68 000.

COMMENT Osmometry is a very important technique for determining the molar masses of biological and synthetic macromolecules. It is widely used in modern laboratories. Note that an osmotic pressure of 395 Pa is sufficient to support a column of water 4 cm high, and so it is readily measurable in an apparatus like that shown in Figure 7.23. In the **Berkeley–Hartley osmometer** an excess pressure is exerted on the concentrated solution so as to oppose the net flow of the solvent into it. The pressure required to stop the net flow is equal to Π. This method has the significant advantage of leaving concentrations unchanged, and the results are more accurate.

Figure 7.24 The plant under construction at Yuma, Arizona, will be the world's largest reverse osmosis desalination facility

Osmosis is important in many contexts, one of them being life-saving. If fresh water is gulped into the lungs, it undergoes osmosis into the more saline blood and causes heart failure in about three minutes. Salt water is more saline than blood, and if it is gulped into the lungs osmosis transfers water *from* the blood. This takes longer, about twelve minutes, to cause death, and so rescuers have slightly longer to work in salt water.

Reverse osmosis is the opposite of osmosis: pressure applied to a solution causes pure solvent to flow out of the solution through a membrane, provided that the pressure applied is greater than the osmotic pressure of the solution. This technique can be used to **desalinate** sea water (see Figure 7.24), the principal (but largely solved) technical difficulty being to find membranes, such as cellulose acetate, that can withstand the high pressures required.

BOX 7.1

Colloids

A colloid consists of very tiny particles (with diameters approaching the wavelength of light) of one substance dispersed through another. They may be distinguished from solutions by their ability to scatter light, like dust particles in a sunbeam. This is called the **Tyndall effect.** A familiar colloidal system is milk, in which tiny oily globules are dispersed through an aqueous phase that consists largely of water and some water-soluble compounds. Liquids dispersed in liquids, such as milk, are called **emulsions**. A fog is also colloidal, the **disperse phase** being microscopic droplets of water, and the **support** (or **dispersion medium**) being air. Liquids dispersed in gases, like fog, are called **aerosols**.

Colloids occur widely in everyday life, as these examples suggest. Some ruby glass in stained windows consists of colloidal gold particles dispersed in transparent glass. Many paints are colloidal emulsions, and coat the surface with an adhering film as the water support evaporates. The cosmetic preparation called 'cold cream' is an emulsion of oil and water; its water evaporates, cooling the skin.

Some colloids are a nuisance, even when they have been precipitated. This is the case with river mud, which is carried downstream as colloidal particles. When the fresh water mingles with salt water at estuaries, the higher ion concentration there causes the colloidal particles to stick together and precipitate (see Figure 7.25). This precipitation is a major source of silting in estuaries.

Figure 7.25 The silting of river estuaries is largely due to colloidal mud particles precipitating out where the fresh water meets the sea (the white flecks in this estuary are small boats at their moorings)

Summary

- ☐ A phase diagram is a plot of the pressures and compositions at constant temperature, or temperatures and compositions at constant pressure, at which phases coexist in equilibrium.
- ☐ A liquid/solid equilibrium line denotes the solubility of one component in the other. The composition at a point on the line is the composition of the saturated solution at that temperature.
- ☐ The eutectic composition is the composition corresponding to the lowest melting point of a mixture of solids, or the lowest freezing point of a liquid mixture. A mixture with the eutectic composition melts and freezes at a single temperature (at a stated pressure) without change of composition.
- ☐ Partially miscible liquids are pairs of liquids that do not mix (to give a single liquid phase) in all proportions at all temperatures in the range of interest.
- ☐ Raoult's law states that $p_A = x_A p_A^*$ (equation 7.1.1).
- ☐ Ideal solutions are solutions that obey Raoult's law.
- ☐ When a liquid mixture is boiling, the equilibrium compositions of the liquid and the vapour are given by the two points at the opposite ends of the tie-line on the liquid/vapour phase diagram.
- ☐ Fractional distillation is a series of distillations and condensations; it is used to separate mixtures of liquids of similar boiling points.
- ☐ Solutions having vapour pressures higher than predicted by Raoult's law are said to show positive deviations from the law; when the deviations are large enough such systems form minimum boiling azeotropes.
- ☐ Solutions having vapour pressures lower than predicted by Raoult's law show negative deviations and, when these are large enough, form maximum boiling azeotropes.
- ☐ An azeotrope is a mixture that boils without change in composition. A maximum boiling azeotrope boils at a higher temperature than either component; a minimum boiling azeotrope boils at a lower temperature than either component.
- ☐ Azeotropic liquid mixtures cannot be completely separated by distillation.
- ☐ Pairs of immiscible liquids boil at lower temperatures than either component alone, the total vapour pressure being the sum of the individual vapour pressures.
- ☐ Steam distillation is used to separate temperature-sensitive materials.
- ☐ Colligative properties are properties that depend on the number but not on the nature of solute particles.
- ☐ Colligative properties include the lowering of vapour pressure, the elevation of boiling point, the depression of freezing point and osmosis.
- ☐ The presence of non-volatile solutes lowers the vapour pressure of a solvent, and therefore raises its boiling point: this is the basis of ebullioscopy, which is used for the determination of molar mass.
- ☐ The presence of a solute that does not dissolve in the solid solvent leads to a lowering of the freezing point of the solvent. This is the basis of cryoscopy, which is also used for the determination of molar mass.
- ☐ Osmosis is the tendency for solvent to flow from a less to a more concentrated solution when the two are separated by a semi-permeable membrane.
- ☐ The osmotic pressure is the extra pressure that must be exerted on the concentrated solution in order to stop the net flow by osmosis; its magnitude is given by the van't Hoff equation (equation 7.2.4).

PROBLEMS

1 Explain the term *mole fraction*. Calculate the mole fractions of ethanol and water in vodka, taken to consist of 35 per cent by mass of ethanol and 65 per cent by mass of water.

2 Explain the terms *eutectic*, *azeotrope*, *minimum boiling azeotrope* and *tie-line*.

3 Show how a temperature–composition phase diagram may be used to describe the fractional distillation of a mixture of two liquids.

4 The separation of methanol and ethanol is an important problem, not only on account of their different physiological effects. Their vapour pressures at 25 °C are 11.730 kPa and 5.866 kPa, respectively. On the basis that 40 g of methanol and 200 g of ethanol mix to give an ideal solution, calculate (a) the partial vapour pressure exerted by each component, (b) the total vapour pressure of the mixture, (c) the composition of the vapour.

5 At 95 °C the vapour pressure of water is 84.52 kPa and that of bromobenzene (C_6H_5Br) is 15.48 kPa. Calculate the mass of water needed to steam distil 100 g of bromobenzene when the atmospheric pressure is 100.0 kPa.

6 What is the effect on the vapour pressure of water and the total vapour pressure of the system when small amounts of the following materials are added (separately) to water at 25 °C: (a) sodium chloride, (b) propanone, (c) tetrachloromethane?

7 Explain why the presence of a non-volatile solute raises the boiling point of a solvent. Explain why the presence of a solute that is insoluble in the solid solvent lowers the freezing point of the solvent.

8 What is the molar mass of a substance which, when 0.2 g is dissolved in 10 g of water, lowers the freezing point to −0.62 °C?

9 Calculate the boiling point under a pressure of 1 atm of 100 g of water sweetened with 1 lump (3.6 g) of sucrose. What is the freezing point?

10 Calculate the osmotic pressure of the sucrose solution in the last question, at 25 °C (take the density of water as 1.00 g cm^{-3}).

11 Suggest an explanation for the similarity between the van't Hoff equation $\Pi V = nRT$ and the perfect gas equation $pV = nRT$.

12 What do you understand by the terms *allotropy*, *alloy* and *eutectic mixture*? How could you distinguish between a eutectic mixture and a pure substance?

A mixture of 27 per cent gold and 73 per cent thallium melts and freezes at 131 °C as if it were a pure substance. Draw on graph paper the phase diagram for the gold–thallium system, labelling each of the main areas.

Sketch the temperature–time curves that would be obtained on allowing the following mixtures to cool to room temperature from 1000 °C:

	(a)	(b)
gold	60%	27%
thallium	40%	73%

Describe what happens in the system during the cooling of mixture (a). *(Oxford and Cambridge)*

13 (a) Draw and suitably label a phase diagram for water. Use the diagram to show the effect of pressure on the freezing point of pure water.

Explain how the vapour pressure changes when an involatile solute is added to water. Add a further part to your diagram to illustrate the effect of this addition on the boiling point and freezing point of pure water.

(b) When an immiscible volatile liquid is added to water explain how the total vapour pressure is affected. Use this concept to explain the basis of steam distillation and give one practical application of steam distillation.

(c) Hexane and heptane are totally miscible and form an ideal two-component system. If the vapour pressures of the pure liquids are 56 000 and 24 000 N m^{-2} at 51 °C calculate (i) the total vapour pressure and (ii) the mole fraction of heptane in the vapour above an equimolar mixture of hexane and heptane. *(Welsh)*

14 State Raoult's law as applied to a liquid binary mixture and describe how it can be explained in terms of a simple molecular model.

Show how binary liquid systems which deviate from ideality may form either a maximum or a minimum boiling mixture.

Discuss the principles involved in the separation, by fractional distillation, of a mixture of two volatile liquids (which do not form a constant boiling mixture). Give brief details of the apparatus and methods you would use to carry out this separation in the laboratory. *(Oxford and Cambridge)*

15 (a) Explain the meaning of the term *osmotic pressure*. Describe with the aid of a sketch an apparatus for the measurement of the osmotic pressure of an aqueous solution.

(b) At 290 K, the osmotic pressure of a solution of 24.3 g of a sugar in 1 dm^3 of water is 9.85 × 10^4 Pa. Estimate the relative molecular mass (molecular weight) of this sugar. (1 Pa = 1 N m^{-2}.)

(c) When the relative molecular mass of a compound is comparatively low, it is far more often measured by determination of freezing-point depression or boiling-point elevation than by osmotic pressure measurements. For macromolecules such as polymers, the reverse is true. Why is this? *(Oxford)*

16 Describe how measurements of freezing-point depression can be used to determine relative molecular masses (molecular weights). It was found that 1 mole of added solute depressed the freezing point of 1 kg of benzene by 5.10 K, while 0.500 g of a hydrocarbon M dissolved in 100 g of benzene depressed the freezing point by 0.212 K. Quantitative analysis indicated that the composition was 90.0 per cent carbon and 10.0 per cent hydrogen by mass. A spectroscopic method established that M possessed a threefold axis of symmetry. Suggest a possible structure for M. What physical method might be used to provide evidence about the symmetry of a molecule? *(Oxford Entrance)*

17 (a) Sketch the boiling point–composition diagram for a simple binary mixture which *does not* show a boiling point maximum, and use it to illustrate the principle of fractional distillation.

(b) Explain *concisely* how steam distillation works.

(c) Methanol and ethanol form a solution which is almost ideal, and the vapour pressures of the pure components at 300 K are 11.82 × 10^3 N m^{-2} and 5.93 × 10^3 N m^{-2}, respectively. Calculate the mole fraction of methanol in the vapour above a solution made by mixing equal *weights* of the two components.

(d) The vapour pressure of the *immiscible* liquid system diethylaniline–water is 10.13 × 10^4 N m^{-2} at 372.5 K. The vapour pressure of water at this temperature is 9.92 × 10^4 N m^{-2}. How many kg of steam are required to distil over 0.1 kg of diethylaniline at a total pressure of 10.13 × 10^4 N m^{-2}? *(Cambridge Entrance)*

8 Ions in solution

In this chapter we examine the properties of a special class of solution, one containing ions. The principal difference between these and the solutions considered so far is that the solute particles carry charges, and so the solutions conduct electricity. We see how the presence of ions affects the properties of solutions, and pay special attention to the description of their electrical properties.

A special type of solution is formed by dissolving an electrolyte, a substance able to act as a supply of ions. Solutions of electrolytes show all the colligative properties discussed in Chapter 7, and in addition they have the very important property of being able to conduct electricity.

8.1 Solutions of electrolytes

When the view was expressed by Svante Arrhenius (in 1884) that many materials simply fell apart into ions when they dissolved in water, his contemporaries were very sceptical (see Box 8.1). After all, they argued, chemical bonds are strong, and the gentle act of dissolution could not be expected to break them. We now know that there are two errors in this argument.

BOX 8.1

The development of the concept of ions

Baron von Grotthuss took the view that the 'molecules' of electrolytes were polar, and that the application of an electric field caused them to orientate in chains and the ones at the ends of the chains to give up their electric charge to the electrodes. There was a shift of opinion with Rudolph Clausius (born in 1822 in the part of Prussia that is now Poland) who thought that electrolytes were partly dissociated even in the absence of an electric field. Svante Arrhenius (born in Sweden in 1859) put forward the suggestion that in solution electrolytes were completely dissociated into **ions** (a term that had been coined by Michael Faraday). He did so in his doctoral thesis; it was greeted with so much disbelief that he was awarded a mere fourth-class pass. Nevertheless, evidence accumulated in its favour with the work of Jacobus van't Hoff (born in 1852 in Rotterdam), who later received the first Nobel prize in chemistry (in 1901) for his work on equilibria and osmosis. The theory of complete dissociation was taken up and supported by the German chemist Wilhelm Ostwald, who can be said to be the father of physical chemistry because he did so much to organize it into a coherent subject. The terms ion, cation, anion, cathode, anode and electrode were all coined by Faraday who began life (in Surrey and in 1791) as the son of a blacksmith, became the assistant of Humphry Davy (see Box 17.1) and in due course came to be regarded as perhaps the greatest experimental scientist of all time. The first unambiguous evidence for the existence of ions, even in the solid state, was obtained by Sir Lawrence Bragg in the course of his X-ray diffraction studies of sodium chloride.

The first error is the assumption that materials do not exist as ions even before they dissolve. The X-ray diffraction of crystalline solids, which we discussed in Section 5.2, shows that many solids are collections of ions. There is clearly no force in the argument that 'NaCl molecules' are unlikely to break up in water, if solid sodium chloride is itself a collection of Na^+ and Cl^- ions.

The second error is to suppose that there are no energy advantages arising from going into solution as ions. For instance, we know (from X-ray studies) that solid citric acid is not ionic, yet when it dissolves in water it dissociates (at least partially) into ions. There must therefore be another contribution to the energy that is strong enough to overcome the forces that hold the molecules together in the solid and the force that keeps the H^+ ion attached to its parent molecule.

We shall find it useful to use the following language. An **ion** is an electrically charged particle. A positively charged ion is called a **cation**; a negatively charged ion is called an **anion**. An **electrolyte** is a substance in which ions conduct the current, and may be a molten salt or a solution. A **strong electrolyte** is a substance (such as sodium chloride or hydrogen chloride) that is essentially completely ionized in solution. Whether or not the compound is ionic in the undissolved form is irrelevant (sodium chloride is ionic in the solid, whereas hydrogen chloride is a gas of discrete covalent molecules, yet both are found experimentally to be strong electrolytes in water). A **weak electrolyte** is a substance that is only partially ionized in solution (such as ethanoic acid in water).

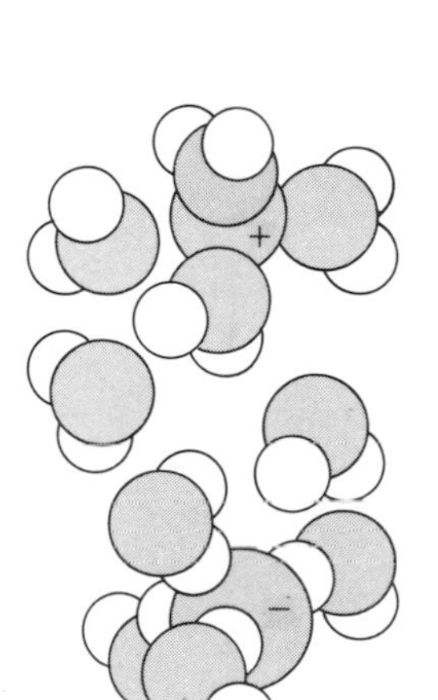

Figure 8.1 The hydration of ions

Dissolving

If an ionic solid is to dissolve there must be energetically favourable interactions between the solvent and the dissolved ions to compensate for losing the favourable interactions in the ionic solid. This new favourable interaction arises from the presence of the dipoles (see Section 2.6) of the solvent molecules: the positive end of the dipole can interact favourably with anions, and the negative end with cations. This **ion–dipole interaction** is illustrated in the case of water as solvent in Figure 8.1. The grouping of solvent molecules around ions is called **solvation** in general, and **hydration** when the solvent is water.

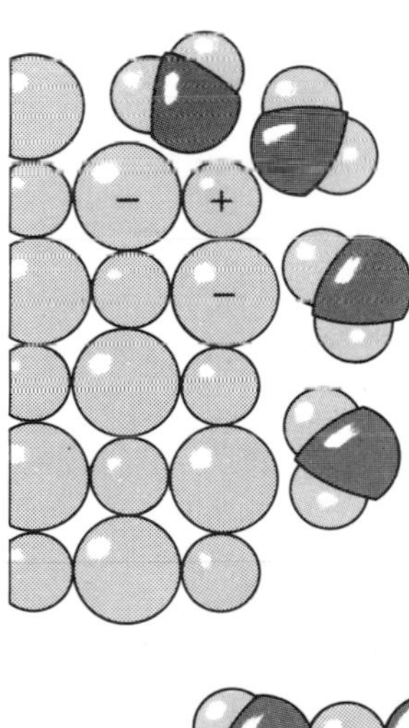

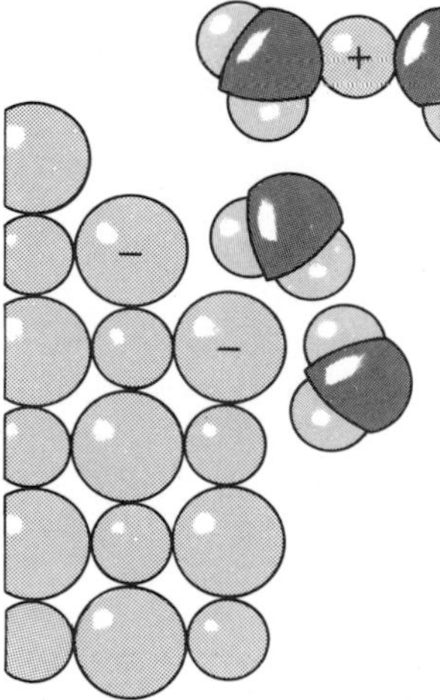

Figure 8.2 Dissolving in water

Dissolving can be pictured as in Figure 8.2. The water molecules nibble at the edge of the crystal, and the removal of the ions is aided by the favourable energy change that accompanies each act of hydration, the **hydration energy**. Dissolution is the continuation of this process of dislodging ions from the crystal under the influence of the electrostatic interactions with the dipoles of the polar solvent molecules. This description accounts qualitatively for the very low solubility of calcium carbonate (and many other carbonates) in water, because the ions in these compounds are so strongly bound that the hydration energy is insufficient to compensate for the loss of the lattice energy (see Section 9.4). It also accounts for the insolubility of ionic crystals in non-polar solvents, for molecules of the latter have no dipole and the solvation energy, arising only from van der Waals forces, is far too small to compensate for the loss of lattice energy.

Although qualitatively correct, the picture is incomplete. We can see this by examining the energy changes that occur when sodium chloride dissolves in water. It turns out that sodium chloride (like many other salts) needs energy to go into solution (in the language of thermochemistry, the dissolution is *endothermic* to the extent of about $4\,kJ\,mol^{-1}$). In other words, instead of the dissolution of the solid being accompanied by a *decrease* in energy, there is an *increase*! In practice this means that the solution cools slightly when salt dissolves in water, because the energy requirement is taken from the motion of the water molecules and from the walls of the vessel.

Why, then, does salt dissolve if there is an energy *disadvantage* in so doing? The explanation is provided by thermodynamics (Chapter 11), but the qualitative reason is easy enough to understand. We have stressed already (in Section 7.2) that objects have a natural tendency to become chaotic. A crystal is an ordered collection of ions; a solution is a widely dispersed random collection. If there were no forces holding the ions together, they would tend to drift off into any solvent in contact with them (like a gas expanding into a vacuum). The direction of natural change is dissolution, because it corresponds to an increase in chaos. (In the more precise language of Chapter 11, the dissolution produces an increase in entropy.) The forces between the ions oppose this tendency. In the presence of a polar solvent, however, the effectiveness of their opposition is reduced because there is now very little energy difference between the crystal and the solution. In many cases the tendency to chaos succeeds in spreading the ions through the solution. In the case of sodium chloride in water (and of other materials that dissolve with a small increase in energy) the tendency to disperse through the solvent overcomes the small energy disadvantage. In contrast, in the case of sodium chloride in benzene, or calcium carbonate in water, the tendency to chaos is still present, but cannot lead to dispersal because the energy change opposes it too strongly.

8.2 Properties of ionic solutions

The direct evidence for the existence of ions in solids (and, by implication, in solution too) comes from X-ray crystallography (see Figure 2.4 and Section 5.2). The indirect evidence for their existence in solution comes from various observations, including electrolysis, the measurement of colligative properties and electric conduction.

Electrolysis

The general term for chemical change brought about by an electric current is **electrolysis** (lysis means 'loosening'). At the electrodes, electrolysis causes deposition of metals from solution (see Figure 8.3), evolution of a gas, and other processes. Electrolysis is widely used industrially for the manufacture of metals, including sodium and aluminium, for preparing chlorine and for electroplating (see Figure 8.4 and Sections 17.7 and 19.7). In each case, the cations in the electrolyte (which may be an ionic solution or a molten ionic compound) move towards the negatively charged cathode where each one gains one or more electrons and is reduced. The anions in the electrolyte move towards the positively charged anode where each one loses one or more electrons and hence is oxidized (reduction and oxidation are discussed in Section 13.2).

Figure 8.3 The interior of an aluminium production plant, illustrating the massive scale of the electrolysis operation: the technician is about to extract molten aluminium from the electrolysis cells below

Figure 8.4 Attractive tree-like shapes form as this nickel crystal grows during electrolysis

The earliest quantitative experimental observations were made by Michael Faraday, and his experimental observations are summarized in two laws.

Faraday's first law: The mass of a substance liberated at an electrode is proportional to the charge passed.

Charge (Q) is measured by noting the current (I) and the time (t) for which it is passed: $Q = It$. With current in amperes (A) and time in seconds (s), charge is expressed in coulombs (C) using $1\,\text{A s} = 1\,\text{C}$. A charge stated in coulombs corresponds to a definite number of electrons, because each electron carries a charge e of magnitude $1.602 \times 10^{-19}\,\text{C}$. The charge per mole of electrons is this charge multiplied by Avogadro's constant, L. The resulting product is called **Faraday's constant**, F:

$$F = eL = 9.648 \times 10^{4}\,\text{C mol}^{-1}$$

It follows that a charge Q corresponds to a definite amount of electrons:

$$n(\text{e}^-) = Q/F$$

and so the law is readily explained on the grounds that electrolysis depends on the supply of electrons to cations or from anions in solution, and the mass liberated is proportional to the amount of electrons supplied.

Faraday's second law: In order to liberate 1 mole of a substance, 1, 2, 3 (or some other whole number) moles of electrons must be supplied.

(This is a modern version of the law.) For instance, it is found that to liberate 1 mole of zinc atoms, 2 moles of electrons must be supplied. This gives an easy way of determining the charges on the ions that are shown to exist by the first law. We conclude, for instance, that zinc ions are Zn^{2+}. Similarly, silver ions are Ag^+, as shown in the following Example.

EXAMPLE

Measurements of the masses of metals deposited by electrolysis gave the following results. When a current of 1.47 A was passed through aqueous copper(II) sulphate for 525 s, 254 mg of copper were deposited. The same current passed for the same time through aqueous silver nitrate led to the deposition of 863 mg of silver. What are the charges on the copper and silver ions in the solutions?

METHOD Determine the charge passed (current × time) and express it as an amount of electrons by noting that the magnitude of the charge on 1 mol of electrons is Faraday's constant. Express the masses of the deposited metals as amounts of substance, and note that a singly charged ion requires one electron whereas a doubly charged ion requires two.

ANSWER The charge passed through each solution is $1.47\,\text{A} \times 525\,\text{s} = 772\,\text{A s} = 772\,\text{C}$ (since $1\,\text{A s} = 1\,\text{C}$). The amount of electrons passed is therefore

$$n(\text{e}^-) = \frac{Q}{F} = \frac{772\,\text{C}}{9.648 \times 10^{4}\,\text{C mol}^{-1}} = 8.00 \times 10^{-3}\,\text{mol}$$

The amount of substance of Cu deposited is

$$n(\text{Cu}) = \frac{254 \times 10^{-3}\,\text{g}}{63.55\,\text{g mol}^{-1}} = 4.00 \times 10^{-3}\,\text{mol}$$

Therefore, one atom is deposited for every two electrons passed, and so copper ions are Cu^{2+}. The amount of substance of Ag deposited is

$$n(\text{Ag}) = \frac{863 \times 10^{-3}\,\text{g}}{107.9\,\text{g mol}^{-1}} = 8.00 \times 10^{-3}\,\text{mol}$$

Therefore, one atom of Ag is deposited for every electron passed, and so silver ions are Ag^+.

COMMENT The silver **coulometer** can be used to measure the charge passed through a circuit. It is basically an electrolysis cell. The mass of silver deposited is measured and interpreted as an amount of electrons passed using a calculation of the type set out in this Example, but with the charge type, Ag^+, known.

Colligative properties of ionic solutions

Colligative properties (see Section 7.2) depend on the *number* of solute particles present in solution and not on their chemical identity. For a solution of sodium chloride, the osmotic pressure, the depression of freezing point and the elevation of boiling point are all twice as great as would be expected on the basis that the solute dissolves as 'NaCl molecules'. This is readily accounted for if the particles present are Na^+ and Cl^- ions, because then there are twice as many solute particles.

The usual way of expressing the increase in the magnitude of the colligative properties is in terms of the **van't Hoff *i*-factor**. If m_A is the molality of solute A, we write:

$$\left.\begin{array}{ll} \textit{Elevation of boiling point:} & \delta T = iK_b m_A \\ \textit{Depression of freezing point:} & \delta T = iK_f m_A \\ \textit{Osmotic pressure:} & \Pi V = inRT \end{array}\right\} \qquad (8.2.1)$$

For very dilute solutions i can be interpreted as the number of ions each formula unit produces in solution (thus $i = 2$ for NaCl, $i = 3$ for Na_2SO_4). An example will make this clear. Consider the effect of concentrated nitric acid on the freezing point of concentrated sulphuric acid. This potent mixture is used in the laboratory for the nitration of aromatic compounds, discussed in Section 34.3. When the freezing point of a given dilute solution of nitric acid in sulphuric acid is measured, about four times the expected depression is observed; hence $i = 4$. This result suggests that nitric acid ionizes:

$$HONO_2 + 2(HO)_2SO_2 \rightarrow NO_2^+ + H_3O^+ + 2HOSO_3^-$$

there now being four ions present for each nitric acid molecule added. This observation was used as evidence for the role of the NO_2^+ ion in nitration.

EXAMPLE

Common salt is spread on roads to prevent the formation of ice. What depression of freezing point of water results from the addition of 100 kg of salt to 1 tonne of water?

METHOD Use equation 8.2.1. The cryoscopic constant is given in Table 7.1 as $1.86\ \mathrm{K\,kg\,mol^{-1}}$; the van't Hoff *i*-factor for sodium chloride is approximately 2 as Na^+ and Cl^- ions go into solution for every NaCl formula unit added. Calculate the molality of the solute m_A, using $M_r(\mathrm{NaCl}) = 22.99 + 35.45 = 58.44$ (see page 518).

ANSWER The molality of the solute is

$$m_A = \frac{(100 \times 10^3\ \mathrm{g})/(58.44\ \mathrm{g\,mol^{-1}})}{1000\ \mathrm{kg}} = 1.71\ \mathrm{mol\,kg^{-1}}$$

The depression of freezing point is therefore

$$\delta T = iK_f m_A = 2 \times 1.86\ \mathrm{K\,kg\,mol^{-1}} \times 1.71\ \mathrm{mol\,kg^{-1}} = 6.4\ \mathrm{K}$$

COMMENT The solution will therefore not freeze until the temperature has fallen below $-6\,°\mathrm{C}$. Note that the molality of the solute is quite high, and so the solution is not ideal. Nevertheless, the measured freezing-point depression ($-6\,°\mathrm{C}$) is almost exactly what this calculation predicts. The *i*-factor is not exactly 2 because ion–ion interactions are significant in solution.

8.3 The conductivities of ions

In metals the cations are in a rigid array and charge is transported by the mobile electrons. In solutions of electrolytes the anions and the cations are free to move, and so both transport charge. The ability of a solution of an electrolyte to carry current, its **conductivity**, depends on the numbers of anions and cations present and on how readily they move (their **mobilities**).

The measurement of conductivity

The **resistance**, R, of a conductor is proportional to its length, l, and inversely proportional to its cross-sectional area, A:

$$R \propto l/A \quad \text{or} \quad R = \rho l/A \tag{8.3.1}$$

ρ (rho) is called the **resistivity** of the substance. The **electrolytic conductivity** (κ, kappa) is the reciprocal of the resistivity:

$$\textit{Electrolytic conductivity: } \kappa = 1/\rho = l/RA \tag{8.3.2}$$

Resistance is normally expressed in ohms (symbol: Ω, omega). The dimensions of conductivity are length/(resistance × area), which is normally expressed in cm/($\Omega \times$ cm^2), or $\Omega^{-1}\,\text{cm}^{-1}$. The internationally recommended unit for Ω^{-1} is the siemens (symbol: S) so that conductivity is usually expressed in S cm^{-1}.

The resistance of an electrolyte solution is measured by connecting a conductivity cell, like the one illustrated in Figure 8.5, to one arm of a bridge

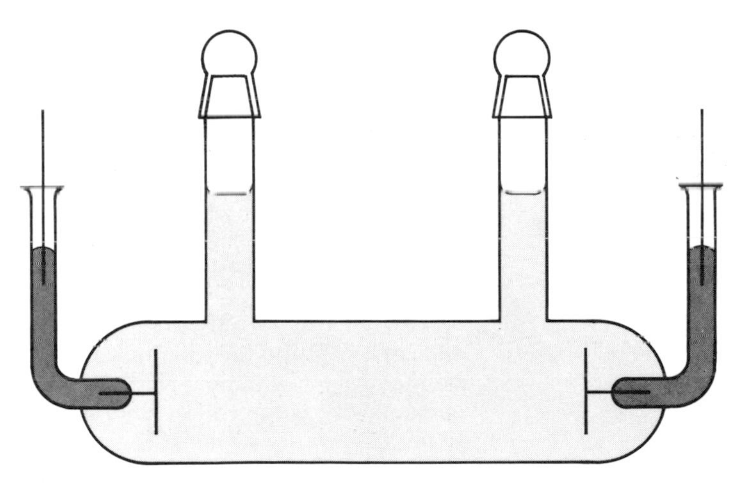

Figure 8.5 A conductivity cell

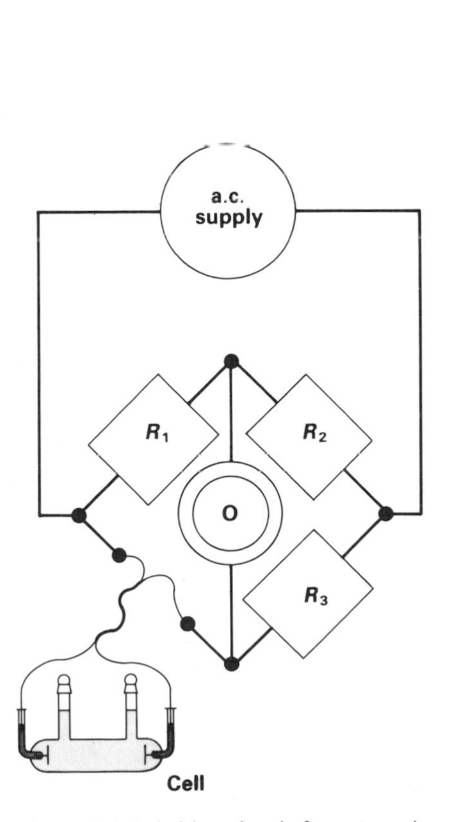

Figure 8.6 A bridge circuit for measuring conductances (R_1, R_2 and R_3 are variable resistances, and O is an oscilloscope)

(a network of resistances, shown in Figure 8.6). In practice, an **a.c. bridge** has to be used to avoid the electrolysis that would occur if current flowed continuously in a single direction. The conductivity of the solution is obtained from the measured resistance by evaluating $\kappa = l/RA$. In practice it proves to be easier to calibrate the cell using a solution of known conductivity, and then to measure the unknown conductivity by comparison. The ratio l/A is called the **cell constant**, and is determined once and for all for a particular cell by measuring the resistance of a solution of known conductivity (often $0.10\,\text{mol dm}^{-3}$ KCl(aq)).

Molar conductivity

The conductivity of a sample depends on the concentration of the ions. The normal practice is to express the conductivity as a **molar conductivity** (Λ, the Greek capital letter lambda), by dividing the measured electrolytic conductivity by the concentration of added electrolyte, c:

$$\textit{Molar conductivity: } \Lambda = \kappa/c \tag{8.3.3}$$

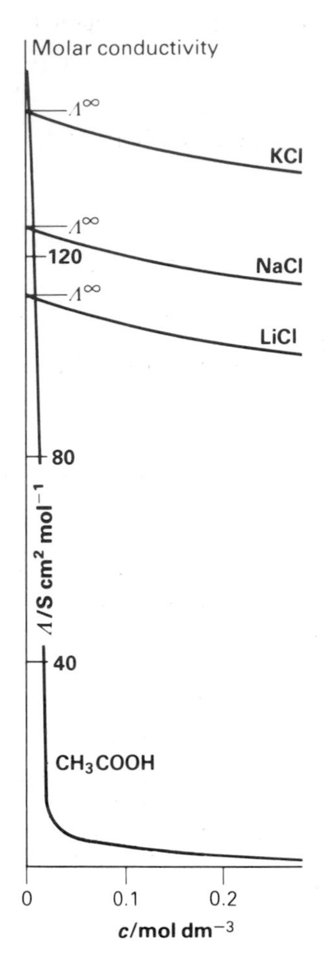

Figure 8.7 The molar conductivities of weak and strong electrolytes

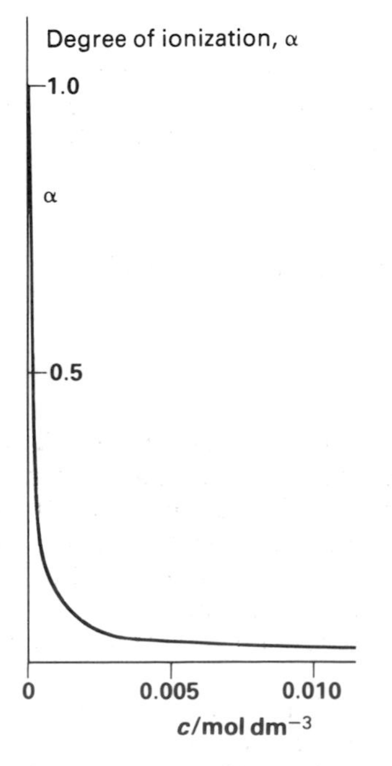

Figure 8.8 The degree of ionization and the concentration of a weak electrolyte (ethanoic acid in water at 25 °C)

EXAMPLE

The electrolytic conductivity of aqueous potassium chloride at a concentration of 0.100 mol dm^{-3} is 0.0129 S cm^{-1} at 25 °C. What is its molar conductivity?

METHOD Use equation 8.3.3. The units may be manipulated on the basis that 1 dm^3 = 10^3 cm^3.

ANSWER From equation 8.3.3, with $\kappa = 0.0129\ \mathrm{S\,cm^{-1}}$ and $c = 0.100\ \mathrm{mol\,dm^{-3}}$,

$$\Lambda = \kappa/c = \frac{0.0129\ \mathrm{S\,cm^{-1}}}{0.100\ \mathrm{mol\,dm^{-3}}} = 0.129\ \mathrm{S\,cm^{-1}/(mol\,dm^{-3})}$$

$$= 0.129\ \mathrm{S\,dm^3\,cm^{-1}\,mol^{-1}} = (0.129\ \mathrm{S})(1000\ \mathrm{cm^3})\,\mathrm{cm^{-1}\,mol^{-1}}$$

$$= 129\ \mathrm{S\,cm^2\,mol^{-1}}$$

COMMENT The advantage of leaving the molar conductivity in the form with mol dm^{-3} written explicitly is that it is easy to go from a molar conductivity to a conductivity simply by multiplying by the concentration. The advantages of going on to express it in the final form are mainly those of tidiness and brevity.

Strong electrolytes have molar conductivities that depend only weakly on the concentration, at least at low concentrations. This is illustrated in Figure 8.7 for solutions of various chlorides. The measured molar conductivities decrease slightly with increasing concentration of added salt, but there is no dramatic change. So as to be precise, we tabulate the **limiting molar conductivity**, the molar conductivity in the limit of zero concentration (formerly called **infinite dilution**). This is denoted Λ^∞ and is marked on the diagram.

The molar conductivities of **weak electrolytes** show a marked dependence on the concentration (this identifies them). Figure 8.7 shows that the molar conductivity of ethanoic acid is quite low at moderate concentrations and remains low as the concentration is reduced; when the concentration is extremely low, however, it rises to values similar to the molar conductivities of strong electrolytes. The explanation is that the proportion of dissociated molecules of weak electrolytes is low at ordinary concentrations, but increases steeply at very low concentration. This behaviour is normally expressed in terms of the **degree of ionization** or **degree of dissociation**, α, defined so that for an electrolyte AB added at a concentration c the concentrations of A^+ and of B^- are each αc, and the concentration of non-ionized AB itself in solution is $(1 - \alpha)c$. The value of α depends on c itself, and its calculation is described in Appendix 8.1. All we need at this stage is the form of the result, shown in Figure 8.8: α is close to unity (complete ionization) at very low concentrations of added electrolyte, but at higher ('typical') concentrations α is very low. This is exactly the kind of behaviour needed to explain the conductivities of solutions of weak electrolytes. This point will be developed after another feature has been introduced.

The contributions of individual ions

Extensive studies of the conductivities of strong electrolytes were made by Friedrich Kohlrausch. He found that a given type of ion makes a characteristic contribution to the limiting molar conductivity of an electrolyte.

Kohlrausch's law: The limiting molar conductivity is the sum of the limiting molar conductivities of the ions present.

The mathematical expression of this law for an AB electrolyte producing singly charged ions is

$$\textit{Kohlrausch's law: } \Lambda^{\infty}(\mathrm{AB}) = \lambda^{\infty}(\mathrm{A}^+) + \lambda^{\infty}(\mathrm{B}^-) \qquad (8.3.4)$$

where $\lambda^{\infty}(\mathrm{A}^+)$ and $\lambda^{\infty}(\mathrm{B}^-)$ are the individual limiting molar conductivities. Some values are listed in Table 8.1.

Table 8.1 Limiting molar conductivities of individual ions in water at 25 °C, $\lambda^{\infty}/\mathrm{S\,cm^2\,mol^{-1}}$

H^+	350	Mg^{2+}	106	Al^{3+}	189	OH^-	199	NO_3^-	71.4
Li^+	38.7	Ca^{2+}	119			F^-	55.4	$CH_3CO_2^-$	40.9
Na^+	50.1	Ba^{2+}	127			Cl^-	76.4	SO_4^{2-}	160
K^+	73.5	Cu^{2+}	107			Br^-	78.1		
Ag^+	61.9					I^-	76.8		

The data in Table 8.1, together with Kohlrausch's law, make it possible to predict the limiting molar conductivity of a dilute solution of a strong electrolyte. For instance, the limiting molar conductivity of aqueous sodium sulphate is

$$\Lambda^{\infty}(\mathrm{Na_2SO_4}) = 2\lambda^{\infty}(\mathrm{Na^+}) + \lambda^{\infty}(\mathrm{SO_4^{2-}}) = 260.2\,\mathrm{S\,cm^2\,mol^{-1}}$$

The law also accounts for the concentration dependence of the molar conductivities of solutions of weak electrolytes. Consider the AB electrolyte treated above. The electrolytic conductivity when the added electrolyte concentration is c is the sum of the conductivities due to the A^+ ions (at a concentration αc) and the B^- ions (also at a concentration αc). In other words, the total electrolytic conductivity is

$$\kappa = \alpha c\lambda(\mathrm{A}^+) + \alpha c\lambda(\mathrm{B}^-) = \alpha c\{\lambda(\mathrm{A}^+) + \lambda(\mathrm{B}^-)\}$$

Since there is only a small degree of ionization, the individual ion conductivities can be replaced by their limiting values, λ^{∞}. It follows that the molar conductivity, κ/c, is

$$\Lambda \approx \alpha\{\lambda^{\infty}(\mathrm{A}^+) + \lambda^{\infty}(\mathrm{B}^-)\} \qquad (8.3.5)$$

In other words, the molar conductivity is proportional to α, and therefore rises sharply from a low value when c is large (α much less than 1) to a high value when c is very small (and $\alpha \approx 1$), just as in Figure 8.7. Note that since the last equation can be written as

$$\alpha \approx \Lambda/\Lambda^{\infty}$$

and Λ^{∞} can be obtained from tables, the degree of ionization can be determined by measuring Λ.

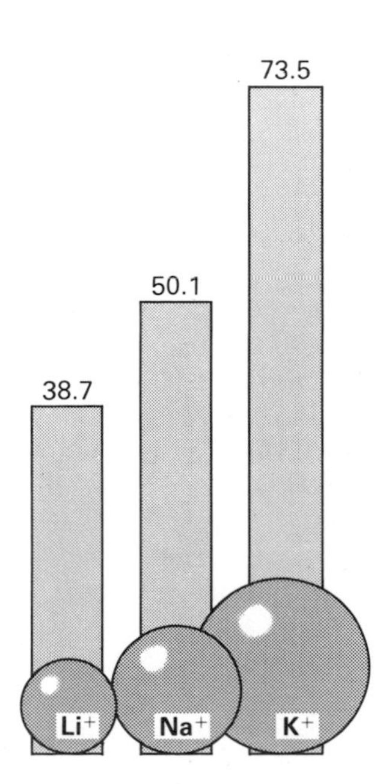

Figure 8.9 Ionic molar conductivities and radii

Ion mobilities

The values in Table 8.1 show that the molar conductivity of cations is larger the greater the charge on the ion, as each individual ion transports more charge. Casual thinking about the relation between the sizes of ions and their molar conductivities would probably suggest that small ions should be more mobile and therefore have greater molar conductivities. A glance at Table 8.1 shows that the opposite is true. Figure 8.9 shows how the limiting molar conductivities of some alkali metals vary with ionic radii (which are obtained from X-ray studies of crystals).

The missing feature is the effect of solvation. When an ion moves through water it carries its **hydration sphere**, the water molecules attached to it, along with it, and so its *effective* size is much larger than the size of the ion itself. A small ion gives rise to a higher electric field at its surface than a big one of the same charge, and so grips its hydration sphere more strongly. The outcome is that a small ion like Li^+ has a *greater* effective size than K^+ has, and consequently is less mobile. The increase in the molar conductivities of ions on going down a given Group of the Periodic Table can therefore be explained in terms of their decreasing effective sizes when hydration is taken into account.

There is, however, a glaring discrepancy. The hydrogen ion, H^+, is extremely small and can therefore be expected to be strongly hydrated, which is why it is normally denoted $H_3O^+(aq)$ or $H^+(aq)$. It is therefore expected to have a very low molar conductivity. Its actual molar conductivity is in fact the highest for any ion. The explanation lies in the fact that H^+ conducts by an entirely different mechanism, called **Grotthuss transfer**. It is illustrated in Figure 8.10. There is only an *effective* transfer of protons through the solution, and the charge of the proton on the left in Figure 8.10(a) is carried to the other end of the chain of water molecules by a small motion of the bonds and hydrogen bonds. This produces the same effect as if the proton had actually moved, but it takes place much more quickly. The OH^- ion, which also has a high molar conductivity in water, is 'transported' in a similar way.

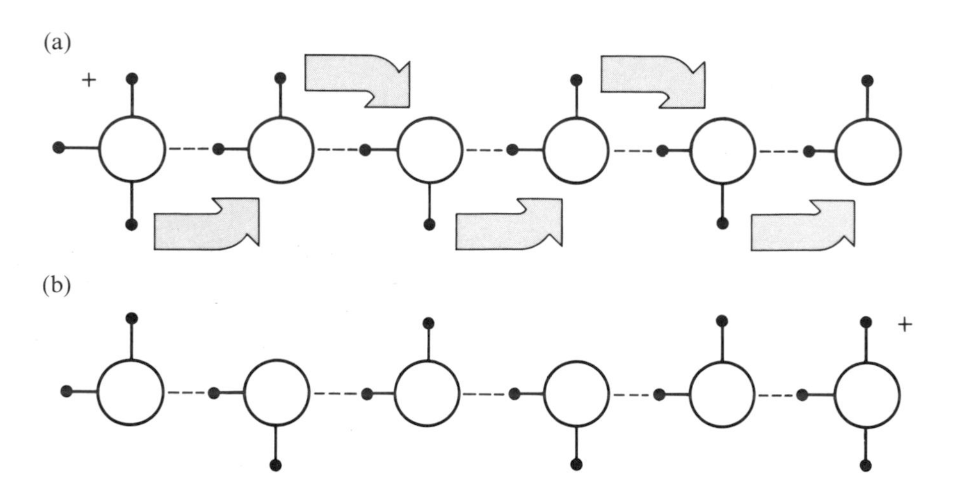

Figure 8.10 The mechanism of conduction by hydrogen ions in water

Applications of conductivity measurements

Conductivity measurements are used to distinguish between weak and strong electrolytes, to measure the degree of ionization, to determine solubilities of sparingly soluble salts (as described in Section 12.1), and to follow acid–base titrations. The apparatus for a **conductometric titration** consists of a conductivity cell dipping into the beaker where the titration is taking place. The important point to remember throughout the following discussion is that the molar conductivities of $H^+(aq)$ and $OH^-(aq)$ are very high.

Consider the changes in conductivity when a strong acid (such as aqueous hydrochloric acid) is titrated against a strong base (such as aqueous sodium hydroxide). Initially the solution contains Na^+ and OH^- ions, and so the conductivity is high. As the titration proceeds some of the OH^- ions are replaced by Cl^- ions. As a result, the conductivity falls sharply, as shown in Figure 8.11(a). At the **equivalence-point** (or **end-point**), when exactly enough acid has been added to neutralize the base, all the OH^- ions have been replaced by Cl^- ions and the solution has a conductivity typical of a solution of sodium chloride in water. After the equivalence-point, the further addition of acid results in the presence of $H^+(aq)$ ions in the solution, and the conductivity rises sharply. The equivalence-point is indicated by the clear minimum in the conductivity plot. This titration technique is useful if the solutions are coloured and indicators cannot be used.

Now consider the case of the titration of a weak base with a strong acid, shown in Figure 8.11(b). Initially the conductivity is low because the base is only slightly ionized. Up to the equivalence-point the addition of acid results in the formation of the fully ionized salt, and so the conductivity rises steadily.

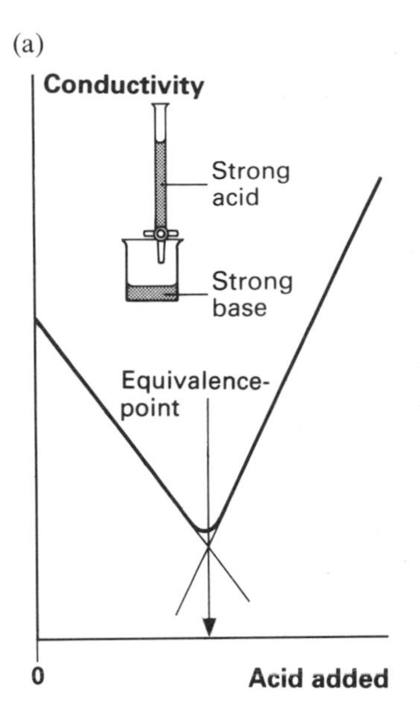

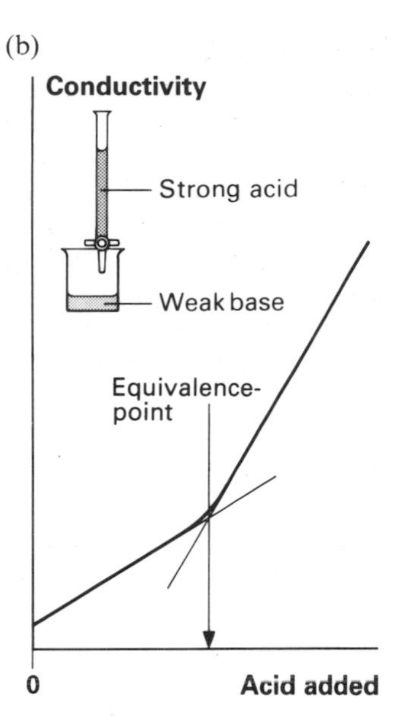

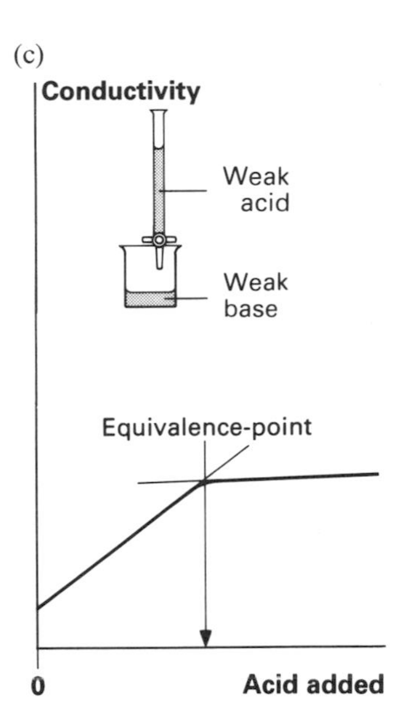

Figure 8.11 Conductometric titrations:
(a) strong acid/strong base,
(b) strong acid/weak base,
(c) weak acid/weak base

After the equivalence-point, when $H^+(aq)$ ions become available, the conductivity rises much more sharply.

A conductometric titration is one of the principal ways of detecting the equivalence-points of weak acid/weak base titrations (for example, for ethanoic acid and aqueous ammonia titrations). Figure 8.11(c) illustrates such a titration: up to the equivalence-point, the change is similar to that in Figure 8.11(b). After the equivalence-point, however, the conductivity barely changes as the acid is added, because the addition of the weakly ionized acid does not result in a significant increase in the number of ions present. The pronounced change in slope at the equivalence-point makes it quite easy to locate.

Appendix 8.1 The degree of ionization

The calculation of α depends on knowing that the equilibrium $AB \rightleftharpoons A^+(aq) + B^-(aq)$ is characterized by the constant

$$K_c = \{[A^+][B^-]/[AB]\}_{eq}$$

This concept is discussed in Chapter 10. In the present case $[A^+] = \alpha c$, $[B^-] = \alpha c$, and $[AB] = (1 - \alpha)c$; therefore

$$K_c = \frac{(\alpha c)(\alpha c)}{(1-\alpha)c} = \frac{\alpha^2 c}{1-\alpha} \approx \alpha^2 c \text{ (for } \alpha \ll 1)$$

The exact expression for α rearranges into

$$(1-\alpha)K_c = \alpha^2 c \quad \text{or} \quad \alpha^2 + (K_c/c)\alpha - (K_c/c) = 0$$

This is a quadratic equation with the (acceptable) root

$$\alpha = (K_c/2c)\{-1 + \sqrt{[1 + (4c/K_c)]}\}$$

which is plotted in Figure 8.8.

Summary

- ☐ An electrolyte is a substance in which ions conduct current, and may be a molten salt or a solution.
- ☐ A strong electrolyte is essentially completely ionized in solution; a weak electrolyte is only partially ionized in solution (except at extreme dilutions). The classification depends on the nature of the solvent.
- ☐ A cation is a positively charged ion; an anion is a negatively charged ion.
- ☐ Solvation (hydration when water is the solvent) is the surrounding of a species in solution by a cluster of solvent molecules.
- ☐ Many dissolutions are endothermic. The driving force for dissolution is the natural tendency towards chaos of ordered structures; this is successful provided little energy has to be supplied to the system from the surroundings.
- ☐ Electrolysis is chemical change brought about by an electric current.
- ☐ Faraday's first law is 'mass liberated is proportional to charge passed'. It indicates the presence of ions in the solution.
- ☐ Faraday's second law is 'to liberate 1 mole of a substance, 1, 2, 3 (or other whole number) moles of electrons must be supplied'. It provides a way of determining the charge type of ions in solution.
- ☐ Faraday's constant is the magnitude of the charge carried per unit amount of electrons, $F = eL$.
- ☐ In very dilute solutions the van't Hoff *i*-factor (equation 8.2.1) can be interpreted as the number of ions which each formula unit produces in solution.
- ☐ The principal evidence for the existence of ions comes from X-ray crystallography, the observation that solutions of electrolytes conduct electricity, Faraday's laws and colligative properties.
- ☐ The colligative properties of ionic solutions suggest that ions exist in solution because these properties are proportional to the number of particles.
- ☐ Electrolytic conductivity (κ) is related to resistance R through $\kappa = l/RA$, where l is the length of the sample and A its cross-sectional area. The electrolytic conductivity is normally measured with the cell acting as the unknown resistance in one arm of an a.c. bridge.
- ☐ The units of electrolytic conductivity are usually $\mathrm{S\,cm^{-1}}$ (S is siemens, $1\,\mathrm{S} \equiv 1\,\Omega^{-1}$, Ω is ohm).
- ☐ The molar conductivity is denoted Λ and defined as κ/c, where c is the concentration of electrolyte added. Its units are $\mathrm{S\,cm^2\,mol^{-1}}$; Λ^{∞} is the limiting molar conductivity.
- ☐ In the case of strong electrolytes, the molar conductivity is only slightly dependent on concentration.
- ☐ In the case of weak electrolytes, as the concentration decreases the molar conductivity increases sharply from a low value to a value typical of strong electrolytes.
- ☐ The behaviour of weak electrolytes is expressed in terms of the degree of ionization α, which is very small except at very low concentrations, when it rises to unity.
- ☐ Kohlrausch's law states that the limiting molar conductivity is the sum of the limiting molar conductivities of the ions present.
- ☐ The degree of ionization of a weak electrolyte can be found from $\alpha \approx \Lambda/\Lambda^{\infty}$.
- ☐ Small ions are more strongly solvated than large ions are; as a consequence their effective sizes are larger and their conductivities lower; $H^+(aq)$ and $OH^-(aq)$ ions have unusually high conductivities.
- ☐ Conductivity measurements are used in conductometric titrations.

PROBLEMS

1 Describe the evidence for the existence of ions in solution.

2 Account for the following facts: (a) many ionic compounds are readily soluble in water, (b) iodine is more soluble in tetrachloromethane than in water, (c) all nitrates are soluble in water but most hydroxides and carbonates are insoluble.

3 How much charge is transported when a current of 2 A flows through an electric light for 30 s? How many electrons are transported? What amount of electrons (that is, how many moles of e^-) does that correspond to?

4 For how long would a 100 A current need to flow through an industrial electrolytic cell in order to generate (a) 1 mol Cl_2 from molten sodium chloride, (b) 3 mol Cu from aqueous copper(II) sulphate?

5 Which would bring about the greater depression of freezing point, (a) 0.20 g sodium chloride, (b) 0.22 g calcium chloride, when dissolved in the same volume of water?

6 Define the terms *resistance, resistivity, electrolytic conductivity* and *molar conductivity*.

7 Identify the errors in the following remarks: (a) the conductivity of sodium chloride in water increases as its concentration decreases, (b) the molar conductivity of sodium chloride in water is independent of concentration.

8 Estimate the molar conductivities of dilute aqueous (a) copper(II) sulphate, (b) potassium nitrate. What will be the resistance of a solution of cross-sectional area 1 cm^2, length 10 cm of 0.01 mol dm^{-3} aqueous potassium nitrate?

9 The limiting molar conductivities of three of the alkali metal halides were measured and found to be as follows: Λ^∞(LiCl) = 115, Λ^∞(NaCl) = 126, Λ^∞(KCl) = 150, all in S cm^2 mol^{-1}. Account for the trend in these values. Would you expect the molar conductivities of the bromides to be uniformly larger or smaller? Confirm your conclusion on the basis of the data in Table 8.1.

10 The molar conductivity of 0.05 mol dm^{-3} aqueous ethanoic acid is 7.36 S cm^2 mol^{-1}. What is its degree of ionization at this concentration?

11 (a) When one mole of a non-ionized solute is dissolved in 1000 g of water, the freezing point of water falls to −1.86 °C at 101 kPa pressure. 4.84 g of a non-ionized solute dissolved in 250 g of water gives a freezing point of −0.20 °C at 101 kPa pressure.
(i) Calculate the number of moles of solute used.
(ii) Calculate the relative molecular mass of the solute.

(b) A solute AB_2 forms A^{2+} and B^- ions in solution and the freezing point depression depends on the total number of particles present in solution.
(i) Write an equation showing the dissociation of AB_2 into ions.
(ii) If the degree of dissociation of AB_2 is α at a particular concentration, how many moles of ions will be present if one mole of AB_2 is in solution at that concentration?
(iii) How many moles of non-ionized AB_2 will remain?
(iv) Use your answers to (ii) and (iii) to write an expression for the ratio

$$\frac{\text{freezing point depression observed}}{\text{freezing point depression calculated if no ionization}}$$

(v) Using the information in (a), calculate the freezing point depression which would have been caused if 0.01 mol of AB_2 were dissolved in 100 g of water and no ionization occurred.
(vi) If 0.01 mol of AB_2 in 100 g of water is observed to give a freezing point depression of 0.487 °C, calculate the degree of dissociation, α, of the solute AB_2. *(AEB 1985)*

12 (a) Describe with essential detail the experiments you would carry out to obtain a value for the molar conductivity at infinite dilution [Λ^∞] of hydrochloric acid in aqueous solution at 25 °C.
(b) State what information you would need, and explain how you would use it, to evaluate Λ^∞ for ethanoic acid (acetic acid) in aqueous solution at 25 °C. [Details of how the necessary data are obtained are not required.]
(c) At 25 °C, Λ^∞ for ethanoic acid is 391 ohm^{-1} cm^2 mol^{-1} [S cm^2 mol^{-1}], and the dissociation constant of this acid is 1.76×10^{-5} mol dm^{-3}. Calculate to two significant figures the electrolytic conductivity (specific conductance) of a 0.100 molar [mol dm^{-3}] solution of ethanoic acid at 25 °C. *(Oxford)*

9 Thermochemistry

In this chapter we investigate the role of energy in chemical reactions. We meet the First law of thermodynamics, and see some of its applications in chemistry. We see how to measure energy changes occurring in the course of reactions, and how to manipulate thermochemical data so that we can make predictions about the energy the reactions release or require. The kind of information described in this chapter lies at the heart of discussions of the properties of fuels, of industrial syntheses and of the energies involved in the activity of biological cells.

Energy is the motive power of all chemical reactions, and we will see its role emerge in this and the following chapters. Without energy, nothing could happen in the world. Yet it is not merely the quantity of energy that matters to us, but also its economic availability. In Chapter 11 we will see exactly what 'availability' means, and come close to understanding the scientific basis of the current concerns about energy. The essential feature of the present discussion is that energy is neither created nor destroyed, but is only *transformed* from one form to another. For instance, burning a fuel in a rocket transforms the energy stored in the fuel and oxygen molecules into the kinetic energy of the space vehicle. Burning coal, gas or oil in a power station transforms the energy stored in these fuels into electrical energy, which can be transported over great distances and then used for doing work or heating.

Thermochemistry, the study of the energy produced or absorbed during chemical reactions, provides a way of assessing the suitability of fuels and of helping a chemical engineer to design a plant so that it operates economically. It is a branch of **thermodynamics**, the science of the transformations of energy. Thermodynamics grew out of the study of steam engines when there was intense interest, during the nineteenth century, in the way their efficiency could be improved. These origins have been left far behind (see Box 9.1), and modern thermodynamics has become the subject of wide applicability. It can tell us nothing about *how fast* a reaction occurs, however; we shall deal with that aspect of chemistry in Chapter 14.

BOX 9.1

The development of thermodynamics

Thermodynamics grew out of the Industrial Revolution, when engineers and scientists turned their attention to the improvement of the efficiencies of steam engines. Sadi Carnot (born in 1796 in Paris) was convinced that France's military defeat was due to the inferiority of her steam power, and investigated ways of improving it. This led to the formulation of the concept of **entropy** (to be discussed in Chapter 11). James Joule (born in Salford in 1818) studied under Dalton, and then turned to a precise study of the properties of heat and work and their interrelation. He was able to demonstrate their interconvertibility, and hence the concept of **energy** entered physical science. Rudolph Clausius (mentioned in Box 8.1) is generally given the credit for turning the subject into an exact science by sharpening the definitions of the quantities involved and showing how to manipulate them mathematically. One of the giants of the subject is William Thomson, later Lord Kelvin (born in 1824 in Belfast). He was a professor at the University of Glasgow at the age of twenty-two, and remained there throughout his life. In the course of that life he became wealthy and famous largely through his inventions relating to telegraphy and submarine cables, but his enduring fame comes from his contribution to thermodynamics. His name is commemorated in the temperature scale, for another major contribution that he made to thermodynamics was the exact definition of the **temperature** of a body. The most important contributor to the application of thermodynamics to chemical problems (it had emerged out of engineering and moved into physics) was Josiah Willard Gibbs. He was born in Connecticut in 1839 and was described by a contemporary historian as 'the greatest of Americans, judged by his rank in science'. In a series of papers between 1876 and 1878 he advanced the very useful concept of **free energy**, which we now call Gibbs energy (see Section 11.3). The man who formulated the connection between the properties of individual molecules and the bulk properties of samples, the branch of physical chemistry known as **statistical thermodynamics**, was Ludwig Boltzmann (born in 1844 in Vienna) who committed suicide, ill and depressed, in 1906. His equation relating entropy with disorder is engraved on his tombstone.

9.1 The First law

The First law of thermodynamics summarizes the fact that no one has ever succeeded in constructing a perpetual motion machine, one that will generate work without consuming fuel. The hope of achieving this has been abandoned, and the failure is now expressed as a law:

First law of thermodynamics: Energy can be neither created nor destroyed.

(Now that we are familiar with nuclear energy, energy must be interpreted as including mass through the relationship $E = mc^2$.)

Work and heat

When interpreting the First law, we must keep in mind several points about the ways in which energy may be transferred between a **system**, the part of the world we are investigating, and its **surroundings**, the rest of the world.

One way of transferring energy is by **doing work**. For instance, energy can be transferred to a vessel of water (with the effect of raising its temperature) by stirring the water vigorously. Another way of transferring energy is by **heating** the water, by setting up a temperature difference between it and its surroundings.

Doing work and heating are the two fundamental ways of transferring energy to or from a system. In that respect, the system can be thought of as being like a bank for energy. Transactions can be made in two currencies: by doing work or by heating. Once the transaction has been made, the reserves of the bank are stored as energy, and may be withdrawn in either currency. An **isolated system** is like a bank when it is shut: no transactions can be made. A stoppered Dewar flask or thermos flask is an example.

The quantity of energy transferred by doing work during a transaction with the system is written w (and normally expressed in joules, J, or kilojoules, kJ). If we transfer 10 kJ of energy by doing work *on* the system, we write $w = +10\,\text{kJ}$. If the system itself *does* 10 kJ of work, we say that $w = -10\,\text{kJ}$, because the system has *lost* 10 kJ of energy by doing work on the outside world.

The energy transferred by heating is written q, and is also normally expressed in joules or kilojoules. If 10 kJ of energy are transferred *to* the system by heating it, we write $q = +10\,\text{kJ}$. If the same quantity of energy is transferred *from* the system, we write $q = -10\,\text{kJ}$, the negative sign indicating loss of energy from the system.

The First law can now be expressed in terms of w and q. We denote the energy possessed by the system, its **internal energy**, as U; the *change* of internal energy when energy is transferred is written ΔU. It follows that the change of internal energy is

$$\Delta U = q + w \qquad (9.1.1)$$

(a)

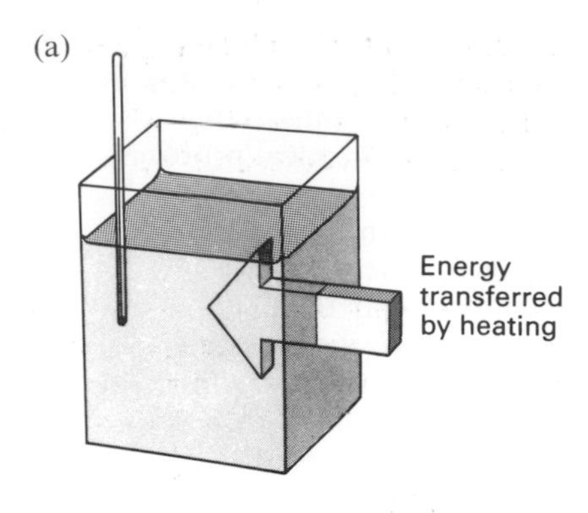

(b)

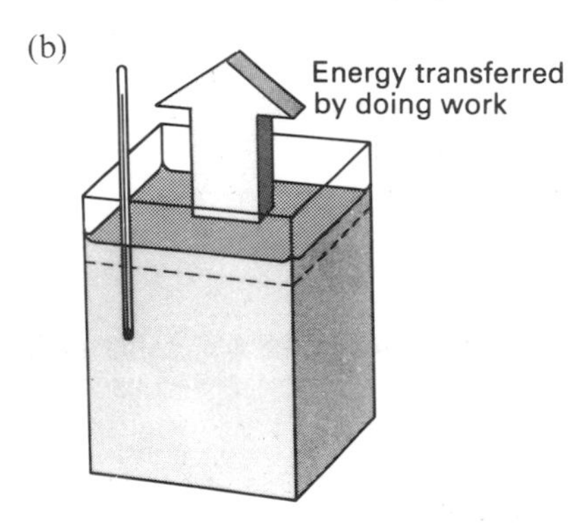

Figure 9.1 Energy can be transferred to or from a system either (a) by heating or (b) by doing work

Therefore, if 250 kJ of energy are transferred to the system by doing work ($w = +250\,\text{kJ}$) and at the same time 150 kJ are transferred back as the system heats its environment ($q = -150\,\text{kJ}$), the overall change in the internal energy is $\Delta U = +100\,\text{kJ}$; the reserves of the energy bank have risen by 100 kJ as a result of the transfers, and are stored until a later process requires them.

Energy and enthalpy

Suppose energy is transferred by heating 500 g of water in an open vessel. As well as getting 'hotter' (that is, reaching a higher temperature) the water expands a little. When the water expands, it pushes back the atmosphere, as shown in Figure 9.1. That pushing means that the system is doing work on its surroundings. In other words, when a quantity of energy, q, is used to heat a system open to the atmosphere, the change in internal energy is less than q because the system does work by pushing back the atmosphere. If the constant external pressure is p, and the system increases volume by ΔV, the work the system does as it expands is $p\Delta V$, and so the internal energy changes by q less this quantity of work, that is, $\Delta U = q - p\Delta V$. If we define the **enthalpy change** at constant pressure as

$$\Delta H = \Delta U + p\Delta V$$

then

$$\Delta H = q - p\Delta V + p\Delta V = q$$

and the energy transferred by heating causes an exactly equal change in enthalpy. (The word enthalpy comes from the Greek for 'inner warmth'.) That is,

$$\Delta H = q \text{ (at constant } \textit{pressure}) \qquad (9.1.2)$$

If the same quantity of energy is transferred by heating at constant volume, no energy leaks away since the system does no work, and so the internal energy changes by

$$\Delta U = q \text{ (at constant } \textit{volume}) \qquad (9.1.3)$$

The difference between ΔH and ΔU is illustrated in Figure 9.2.

Usually the change of volume that occurs when energy is transferred at constant pressure is small; the energy that leaks away is therefore also small and ΔH and ΔU are almost the same. Reactions that involve gases, however, often lead to big changes of volume, and therefore do a lot of work on the surroundings. For them ΔH and ΔU can be significantly different (by several kilojoules per mole).

In chemistry we are mainly concerned with energy transfers at constant pressure, and so we concentrate on the enthalpy. Once we know the enthalpy changes that occur during reactions (using, for example, the information given in the tables later in this chapter), we can state how much energy that reaction requires or liberates when it takes place at constant pressure.

(a)

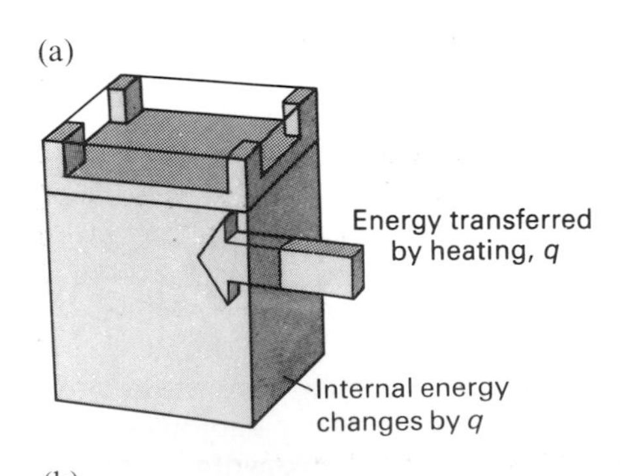

(b)

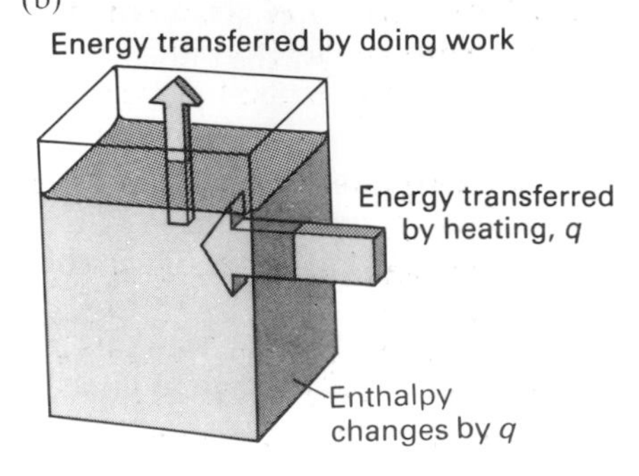

Figure 9.2 $\Delta U = q$ at constant volume (a), but $\Delta H = q$ at constant pressure (b)

9.2 Enthalpies of transition

The easiest type of process to consider is a change of state, such as melting or evaporation. In a kettle, water boils under conditions of constant pressure (usually a pressure of around 1 atm), and the energy required to vaporize a given amount of water is the enthalpy change for the process liquid water → water vapour. The **vaporization enthalpy** $\Delta_v H$ of water is the enthalpy change at the specified temperature per unit amount of substance – for example, per mol H_2O in the process

$$H_2O(l) \rightarrow H_2O(g) \quad \Delta_v H(373\,\text{K}) = +40.7\,\text{kJ mol}^{-1}$$

Every time we use a symbol of the form $\Delta_p H$ for a process p (such as $\Delta_v H$ for the process of vaporization) we mean a **molar enthalpy change**, that is, an enthalpy change per unit amount of substance.

The numerical value shows that in order to vaporize 18 g of water at 100 °C, 40.7 kJ of energy must be provided. Standard vaporization enthalpies of several substances are listed in Table 9.1 (the term 'standard' will be explained in Section 9.4). The value for water is anomalously high; this is another consequence of the presence of hydrogen bonds, which have the effect of strapping the molecules together and making vaporization an energetically more demanding process.

Table 9.1 Standard melting and vaporization enthalpies

	T_f/K	$\Delta_m H^\ominus$/kJ mol^{-1}	T_b/K	$\Delta_v H^\ominus$/kJ mol^{-1}
H_2	13.96	0.12	20.38	0.92
N_2	63.15	0.72	77.35	5.59
O_2	54.36	0.44	90.18	6.82
H_2O	273.15	6.01	373.15	40.66 (44.02 at 25 °C)
NH_3	195.40	5.65	239.73	23.35
CH_4	90.68	0.94	111.66	8.18
C_6H_6	278.65	9.83	353.25	30.8

Values relate to the melting or boiling point

EXAMPLE

Calculate how long it would take for a 2 kW heater (such as in an electric kettle) to evaporate 1.00 kg of water at 100 °C. The vaporization enthalpy of water at that temperature is 40.66 kJ mol^{-1}.

METHOD Calculate the total quantity of energy $n\Delta_v H$ that has to be supplied by heating. To find the amount of substance divide the mass by the molar mass. Calculate the time it takes for a 2 kW heater to supply that quantity of energy from power = energy/time.

ANSWER The amount of substance of H_2O to be evaporated is

$$n = \frac{1.00 \times 10^3\,\text{g}}{18.02\,\text{g mol}^{-1}} = 55.5\,\text{mol}$$

The quantity of energy to be supplied at constant pressure is

$$q = n\Delta_v H = 55.5\,\text{mol} \times 40.66\,\text{kJ mol}^{-1} = 2.26 \times 10^3\,\text{kJ}$$

The time it takes for this quantity of energy to be supplied by a 2 kW heater (that is, a heater supplying energy at a rate of $2 \times 10^3\,\text{J s}^{-1}$) is

$$t = \frac{2.26 \times 10^6\,\text{J}}{2 \times 10^3\,\text{J s}^{-1}} = 1.13 \times 10^3\,\text{s}$$

The kettle will boil to dryness in about 19 minutes.

COMMENT The time to heat the water from room temperature to the boiling point can be calculated as the molar heat capacity of the water is known (and equal to 75 J K^{-1} mol^{-1}). The calculation could be continued by finding the total time it takes to evaporate water starting at room temperature.

Solids require energy in order to melt. Melting (or **fusion**) normally takes place under constant pressure, and so the energy absorbed from the heater can be identified with the change of the system's enthalpy. We therefore speak of the **melting enthalpy** $\Delta_m H$ at the specified temperature. For instance, the melting enthalpy of ice at 0 °C is the enthalpy change per mol H_2O for the process

$$H_2O(s) \rightarrow H_2O(l) \quad \Delta_m H = +6.01\,\text{kJ mol}^{-1}$$

that is, 6.01 kJ are required to melt 18 g of ice. Some other melting enthalpies are given in Table 9.1.

9.3 Measurement of enthalpy changes

By far the most important type of enthalpy change for the chemist is the change that accompanies a chemical reaction. It is measured using a **calorimeter**, a device for monitoring the energy transferred by heating as the reaction takes place.

Calorimetry

The energy transaction in a calorimeter is measured from the change of temperature it causes. A change of temperature, δT, is converted to the energy supplied, q, using the **heat capacity**, C, of the calorimeter and the relation

$$q = C\delta T \tag{9.3.1}$$

For instance, if the heat capacity of a lump of metal is $231\,\mathrm{J\,K^{-1}}$ (as would be the case for a 600 g block of copper), and during a reaction we observe a temperature rise of 5.2 K, we know that 1.2 kJ of energy has been transferred. (The **specific heat capacity**, c, is often listed in reference books; this is the heat capacity per unit mass of material.) The actual heat capacity of the calorimeter (and its contents) can then be calculated if the masses of its parts are known.

There are several ways of putting these ideas into practice. The simplest type of calorimeter is an insulated polystyrene cup or vacuum flask fitted with a thermometer. The reaction is started by mixing the reactants, and the temperature change is recorded. The heat capacity of the system (the calorimeter plus its contents) is measured in a separate experiment (for instance, by heating it electrically) or calculated from specific heat capacities.

A more elaborate piece of apparatus is the **flame calorimeter** illustrated in Figure 9.3(a). It is used for measuring the enthalpies of combustion of gases and volatile liquids. In a typical experiment, oxygen is passed through the combustion chamber and the energy evolved by the reaction heats the surrounding water-bath. The temperature rise produced by the combustion of a known mass of the compound is measured.

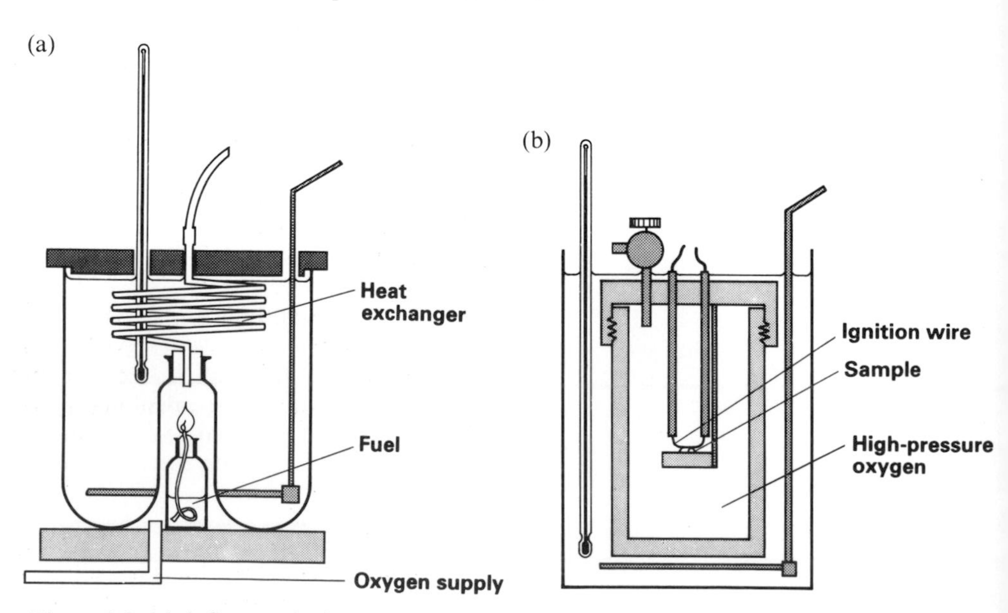

Figure 9.3 (a) A flame calorimeter, (b) a bomb calorimeter

The third basic type of calorimeter is the **bomb calorimeter**, shown in Figure 9.3(b). This is a strong, closed vessel immersed in an insulated water-bath. The solid or liquid sample in the holder is ignited electrically in oxygen (for a combustion) at about 25 atm. The temperature rise of the entire assembly is recorded. The calorimeter is calibrated using benzoic acid, $C_6H_5COOH(s)$, which has an accurately known combustion enthalpy. Modern bomb calorimeters are so sensitive that very small amounts of material are sufficient. This is important when the substance is in short supply and expensive, as is often the case with biological materials. A significant point about a bomb calorimeter is that it operates at constant *volume*, and so the internal energy change ΔU, not the enthalpy change ΔH, is measured.

EXAMPLE

The combustion of 0.500 g of ethanol, C_2H_5OH, in a flame calorimeter produced a temperature rise of 5.01 K. The same rise in temperature was brought about electrically using a 10.0 V supply passing a current of 2.50 A for 600 s (10 min). Find the combustion enthalpy, $\Delta_c H$, of ethanol.

METHOD The energy supplied by the combustion of ethanol is the same as that supplied in the electrical heating experiment because the temperature rise is the same in each case. Electrical energy is given by (current) × (time) × (potential difference), or ItV for short. Note that $1\,\text{A s V} = 1\,\text{J}$. The molar mass of ethanol is $46.07\,\text{g mol}^{-1}$; use this to convert q to a molar quantity by dividing by the amount of substance, n.

ANSWER The energy supplied is

$$q = 2.50\,\text{A} \times 600\,\text{s} \times 10.0\,\text{V} = 15\,000\,\text{A s V} = 15.0\,\text{kJ}$$

The amount of substance of C_2H_5OH in the sample is

$$n = \frac{0.500\,\text{g}}{46.07\,\text{g mol}^{-1}} = 1.09 \times 10^{-2}\,\text{mol}$$

Therefore, since 0.500 g produces 15.0 kJ, the combustion enthalpy is

$$\Delta_c H = \frac{-15.0\,\text{kJ}}{1.09 \times 10^{-2}\,\text{mol}} = -1380\,\text{kJ mol}^{-1}$$

($\Delta_c H$ is negative because energy leaves the ethanol.)

COMMENT Instead of arranging for the electrical heating to supply the same quantity of energy as the combustion, it is possible to determine the heat capacity, C, of the calorimeter from C = (energy supplied by heating)/(temperature rise): in the present case the electrical heating experiment gives $C = 2.99\,\text{kJ K}^{-1}$. Then if in the combustion experiment a temperature rise δT is observed, $q = C\delta T$. In this way the temperature rises do not have to be precisely matched.

Not all reactions are combustions. For instance, we might be interested in the **solution enthalpy** of a salt such as iron(III) chloride in water. The procedure outlined above may be used: the salt is added to water in an insulated vessel, and the change of temperature is recorded and converted to a molar enthalpy change. The **neutralization enthalpy** is another quantity of interest: it is the enthalpy change per mol H_2O that accompanies the neutralization of an acid by a base. Solution and neutralization enthalpies are usually quoted for the formation of infinitely dilute solutions to avoid complications arising from the interactions of ions.

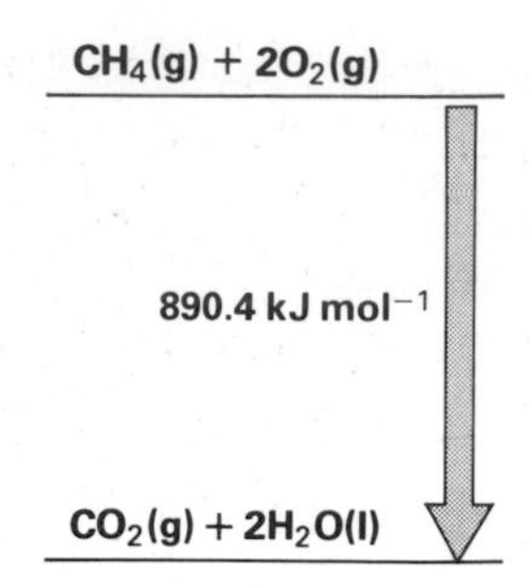

Figure 9.4 The standard combustion enthalpy of methane

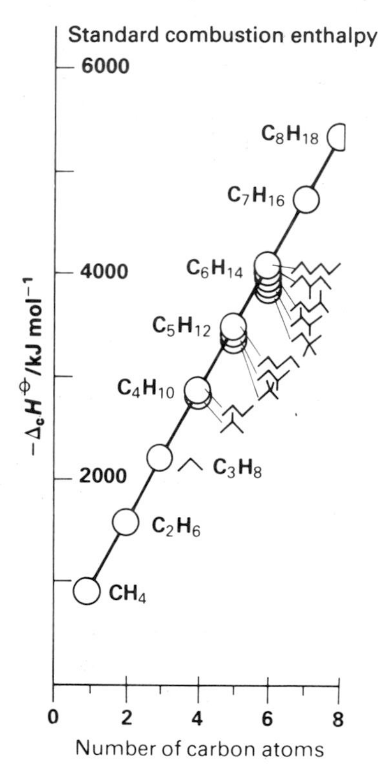

Figure 9.5 The standard combustion enthalpies of some alkanes

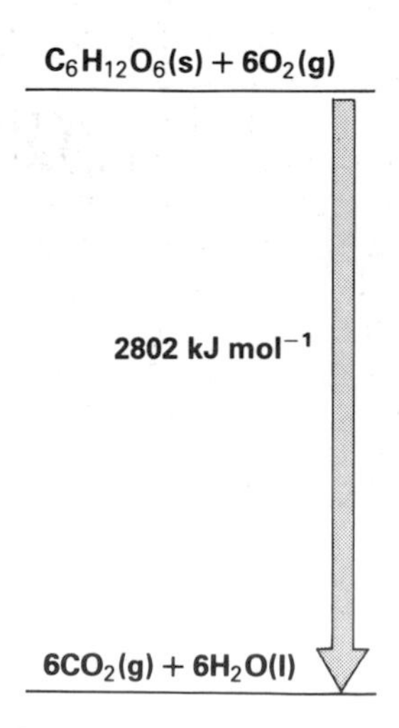

Figure 9.6 The standard combustion enthalpy of glucose

9.4 Enthalpy in action

We now consider how to use the enthalpy changes measured experimentally.

Enthalpies of combustion

One very important type of reaction is combustion, especially the combustion of organic materials to carbon dioxide and water. Information on this type of reaction is central to an understanding of the metabolism of organisms and the utilization of fuels.

Consider the combustion of methane:

$$CH_4 + 2O_2 \rightarrow CO_2 + 2H_2O$$

This reaction takes place when natural gas is burnt. When the reaction takes place in a flame calorimeter similar to the kind illustrated in Figure 9.3(a), the energy evolved heats its surroundings by 890.4 kJ per mole of CH_4 consumed. The combustion enthalpy of methane is therefore $-890.4\,\mathrm{kJ\,mol^{-1}}$ (the negative sign indicating that energy is released in the combustion).

The last statement must be made more precise. What is normally reported is the **standard combustion enthalpy**. This is the enthalpy change per unit amount of substance, all the reactants and products being in their standard states at the specified temperature. A **standard state** is **the pure substance at a pressure of 1 bar** (100 kPa, see Section 4.1). Normally each substance is taken to be in its **reference phase**, the most stable phase at the temperature specified. At 298 K that means gaseous methane, oxygen and carbon dioxide, but liquid water. This, however, is not a necessary part of the definition, and any phase could be specified: there are values for both graphite and diamond in Table 9.2.

Table 9.2 Standard combustion enthalpies at 25 °C, $\Delta_c H^{\ominus}/\mathrm{kJ\,mol^{-1}}$

H_2(g)	−285.8	CH_4(g)	−890.4	C_6H_6(l)	−3268
C(graphite)	−393.5	C_2H_6(g)	−1560	C_2H_5OH(l)	−1367
C(diamond)	−395.4	C_8H_{18}(l)	−5512	CH_3CHO(l)	−1167
S(rhombic)	−296.9	C_2H_4(g)	−1411	CH_3COOH(l)	−875
S(monoclinic)	−297.2	C_2H_2(g)	−1300	$C_6H_{12}O_6$(s) (glucose)	−2802

A standard enthalpy change is denoted by a superscript ⊖, which is read 'standard'; the temperature must also be specified. For example, the standard combustion enthalpy of methane at 25 °C is written as follows (if the product is liquid water):

$$CH_4(g) + 2O_2(g) \rightarrow CO_2(g) + 2H_2O(l) \quad \Delta_c H^{\ominus}(298\,\mathrm{K}) = -890.4\,\mathrm{kJ\,mol^{-1}}$$

The negative sign indicates a *release* of energy in the combustion (see Figure 9.4). The methane and oxygen are originally pure and each at 1 bar pressure, and the products are pure carbon dioxide at 1 bar and pure water at that pressure.

Many other combustion reactions are of importance, and some standard combustion enthalpies are listed in Table 9.2. The values for the alkanes increase with the number of CH_2 units and depend on the degree of chain branching (see Figure 9.5; these compounds are described in Chapter 27). The combustion of glucose at 25 °C is the reaction

$$\underset{\text{glucose}}{C_6H_{12}O_6(s)} + 6O_2(g) \rightarrow 6CO_2(g) + 6H_2O(l) \quad \Delta_c H^{\ominus} = -2802\,\mathrm{kJ\,mol^{-1}}$$

(see Figure 9.6), and this immense quantity of energy per mole (which corresponds to 15.6 kJ per gram of glucose) is used by reactions inside living cells. It is the source of energy for animals and plants, which use respiration to tap energy resources and to power metabolic processes (see Figure 9.7).

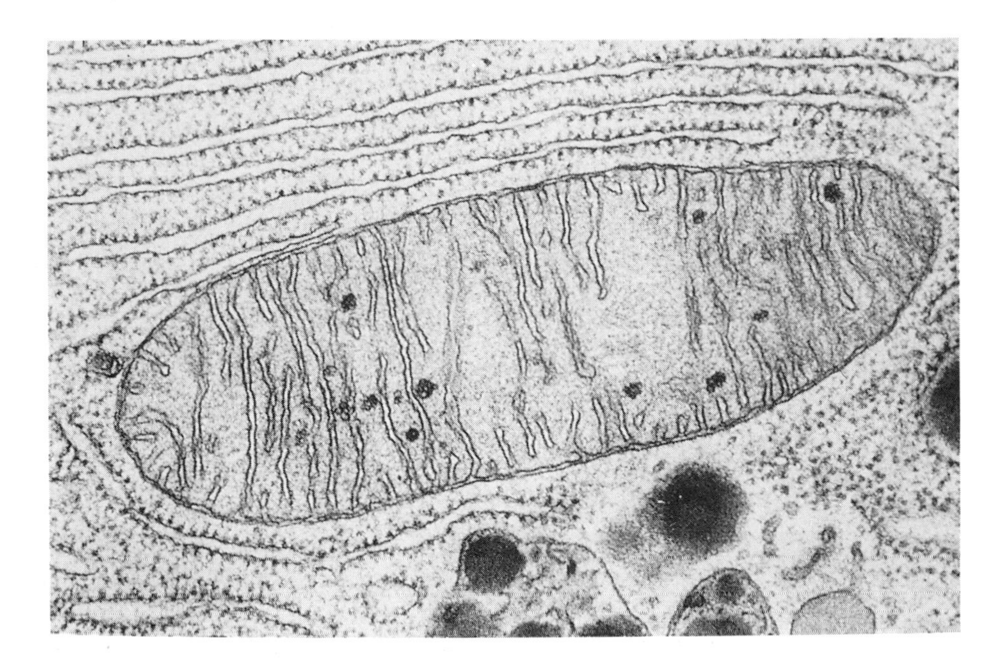

Figure 9.7 A mitochondrion, the site of respiration within a cell. Magnification approx. ×40 000

Enthalpies of formation

The enthalpy of formation of a compound is the enthalpy change that accompanies its formation from its elements. When unit amount of a compound is formed from its elements, and every substance involved is in its reference phase at the temperature specified, the enthalpy change is called the **standard formation enthalpy** of the compound, $\Delta_f H^\ominus$. Standard formation enthalpies of the elements are therefore zero, when they are in their reference phases.

As an example, consider the formation of methane at 25 °C:

$$C(s) + 2H_2(g) \rightarrow CH_4(g) \quad \Delta_f H^\ominus(298\,K) = -74.8\,kJ\,mol^{-1}$$

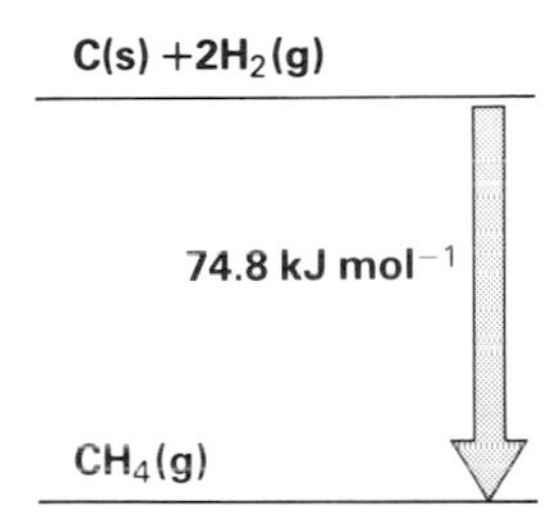

Figure 9.8 The standard formation enthalpy of methane

The reference phases at this temperature (25 °C) are graphite, gas and gas respectively, all at 1 bar. Note that the standard formation enthalpy of methane is negative, as shown in Figure 9.8, which indicates that 74.8 kJ of energy is liberated to heat its surroundings for each mole of CH_4 formed under standard conditions.

The formation of benzene

$$6C(s) + 3H_2(g) \rightarrow C_6H_6(l) \quad \Delta_f H^\ominus(298\,K) = +49.0\,kJ\,mol^{-1}$$

(see Figure 9.9) illustrates two points. The first is that *standard formation enthalpies may be either positive or negative.* A positive standard formation enthalpy indicates that energy has to be provided as the reaction proceeds. In the case of the formation of benzene, 49.0 kJ of energy is needed to produce 1 mol $C_6H_6(l)$ in its standard state at 25 °C. The second point is that the formation reaction need not be an actual, feasible reaction: in this instance, heating carbon and hydrogen together does not in fact generate benzene. Likewise, we can speak of the formation enthalpy of complex molecules like glucose even though the direct synthesis is impracticable.

Standard formation enthalpies are very important in thermochemistry, as we shall now see. They have been determined (often from combustion reactions as the following Example shows) for many substances over a wide range of temperatures. Some values at 25 °C are listed in Table 9.3.

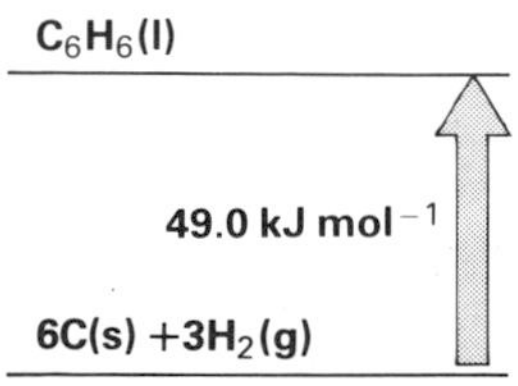

$C_6H_6(l)$

49.0 kJ mol^{-1}

$6C(s) + 3H_2(g)$

Figure 9.9 The standard formation enthalpy of benzene

Table 9.3 Standard formation enthalpies at 25 °C, $\Delta_f H^\ominus/\mathrm{kJ\,mol^{-1}}$

$H_2O(l)$	−285.8	$Fe_2O_3(s)$	−824.2	$Na^+(aq)$	−240.1
$H_2O_2(l)$	−187.8	$SiO_2(s)$	−910.9	$K^+(aq)$	−252.4
$NH_3(g)$	−46.1	$CaCO_3(s)$	−1207	$Mg^{2+}(aq)$	−466.9
$NH_3(aq)$	−80.3	$BaCO_3(s)$	−1219	$Ca^{2+}(aq)$	−542.8
$N_2H_4(l)$	50.6	$NaOH(s)$	−425.6	$Al^{3+}(aq)$	−524.7
$HF(g)$	−271.1	$KOH(s)$	−424.8	C(s, diamond)	1.9
$HCl(g)$	−92.3	$HONO_2(l)$	−174.1	$CH_4(g)$	−74.8
$HCl(aq)$	−167.2	$HONO_2(aq)$	−207.4	$C_2H_2(g)$	226.7
$HBr(g)$	−36.4	$(HO)_2SO_2(l)$	−814.0	$C_2H_4(g)$	52.3
$HI(g)$	26.5	$(HO)_2SO_2(aq)$	−909.3	$C_2H_6(g)$	−84.7
$H_2S(g)$	−20.6	$NH_4Cl(s)$	−314.4	$C_3H_8(g)$	−103.9
$CO(g)$	−110.5	$NaCl(s)$	−411.2	$C_4H_{10}(g)$	−126.2
$CO_2(g)$	−393.5	$NaBr(s)$	−361.1	$C_6H_6(l)$	49.0
$NO(g)$	90.3	$NaI(s)$	−287.8	$C_6H_5CH_3(l)$	12.1
$NO_2(g)$	33.2	$KCl(s)$	−436.8	$C_8H_{18}(l)$	−249.9
$N_2O_4(g)$	9.2	$KBr(s)$	−393.8	$CH_3OH(l)$	−238.7
$SO_2(g)$	−296.8	$KI(s)$	−327.9	$C_2H_5OH(l)$	−277.7
$SO_3(g)$	−395.7	$AgCl(s)$	−127.1	$CH_3CHO(l)$	−192.3
$MgO(s)$	−601.7	$CaCl_2(s)$	−795.8	$CH_3COCH_3(l)$	−248.1
$CaO(s)$	−635.1	$AlCl_3(s)$	−704.2	$CH_3COOH(l)$	−484.5
$BaO(s)$	−553.5	$Cl^-(aq)$	−167.2	$CH_3COOC_2H_5(l)$	−479.0
$ZnO(s)$	−348.3	$Br^-(aq)$	−121.6	$C_6H_{12}O_6(s)$	−1274
$Al_2O_3(s)$	−1675.7	$I^-(aq)$	−55.2	$C_{12}H_{22}O_{11}(s)$	−2222

EXAMPLE

The standard combustion enthalpies of carbon, hydrogen and methane are $-393.5\,\mathrm{kJ\,mol^{-1}}$, $-285.8\,\mathrm{kJ\,mol^{-1}}$ and $-890.4\,\mathrm{kJ\,mol^{-1}}$ respectively, at 25 °C. Deduce the value of the standard formation enthalpy of methane at that temperature.

METHOD The reaction is $C(s) + 2H_2(g) \rightarrow CH_4(g)$ where the reference phase of carbon is graphite (at 25 °C). The three combustion reactions are

(a) $C(s) + O_2(g) \rightarrow CO_2(g) \quad \Delta_c H^\ominus = -393.5\,\mathrm{kJ\,mol^{-1}}$

(b) $H_2(g) + \tfrac{1}{2}O_2(g) \rightarrow H_2O(l) \quad \Delta_c H^\ominus = -285.8\,\mathrm{kJ\,mol^{-1}}$

(c) $CH_4(g) + 2O_2(g) \rightarrow CO_2(g) + 2H_2O(l) \quad \Delta_c H^\ominus = -890.4\,\mathrm{kJ\,mol^{-1}}$

The reaction of interest can formally be constructed as follows: burn C(s); burn $2H_2(g)$; 'unburn' (that is, run the combustion reaction in reverse) $CO_2(g) + 2H_2O(l)$ to form $CH_4(g)$. The overall result is the same as the required formation reaction. Add the enthalpies for each step.

ANSWER The standard formation enthalpy is the sum of (a) + 2(b) − (c):

$$\Delta_f H^\ominus = (-393.5\,\mathrm{kJ\,mol^{-1}}) + 2 \times (-285.8\,\mathrm{kJ\,mol^{-1}}) - (-890.4\,\mathrm{kJ\,mol^{-1}})$$

$$= -74.7\,\mathrm{kJ\,mol^{-1}}$$

Check that the overall reaction is indeed the formation reaction:

$$
\begin{array}{lll}
(a) & C + O_2 & \rightarrow CO_2 \\
+2(b) & 2H_2 + O_2 & \rightarrow 2H_2O \\
-(c) & CO_2 + 2H_2O & \rightarrow CH_4 + 2O_2 \\
\hline
& C + 2H_2 + CO_2 + 2H_2O + 2O_2 & \rightarrow CH_4 + CO_2 + 2H_2O + 2O_2 \\
& C + 2H_2 & \rightarrow CH_4
\end{array}
$$

which is the reaction required.

> COMMENT Note that this kind of calculation is independent of whether or not the individual reaction steps can be carried out in practice; only *formally* possible steps are required. The standard combustion enthalpy of methane ($-890.4\,kJ\,mol^{-1}$) is high, which is an advantage for its applications in domestic and industrial heating.

Reaction enthalpies

We now show that the data such as those in Tables 9.2 and 9.3 can be used in the discussion of any reaction, not only combustion and formation, and that from them we can predict the enthalpy change of any reaction, including reactions that have not been studied experimentally.

Suppose we were interested in some fairly complicated reaction such as the production of chloroethane (see Section 28.5):

$$HCl(g) + CH_2{=}CH_2(g) \rightarrow CH_3CH_2Cl(g)$$

and we wanted to know the enthalpy change. The First law tells us that whatever route we select between reactants and products, the same change occurs. The path-independence of standard reaction enthalpies was noticed by Germain Hess in his extensive series of measurements on 'heats of reaction' (which we now call standard reaction enthalpies, $\Delta_r H^\ominus$) before thermodynamics had been fully established. In 1840 he summarized his conclusions in a law that now bears his name. In modern language it reads:

Hess's law: The standard reaction enthalpy is the sum of the standard reaction enthalpies of each step into which the reaction can formally be divided.

As an illustration, consider the CH_3CH_2Cl reaction. We can *formally* regard it (that is, write it on paper, not necessarily actually carry it out in the laboratory) as occurring in the following three steps, illustrated in Figure 9.10:

(a) $HCl(g) \rightarrow \frac{1}{2}H_2(g) + \frac{1}{2}Cl_2(g)$ $\quad \Delta_r H^\ominus = +92.3\,kJ\,mol^{-1}$

(b) $CH_2{=}CH_2(g) \rightarrow 2C(s) + 2H_2(g)$ $\quad \Delta_r H^\ominus = -52.3\,kJ\,mol^{-1}$

(At this stage we have reversed the formation reactions, and so have produced the elements C(s), $H_2(g)$ and $Cl_2(g)$, the values are taken from Table 9.3 by changing the signs of the standard formation enthalpies listed there; they refer to 25 °C.) Now form the elements into the final product:

(c) $2C(s) + \frac{5}{2}H_2(g) + \frac{1}{2}Cl_2(g) \rightarrow CH_3CH_2Cl(g)$ $\quad \Delta_r H^\ominus = -106.7\,kJ\,mol^{-1}$

The enthalpy change is the standard formation enthalpy of CH_3CH_2Cl. According to Hess's law, it follows that the standard reaction enthalpy is

$$\Delta_r H^\ominus = (+92.3\,kJ\,mol^{-1}) + (-52.3\,kJ\,mol^{-1}) + (-106.7\,kJ\,mol^{-1}) = -66.7\,kJ\,mol^{-1}$$

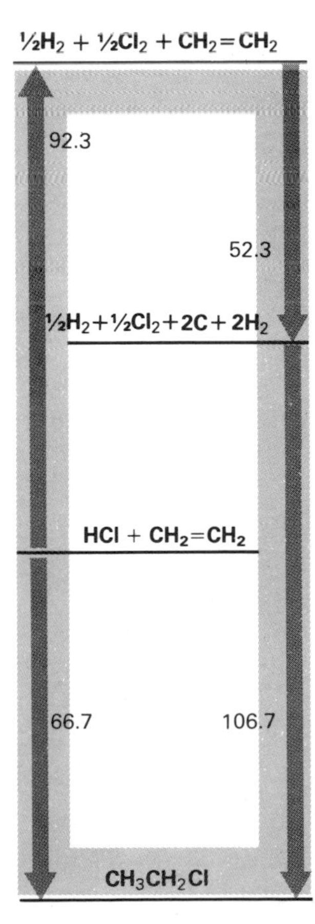

Figure 9.10 Hess's law and standard reaction enthalpy; the circuit of reactions is called a cycle

The classification of reactions

Reactions can be classified according to the sign of their standard reaction enthalpy:

Exothermic reactions are those for which $\Delta_r H^\ominus < 0$
Endothermic reactions are those for which $\Delta_r H^\ominus > 0$

All the combustion reactions in Table 9.2 are exothermic and liberate energy under standard conditions. Some of the values in Table 9.3 are positive and some are negative. Methane, with a negative standard formation enthalpy, is produced by an exothermic formation reaction and is called an **exothermic compound**. Benzene is an **endothermic compound**.

How the exothermicity or endothermicity shows itself in practice depends on the reaction conditions. For instance, if the reaction is carried out in a thermally insulated container, then an endothermic reaction draws its energy from the contents of the vessel; as a result, they cool. In contrast, an exothermic reaction in a thermally insulated vessel heats the contents of the vessel, and so their temperature rises – that is why water gets hot when concentrated sulphuric acid is added, because the solution process is exothermic. If an endothermic reaction occurs in a vessel in a water-bath that maintains constant temperature, then it draws in energy from the bath; conversely, an exothermic reaction releases energy into the bath.

Dissociation, atomization and sublimation enthalpies

The **standard dissociation enthalpy** of a diatomic molecule is the enthalpy change accompanying the break-up into atoms of a mole of *molecules*, all substances being in their standard states. For example,

$$Cl_2(g) \rightarrow 2Cl(g) \quad \Delta_d H^\ominus(298\,K) = +242\,kJ\,mol^{-1}$$

The **standard atomization enthalpy** of an element is the enthalpy change accompanying the break-up into atoms, per mole of *atoms*, all substances being in their standard states:

$$\tfrac{1}{2}Cl_2(g) \rightarrow Cl(g) \quad \Delta_a H^\ominus(298\,K) = +121\,kJ\,mol^{-1}$$

For a solid element that evaporates as a monatomic vapour, the standard atomization enthalpy is equal to its **standard sublimation enthalpy**:

$$Na(s) \rightarrow Na(g) \quad \Delta_a H^\ominus(298\,K) = +108\,kJ\,mol^{-1}$$

Some elements form a molecular vapour – phosphorus, for instance, forms $P_4(g)$ – and for them the two values are different.

Lattice enthalpies and the Born–Haber cycle

The **lattice enthalpy** is a measure of the strength of an ionic crystal lattice (see Box 9.2); for sodium chloride, for instance, it is the standard reaction enthalpy

BOX 9.2

Lattice enthalpy

The formal name for 'lattice enthalpy' is the **standard lattice-breaking enthalpy**. It is the standard reaction enthalpy for the process

$$MX(s) \rightarrow M^+(g) + X^-(g)$$

for singly charged ions. For MX = NaCl, $\Delta_l H^\ominus(298\,K) = +787\,kJ\,mol^{-1}$. All lattice enthalpies are endothermic. The size of the lattice enthalpy, calculated on the ionic model, depends on three main factors. The first is the charges on the ions; the larger their charges the greater the lattice enthalpy. The second is the radii of the ions, smaller radii leading to larger lattice enthalpies (because ions interact more strongly when they are close together). The third is the arrangement of ions in the crystal.

Any one ion interacts with all the other ions in the crystal, and the total electrostatic energy is the sum of a very large number of positive (repulsive) and negative (attractive) terms. Fortunately, it turns out that the total energy is equal to the interaction energy between a single neighbouring cation and anion multiplied by a numerical constant called the **Madelung constant**, after the scientist who calculated some of their values, which is characteristic of the crystal structure. The Madelung constant for the rock-salt structure is 1.748, which means that the electrostatic energy of the entire crystal of N formula units is equal to $1.748 \times N$ times the energy of interaction of a sodium ion and a chloride ion at their actual separation in the crystal. The constant for the caesium chloride structure is 1.763.

The calculated and experimental lattice enthalpies agree very well for the alkali metal halides, but not for the silver halides. This discrepancy is ascribed to the importance of covalent character in the silver halides and the consequent failure of the purely ionic model.

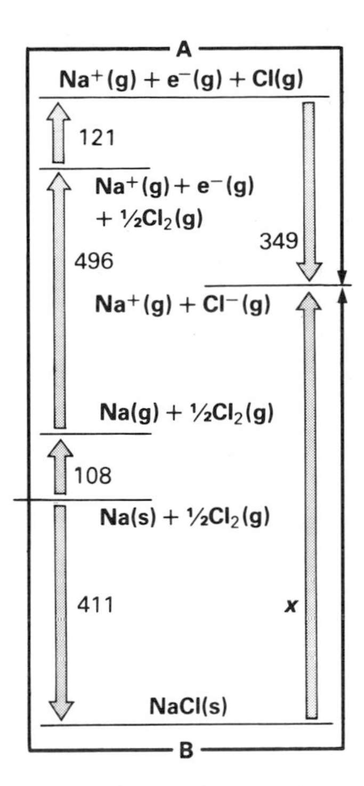

Figure 9.11 A Born–Haber cycle

for the process in which the lattice is broken up into gaseous ions:

$$NaCl(s) \rightarrow Na^+(g) + Cl^-(g) \quad \Delta_l H^\ominus = +787\,kJ\,mol^{-1}$$

The lattice enthalpy can be calculated from other data by using an adaptation of Hess's law known as the **Born–Haber cycle**. Consider the case of sodium chloride. We know (from calorimetric measurements) that its standard formation enthalpy is $-411\,kJ\,mol^{-1}$. This formation reaction can be thought of as taking place in several steps, one of which is the break-up of the lattice. This complete sequence of reactions, shown in Figure 9.11, is called a **cycle**. It should be remembered that a cycle is only a formal concept, and need not involve reactions that actually happen.

The first step along route A is the atomization of sodium metal; the second is the ionization of its gaseous atoms; the third is the atomization of chlorine gas. These are all endothermic, and in Figure 9.11 they are drawn leading upwards. The fourth step is the attachment to the chlorine atom of the electron released by the sodium atom; this stage is exothermic, and steps downwards. In Section 1.2, we used the term 'electron-gain energy' for the energy change that takes place when an electron attaches to a gas-phase atom. It would be more precise to call it the **standard electron-gain enthalpy**.

The alternative route B to the gaseous ions begins with the exothermic formation of solid sodium chloride, and is followed by a step upwards equal in size to the lattice enthalpy. If the size of any one step is unknown, it may be determined by noting that the two routes must lead to the same level on the diagram.

EXAMPLE

Use thermochemical data to find the lattice enthalpy of sodium chloride.

METHOD Construct a Born–Haber cycle which includes the NaCl formation reaction and the break-up of the lattice into a gas of Na^+ and Cl^- ions. The ionization energy of Na, $+496\,kJ\,mol^{-1}$, and the electron-gain energy of Cl, $-349\,kJ\,mol^{-1}$, are given in Table 1.2. The atomization enthalpy of Na(s), $+108\,kJ\,mol^{-1}$, and the atomization enthalpy of $Cl_2(g)$, $+121\,kJ\,mol^{-1}$, are given in the text.

ANSWER A suitable Born–Haber cycle is shown in Figure 9.11. The unknown quantity is x, and on the basis that we have:

$$\text{Route A}/kJ\,mol^{-1} = 108 + 496 + 121 - 349 = 376$$
$$\text{Route B}/kJ\,mol^{-1} = -411 + x$$

Therefore $x = 787$. It follows that the lattice enthalpy, $\Delta_l H^\ominus$, is $+787\,kJ\,mol^{-1}$.

COMMENT The electron-gain energy of atoms is a difficult quantity to measure, and it is therefore often taken as the unknown in a cycle of the kind treated here. This is how some of the values listed in Table 1.2 were obtained. Others were obtained by calculation using quantum mechanics.

Bond enthalpies

Scientists are always on the alert for patterns emerging in the data they collect. Patterns are found in thermochemistry, and help in the applications of the data and in the understanding of the underlying molecular properties.

A reaction enthalpy arises from the changes that accompany the breaking of old bonds and the formation of new ones. Therefore, if we could compile a table of **bond enthalpies** we would be able to predict the reaction enthalpy for organic reactions by noting which bonds are changed. The procedure runs into difficulty as soon as we start, however. Consider the case of water. Calculating the standard formation enthalpy requires a number of steps, as

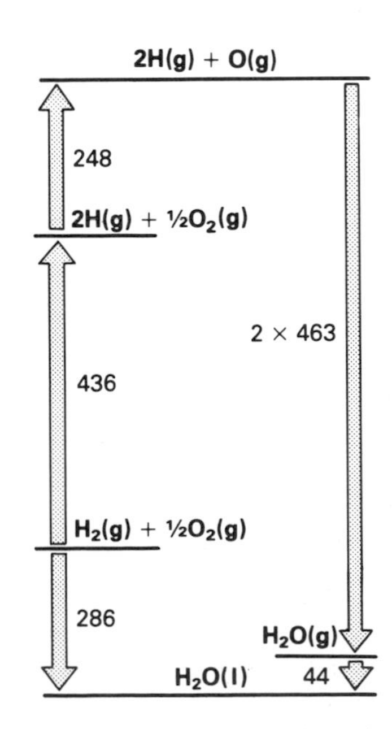

Figure 9.12 A thermodynamic cycle

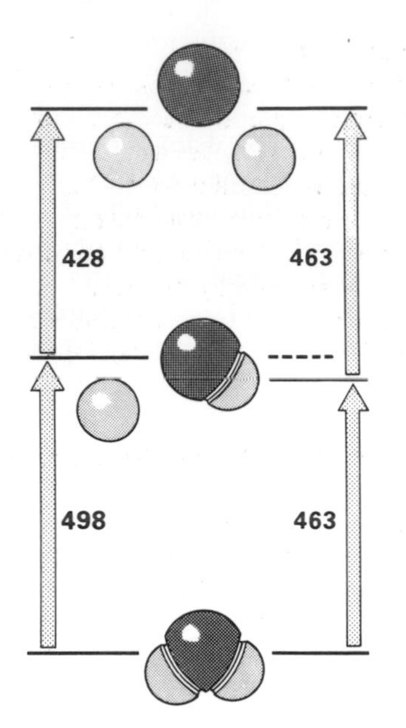

Figure 9.13 Bond enthalpies and mean bond enthalpy

indicated in Figure 9.12. However, it is known from spectroscopy that the strength of an HO—H bond is different from that of an O—H bond, their standard dissociation enthalpies being +498 kJ mol^{-1} and +428 kJ mol^{-1} respectively (see Figure 9.13), and so we cannot accurately list a bond enthalpy without specifying what other atoms are already present.

The way round this problem is to draw up a list of **mean bond enthalpies**, and to use them in *approximate* calculations of reaction enthalpies. For example, the mean O—H bond enthalpy in water is +463 kJ mol^{-1}. A list of mean bond enthalpies is given in Table 9.4. By adding together the appropriate values, reaction enthalpies may be estimated.

Table 9.4 Mean bond enthalpies/kJ mol^{-1}

H—H	436*	F—F	158*	H—O	463
C—C	348	Cl—Cl	242*	C—O	360
C=C	612	Br—Br	193*	C=O	743
Si—Si	176	I—I	151*	C—N	305
N—N	163	H—F	562*	C=N	613
N=N	409	H—Cl	431*	C≡N	890
N≡N	944*	H—Br	366*	C—F	484
O—O	146	H—I	299*	C—Cl	338
O=O	496*	H—C	413	C—Br	276
S—S	264	H—N	388	C—I	238

The asterisked values are exact values for the appropriate molecules; the others are mean values over a variety of compounds containing the bond

EXAMPLE

Estimate the standard formation enthalpy of methane on the basis of mean bond enthalpies.

METHOD The formation reaction is

$$C(s) + 2H_2(g) \rightarrow CH_4(g)$$

In order to estimate the overall enthalpy change we have first to atomize carbon ($\Delta_a H^\ominus$ = +715 kJ mol^{-1}) and $2H_2$ ($\Delta_a H^\ominus$ = +218 kJ mol^{-1}), and then form four C—H bonds (mean bond enthalpies in Table 9.4). The overall process is represented by the cycle shown in Figure 9.14.

ANSWER Route A/kJ mol^{-1} = 715 + 4 × 218 − 4 × 413. Therefore the standard formation enthalpy of methane, B in the diagram, is −65 kJ mol^{-1}.

COMMENT The experimental value is −74.8 kJ mol^{-1} (Table 9.3). The use of mean bond enthalpies is only an approximation, but it is a useful and easy way of estimating data that may not be to hand.

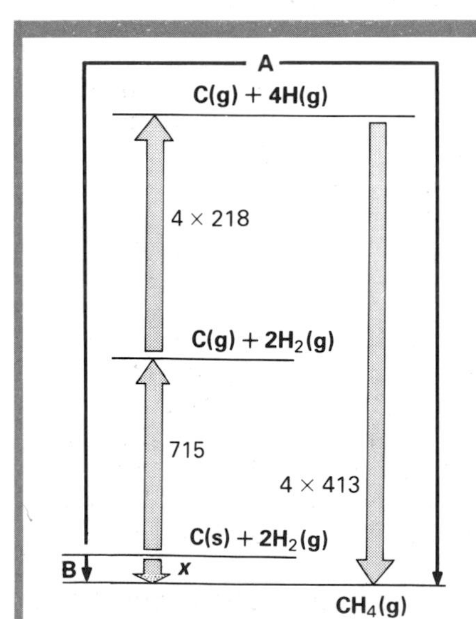

Figure 9.14 Estimating the standard formation enthalpy of methane

Delocalization enthalpy

When thermochemical arguments are applied to benzene (and to other aromatic molecules, discussed in Chapters 34 and 35), unexpected results are obtained. At least, they were unexpected at the time: our familiarity with the stability of aromatic rings has removed the striking quality of the results obtained by early chemists.

The standard formation enthalpy of benzene can be predicted by adding together the appropriate mean bond enthalpies; this is done in Figure 9.15 on the basis that it is a ring of alternating double and single bonds. The result is +209 kJ mol^{-1}. The experimental value, however, is +49 kJ mol^{-1}, indicating that the calculation has *underestimated* the compound's thermochemical

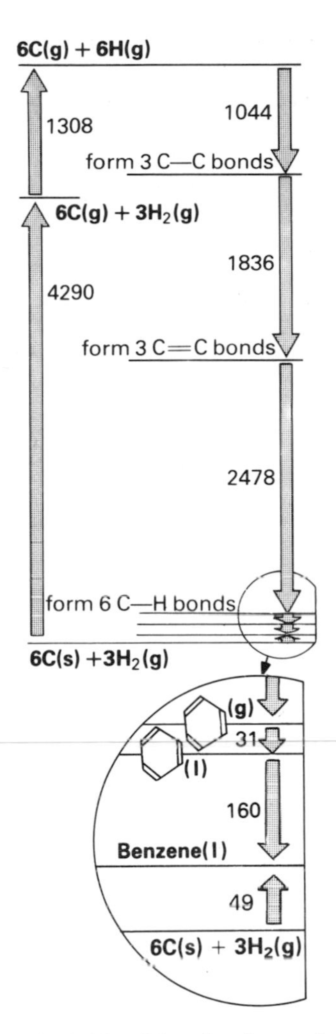

Figure 9.15 The delocalization enthalpy of benzene

stability by $160\,kJ\,mol^{-1}$, a chemically very significant amount.

We now know, as we saw in Section 2.5, that it is an oversimplification to regard a benzene molecule as a ring of alternating single and double bonds; in fact, the molecule is stabilized by the delocalization of the electrons. Thermochemistry provides a quantitative measure of this delocalization stabilization, and has shown that the delocalization reduces the molar enthalpy of the molecule by as much as $160\,kJ\,mol^{-1}$. This quantity is called the **delocalization enthalpy** of benzene. A similar discrepancy between observation and prediction is provided by measurements of the enthalpies of hydrogenation of benzene and cyclohexene, as demonstrated in the following Example.

EXAMPLE

The standard hydrogenation enthalpy ($\Delta_h H^\ominus$) of cyclohexene is $-120\,kJ\,mol^{-1}$; the standard hydrogenation enthalpy of benzene is $-208\,kJ\,mol^{-1}$. Calculate the delocalization enthalpy of benzene.

METHOD The two reactions are

$$+ H_2(g) \longrightarrow \qquad \Delta_h H^\ominus = -120\,kJ\,mol^{-1}$$

$$+ 3H_2(g) \longrightarrow \qquad \Delta_h H^\ominus = -208\,kJ\,mol^{-1}$$

Calculate the standard hydrogenation enthalpy of benzene on the basis that it has three localized double bonds. The difference between this value and the observed value is the delocalization enthalpy.

ANSWER The calculated value is $3 \times (-120\,kJ\,mol^{-1}) = -360\,kJ\,mol^{-1}$. The delocalization enthalpy is therefore the difference $\Delta_{del}H^\ominus = (-208\,kJ\,mol^{-1}) - (-360\,kJ\,mol^{-1}) = 152\,kJ\,mol^{-1}$.

COMMENT When benzene is hydrogenated, a smaller enthalpy change occurs than would be expected on the basis that it has three double bonds; we therefore conclude that it is more stable than the latter structure suggests. The value $152\,kJ\,mol^{-1}$ represents the thermochemical stabilization brought about by delocalization of the bonds. Note that slightly different values of $\Delta_{del}H^\ominus$ are obtained depending on the basis of the calculation: this shows that the delocalization enthalpy is not a well-defined characteristic of the molecule.

Summary

- ☐ Thermodynamics is the study of the transformations of energy; thermochemistry is the branch of thermodynamics concerned with transformations of energy in the course of chemical reactions.
- ☐ The First law of thermodynamics states that energy can be neither created nor destroyed.
- ☐ Energy can be transferred from one system to another by heating or by doing work.
- ☐ If a quantity of energy, w, is transferred to a system by doing work and a quantity of energy, q, is transferred to the system by heating, the transfers change the internal energy, U, by $\Delta U = q + w$.
- ☐ When a quantity of energy, q, is transferred to a constant volume system by heating, the internal energy changes by $\Delta U = q$.
- ☐ The enthalpy, H, of a system is a modification of the internal energy which automatically takes into account the work of expansion when a system is heated at constant pressure.

- ☐ When a quantity of energy, q, is transferred to a constant pressure system by heating, the enthalpy changes by $\Delta H = q$.
- ☐ The vaporization enthalpy is the enthalpy change that occurs when unit amount of substance of a liquid evaporates; it is the energy that has to be supplied by heating when vaporization occurs at constant pressure.
- ☐ The melting enthalpy is the enthalpy change that occurs when unit amount of substance of a solid melts; it is the energy that has to be supplied by heating when melting occurs at constant pressure.
- ☐ Vaporization and melting enthalpies are positive.
- ☐ The flame calorimeter operates at constant pressure, and so gives the reaction enthalpy directly (via the temperature change and the heat capacity of the calorimeter). The bomb calorimeter operates at constant volume, and so gives the internal energy change during the reaction.
- ☐ The solution enthalpy is the enthalpy change accompanying dissolution of a salt to give an infinitely dilute solution; the neutralization enthalpy is the enthalpy change accompanying neutralization of an acid by a base in infinitely dilute solution.
- ☐ The standard state of a substance is the pure substance at a pressure of 1 bar (100 kPa) and the temperature specified. The reference phase of a substance is its most stable form at a pressure of 1 bar and the temperature specified.
- ☐ The standard reaction enthalpy, $\Delta_r H^\ominus$, is the enthalpy change per unit amount of substance when all the reactants and all the products are in their standard states.
- ☐ The standard combustion enthalpy, $\Delta_c H^\ominus$, is the enthalpy change accompanying the complete combustion of unit amount of substance of a sample, all substances being in their standard states.
- ☐ The standard formation enthalpy, $\Delta_f H^\ominus$, is the enthalpy change accompanying formation of unit amount of the compound from its elements, all substances being in their standard states.
- ☐ The standard atomization enthalpy, $\Delta_a H^\ominus$, is the enthalpy change accompanying the decomposition of a substance into unit amount of atoms, all substances being in their standard states.
- ☐ Hess's law states that the overall standard reaction enthalpy is the sum of the standard reaction enthalpies of each step into which the reaction may formally be divided.
- ☐ A Born–Haber cycle is a complete sequence of reactions from an initial set of substances and back to them through a different route.
- ☐ If $\Delta_r H^\ominus < 0$, a reaction is exothermic; if $\Delta_r H^\ominus > 0$, a reaction is endothermic.
- ☐ The mean A—B bond enthalpy is the average of the enthalpy changes accompanying the breaking of the A—B bond in a series of compounds.
- ☐ The thermochemical stabilization of aromatic compounds is indicated by their delocalization enthalpies.

PROBLEMS

1 Explain the terms *work*, *heat* and *internal energy*, and state the First law of thermodynamics.

2 Explain why enthalpy changes are so important in chemistry.

3 The standard vaporization enthalpy of benzene at its boiling point is 30.8 kJ mol^{-1}: for how long would a 100 W electric heater have to operate in order to vaporize a 100 g sample at that temperature?

4 Suppose you were caught on a mountain with soaking wet clothes; what energy would your body have to supply if the wind caused the evaporation of 500 g of water?

5 Foodstuffs are often classified as having 'so many Calories': given that 1 Calorie (as used by dieticians) corresponds to 4.18 kJ, calculate (to one significant figure) the mass of cheese that must be consumed in order to offset the energy loss in the last question, given that the calorific value of cheese is 5.0 Calories per gram.

6 The calorific value of milk is about 2.9 kJ g^{-1}. To what height could a person (mass 70 kg) climb above the surface of the Earth in the course of using the entire energy one pint can supply?

7 Explain the terms *standard state* of a substance and *standard formation enthalpy*.

8 Explain the connection between Hess's law and the First law of thermodynamics. What is the evidence for each law?

9 Use the standard formation enthalpies in Table 9.3 to calculate the standard enthalpy changes accompanying the following reactions:
(a) $CH_4(g) + 2O_2(g) \rightarrow CO_2(g) + 2H_2O(l)$
(b) $4Al(s) + 3O_2(g) \rightarrow 2Al_2O_3(s)$

10 Calculate the dissolution enthalpy of sodium chloride (the enthalpy change for 1 mol $NaCl(s) \rightarrow NaCl(aq)$) on the basis that the solvation enthalpies of $Na^+(g)$ and $Cl^-(g)$ are $-406\,kJ\,mol^{-1}$ and $-377\,kJ\,mol^{-1}$, respectively. (Recall the discussion in Chapter 8.)

11 Present a thermochemical argument to explain why sodium chloride does not dissolve in benzene.

12 (a) State Hess's law of constant heat summation (law of conservation of energy).
(b) Define precisely the terms
(i) *enthalpy of formation* (*heat of formation*);
(ii) *enthalpy of neutralization* (*heat of neutralization*).
(c) Using the following data, collected at 25 °C and standard atmospheric pressure, in which the negative sign indicates heat evolved, calculate the enthalpy of formation of potassium chloride, KCl(s):

	$\Delta H(298\,K)/kJ\,mol^{-1}$
(i) $KOH(aq) + HCl(aq) \rightarrow KCl(aq) + H_2O(l)$	−57.3
(ii) $H_2(g) + \frac{1}{2}O_2(g) \rightarrow H_2O(l)$	−285.9
(iii) $\frac{1}{2}H_2(g) + \frac{1}{2}Cl_2(g) + aq \rightarrow HCl(aq)$	−164.2
(iv) $K(s) + \frac{1}{2}O_2(g) + \frac{1}{2}H_2(g) + aq \rightarrow KOH(aq)$	−487.0
(v) $KCl(s) + aq \rightarrow KCl(aq)$	+18.4

(d) The enthalpy of neutralization of ethanoic acid (acetic acid) is $-55.8\,kJ\,mol^{-1}$ while that of hydrochloric acid is $-57.3\,kJ\,mol^{-1}$, both reactions being with potassium hydroxide solution. Explain the difference in these two values and make what deductions you can.
(e) Describe briefly an experiment by which the enthalpy change of a chemical reaction *of your own choice* may be determined, and outline how the various measurements would be used to calculate the final value.
(f) When solid potassium chloride is dissolved in water, heat is absorbed.

$$KCl(s) + aq \rightarrow KCl(aq) \quad \Delta H = +18.4\,kJ\,mol^{-1}$$

Use the ideas formulated in the kinetic-molecular theory of matter to discuss what happens during the dissolution of a crystalline salt in water, and explain why heat is absorbed in the above reaction. (*SUJB*)

13 (a) Use the following information to calculate (i) the enthalpy of formation of methane from graphite and hydrogen in their standard states, (ii) the C—H bond energy [enthalpy] in the methane molecule. (All data refer to 298 K.)

$CH_4(g) + 2O_2(g) \rightarrow CO_2(g) + 2H_2O(l)$	$\Delta H = -890\,kJ\,mol^{-1}$
$C(graphite) + O_2(g) \rightarrow CO_2(g)$	$\Delta H = -394\,kJ\,mol^{-1}$
$H_2(g) + \frac{1}{2}O_2(g) \rightarrow H_2O(l)$	$\Delta H = -286\,kJ\,mol^{-1}$
$C(graphite) \rightarrow C(g)$	$\Delta H = 717\,kJ\,mol^{-1}$
$H_2(g) \rightarrow 2H(g)$	$\Delta H = 436\,kJ\,mol^{-1}$

(b) State what information you would need about ethane such that, together with the above data, you could make an estimate of the C—C bond energy in the ethane molecule. (An explanation of your answer is not required.) (*Oxford*)

14 State the First law of thermodynamics in terms of *heat* and *work*, defining these quantities carefully.
Use the bond enthalpies [in Table 9.4] to calculate the standard enthalpy change for the reaction

$$C_2H_4(g) + HBr(g) \rightarrow C_2H_5Br(g)$$

Given that the molar enthalpy of vaporization of bromoethane is $27.0\,kJ\,mol^{-1}$, compare your calculated value of $\Delta H^{\ominus}(298\,K)$ with the one calculated from the standard enthalpies of formation [$-85.4\,kJ\,mol^{-1}$ for $C_2H_5Br(l)$] and comment on your answer.
(*Oxford and Cambridge S part question*)

15 Two possible ways of fixing atmospheric nitrogen may be represented by the equations:

$$\tfrac{1}{2}N_2(g) + \tfrac{3}{2}H_2O(g) \rightarrow NH_3(g) + \tfrac{3}{4}O_2(g)$$
$$\tfrac{1}{2}N_2(g) + \tfrac{1}{2}H_2O(g) + \tfrac{5}{4}O_2(g) \rightarrow HNO_3(l)$$

Calculate the value of $\Delta H^{\ominus}$ for each of these processes at 298 K.
Why do you think that the process represented by the second equation does not provide a suitable method of fixing nitrogen? (*Oxford and Cambridge S part question*)

16 (a) (i) Define enthalpy of formation $\Delta_f H$ of a compound.
(ii) What extra conditions must be imposed to specify the standard enthalpy of formation $\Delta_f H^{\ominus}$ of a compound?
(b) When ethanol burns in oxygen, carbon dioxide and water are formed.
(i) Write the equation which describes this reaction.
(ii) Using the data

$\Delta_f H^{\ominus}$ for ethanol (l) $= -277.0\,kJ\,mol^{-1}$
$\Delta_f H^{\ominus}$ for carbon dioxide (g) $= -393.7\,kJ\,mol^{-1}$
$\Delta_f H^{\ominus}$ for water (l) $= -285.9\,kJ\,mol^{-1}$

calculate the value of $\Delta H^{\ominus}$ for the combustion of ethanol. (*JMB*)

17 What factors control the magnitude of lattice energies [enthalpies]? What information can be obtained from a comparison of calculated lattice energies with those obtained from a Born–Haber cycle?
Given the following quantities (in $kJ\,mol^{-1}$) for rubidium chloride, obtain a value for the electron affinity [negative of electron-gain energy] of the Cl atom:

Lattice energy of RbCl	665
Dissociation energy of Cl_2 (gas)	226
Heat [enthalpy] of sublimation of Rb metal	84
Ionization energy of the Rb atom	397
Standard heat [enthalpy] of formation of solid RbCl	−439

(*Cambridge Entrance*)

10 Equilibrium

One of the most important pieces of information in chemistry is whether a reaction will run in one direction or another. We first describe what it means to say that a chemical reaction is at equilibrium, and see that the composition at equilibrium can be specified in terms of a single quantity, the equilibrium constant. We also see how it is possible to predict how the position of equilibrium responds to changes in the conditions, and go on to show how it is also possible to make quantitative predictions.

Synthesis of ammonia

Some reactions seem to go just so far and then stop, even though there are plenty of reactants left. Others seem to go all the way to products. An example of the first type is the synthesis of ammonia; a mixture of hydrogen and nitrogen reacts under suitable conditions and forms ammonia, but the production of ammonia comes to a halt at some stage even though there is plenty of unreacted hydrogen and nitrogen. As an example of the second type, take the closely related reaction of hydrogen with oxygen to give water. When they are induced to react (by a spark, for instance), they go completely to the product.

These examples raise several questions. In the first place, are there really two sorts of reaction, one going to completion and the other not? We shall see that in fact all reactions are similar in kind, but differ in degree. In the second place, how can we describe the composition at chemical equilibrium? We shall see that this can be done through a quantity called the equilibrium constant of the reaction. Once the value of the equilibrium constant is known we can discuss the chemical composition of reaction mixtures and their response to the conditions, such as changes in the temperature and the pressure. This aspect of chemistry is developed in the following three chapters.

10.1 The description of equilibrium

When a reaction reaches **chemical equilibrium**, that is, when the composition of the reaction mixture is steady and has no further tendency to change, the mixture is not just dead. In the ammonia synthesis at equilibrium the reaction mixture consists of a mixture of hydrogen, nitrogen and ammonia with constant concentrations of each substance, but individual molecules continue to react in such a way that the rate of formation of the product is exactly balanced by the rate of its disappearance. In other words, chemical equilibrium is *dynamic* equilibrium. This can be demonstrated using isotopes: if a small quantity of ND_3 (where D is deuterium, ^2H – see Chapter 16) is added to the equilibrium mixture in place of some NH_3, the equilibrium composition of the mixture is not significantly altered, yet in due course the mass spectrum of a sample will show that HD, D_2 and molecules such as NDH_2 are present.

Furthermore, the equilibrium composition (at a given temperature) is the same whether we start with pure ammonia or with hydrogen and nitrogen. Similarly, a mixture of hydrogen and iodine reaches an equilibrium composition in which hydrogen iodide is present, and the same equilibrium composition is reached if we start from pure hydrogen iodide and allow it to decompose (see Figure 10.1).

When we turn to a reaction that appears to 'go to completion' we find exactly the same type of behaviour. Such reactions also reach an equilibrium, a stationary but dynamic condition in which the forward and backward reaction rates balance, but for them the equilibrium composition lies strongly in favour of the products. Even in a reaction that apparently 'goes to completion' there is always a tiny amount of the reactants in the equilibrium mixture.

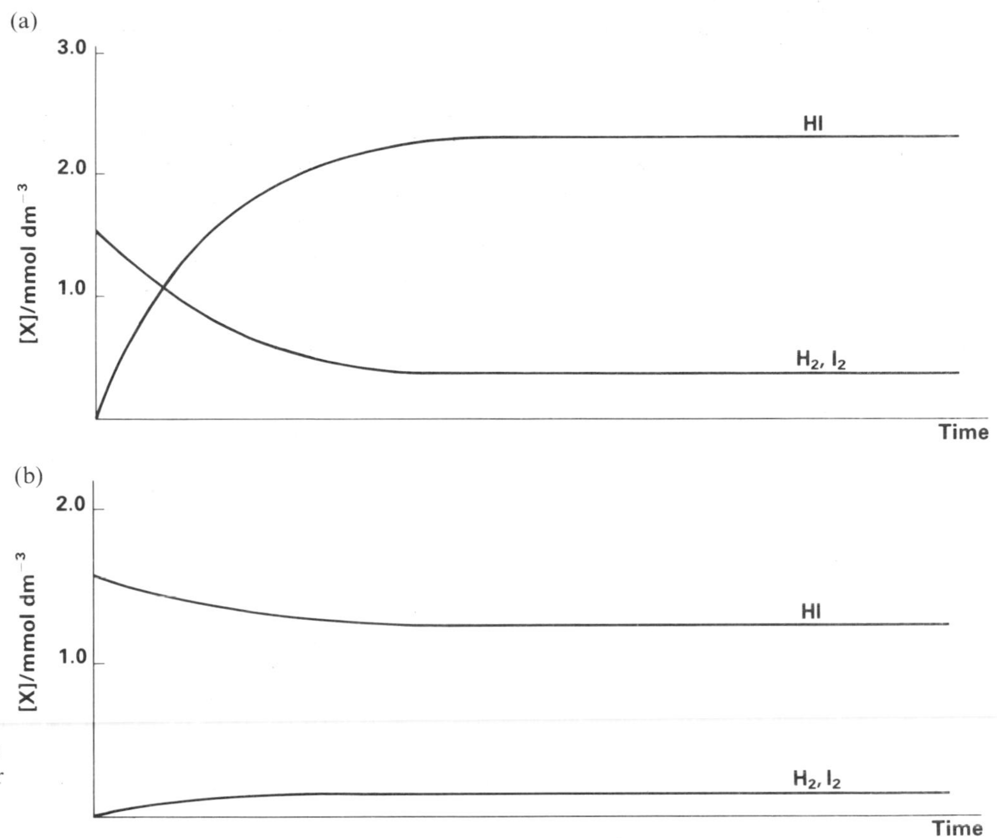

Figure 10.1 The same equilibrium (ratio of concentrations) is reached from either (a) pure H_2 and I_2 or (b) pure HI (at 764 K)

All reactions are the same in kind but different in degree; all reach dynamic equilibrium with both reactants and products present, but in some the equilibrium lies so strongly in favour of the products that we fail to notice the presence of the minute amounts of reactants.

The equilibrium constant

Many experiments have been carried out to measure the compositions of reaction mixtures at equilibrium. It has been discovered that the equilibrium composition for each reaction can be characterized by a single constant K. This **equilibrium constant** can be measured experimentally, and its value can also be predicted using thermodynamics (see Section 11.3). Once its value is known for a reaction, it is possible to do a variety of useful calculations.

Consider first a special case, the reaction of ethanoic acid with ethanol (this kind of reaction is called esterification – see Section 31.4). The reaction is illustrated in Figure 10.2, and is written

$$CH_3COOH + C_2H_5OH \rightleftharpoons CH_3COOC_2H_5 + H_2O$$

where the symbol $\rightleftharpoons$ signifies an equilibrium.

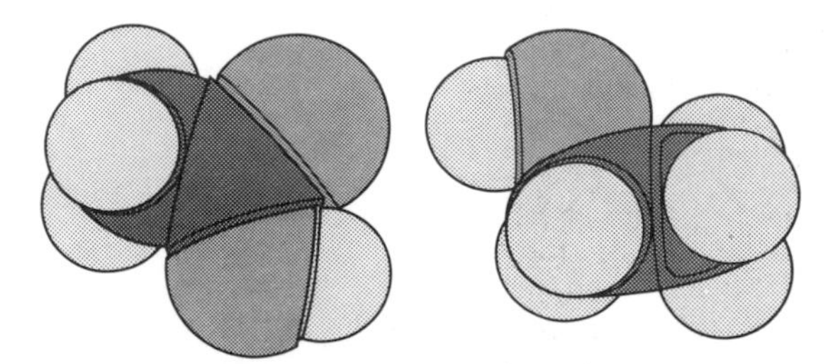

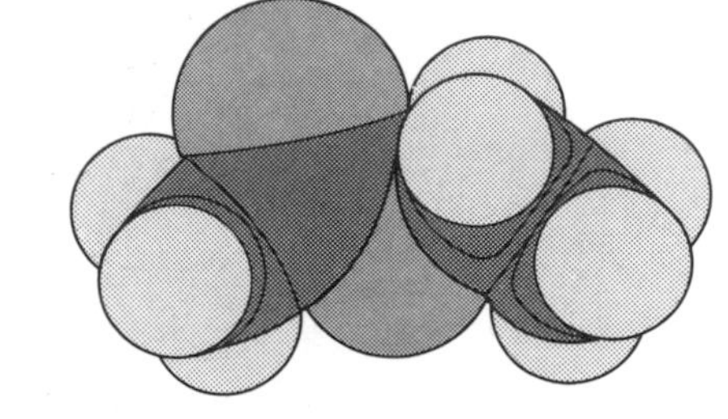

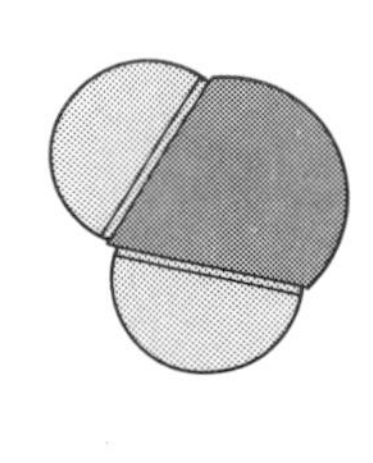

Figure 10.2 A molecular model of the esterification of ethanoic acid

This is an example of a **homogeneous equilibrium**, since all the substances are in the same phase (liquid). From the data in Table 10.1 it can be seen how the equilibrium composition of the reaction mixture depends on the relative amounts of acid and alcohol mixed initially. Although the information may seem to have no pattern, one does emerge if we look more closely. The fifth column in Table 10.1 gives the value of

$$K_c = \left\{\frac{[CH_3COOC_2H_5][H_2O]}{[CH_3COOH][C_2H_5OH]}\right\}_{eq} \tag{10.1.1}$$

where [X] is the concentration of substance X and the subscript 'eq' denotes that the concentrations are those at equilibrium. The subscript 'c' on K_c is used when the constant is expressed in terms of concentrations (as distinct from partial pressures). Within experimental error, K_c has about the same value, 4.0 at 100 °C, whatever the initial composition of the mixture. K_c therefore characterizes the equilibrium composition of this reaction at this temperature.

Table 10.1 The esterification of ethanoic acid at 100 °C

		Amounts of substance		
n_{acid}/mol at start	$n_{alcohol}$/mol at start	n_{ester}/mol at equilibrium	n_{water}/mol at equilibrium	K_c
1.00	0.18	0.171	0.171	3.9
1.00	0.50	0.420	0.420	3.8
1.00	1.00	0.667	0.667	4.0
1.00	2.00	0.858	0.858	4.5
1.00	8.00	0.966	0.966	3.9

The experimental measurement of the equilibrium constant thus depends on the measurement of concentrations. For the esterification reaction, measured quantities of ethanoic acid and ethanol are mixed together and a measured quantity of concentrated sulphuric acid is added to catalyse the reaction. The reactants are left to equilibrate (that is, to reach equilibrium) in a constant-temperature bath (because the equilibrium constant depends on the temperature), and then the acid concentration is measured by titration against standard ('of known concentration') sodium hydroxide with phenolphthalein as indicator (see Section 12.3). When allowance has been made for the sulphuric acid catalyst, the ethanoic acid concentration can be deduced. The concentrations of the other substances are then calculated.

EXAMPLE

The analysis of the product composition of the equilibrium mixture of the esterification reaction showed that it contained 0.67 mol CH_3COOH, 0.82 mol C_2H_5OH and 0.73 mol H_2O. What amount of substance of the ester $CH_3COOC_2H_5$ is present?

METHOD At equilibrium the concentrations of the four substances are related by equation 10.1.1 with $K_c = 4.0$ at 100 °C. The volumes in the concentration terms all cancel, and so the amount of ester, n, may be obtained by using the expression in the form

$$n = \left\{\frac{n(CH_3COOH) \times n(C_2H_5OH)}{n(H_2O)}\right\} \times K_c$$

ANSWER Inserting the data into the expression just quoted gives

$$n = \left\{\frac{0.67\,\text{mol} \times 0.82\,\text{mol}}{0.73\,\text{mol}}\right\} \times 4.0 = 3.0\,\text{mol}$$

COMMENT If the reaction had been run in a large quantity of water so that the equilibrium composition corresponded to 0.67 mol CH_3COOH and 0.82 mol C_2H_5OH in 1 dm^3 water (55.5 mol H_2O), the same calculation leads to $n = 0.040$ mol, a huge decrease in ester (as a result of hydrolysis – see Section 31.4). The equilibrium constant is the same, but the individual compositions have adjusted. The equilibrium constant for this reaction barely changes with temperature between 25 °C and 100 °C. We will develop this Example later.

The general definition of the equilibrium constant, valid for any reaction, can be introduced in two steps. First, consider a homogeneous reaction where all the substances are in solution (the esterification reaction is an example). Then it has the form

$$aA + bB \rightleftharpoons cC + dD \qquad K_c = \left\{\frac{[C]^c[D]^d}{[A]^a[B]^b}\right\}_{eq} \tag{10.1.2}$$

For the esterification reaction, $a = b = c = d = 1$. We shall see more examples later.

Now consider a homogeneous reaction in the gas phase (such as the synthesis of ammonia). Although the equilibrium constant could be expressed in terms of the concentrations of the gases present at equilibrium (using the last equation), it is more natural to express it in terms of the equilibrium partial pressures (see Section 4.1) of the gases (and to denote it K_p):

$$aA(g) + bB(g) \rightleftharpoons cC(g) + dD(g) \qquad K_p = \left\{\frac{p_C{}^c p_D{}^d}{p_A{}^a p_B{}^b}\right\}_{eq} \tag{10.1.3}$$

As an example, consider the ammonia synthesis:

$$N_2(g) + 3H_2(g) \rightleftharpoons 2NH_3(g) \qquad K_p = \left\{\frac{p_{NH_3}{}^2}{p_{N_2}p_{H_2}{}^3}\right\}_{eq} \tag{10.1.4}$$

since $a = 1$, $b = 3$, $c = 2$ and $d = 0$. At 400 °C the value of this equilibrium constant is $1.8 \times 10^{-4}\ bar^{-2}$, and so we can begin to make predictions about the composition of the reaction mixture at equilibrium.

Using the equilibrium constant

The principal feature of the equilibrium constant is that it is *independent of the initial composition* of the reaction mixture, so that, at a fixed temperature, the same value of the constant describes the equilibrium composition whatever the initial mixture. However, because the equilibrium constant specifies the composition in the form of a ratio of concentrations or pressures, it is not always immediately obvious what the equilibrium composition will be. We will therefore illustrate how to extract the information by working through two examples.

Consider first the esterification reaction at equilibrium. If the amount of ester present is n, then the initial amounts of acid and alcohol must each have been reduced by an amount n. This can be set out clearly by drawing up the following table:

	Acid CH_3COOH	Alcohol C_2H_5OH	Ester $CH_3COOC_2H_5$	Water H_2O
Initial amounts of substance	n_{ac}	n_{al}	0	0
Equilibrium amounts of substance	$n_{ac} - n$	$n_{al} - n$	n	n
Concentration at equilibrium	$\frac{n_{ac} - n}{V}$	$\frac{n_{al} - n}{V}$	$\frac{n}{V}$	$\frac{n}{V}$

Substituting the equilibrium concentrations into the expression for the equilibrium constant (equation 10.1.1) gives

$$K_c = \frac{(n/V)(n/V)}{[(n_{ac}-n)/V][(n_{al}-n)/V]} = \frac{n^2}{(n_{ac}-n)(n_{al}-n)} \qquad (10.1.5)$$

This can be solved for n (it can be turned into a quadratic equation), and so the equilibrium composition can be predicted simply by feeding in the appropriate values of the initial composition and the equilibrium constant.

EXAMPLE

50.0 cm³ of ethanoic acid is mixed with 50.0 cm³ of ethanol. What is the equilibrium composition of the mixture at 100 °C?

METHOD Use equation 10.1.5. Find n_{ac} and n_{al} by converting the volumes to masses, using the densities 1.049 g cm^{-3} (acid) and 0.789 g cm^{-3} (alcohol), and then to amounts of substance using $M = 60.05\ \text{g mol}^{-1}$ (acid) and 46.07 g mol^{-1} (alcohol). Rearrange equation 10.1.5 into a quadratic equation in n, and solve using the standard formula for quadratic equations of the form $an^2 + bn + c = 0$; that is, $n = (1/2a)\{-b \pm \sqrt{(b^2 - 4ac)}\}$ and select the correct root. Work out the amounts of substance by calculating $n_{ac} - n$, $n_{al} - n$, n, n for the amounts of acid, alcohol, ester and water present at equilibrium. Use $K_c = 4.0$.

ANSWER $$n_{ac} = \frac{50.0\ \text{cm}^3 \times 1.049\ \text{g cm}^{-3}}{60.05\ \text{g mol}^{-1}} = 0.873\ \text{mol}$$

$$n_{al} = \frac{50.0\ \text{cm}^3 \times 0.789\ \text{g cm}^{-3}}{46.07\ \text{g mol}^{-1}} = 0.856\ \text{mol}$$

Equation 10.1.5 rearranges into

$$n^2 - (n_{ac} - n)(n_{al} - n)K_c = 0$$

or

$$(1 - K_c)n^2 + (n_{ac} + n_{al})K_c n - n_{ac}n_{al}K_c = 0$$

Therefore, for an equation of the form $an^2 + bn + c = 0$

$$a = 1 - K_c = 1 - 4.0 = -3.0$$

$$b = (n_{ac} + n_{al})K_c = (0.873\ \text{mol} + 0.856\ \text{mol}) \times 4.0 = 6.92\ \text{mol}$$

$$c = -n_{ac}n_{al}K_c = -(0.873\ \text{mol}) \times (0.856\ \text{mol}) \times 4.0 = -2.99\ \text{mol}^2$$

$$n = (1/2a)\{-b \pm \sqrt{(b^2 - 4ac)}\}$$

$$= \frac{(-6.92\ \text{mol}) \pm \sqrt{\{(6.92\ \text{mol})^2 - 4 \times (-3.0) \times (-2.99\ \text{mol}^2)\}}}{2 \times (-3.0)}$$

$$= \frac{-6.92\ \text{mol} \pm 3.47\ \text{mol}}{-6.0} = 0.58\ \text{mol or } 1.73\ \text{mol}$$

The second answer is impossible (because no more than 0.856 mol of product can be generated with the specified starting amounts), and so we conclude that at equilibrium $n = 0.58$ mol. It follows that the equilibrium composition is 0.58 mol $CH_3COOC_2H_5$, 0.30 mol CH_3COOH, 0.28 mol C_2H_5OH and 0.58 mol H_2O.

COMMENT The amount of acid is found by evaluating n to *three* significant figures, and subtracting from n_{ac} before returning the final answer to two significant figures.

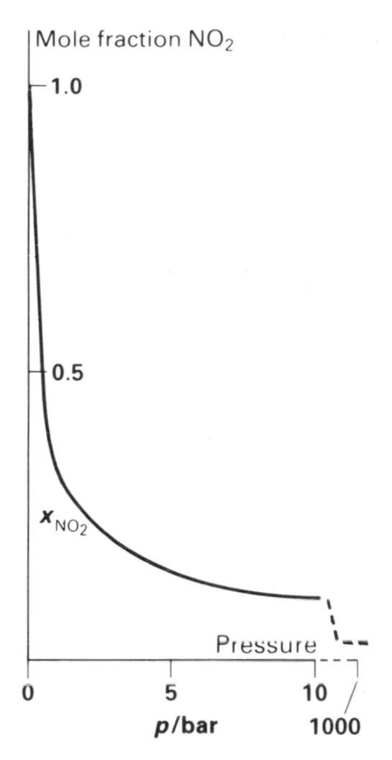

Figure 10.3 The mole fraction of NO_2 in the $N_2O_4 \rightleftharpoons 2NO_2$ equilibrium at 25 °C

As a second example, consider the gas-phase equilibrium $N_2O_4(g) \rightleftharpoons 2NO_2(g)$. Since NO_2 is strongly coloured (brown) and N_2O_4 is colourless (see Section 21.3), it is easy to follow the composition of the vapour spectroscopically. The reaction equilibrium is

$$N_2O_4(g) \rightleftharpoons 2NO_2(g) \qquad K_p = \left\{\frac{p_{NO_2}{}^2}{p_{N_2O_4}}\right\}_{eq} \tag{10.1.6}$$

It follows from the equation that for each N_2O_4 molecule that dissociates two NO_2 molecules are formed. The same kind of calculation used for the esterification equilibrium leads to the result that the mole fraction of NO_2 depends on the total pressure, as shown in Figure 10.3. This shows that as the total pressure is increased, the mole fraction of NO_2 decreases, and at very high pressures there is very little present at all. It is as though the NO_2 molecules have been squashed together into N_2O_4 molecules.

EXAMPLE

Calculate the partial pressure of nitrogen dioxide when 1 mol N_2O_4 reaches equilibrium in the reaction $N_2O_4(g) \rightleftharpoons 2NO_2(g)$ at 25 °C at a total pressure of 1.000 bar, given that $K_p = 0.148$ bar.

METHOD Express the equilibrium constant in terms of partial pressures. Calculate the mole fractions of the two components by finding the amount of substance of NO_2 formed when y mol N_2O_4 dissociates, and hence find the mole fractions of NO_2 and N_2O_4 present in a mixture in which that extent of dissociation has occurred. Use the mole fractions, expressed in terms of y, to relate the partial pressures to the total pressure, express K_p in terms of y, and solve for y at equilibrium. Use this equilibrium y in the expression for the partial pressure of NO_2.

ANSWER The equilibrium constant is

$$K_p = \left\{\frac{p_{NO_2}{}^2}{p_{N_2O_4}}\right\}_{eq}$$

Since y mol N_2O_4 results in $2y$ mol NO_2, the mole fractions present when 1 mol N_2O_4 dissociates are

$$x_{NO_2} = \frac{2y}{1 - y + 2y} = \frac{2y}{1 + y}$$

$$x_{N_2O_4} = \frac{1 - y}{1 + y}$$

Hence the partial pressures (see Section 4.1) are

$$p_{NO_2} = x_{NO_2}p = \frac{2yp}{1 + y}$$

$$p_{N_2O_4} = x_{N_2O_4}p = \frac{(1 - y)p}{1 + y}$$

The equilibrium equation is therefore

$$K_p = \left\{\frac{p_{NO_2}{}^2}{p_{N_2O_4}}\right\}_{eq} = \left\{\frac{4y^2p}{1 - y^2}\right\}_{eq}$$

On rearranging for y^2 and taking the square root,

$$y_{eq} = \sqrt{\left(\frac{K_p}{K_p + 4p}\right)}$$

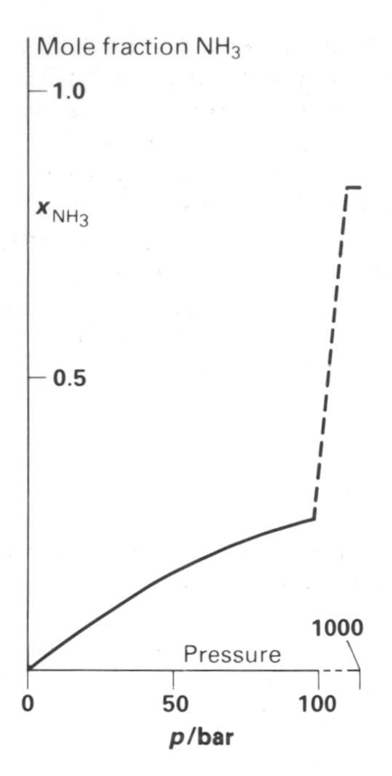

Figure 10.4 The mole fraction of NH_3 in the $N_2 + 3H_2 \rightleftharpoons 2NH_3$ equilibrium at 400 °C

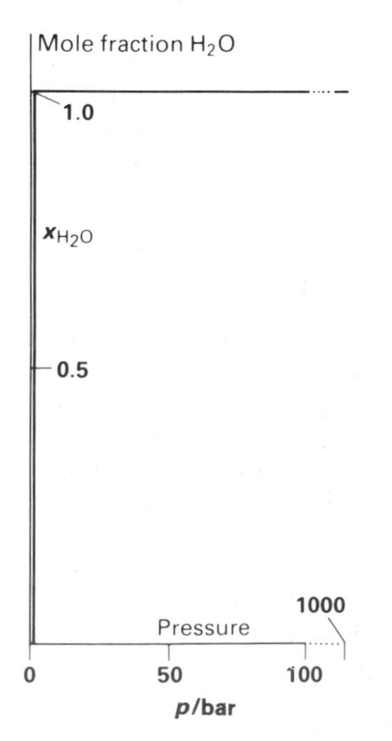

Figure 10.5 The mole fraction of H_2O in the $O_2 + 2H_2 \rightleftharpoons 2H_2O$ equilibrium at 400 °C

and on substituting the data,

$$y_{eq} = \sqrt{\left(\frac{0.148}{4.148}\right)} = 0.189$$

Hence, at equilibrium,

$$\{p_{NO_2}\}_{eq} = \frac{2 \times 0.189 \times 1.000\,\text{bar}}{1 + 0.189} = 0.318\,\text{bar}$$

COMMENT Although the equilibrium constant is independent of the total pressure, the individual partial pressures do depend on the total pressure. This can be confirmed by substituting different values of p into the expression derived above.

The similar but contrasting cases of the syntheses of ammonia and water can be dealt with in the same way. If nitrogen and hydrogen are present in the mole ratio 1:3 and this mixture is allowed to come to equilibrium, then the mole fraction of ammonia depends on the total pressure of the mixture, as shown in Figure 10.4. In contrast to the $N_2O_4 \rightleftharpoons 2NO_2$ equilibrium, an increase in pressure drives the ammonia synthesis equilibrium to the right. A pattern is beginning to emerge.

When the technique is applied to the water synthesis reaction under conditions such that all the water produced is a vapour, its mole fraction in the equilibrium mixture depends on the total pressure, as shown in Figure 10.5.

These results show that the differences between the equilibrium compositions of the ammonia and water synthesis reactions are of degree rather than of kind. First, the pressure dependence of the equilibrium is similar in the two cases; the curves drawn in Figures 10.4 and 10.5 differ, but both rise with increasing pressure. Quantitatively, though, at any given pressure the equilibrium compositions are very different. When the total pressure is 100 bar, the mole fraction of ammonia at equilibrium is 0.25, a readily measurable but not a dominating quantity. In contrast, under the same conditions, the mole fraction of water is so close to 1 that, for all practical purposes, only water vapour is present. In other words, the synthesis equilibrium lies slightly in favour of the reactants in the case of ammonia, but overwhelmingly in favour of the product in the case of water: qualitatively the reactions are similar, but quantitatively they are very different.

Equilibria involving solids

When a chemical reaction involves substances in different phases, such as a solid decomposing into another solid plus a gas, the stationary composition is called a **heterogeneous equilibrium**. An example is the decomposition of calcium carbonate into calcium oxide plus carbon dioxide:

$$CaCO_3(s) \rightleftharpoons CaO(s) + CO_2(g) \qquad K = \left\{\frac{[CaO]p_{CO_2}}{[CaCO_3]}\right\}_{eq}$$

The equilibrium constant is expressed in terms of the concentrations of the solids and the pressure of the gas. But what is meant by the concentration of a pure solid? We express concentration as *amount of substance per unit volume* (which in practice means mol dm^{-3}), but for a pure solid, amount of substance per unit volume is proportional to mass per unit volume, or density. The density of a solid is independent of how much of it is present, therefore the 'concentration' of a pure solid is also a constant. It is obviously sensible to combine any constants with K itself, and so from now on we will write

$$CaCO_3(s) \rightleftharpoons CaO(s) + CO_2(g) \qquad K_p = \{p_{CO_2}\}_{eq} \qquad (10.1.7)$$

Similar arguments apply to any heterogeneous equilibrium involving solids or pure liquids.

At 800 °C, $K_p = 22$ kPa. Therefore when calcium carbonate is heated to 800 °C in a *closed* vessel and allowed to come to equilibrium the pressure of carbon dioxide is 22 kPa. In this case K_p is the **decomposition pressure** of calcium carbonate at 800 °C. This shows that we can easily measure the equilibrium constant by noting the pressure exerted by a sample heated to the temperature of interest. When calcium carbonate is heated to 800 °C in an *open* vessel it will continue to decompose until the partial pressure of carbon dioxide reaches its equilibrium pressure, 22 kPa. However, since the gas is free to escape, the local partial pressure will never reach that value, because the carbon dioxide simply disperses into the surroundings. Therefore more carbonate decomposes, and the reaction continues until only calcium oxide remains.

10.2 The effect of the conditions

In this section we look at some of the ways of controlling the composition at equilibrium. This is enormously important when thinking about the design of an industrial plant or how a biological cell contributes to the activity of an organism. For instance, what is the pressure of oxygen in equilibrium with blood cells inside the lung? Is it better to run the ammonia synthesis reaction at a high temperature? Since the Haber–Bosch synthesis of ammonia (see Section 21.7 and Box 10.1) is now the starting point of the fertilizer industry, there is obviously great human as well as economic benefit to be obtained from maximizing the yield.

BOX 10.1

The Haber–Bosch process: a landmark in chemical technology

The first industrial plant for the synthesis of ammonia went on stream at the works of the Badische Anilin und Soda Fabrik (BASF) in the German town of Ludwigshafen-Oppau, near Mannheim, in 1913 and went out of operation in 1982. It had a capacity of 30 tonnes a day, and used the then technologically novel process of passing high-pressure gases over a heated catalyst. Synthetic ammonia plants have since been built all over the world, but the capacity of a plant is now typically 1500 tonnes a day. The source of the hydrogen has also changed: originally it was obtained from the electrolysis of water, but now it is commonly obtained from natural gas (see Section 16.4). Haber used osmium as a catalyst, but now iron is used, as it is much cheaper.

This crucially important chemical process is named after the two people whose collaboration made it possible. Fritz Haber (1868–1934) was an academic chemist who saw the achievement of the synthesis as an intellectual challenge. He was encouraged by Walther Nernst, who already had a reputation in thermodynamics, and despite many setbacks his high-pressure system finally yielded 100 g of ammonia in 1909. The task of scaling up this laboratory process fell to Carl Bosch (1874–1940), a chemical engineer employed by BASF. Both Haber and Bosch were awarded Nobel prizes for chemistry, Haber in 1918 for his fundamental contribution, and Bosch in 1931 for his invention and development of high-pressure technology.

The Haber–Bosch process came just in time for Germany, for the country was shortly to be cut off by the First World War from the main natural source of nitrogen compounds, which at that time was sodium nitrate from Chile. With synthetic ammonia Germany could make fertilizer to increase food production and negate the Allied blockade, and nitric acid for the production of explosives. Now the main uses of ammonia are for fertilizers and nitric acid, and as a raw material for the production of nylon.

Le Chatelier's principle

A simple rule lets us predict, with reasonable confidence, what happens to the position of chemical equilibrium when the conditions are changed. It was formulated by Henri Le Chatelier (in 1888) on the basis of experimental observations.

Le Chatelier's principle: When a reaction at equilibrium is subjected to a change in conditions, the composition tends to adjust so as to minimize the change.

We shall apply the principle to the three basic types of change: changes of concentration, of pressure and of temperature. We first show how the principle lets us make *qualitative* predictions, and then go on to justify the principle in terms of the equilibrium constant. This second step also allows us to make *quantitative* predictions. This is a principal feature of science: it is basically common sense equipped with numbers.

How equilibria respond to changes of concentration

The central point to remember is that *the value of the equilibrium constant is independent of the initial concentrations.* This means that if the conditions are changed, the composition will always adjust so as to keep the value of K_c constant.

Consider the effect of adding water on the esterification equilibrium

$$CH_3COOH + C_2H_5OH \rightleftharpoons CH_3COOC_2H_5 + H_2O$$

According to Le Chatelier's principle, the equilibrium tends to shift so as to minimize the increase in the water concentration. Therefore it predicts that the equilibrium will tend to shift to the left, and the ester will tend to be hydrolysed.

This prediction is made quantitative using the equilibrium constant,

$$K_c = \left\{\frac{[CH_3COOC_2H_5][H_2O]}{[CH_3COOH][C_2H_5OH]}\right\}_{eq} = 4.0 \quad \text{at } 100\,°C$$

In order to maintain the ratio of concentrations constant when the concentration of water is increased (so that the numerator increases), the ester concentration must fall, and the denominator must increase. In other words, the equilibrium tends to shift back towards the reactants, in accord with Le Chatelier's principle. Since the ratio of concentrations must continue to equal 4.0, we can also state *quantitatively* the effect of the addition of a known amount of water.

How equilibria respond to pressure

In this connection it is important to remember that, since increasing the total pressure is equivalent to increasing the initial concentrations of the gases, *the value of the equilibrium constant is also independent of the total pressure.* We limit the discussion to reactions involving gases, because only they respond significantly to changes of pressure (except at very high pressures, as under geological conditions). According to Le Chatelier's principle, a reaction at equilibrium responds to an increase of pressure by minimizing the increase.

Take as an example the synthesis of ammonia:

$$N_2(g) + 3H_2(g) \rightleftharpoons 2NH_3(g) \qquad K_p = \left\{\frac{p_{NH_3}{}^2}{p_{N_2}p_{H_2}{}^3}\right\}_{eq} \qquad (10.2.1)$$

Two molecules of ammonia are produced from four molecules of gaseous reactants (one N_2 and three H_2). If the reaction is in a constant-volume vessel, the reduction of the number of molecules caused by the forward reaction results in a decrease in the pressure. Therefore (according to Le Chatelier's principle) if the pressure is increased, the equilibrium responds by tending to shift towards ammonia. It is important to stress, however, that the equilibrium constant itself does not change with pressure: the individual partial pressures adjust to keep the value of K_p constant.

This can be demonstrated as follows. From Dalton's law (equation 4.1.9), the partial pressure of each gas, p_A, is equal to $x_A p$, where x_A is the mole fraction of the gas in the reaction mixture. Hence,

$$K_p = \left\{\frac{(x_{NH_3}p)^2}{(x_{N_2}p)(x_{H_2}p)^3}\right\}_{eq}$$

$$= \left\{\frac{x_{NH_3}{}^2}{x_{N_2}x_{H_2}{}^3}\right\}_{eq} \times \frac{1}{p^2}$$

This shows that when the total pressure p is increased, the mole fraction term must also increase in order to keep K_p constant. That is, x_{NH_3} increases and x_{N_2} and x_{H_2} decrease when the total pressure is increased.

Figure 10.4 expresses this conclusion quantitatively. It shows that the gas becomes richer in ammonia as the pressure is increased. Therefore, the yield from an ammonia synthesis plant can be improved by operating it at high pressures. The reverse applies to the N_2O_4 dissociation, for the forward reaction increases the number of molecules, and therefore corresponds to an increase in pressure. Le Chatelier's principle therefore predicts that an increase of pressure favours a shift of the equilibrium position to the left, in favour of N_2O_4.

A general reaction can now be treated very easily. If the number of gas-phase molecules on the right of the equation is smaller than the number on the left, then an increase of pressure results in the equilibrium tending to shift to the right, because in that way the system tends to minimize the increase of pressure. If there are fewer molecules on the left an increase in pressure tends to shift the equilibrium to the left. This can be summarized as follows:

When the pressure is increased, an equilibrium involving gases tends to shift as shown:

$$\text{reactants} \underset{\text{fewer on left}}{\overset{\text{fewer on right}}{\rightleftharpoons}} \text{products}$$

How equilibria respond to temperature

According to Le Chatelier's principle, an increase in the temperature of an equilibrium mixture tends to shift the position of equilibrium in the direction that minimizes the increase. The explanation is different in this case because the equilibrium constant *does* depend on the temperature, and so the concentrations adjust until they reach its new value.

Only changes in *temperature* affect the value of the equilibrium constant. Changes in concentration and pressure affect only the composition at equilibrium (the individual values of the concentration or partial pressure), not the value of the equilibrium constant itself, and so they bring about compensating changes in the numerator and the denominator – see Table 10.2.

Table 10.2 The effect of conditions

Condition altered	Equilibrium constant	Equilibrium composition
concentration	unchanged	changed
pressure	unchanged	changed
temperature	changed	changed

Suppose the reaction is endothermic ($\Delta_r H^\ominus$ positive, see Section 9.4), then an increase in the temperature is predicted to tend to shift the equilibrium towards products, because that results in the absorption of energy and tends to reduce the temperature rise. On the other hand, if the reaction is exothermic, an increase in temperature is predicted to favour the formation of reactants, because this involves an absorption of energy and therefore tends to

(a)

(b)

Figure 10.6 Effect of temperature on NO_2 dimerization: at room temperature (a), the proportion of NO_2 is larger than when the flask is cooled in ice (b), and the sample is darker

reduce the temperature rise. This behaviour can be summarized as follows:

Endothermic reactions: a rise in temperature favours products (*K increases*).

Exothermic reactions: a rise in temperature favours reactants (*K decreases*).

The $N_2O_4 \rightleftharpoons 2NO_2$ reaction is endothermic in the forward direction (like all dissociations, it requires energy). Therefore we can predict that on increasing the temperature the dissociation equilibrium will shift to the right in favour of nitrogen dioxide. In practice this means that the colour of the sample deepens as the temperature is raised, because there is then a higher proportion of the brown NO_2 molecules (see Figure 10.6).

The ammonia synthesis (equation 10.2.1), is exothermic ($\Delta_r H^\ominus = -92.2\ \mathrm{kJ\,mol^{-1}}$ at 25 °C). We can therefore predict that the equilibrium will tend to shift in favour of the *reactants* when the temperature is raised. It follows that the equilibrium yield from an ammonia synthesis plant can be maximized by working at low temperatures (and, as we saw in the last section, at high pressures). This information immediately presents a problem: the synthesis reaction is very slow at low temperatures, and so although the equilibrium then favours the product, it is formed extremely slowly. The answer is to use a **catalyst** (a substance that speeds up a chemical reaction without being used up in the process; we deal with them in more detail in Section 14.5). *Catalysts do not affect the position of equilibrium, but they do affect the rate at which it is attained.* In the Haber–Bosch ammonia synthesis the catalyst is iron, together with a mixture of metal oxides. The synthesis is run at a moderate temperature (about 450 °C) so that the rate is acceptable and the equilibrium yield is not too low, and at about 250 atm, which favours ammonia production but is not so high as to require expensive very-high-pressure plant.

The quantitative dependence of the equilibrium constant on temperature is calculated using the **van't Hoff isochore**:

$$\ln K = \ln K' + \frac{\Delta_r H^\ominus}{R}\left(\frac{1}{T'} - \frac{1}{T}\right) \tag{10.2.2}$$

where K is the equilibrium constant at a temperature T and K' its value at T'. This equation also shows that the standard reaction enthalpy can be obtained by plotting $\ln K$ against $1/T$, for the slope of this line is $-\Delta_r H^\ominus/R$.

10.3 Equilibria between phases

Closely related to chemical equilibria are **distribution equilibria**, in which a substance dissolves to different extents in two or more phases. These can also be described in terms of Le Chatelier's principle and, for quantitative predictions, in terms of an equilibrium constant.

Gases in liquids

Consider the distribution of a gas between a solvent and the gas phase itself. We limit attention to cases where the gas does not react with the solvent, and so while we include oxygen dissolving in water or in animal fat, we exclude chlorine dissolving in water (see Section 23.2). Gas solubilities usually decrease with increasing temperature; at higher temperatures the solvent molecules in effect shake the gas out of solution. Helium has a very low solubility; that is one of the reasons why it is used instead of nitrogen in deep-sea diving, for less dissolves in the bloodstream and there is less likelihood of the 'bends', the formation of small bubbles of gas in the blood as the diver comes to the surface.

The mass of gas that dissolves depends on its partial pressure. Le Chatelier's principle predicts that increased pressure should push the gas into solution. The behaviour observed for sparingly soluble gases is summarized by a law formulated by William Henry in the earliest days of physical chemistry; *Henry's law* (1803) states that *the mass of gas dissolved by unit volume of solvent is proportional to its partial pressure:*

$$\text{Henry's law: } m = K_{\text{H}}p \qquad (10.3.1)$$

K_{H}, a kind of equilibrium constant, depends on both the gas and the solvent. Once values of K_{H} have been measured, they are useful for discussing the dissolution of air and other gases in ponds, and in fluids such as blood plasma; some are listed in Table 10.3.

Table 10.3 Henry's law constants at 25 °C, $K_{\text{H}}/\text{g dm}^{-3}\,\text{Pa}^{-1}$

	Water	Benzene
N_2	1.8×10^{-7}	1.3×10^{-6}
O_2	4.0×10^{-7}	2.3×10^{-6}
CO_2	1.5×10^{-5}	4.3×10^{-5}

Partition equilibria

When a solute B is shaken with two immiscible liquids it dissolves in them to different extents. At equilibrium, which is called a **partition equilibrium**, the concentration of the solute will be $[\text{B}]_1$ in one liquid and $[\text{B}]_2$ in the other. The ratio of concentrations at equilibrium is called the **partition coefficient** (or **distribution coefficient**) for the system:

$$K_{\text{part}} = \left\{\frac{[\text{B}]_1}{[\text{B}]_2}\right\}_{\text{eq}} \qquad (10.3.2)$$

(The solute must be in the same molecular form in both solvents.) Polar solutes favour polar solvents. The partition coefficient reflects the energetics of the dissolution – energetically favourable interactions weight the partition equilibrium in favour of that solvent.

The preference of organic solutes for organic solvents is put to good use in **solvent extraction** (see Figure 10.7). A solvent such as ethoxyethane ('ether') is shaken with the aqueous reaction mixture in a **separating funnel**, and it extracts most of the organic material. On standing, two layers form, and the organic layer can be separated. The solvent is then distilled off or simply allowed to evaporate, leaving the organic material. Penicillin is extracted like this from aqueous solution using trichloromethane, and a similar technique is used to de-caffeinate coffee.

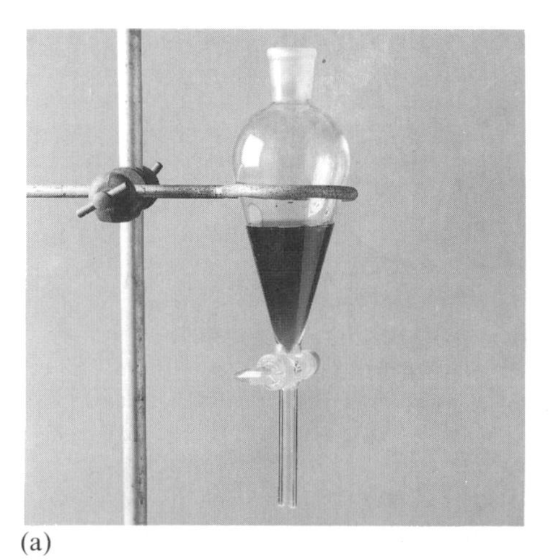

(a)

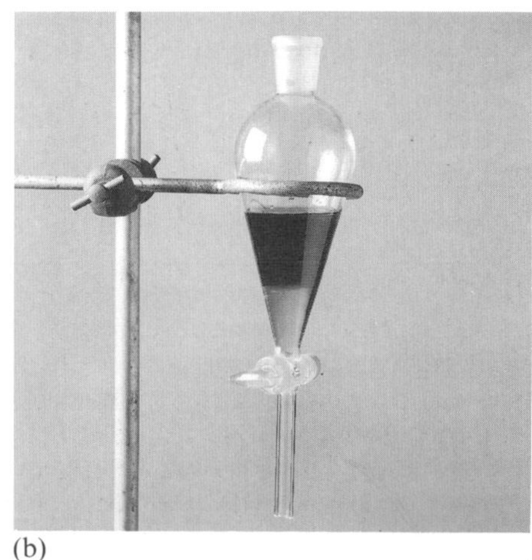

(b)

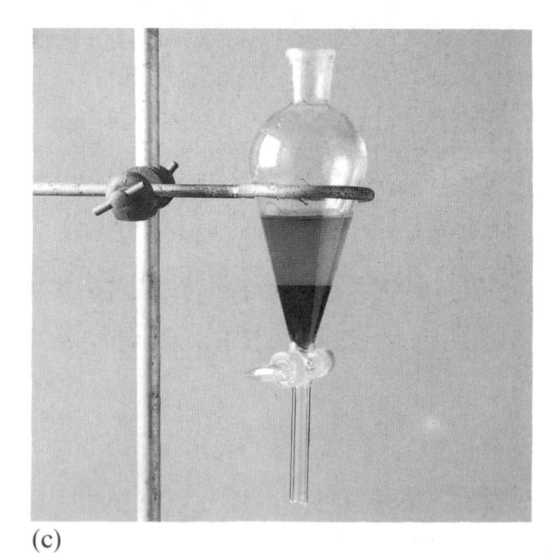

(c)

Figure 10.7 Solvent extraction in progress. The original solution (a) consists of iodine in dilute aqueous potassium iodide. A small volume of tetrachloromethane is carefully added, to give (b), and forms the lower layer. After the flask has been shaken (c), the iodine is mainly in the tetrachloromethane layer

EXAMPLE

Iodine was extracted from an aqueous solution by shaking with one-third its volume of tetrachloromethane. What proportion of the iodine will be present in the organic solvent?

METHOD The partition coefficient K_{part} refers to the ratio of concentrations. Let the total amount of iodine be n; then the amounts at equilibrium will be $(1-\alpha)n$ in water (1) and αn in the organic solvent (2). The volumes are V and $\frac{1}{3}V$, respectively. The concentrations at equilibrium are therefore $(1-\alpha)n/V$ and $3\alpha n/V$. Insert these values into equation 10.3.2 and solve for α. $K_{part} = 0.012$.

ANSWER The equilibrium is described by

$$K_{part} = \left\{\frac{[I_2]_1}{[I_2]_2}\right\}_{eq} = \frac{(1-\alpha)n/V}{3\alpha n/V} = \frac{1-\alpha}{3\alpha}$$

This rearranges to

$$\alpha = \frac{1}{1+3K_{part}} = \frac{1}{1+0.036} = 0.97$$

Therefore, 97 per cent of the iodine is in the tetrachloromethane.

COMMENT Oxidizing agents are often detected by adding acidified potassium iodide and looking for the formation of iodine by extracting it with tetrachloromethane, as in this Example. Notice the efficiency of the solvent extraction procedure; this is on account of the favourable van der Waals interactions between the solvent and the iodine. The extraction is less efficient from KI(aq) due to the formation of the I_3^-(aq) complex ion (see Section 23.4).

Chromatography

The important technique of chromatography depends on the distribution equilibrium between two phases, one stationary, the other mobile. The different partition coefficients, or adsorption characteristics, of substances are then exploited in order to separate and identify the components of a mixture.

In **paper chromatography** the stationary phase is water held on paper (see Figure 10.8) and the mobile phase is some solvent; the latter runs up or down the paper and the solutes are partitioned between it and the water. The solutes separate out according to their partition coefficients, those that favour the mobile phase being carried along the paper more quickly. The pigments in chlorophyll can be separated in this way. The paper acts only as a support for the stationary phase, and is often replaced by some other medium, such as a column filled with aluminium oxide or a similar material.

An important recent development in chromatography has been the introduction of **high-performance liquid chromatography** (h.p.l.c.). This is a variant of the older column chromatography, but with the diameter of the column so small (about 5 mm) that the solvent must be forced through it by pumping. Separations of enantiomers (see Section 32.6) have been achieved on optically active h.p.l.c. columns.

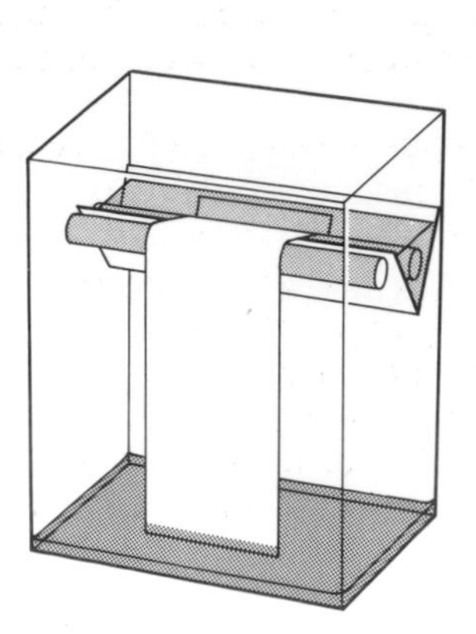

Figure 10.8 Paper chromatography

Thin-layer chromatography (t.l.c.) is a technique in which the stationary phase is trapped in a thin layer of the support medium, which is often aluminium oxide (Al_2O_3) spread on a glass plate or plastic film. As the solvent advances up the plate, the solutes are separated because of their differing preferences for the mobile or the stationary phases (see Colour 2). An important application of t.l.c. is to the detection of certain hormones in the urine as a sign of pregnancy.

In two-dimensional t.l.c. the chromatogram is first developed in one direction, then the solvent is changed and the chromatogram is developed in

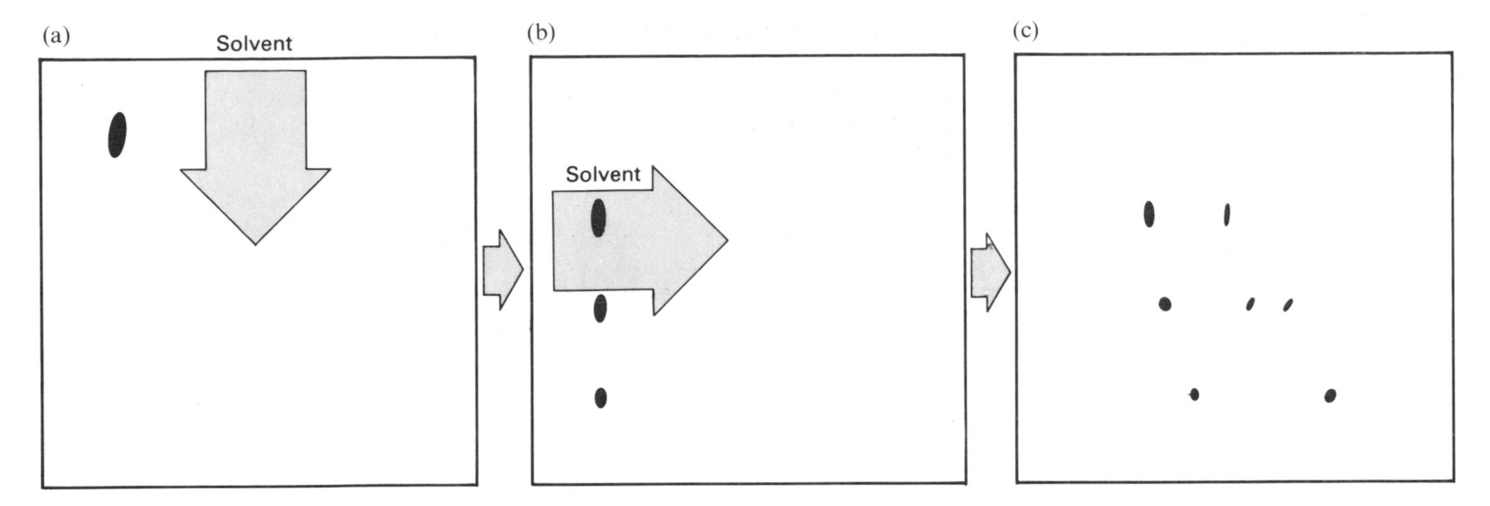

Figure 10.9 Two-dimensional thin-layer chromatography: (a) the first solvent partly separates, then (b) the second solvent produces (c) clearly separated spots

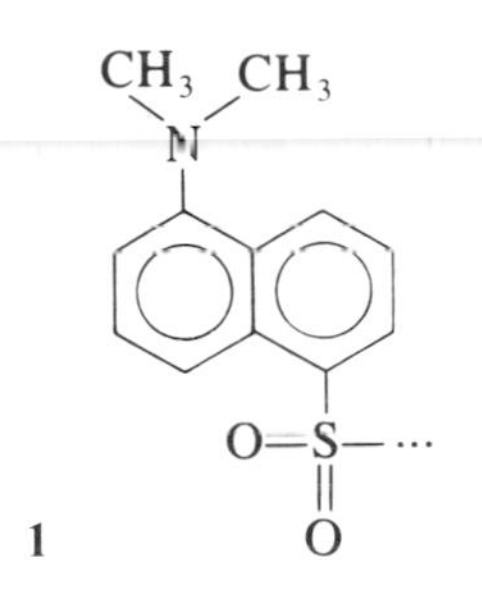

the perpendicular direction (see Figure 10.9). This technique is very useful for separating amino-acids (see Section 32.5), and with a clever choice of solvents the twenty naturally occurring amino-acids can be separated and identified. The technique is made more sensitive (as little as 1 nmol being detectable) in the **dansyl procedure** for sequencing proteins. The dansyl (**d**imethyl**amino**-**n**aphthalene**sulphonyl**) group (**1**) attaches to the free H_2N— group of the end amino-acid of a peptide or protein molecule, and fluoresces in ultraviolet light. The end amino-acid's identity can thus be found after hydrolysis of the protein and t.l.c. of the hydrolysate.

In **gas–liquid chromatography** (g.l.c.), the stationary phase is a liquid trapped on some finely divided inert support held in a long thin tube (see Figure 10.10). The mobile phase is an unreactive carrier gas such as nitrogen or helium. The sample to be analysed (about 1 mm^3 is sufficient) is injected into the gas stream just before it enters the stationary phase, and is swept through by the stream. The components of the mixture are distributed to different extents between the two phases, and so emerge at different times. In sophisticated instruments the detector on the outlet tube is a mass spectrometer (see Section 3.1 and Figure 10.11), and the sample can be analysed directly and almost automatically. The technique is very sensitive, and is used to detect small quantities of alcohol in blood and urine, or small quantities of explosives, as well as to separate and identify mixtures of similar compounds. Figure 10.12, for example, shows the **gas chromatograms** of lemon oil and lime oil; the small differences of composition account for the subtle shift of odour between the two oils. Whisky has been analysed by gas chromatography (Figure 10.13) but mixing the appropriate chemicals has not yet reproduced the taste, probably because the numerous minor components play an important role.

Ion-exchange resins are synthetic materials, often artificial zeolites (which are aluminosilicates with an open structure – see Section 20.3), which have the property of retaining different ions more or less strongly. When they are used in an **ion-exchange chromatography** column they can separate similar substances. Simply by washing the sample through the column with a solvent, the different substances are made to emerge at different times and may be collected separately. The time for a substance to be washed through a column is called its **elution time**; the washing process itself is **elution**. This technique is especially useful for the separation of the rare-earth (lanthanoid) elements and also of the actinoid elements.

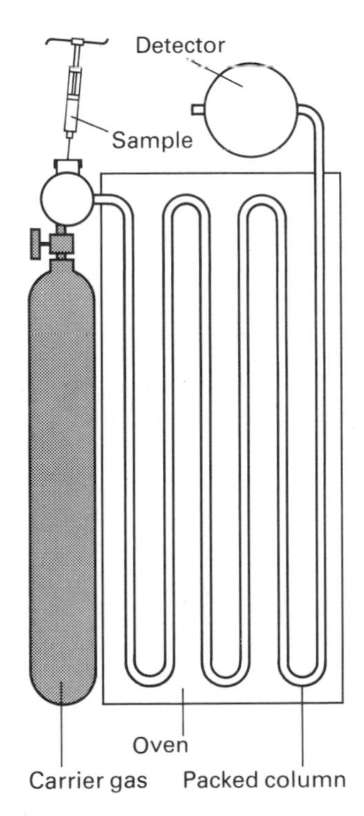

Figure 10.10 Gas–liquid chromatography

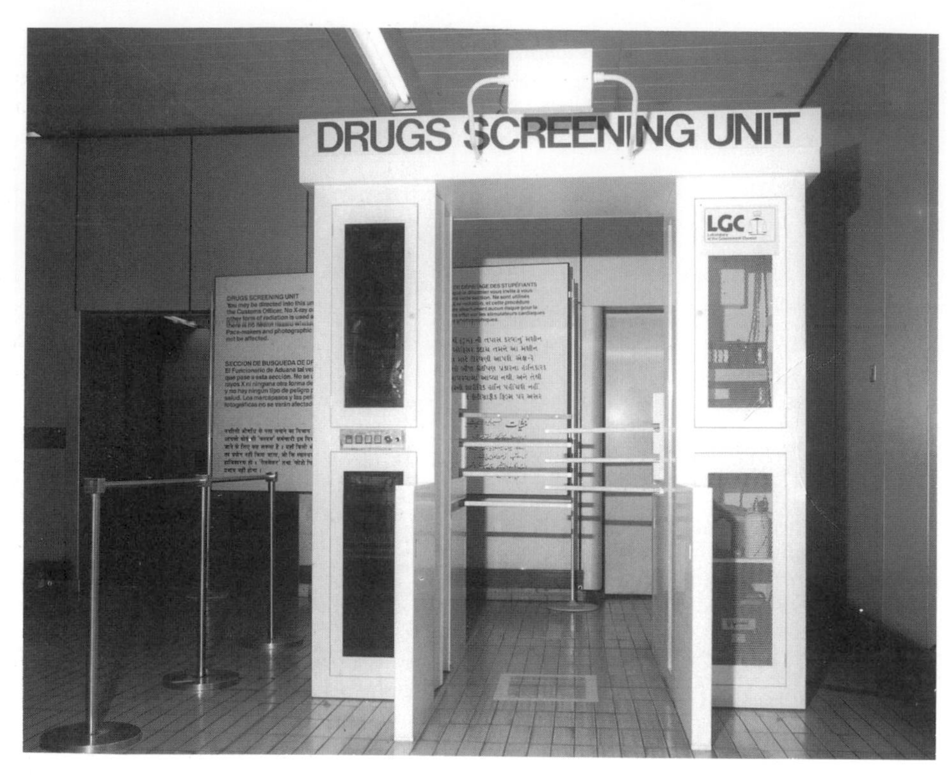

Figure 10.11 This airport 'drug sniffer' is based on a gas chromatograph connected to a mass spectrometer, and can detect tiny quantities of drugs

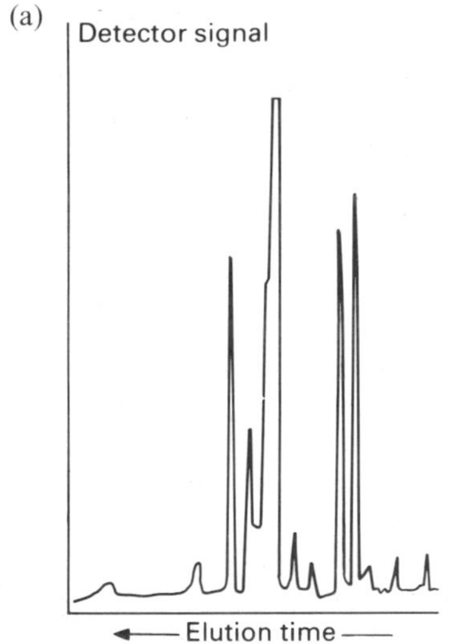

Figure 10.12 The gas chromatograms of (a) lemon oil, (b) lime oil

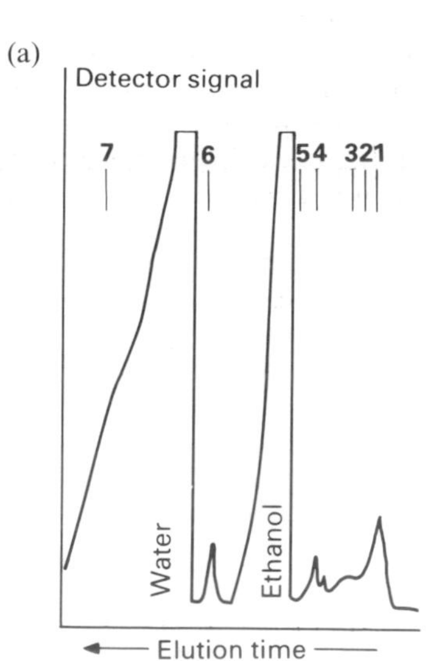

Figure 10.13 The gas chromatograms of (a) Scotch whisky, (b) bourbon whiskey (key: **1**, ethanal; **2**, methanal; **3**, ethyl methanoate; **4**, ethyl ethanoate; **5**, methanol; **6**, propan-1-ol; **7**, 3-methylbutan-1-ol)

Summary

- □ A chemical reaction is at equilibrium when the composition is stationary and has no further tendency to change.
- □ Chemical equilibria are dynamic equilibria; the forward and backward reactions continue, but there is no net change.
- □ The equilibrium constant characterizes the equilibrium composition of a reaction. K_c is defined in equation 10.1.2 and K_p, for gas-phase reactions, in equation 10.1.3.
- □ Equilibrium constants are independent of the concentrations and pressures of the substances involved in the reaction, but depend on the temperature.
- □ In homogeneous reactions all the substances are in the same phase; in heterogeneous reactions the substances are in more than one phase. The corresponding equilibria are called homogeneous equilibria and heterogeneous equilibria, respectively.
- □ The concentrations of pure solids are constant, and may be absorbed into the definition of the equilibrium constant for a heterogeneous reaction.
- □ The equilibrium constant for the decomposition of a solid to form a gas is the decomposition pressure of the solid at the temperature specified.
- □ Le Chatelier's principle states that when a reaction at equilibrium is subjected to a change in conditions, the composition tends to adjust so as to minimize the change.
- □ Le Chatelier's principle is used to make qualitative predictions; they can be interpreted and expressed quantitatively in terms of the equilibrium constant.
- □ If the concentration of a substance involved in an equilibrium is artificially increased, the reaction tends to adjust the composition so as to minimize the increase: K_c remains unchanged.
- □ If a reaction involving gases is subjected to an increase in pressure the reaction tends to adjust the composition so as to reduce the number of gas-phase molecules (and thereby to minimize the increase in pressure): K_p remains unchanged.
- □ Equilibria that do not involve gases are insensitive to moderate changes of pressure.
- □ An increase of temperature shifts the position of equilibrium in favour of products in the case of endothermic reactions, and in favour of reactants for exothermic reactions.
- □ The yield in the ammonia synthesis is increased by increasing the pressure and decreasing the temperature. Decreasing the temperature also has the result of slowing the rate at which equilibrium is reached. Therefore a catalyst is added.
- □ Catalysts affect the rates of reactions, not their positions of equilibrium.
- □ Henry's law states that the mass of gas dissolved by unit volume of solvent is proportional to its partial pressure (equation 10.3.1).
- □ The partition coefficient expresses the partition equilibrium of a solute between two solvents (equation 10.3.2).
- □ Chromatography (paper, high-performance liquid, thin-layer, gas–liquid and ion-exchange) is a technique for separating the components of a mixture making use of their different distribution equilibria between mobile and stationary phases.

PROBLEMS

1 Explain the nature of chemical equilibrium and describe the evidence showing that it is dynamic.

2 Write expressions for the equilibrium constants for the following reactions:

(a) $CdCO_3(s) \rightleftharpoons CdO(s) + CO_2(g)$
(b) $2SO_2(g) + O_2(g) \rightleftharpoons 2SO_3(g)$
(c) $2SO_3(g) \rightleftharpoons 2SO_2(g) + O_2(g)$
(d) $2H_2(g) + O_2(g) \rightleftharpoons 2H_2O(l)$
(e) $C_2H_5COOH(l) + CH_3OH(l) \rightleftharpoons C_2H_5COOCH_3(l) + H_2O(l)$

3 Explain why the concentration of a pure solid is a constant.

4 Explain why the equilibrium constant for the decomposition of a pure solid to a gas is equal to the decomposition pressure of the solid.

5 The equilibrium constant for the decomposition of cadmium carbonate, reaction (a) in Question 2, has the values 0.5 kPa at 550 K and 25 kPa at 600 K. Describe what happens when a sample of the solid is heated to 600 K in a vessel fitted with a piston which exerts a pressure of 10 kPa but which cannot be pushed out beyond some point, and then allowed to cool back to room temperature.

6 State Le Chatelier's principle. Why is it limited to states of dynamic equilibrium and does not apply to static equilibria (for example, a pencil balanced on its point)?

7 State what direction is favoured when the reactions in Question 2 are subjected to an increase of pressure.

8 Is the water synthesis equilibrium shifted in favour of products by an increase of temperature?

9 An inventor files a patent for a substance he calls *Anticat* which, he claims, eliminates the tendency of hydrogen and oxygen to combine to form water. Comment on the plausibility of the claim. In what sense might he be correct, and how should he reword his claim?

10 A steel vessel containing ammonia, hydrogen and nitrogen is at equilibrium at 1000 K. Analysis of the contents shows that the concentrations are $[NH_3] = 0.142\ mol\ dm^{-3}$, $[H_2] = 1.84\ mol\ dm^{-3}$ and $[N_2] = 1.36\ mol\ dm^{-3}$. Calculate the value of K_c for the reaction $N_2(g) + 3H_2(g) \rightleftharpoons 2NH_3(g)$ at this temperature.

11 Calculate the value of K_p for the reaction in the last question, on the assumption that the gases behave perfectly.

12 The reaction $2H_2S(g) \rightleftharpoons 2H_2(g) + S_2(g)$ has $K_c = 1.06 \times 10^{-6}\ mol\ dm^{-3}$ at 750 °C. What amount of substance of $S_2(g)$ (expressed as moles of S_2) will be present at equilibrium in a 5.00 dm^3 vessel in which there are known to be 2.21 mol $H_2S(g)$ and 1.17 mol $H_2(g)$?

13 At 2200 K and a total pressure of 1 atm, steam is 1.18 per cent dissociated into hydrogen and oxygen. Calculate K_p for the reaction $2H_2O(g) \rightleftharpoons 2H_2(g) + O_2(g)$.

14 When molecular iodine is dissolved in an aqueous solution of potassium iodide, the following equilibrium is established:

$$I_3^- \rightleftharpoons I_2 + I^-$$

[The equilibrium constant is 0.0015 mol dm^{-3} at 298 K.]
0.02 mole of iodine is dissolved in 500 cm^3 of a solution containing 0.2 mol dm^{-3} of potassium iodide at 298 K. Calculate the concentration of each ion present at equilibrium.
The above equilibrium may be investigated by partition between two solvents, a method which can sometimes be used to determine the molecularity of a compound in solution. Benzenecarboxylic acid (benzoic acid) is more soluble in benzene than in water. In benzene it exists as double molecules. Explain how you could confirm this fact experimentally using a partition method, giving important experimental details and the relevant theoretical background. *(Welsh S)*

15 Write a *short account* of the factors affecting the position of equilibrium of a balanced reaction, the rate at which equilibrium is attained and the value of the equilibrium constant.
(a) Using partial pressures, show that for gaseous reactions of the type

$$XY(g) \rightleftharpoons X(g) + Y(g)$$

at a given temperature, the pressure at which XY is exactly *one-third* dissociated is numerically equal to *eight* times the equilibrium constant at that temperature.
(b) When one mole of ethanoic acid (acetic acid) is maintained at 25 °C with 1 mole of ethanol, one-third of the ethanoic acid remains when equilibrium is attained. How much would have remained if one-half of a mole of *ethanol* had been used instead of one mole at the same temperature? *(SUJB)*

16 This question concerns the reaction of nitrogen and hydrogen to form ammonia which has an equilibrium constant (K_p) value of $5 \times 10^5\ atm^{-2}$ at 25 °C and one atmosphere pressure. Under these conditions the bond energies [enthalpies] (in kJ mol^{-1}) are N≡N, 945; H—H, 436; N—H, 391.
In the Haber–Bosch industrial process, gaseous nitrogen and hydrogen in a mole ratio of 1:3 are reacted, typically at a temperature of 500 °C, at a pressure of 200 atm and in the presence of finely divided iron when the equilibrium mole percentage of ammonia in the product mixture is 18.
(a) Give the equation for the reaction.
(b) Calculate a value for the standard enthalpy change for the reaction ($\Delta_r H^\ominus$) at 25 °C.
(c) Give an expression for K_p in terms of the partial pressures of the gases.
(d) (i) Comment on the effect of increasing the pressure on the equilibrium constant.
(ii) Explain why high pressures are used in the industrial process.
(iii) Explain whether the high pressures in the industrial process could be achieved by adding an excess of an inert gas (e.g., argon).
(e) Outline the reasons for using a temperature of 500 °C for the industrial process.
(f) Explain the use of 'finely divided iron'.
(g) Calculate the mole percentage conversion of nitrogen to ammonia for the industrial process. *(Welsh)*

17 (a) State one essential characteristic of a chemical equilibrium.
(b) State two ways in which the time taken to establish such an equilibrium may be altered.

(c) The following data indicate the effect of temperature and pressure on the equilibrium concentration of the product, X, of the forward reaction in a gaseous equilibrium.

Temperature/°C	Percentage of X present in the equilibrium mixture at 1 atm	100 atm	200 atm
550	0.077	6.70	11.9
650	0.032	3.02	5.71
750	0.016	1.54	2.99
850	0.009	0.87	1.68

(i) Use the above data to deduce whether the production of X is accompanied by an increase or a decrease in volume and explain your answer.
(ii) Use the above data to determine whether the production of X is an exothermic or an endothermic process and explain your answer.
(iii) State qualitatively the theoretical optimum conditions of temperature and pressure suggested by the above data for the commercial production of X. (*JMB*)

18 (a) State what is meant by (i) dynamic equilibrium, (ii) the equilibrium law.
(b) For the reaction

$$N_2(g) + 3H_2(g) \rightleftharpoons 2NH_3(g)$$

(i) calculate the mole percentage of ammonia in the equilibrium mixture formed at 400 °C and 3×10^7 Pa pressure, when gaseous hydrogen and nitrogen are mixed in a 3:1 mole ratio, and there is 61 per cent conversion of nitrogen to ammonia;
(ii) write an expression for K_p in terms of the partial pressures of the three gases in equilibrium;
(iii) given that the value of K_p at 400 °C is 2.0×10^{-14} Pa^{-2}, calculate the pressure at which ammonia is 95 per cent dissociated into its elements at 400 °C.
(*AEB 1985*)

19 The distribution (partition) coefficient of a compound X between trichloromethane and water is 12, X being more soluble in trichloromethane than in water. What mass of X will be extracted from 200 cm^3 of an aqueous solution containing 8 g of X by shaking it with 50 cm^3 of trichloromethane? (*Oxford*)

20 Discuss briefly the principles of chromatography.
Give brief details of the experimental methods used for column chromatography, thin-layer chromatography and paper chromatography giving a specific example of the use of each specialized technique.
Suggest analytical techniques (not necessarily chromatographic) which would be most appropriate for quantitative analysis of the major components of two of the following samples:
(a) a mixture of nitrogen(II) oxide and ethane,
(b) a sample of North Sea oil,
(c) a mineral sample *brought back* to Earth from the Moon,
(d) a mineral sample *on* the planet Mars.
(*Oxford and Cambridge S*)

21 If lead(II) chloride is precipitated in the presence of thorium nitrate, using an aqueous solution of lead(II) nitrate and dilute hydrochloric acid, the lead(II) chloride contains radioactive ^{212}Pb atoms. (These radioactive lead atoms are a daughter product of the thorium.)
Show how you could experimentally use this information to establish that lead(II) chloride and its saturated solution are in dynamic equilibrium.
(*SUJB; continued from Question* 1.13)

22 When a substance is added to a two-phase liquid system it is generally distributed with different equilibrium concentrations in the two phases. Why is this so?
A certain amount of iodine was shaken with CS_2 and an aqueous solution containing 0.3 mol dm^{-3} of potassium iodide. By titrating with thiosulphate solution, it was found that the CS_2 phase contained 32.3 g dm^{-3} and the aqueous phase 1.14 g dm^{-3} of iodine. The distribution coefficient for iodine distributed between CS_2 and water is 585. Calculate the equilibrium constant for the reaction $I_2 + I^- \rightleftharpoons I_3^-$ at the prevailing temperature. (*Cambridge Entrance*)

23 Discuss the importance of the concept of equilibrium constant in chemistry, including examples from as wide a range of applications as possible. (*Cambridge Entrance*)

11 The direction of natural change

It is possible to account for all natural processes in terms of a single idea, and in order to express this idea precisely we are led to the concept of entropy. We see that for chemical applications it is helpful to introduce a related quantity, the Gibbs energy. The latter is extremely important because it lets us predict the values of equilibrium constants. Moreover, it also lets us judge the energy resources of chemical reactions taking place in biological systems and in sources of electrical energy, such as fuel cells.

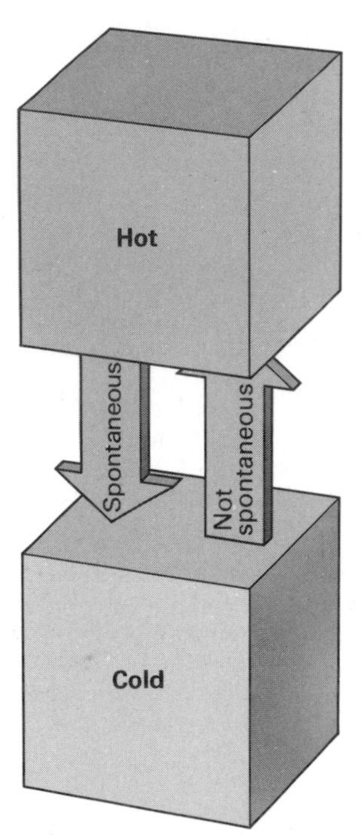

Figure 11.1 Cooling is a spontaneous process

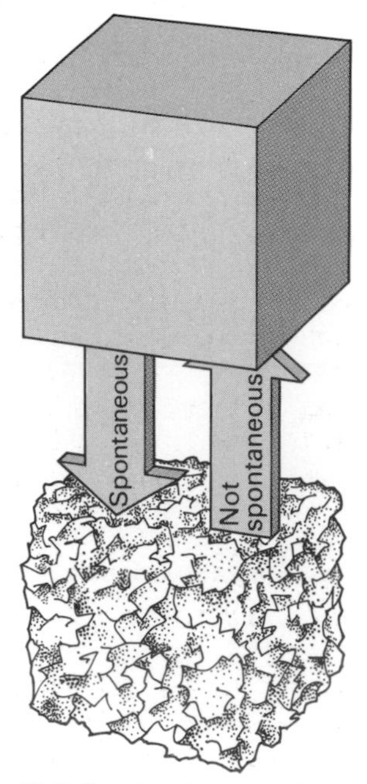

Figure 11.2 Rusting is a spontaneous process

So far, we have treated the equilibrium constant only as something measured experimentally. In this chapter, the world of the **Second law of thermodynamics**, we look into what determines the value of the equilibrium constant. The Second law deals with the direction of natural change, with questions such as why a reaction runs in one direction rather than another, and with what determines the composition of a reaction mixture at equilibrium.

The Second law has such a simple interpretation that, even without going into detailed thermodynamic calculations, we are led to the heart of understanding why reactions take place in one direction and not in another. It also lets us construct tables of data that can be used to predict the values of equilibrium constants, to decide how they depend on the temperature and to assess the quantity of electrical energy available from a given reaction.

11.1 Why things change

A **natural change** (which is also called a **spontaneous change**) is one that has a tendency to occur without needing to be driven. At the outset, however, it must be emphasized that neither 'natural' nor 'spontaneous' means 'fast'. Thermodynamics deals only with *tendencies* to change, and although there may be a strong tendency for a change to occur, that change may be extremely slow.

We all know that some changes are natural whereas others are not. A hot block of metal cools, but a cool block does not tend to become hot (Figure 11.1). Iron tends to rust; rust does not tend to turn into iron and release oxygen (Figure 11.2). What is the reason?

The dispersal of energy

There is a feature common to all types of natural change: *they are all accompanied by an increase in the dispersal of energy and matter*. At this stage, we may take the word 'dispersal' to have its everyday meaning of spreading. Energy and matter that is not dispersed can be thought of as confined to a small, well-defined location. Energy that has dispersed can be thought of as being spread more widely, such as into the motion of the particles in the surroundings of some object.

As a first example of how dispersal governs change, consider a hot block of metal in contact with cooler surroundings (your hand or the atmosphere). The block is a collection of electrons and vigorously vibrating ions, as shown diagrammatically in Figure 11.3(a). At this stage the energy is concentrated within the volume occupied by the block. The ions in the block jostle their neighbours, and those at the edge jostle the particles in the surroundings. The latter pick up energy as a result of the jostling, and in turn pass it on to their neighbours, as illustrated in Figure 11.3(b). There are very many particles in the surroundings, and so the energy spreads away: the direction of natural change is in the direction of the dispersal of energy. The reason why we never see a block in a cool room spontaneously becoming hot is that it is very

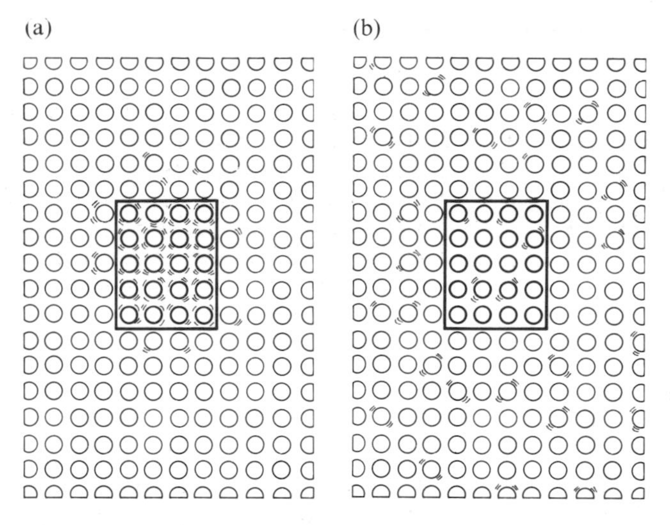

Figure 11.3 Energy initially localized (a) tends to disperse (b)

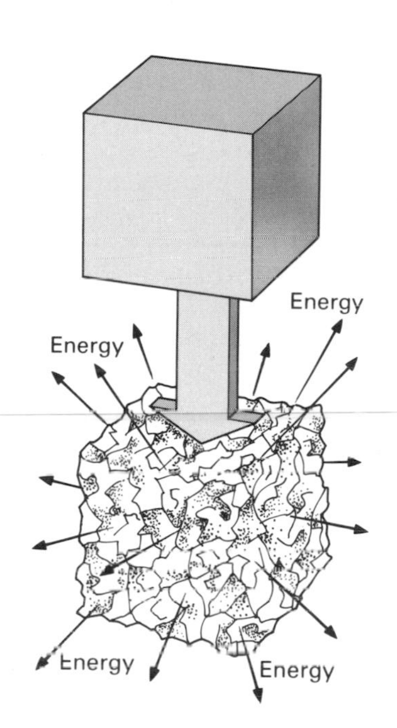

Figure 11.4 The (exothermic) rusting of iron disperses energy into the surroundings

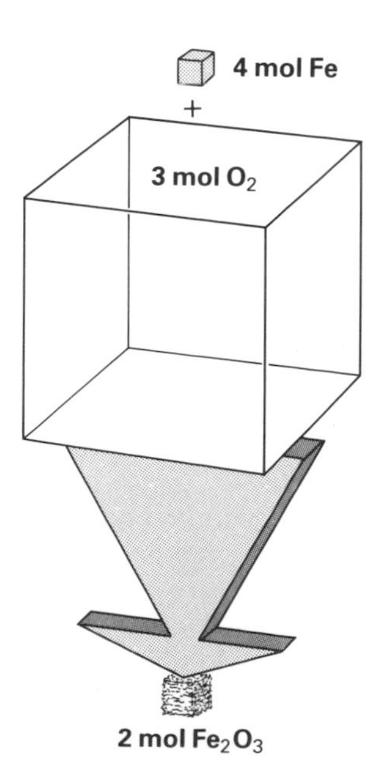

Figure 11.5 The reactants become more localized

improbable that by random jostling a lot of energy in the surroundings will return to the block at the same time.

Now consider the expansion of a gas into a vacuum, or the spreading of perfume through a room. The random motion of the molecules results in them spreading throughout the available space, and it is extremely unlikely that most of them will congregate back in one small region of the container or room.

Every type of natural change can be explained along similar lines. Chemists are interested in the natural direction of reactions. Since a reaction changes the nature of the participating substances, accounting for its natural direction is trickier than in the case of a physical change like cooling. Nevertheless it can be done, as we shall now see.

Chemical reactions as dispersers of energy

Consider the changes that occur when iron rusts. A sufficiently close model reaction is:

$$4Fe(s) + 3O_2(g) \rightarrow 2Fe_2O_3(s) \quad \Delta_r H^\ominus(298\ K) = -1648\ kJ\ mol^{-1} \qquad (11.1.1)$$

The oxidation is exothermic, so that energy is released when it takes place. This released energy disperses into the surroundings (Figure 11.4). The direction of natural change is from the reactants, with the energy locked up in the bonds holding the metal together and in the molecules of oxygen, to the product, where some energy remains in the highly localized solid (the heap of rust) but 1648 kJ for every 2 mol Fe_2O_3 has jostled away and is spread over the surroundings.

Once we stop to think about the reaction, however, we see that there is more to it than this discussion suggests. The oxygen is initially present as a gas and so its energy is spread over a large volume (for example, it is spread over about 72 dm^3 if we are thinking of 3 mol O_2 at 1 atm pressure and 20 °C). When the reaction has taken place the products are entirely solid, and 2 mol Fe_2O_3 occupies only about 0.06 dm^3: all the energy possessed by the compound is now confined to that volume (see Figure 11.5). We see that there is some kind of competition between the *localization* of energy and matter that occurs when the widely dispersed oxygen gas reacts to produce a highly localized solid and the *dispersal* of the energy that is released into the surroundings as a result of the reaction.

Assessing the competition between the localization of energy and matter by the elimination of a gas and the dispersal of the liberated reaction energy looks as though it could be quite complex. In fact it is simple once we have introduced another thermodynamic quantity, the **entropy**.

11.2 Entropy and the Second law

The formal definition of entropy is set out in Appendix 11.1. For our purposes its interpretation is what is important. *The entropy is a measure of the chaotic dispersal of energy and matter.* As the dispersal increases, so too does the entropy. Whenever a system gets more chaotic, like an ordered pack of cards being shuffled, then the entropy increases. We now develop this idea.

The Second law of thermodynamics

We have already seen that the direction of natural change is towards increasing dispersal. In terms of the entropy, natural change corresponds to the direction of increasing entropy. This statement is nothing other than the Second law of thermodynamics (although that law was built up on the basis of much more rigorous arguments than we have expressed here):

Second law of thermodynamics: The entropy of the universe increases in the course of every natural change.

Bear in mind when reading this chapter the present concern about the world's energy resources. It should be clear that the First law of thermodynamics eliminates the conventional worry; if energy cannot be destroyed, there is no need to worry about using it up. It will become clear that it is the *quality* of the energy (in the sense of its dispersal) which should be the true object of concern. Human beings are conserving energy but producing too much entropy. That is, highly localized energy (such as the energy stored in the bonds between atoms in fossil fuels) is being dispersed, and its economically viable sources are being depleted. We have an *entropy crisis*, not an energy crisis.

Before we go on there are two words of warning. The Second law refers in a grand way to the 'universe'. In practice that means the **system**, the contents of the reaction vessel, and its **surroundings**, such as the atmosphere or a water-bath. The first warning therefore is that the calculation of the direction of natural change must take into account the entropies both of the system itself *and* of its surroundings. The entropy of the system itself may decrease, yet the corresponding process will still be a natural, spontaneous change if that decrease is accompanied by a larger increase of entropy in the surroundings. Second, it is most important to remember the limitation of thermodynamics mentioned above: thermodynamics is silent on the *rates* at which changes occur, and a change may be spontaneous but slow. Thermodynamics deals only with the natural *direction* of change, not with the *rate* of change.

Entropies of substances

Entropy is always denoted by the letter S, and molar entropy, the entropy per unit amount of substance, by S_m. **Standard molar entropy**, the entropy per unit amount of substance in its standard state at the temperature specified, is denoted $S_m^{\ominus}$. The precise thermodynamic definition of the entropy (Appendix 11.1) leads to a way of measuring entropies by a straightforward laboratory experiment using a calorimeter, and so tables of values, such as Table 11.1, can be compiled.

The first point to notice about the values in Table 11.1 is their units. Entropies are normally expressed as so many joules per kelvin, $J\,K^{-1}$, so that molar entropies are normally expressed in $J\,K^{-1}\,mol^{-1}$. (The molar entropy has the same units as the molar gas constant R.)

The second point to notice is the relative magnitudes of the entries in the table. The standard molar entropies of gases at 1 bar pressure and 25 °C are all similar and are much larger than the standard molar entropies of most simple solids. This is because in a gas the energy is due to the motion of the particles, and the particles themselves (with their energies) are dispersed over a large volume. On the other hand, the energy of a solid is confined to a small region, the volume it occupies, and the particles can only vibrate. Note, however, that the standard molar entropies of solids composed of complex molecules, such as sucrose, can be very high. This is because the molecules have many atoms and the energy can be shared among them.

We can also see from Table 11.1 that the standard molar entropies of liquids are intermediate between those of gases and solids. This is in line with their intermediate structures. Notice how the value for water is quite low in comparison with those of other liquids at the same temperature. This is yet another way in which hydrogen bonds affect its properties. In this case they are holding the molecules together into an orderly structure, and therefore water is more 'solid-like' in its structure than are other liquids.

Table 11.1 Standard molar entropies at 25 °C, $S_m^{\ominus}/\mathrm{J\,K^{-1}\,mol^{-1}}$

Solids		Liquids		Gases	
C(graphite)	5.7	Hg	76.0	H_2	130.6
C(diamond)	2.4	H_2O	69.9	N_2	191.6
Fe	27.3	C_2H_5OH	160.7	O_2	205.0
Cu	33.1	C_6H_6	173.3	CO_2	213.6
AgCl	96.2	CH_3COOH	159.8	NO_2	240.1
Fe_2O_3	87.4			N_2O_4	304.3
$CuSO_4.5H_2O$	300.4			NH_3	192.3
sucrose	360.2			CH_4	186.2

Changes of entropy

The change of entropy when reactants change completely into products is calculated by taking the appropriate combinations of the entropies listed in Table 11.1 (for 25 °C). Note, however, that the values listed there are the *standard* molar values, and so for a reaction such as the oxidation of iron (equation 11.1.1) the standard molar entropy change obtained from the data is the value when pure oxygen gas at 1 bar combines with solid iron and reacts *completely* to form iron(III) oxide. In general, the **standard reaction entropy**, $\Delta_r S^{\ominus}$, of a reaction

$$aA + bB \rightarrow cC + dD$$

is defined as the difference

$$\Delta_r S^{\ominus} = (cS^{\ominus}_{m,C} + dS^{\ominus}_{m,D}) - (aS^{\ominus}_{m,A} + bS^{\ominus}_{m,B})$$

where the individual entropies are the standard molar entropies of each pure substance at the temperature of the reaction. For the oxidation of iron,

$$4Fe(s) + 3O_2(g) \rightarrow 2Fe_2O_3(s)$$

we have

$$\begin{aligned}\Delta_r S^{\ominus} &= 2 \times 87.4\,\mathrm{J\,K^{-1}\,mol^{-1}} \\ &\quad - (4 \times 27.3\,\mathrm{J\,K^{-1}\,mol^{-1}} + 3 \times 205.0\,\mathrm{J\,K^{-1}\,mol^{-1}}) \\ &= -549.4\,\mathrm{J\,K^{-1}\,mol^{-1}}\end{aligned}$$

As expected, there is a large *decrease* in entropy in this reaction, partly because the highly dispersed oxygen gas reacts to form a compact solid and partly because, since the reaction is exothermic, less energy is dispersed over the products than was initially dispersed over the reactants. The interesting point about the calculation is that it gives a *numerical* value for the change in dispersal brought about by the conversion of pure reactants into pure products.

But if iron oxidation is accompanied by a decrease in entropy, why is it spontaneous? The missing feature that we have not yet taken into account is the change of entropy in the surroundings. This emphasizes how important it is to consider both contributions to the change of entropy, and we will return to this point shortly.

Using the same technique as for the oxidation of iron, we can also find the change of entropy during the complete dissociation of N_2O_4 at 25 °C and

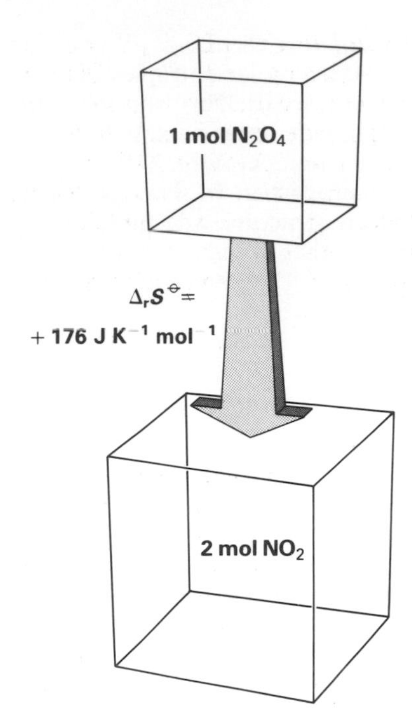

Figure 11.6 The standard reaction entropy of the system

1 bar pressure. We can expect the entropy to *increase* because there are 2 mol NO_2 for each 1 mol N_2O_4, so that at constant pressure complete reaction doubles the volume over which the energy is dispersed (see Figure 11.6). Moreover, since the reaction is endothermic, there is more energy to disperse over the reactants. Application of the data in Table 11.1 gives a value of $(2 \times 240.1 - 304.3)\,\mathrm{J\,K^{-1}\,mol^{-1}} = +175.9\,\mathrm{J\,K^{-1}\,mol^{-1}}$ for the standard reaction entropy for the reaction

$$N_2O_4(g) \rightarrow 2NO_2(g)$$

in agreement with this qualitative prediction.

Changes of entropy in the surroundings

Before making use of the information just derived, it is necessary to assess the change of entropy in the surroundings of the reaction vessel. This turns out to be very easy, because thermodynamic arguments show that it can be expressed in terms of the standard reaction enthalpy, $\Delta_r H^\ominus$. At a temperature T, the standard reaction entropy for the *surroundings* is

$$\Delta_r S^\ominus(\text{surroundings}) = -\frac{\Delta_r H^\ominus}{T} \qquad (11.2.1)$$

The T in the denominator comes from the thermodynamic definition of entropy (see Appendix 11.1). The negative sign is justified as follows. Suppose the reaction is exothermic ($\Delta_r H^\ominus < 0$). Intuitively we then expect the energy released to increase the entropy of the surroundings. This is as predicted by equation 11.2.1, because $-\Delta_r H^\ominus$ is positive and so $\Delta_r S^\ominus$(surroundings) is also positive, corresponding to an increase from its initial value.

The changes of entropy in the surroundings accompanying a chemical reaction are obtained very easily from the information about standard reaction enthalpies discussed in Chapter 9 and in particular the data in Table 9.3. For instance, the standard reaction enthalpy of the iron oxidation reaction (equation 11.1.1) is $-1648\,\mathrm{kJ\,mol^{-1}}$ and so the standard reaction entropy for the surroundings at 25 °C (298.15 K) is

$$\Delta_r S^\ominus(\text{surroundings}) = \frac{-(-1648 \times 10^3\,\mathrm{J\,mol^{-1}})}{298.15\,\mathrm{K}} = +5527\,\mathrm{J\,K^{-1}\,mol^{-1}}$$

This is a massive *increase* in entropy, and arises because the reaction releases energy into the surroundings.

The dissociation of N_2O_4 is endothermic and $\Delta_r H^\ominus = +57.2\,\mathrm{kJ\,mol^{-1}}$ (see Figure 11.7); hence

$$\Delta_r S^\ominus(\text{surroundings}) = \frac{-(+57.2 \times 10^3\,\mathrm{J\,mol^{-1}})}{298.15\,\mathrm{K}} = -192\,\mathrm{J\,K^{-1}\,mol^{-1}}$$

All endothermic reactions decrease the entropy of their surroundings.

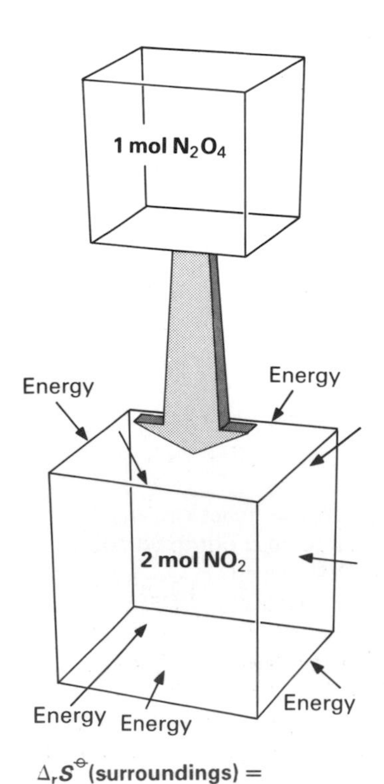

Figure 11.7 The standard reaction entropy of the surroundings during an endothermic reaction

Total entropy changes

The total standard reaction entropy is the sum of the changes that take place in the system and in the surroundings:

$$\begin{aligned}\Delta_r S^\ominus(\text{total}) &= \Delta_r S^\ominus(\text{surroundings}) + \Delta_r S^\ominus(\text{system}) \\ &= -\Delta_r H^\ominus/T + \Delta_r S^\ominus(\text{system}) \qquad (11.2.2)\end{aligned}$$

From now on we drop the label 'system' and remember that whenever the unlabelled term $\Delta_r S^\ominus$ appears it refers to the system itself. In order to find $\Delta_r S^\ominus$(total) we simply have to add together the values found in the two preceding sections.

For the oxidation of iron, the total standard reaction entropy is

$$\begin{aligned}\Delta_r S^\ominus(\text{total}) &= (+5527\,\mathrm{J\,K^{-1}\,mol^{-1}}) + (-549\,\mathrm{J\,K^{-1}\,mol^{-1}}) \\ &= +4978\,\mathrm{J\,K^{-1}\,mol^{-1}}\end{aligned}$$

Figure 11.8 Iron tends to rust spontaneously: this badly corroded ship's hull is that of the *Titanic*

That is, for each 4 mol Fe and 3 mol O_2 converted to 2 mol Fe_2O_3 there is an overall *increase* in the entropy of the universe of almost $5000\,J\,K^{-1}$. This is a large positive quantity: the reaction is spontaneous. That is the thermodynamic reason why iron corrodes. It may be possible to slow the rate at which the corrosion takes place (see Section 13.5), and that is one of the reasons for electroplating or painting bridges, ships and cars. Nevertheless, we can see from this result that iron (and steel) structures are fundamentally unstable in air and have a built-in tendency to decay (see Figure 11.8).

Similarly, the total standard reaction entropy accompanying the complete dissociation of N_2O_4 at 1 bar and 25 °C is

$$\begin{aligned}\Delta_r S^{\ominus}(\text{total}) &= (-192\,J\,K^{-1}\,mol^{-1}) + (+176\,J\,K^{-1}\,mol^{-1})\\ &= -16\,J\,K^{-1}\,mol^{-1}\end{aligned}$$

This is small and negative: the reaction is not spontaneous at 25 °C.

So far, we have ignored the effect on the system's entropy of having a *mixture* of substances at intermediate stages of the reaction, and the effect this has of contributing an additional positive term to the system's entropy (from the extra chaos present). This additional entropy contribution is zero for either pure N_2O_4 or pure NO_2, and a maximum when both are present in equal amounts. Therefore there are two contributions to the entropy change that occurs *in the system*: the substances *change* and they *mix*. Overall there are *three* contributions to the entropy change: the two contributions in the system, and the change of entropy in the surroundings.

The sum of the three contributions is shown in Figure 11.9. The total entropy goes through a *maximum* at a composition that corresponds exactly to the equilibrium composition as found from the equilibrium constant. Pure N_2O_4 tends to dissociate until the universe (the system plus its surroundings) reaches the condition of greatest chaos (maximum total entropy); similarly, pure NO_2 tends to dimerize until the same condition of maximum total entropy is reached.

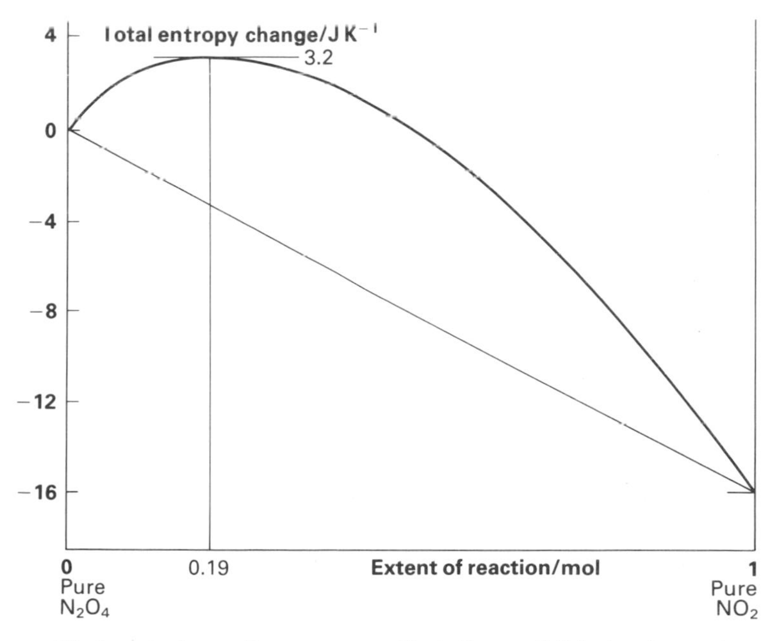

Figure 11.9 Total entropy change during the course of reaction at 1 bar and 25 °C: at equilibrium, 0.19 mol N_2O_4 has dissociated to produce 0.38 mol NO_2, hence $x_{NO_2} = 0.38\,mol/1.19\,mol = 0.32$, as in the Example in Section 10.1

All chemical reactions are manifestations of this tendency towards maximum total entropy. *All* equilibrium constants are shorthand ways of denoting the composition corresponding to maximum total entropy. *All* initial compositions tend to move spontaneously towards *equilibrium, the condition of maximum total entropy.*

11.3 The Gibbs energy

We have emphasized that whether or not a reaction has a natural tendency to occur depends on the value of $\Delta_r S^\ominus$(total). That quantity is composed of two parts. One is the standard reaction entropy in the system itself. The other is an indication of the extent of dispersal in the surroundings, and can always be calculated very simply in terms of the standard reaction enthalpy. The total standard reaction entropy can therefore be expressed entirely in terms of quantities related to the reaction:

$$\Delta_r S^\ominus(\text{total}) = -\frac{\Delta_r H^\ominus}{T} + \Delta_r S^\ominus \qquad (11.3.1)$$

In order to use this expression we need to refer to two tables of data, one for the standard molar entropies of species (Table 11.1) and the other for the standard formation enthalpies (Table 9.3). It is very helpful to combine the two types of data into a single table.

As a first step we reorganize equation 11.3.1 by multiplying through by $-T$, giving

$$-T\Delta_r S^\ominus(\text{total}) = \Delta_r H^\ominus - T\Delta_r S^\ominus$$

The new quantity $-T\Delta_r S^\ominus$(total) is now given a new symbol and a new name. It is called the **standard reaction Gibbs energy**, and denoted $\Delta_r G^\ominus$. Hence the last equation becomes

$$\Delta_r G^\ominus = \Delta_r H^\ominus - T\Delta_r S^\ominus \qquad (11.3.2)$$

The new quantity is named after Josiah Willard Gibbs (see Box 9.1), who contributed greatly to the development of thermodynamics.

Some of the features of the Gibbs energy are explained below, and its precise significance is explained in Appendix 11.2. At this stage it is enough to know that although $\Delta_r G^\ominus$ is composed of two parts, one an enthalpy change and the other an entropy change, it can be thought of as the change that takes place in a single property – the **Gibbs energy**, G – when reactants convert to products. For the reaction $a\text{A} + b\text{B} \rightarrow c\text{C} + d\text{D}$, $\Delta_r G^\ominus$ is the change in Gibbs energy when a mol of pure A and b mol of pure B become c mol of pure C and d mol of pure D, with all the substances in their standard states.

Using the Gibbs energy

The main point about the standard reaction Gibbs energy is that it shows at a glance whether or not a reaction has a tendency to occur spontaneously. We know already that a reaction is spontaneous (under standard conditions) if $\Delta_r S^\ominus$(total) is positive. Therefore, in terms of the standard reaction Gibbs energy, which is proportional to $-\Delta_r S^\ominus$(total), *a reaction is spontaneous under standard conditions if* $\Delta_r G^\ominus$ *is negative.*

Similar considerations apply if the conditions are not standard (if, for instance, the pressures of the gases present are not 1 bar), and in all cases *the reaction is spontaneous in the direction of decreasing Gibbs energy.* This is illustrated for the $N_2O_4 \rightleftharpoons 2NO_2$ equilibrium in Figure 11.10, which shows how the Gibbs energy changes with composition. It is nothing other than Figure 11.9 inverted (because of the negative sign in $\Delta_r G^\ominus = -T\Delta_r S^\ominus$(total)) and with a change of scale (on account of the factor T). Instead of saying that any initial composition shifts towards maximum total entropy (as illustrated in Figure 11.9), we can now say that any initial composition shifts towards minimum Gibbs energy (at constant temperature and pressure).

For many reactions $\Delta_r H^\ominus$ is quite a lot bigger than $T\Delta_r S^\ominus$, and so in these cases we can write $\Delta_r G^\ominus \approx \Delta_r H^\ominus$ (as a very crude approximation). Then an exothermic reaction, one corresponding to a negative $\Delta_r H^\ominus$, has a negative $\Delta_r G^\ominus$, and is therefore spontaneous. That is why we are so familiar with spontaneous exothermic reactions. An endothermic reaction (with $\Delta_r H^\ominus$ positive) is spontaneous only if $T\Delta_r S^\ominus$ is large enough, because only then is $\Delta_r H^\ominus - T\Delta_r S^\ominus$ negative. Spontaneous exothermic reactions are driven by the large quantity of entropy they generate in the surroundings; spontaneous

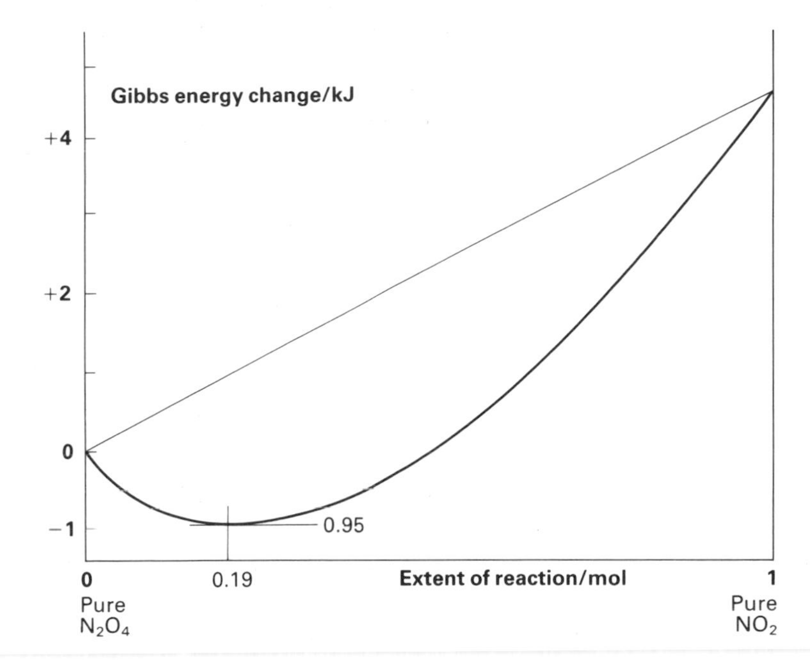

Figure 11.10 Gibbs energy change during the course of reaction at 1 bar and 25 °C: the spontaneous direction is from either pure N_2O_4 or pure NO_2 to the composition corresponding to minimum Gibbs energy

endothermic reactions are driven by the large quantity of entropy generated in the reaction mixture itself. A reaction with a large positive entropy change is that between solid hydrated iron(III) chloride and sulphur dichloride oxide (thionyl chloride, SCl_2O). The copious release of gas is obvious; so too is the endothermic character of the reaction, because the reaction mixture cools by over 10 °C.

The essential information for calculating $\Delta_r G^{\ominus}$ is carried in tables of **standard formation Gibbs energies**, $\Delta_f G^{\ominus}$, such as Table 11.2. These are the Gibbs energy changes per unit amount of compound in its standard state when it is formed from its elements in their reference phases at the temperature specified (which is usually but not necessarily 298 K). This is similar to the definition of standard formation enthalpy (see Section 9.4). To calculate the standard reaction Gibbs energy of any reaction, the appropriate combinations of the standard formation Gibbs energies are taken – just as standard formation enthalpies are combined to calculate standard reaction enthalpies. This is illustrated in the following Example.

Table 11.2 Standard formation Gibbs energies at 25 °C, $\Delta_f G^{\ominus}$/kJ mol^{-1}

Solids		Liquids		Gases	
NaCl	−384.0	H_2O	−237.2	NH_3	−16.5
KCl	−408.3	H_2O_2	−120.4	NO_2	51.3
NH_4Cl	−203.0	CH_3OH	−166.4	N_2O_4	97.9
Al_2O_3	−1582.4	C_2H_5OH	−174.8	CO	−137.2
Fe_2O_3	−742.2	C_6H_6	124.3	CO_2	−394.4
SiO_2	−856.6	HCl(aq)	−131.3	CH_4	−50.8

EXAMPLE

A fuel cell (a device for producing electricity directly from a chemical reaction) uses the oxidation of methanol. Calculate the standard reaction Gibbs energy.

METHOD The reaction is

$$CH_3OH(l) + \tfrac{3}{2}O_2(g) \rightarrow CO_2(g) + 2H_2O(l)$$

The standard reaction Gibbs energy is calculated by taking the difference of the sums of the standard formation Gibbs energies for the products and the reactants. Use the data in Table 11.2. Standard formation Gibbs energies of elements are zero.

ANSWER The overall standard reaction Gibbs energy is

$$\begin{aligned}\Delta_r G^\ominus &= \{\Delta_f G^\ominus(CO_2, g) + 2\Delta_f G^\ominus(H_2O, l)\} \\ &\qquad - \{\Delta_f G^\ominus(CH_3OH, l) + \tfrac{3}{2}\Delta_f G^\ominus(O_2, g)\} \\ &= \{(-394.4) + 2(-237.2)\}\,\mathrm{kJ\,mol^{-1}} - \{(-166.4) + \tfrac{3}{2}(0)\}\,\mathrm{kJ\,mol^{-1}} \\ &= -702.4\,\mathrm{kJ\,mol^{-1}}\end{aligned}$$

COMMENT We shall go on to see that the value of $\Delta_r G^\ominus$ gives the electrical work that may be produced by a reaction when it is coupled to some kind of electrical device, such as an electric motor. In the present case, therefore, we know that for every 32 g of methanol (1 mol CH_3OH) consumed, 702.4 kJ of electrical work can be obtained. The standard combustion enthalpy of methanol is $-726\ \mathrm{kJ\,mol^{-1}}$, and so direct conversion to electrical energy provides 97 per cent of the energy the reaction can provide on combustion.

Gibbs energy and chemical equilibrium

At this stage all the Gibbs energy appears to do is to reproduce the information that we worked through in a more laborious fashion in the earlier sections. Instead of arriving at the conclusion that the complete dissociation of N_2O_4 is not spontaneous, on the grounds that it is accompanied by a net decrease in the total entropy, we have now come to the same conclusion on the basis that it is accompanied by an increase in the Gibbs energy. The great advantage of the Gibbs energy now begins to emerge, however. This is because thermodynamic arguments show us how to use $\Delta_r G^\ominus$ to predict the composition at which the reaction mixture has no further natural tendency to change. In other words, if we know $\Delta_r G^\ominus$ we are able to predict the equilibrium constant for a reaction.

The connection between the standard reaction Gibbs energy and the reaction's equilibrium constant comes from thermodynamics and is as follows:

$$\Delta_r G^\ominus = -RT \ln K \qquad (11.3.3)$$

This very important relation links the work of this chapter to that of the last. On the left of the equation there is a quantity that we can obtain from Table 11.2 by taking appropriate combinations of the entries. On the right we have the equilibrium constant, the quantity at the centre of the stage in the last chapter. Therefore we now have a way of predicting the equilibrium properties of a reaction even if it has not been investigated experimentally. The way this can be used to discuss equilibrium constants is illustrated in the following Example (we shall also see other examples in the next two chapters).

EXAMPLE

Predict the value of the equilibrium constant for the reaction $C(s) + H_2O(g) \rightleftharpoons CO(g) + H_2(g)$ on the basis that at 1000 K the standard reaction Gibbs energy is $-8.1\ \mathrm{kJ\ mol^{-1}}$.

METHOD Use equation 11.3.3 in the form $\ln K = -\Delta_r G^\ominus/RT$. The specification of the equilibrium constant is

$$K_p = \left\{\frac{p(CO)p(H_2)}{p(H_2O)}\right\}_{eq}$$

and its units are therefore those of pressure (e.g., bar).

ANSWER At 1000 K the value of RT is

$$RT = 8.314\ \mathrm{J\ K^{-1}\ mol^{-1}} \times 1000\ \mathrm{K} = 8.314\ \mathrm{kJ\ mol^{-1}}$$

Therefore the equilibrium constant has the value

$$\ln K = \frac{-(-8.1\ \mathrm{kJ\ mol^{-1}})}{8.314\ \mathrm{kJ\ mol^{-1}}} = 0.97, \text{ implying } K_p = 2.6\ \mathrm{bar}.$$

COMMENT The reaction is part of the water gas reaction (see Section 16.4), which is used in the industrial reduction of steam to hydrogen. Hydrogen is used widely commercially, for example in the Haber Bosch process for the synthesis of ammonia, and hence of nitrogenous fertilizers.

Work and the Gibbs energy

There is a second aspect of $\Delta_r G^\ominus$ that opens up even wider applications than the prediction of equilibrium constants; in some ways this is even more important, because it is the basis of the application of physical chemistry to biological systems, fuel cells and electrochemistry.

The central lesson of the Second law is that spontaneous changes are those that are accompanied by an increase in total entropy. This implies, as we shall now show, that it is not always possible to convert all the energy change of a system into work.

Consider, for example, an exothermic reaction with a negative standard reaction entropy (such as the oxidation of iron). If the forward reaction is to occur naturally, there must be some way of producing enough entropy in the surroundings to compensate for the reduction of entropy in the system. Entropy is generated in the surroundings only if the system heats them, and not if it merely does work (because the latter involves only the *orderly* motion of particles whereas entropy is a measure of chaos). The reaction is therefore not spontaneous unless part of the standard reaction enthalpy is used to heat the surroundings at the same time as it is causing work to be done: an attempt to extract all the standard reaction enthalpy as work will fail because such a process is not spontaneous. When the thermodynamic arguments are carried through it turns out that the minimum amount of energy that must be 'wasted' in order to achieve spontaneity is $-T\Delta_r S^\ominus$. Hence the maximum work available per mole of reaction under standard conditions is equal to $-\Delta_r G^\ominus$ (this is why the Gibbs energy is also often called the **free energy**).

These very important conclusions can be illustrated as follows. The standard reaction enthalpy for equation 11.1.1 (the oxidation of iron) is $-1648\ \mathrm{kJ\ mol^{-1}}$, and the standard reaction entropy is $-549.4\ \mathrm{J\ K^{-1}\ mol^{-1}}$. Hence $1648\ \mathrm{kJ\ mol^{-1}}$ can be used to heat the surroundings (at constant temperature and pressure); if the reaction is used to do work, however, at least

$$-T\Delta_r S^\ominus = -298\ \mathrm{K} \times -549.4\ \mathrm{J\ K^{-1}\ mol^{-1}} = 164\ \mathrm{kJ\ mol^{-1}}$$

must still be discarded to heat the surroundings. Hence, the energy free to do

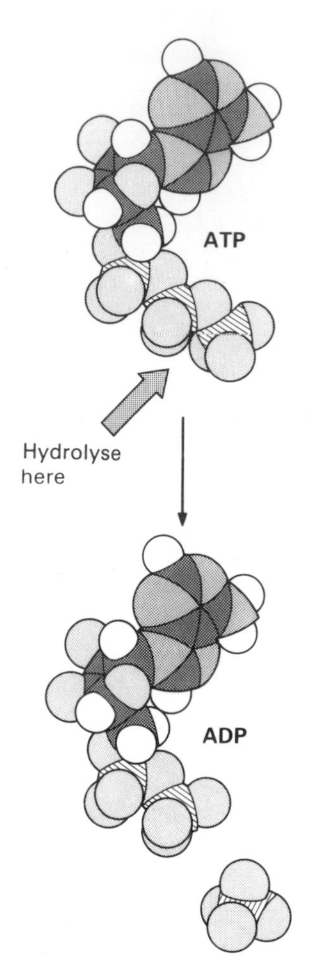

Figure 11.11 The hydrolysis of ATP to ADP

work (such as driving an electric motor) is only $-\Delta_r G^\ominus$, that is, ($1648\,kJ\,mol^{-1} - 164\,kJ\,mol^{-1}$), or $1484\,kJ\,mol^{-1}$.

The importance of this result should be obvious. In the first place, electrical cells and fuel cells, which we shall discuss in Chapter 13, are producers of electricity and as such they are sources of electrical work. Therefore, if we know the standard reaction Gibbs energy for the reaction going on inside them, we can state the maximum quantity of electrical work that can be obtained. Moreover, the processes going on inside our bodies, such as thinking, are basically electrochemical, and so when we assess the energy resources of molecules involved in metabolism we should really do so in terms of the Gibbs energy.

Once again, consider the oxidation of iron. The oxidation of 4 mol Fe corresponds to a change of Gibbs energy of -1484 kJ (see Table 11.2). Therefore, if we can devise a way of tapping that energy electrically, we can produce 1484 kJ of electrical work for every 223 g of iron consumed. In some sense the world is wasting a vast quantity of energy simply by letting steel objects corrode; it is like burning the iron and discarding the energy.

When methane is burnt the standard reaction Gibbs energy at 1 bar and 25 °C is $-818.0\,kJ\,mol^{-1}$, and so 818 kJ of electrical energy can be produced for each 16 g of methane that is oxidized in a fuel cell (a device for converting the energy of chemical reactions directly into electricity).

The biologically important ATP (adenosine triphosphate) molecule is shown in Figure 11.11. Its crucial role in metabolism is as a store of energy obtained from food. ATP makes energy available by being able to lose the terminal phosphate group, forming the diphosphate (ADP) and releasing energy in the process. The basic reaction is the hydrolysis of ATP according to

$$ATP^{4-}(aq) + H_2O(l) \rightarrow ADP^{3-}(aq) + HOPO_3{}^{2-}(aq) + H^+(aq)$$

$$\Delta_r G^\ominus \approx -30\,kJ\,mol^{-1} \quad \text{at } 37\,°C \text{ and pH} = 7$$

(37 °C is blood temperature). This indicates that the hydrolysis reaction can provide 30 kJ of useful work for every mole of ATP molecules consumed, which can be used to drive other reactions that are themselves not spontaneous. For instance, the standard reaction Gibbs energy of the reaction in which glucose and fructose are combined to form sucrose is $\Delta_r G^\ominus \approx +23\,kJ\,mol^{-1}$. It is therefore not spontaneous. But if the energy of the ATP hydrolysis can be channelled to it, then since 30 kJ are available from 1 mol ATP molecules, the overall change is $-7\,kJ\,mol^{-1}$ and the coupled reactions have a natural tendency to occur. (Whether or not they actually do so depends on the availability of suitable enzymes in the cell.)

Some idea of the energy resources required to build proteins can be obtained from the fact that the work from about three ATP hydrolysis reactions is needed to form one peptide link. Even as small a protein as myoglobin contains about 150 peptide linkages, and so every molecule requires for its construction the energy resources of around 450 ATP molecules.

Ellingham diagrams

Gibbs energies also have a value in industrial chemistry, for they can be used to explain (and choose) the techniques for extracting metals from their ores. An **Ellingham diagram** shows how the standard reaction Gibbs energy for the conversion of elements to their oxides (expressed per mole $O_2(g)$) varies with temperature. As

$$\Delta_r G^\ominus = \Delta_r H^\ominus - T\Delta_r S^\ominus$$

the slope of each line is $-\Delta_r S^\ominus$ (see Figure 11.12), provided that neither the standard reaction enthalpy nor the standard reaction entropy depends strongly on the temperature (as is usually the case).

In the reaction

$$2M(s) + O_2(g) \rightarrow 2MO(s)$$

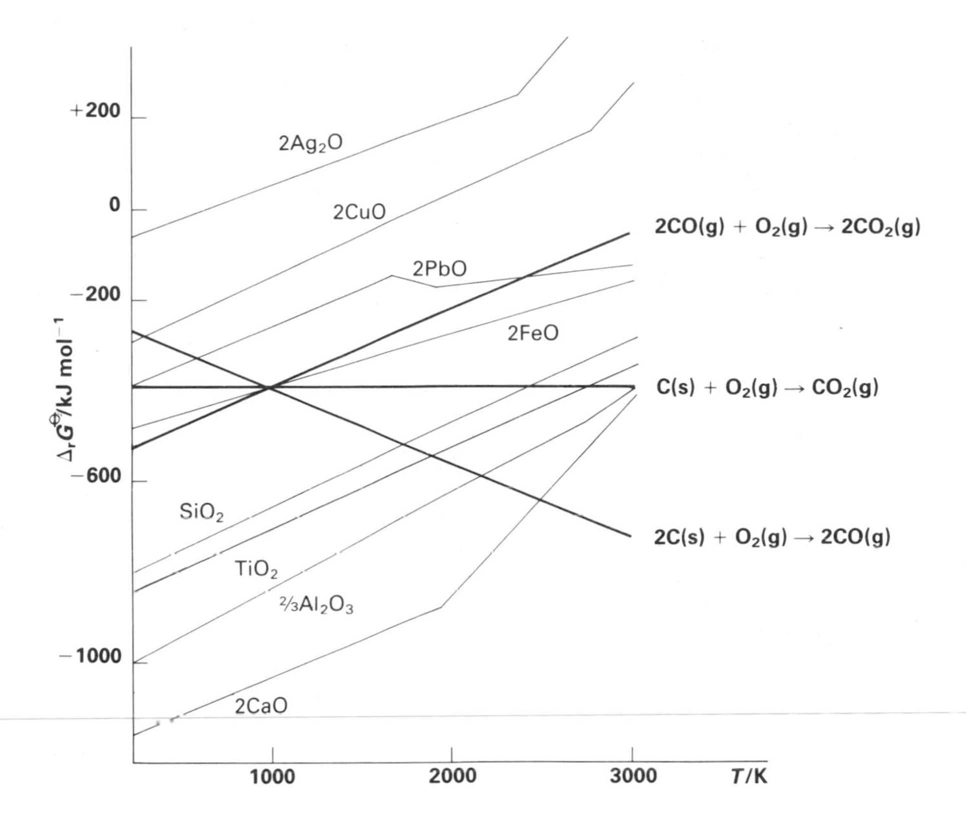

Figure 11.12 An Ellingham diagram

1 mol O_2(g) is consumed and the only product is a solid: hence the standard reaction entropy is negative and largely independent of the identity of the metal M. Thus, most of the lines have positive slopes and are approximately parallel. The slope of each line changes abruptly at the boiling point of the metal (and at the temperature of other phase changes), however. Above the boiling point the standard reaction entropy is even more negative, as 3 mol gas particles are replaced by solid oxide.

The slopes of the carbon lines are very different. Oxidation of carbon to carbon dioxide does not change the number of gas molecules, and so the standard reaction entropy is small and the line is almost horizontal. The line for the oxidation of carbon to carbon monoxide slopes *downwards*, because 1 mol O_2(g) is replaced by 2 mol CO(g) and the standard reaction entropy is positive.

Only if the standard reaction Gibbs energy is positive will the oxide spontaneously decompose to the metal under standard conditions – silver, for instance, is found native (i.e. uncombined) in nature. Mercury can be extracted at normal industrial temperatures (typically below 2000 K) simply by heating. Carbon can be used to extract a number of the more reactive metals (compare Figure 11.13). For example, the iron/iron(II) oxide line crosses the carbon monoxide/carbon dioxide line at 1000 K; below that temperature the standard reaction Gibbs energy for the conversion of carbon monoxide to carbon dioxide is more negative than that for the conversion of iron to iron(II) oxide. Therefore iron(II) oxide can be reduced by carbon monoxide below 1000 K. The carbon monoxide, however, is produced *lower down* the furnace (see Section 25.9): carbon reduces carbon dioxide to carbon monoxide only *above* 1000 K. Similarly, zinc oxide can be reduced by carbon above 1300 K, a method used in practice in industry. The line for aluminium lies below all three carbon lines in the industrially economical range of temperatures, however: aluminium cannot be obtained from its oxide by reduction with carbon. It is in fact obtained electrolytically (see Section 19.7).

Figure 11.13 Lead(II) oxide, which is yellow, can be reduced by carbon to form globules of metallic lead

Appendix 11.1 The background to entropy

Entropy was described earlier in terms of the energy transferred by heating the system at a temperature T. An entropy change is *defined* as

$$\Delta S = \frac{q}{T} \quad q \text{ being transferred reversibly.}$$

ΔS is the change of entropy brought about by a transfer of a quantity of energy q by heating at the temperature T. The term 'reversible' refers to the way the transfer is carried out; it must be carried out extremely carefully, and differences of temperature must not be allowed to develop either between the sample and its surroundings or within the sample itself. We can see that there is a smaller change of entropy when a given quantity of energy is transferred to an object at high temperature than at low: more chaos is introduced when the sample is cool than when it is already hot. A sneeze in a library creates more disorder than one in a crowded street.

The ability of a substance to distribute energy over its particles is related to its heat capacity (this is the reason why heat capacity measurements are used to determine entropies). The usual calorimetric measurement of entropy involves measuring heat capacities from very low temperatures all the way up to the temperature of interest. The entropy can also be calculated if the number of ways that the energy can be distributed is known. The formula for the calculation was deduced by Ludwig Boltzmann, and is

$$S = k \ln W$$

where W is the number of ways in which the energy can be distributed and k is **Boltzmann's constant**, which is equal to the gas constant R divided by Avogadro's constant L (in other words: $R = kL$) and has the value $1.38 \times 10^{-23}\,\mathrm{J\,K^{-1}}$.

Appendix 11.2 Reaction Gibbs energy and standard reaction Gibbs energy

The reaction Gibbs energy $\Delta_r G$ and the standard reaction Gibbs energy $\Delta_r G^\ominus$ are distinct, and the two should not be confused.

The **standard reaction Gibbs energy** $\Delta_r G^\ominus$ is the change of Gibbs energy per mole of reaction that would occur if *pure* reactants in their standard states changed to *pure* products in theirs. Thus it takes no account of mixing, and refers to an artificial *Pure* → *Pure* 'book' reaction.

The **reaction Gibbs energy** $\Delta_r G$ is the change of Gibbs energy per mole of reaction *at some specified, fixed composition of the reaction system.* It can be imagined as the change in Gibbs energy that occurs when 1 mol of products is formed in a reaction vessel the size of a swimming pool, so that the new products (and the destroyed reactants) have negligible effect on the concentrations of the mixture.

The two are related by

$$\Delta_r G = \Delta_r G^\ominus + RT \ln Q$$

where Q is the **reaction quotient**, a quantity defined like the equilibrium constant, but with the concentrations having their specified, not necessarily equilibrium, values:

$$Q = \frac{[\mathrm{C}]^c[\mathrm{D}]^d}{[\mathrm{A}]^a[\mathrm{B}]^b}$$

Thermodynamics shows that the condition for equilibrium at constant temperature and pressure is formally $\Delta_r G = 0$. Since $Q = K$ when the concentrations have their equilibrium values, substituting these into the expression above gives

$$0 = \Delta_r G^\ominus + RT \ln K \quad \text{or} \quad \Delta_r G^\ominus = -RT \ln K$$

the expression used in the text.

Summary

- ☐ The direction of natural change is the direction of increasing dispersal of energy and matter.
- ☐ The entropy is a measure of the chaotic dispersal of energy and matter; as the dispersal increases so does the entropy.
- ☐ The Second law of thermodynamics states that the entropy of the universe increases in the course of every natural change. The 'universe' means the system and its surroundings.
- ☐ The Second law refers to the natural tendency to change, not the rate of change.
- ☐ The standard molar entropy, $S_m^{\ominus}$, is the entropy of unit amount of substance when it is in its standard state at the temperature specified. The units are normally $\mathrm{J\,K^{-1}\,mol^{-1}}$.
- ☐ Entropies can be measured using a calorimeter (by measuring the heat capacity over a range of temperatures).
- ☐ The standard reaction entropy, $\Delta_r S^{\ominus}$, is the standard molar entropy change accompanying complete reaction, all substances being in their standard states at the temperature specified.
- ☐ The standard reaction entropy for the surroundings when a reaction with standard reaction enthalpy $\Delta_r H^{\ominus}$ occurs is $\Delta_r S^{\ominus}(\text{surroundings}) = -\Delta_r H^{\ominus}/T$ (equation 11.2.1).
- ☐ A spontaneous process is one that may occur without needing to be driven. 'Spontaneous' does not mean 'fast'.
- ☐ A process is spontaneous if the total entropy change, the sum of the changes in the system and in the surroundings, is positive. Equilibrium corresponds to the position of maximum total entropy.
- ☐ The standard reaction Gibbs energy, $\Delta_r G^{\ominus}$, is given by $\Delta_r G^{\ominus} = \Delta_r H^{\ominus} - T\Delta_r S^{\ominus}$ if the temperature is constant (equation 11.3.2).
- ☐ A process is spontaneous if it corresponds to a negative value of $\Delta_r G^{\ominus}$.
- ☐ The standard formation Gibbs energy, $\Delta_f G^{\ominus}$, is the Gibbs energy change when unit amount of substance of the compound in its standard state is formed from its elements in their reference phases at the temperature specified.
- ☐ The standard reaction Gibbs energy is related to the equilibrium constant by $\Delta_r G^{\ominus} = -RT\ln K$ (equation 11.3.3). When $\Delta_r G^{\ominus}$ is large and negative, the equilibrium lies strongly in favour of the products; when it is large and positive, the equilibrium lies strongly in favour of the reactants.
- ☐ The maximum quantity of non-expansion work per mole of reaction, under standard conditions, equals $-\Delta_r G^{\ominus}$.
- ☐ Ellingham diagrams describe the thermodynamics behind metal extraction.

PROBLEMS

1 Write an account of the direction of natural change. Include as examples the cooling of hot objects, and chemical reactions.

2 Why does a bouncing ball come to rest?

3 State the Second law of thermodynamics (a) in terms of entropy, (b) without using the word 'entropy'.

4 Why do endothermic reactions occur?

5 Explain, in terms of the changes in the entropy of the system and of the surroundings, why endothermic reactions are favoured by an increase of temperature. (Investigate the effect of the presence of T in the entropy change of the surroundings.)

6 Explain, in terms of entropy and the natural tendency of energy to disperse, why chemical reactions take place spontaneously in the direction corresponding to a decrease in the Gibbs energy.

7 Explain why the Gibbs energy may also be called the Gibbs free energy.

8 Calculate the standard reaction entropy at 25 °C for the following reactions:
(a) $2H_2(g) + O_2(g) \rightarrow 2H_2O(l)$
(b) $N_2(g) + 3H_2(g) \rightarrow 2NH_3(g)$
(c) $12C(s) + 11H_2O(l) \rightarrow C_{12}H_{22}O_{11}(s)$, sucrose

9 Calculate the standard reaction enthalpies at 25 °C for the reactions in the last question, and then find the values of the standard reaction Gibbs energy for the reactions.

10 Calculate the standard reaction Gibbs energy for reactions (a) and (b) in Question 8 directly from the data in Table 11.2.

11 The standard molar entropy of sucrose is large, but that is partly because each molecule contains so many atoms. What are the standard entropies of (a) sucrose, (b) copper per unit amount of *atoms* present?

12 The standard reaction Gibbs energies

$$C + O_2 \rightarrow CO_2 \qquad \Delta_r G^\ominus = -380\,\text{kJ mol}^{-1}$$

$$2C + O_2 \rightarrow 2CO \qquad \Delta_r G^\ominus = -500\,\text{kJ mol}^{-1}$$

are given for 1500 °C. On the basis of the following information, discuss the possibility of reducing Al_2O_3, FeO, PbO and CuO with carbon at this temperature.

$$4Al + 3O_2 \rightarrow 2Al_2O_3 \qquad \Delta_r G^\ominus = -2250\,\text{kJ mol}^{-1}$$

$$2Fe + O_2 \rightarrow 2FeO \qquad \Delta_r G^\ominus = -250\,\text{kJ mol}^{-1}$$

$$2Pb + O_2 \rightarrow 2PbO \qquad \Delta_r G^\ominus = -120\,\text{kJ mol}^{-1}$$

$$2Cu + O_2 \rightarrow 2CuO \qquad \Delta_r G^\ominus \approx 0$$

Calculations like this are used in the discussion of metal extraction.

13 Explain why ΔG, the change in the Gibbs free energy, is a better guide to whether a reaction will occur spontaneously than is ΔH, the enthalpy change.

The standard free energies of formation of propene and propane are $+62.7$ and $-23.5\,\text{kJ mol}^{-1}$. Calculate the free energy change for the reaction of propene with molecular hydrogen to give propane, indicate whether the reaction is favourable thermodynamically, and comment on the need for a catalyst when the hydrogenation of propene is carried out in practice.

The addition of an excess of sulphur dichloride oxide ($SOCl_2$, the acid chloride of sulphurous acid) to the hydrated chloride of a transition metal leads to a brisk reaction attended by gas evolution and a colour change, and the temperature of the mixture drops markedly. Comment on the reaction which occurs and on the changes in enthalpy and entropy which accompany it.

(Oxford and Cambridge S)

12 Acids, bases and salts

Now we apply the discussion of the last two chapters to the case of ionic substances in water. First we see how solubilities depend on the conditions, including the presence of other salts. We then turn to a particularly important ion, the hydrogen ion. This ion is responsible for the properties of acids, and is therefore of central importance in chemistry. Once we know about pH we can discuss acid–base titrations, hydrolysis and indicators. We can also discuss the chemically and biologically important buffer solutions.

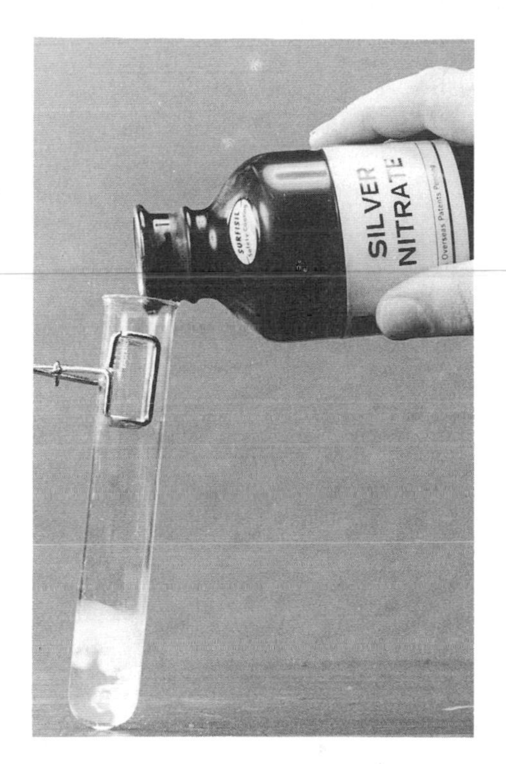

Figure 12.1 Silver chloride is precipitated when aqueous silver nitrate is added to aqueous sodium chloride

Many chemical reactions involve the reactions of ions in solution, including precipitation, as when barium sulphate or silver chloride is precipitated during analysis. An ion second only to the electron in its importance in chemistry is the hydrogen ion, the proton. Its properties in solution govern the behaviour of acids and bases – substances that play a role in all aspects of daily life, including life itself. Proteins, for instance, can be thought of as being complicated acids and bases and their function in a cell is often merely a very elaborate form of neutralization.

12.1 Dissolving

Compounds dissolve until the solution is **saturated**. Their concentration in the saturated solution is termed their **solubility**. In some cases the solubility is very high and a large quantity of the solid may dissolve before the solution is saturated. The solubility of common salt, for instance, is 357 g in 1 kg of water at 20 °C. In other cases the solubility is very low. For instance, the solubility of silver chloride is only 2×10^{-3} g in 1 kg of water at the same temperature.

The solubility product

The link with the ideas in the last two chapters is that a saturated solution is in *dynamic equilibrium* with the excess solid present. The dissolution equilibrium can be expressed in terms of an equilibrium constant. For instance, in the case of silver chloride,

$$AgCl(s) \rightleftharpoons Ag^+(aq) + Cl^-(aq) \qquad K_c = \left\{\frac{[Ag^+][Cl^-]}{[AgCl]}\right\}_{eq}$$

As explained in Section 10.1, the concentration of a solid is a constant, and so [AgCl] may be absorbed into K_c. When this is done it is conventional to refer to the resulting quantity as the **solubility product**, and to denote it K_{sp}:

$$K_{sp} = \{[Ag^+][Cl^-]\}_{eq}$$

Likewise, the solubility product for a salt M_xA_y composed of M^{m+} cations and A^{a-} anions is

$$K_{sp} = \{[M^{m+}]^x[A^{a-}]^y\}_{eq} \qquad (12.1.1)$$

That the equilibrium is *dynamic* can be demonstrated by adding some solid silver chloride, labelled with radioactive ^{112}Ag, to a saturated solution. After an hour, some of the radioactivity can be detected in the aqueous phase.

The solubility product is most useful for the description of the properties of sparingly soluble salts (which means salts with solubilities up to about 10^{-3} mol per kg of water) and therefore we will confine our discussion to them.

The solubility products of sparingly soluble salts can be determined in a variety of ways. The aim is to measure the concentrations of the cations and the anions in the saturated solution in contact with excess undissolved solid. The principal techniques involve evaporating a known volume of saturated solution to dryness and then weighing the residue, titration, ion exchange (see Section 10.3 – the cations are replaced by protons, which are then estimated by titration) and conductivity measurement, as in the Example below. We shall see in the next chapter that an electrochemical cell can also be used.

EXAMPLE

In an experiment to measure the solubility product of silver chloride the electrolytic conductivity of a saturated aqueous solution was measured and found to be 1.96×10^{-6} S cm^{-1} at 25 °C. The water itself had an electrolytic conductivity of 0.12×10^{-6} S cm^{-1} at that temperature. Find the value of K_{sp}.

METHOD The aim is to find the concentrations of the Ag^+ and Cl^- ions in the saturated solution; then use $K_{sp} = \{[Ag^+][Cl^-]\}_{eq}$. Find the electrolytic conductivity due to the ions by taking the difference between the measured values for the solution and the water. Find the molar conductivities for the ions from Table 8.1. Find the concentration by rearranging $\Lambda = \kappa/c$. Take the molar conductivity as having its limiting value, since the solution is so dilute.

ANSWER The observed electrolytic conductivity due to the dissolved ions is

$$\begin{aligned}\kappa &= 1.96 \times 10^{-6}\,\mathrm{S\,cm^{-1}} - 0.12 \times 10^{-6}\,\mathrm{S\,cm^{-1}} \\ &= 1.84 \times 10^{-6}\,\mathrm{S\,cm^{-1}}\end{aligned}$$

The limiting molar conductivity is

$$\begin{aligned}\Lambda^{\infty} &= \lambda^{\infty}(\mathrm{Ag^+}) + \lambda^{\infty}(\mathrm{Cl^-}) = (61.9 + 76.4)\,\mathrm{S\,cm^2\,mol^{-1}} \\ &= 138.3\,\mathrm{S\,cm^2\,mol^{-1}}\end{aligned}$$

Therefore the concentration of ions is

$$\begin{aligned}c &= \frac{1.84 \times 10^{-6}\,\mathrm{S\,cm^{-1}}}{138.3\,\mathrm{S\,cm^2\,mol^{-1}}} = 1.33 \times 10^{-8}\,\mathrm{mol\,cm^{-3}} \\ &= 1.33 \times 10^{-5}\,\mathrm{mol\,dm^{-3}}\end{aligned}$$

(since $10^3\,\mathrm{cm^3} = 1\,\mathrm{dm^3}$). Since $[Ag^+] = [Cl^-]$ in the electrically neutral solution, we have $[Ag^+] = [Cl^-] = c$. Therefore

$$K_{sp} = (1.33 \times 10^{-5}\,\mathrm{mol\,dm^{-3}})^2 = 1.8 \times 10^{-10}\,\mathrm{mol^2\,dm^{-6}}$$

COMMENT Note that the solubility product depends on the temperature, and so in all measurements of K_{sp} the experiment has to be carried out under careful temperature control.

Values of solubility products for some sparingly soluble compounds are listed in Table 12.1. A knowledge of K_{sp} lets us make both qualitative and quantitative predictions, exactly as in the case of the equilibrium constants for chemical reactions. The point to remember is that at a given temperature the

Table 12.1 Solubility products in water at 25 °C

$CaCO_3$	4.8×10^{-9}	$BaSO_4$	1.3×10^{-10}	$Fe(OH)_3$	2.0×10^{-39}
$BaCO_3$	5.1×10^{-9}	FeS	6.3×10^{-18}	$Fe(OH)_2$	7.9×10^{-16}
$PbCl_2$	1.6×10^{-5}	CuS	6.3×10^{-36}	$Zn(OH)_2$	2.0×10^{-17}
AgCl	1.8×10^{-10}	PbS	1.3×10^{-28}	$Mg(OH)_2$	1.1×10^{-11}
AgI	8.3×10^{-17}	$Al(OH)_3$	1.0×10^{-33}	$Ca(OH)_2$	5.5×10^{-6}

Note that these are the *numerical* values of K_{sp} with concentrations expressed in mol dm^{-3}

solubility product is a *constant*, independent of the individual ion concentrations. The ion concentrations therefore always tend to adjust so that their product is equal to the value of K_{sp}.

Experimental measurements of the concentrations of silver and chloride ions in a saturated solution of silver chloride in water at 25 °C give the solubility product at this temperature as

$$K_{sp} = \{[Ag^+][Cl^-]\}_{eq} = 1.8 \times 10^{-10}\,mol^2\,dm^{-6}$$

Consider what happens when we add to water more silver chloride than would be needed to form a saturated solution. At first the rate of dissolution is greater than that of precipitation. When the concentration has risen sufficiently, a dynamic equilibrium is established and there is no more net dissolution. What is the concentration of salt in solution when this stage is reached? We know the value of the product of the ion concentrations. The concentrations of chloride and silver ions are equal (because whenever an AgCl formula unit goes into solution it produces one Ag^+ ion and one Cl^- ion). Therefore $[Ag^+] = [Cl^-]$ at every stage of dissolution, so that at equilibrium

$$K_{sp} = \{[Ag^+][Cl^-]\}_{eq} = [Ag^+]^2_{eq}$$

The concentration of silver ions in the saturated solution is therefore

$$[Ag^+]_{eq} = \sqrt{K_{sp}} = \sqrt{(1.8 \times 10^{-10}\,mol^2\,dm^{-6})} = 1.3 \times 10^{-5}\,mol\,dm^{-3}$$

We can conclude that the solubility of silver chloride is $1.3 \times 10^{-5}\,mol\,dm^{-3}$. This result also applies to other salts of the type MA:

The solubility of a salt of the type MA *is equal to* $\sqrt{K_{sp}}$ *at the temperature specified.*

The effects of added salts

The solubility product also lets us predict what will happen when the concentration of *one* of the ions present in the equilibrium is increased, for example when sodium chloride is added to a solution already saturated with silver chloride.

The *qualitative* prediction comes from an application of Le Chatelier's principle (see Section 10.2): increasing the concentration of chloride ions results in the equilibrium tending to shift so as to minimize the increase. In other words, the sparingly soluble salt precipitates, and hence is less soluble in the presence of a common ion. This is called the **common-ion effect**.

The same conclusion follows from the solubility product, for if $[Cl^-]$ is increased by the addition of chloride ions, the value of $[Ag^+]$ must decrease in order to preserve the value of K_{sp}. We can go on to make a *quantitative* prediction about the size of the effect. Suppose we add enough sodium chloride to the solution to make it $1.0 \times 10^{-4}\,mol\,dm^{-3}$ in NaCl (5.8 mg in $1\,dm^3$ – a few grains). Then $[Cl^-]_{eq} \approx 1.0 \times 10^{-4}\,mol\,dm^{-3}$, as the chloride ion concentration is dominated by the contribution from the sodium chloride. In order for the solubility product to remain at $1.8 \times 10^{-10}\,mol^2\,dm^{-6}$ the silver ion concentration must fall to

$$[Ag^+]_{eq} = \frac{K_{sp}}{[Cl^-]_{eq}} = \frac{1.8 \times 10^{-10}\,mol^2\,dm^{-6}}{1.0 \times 10^{-4}\,mol\,dm^{-3}} = 1.8 \times 10^{-6}\,mol\,dm^{-3}$$

In other words, whereas the solubility of silver chloride in pure water is $1.3 \times 10^{-5}\,mol\,dm^{-3}$, its solubility in $1.0 \times 10^{-4}\,mol\,dm^{-3}$ NaCl(aq) is only about $1.8 \times 10^{-6}\,mol\,dm^{-3}$.

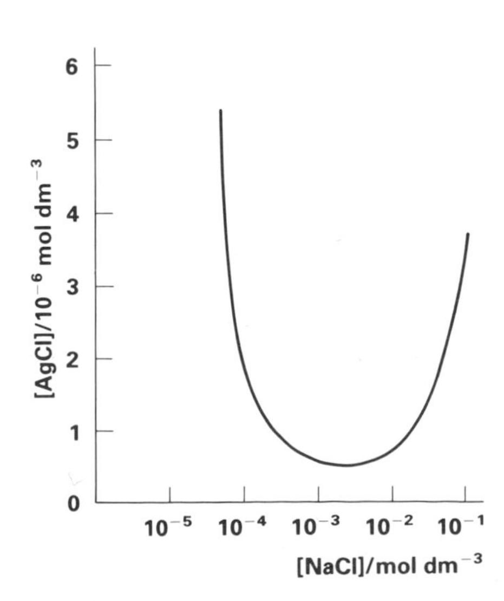

Figure 12.2 The solubility of silver chloride in the presence of sodium chloride

The experimentally observed dependence of the solubility on the concentration of added salt is shown in Figure 12.2. The solubility decreases in line with the prediction just made. Yet something strange happens when the added salt concentration exceeds about $2.5 \times 10^{-3}\,mol\,dm^{-3}$; the solubility of silver chloride begins to increase!

The explanation for this behaviour lies in the chemistry of silver. In the presence of excess chloride ions, Ag^+ forms complex ions (the term is

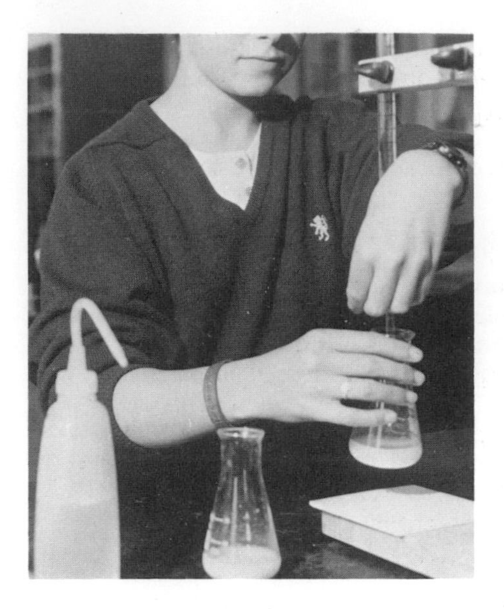

Figure 12.3 Titrations are usually performed between acids and bases, but precipitation titrations are also used; here the concentration of chloride ions is being measured by reaction with aqueous silver nitrate

explained in Section 25.7) such as $AgCl_2^-$, which are soluble. Not only are they soluble, but as a result of their formation the concentrations of both free Ag^+ and Cl^- ions decrease in the solution; therefore more of the solid goes into solution in order to maintain the constancy of the solubility product, and so there is an increase in solubility. Similar **complexation reactions** occur with many other ions. For instance, such a reaction accounts for the dissolution of amphoteric hydroxides such as aluminium hydroxide by the addition of excess sodium hydroxide (which complexes aluminium to form the soluble $[Al(OH)_4]^-$ ion – see Section 19.3).

12.2 Acids and bases

The definitions of acids and bases have increased in sophistication since Robert Boyle ascribed the name 'acid' to substances with a sharp taste and interpreted this property in terms of spikes on the atoms.

Brønsted acids and bases

One definition of acids and bases is due to Johannes Brønsted and Thomas Lowry:

a **Brønsted acid** is a proton donor,
a **Brønsted base** is a proton acceptor.

The reaction between a Brønsted acid and a Brønsted base is a **neutralization reaction**.

Not only do these definitions include 'conventional' acids and bases, but they also go beyond by capturing less obvious substances and labelling them as acids or bases. This is typical of chemistry: chemists recognize that superficially unrelated species may behave in the same way and then generalize definitions so that they become more widely applicable. In that way the ideas behind the subject become unified and concepts become simplified.

Consider, for example, the character of ethanoic acid. In water ethanoic acid is partially ionized according to the equilibrium

$$CH_3COOH(aq) + H_2O(l) \rightleftharpoons H_3O^+(aq) + CH_3CO_2^-(aq)$$

where the ethanoate ion is written $CH_3CO_2^-$ to emphasize the delocalization of the charge over both oxygen atoms, as explained in Section 2.5. $H_3O^+(aq)$ is called the **oxonium ion**, and is an attempt to represent the hydrated proton, $H^+(aq)$, more realistically. The proton is highly hydrated in water, however, and even $H_3O^+(aq)$ is a simplification, several hydrated species including $H_9O_4^+(aq)$ being present as well. For most purposes, the precise number of strongly attached water molecules is not very important, and we will often denote the hydrated hydrogen ion simply as $H^+(aq)$. Clearly, the CH_3COOH molecule is a proton donor in water, and so is properly classified as an acid. We can also begin to see how the Brønsted–Lowry definition captures more than just the obvious and the trivial. Note, for instance, that the CH_3COOH molecule donates a proton to an H_2O molecule. That shows that H_2O is a proton acceptor. According to the Brønsted–Lowry definition water is therefore a Brønsted *base* in this reaction.

Now consider the product side of the equilibrium. Since there is an equilibrium (which, like all chemical equilibria, is dynamic – see Section 10.1) the reverse reaction, the formation of CH_3COOH, occurs when H_3O^+ donates a proton to the $CH_3CO_2^-$ ion. Therefore, H_3O^+ is a Brønsted acid and $CH_3CO_2^-$ is a Brønsted base. The ethanoate ion, $CH_3CO_2^-$, is called the **conjugate base** of ethanoic acid, CH_3COOH. Similarly, H_3O^+ is called the **conjugate acid** of the base H_2O. Since the ethanoic acid is only weakly ionized at moderate concentrations, the equilibrium lies in favour of CH_3COOH. One way of expressing this is to regard H_3O^+ as a stronger acid than CH_3COOH, as it has a stronger tendency to donate a proton. Alternatively, $CH_3CO_2^-$ can be regarded as a stronger base than H_2O, so that it dominates the accepting of any protons that are available.

Similar considerations apply to substances we conventionally regard as bases, such as ammonia. In water the equilibrium established is

$$NH_3(aq) + H_2O(l) \rightleftharpoons NH_4^+(aq) + OH^-(aq)$$

NH_3 is clearly a proton acceptor, and is therefore a Brønsted base. In this reaction, however, the H_2O molecule is the donor of the proton, not the acceptor. Therefore it is acting as a Brønsted *acid* in this reaction. Note how *water can act both as an acid and as a base*; that is of central importance. The reverse reaction in the equilibrium is the donation of a proton by the ammonium ion, NH_4^+, to a hydroxide ion, OH^-. It follows that NH_4^+ is a Brønsted acid, the conjugate acid of the base NH_3, while OH^- is the conjugate base of the acid H_2O. Since the equilibrium lies to the left, OH^- is a stronger base than NH_3; alternatively, NH_4^+ is a stronger acid than H_2O.

We have emphasized that water can act either as an acid or as a base, depending on the demands of the other substances present. It is not alone in this property. Consider, for instance, what happens when hydrogen chloride dissolves in pure ethanoic acid. The following equilibrium is established:

$$HCl + CH_3COOH \rightleftharpoons CH_3C(OH)_2^+ + Cl^-$$

HCl is a strong proton donor (which is why it behaves as an acid in water); CH_3COOH is a relatively weak donor. Therefore, in this system, HCl acts as the donor while CH_3COOH acts as the acceptor; that is, 'ethanoic acid' is a base! The important conclusion to draw is that *whether a substance behaves as an acid or a base depends on the other substances present*. A significant part of modern chemistry is concerned with **non-aqueous solvents**, and a correct understanding of the roles of substances as acids or as bases is essential.

Finally, consider the case of water alone. In the pure liquid there is the equilibrium

$$H_2O(l) + H_2O(l) \rightleftharpoons H_3O^+(aq) + OH^-(aq)$$

Here we see H_2O acting simultaneously as a Brønsted acid (in donating a proton leaving OH^-, its conjugate base) and as a Brønsted base (in accepting a proton to form H_3O^+, its conjugate acid). In pure water the concentrations of hydrogen ions and hydroxide ions are equal and the liquid is *neutral*. If a proton donor is present $[H_3O^+]$ exceeds $[OH^-]$, and the solution is *acidic*. When a proton acceptor is present $[OH^-]$ exceeds $[H_3O^+]$, and the solution is *alkaline*. (An **alkali** is a water-soluble base.)

Ionization constants

All the equations we have written in this section are equilibria. Therefore we can use the techniques developed in Chapter 10 to describe them; we introduce equilibrium constants, and then explore their implications.

Consider the equilibrium established in aqueous ethanoic acid:

$$CH_3COOH(aq) + H_2O(l) \rightleftharpoons H_3O^+(aq) + CH_3CO_2^-(aq)$$

$$K_c = \left\{\frac{[H_3O^+][CH_3CO_2^-]}{[CH_3COOH][H_2O]}\right\}_{eq}$$

The concentration of water is almost constant on account of the relatively small amount involved in the ionization. Therefore we absorb $[H_2O]$ into the equilibrium constant and write the new constant as the **acid ionization constant** or **acid dissociation constant**, K_a:

$$K_a = \left\{\frac{[H_3O^+][CH_3CO_2^-]}{[CH_3COOH]}\right\}_{eq} \qquad (12.2.1)$$

In many applications it is much more convenient to report and use the **pK_a value** rather than K_a itself. This is defined* as follows:

$$pK_a = -\lg K_a \qquad (12.2.2)$$

* *In the definition here and the analogous ones to come, the definition formally requires the* numerical values *of K_a, that is, the definition is really $pK_a = -\lg\{K_a/\text{mol dm}^{-3}\}$, as it is impossible to take the logarithm of a quantity with dimensions.*

The value of K_a for ethanoic acid at 25 °C is 1.76×10^{-5} mol dm^{-3}, and so its pK_a is 4.75. This shows one advantage of pK_a: the values are much less cumbersome than those of K_a itself. Remember that, because of the negative sign defining pK_a, *the larger the value of* pK_a, *the smaller the acid ionization constant*:

$$K_a = 10^{-pK_a} \text{ mol dm}^{-3}$$

The small value of K_a for ethanoic acid classifies it as a weak acid. In general, the terms 'strong' and 'weak' indicate the extent to which an acid is ionized in solution, just as for electrolytes in general (see Section 8.3):

a **strong acid** is one that is almost fully ionized in solution;
a **weak acid** is one that is partially ionized in solution.

Table 12.2 Acid ionization constants in water at 25 °C, K_a/mol dm^{-3} and pK_a

	K_a/mol dm^{-3}	pK_a
CH_3COOH	1.76×10^{-5}	4.75
HF	5.62×10^{-4}	3.25
HOCl	3.72×10^{-8}	7.43
$(HO)_2SO_2$	1.20×10^{-2}	1.92 (pK_{a2})
$(HO)_3PO$	7.52×10^{-3}	2.12 (pK_{a1})
	6.17×10^{-8}	7.21 (pK_{a2})
	4.37×10^{-13}	12.36 (pK_{a3})

Hydrochloric and nitric acids ionize almost completely in water, and so in aqueous solution they are classified as strong acids. (They may be either dilute or concentrated: 'strong' and 'weak' are *types* of acid and must not be confused with measures of concentration.) Ethanoic acid is only partially ionized at normal concentrations in water, and so it is a weak acid (whether the solution is dilute or concentrated). Sulphuric acid, $(HO)_2SO_2$ (see Section 15.7), is interesting: although $(HO)_2SO_2$ itself is strong (in the sense that a typical aqueous solution consists of H^+(aq) and $HOSO_3^-$(aq) ions), the hydrogensulphate ion, $HOSO_3^-$, is only a weak acid in water. This decrease in strength on successive ionizations is a common feature of **polyprotic acids** (or **polybasic acids**), acids that can donate more than one proton. It is shown in the decrease of their acid ionization constants, typically by three to five orders of magnitude for each proton lost (so that pK_a increases in steps of from 3 to 5). For example, phosphoric acid is a weak triprotic acid with the following acid ionization constants:

$$(HO)_3PO(aq) + H_2O(l) \rightleftharpoons H_3O^+(aq) + (HO)_2PO_2^-(aq) \qquad pK_{a1} = 2.1$$

$$(HO)_2PO_2^-(aq) + H_2O(l) \rightleftharpoons H_3O^+(aq) + HOPO_3^{2-}(aq) \qquad pK_{a2} = 7.2$$

$$HOPO_3^{2-}(aq) + H_2O(l) \rightleftharpoons H_3O^+(aq) + PO_4^{3-}(aq) \qquad pK_{a3} = 12.4$$

The decrease results from the increasing negative charge of the ions and the greater difficulty of removing the positively charged proton.

The most important characteristic of an aqueous solution of an acid is the hydrogen ion concentration $[H_3O^+]$. For weak acids an approximate way of relating $[H_3O^+]$ to K_a is to suppose that the weak acid ionizes so little that the concentration of the non-ionized form (CH_3COOH, for example) is equal to the concentration originally added. We shall write $[CH_3COOH]_{eq} \approx [CH_3COOH]_{added} = A$. Moreover, if we ignore the relatively slight ionization of water, the hydrogen ion and ethanoate ion concentrations are equal, because one ion of each type is formed whenever an acid molecule ionizes. We therefore write $[CH_3CO_2^-]_{eq} \approx [H_3O^+]_{eq}$. These two approximations turn equation 12.2.1 into

$$K_a \approx \frac{[H_3O^+]_{eq}^2}{A} \quad \text{or} \quad [H_3O^+]_{eq} \approx \sqrt{(K_a A)} \qquad (12.2.3)$$

If we have an independent source of information about the value of K_a, or access to tables of data (such as those in Table 12.2), it is easy to predict the hydrogen ion concentration in a solution of known acid concentration.

Acids and pH

Even in quite ordinary chemistry the value of $[H_3O^+]$ can vary over many orders of magnitude. It therefore turns out to be sensible to introduce a quantity called the **pH** of a solution (it was originally introduced by Søren Sørensen in an attempt to improve the quality control of beer). pH is defined as the *negative* of the logarithm of the hydrogen ion concentration:

$$\textit{pH of a solution: } \mathrm{pH} = -\lg[H_3O^+] \tag{12.2.4}$$

The log is to the base 10 (common logarithms). Since it is impossible to take the log of a quantity with dimensions we must interpret $[H_3O^+]$ in this expression as the numerical value of the concentration when it is expressed in mol dm^{-3}. A better definition, which is explicit but clumsy to keep writing, is therefore

$$\mathrm{pH} = -\lg\{[H_3O^+]/\mathrm{mol\,dm^{-3}}\} \tag{12.2.5}$$

We use this form in the Examples because it eliminates any worries about units.

Consider a solution of ethanoic acid which is so dilute that we can use the simple expression in equation 12.2.3. The pH is given by

$$\mathrm{pH} = -\lg[H_3O^+] = -\lg\sqrt{(K_aA)} = -\tfrac{1}{2}\lg K_aA = -\tfrac{1}{2}\lg K_a - \tfrac{1}{2}\lg A$$

(since $\lg\sqrt{x} = \frac{1}{2}\lg x$ and $\lg ab = \lg a + \lg b$). The term $-\lg K_a$ equals $\mathrm{p}K_a$ (equation 12.2.2), and so

$$\mathrm{pH} = \tfrac{1}{2}\mathrm{p}K_a - \tfrac{1}{2}\lg A \tag{12.2.6}$$

The pH of 1 mol dm^{-3} ethanoic acid, using the $\mathrm{p}K_a$ value from Table 12.2, is given by

$$\mathrm{pH} = \tfrac{1}{2} \times 4.75 - \tfrac{1}{2} \times \lg 1 = 2.4$$

EXAMPLE

Calculate the pH of the following aqueous solutions at 25 °C: (a) 2.0 mol dm^{-3} HCl(aq), (b) 0.1 mol dm^{-3} HCl(aq), (c) 0.1 mol dm^{-3} CH_3COOH(aq).

METHOD Base the answer on equation 12.2.5 for the strong acid (HCl) and on equation 12.2.6 for the weak acid (CH_3COOH). For (c) it is necessary to know the value of $\mathrm{p}K_a$: use Table 12.2.

ANSWER Since hydrochloric acid is almost completely ionized in aqueous solution we use

(a) $[H_3O^+] = [HCl] = 2.0\ \mathrm{mol\,dm^{-3}}$ $\quad \mathrm{pH} = -\lg 2.0 = -0.30$

(b) $[H_3O^+] = [HCl] = 0.1\ \mathrm{mol\,dm^{-3}}$ $\quad \mathrm{pH} = -\lg 0.1 = 1.0$

Since ethanoic acid is only partially ionized in aqueous solution, $\mathrm{p}K_a = 4.75$ at 25 °C, and $A = 0.1\ \mathrm{mol\,dm^{-3}}$, we have

(c) $\mathrm{pH} = \frac{1}{2}\mathrm{p}K_a - \frac{1}{2}\lg A = \frac{1}{2}(4.75) - \frac{1}{2}\lg(0.1) = 2.9$

COMMENT Note that at a concentration of 0.1 mol dm^{-3}, ethanoic acid has a higher pH than hydrochloric acid; this arises from its incomplete ionization and therefore lower hydrogen ion concentration. Vinegar is about 0.7 mol dm^{-3} CH_3COOH(aq), corresponding to pH ≈ 2.5; Coca-Cola is only slightly less acidic, having pH ≈ 3.

Alkalis and pH

The next step is to apply similar arguments to the ionization equilibria of alkalis, and in particular weak alkalis such as ammonia in water. The equilibrium when ammonia dissolves in water is

$$NH_3(aq) + H_2O(l) \rightleftharpoons NH_4^+(aq) + OH^-(aq) \qquad K_c = \left\{\frac{[NH_4^+][OH^-]}{[NH_3][H_2O]}\right\}_{eq}$$

As in the case of acids, it is sensible to absorb the water concentration into the constant, and to define the **base ionization constant** and its negative logarithm:

$$K_b = \left\{\frac{[NH_4^+][OH^-]}{[NH_3]}\right\}_{eq} \qquad pK_b = -\lg K_b \qquad (12.2.7)$$

Some values are listed in Table 12.3.

Table 12.3 Base ionization constants in water at 25 °C, K_b/mol dm^{-3} and pK_b

	K_b/mol dm^{-3}	pK_b
NH_3	1.77×10^{-5}	4.75
$C_2H_5NH_2$	5.62×10^{-4}	3.25
$(C_2H_5)_2NH$	9.55×10^{-4}	3.02
$C_6H_5NH_2$	3.80×10^{-10}	9.42

The OH^- ion concentration can be calculated from the value of K_b and the concentration of base added initially, but it turns out to be much more useful to express the properties of aqueous solutions of bases in terms of the *hydrogen ion* concentration. This may seem curious, for we do not refer to hydrogen ions in the base ionization equilibrium. The problem is resolved once it is realized that the base is present in water, and that water itself ionizes slightly to produce hydrogen ions:

$$H_2O(l) + H_2O(l) \rightleftharpoons H_3O^+(aq) + OH^-(aq) \qquad K_c = \left\{\frac{[H_3O^+][OH^-]}{[H_2O]^2}\right\}_{eq} \qquad (12.2.8)$$

As in the case of weak acids and bases, the ionization is so slight that $[H_2O]$ is approximately constant, and can be absorbed into the equilibrium constant. This gives the **ionic product for water**, K_w, and its negative logarithm:

$$\textit{Ionic product for water: } K_w = \{[H_3O^+][OH^-]\}_{eq} \qquad pK_w = -\lg K_w \quad (12.2.9)$$

At 25 °C $K_w = 1.0 \times 10^{-14}$ mol^2 dm^{-6} ($pK_w = 14.00$) reflecting the very low degree of ionization of water.

The water ionization equilibrium has several consequences. In the first place we can immediately state the hydrogen ion concentration in pure water from a knowledge of the value of K_w. Since both an OH^- and an H_3O^+ ion are formed in equation 12.2.8, in the neutral solution $[H_3O^+] = [OH^-]$; hence

$$[H_3O^+] = \sqrt{K_w} = 1.0 \times 10^{-7}\ \text{mol dm}^{-3} \quad \text{at } 25\ °\text{C}$$

It follows that the pH of pure water at 25 °C is

$$pH = -\lg[H_3O^+] = -\lg(1.0 \times 10^{-7}) = 7.0$$

In other words, *the pH of a neutral solution at 25 °C is 7.0.* This is an extremely important point, and will recur throughout the remainder of this discussion.

Since K_w is a constant, it follows that when $[OH^-]$ increases in the presence of a base, $[H_3O^+]$ must decrease, and vice versa. In particular the two concentrations are related by equation 12.2.9. Therefore equilibria involving alkalis can be expressed in terms of the hydrogen ion concentration. This is a very important chain of argument, and we shall now show what it means in practice.

When a base dissolves in water the ionization equilibrium is established and OH^- ions are generated. In the case of ammonia, for instance, since the ionization is so small, the value of $[NH_3]$ in the expression for K_b is almost equal to the concentration of ammonia added ($[NH_3]_{eq} \approx [NH_3]_{added} = B$) and the concentrations of OH^- and NH_4^+ are approximately the same; therefore equation 12.2.7 gives

$$K_b \approx \frac{[OH^-]^2_{eq}}{B} \quad \text{or} \quad [OH^-]_{eq} \approx \sqrt{(K_b B)}$$

The water ionization equilibrium adjusts so as to maintain $[H_3O^+] = K_w/[OH^-]$. Then following the same argument as before, we find

$$pH = pK_w - \tfrac{1}{2}pK_b + \tfrac{1}{2}\lg B \qquad (12.2.10)$$

and we can calculate the pH of any weak alkali.

The pK_b of a base is related to the pK_a of its conjugate acid. This can be shown as follows. Consider the base B and its ionization equilibrium

$$B(aq) + H_2O(l) \rightleftharpoons BH^+(aq) + OH^-(aq) \qquad K_b = \left\{\frac{[BH^+][OH^-]}{[B]}\right\}_{eq}$$

Now consider the ionization equilibrium of the conjugate acid, BH^+:

$$BH^+(aq) + H_2O(l) \rightleftharpoons H_3O^+(aq) + B(aq) \qquad K_a = \left\{\frac{[H_3O^+][B]}{[BH^+]}\right\}_{eq}$$

We can express K_a in terms of K_b. First, note that

$$K_a K_b = \left\{\frac{[H_3O^+][B][BH^+][OH^-]}{[BH^+][B]}\right\}_{eq} = K_w$$

Then taking logs and changing the sign results in:

$$pK_a + pK_b = pK_w \qquad (12.2.11)$$

The strength of a base can therefore be expressed in terms of the pK_a value of its conjugate acid. Thus instead of reporting that $pK_b(NH_3) = 4.75$, it can be reported that $pK_a(NH_4^+) = 14.00 - 4.75 = 9.25$.

EXAMPLE

Calculate the pH at 25 °C of (a) 0.1 mol dm^{-3} NH_3(aq), (b) the same concentration of NaOH(aq).

METHOD For the partially ionized ammonia, use equation 12.2.10 with pK_b taken from Table 12.3. The pH of a strong alkali can be obtained on the basis that it is completely ionized, and that $[OH^-] = [NaOH]$; then express $[H_3O^+]$ in terms of $[OH^-]$ through equation 12.2.9.

ANSWER (a) Since $pK_b = 4.75$, $pK_w = 14.00$, and $B = 0.1$ mol dm^{-3} use of equation 12.2.10 gives

$$pH = pK_w - \tfrac{1}{2}pK_b + \tfrac{1}{2}\lg B = 14.00 - \tfrac{1}{2} \times 4.75 + \tfrac{1}{2}\lg 0.1 = 11.1$$

(b) For the strong alkali at the same concentration,

$$[H_3O^+] = K_w/[OH^-] \text{ so that } pH = pK_w + \lg [OH^-] = pK_w + \lg B$$

where B is the concentration of base added. In the present case,

$$pH = 14.00 + \lg 0.1 = 13.0$$

COMMENT Both aqueous ammonia and aqueous sodium hydroxide are key industrial and laboratory alkalis. Note how the pH of the strong alkali is higher than that for ammonia even though the nominal concentrations are the same; this is because ammonia is only partially ionized, the OH^- ion concentration is lower, so that (on account of the water ionization equilibrium) the hydrogen ion concentration is higher.

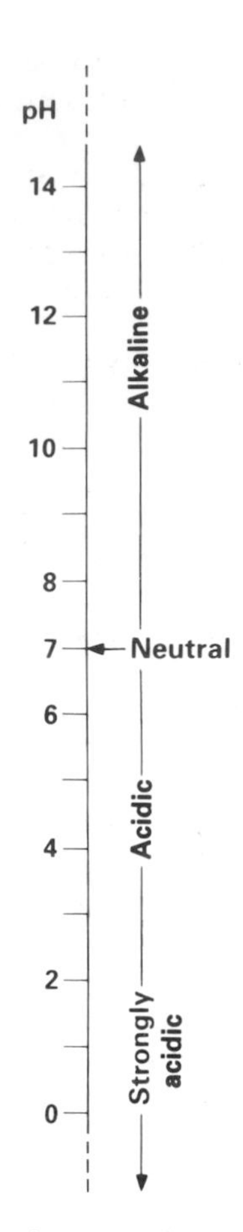

Figure 12.4 The pH scale

12.3 Applications of pH

The pH of a solution expresses the hydrogen ion concentration. From its definition, and as we have seen in the Examples, at 25 °C:

the pH of an acidic solution is less than 7.0,
the pH of a neutral solution is 7.0,
the pH of an alkaline solution is greater than 7.0.

There are no upper or lower limits to the values of pH, but for most applications it has values in the range 0 to 14 (see Figure 12.4), and so most devices for measuring pH (the most important of which we meet in the next chapter) are calibrated in that range.

Salt hydrolysis

We can now account for the behaviour of salts of weak acids and strong bases, and of strong acids and weak bases, when they are dissolved in water. All we have to bear in mind is that the hydrogen ion concentration adjusts so as to maintain the constancy *both* of the acid or base ionization constant *and* of the ionic product for water. The modification of the pH of a solution by a salt is called **salt hydrolysis**: the water is ionized as a result of responding to equilibria involving the salt's ions.

Consider sodium ethanoate in water. This is the salt of a weak acid (ethanoic acid, $pK_a = 4.75$) and a strong base (sodium hydroxide; almost fully ionized). When it dissolves it gives rise to ethanoate ions, $CH_3CO_2^-$. In aqueous solution there are two important equilibria:

$$H_2O(l) + H_2O(l) \rightleftharpoons H_3O^+(aq) + OH^-(aq) \qquad K_w = \{[H_3O^+][OH^-]\}_{eq}$$

$$CH_3COOH(aq) + H_2O(l) \rightleftharpoons H_3O^+(aq) + CH_3CO_2^-(aq)$$

$$K_a = \left\{\frac{[H_3O^+][CH_3CO_2^-]}{[CH_3COOH]}\right\}_{eq}$$

The second equilibrium lies strongly in favour of the non-ionized ethanoic acid. Therefore, when ethanoate ions are first added, it shifts to the left, using hydrogen ions provided by the first equilibrium. This takes place until enough CH_3COOH has been formed and the value of K_a is reached. In order to keep K_w constant, the first equilibrium must shift to the right. Since the H_3O^+ ions are consumed, an excess of OH^- ions is left in the solution. As a result of this hydrolysis, the solution of the salt is slightly alkaline, that is, its pH is greater than 7.0. A 0.1 mol dm^{-3} solution of sodium ethanoate, for example, is observed to have pH = 8. A particularly important example of a salt formed from a weak acid and a strong base is sodium carbonate, which is commonly used in the laboratory as an alkali.

A solution of a salt of a strong acid and a weak base, such as ammonium chloride, shows the opposite behaviour, and its hydrolysis leads to a slightly acidic solution. As a result of the presence of ammonium ions in the solution, the system establishes the following equilibria:

$$H_2O(l) + H_2O(l) \rightleftharpoons H_3O^+(aq) + OH^-(aq)$$

$$NH_3(aq) + H_2O(l) \rightleftharpoons NH_4^+(aq) + OH^-(aq) \qquad K_b = \left\{\frac{[NH_4^+][OH^-]}{[NH_3]}\right\}_{eq}$$

Since ammonia is a weak base ($pK_b = 4.75$) the equilibrium lies strongly to the left, and so the OH^- ions provided by the water ionization are removed. The hydrogen ion concentration rises in order to keep K_w constant, and as a result the pH of the solution falls below 7.0. A solution of a salt of a strong acid and a weak base is therefore slightly acidic: a 0.1 mol dm^{-3} solution of ammonium chloride in water has pH = 6 at 25 °C.

Acid–base titrations

The importance of hydrolysis equilibria lies in their applications, for example to the **equivalence-point** (or **end-point**) **of titrations**. At the equivalence-point of

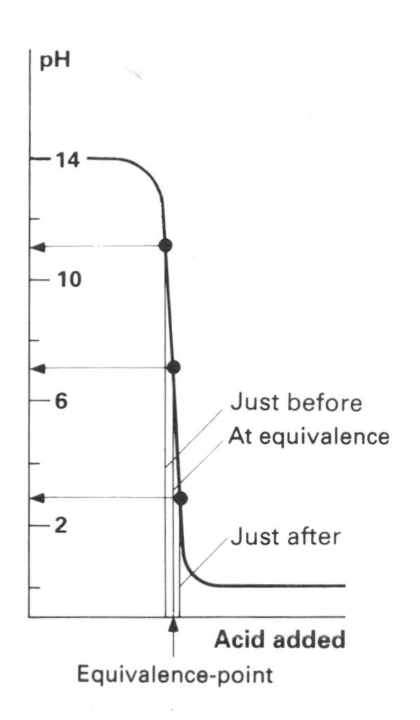

Figure 12.5 The rapid change of pH at the equivalence-point

an acid–base titration, precisely enough acid has been added to turn all the alkali initially present into a salt. In other words, at the exact equivalence-point there is a solution of a salt.

In the titration of a strong acid with a strong base, the salt produced is not hydrolysed and the only significant equilibrium in solution is the ionization of water. The solution is neutral, and its pH is 7. Hence, by measuring the pH of the solution during the titration, the equivalence-point is detected by noting when the pH reaches 7.

The detection of the equivalence-point of this titration is helped by the fact that pH changes very rapidly there. Suppose the titration is a few drops away from the equivalence-point, and the remaining alkali concentration is only 0.001 mol dm^{-3}; its pH is then 11 (from $[H_3O^+] = K_w/[OH^-]$). At the equivalence-point it is 7. Just after the equivalence-point, when acid is in slight excess – of the order of 0.001 mol dm^{-3}, for instance – the pH is about 3 (see Figure 12.5). Therefore, that very small shift of composition drives the pH rapidly from 11 down to 3. The sharp fall of pH allows the easy detection of the equivalence-point.

For the titration of a strong acid with a weak base, the equivalence-point also corresponds to a solution of the salt, but as a result of hydrolysis the solution is acidic. The equivalence-point is at a pH less than 7, as indicated in Figure 12.6(a). The actual position can be detected by noting where the pH changes most rapidly. The titration of a weak acid with a strong base is similar except that the equivalence-point lies to higher, more alkaline values of pH, as indicated in Figure 12.6(b).

The titration of a weak acid with a weak base is shown in Figure 12.6(c). This is more complicated to deal with because there are now three equilibria involved at the equivalence-point, and the sluggish variation of the pH makes the equivalence-point difficult to detect (and a conductometric titration, as described in Section 8.3, is used instead).

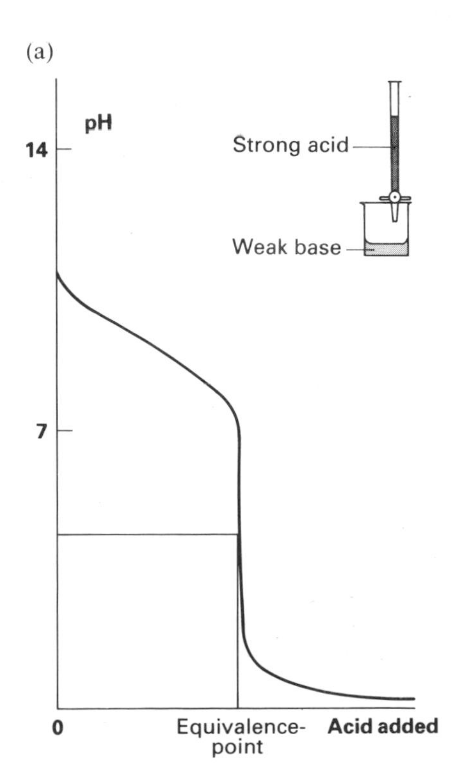

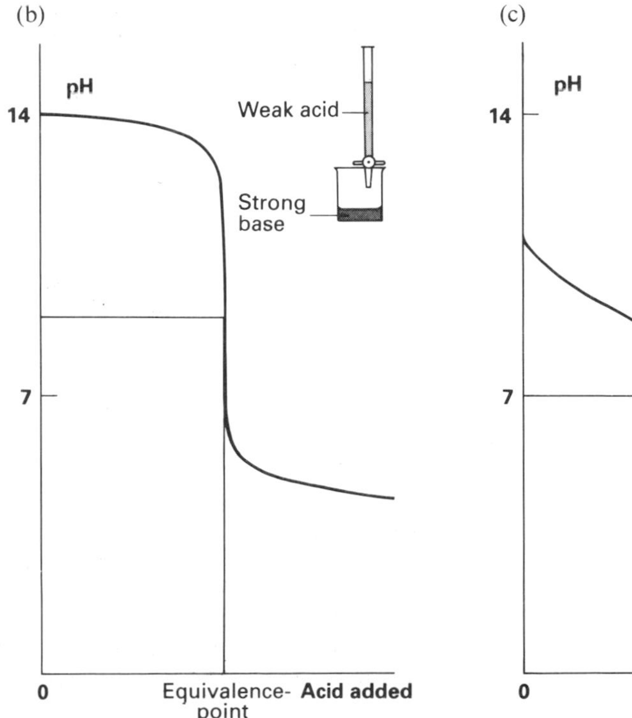

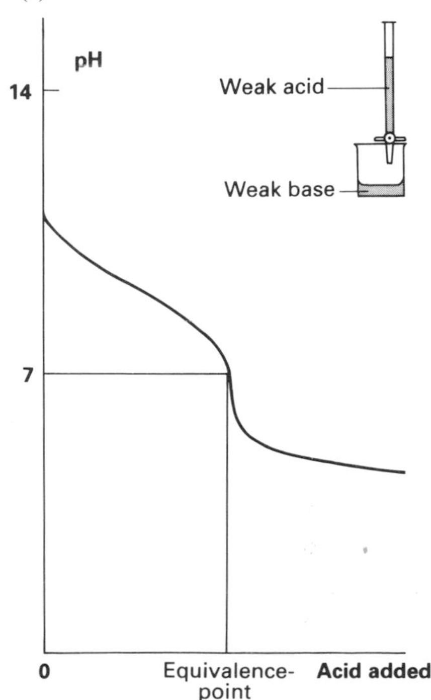

Figure 12.6 Titrations: (a) strong acid/weak base, (b) weak acid/strong base, (c) weak acid/weak base

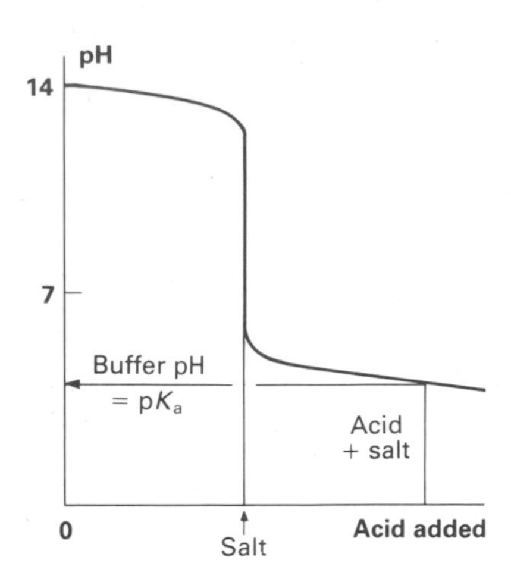

Figure 12.7 An acidic buffer region

Buffers

We have seen that at the equivalence-point of a titration there is a rapid change of pH for even small changes of concentration. In contrast, in the regions where there are approximately equal concentrations of weak acid and salt, the pH is fairly insensitive to the acid added. This **buffer region** is indicated in Figure 12.7.

Buffer solutions – solutions that resist changes of pH – are of immense importance in living systems because most metabolic processes have to occur at strictly controlled pH. A part of the reason is that the function of proteins depends on their structures, and structures depend on pH. This is because proteins possess acidic and basic groups and can exchange protons with the medium around them. When they do so they acquire electric charges, and electrostatic interactions then force the protein into a new shape. Only if that shape is correct will the metabolic process occur; otherwise the organism may die (see Figure 12.8). For instance, blood pH has to be maintained near 7.4, saliva functions at a pH near 6.6 and the enzymes acting in the stomach require an acidic environment with a pH in the region of 1.6 to 2.0. All these conditions are buffered in the way we shall now describe.

(a)

(b)

Figure 12.8 The consequences of acid rain; these photographs, of a forest site in West Germany, were taken in 1970 (a) and 1986 (b). Air pollution has resulted in an environment of such low pH that the buffer systems in the tree tissues can no longer cope

Suppose a solution consists of approximately equal concentrations of a weak acid (ethanoic acid, for instance) and its salt with a strong base (such as sodium ethanoate). There is an abundant supply both of ethanoate ions (from the salt) and of the non-ionized acid. The equilibrium

$$CH_3COOH(aq) + H_2O(l) \rightleftharpoons H_3O^+(aq) + CH_3CO_2^-(aq)$$

$$K_a = \{[H_3O^+][CH_3CO_2^-]/[CH_3COOH]\}_{eq}$$

is established. When a little more acid is added to this solution, the increase in hydrogen ion concentration is opposed (by Le Chatelier's principle, the equilibrium tends to minimize the change). This is possible because there are many ethanoate ions and the equilibrium can shift strongly to the left. The change in H_3O^+ ion concentration is therefore not as marked as when acid is added to water alone. Likewise, if a base is added, the equilibrium responds by shifting to the right and opposing the loss of the hydrogen ions. This is possible because there are plenty of non-ionized acid molecules. Hence the hydrogen ion concentration responds only slightly. Furthermore, we are interested in pH, the *logarithm* of a concentration. Whenever a logarithm is taken it tends to flatten out the change (for instance, on changing x from 1 to 1 000 000, lg x changes only from 0 to 6). Therefore although $[H_3O^+]$ might change slightly, the pH barely changes at all.

Buffering can be expressed quantitatively in terms of the acid ionization constant. Since there is added salt, we shall assume that the ethanoate ion concentration is due entirely to it, and therefore write $[CH_3CO_2^-]_{eq} \approx [Salt]_{added} = S$. We shall also assume that the acid is so little ionized that we can replace $[CH_3COOH]_{eq}$ by the concentration added, and write $[CH_3COOH]_{eq} \approx [Acid]_{added} = A$. The expression for K_a above then lets us write

$$[H_3O^+] = K_a \left\{ \frac{[CH_3COOH]}{[CH_3CO_2^-]} \right\}_{eq} \approx \frac{K_a A}{S}$$

Then, by taking logs and changing the sign,

$$pH = pK_a + \lg (S/A) \qquad (12.3.1)$$

This is called the **Henderson–Hasselbalch equation**. Since the logarithm is zero when $S = A$, *the pH of a solution containing equal concentrations of acid and salt is equal to the* pK_a *of the acid.* If the ratio S/A is not too far from unity, its logarithm changes only slowly as S/A varies. Consequently, *the pH of the buffer is close to the* pK_a *of the acid.* For example, since for phosphoric acid $pK_{a2} = 7.2$ (see Section 12.2), a solution containing $(HO)_2PO_2^-$ and $HOPO_3^{2-}$ ions in equal concentrations should be a suitable buffer to use near a pH of 7.

EXAMPLE

What is the pH of a buffer solution containing 0.1 mol dm^{-3} ethanoic acid and 0.1 mol dm^{-3} sodium ethanoate? What is its pH after the addition of 10 cm^3 of 1.0 mol dm^{-3} HCl(aq) to 1 dm^3 of the solution?

METHOD Use equation 12.3.1 with $S = A = 0.1$ mol dm^{-3}. Take the value of pK_a from Table 12.2. For the second part, the addition of strong acid results in a change in the concentrations of weak acid and salt; therefore, find the new values of A and S following the addition of acid, and use equation 12.3.1 again.

ANSWER From Table 12.2 we have $pK_a = 4.75$. Since initially $S = A$, $S/A = 1$, $\lg 1 = 0$, and so from equation 12.3.1, $pH = pK_a = 4.75$. This is the pH of the buffer solution. 10 cm^3 of 1.0 mol dm^{-3} HCl(aq) contains $(10 \times 10^{-3}\ dm^3) \times (1.0\ mol\ dm^{-3}) = 0.01\ mol\ H_3O^+(aq)$. 1 dm^3 of the buffer solution contains 0.1 mol of $CH_3CO_2^-$ and 0.1 mol of CH_3COOH. When the acid is added, the amount of substance of salt in the buffer solution is reduced from 0.10 mol to 0.09 mol, and the amount of substance of acid is increased from 0.10 mol to 0.11 mol. The new concentrations are $A = 0.11$ mol dm^{-3}, $S = 0.09$ mol dm^{-3}. Therefore, using equation 12.3.1,

$$pH = pK_a + \lg(S/A) = 4.75 + \lg(0.09/0.11) = 4.66$$

COMMENT If the same quantity of strong acid had been added to 1 dm^3 of pure water, the pH would have changed from 7.0 to 2.0; the stabilizing effect of the buffer on the pH of the medium is very evident.

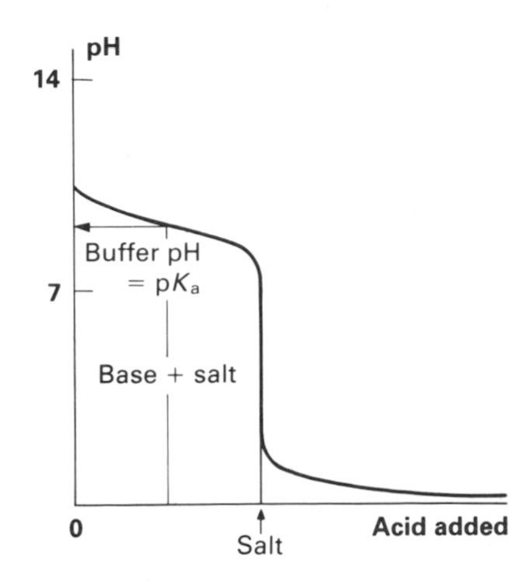

Figure 12.9 An alkaline buffer region

A weak base and its salt can buffer at alkaline pH; a typical example is aqueous ammonia/ammonium chloride, which buffers close to a pH of 9.25 (the pK_a of ammonia's conjugate acid – see Figure 12.9).

Indicators

A common way of detecting the equivalence-point of an acid–base titration is to use an **indicator** – a substance (a dye) that changes colour according to the pH of the medium, such as methyl orange (Figure 12.10). The colour change arises from the attachment or removal of a proton from the substance, which

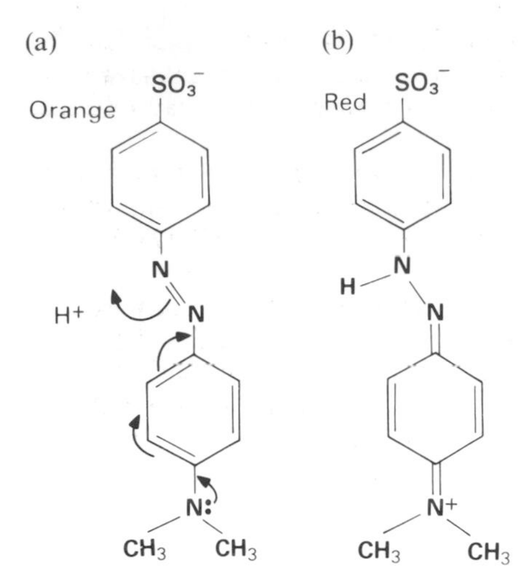

Figure 12.10 Methyl orange: (a) in its base (orange) form, and (b) in its conjugate acid (red) form

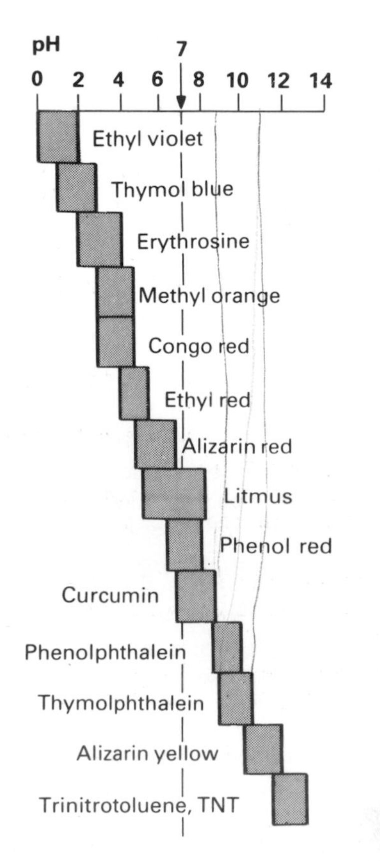

Figure 12.11 The pH range for the colour changes of various indicators

changes the energy levels of the molecule. We denote the protonated form as HIn and the unprotonated form as In^-. Then HIn, a weak acid, is in equilibrium with its conjugate base In^-:

$$HIn(aq) + H_2O(l) \rightleftharpoons H_3O^+(aq) + In^-(aq)$$

$$K_{In} = \left\{\frac{[H_3O^+][In^-]}{[HIn]}\right\}_{eq} \qquad pK_{In} = -\lg K_{In}$$

As usual, we have absorbed the water concentration into the equilibrium constant and introduced the **indicator ionization constant,** K_{In}.

The interpretation of how an indicator responds to pH follows the usual arguments about equilibria. For instance, suppose the pH is in the strongly acidic region (pH near 0), then there are many hydrogen ions in the solution. Therefore, in order to maintain the constancy of K_{In}, $[In^-]$ must fall and [HIn] must rise. Consequently, in acidic solution the indicator is mainly in the protonated form and shows its characteristic 'acidic' colour. In contrast, in alkaline solution, when the concentration of hydrogen ions is low, the indicator equilibrium requires $[In^-]$ to increase and [HIn] to decrease. Therefore the indicator is then mainly present as In^-, and shows its characteristic 'alkaline' colour. It follows that when the solution swings from low pH to high (or vice versa) the indicator swings from being predominantly HIn to In^- (or vice versa), and we see the corresponding change of colour. The colour begins to change noticeably when $[HIn] \approx 10[In^-]$, i.e. at about $pH = pK_{In} - 1$ and is nearly complete when $[HIn] \approx [In^-]/10$, at about $pH = pK_{In} + 1$.

The precise range over which an indicator changes predominantly from one form to another depends on its pK_{In} (see Figure 12.11). This is valuable when seen in relation to the discussion about the equivalence-points of different types of acid–base titration. For instance, we have seen that as a result of salt hydrolysis the titration of a strong acid and a weak base has an equivalence-point in the acidic region. Therefore an indicator is needed that changes colour at around pH = 4. Figure 12.11 shows that methyl orange is suitable. In contrast, for the alkaline equivalence-point of a weak acid/strong base titration, an indicator with a colour change around pH = 9 must be chosen; phenolphthalein, which is red above pH = 10 and colourless below pH = 8, is suitable.

The colours of several of these indicators are familiar in everyday life, as they are some of the permitted colourings of foodstuffs; erythrosine, for example, is the additive coded E127. Colour 3 illustrates the dependence on pH of the colours of certain plant pigments.

12.4 Lewis acids and bases

We have concentrated on the Brønsted–Lowry definition of acids and bases. This, it should be recalled, is a generalization of a primitive definition of acids and bases. There is, however, an even more general definition. It captures everything we have done so far but is much broader in scope. Breadth of scope, as we have stressed before, means simplification because it unifies what may seem to be quite different concepts.

The Lewis definition was introduced by G. N. Lewis (Box 2.1) in the same year in which the Brønsted–Lowry definition appeared, and runs as follows:

a **Lewis acid** is an electron pair acceptor,
a **Lewis base** is an electron pair donor.

Lewis acids (and bases) include Brønsted acids (and bases), but with a slight shift of emphasis. Thus because protons attach to pairs of electrons (as in NH_3, which becomes NH_4^+), an electron pair donor is also a proton acceptor, and for bases the two definitions are essentially the same. Likewise, the outstanding example of an electron pair acceptor is the proton; the proton is therefore a Lewis acid and, as we have seen, substances that can provide protons are Brønsted acids. However, Lewis acids include many substances other than proton donors. Thus boron trifluoride, BF_3, cannot be classified under the Brønsted–Lowry definition, yet it forms a compound with ammonia (see Section 2.6): the BF_3 molecule is an electron pair acceptor, a Lewis acid, and the NH_3 molecule is a Lewis base.

The Lewis definition goes much further than the Brønsted–Lowry definition because it also includes reactions such as the formation of complex ions with ligands (see Section 15.5), including water molecules to form $[Fe(H_2O)_6]^{2+}$ (Section 2.6), and halide ions to form $AgCl_2^-$ (Section 12.1).

Under the Lewis definition, the whole of the chemistry of acids and bases can be regarded in terms of the availability of electron pairs, and this fits much more naturally into the general view that chemistry is principally concerned with the behaviour of electrons. It does not entirely supersede the Brønsted–Lowry definition, however, because the Lewis theory is essentially qualitative whereas the Brønsted–Lowry theory can be discussed quantitatively in terms of pH and pK_a.

Summary

- ☐ A saturated solution is one in which no more of the solute will dissolve; the solution is in dynamic equilibrium with undissolved solid.
- ☐ The solubility product for a salt M_xA_y is $K_{sp} = \{[M^{m+}]^x[A^{a-}]^y\}_{eq}$. Solubility products are important for sparingly soluble salts, and can be found by measuring the ion concentrations in the saturated solutions.
- ☐ The solubility of a sparingly soluble salt MA is equal to $\sqrt{K_{sp}}$ at the temperature specified.
- ☐ Addition of a common ion causes precipitation of a sparingly soluble salt. Increased solubility when a common ion is added indicates the formation of a complex ion.
- ☐ A Brønsted acid is a proton donor. A Brønsted base is a proton acceptor. Brønsted acid–base reactions are neutralization reactions. An alkali is a water-soluble base.
- ☐ To every acid there is a conjugate base, formed by loss of a proton. To every base there is a conjugate acid, formed by gain of a proton.
- ☐ The classification of a substance as an acid or as a base depends on the other substance present in the solution.
- ☐ A strong acid is almost fully ionized. A weak acid is only partially ionized. The acid ionization constant of an acid HA is $K_a = \{[H_3O^+][A^-]/[HA]\}_{eq}$; by definition, $pK_a = -\lg K_a$.
- ☐ The pH of a solution is defined as $pH = -\lg\{[H_3O^+]/\text{mol dm}^{-3}\}$. The pH of neutral water is 7.0 (at 25 °C). Acidic solutions have pH less than 7.0 while alkaline solutions have pH greater than 7.0.
- ☐ Polyprotic acids can donate more than one proton.
- ☐ The base ionization constant of a base B is $K_b = \{[BH^+][OH^-]/[B]\}_{eq}$; by definition, $pK_b = -\lg K_b$.
- ☐ The ionic product for water is $K_w = \{[H_3O^+][OH^-]\}_{eq}$; the concentrations of hydrogen and hydroxide ions are linked by this equilibrium.
- ☐ The salts of weak acids with strong bases are hydrolysed in water and are mildly alkaline. The salts of strong acids with weak bases are also hydrolysed, and are mildly acidic.
- ☐ The equivalence-point of an acid–base titration, except for titrations of weak acids with weak bases, can be determined using pH measurement; the pH changes sharply at the equivalence-point.
- ☐ A buffer solution resists changes of pH; an acidic buffer consists of a solution of a weak acid and one of its salts in approximately equal concentrations.
- ☐ Titration equivalence-points are frequently determined using indicators, dyes with colours that depend on pH. The selection of the indicator must take into account the effects of salt hydrolysis.
- ☐ A Lewis acid is an electron pair acceptor; a Lewis base is an electron pair donor. This represents a generalization of acid–base classification.
- ☐ Lewis acid-base reactions include neutralization and complexation.

PROBLEMS

1 Describe what happens when common salt is added to a saturated solution of silver chloride in water.

2 Give *three* examples of complexation reactions.

3 Discuss the effect of temperature on the solubility of a sparingly soluble salt in the same terms as were used for the effect of temperature on an equilibrium (see Chapter 10). The standard dissolution enthalpy of silver chloride is $+65\,kJ\,mol^{-1}$. Is silver chloride more or less soluble at higher temperatures?

4 The solubility product of zinc hydroxide is given in Table 12.1. What is its solubility? What is its solubility in $1.0 \times 10^{-4}\,mol\,dm^{-3}$ sodium hydroxide solution?

5 The conductivity of a $0.0634\,mol\,dm^{-3}$ solution of 2-hydroxypropanoic acid (lactic acid, the acid responsible for 'cramp') was measured as $1.138 \times 10^{-3}\,S\,cm^{-1}$. Its limiting molar conductivity is $388.5\,S\,cm^2\,mol^{-1}$. What is (a) the pH of the solution, (b) the acid ionization constant?

6 Calculate the pH of $0.1\,mol\,dm^{-3}$ sulphuric acid.

7 Explain why pK_w increases with temperature.

8 Define the terms *Brønsted acid* and *Brønsted base*. Classify as acids or bases, where possible, the following species: (a) hydrogen chloride in water, (b) sodamide ($NaNH_2$) in liquid ammonia, (c) iron(III) chloride in water.

9 Outline the Lewis definition of acids and bases, and classify, where possible, the three substances in the last question, giving reasons.

10 (a) (i) What is a buffer solution?
(ii) What is meant by a weak acid?
(b) Ethanoic acid is a weak acid with a pK_a value of 4.76 at 298 K.
(i) Write an expression for K_a for aqueous ethanoic acid.
(ii) Calculate the value of K_a.
(iii) Calculate the pH of an aqueous solution of ethanoic acid, CH_3COOH, containing $0.25\,mol\,dm^{-3}$.
(c) Describe how an aqueous solution of ethanoic acid and sodium ethanoate behaves as a buffer solution.
(AEB 1985)

11 At 20 °C, the solubility product of strontium sulphate is $4.0 \times 10^{-7}\,mol^2\,dm^{-6}$, and that of magnesium fluoride is $7.2 \times 10^{-9}\,mol^3\,dm^{-9}$. Estimate to two significant figures the solubility at 20 °C in $mol\,dm^{-3}$ of
(a) strontium sulphate in a 0.1 M solution of sodium sulphate,
(b) magnesium fluoride in a 0.2 M solution of sodium fluoride. *(Oxford)*

12 Distinguish carefully between the terms *solubility* and *solubility product* as applied to a sparingly soluble electrolyte.
What happens when aqueous solutions of the following substances are added to a saturated aqueous solution of lead chloride containing a small excess of undissolved solid?
(a) sodium chloride,
(b) sodium iodide,
(c) silver nitrate.
Account quantitatively for the fact that copper(II) sulphide can be precipitated by hydrogen sulphide from an acidic solution (assume $0.1\,mol\,dm^{-3}$ of free H^+) whereas iron(II) sulphide cannot be precipitated under these conditions.
(Oxford and Cambridge S part question)

13 (a) In each of the following reactions one of the reactants is behaving as a base. In each case underline the species you believe to be the base.
(i) $HSO_4^- + HNO_2 \rightarrow H_2NO_2^+ + SO_4^{2-}$
(ii) $H_2PO_4^- + HCO_3^- \rightarrow HPO_4^{2-} + H_2O + CO_2$
(iii) $CH_3CO_2H + HNO_3 \rightarrow CH_3CO_2H_2^+ + NO_3^-$
(iv) $HBr + HCl \rightarrow Br^- + H_2Cl^+$
(b) (i) Using any one of the systems given above as an example, give and explain briefly the Brønsted–Lowry theory of acids and bases.
(ii) According to the theory, water is able to function both as an acid and as a base. Give *one* reaction in which water functions as an acid and *another* in which it functions as a base.
(c) Comment on the fact that a mixture of pure nitric acid, HNO_3, and pure perchloric acid, $HClO_4$, contains the ions $H_2NO_3^+$ and ClO_4^-. *(London)*

14 (a) Define the terms *Brønsted acid* and *Brønsted base*.
(b) Explain whether each of the following species normally acts as an acid or a base or both in (i) aqueous solution and (ii) liquid ammonia: NH_4^+, HSO_4^-, NH_2^-, NH_3.
(c) Discuss the factors which have to be taken into account in determining the equivalence-point (end-point) in the titration of an aqueous solution of ethanoic acid (acetic acid) with sodium hydroxide solution. *(Welsh)*

15 (a) Explain why an aqueous solution of ammonium chloride is acidic.
(b) When aqueous solutions of ammonium chloride and ammonia are mixed, a buffer solution can be produced.
(i) What is a buffer solution?
(ii) Explain the buffer action of this mixture.
(c) What is the hydrogen ion concentration of a 0.01 M aqueous solution of methylamine at 25 °C? K_b for methylamine is $4 \times 10^{-4}\,mol\,l^{-1}$ [$mol\,dm^{-3}$] and K_w for water is $10^{-14}\,mol^2\,l^{-2}$ [$mol^2\,dm^{-6}$] both at 25 °C.
(JMB)

16 What is the definition of pH? Describe how you would measure the pH of an unknown solution (*not* using an indicator).
Discuss and evaluate the following statements.
(a) More hydrogen will be evolved when excess zinc is added to $50\,cm^3$ of hydrochloric acid of concentration $0.100\,mol\,dm^{-3}$ than when excess zinc is added to an equal volume of ethanoic (acetic) acid of the same concentration.

(b) The same volume of a standard solution of sodium hydroxide is needed to neutralize equal volumes of hydrochloric acid and ethanoic acid of the same concentration.

Calculate the pH of the following solutions:
(i) nitric acid of concentration 0.025 mol dm^{-3},
(ii) a saturated solution of 4-hydroxybenzoic acid which contains 6.50 g dm^{-3} at 25 °C ($pK_a = 4.58$),
(iii) a buffer solution made by mixing 60 cm^3 of ethanoic acid of concentration 0.100 mol dm^{-3} with 40 cm^3 of sodium hydroxide of the same concentration.

(Oxford and Cambridge)

17 Explain qualitatively why aqueous solutions of sodium ethanoate are not neutral. A hydrolysis constant K_h can be defined as shown below, and has the value given at 298 K.

$$K_h = \frac{[CH_3CO_2H][OH^-]}{[CH_3CO_2^-]} = 5.7 \times 10^{-10}\,\text{mol dm}^{-3}$$

Calculate the concentration of hydroxide ion in an 0.08 M [mol dm^{-3}] aqueous solution of sodium ethanoate at 298 K.

Some information about three titration indicators is shown in the table.

Name	*Colour in 1M acid*	*Colour in 1M alkali*	*Useful pH range*
bromocresol green	yellow	blue	3.8–5.4
phenolphthalein	colourless	pink	8.3–10.0
4-nitrophenol	colourless	yellow	5.6–7.6

(a) Explain why 4-nitrophenol can be used as an indicator for acid-base titrations.
(b) Which of the indicators in the table would be the most appropriate to use in the titration of ethanoic acid with sodium hydroxide?
(c) How are the figures for the 'useful pH range' obtained?

Comment on the following observation. Addition of a small quantity of bromocresol green to 0.1 M aqueous ammonium chloride at 298 K results in a blue coloration. If the solution is warmed to 360 K in the absence of air the colour changes to green; the blue colour is restored on cooling.

(Oxford)

18 (a) Define the terms 'Lewis acid' and 'Lewis base' and give one example of each.
(b) Define the terms 'acid' and 'base' on the Brønsted–Lowry theory and give one example of each.
(c) Explain why
(i) CCl_3COOH is a stronger acid than CH_3COOH.
(ii) the $[Al(H_2O)_6]^{3+}$ ion is a stronger acid than the $[Mg(H_2O)_6]^{2+}$ ion.

(JMB)

19 In a buffered solution

$$pH = -\lg K_a + \lg\frac{[\text{base}]_{eqm}}{[\text{acid}]_{eqm}}$$

Human plasma is buffered mainly by dissolved carbon dioxide which has reacted to form carbonic acid.

$$H_2CO_3(aq) \rightleftharpoons H^+(aq) + HCO_3^-(aq)$$

(a) Explain how carbonic acid can buffer human plasma. Give an example to illustrate your answer.
(b) When the concentrations of carbonic acid and hydrogencarbonate ion are equal, the concentration of hydrogen ions is 7.9×10^{-7} mol dm^{-3}. Calculate the value of $\lg K_a$ for carbonic acid.
(c) Usually the pH of human plasma is about 7.4. Calculate the *ratio* of the concentrations of hydrogencarbonate ion and carbonic acid in plasma.
(d) If the total concentration of hydrogencarbonate ion and carbonic acid was equivalent to 2.52×10^{-2} mol dm^{-3} of carbon dioxide, calculate the separate *concentrations* of the hydrogencarbonate ion and the carbonic acid in plasma.

(Nuffield)

20 Explain what is meant by pH and describe briefly one method of measuring it.

A weak acid HA has an ionization constant $K_a = 2 \times 10^{-5}$ mol dm^{-3}. A solution S is made up with 0.1 mole of HA in 1 dm^3, and a second solution P has 0.1 mole of HA and 0.5 mole of the sodium salt of the weak acid (NaA) in 1 dm^3. Calculate the pH of solutions S and P, stating any approximations that are made. If 0.01 mole of a strong acid HX is added to solution S, and an identical amount is added to solution P, what are the new pH values? Comment on your results.

(Oxford Entrance)

13 Electrochemistry

The generation and storage of electric power in electric cells (including batteries, fuel cells and biological cells) is a feature of everyday life, and the chemistry of these devices is described in this chapter. We see that the transport of electrons from one substance to another in the course of a chemical reaction can be harnessed to do work, and that measurements of the e.m.f.s (voltages) of cells can be interpreted in terms of the thermodynamic characteristics of the reactions taking place within them. Tables of data compiled from e.m.f. measurements can be used to predict the values of equilibrium constants for a wide variety of reactions and applied in many important areas of chemistry, including corrosion, pH and titrations.

Electrochemistry deals with a wide range of important phenomena, many of which depend on the transfer of electrons from one substance to another. It developed from investigations into the production of electrical power by chemical reactions, but is now much more widely applied. It includes the development of transportable, efficient sources of energy, such as the fuel cells that are already used in spacecraft and are under investigation for road vehicles, and devices capable of generating electricity directly from solar radiation (see Figure 13.1). Electrochemistry now includes aspects of physiology, for the nerve action inside our bodies depends on electrochemical effects: even to read and think about electrochemistry involves electrochemistry!

Electrochemical techniques are also available for measuring and predicting the thermodynamic properties of substances and reactions. They can be used, as we will see, for calculating equilibrium constants and reaction Gibbs energies. They can also be used analytically, because they can be adapted to measure pH and to control automatic titrations.

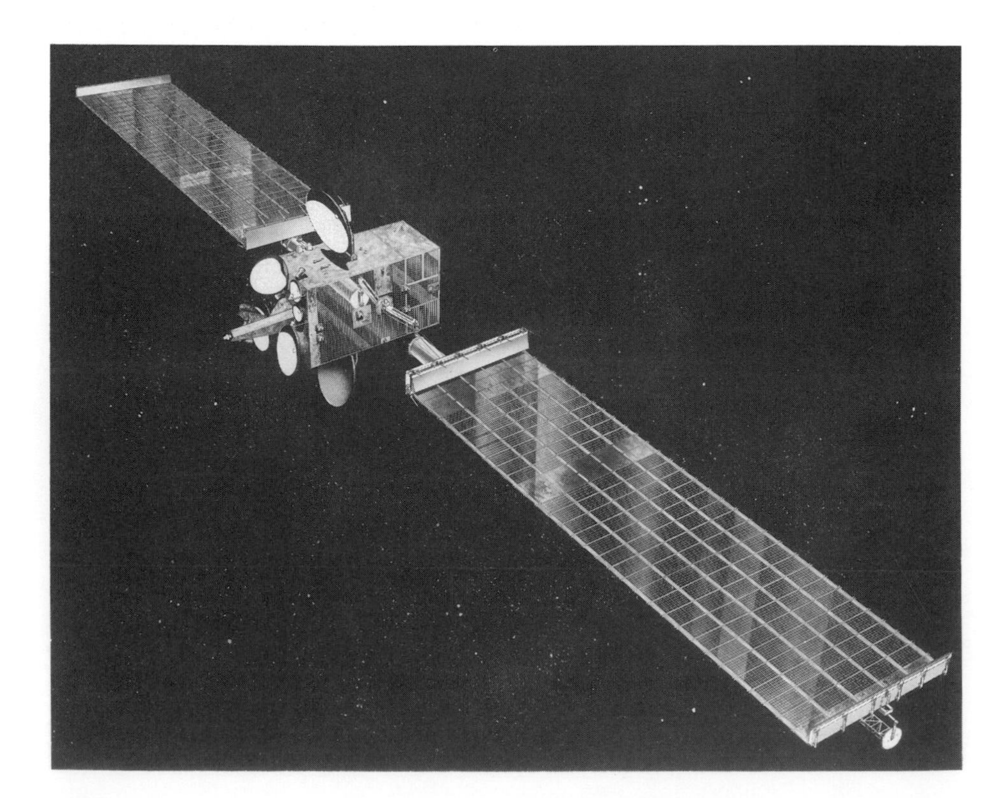

Figure 13.1 The British Aerospace *Olympus 1* communication satellite is powered by solar panels, which have a span of 27 metres

13.1 Electrochemical cells

When a piece of zinc is placed in an aqueous solution of copper(II) sulphate the blue colour fades as a result of the reaction

$$Zn(s) + CuSO_4(aq) \rightarrow ZnSO_4(aq) + Cu(s) \qquad (13.1.1)$$

The sulphate ions remain in the solution unchanged: they are merely **spectator ions**. The net reaction can therefore be expressed more simply as the **net ionic equation**

$$Zn(s) + Cu^{2+}(aq) \rightarrow Zn^{2+}(aq) + Cu(s) \qquad (13.1.2)$$

This equation can be written formally as the sum of the **half-equations** for the **half-reactions**. In one half-reaction zinc donates electrons:

$$Zn(s) \rightarrow Zn^{2+}(aq) + 2e^-$$

In the other half-reaction copper(II) ions accept them:

$$Cu^{2+}(aq) + 2e^- \rightarrow Cu(s)$$

The state of the electrons is not specified because they are regarded as 'in transit'.

The net ionic reaction proceeds in the direction shown in equation 13.1.2, which indicates that zinc has a greater tendency to supply electrons and go into solution than has copper (see Section 13.5). That is, the zinc half-reaction drives the copper half-reaction, and not vice versa; in other words, zinc is more **electropositive** than copper. This suggests that it may be possible to set up a scale of electropositive character, and hence to predict which half-reactions can drive a given half-reaction.

As shown in Figure 13.2, the zinc/copper reaction takes place as electrons are transferred in random directions between the metal and the solution: the reaction enthalpy released heats the solution. Suppose, however, that a way could be found of allowing the zinc to supply electrons to an external circuit and for them to return to the system through an electronic conductor (graphite or a metal), an **electrode**, and attach to the copper(II) ions, as in Figure 13.3. The net ionic reaction is the same as before, but as it takes place electrons flow in an orderly way through an external circuit and hence can be used to do useful work, such as driving an electric motor. That is, the energy liberated by the reaction can be used to do *work* rather than being wasted in heating the solution.

The replacement of a disorderly transfer of electrons by an orderly flow through an external circuit is the basis of all electrochemical power production. **Galvanic cells** (or **voltaic cells**), such as the cells used in portable

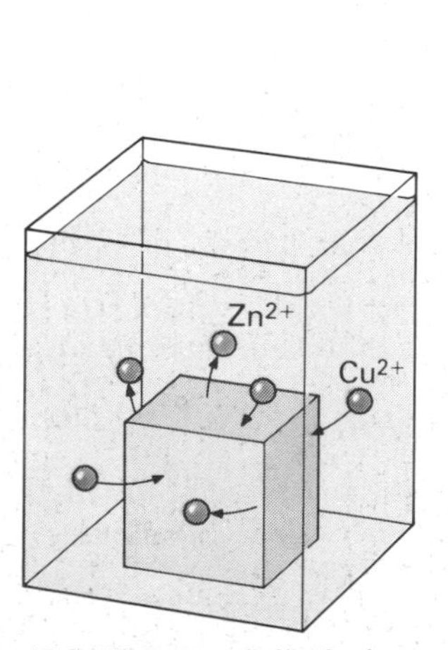

Figure 13.2 Zinc metal displacing copper ions

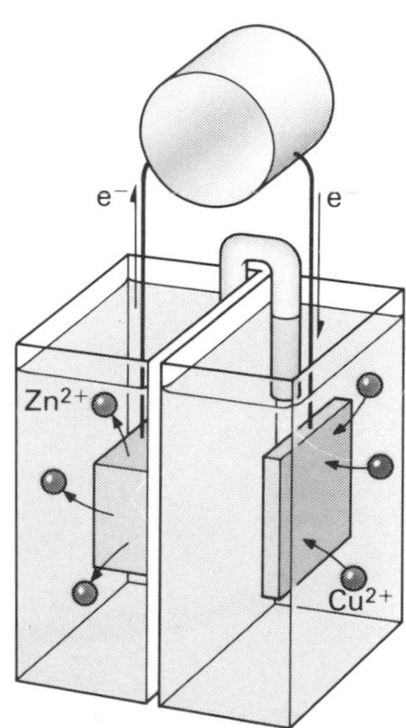

Figure 13.3 Zinc metal displacing copper ions with a flow of electrons through an external circuit

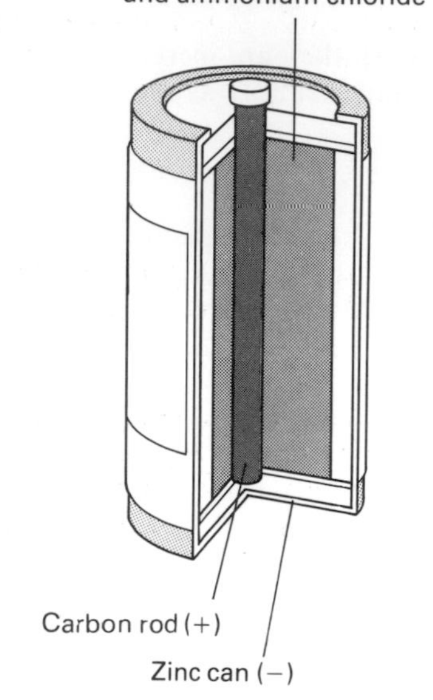

Figure 13.4 A dry cell

electric equipment, are containers in which a chemical reaction produces electricity. They have the general structure

Electrode|Electrolyte|Electrode

where the **electrolyte** is an ionically conducting medium (either an ionic solution or a molten salt). The zinc/copper galvanic cell is called the **Daniell cell** after its inventor. The familiar **dry cell**, illustrated in Figure 13.4, is powered by a reaction in which the zinc can acts as one electrode and supplies electrons through the external circuit and back through the central carbon rod electrode. The electrolyte is a paste of powdered carbon, manganese(IV) oxide and ammonium chloride (see Box 25.1).

BOX 13.1

The car battery

Most vehicles need a rechargeable electric storage system. By far the most common at present is the **lead–acid battery**, the car battery. The battery consists of six cells, each one capable of producing about 2 V, so that overall the battery can produce 12 V.

As in all rechargeable storage cells, the operation of the battery depends on there being *reversible* redox reactions at each electrode. In a lead–acid cell, the electrodes are lead–antimony alloy frameworks holding spongy lead and lead(IV) oxide alternately. The electrolyte is dilute sulphuric acid.

During discharge (current generation), the negative electrode is the spongy lead. In the presence of the sulphuric acid the overall oxidation reaction is

$$Pb(s) + SO_4^{2-}(aq) \rightarrow PbSO_4(s) + 2e^-$$

The electrons released at this anode pass through the external circuit, and reduce the lead(IV) oxide electrode. The cathode is in contact with sulphuric acid, and the reduction reaction is

$$PbO_2(s) + 4H^+(aq) + SO_4^{2-}(aq) + 2e^- \rightarrow PbSO_4(s) + 2H_2O(l)$$

The overall reaction is

$$Pb(s) + PbO_2(s) + 2(HO)_2SO_2(aq) \rightarrow 2PbSO_4(s) + 2H_2O(l)$$

The significance of this reaction is that, during discharge, sulphuric acid is replaced by water, and so the state of charge of each cell can be monitored by measuring the density of the electrolyte with a hydrometer.

The charging process uses an external source of current to drive these reactions in reverse, so that at one electrode lead(II) sulphate is reduced to lead and at the other it is oxidized to lead(IV) oxide.

13.2 Redox reactions

The scope of electrochemistry is considerably broadened by generalizing two concepts that were identified by the early chemists: oxidation and reduction.

Oxidation

The classical meaning of **oxidation** is reaction with oxygen. For example, zinc reacts with oxygen as follows:

$$2Zn(s) + O_2(g) \rightarrow 2ZnO(s)$$

A *formal* way of expressing this reaction (that is, one that leads to the same overall reaction without signifying how it actually takes place) is to write it as the sum of two half-reactions:

$$2Zn(s) \rightarrow 2Zn^{2+}(s) + 4e^-$$
$$O_2(g) + 4e^- \rightarrow 2O^{2-}(s)$$

$$2Zn(s) + O_2(g) \rightarrow 2ZnO(s)$$

Similar pairs of half-equations can be written for the oxidation of other metals and, more significantly, for the reaction of zinc with chlorine:

$$Zn(s) \rightarrow Zn^{2+}(s) + 2e^-$$
$$Cl_2(g) + 2e^- \rightarrow 2Cl^-(s)$$

$$Zn(s) + Cl_2(g) \rightarrow ZnCl_2(s)$$

This suggests that the common unifying feature is the loss of electrons by the zinc. For this reason, the modern definition of an oxidation is as follows:

Oxidation is the loss of electrons.

This definition concentrates on the central unifying feature of a wide variety of reactions, the behaviour of electrons; it makes no mention of oxygen. Oxidation reactions without oxygen are perfectly possible, as in the oxidation of iron(II) ions to iron(III) ions in solution, which is formally the electron loss

$$Fe^{2+}(aq) \rightarrow Fe^{3+}(aq) + e^-$$

Reduction

Early chemists called loss of oxygen, often by reaction with hydrogen, a **reduction**, because typically a metallic oxide was reduced to the metal. For example, black copper(II) oxide becomes coated with shiny pink copper when heated in hydrogen:

$$CuO(s) + H_2(g) \xrightarrow{\text{heat}} Cu(s) + H_2O(g)$$

This reaction may formally be divided into half-reactions, the copper half-equation being

$$Cu^{2+}(s) + 2e^- \rightarrow Cu(s)$$

A similar half-equation describes the reduction of other metal oxides. Therefore the modern definition of a reduction is as follows:

Reduction is the gain of electrons.

This definition extends the meaning of reduction to reactions in which hydrogen is not involved, such as the reverse of the iron(II) oxidation:

$$Fe^{3+}(aq) + e^- \rightarrow Fe^{2+}(aq)$$

which is the reduction of iron(III) ions. Note that the reverse of any reduction half-reaction is an oxidation half-reaction, and vice versa.*

The combination of reduction and oxidation

The zinc/copper reaction described in Section 13.1 can now be seen to be a combination of two half-reactions: the reduction of copper(II) ions and the oxidation of zinc metal. It is therefore an example of a **redox reaction**, the sum of a **red**uction and an **ox**idation half-reaction in which electrons are transferred from one substance to another. The substance that causes reduction (Zn) is the **reducing agent** and is itself oxidized; the substance that causes oxidation (Cu^{2+}) is the **oxidizing agent**, and is itself reduced.

Treating a redox reaction as the sum of half-reactions makes it easy to write the balanced equation: the number of electrons supplied in the oxidation half-reaction must equal the number used in the reduction half-reaction. The following Example explains the technique.

* *One way of remembering the distinction between oxidation and reduction is by the mnemonic OIL RIG: Oxidation Is Loss, Reduction Is Gain (of electrons).*

EXAMPLE

Write the balanced equation for the oxidation of iron(II) ions by acidified dichromate(VI) ions.

METHOD Write the reduction and oxidation half-reactions, and then add them after multiplying by factors that match the numbers of electrons involved in each step.

ANSWER The iron(II) oxidation half-reaction is

$$Fe^{2+}(aq) \rightarrow Fe^{3+}(aq) + e^-$$

The dichromate(VI) reduction half-reaction is

$$Cr_2O_7^{2-}(aq) + 14H^+(aq) + 6e^- \rightarrow 2Cr^{3+}(aq) + 7H_2O(l)$$

Multiply the first half-reaction by 6 to match the numbers of electrons:

$$Cr_2O_7^{2-}(aq) + 14H^+(aq) + 6Fe^{2+}(aq) \rightarrow 2Cr^{3+}(aq) + 6Fe^{3+}(aq) + 7H_2O(l)$$

COMMENT Many useful half-reactions are given in Table 13.1, and can be combined together in a similar way.

It will become increasingly clear that redox reactions are an extremely important part of chemistry, and that together with Lewis acid–base reactions, discussed in Section 12.4, account for most of the chemical properties of substances.

13.3 Electrode potentials

A typical arrangement for a galvanic cell is shown in Figure 13.5. It consists of two **half-cells** in electrical contact, often through an ionically conducting medium (commonly a concentrated solution of potassium chloride or ammonium nitrate) called a **salt bridge**. Each half-cell consists of an electrode, which is normally a metallic conductor, and an electrolyte, which is normally an aqueous solution of ions. When an electric circuit is completed, as in Figure 13.6, the zinc electrode loses electrons and the zinc ions dissolve into the solution. The electrons travel through the external circuit and re-enter the cell at the copper electrode where they attach to the copper(II) ions in the solution; the ions deposit as metallic copper.

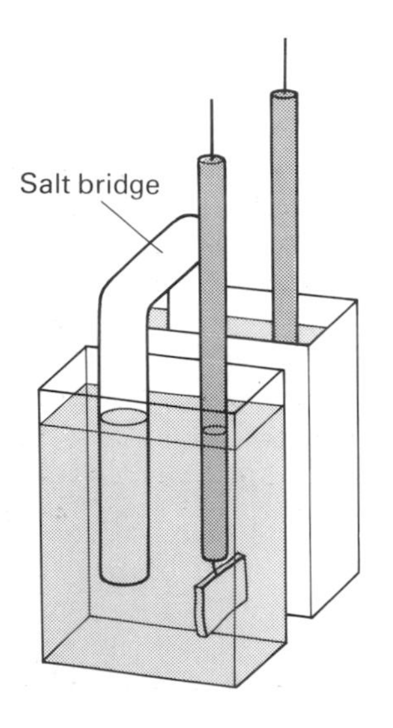

or, more simply :

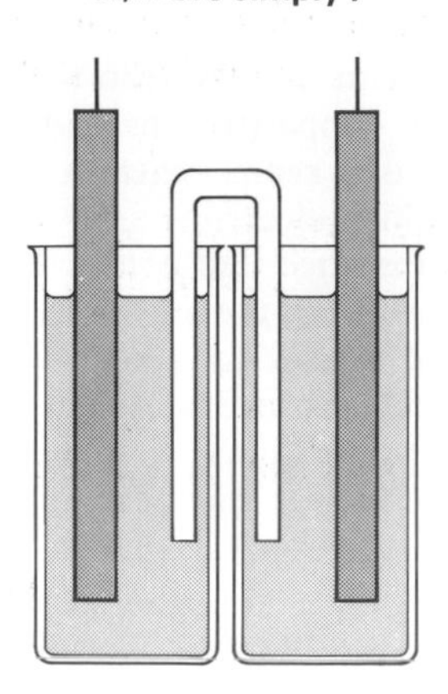

Figure 13.5 An electrochemical cell

Half-cells

There are several other types of half-cell in addition to the **metal/metal ion** type just described. One important example consists of an inert metal electrode (usually platinum) dipping into a solution containing a mixture of the reduced and oxidized ions of another metal, such as a solution of iron(II) and iron(III) ions: this is called a **redox half-cell**. Another is a **gas half-cell**, in which a gas is bubbled over a platinum electrode that dips into a solution containing ions derived from the gas. Such an electrode is called a **gas electrode**; in the **hydrogen electrode** the gas is hydrogen and the solution contains hydrogen ions.

The cell e.m.f.

Whatever the details of the cell, the negatively charged electrons tend to travel through the external circuit to the region of positive potential. A digital voltmeter, which shows the direction of the current, reveals that the electrons travel from the zinc electrode to the copper electrode. The copper is therefore the *positive* electrode in the Daniell cell shown in Figure 13.6.

The 'voltage' of a cell is more precisely a measure of the electrical potential

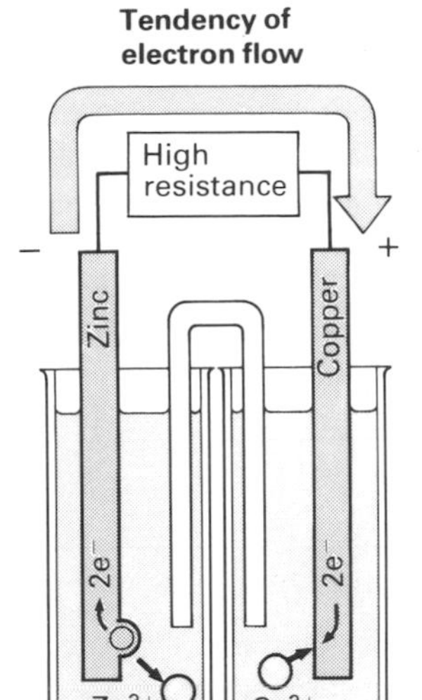

Figure 13.6 The relative electric potentials of electrodes and the tendency of electron flow

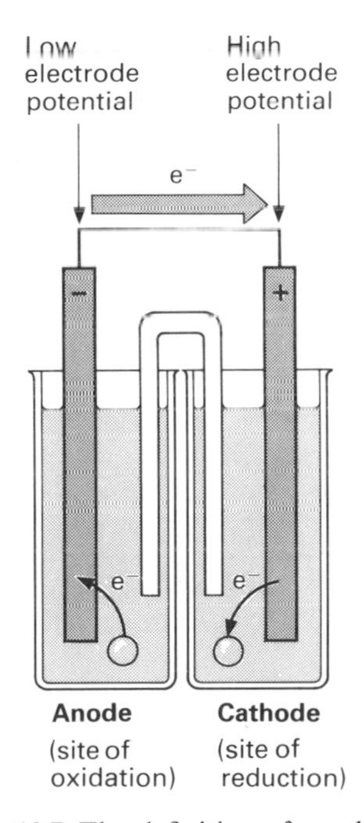

Figure 13.7 The definition of anode and cathode, and the signs of the electrode potentials

difference between its two electrodes. Since the voltage depends on the concentrations of the electrolytes, and these change as the reaction proceeds, care must be taken to ensure that negligible current is drawn when the voltage is measured. In modern measurements, an electrometer or a high-resistance digital voltmeter is used. The potential difference between the electrodes measured while drawing negligible current is called the **electromotive force** (e.m.f.) of the cell, and is expressed in volts (symbol: V).

The sign convention

The e.m.f. of the Daniell cell with a certain concentration of electrolytes may be found to be 1.14 V, with the copper electrode positive. These two pieces of experimental information are combined as follows. First, the structure of the cell is written, using a vertical solid line to separate each distinct phase from the next, and double vertical broken lines for the salt bridge. Using this notation the Daniell cell can be written in either of the following ways:

$$\mathrm{Zn(s)|Zn^{2+}(aq)\,\vdots\vdots\,Cu^{2+}(aq)|Cu(s)} \qquad \text{(A)}$$

$$\mathrm{Cu(s)|Cu^{2+}(aq)\,\vdots\vdots\,Zn^{2+}(aq)|Zn(s)} \qquad \text{(B)}$$

The e.m.f. is then written beside the cell, and *the sign is the sign of the charge of the right-hand electrode as written.* Since the copper electrode is positive, and in cell A it is the right-hand electrode, the e.m.f. is reported as +1.14 V:

$$\mathrm{Zn(s)|Zn^{2+}(aq)\,\vdots\vdots\,Cu^{2+}(aq)|Cu(s)} \qquad E = +1.14\ \mathrm{V}$$

For cell B – the same cell, but written differently – the right hand (negative) electrode is zinc, and the e.m.f. is reported as −1.14 V:

$$\mathrm{Cu(s)|Cu^{2+}(aq)\,\vdots\vdots\,Zn^{2+}(aq)|Zn(s)} \qquad E = -1.14\ \mathrm{V}$$

Cathodes and anodes

A final piece of terminology labels the electrodes as either a cathode or an anode: the **cathode** in any electrochemical cell is the electrode at which reduCtion occurs; the **anode** is the electrode at which oxidAtion occurs. Since in a galvanic cell the positive electrode is the point of entry of electrons into the cell, the positive electrode is the cathode; likewise, the negative electrode is the anode (see Figure 13.7). The + sign on the commercial dry cell therefore labels the cathode, the site of reduction, and the − sign labels the anode, the site of oxidation. The electrons can be imagined as being driven through the external circuit (which may be a radio, a cassette recorder, a torch bulb and so on) from the anode to the cathode as a result of the chemical reaction going on inside the cell.

13.4 The concentration dependence of the e.m.f.

One of the most important aspects of a galvanic cell is the dependence of its e.m.f. on the concentration of the substances taking part in the cell reaction.

The Nernst equation

Thermodynamics can be used to derive an expression relating a cell's e.m.f. to the concentration of the electrolyte: the information needed is given in Chapter 11. Consider a cell reaction of the form

$$a\mathrm{A} + b\mathrm{B} \rightarrow c\mathrm{C} + d\mathrm{D}$$

First, the reaction Gibbs energy depends on the concentration of the substances as follows (see Appendix 11.2):

$$\Delta_r G = \Delta_r G^{\ominus} + RT \ln Q \qquad (13.4.1a)$$

where Q is the reaction quotient $Q = \dfrac{[\mathrm{C}]^c[\mathrm{D}]^d}{[\mathrm{A}]^a[\mathrm{B}]^b}$

Second, from Section 11.3, $-\Delta_r G^\ominus$ is equal to the maximum electrical work per mole of reaction when the reaction takes place at constant temperature under standard conditions. This electrical work can be expressed in terms of the e.m.f. as follows. Suppose that in the reaction z moles of electrons are transferred per mole of reaction (in the Daniell cell, $z = 2$); then the total charge transferred between the electrodes per mole of reaction is zF, where F is Faraday's constant, the charge per mole of electrons (see Section 8.2). As this charge is moved against an electromotive force $E^\ominus$, the electrical work done is $zFE^\ominus$. That is,

$$\Delta_r G^\ominus = -zFE^\ominus \qquad (13.4.1b)$$

The corresponding expression when the substances are not in their standard states is

$$\Delta_r G = -zFE \qquad (13.4.1c)$$

Combining equations 13.4.1*a*, 13.4.1*b* and 13.4.1*c* gives the **Nernst equation**:

$$E = E^\ominus - (RT/zF)\ln Q \qquad (13.4.2)$$

where $E^\ominus$ is the **standard e.m.f.** of the cell. At 25 °C, $RT/F = 25.7$ mV.

For the Daniell cell, $z = 2$ and the reaction quotient for equation 13.1.2 is

$$Q = \frac{[Zn^{2+}(aq)]}{[Cu^{2+}(aq)]}$$

because the concentrations of the pure solids are constant (see Section 10.1). Therefore, the concentration dependence of its e.m.f. is predicted to be

$$E = E^\ominus - (RT/2F)\ln\left\{\frac{[Zn^{2+}(aq)]}{[Cu^{2+}(aq)]}\right\}$$

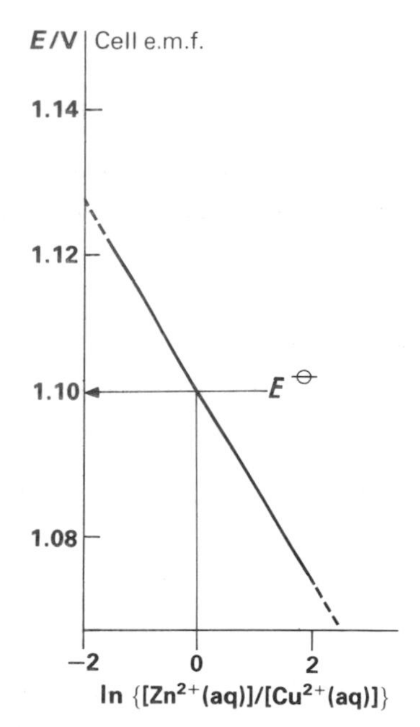

Figure 13.8 A cell e.m.f. depends logarithmically on the ion concentrations

Hence, by plotting E against different values of $\ln\{[Zn^{2+}(aq)]/[Cu^{2+}(aq)]\}$, the standard e.m.f. can be obtained from the intercept at $\ln\{[Zn^{2+}(aq)]/[Cu^{2+}(aq)]\} = 0$, as shown in Figure 13.8. When this is done (in practice, a more complicated procedure is used), it is found that $E^\ominus = +1.10$ V.

Concentration cells and membrane potentials

The Nernst equation shows that a cell can produce an e.m.f. even if its two half-cells differ only in the concentration of the electrolyte. Although the standard e.m.f. of such a **concentration cell** is zero (as the two half-cells are then identical for they have equal concentrations), the $\ln Q$ term can still give a contribution. If the cell is

$$Cu(s)|Cu^{2+}(aq, conc.)\,\vdots\vdots\,Cu^{2+}(aq, dil.)|Cu(s)$$

with the less concentrated solution on the right, then electrons will tend to flow from the right-hand electrode to the left, since in that way the concentration of ions will increase on the right and decrease on the left. In other words, the cell as written will have a negative e.m.f. as a result of the natural tendency for the two concentrations to equalize.

Biological membranes act like salt bridges separating two regions of different ion concentration. For instance, the membrane of a nerve cell (see Figure 13.9) is permeable to potassium ions when it is not activated, and so the potential difference across the membrane is

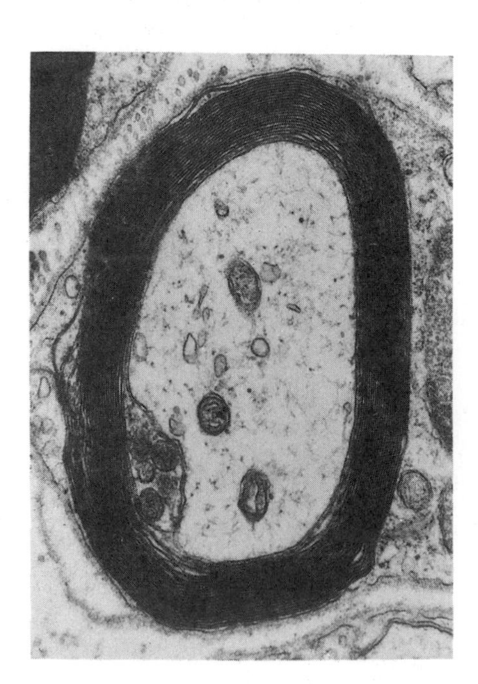

Figure 13.9 Electron micrograph of a section of a nerve cell, showing the cell membrane

$$E = -(RT/F)\ln\left\{\frac{[K^+(aq, inside)]}{[K^+(aq, outside)]}\right\}$$

As the concentration of potassium ions inside the cell is about 25 times greater than that outside, at 37 °C

$$E = -26.7\,\text{mV} \times \ln 25 = -86\,\text{mV}$$

in reasonable agreement with the experimental value of -75 mV for the potential difference across the resting nerve, the interior being negative.

Cells at equilibrium

We now consider what will prove to be a simple but far-reaching question: what does it mean to say that a cell is *exhausted*?

An exhausted cell (a 'flat battery') is one in which there is no net tendency to react: the cell reaction is at equilibrium. Therefore, *when the cell reaction is at equilibrium, the cell's e.m.f. is zero.*

The Nernst equation takes a very useful form when the cell reaction is at equilibrium. We set $E = 0$, and replace Q by the equilibrium constant K (since $Q = K$ at equilibrium). This gives

$$0 = E^{\ominus} - (RT/zF)\ln K$$

which rearranges to

$$\ln K = zFE^{\ominus}/RT \qquad (13.4.3)$$

an equation relating the equilibrium constant to an electrochemical quantity, the standard e.m.f. of the cell. This important relation is exploited in the next section.

13.5 Standard electrode potentials

It is not necessary to measure the standard e.m.f. of a cell in order to calculate the equilibrium constant for the cell reaction: the values can be calculated from tables of **standard electrode potentials.**

The standard e.m.f. of a cell is the difference between the standard electrode potentials of the left and right half-cells:

$$E^{\ominus} = E^{\ominus}(\text{r.h. half-cell}) - E^{\ominus}(\text{l.h. half-cell})$$

Reporting potentials

The potential of a single electrode cannot be measured, but as only differences are ever needed, the electrode of one selected half-cell can be *defined* as having zero potential and all others referred to it. This is like defining sea level as zero altitude, and referring the heights of mountains and depths of oceans to that artificial zero.

The electrode chosen to have $E^{\ominus} = 0$ is the **standard hydrogen electrode** (SHE), which is the electrode of a hydrogen half-cell as described in Section 13.3, with the hydrogen gas in its standard state (1 bar pressure) and the hydrogen ion concentration $1\,\text{mol dm}^{-3}$ (so that $\text{pH} = 0$) (see Figure 13.10).

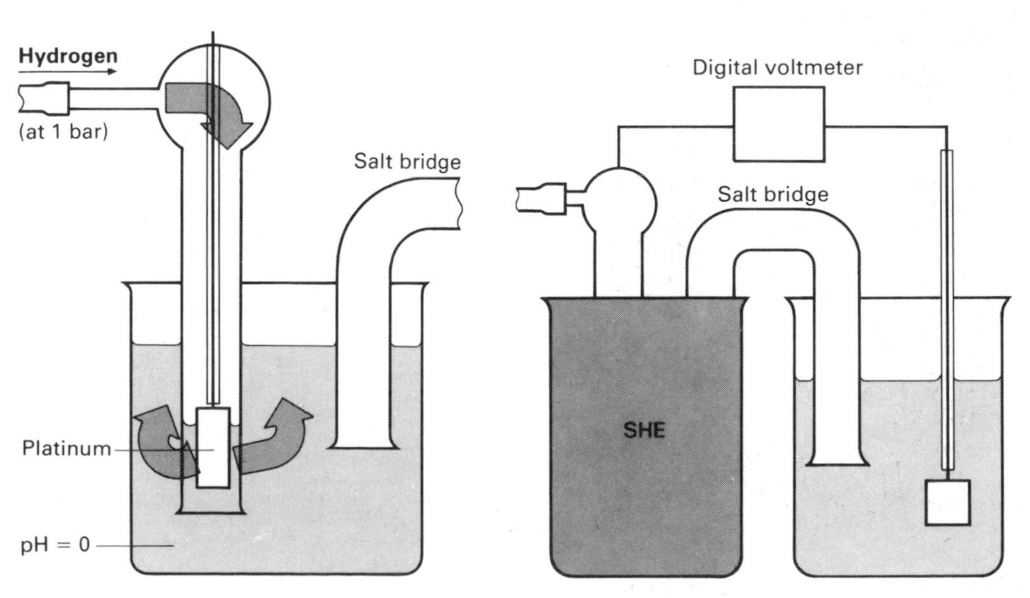

Figure 13.10 The standard hydrogen electrode

Figure 13.11 Measuring a standard electrode potential (of the cell on the right)

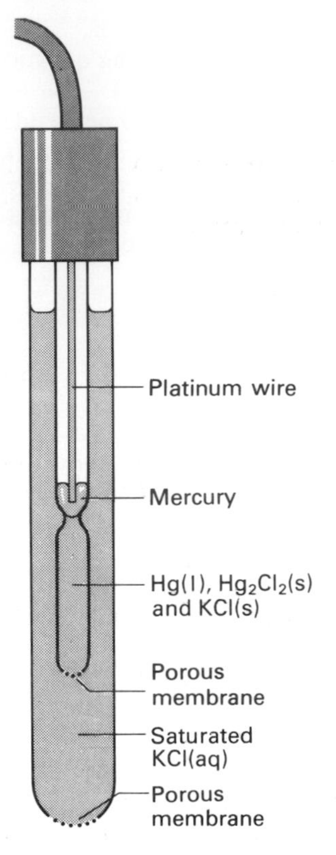

Figure 13.12 A saturated calomel half-cell

The standard potential of any other half-cell is then defined to be the standard e.m.f. of the following cell, shown in Figure 13.11:

SHE¦¦Electrolyte|Electrode

$$E^{\ominus} = E^{\ominus}(\text{r.h. half-cell}) - E^{\ominus}(\text{SHE})$$
$$= E^{\ominus}(\text{r.h. half-cell})$$

as $E^{\ominus}(\text{SHE}) = 0$.

In practice a SHE is difficult to set up, and so **secondary standards** are used: these are half-cells with well-established properties, calibrated against a SHE. The most widely used is the **saturated calomel half-cell** illustrated in Figure 13.12 ('calomel' is the old name for mercury(I) chloride, Hg_2Cl_2), for which $E^{\ominus} = +0.246$ V at 25 °C.

Cell e.m.f. and electrode potential

Some values of standard electrode potentials for common **couples** (that is, the oxidized and reduced forms of substances taking part in a half-reaction) are listed in Table 13.1. They may be used to calculate the standard e.m.f. of any cell by taking the difference:

$\text{Electrode}_L|\text{Electrolyte}_L$¦¦$\text{Electrolyte}_R|\text{Electrode}_R$

$$E^{\ominus} = E^{\ominus}(\text{r.h. half-cell}) - E^{\ominus}(\text{l.h. half-cell})$$

Table 13.1 Standard electrode potentials at 25 °C, $E^{\ominus}$/V

↓ most reducing	
$Li^+ + e^- \rightleftharpoons Li$	−3.04
$K^+ + e^- \rightleftharpoons K$	−2.92
$Ca^{2+} + 2e^- \rightleftharpoons Ca$	−2.87
$Na^+ + e^- \rightleftharpoons Na$	−2.71
$Mg^{2+} + 2e^- \rightleftharpoons Mg$	−2.36
$Be^{2+} + 2e^- \rightleftharpoons Be$	−1.85
$Al^{3+} + 3e^- \rightleftharpoons Al$	−1.66
$Zn^{2+} + 2e^- \rightleftharpoons Zn$	−0.76
$Fe^{2+} + 2e^- \rightleftharpoons Fe$	−0.44
$Sn^{2+} + 2e^- \rightleftharpoons Sn$	−0.14
$Pb^{2+} + 2e^- \rightleftharpoons Pb$	−0.13
$2H^+ + 2e^- \rightleftharpoons H_2$	0
$AgBr + e^- \rightleftharpoons Ag + Br^-$	0.07
$AgCl + e^- \rightleftharpoons Ag + Cl^-$	0.22
$Cu^{2+} + 2e^- \rightleftharpoons Cu$	0.34
$Cu^+ + e^- \rightleftharpoons Cu$	0.52
$I_2 + 2e^- \rightleftharpoons 2I^-$	0.54
$Fe^{3+} + e^- \rightleftharpoons Fe^{2+}$	0.77
$Hg_2{}^{2+} + 2e^- \rightleftharpoons 2Hg$	0.79
$Ag^+ + e^- \rightleftharpoons Ag$	0.80
$2Hg^{2+} + 2e^- \rightleftharpoons Hg_2{}^{2+}$	0.92
$Br_2 + 2e^- \rightleftharpoons 2Br^-$	1.07
$O_2 + 4H^+ + 4e^- \rightleftharpoons 2H_2O$	1.23
$MnO_2 + 4H^+ + 2e^- \rightleftharpoons Mn^{2+} + 2H_2O$	1.23
$Cr_2O_7{}^{2-} + 14H^+ + 6e^- \rightleftharpoons 2Cr^{3+} + 7H_2O$	1.33
$Cl_2 + 2e^- \rightleftharpoons 2Cl^-$	1.36
$Au^{3+} + 3e^- \rightleftharpoons Au$	1.50
$MnO_4{}^- + 8H^+ + 5e^- \rightleftharpoons Mn^{2+} + 4H_2O$	1.51
$Ce^{4+} + e^- \rightleftharpoons Ce^{3+}$	1.61
$MnO_4{}^- + 4H^+ + 3e^- \rightleftharpoons MnO_2 + 2H_2O$	1.69
$F_2 + 2e^- \rightleftharpoons 2F^-$	2.87
↑ most oxidizing	

For the Daniell cell, for instance,

$$E^\ominus = E^\ominus(Cu^{2+},Cu) - E^\ominus(Zn^{2+},Zn) = 0.34\,V - (-0.76\,V) = +1.10\,V$$

as used before.

Cell e.m.f. measurements, and the tables of standard electrode potentials constructed from those measurements, have numerous chemical applications, and we explore some of them in the following paragraphs.

The calculation of equilibrium constants

The key equation is equation 13.4.3:

$$\ln K = zFE^\ominus/RT \quad (13.5.1a)$$
$$= zE^\ominus/(25.7\,mV) \text{ at } 25\,^\circ C \quad (13.5.1b)$$

When $E^\ominus$ for the cell reaction is large and positive, the equilibrium constant is also large, and so the equilibrium lies strongly in favour of the products. If $E^\ominus$ is negative, then the equilibrium constant is less than unity, and the equilibrium lies in favour of the reactants.

EXAMPLE

Predict the value of the equilibrium constant for the reaction $Zn(s) + Cu^{2+}(aq) \rightarrow Zn^{2+}(aq) + Cu(s)$ at 25 °C.

METHOD Write the expression for the equilibrium constant, and calculate its value from equation 13.5.1, using $z = 2$ for the reaction as written (because two moles of electrons are transferred for each mole of reaction) and calculating $E^\ominus$ from $E^\ominus$(r.h. half-cell) – $E^\ominus$(l.h. half-cell). In this case, the couple in the r.h. half-reaction is Cu^{2+},Cu; take the values from Table 13.1.

ANSWER The equilibrium constant of the reaction is

$$K_c = \left\{\frac{[Zn^{2+}(aq)]}{[Cu^{2+}(aq)]}\right\}_{eq}$$

From Table 13.1:

$$E^\ominus = 0.34\,V - (-0.76\,V) = +1.10\,V$$

Then, from equation 13.5.1:

$$\ln K_c = \frac{2 \times 1.10\,V}{0.0257\,V} = 85.6$$

Hence, $K_c = 1.5 \times 10^{37}$.

COMMENT We have avoided the need to perform the actual experiment by using tabulated data deduced from other experiments: this illustrates one of the powers of science, in that it can use data from different sources to make verifiable predictions about other experiments. Note that the values in Table 13.1 apply only to 25 °C, but they vary only slightly for small changes of temperature.

The electrochemical series

The standard electrode potentials of the elements can be used to set up a scale that allows us to state, at a glance, whether one metal has a thermodynamic tendency to displace another from solution: the **electrochemical series**.

A typical displacement reaction is

$$Zn(s) + Cu^{2+}(aq) \rightarrow Zn^{2+}(aq) + Cu(s) \qquad E^\ominus = +1.10\,V$$

The standard e.m.f. is positive, the equilibrium constant for the cell reaction is large (1.5×10^{37}), the products are favoured and zinc displaces copper(II) ions from solution. Since zinc tends to form zinc ions in aqueous solution more readily than copper tends to form copper(II) ions, we report that zinc is more **electropositive** than copper. A more electropositive metal has a thermodynamic tendency to displace a less electropositive metal. In other words:

Under standard conditions, an element with a given standard electrode potential has a tendency to reduce the ions of any element with a more positive standard electrode potential.

Dental fillings, which are often amalgams (solutions in mercury) of silver, copper and tin, illustrate this principle. A filling can be the source of an electric current: if a piece of aluminium foil is bitten a cell is set up in which the electrolyte is saliva, the aluminium acts as the anode, and the filling is the cathode. The e.m.f. of this cell is about 2 V, enough to cause discomfort but not electrocution. A mixture of gold and amalgam fillings in the same mouth can lead to a persistent metallic taste because tin ions are released from the amalgam, which is now the anode.

EXAMPLE

Can aluminium displace copper from aqueous solution?

METHOD The couple with the more negative standard electrode potential reduces the other; reduction corresponds to displacement of the metal involved in the other couple. Therefore, inspect Table 13.1 to see whether the Al^{3+}, Al couple has a more negative standard electrode potential than Cu^{2+}, Cu.

ANSWER From Table 13.1, $E^{\ominus}(Al^{3+}, Al) = -1.66$ V and $E^{\ominus}(Cu^{2+}, Cu) = +0.34$ V; hence, aluminium has a thermodynamic tendency to displace copper from aqueous solution.

COMMENT The difference of standard electrode potentials, 2.0 V, is large. The displacement therefore has a tendency to go to virtual completion. Nevertheless, the equilibrium may be reached very slowly because aluminium left exposed to the atmosphere acquires a protective layer of unreactive oxide.

Figure 13.13 Dilute acid attacks aluminium where a scratch has broken through the surface layer of oxide

Although the electrochemical series is an extremely useful measure of reducing strengths, it suffers from two limitations. The more important is that it can be used to predict only *tendencies* and, like the rest of thermodynamics (of which it is a part), it is silent on the *rate* at which the displacement occurs in practice. Thus, although aluminium is a strongly electropositive element and has a thermodynamic tendency to displace hydrogen from water and acids, a piece of aluminium placed in water (such as an aircraft flying through a cloud) does not react. Only if the tough, unreactive oxide layer on its surface is removed first does the aluminium react at a measurable rate (see Section 19.2 and Figure 13.13).

The second restriction of the electrochemical series is that it deals only with *aqueous solutions* of ions, and reducing powers may be quite different in other solvents or for reactions involving solids and gases.

Redox reactions in solution

Table 13.1 also includes oxoanions and ions of non-metals, and their chemical properties can also be rationalized in terms of their standard electrode potentials. This rationalization of chemistry will become increasingly apparent in the following chapters, but a glimpse of how the table is used is possible even now. For instance, from Table 13.1 it is clear that whereas manganate(VII) ions can oxidize chloride ions in acidic solution under standard conditions at 25 °C (because the Cl_2, Cl^- couple has a more negative standard electrode potential), manganese(IV) oxide cannot. The practical consequence of this difference is that when hydrochloric acid and manganese(IV) oxide are used to prepare chlorine, the temperature must be changed (the mixture heated) and the concentrations must be made significantly different from standard conditions (by using concentrated acid).

If we wish to use an oxidizing agent to test for the presence of a reducing agent in aqueous solution, we should choose a substance with a couple that has a more positive standard electrode potential and undergoes an easily detectable colour change. A substance that fulfils these criteria is acidified aqueous potassium manganate(VII), which forms a purple solution that is usually decolorized on reduction to Mn^{2+}(aq) ions, which are very pale pink:

$$MnO_4^-(aq) + 8H^+(aq) + 5e^- \rightarrow Mn^{2+}(aq) + 4H_2O(l) \qquad E^\ominus = +1.51\ V$$

Alternatively, it may turn brown as a result of its reduction to a precipitate of dark brown MnO_2:

$$MnO_4^-(aq) + 4H^+(aq) + 3e^- \rightarrow MnO_2(s) + 2H_2O(l) \qquad E^\ominus = +1.69\ V$$

Another choice is acidified aqueous potassium dichromate(VI), which is orange; the solution turns green when reduced to Cr^{3+}(aq) ions:

$$Cr_2O_7^{2-}(aq) + 14H^+(aq) + 6e^- \rightarrow 2Cr^{3+}(aq) + 7H_2O(l) \qquad E^\ominus = +1.33\ V$$

In both cases acidification is needed to ensure that the hydrogen ion concentration is close to the standard value of 1 mol dm^{-3}. Substances that have standard electrode potentials more positive than these values cannot bring about the reduction, and so are not detected as reducing agents.

Similar reasoning applies to devising a test for an oxidizing agent, for which we need a moderately strong reducing agent (a less positive standard electrode potential) that undergoes a change of colour when it is oxidized. One common test is based on acidified aqueous potassium iodide:

$$I_2(s) + 2e^- \rightarrow 2I^-(aq) \qquad E^\ominus = +0.54\ V$$

If the colourless iodide ion is oxidized, the brown colour typical of iodine develops.

Corrosion

Corrosion is most familiar in the form of the rusting of iron. It depends on redox reactions and can be discussed electrochemically. The energetics of rusting were mentioned in Section 11.2, but it is now possible to see more clearly the processes that take place, how they are affected by the conditions, and how they may be retarded. Since the annual cost of corrosion in the United Kingdom is over 1 billion pounds, its control is of great economic importance.

One of the iron half-reactions involved in rusting is

$$Fe^{2+}(aq) + 2e^- \rightarrow Fe(s) \qquad E^\ominus = -0.44\ V$$

The oxidation of iron, the reverse of this reaction, is driven by the presence of oxygen; the iron(II) ions are oxidized further to iron(III), and various insoluble hydrated oxides of iron(III) are deposited as the red-brown precipitate known as rust. These oxides lack the rigidity of the metal, and (unlike the oxides of aluminium and zinc) flake off, so that the iron artefact decays: thus cars slowly decay back to an oxide similar to the ore from which the metal was originally refined with such a huge investment of energy.

The half-reaction written above shows that iron is moderately electropositive, and can reduce oxygen (that is, it can be oxidized by oxygen). The precise way in which the reaction occurs depends on the conditions. In the acidic atmosphere of some industrial areas, the reduction half-reaction is

$$O_2(g) + 4H^+(aq) + 4e^- \rightarrow 2H_2O(l) \qquad E^\ominus = +1.23\ V$$

This is more positive than the Fe^{2+}, Fe couple is, and so can drive the latter's oxidation. The standard e.m.f. of the combined half-reactions is

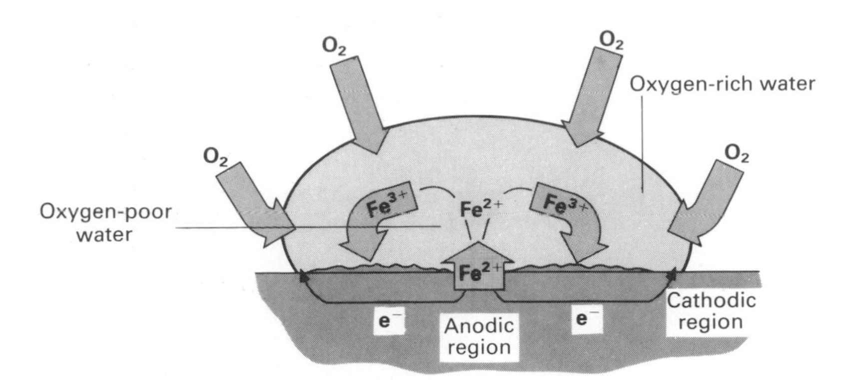

Figure 13.14 Corrosion at a drop of water

1.23 V − (−0.44 V) = +1.67 V; therefore at pH = 0 there is a very strong tendency towards oxidation of the iron (see Figure 13.14).

Corrosion cannot be eliminated, but it can be slowed and diverted. Slowing corrosion often means sealing the surface from attack and is one reason why steel cars, bridges and ships are painted. The tendency to corrode can also be reduced by decreasing the acidity of the environment (because the electrode potential of the oxygen reduction half-reaction depends on the pH); and the corrosion of a car can be reduced by washing off the salt that may settle on surfaces after it has been spread on roads (because that reduces the conductivity of the water that splashes on to the car body, and hence the rate of the corrosion reaction).

The diversion of corrosion into different electrochemical channels deals with the problem more subtly. For example, reference to the electrochemical series shows that magnesium is more electropositive than iron, and hence it has a tendency to reduce any $Fe^{2+}(aq)$ ions present. In **cathodic protection**, a lump of magnesium (or other electropositive metal) is joined to the steel artefact. The cell so set up, with the magnesium the anode and the artefact the cathode, results in a supply of electrons from the magnesium to the iron. The iron object is preserved (because, on account of the supply of electrons, Fe^{2+} no longer forms) while the magnesium decays; it is called a **sacrificial anode**. A lump of magnesium is much less expensive than a battleship (or even an outboard motor), and the corrosion is diverted into the destruction of the relatively cheap sacrificial metal block. It is for this reason that the electrical circuits in cars generally have a negative earth (that is, the steel body is connected to the anode of the battery): the supply of electrons from the anode helps to diminish the oxidation of the iron.

EXAMPLE

Explain the electrochemistry involved in (a) galvanizing (zinc-plating) and (b) tin-plating iron objects.

METHOD Note the relative positions of the metals in the electrochemical series (Table 13.1). If the plating metal is more electropositive than iron it tends to reduce any iron ions that may be present, and hence helps to preserve the metal.

ANSWER (a) Zinc is more electropositive than iron, and hence tends to reduce any iron ions that form. It does not itself corrode, because it is protected by a layer of oxide.
(b) Tin is less electropositive than iron, and so it tends to cause the oxidation of the iron. Tin seals the surface of the iron against corrosion, but once the seal is broken (by scratching) corrosion is *more* likely than for unplated iron.

COMMENT As always in electrochemistry, it is important to consider both thermodynamic (tendency) factors, and then, if these are favourable, rate factors.

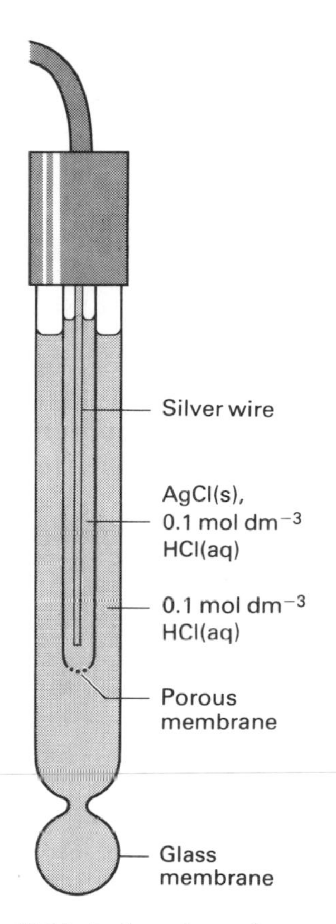

Figure 13.15 A glass electrode

The measurement of pH

The measurement of e.m.f. provides a convenient and simple method of determining pH. The e.m.f. of a cell in which one half-cell contains a hydrogen electrode is proportional to the logarithm of the hydrogen ion concentration, and hence is also proportional to the pH. Therefore, to measure pH (in principle, at least) a cell is set up with a hydrogen electrode (HE) dipping into the solution of interest, the other half-cell being typically a saturated calomel half-cell in contact with the solution:

Calomel¦¦Solution|HE

$$E = E^{\ominus} + (RT/F)\ln[H^+(aq)]$$
$$= E^{\ominus} + (2.303RT/F)\lg[H^+(aq)]$$
$$= E^{\ominus} - (59.2\,\text{mV}) \times \text{pH}$$

$E^{\ominus}$ is the standard e.m.f. of the cell (-0.246 V) and the numerical values in the last line refer to 25 °C. This expression is easy to rearrange to give the pH in terms of the observed value of E.

In practice, the hydrogen electrode is messy and tiresome to set up. The right-hand electrode in the cell described above can be replaced by any electrode sensitive to hydrogen ions, however, and a very convenient one is the **glass electrode** (GE) (see Figure 13.15), which is found experimentally to have a potential proportional to the pH. The second half-cell is normally a saturated calomel half-cell, as before, so that the overall arrangement is

Calomel¦¦Solution|GE

(see Figure 13.16) and the scale of the voltmeter that measures the e.m.f. of the cell is marked to give the pH directly. The entire instrument, a **pH meter**, is calibrated using buffer solutions (see Section 12.3).

Using a glass electrode, pH is represented by an electrical signal. This signal can be used to automate titrations, with the readings carried out on a microcomputer linked to the titration equipment. Apart from being accurate and reliable, automation of this kind releases the operator for more imaginative and interesting investigations.

13.6 Thermodynamics and electrochemistry

Electrochemistry is a branch of thermodynamics, and thermodynamic measurements can be made electrochemically.

The measurement of reaction Gibbs energy

Since the standard e.m.f. of a cell is related to the standard reaction Gibbs energy of the cell reaction, its measurement can be converted directly, using equation 13.4.1*b*:

$$\Delta_r G^{\ominus} = -zFE^{\ominus} \qquad (13.6.1)$$

This equation is very simple to use, because the standard e.m.f. of the cell can be calculated by combining the standard electrode potentials in Table 13.1.

EXAMPLE

Calculate the standard reaction Gibbs energy of $Zn(s) + Cu^{2+}(aq) \rightarrow Zn^{2+}(aq) + Cu(s)$ at 25 °C.

METHOD Use equation 13.6.1. Since two moles of electrons are transferred for each mole of reaction as written, take $z = 2$. Calculate $E^{\ominus}$ from the difference $E^{\ominus}$(r.h. half-cell) – $E^{\ominus}$(l.h. half-cell), taking standard electrode potentials from Table 13.1.

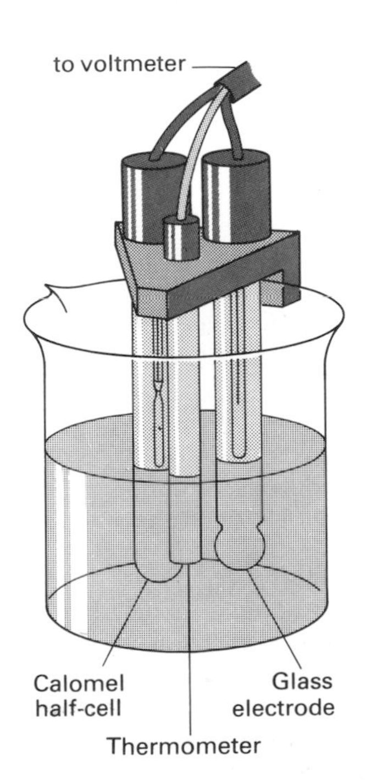

Figure 13.16 The probe of a pH meter

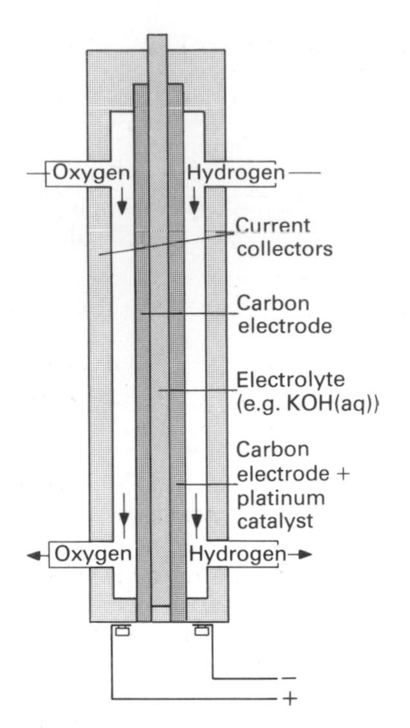

Figure 13.17 A hydrogen/oxygen fuel cell element

ANSWER As in the Example in Section 13.5, $E^{\ominus} = +1.10\,V$. Then, from equation 13.6.1,

$$\begin{aligned}\Delta_r G^{\ominus} &= -2 \times 9.648 \times 10^4\,C\,mol^{-1} \times 1.10\,V \\ &= -2.12 \times 10^5\,C\,V\,mol^{-1} = -212\,kJ\,mol^{-1}\end{aligned}$$

(since $1\,C\,V = 1\,J$).

COMMENT The value is that of the *standard* reaction Gibbs energy at 25 °C, the change in Gibbs energy per mole of reaction as written when zinc metal displaces copper(II) ions from a $1\,mol\,dm^{-3}$ $Cu^{2+}(aq)$ solution and resulting in a $1\,mol\,dm^{-3}$ $Zn^{2+}(aq)$ solution at the same temperature, there being no zinc ions present initially and no copper(II) ions present finally (that is, a hypothetical 'complete reaction'). The value of $\Delta_r G^{\ominus}$ is strongly negative, corresponding to a strong thermodynamic tendency for the reaction to run to the right (for zinc to displace copper from solution).

Fuel cells

In a **fuel cell** the reactants are provided from external reserves rather than being included once and for all when the cell is manufactured (as in a dry cell). The mode of operation is otherwise exactly the same as in any galvanic cell, with electrons emerging from an oxidation reaction at an anode, passing through an external circuit, and then bringing about a reduction at the cathode. A version of a hydrogen/oxygen cell, which is also called a **Bacon cell** after its inventor, is shown in Figure 13.17.

The theoretical maximum work that can be obtained per gram of hydrogen consumed can be calculated from the standard reaction Gibbs energy of the cell reaction, which in the Bacon cell is

$$2H_2(g) + O_2(g) \rightarrow 2H_2O(l)$$

Under standard conditions at 25 °C, Table 11.2 gives

$$\Delta_r G^{\ominus} = 2 \times -237.2\,kJ\,mol^{-1} = -474.4\,kJ\,mol^{-1}$$

Hence, from the consumption of 2 mol H_2, or 4 g of hydrogen, under standard conditions, up to 474 kJ of electrical work is available at 25 °C, which is enough energy to raise a 1000 kg object to a height of about 50 m. In the process 36 g of water is produced as waste product (waste, that is, on Earth, but the passengers in a space vehicle could drink it).

The standard e.m.f. of the Bacon cell can be calculated from the standard reaction Gibbs energy using equation 13.6.1, with $z = 4$:

$$E^{\ominus} = \frac{-(-474.4 \times 10^3\,J\,mol^{-1})}{4 \times 9.65 \times 10^4\,C\,mol^{-1}} = +1.23\,V$$

Calculations such as these are at the heart of modern work on the development of fuel cells such as that in Figure 13.18, in the hope of producing pollution-free, convenient mobile sources of energy.

Figure 13.18 One of the original fuel cells constructed by Francis Bacon, using pure hydrogen and pure oxygen

Summary

- ☐ Oxidation is the loss of electrons from a species; reduction is the gain of electrons.
- ☐ Redox reactions are combinations of oxidation and reduction reactions; the oxidizing agent is itself reduced and the reducing agent is itself oxidized.
- ☐ An electrochemical cell consists of two half-cells; often they are joined through a salt bridge which establishes electrical contact.
- ☐ The electromotive force (e.m.f.) of a cell is the electric potential difference between the electrodes of the cell, E(r.h. half-cell) − E(l.h. half-cell), when no current is drawn. The e.m.f. is measured with a high-resistance digital voltmeter.

- ☐ The e.m.f. of a cell depends on the logarithm of the concentrations of the species present; the e.m.f. of the cell when the concentrations of the solutions are 1 mol dm^{-3} is the standard e.m.f., $E^\ominus$.
- ☐ The standard e.m.f. of a cell is the difference between the standard electrode potentials of the two half-cells.
- ☐ Standard electrode potentials are reported on the basis that the potential of the standard hydrogen electrode is zero. In practice the calomel electrode is used as a secondary standard.
- ☐ When a cell reaction is at equilibrium it produces no e.m.f.
- ☐ The equilibrium constant K and the standard e.m.f. of a reaction are related by $RT \ln K = zFE^\ominus$.
- ☐ Standard electrode potentials determine the electrochemical series; a metal of a given standard electrode potential will be displaced or reduced by a metal higher up the series (with a less positive standard electrode potential).
- ☐ A pH meter makes use of the glass electrode (an electrode with a potential that depends on the hydrogen ion concentration of the medium) in combination with a reference electrode.
- ☐ The standard reaction Gibbs energy $\Delta_r G^\ominus$ and the standard e.m.f. of a reaction are related by $\Delta_r G^\ominus = -zFE^\ominus$.
- ☐ Measurements of the e.m.f. of a cell are used to obtain thermodynamic information, such as the value of $\Delta_r G^\ominus$ for the reaction.
- ☐ The maximum electrical work that can be obtained in the course of a chemical reaction is $-\Delta_r G^\ominus$; thermodynamic data can therefore be used to assess the usefulness of chemical reactions as sources of electrical energy.

PROBLEMS

1 Define the terms *oxidation*, *reduction* and *redox reaction*.

2 State which of the following reactions are redox reactions and, in those cases, identify the species being reduced and oxidized.

(a) $Zn + S \rightarrow ZnS$
(b) $H_2S + Cl_2 \rightarrow S + 2HCl$
(c) $KOH + HCl \rightarrow KCl + H_2O$
(d) $CuO + H_2 \rightarrow Cu + H_2O$
(e) $C_2H_4 + HBr \rightarrow C_2H_5Br$
(f) $2H_2O_2 \rightarrow 2H_2O + O_2$
(g) $MnO_2 + 4HCl \rightarrow MnCl_2 + Cl_2 + 2H_2O$
(h) $Fe_2O_3 + 3CO \rightarrow 2Fe + 3CO_2$

3 Aluminium objects frequently fail to show the reactivity expected on the basis of its position in the electrochemical series. Explain this observation.

4 State which is the positive electrode in the following electrochemical cells:

(a) $Zn|ZnSO_4(aq)⦙CuSO_4(aq)|Cu$
(b) $Pt, H_2|HCl(aq)|AgCl, Ag$
(c) $Pt, O_2|HCl(aq)|H_2, Pt$
(d) $Ag, AgCl|HCl(aq)⦙HBr(aq)|AgBr, Ag$

5 What are the standard e.m.f.s (at 25 °C) of the cells in the last question?

6 Write the cell reactions, together with expressions for the equilibrium constants of the cell reactions, for the cells in Question 4, and deduce the numerical values of the equilibrium constants for the reactions.

7 Arrange the following metals in order of decreasing displacing power: aluminium, copper, iron, magnesium, sodium, zinc.

8 Why should copper and iron pipes not be used in the same domestic water system?

9 A lump of magnesium is sometimes buried near a steel object (such as a pipeline) and connected to it by conducting cables. Why?

10 Given a molar solution of silver nitrate in water and any equipment necessary, describe how *you* would determine the standard electrode potential of silver.

Potassium manganate(VII) ($KMnO_4$) and hydrogen peroxide react in the presence of acid with the evolution of oxygen to give a clear solution. Show how to use the following information to determine the stoichiometric (balanced) ionic equation of the reaction.

$$MnO_4^- + 8H^+ + 5e^- \rightleftharpoons Mn^{2+} + 4H_2O \qquad E^\ominus = +1.52\,V$$
$$O_2 + 2H^+ + 2e^- \rightleftharpoons H_2O_2 \qquad E^\ominus = +0.68\,V$$

What volume of potassium manganate(VII) of concentration 0.200 mol dm^{-3} would be required to react with 100 cm^3 of hydrogen peroxide of concentration 0.0100 mol dm^{-3} (with excess acid present), and what volume of oxygen (at 20 °C and 1 atm pressure) would be evolved?

(Oxford and Cambridge part question)

11 (a) Explain the processes of (i) *oxidation*, (ii) *reduction*, in terms of electron transfer.

(b) For each of the following redox reactions, write two balanced equations each involving electron transfer, one for the oxidation process and one for the reduction process (stating in each case which equation describes oxidation and which reduction):
(i) reaction between $Cr_2O_7^{2-}$ and Fe^{2+} in aqueous solution in the presence of H^+;
(ii) reaction between Cl_2 and I^- in aqueous solution;
(iii) reaction between Li and H_2. *(Oxford)*

12 An electrochemical cell

$$Pt(H_2, 1\,atm)|HNO_3(m = 1), AgNO_3(m = 1)|Ag$$

is set up at 298 K. State the e.m.f. of the cell and its polarity on open circuit. What chemical changes begin to occur if it is connected across a high resistance?

The following cell is set up to observe electrolysis at 298 K:

$$Pt(H_2, 1\,atm)|HNO_3(m = 1), Cu(NO_3)_2(m = 1)|Cu$$

What is the minimum voltage which must be applied, and of what polarity, to cause deposition of copper on the copper electrode?

How does the situation differ if:

(a) the left-hand electrode is not as shown but is made of Pt with no supply of hydrogen,
(b) the left-hand electrode is made of silver with no supply of hydrogen,
(c) the electrodes are as shown but the electrolyte is diluted by a factor of ten?

State briefly with reasons whether you expect standard electrode potentials for metal ion/metal electrodes to vary much with temperature.

$Cu^{2+}(aq)|Cu$, $E^{\ominus} = 0.337$ V; $Ag^{+}(aq)|Ag$, $E^{\ominus} = 0.799$ V.

(Cambridge Entrance)

14 Chemical kinetics

We have seen what holds atoms together in compounds and what determines the tendency of these compounds to react. Now we investigate how fast these reactions occur. We see how to define and measure the rate of reaction, and how to account for the dependence of the reaction rate on the composition and temperature. Some reactions of great importance in industry normally take place far too slowly for commercial use, and we see how they can be made to go faster.

Chemical reactions take place at different rates. Some are very slow, like fermentation (which might require several weeks to produce enough product). Others are moderately fast, like the reactions that contract muscles, transmit impulses along nerves and record photographic images. Others are explosively fast, like the reactions inside a car engine.

A knowledge of reaction rates is central to chemistry. In industry it is important to be able to predict how fast a reaction will go under various conditions. Many reactions take place in a series of steps, and we can only truly say that we understand an overall process when the steps have been identified and investigated: as we will see, the study of reaction rates leads to insight into these individual steps.

14.1 The rates of reactions

Rate in chemistry is defined like speed in physics: it is the change in some property divided by the time it takes for the change to take place. The property in this instance is the concentration (or partial pressure) of a substance, and so a preliminary definition of reaction rate (one made more precise in Appendix 14.1) is

$$rate = \frac{change\ of\ concentration\ of\ a\ substance}{time\ taken\ for\ the\ change}$$

Rate is expressed as 'concentration per unit time'; when concentrations are reported in $mol\,dm^{-3}$, rates are reported in $mol\,dm^{-3}\,s^{-1}$.

The measurement of rate

Just as the speed of a car may change during a journey, so the rate may change during a reaction. The concentration of the selected substance is therefore followed and plotted against time, and the rate at the time of interest is found by drawing the tangent to the graph and calculating its slope (see Figure 14.1).

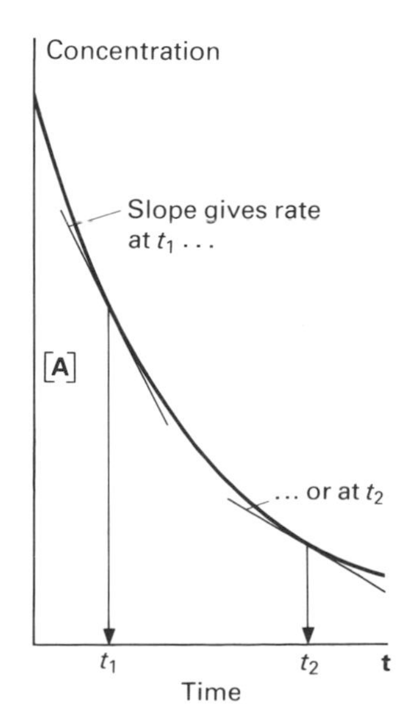

Figure 14.1 The reaction rate and the slope of a concentration/time graph

The choice of an experimental technique used to follow the concentration depends on how quickly it changes: there is not much point in using a titration method if the reaction is over in a millisecond! For moderately slow reactions (those in which there is little change in concentration over several minutes), or for reactions that can be **quenched** – brought to a virtual standstill, by rapid dilution or by cooling – titrations can be used. Sometimes this is done by withdrawing a small sample of the reaction mixture (an **aliquot**) and quenching it at a definite time after the start of the reaction. If the reaction produces or removes ions the electrical conductivity of the sample can be monitored and interpreted in terms of ion concentrations. If a reactant or a product (bromine, for example) has a pronounced colour, or a characteristic absorption in a convenient spectral region, then spectroscopic techniques may be used to follow the reaction. Gas-phase reactions and reactions producing gases can also often be followed by measuring the pressure of the gas. The last technique was used in the early studies of the biochemical Krebs cycle, a series of reactions important in respiration, in which carbon dioxide is released.

EXAMPLE

Suggest methods for the measurement of the rates of (a) the hydrolysis of ethyl ethanoate and (b) the reaction between hydriodic acid and hydrogen peroxide.

METHOD Look for methods by which the concentrations of at least one substance (either a reactant or a product) may be followed. Unless informed otherwise, assume that the reactions are sufficiently slow for classical techniques (typically titration) to be suitable. Rates are determined from concentration *v.* time plots by taking the slope of the graph at the time of interest. Devise a way of terminating the reaction at each time of interest so that the composition of the mixture remains unchanged while the concentrations are measured.

ANSWER (a) The hydrolysis is

$$CH_3COOC_2H_5 + H_2O \rightarrow CH_3COOH + C_2H_5OH$$

The rate can be followed by monitoring the growth of the concentration of ethanoic acid. Begin the reaction by mixing the reactants, and add a measured quantity of dilute hydrochloric acid as catalyst (the amount added must be allowed for in the subsequent steps). At measured time intervals, use a pipette to withdraw measured quantities (aliquots) and quench the reaction by dilution in ice-cold water. Titrate the diluted aliquot against standard aqueous sodium hydroxide solution with phenolphthalein as indicator. Calculate the amount of ethanoic acid in the aliquot, and convert it to concentration in the reaction solution. Plot the concentration against time, and evaluate the rate from the slope of the graph.

(b) The second reaction, often called the **Harcourt–Esson reaction**, is

$$2HI(aq) + H_2O_2(aq) \rightarrow I_2(aq) + 2H_2O(l)$$

The rate can be measured titrimetrically, as in (a). Alternatively, the reaction could be carried out in the presence of a starch indicator and a measured amount of sodium thiosulphate; this reacts very rapidly with the iodine, reducing it back to iodide ions, and so keeping their concentration constant. The time when all the thiosulphate has been used is indicated by the solution turning blue as iodine survives. Another measured amount of thiosulphate is then added, and the observation is repeated until all the hydrogen peroxide has reacted.

COMMENT Each reaction should be run in a thermostatted water-bath so that the temperature (which affects the rate) does not vary.

In modern chemical kinetics, there is a great deal of interest in fast reactions – those reaching equilibrium in less than about a second. One important technique for following such reactions is the **stopped-flow method**, in which the reactants are made to flow very quickly into a mixing chamber as a piston moves back to a stop (see Figure 14.2). The time-dependence of the composition of the solution inside the chamber is followed, often spectroscopically. The fastest reactions, those complete within nanoseconds ($1\ ns = 10^{-9}\ s$), are followed using **laser flash photolysis** (see Figure 14.3). In this technique, a very brief, intense flash from a laser (the **photoflash**) is used to start the reaction, and then the composition at later times is followed spectroscopically with a second flash (the **specflash**). Laser flashes can be very brief (see Box 14.2): nanosecond flash photolysis is now commonplace, and special techniques have extended this scale down to picoseconds ($1\ ps = 10^{-12}\ s$).

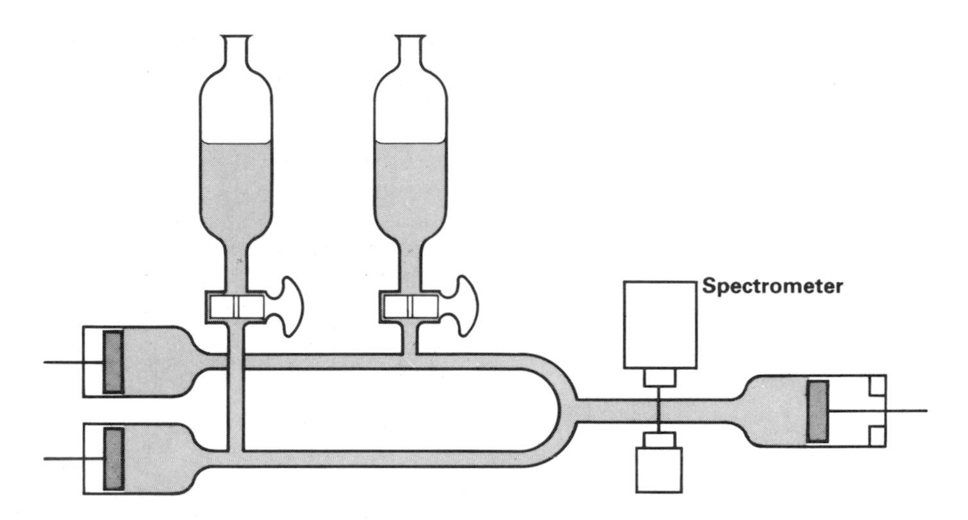

Figure 14.2 The stopped-flow technique

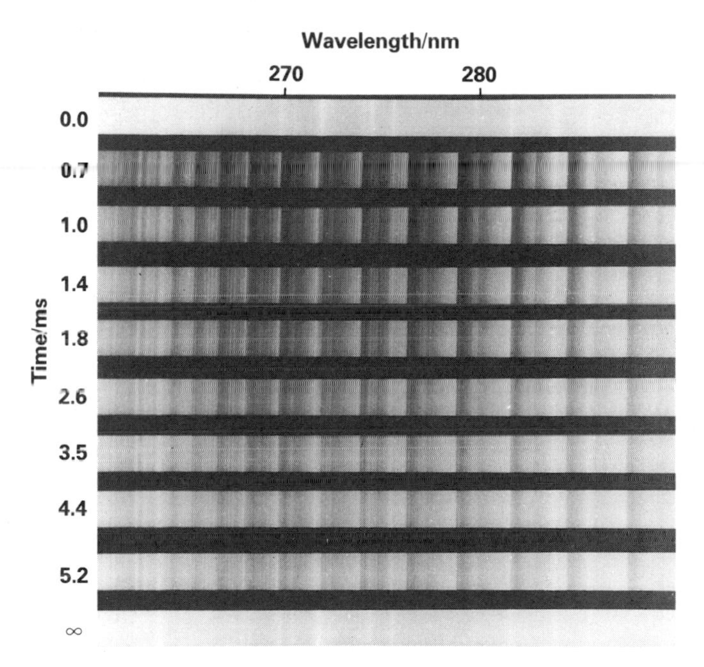

Figure 14.3 The spectrum of ClO following flash photolysis of a mixture of chlorine and oxygen: there is virtually no ClO present initially, but its concentration rises very rapidly following the flash and then decays as it reacts

BOX 14.1

The development of chemical kinetics

The first measurement of the rate of a chemical reaction was by L. Wilhelmy (in 1850) who observed the rate at which the optical activity (see Section 32.6) of sucrose solution changed. The concepts of rate law and order of reaction grew out of the work of C. M. Guldberg and P. Waage (in about 1865) who formulated the **law of mass action**, that 'the rate of a chemical reaction is proportional to the active masses (concentrations) of the reacting substances' (which is what we would now write as $rate = k[A][B]$). The analysis of many gas-phase reactions was carried out by C. N. Hinshelwood (who received the Nobel prize, and was uniquely the President of the Royal Society and of the Classical Association), but the delicacy of the problem of working out reaction mechanisms is illustrated by the fact that the mechanism of the decomposition of hydrogen iodide proposed by M. Bodenstein in 1899 survived until it was disproved by J. H. Sullivan in 1967. The development of modern chemical kinetics has been due to the application of electronic methods for following concentrations, and the resolution of changes on ever shorter time scales. Fast reaction techniques include ultrasonic methods, flash photolysis, magnetic resonance and shock tubes. The introduction of lasers has extended the time-scale of observations down to about 1 ps (10^{-12} s), and extremely detailed information about reactions can now be obtained. Detail of another kind can also be obtained from experiments using **molecular beams**, in which streams of molecules are shot at each other and the products are analysed. In this way, the individual events involved in reactions can be understood, and we can now deduce how bonds are made and broken and how molecules are ripped apart or formed during collisions.

BOX 14.2

Lasers

The word LASER is an acronym for Light Amplification by Stimulated Emission of Radiation. The action of lasers depends on two properties. The first is the ability of atoms, ions and molecules to be excited to a higher energy state and to remain in that excited condition for a short time. The second is the possibility that these excited entities can be stimulated to discard their energy as light when they are exposed to light of the same frequency. If the laser medium is contained in a cavity between two parallel mirrors, the light reflects backwards and forwards between them, and on each passage it stimulates more excited entities to emit their energy. The intensity of the light grows very rapidly, and can be allowed to escape from the cavity in a burst, or to escape continuously if one of the mirrors is half-silvered.

The first laser was constructed (in 1960) from a single crystal ruby rod. Ruby is aluminium oxide with some chromium(III) impurities. The chromium ions can be excited by a bright burst of radiation from a xenon discharge tube, and will then emit laser radiation. Since that time, numerous other substances have been used for lasers, including gases (particularly carbon dioxide, neon, argon and nitrogen), liquids (solutions of organic dyes) and semiconductors (particularly gallium arsenide).

The properties of laser radiation that are particularly useful in chemistry include its high intensity, its ability to be produced in very short (picosecond) pulses, and its well-defined frequency (its 'monochromaticity'). The high intensity is useful for inducing photochemical reactions efficiently; the pulses are useful for observing the details of reactions on a very short time scale; and the monochromaticity is useful for spectroscopy and isotope separation.

At this stage it might appear that all these experiments are doing is providing tables of data, and that the only way of discussing reaction rates is to quote their values for the conditions and times of interest. Happily this is not so. The rates of many reactions depend on the concentrations of the reactants in a simple way. For instance, in some reactions it is found that the rate of disappearance of a reactant A is proportional to its concentration, so that $rate \propto [A]$. The discovery of relations like this is a valuable simplification, and makes the study of reaction rates of great practical use.

14.2 Rate laws

Reaction rates usually depend on the current concentrations of the reactants. For example, the rate at which magnesium reacts with acid depends on the pH; the higher the concentration of hydrogen ions the more quickly the reaction produces hydrogen gas, and as the acid is consumed the reaction

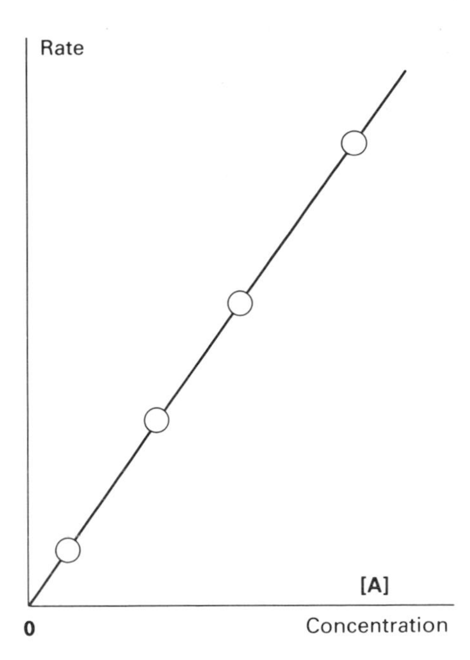

Figure 14.4 Rate is proportional to concentration for a first order reaction

slows. When there are several reactants, the dependence of the rate on each one can be determined by ensuring that all the others are in such great excess that their concentrations are virtually unchanged throughout the reaction. This is called the **Ostwald isolation method** for measuring rates, because the effect of each reactant is isolated in turn. An alternative method is to measure the **initial rate**, that is, the rate at a time when the reaction has not proceeded far, and the concentrations of all the substances are virtually unchanged from their initial values.

Consider a reaction in which A is destroyed at a rate found experimentally to be proportional to its concentration. An example is the decomposition of dinitrogen pentoxide:

$$2N_2O_5(g) \rightarrow 2N_2O_4(g) + O_2(g)$$

At any stage of the reaction the rate is given by the **rate law** (or **rate equation**):

$$rate = k[A] \tag{14.2.1}$$

(see Figure 14.4). The factor k, which is a constant for a particular reaction and temperature, is the **rate constant** (or **rate coefficient**) for the reaction at that temperature. Because the reaction rate is proportional to the first power of [A], it is an example of a **first-order rate law**. Reactions showing this behaviour are called **first-order reactions** and are said to follow **first-order kinetics**. As the rate itself has the dimensions of concentration/time, the rate constant has the dimensions 1/time and is normally expressed in s^{-1}. It is most important to appreciate that in general *the rate law cannot be predicted from the chemical equation, but must be determined experimentally*.

First-order reactions

If it is found by experiment that a reaction is first-order, the consequence is that its rate at *any* concentration of A can be summarized by stating the value of its rate constant. For instance, if k is reported as $2.4 \times 10^{-3}\ s^{-1}$, we would know that when the concentration of A is $1.0\ mol\ dm^{-3}$, the rate of consumption of A at that temperature would be $2.4 \times 10^{-3}\ mol\ dm^{-3}\ s^{-1}$; it would be $1.2 \times 10^{-3}\ mol\ dm^{-3}\ s^{-1}$ when the concentration has fallen to $0.50\ mol\ dm^{-3}$, and so on.

The rate law also provides a simple way of predicting the concentration of the reactants and the products at any time after the start of the reaction because it can be solved for the time dependence of [A]. The solution is described in Appendix 14.1; here only the result is needed. If the initial concentration of A is $[A]_0$, then at a later time t it will have fallen to [A], where

$$[A] = [A]_0 e^{-kt} \tag{14.2.2a}$$

which shows that the concentration follows an exponential decay (see Figure 14.5).

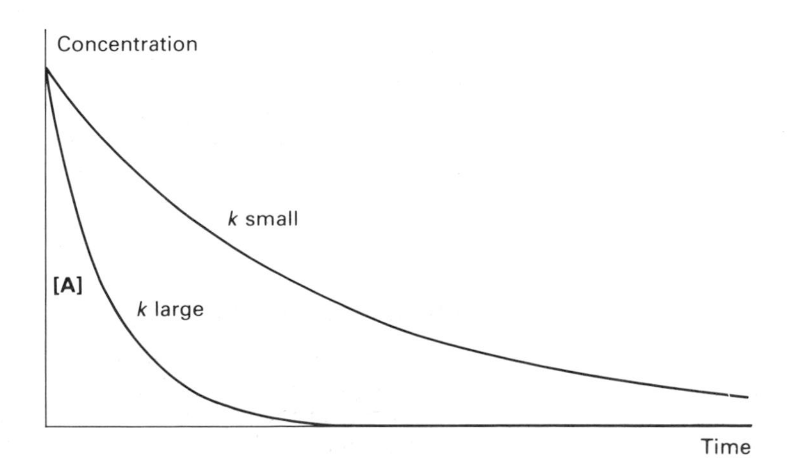

Figure 14.5 The time dependence of the concentration of a reactant in a first-order reaction

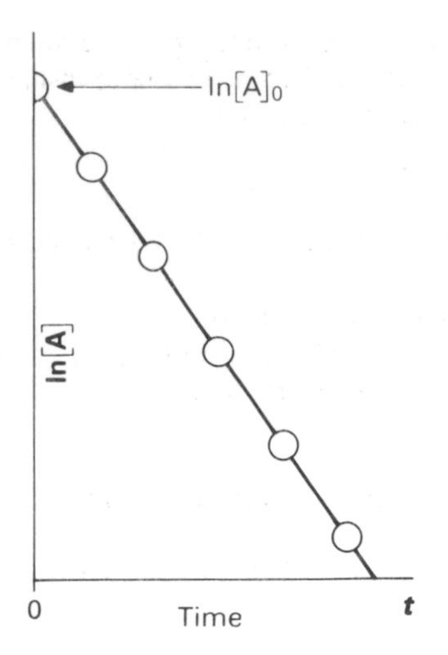

Figure 14.6 Investigating a first-order reaction

On taking natural logarithms:

$$\ln[\mathrm{A}] = \ln[\mathrm{A}]_0 - kt \qquad (14.2.2b)$$

or

$$\ln\frac{[\mathrm{A}]}{[\mathrm{A}]_0} = -kt \qquad (14.2.2c)$$

Figure 14.6 shows a graph of the (natural) logarithm of the concentration of a reactant in a first-order reaction against time: it is a straight line, as the rate law predicts. A reaction can be tested for first-order character by plotting $\ln[\mathrm{A}]$ against time. If the line obtained is straight, then the reaction is first-order and the slope (which is equal to $-k$) gives the rate constant.

Equation 14.2.2 can be used to find the **half-life** ($t_{\frac{1}{2}}$) of a reactant, that is, the time needed for its concentration to fall to half its initial value. When $t = t_{\frac{1}{2}}$, the concentration of A is $[\mathrm{A}] = \frac{1}{2}[\mathrm{A}]_0$; substituting these values into equation 14.2.2*c* gives

$$\ln\frac{[\mathrm{A}]_0}{2[\mathrm{A}]_0} = -kt_{\frac{1}{2}} \quad \text{or} \quad \ln\tfrac{1}{2} = -kt_{\frac{1}{2}}$$

which rearranges (using $\ln\frac{1}{2} = -\ln 2$) to

$$t_{\frac{1}{2}} = \frac{\ln 2}{k} = \frac{0.693}{k} \qquad (14.2.3)$$

For a first-order reaction in which $k = 1.00 \times 10^{-3}\,\mathrm{s}^{-1}$, it follows that the half-life is 693 s.

A unique feature of first-order reactions is that *the half-life is independent of the initial concentration.* Whatever the initial concentration of A, it falls to half that value in a time $0.693/k$ (see Figure 14.7). This behaviour is the basis of the radiocarbon dating technique discussed in Section 1.4, because nuclear disintegration is a first-order process and carbon-14 nuclei have the same half-life whatever their abundance.

EXAMPLE

The hydrolysis of sucrose is first-order in sucrose, and the half-life is 80 min at 20 °C. Calculate the proportion of the initial sucrose concentration that remains (a) after 160 min, (b) after 240 min.

METHOD Since the half-life of a first-order reaction is independent of the concentration, each period of 80 min results in a halving of the concentration at the start of that period.

ANSWER (a) 160 min is two consecutive half-lives; and so the initial concentration is halved and then halved again. The final concentration is therefore one-quarter its initial value.
(b) 240 min is three consecutive half-lives; and so the initial concentration is halved, halved again, and then halved again. The final concentration is reduced to one-eighth its initial value.

COMMENT The hydrolysis reaction is a first step in the fermentation of sugar and the production of alcohol, described in Section 29.6. Brewers' yeast contains the enzyme sucrase, which catalyses the reaction.

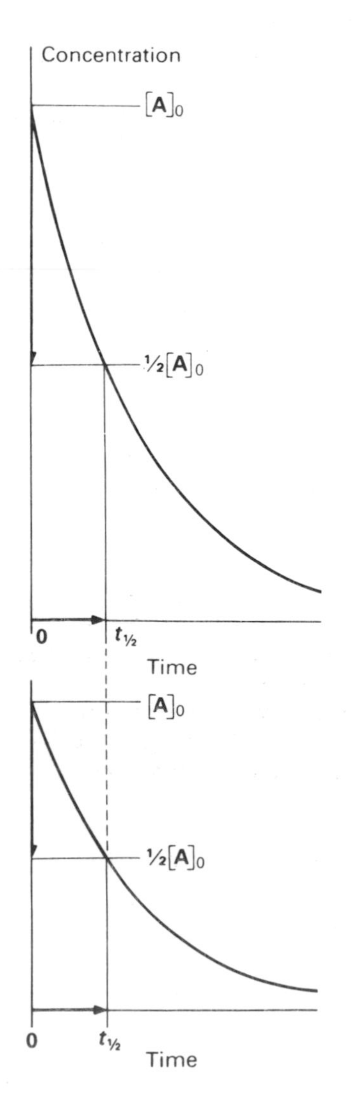

Figure 14.7 The half-life of a first-order reaction is independent of the initial concentration

Second-order reactions

For some reactions the plot of $\ln[\mathrm{A}]$ against time is found not to be a straight line, and so they are not first-order. They may be **second-order**, meaning that the rate depends on the product of two concentrations: the rate law may have the form

$$rate = k[\mathrm{A}]^2 \qquad (14.2.4)$$

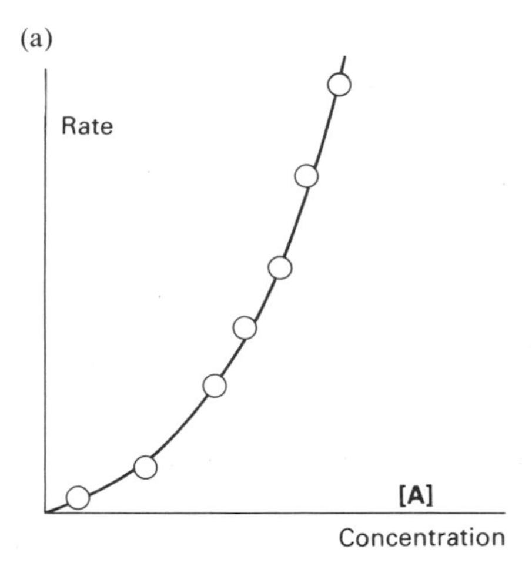

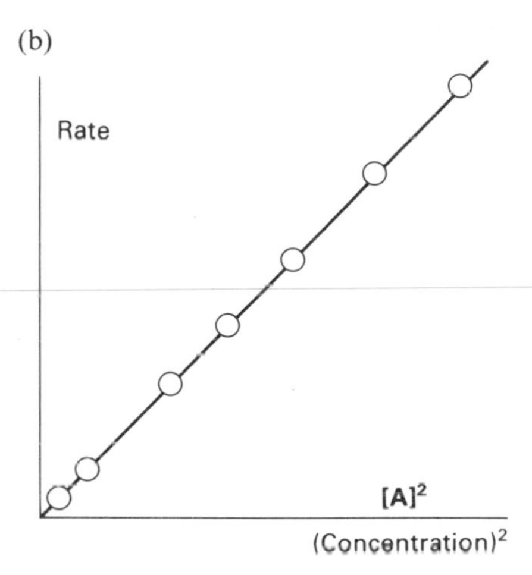

Figure 14.8 For a second-order reaction, rate (a) is not proportional to concentration but (b) is proportional to the square of the concentration

k is the **second-order rate constant**, and the equation is an example of a **second-order rate law** (see Figure 14.8). Such reactions are said to follow **second-order kinetics**. An example is the decomposition of nitrogen dioxide:

$$2NO_2(g) \rightarrow 2NO(g) + O_2(g) \qquad rate = k[NO_2]^2$$

EXAMPLE

The gas-phase reaction $H_2 + I_2 \rightarrow 2HI$ obeys overall second-order kinetics and is first-order in each reactant. At 400 °C the rate constant is $2.42 \times 10^{-2}\,dm^3\,mol^{-1}\,s^{-1}$. Calculate the reaction rate when the concentration of each reactant is $0.50\,mol\,dm^{-3}$.

METHOD Write the rate law for the reaction and substitute the data.

ANSWER The rate law is $rate = k[H_2][I_2]$; therefore, under the stated conditions,

$$\begin{aligned} rate &= 2.42 \times 10^{-2}\,dm^3\,mol^{-1}\,s^{-1} \times 0.50\,mol\,dm^{-3} \times 0.50\,mol\,dm^{-3} \\ &= 6.1 \times 10^{-3}\,mol\,dm^{-3}\,s^{-1} \end{aligned}$$

COMMENT In this case the reactant concentrations are the same; that need not be so, however, and the rate law is a *universal* summary of the reaction rate whatever the specified concentrations. The units are always chosen so that the rate is expressed in terms of concentration units per unit time ($mol\,dm^{-3}\,s^{-1}$, for instance).

One way of testing whether equation 14.2.4 describes a reaction is to measure the rate at several concentrations and then to plot them against $[A]^2$: a straight line would confirm second-order kinetics.

EXAMPLE

A substance A reacts at a rate r which is found to vary with the concentration of A as follows:

$[A]/mol\,dm^{-3}$	0.090	0.15	0.30	0.54
$r/mol\,dm^{-3}\,s^{-1}$	6.2×10^{-5}	1.7×10^{-4}	6.9×10^{-4}	2.2×10^{-3}

Find the order of the reaction and its rate constant.

METHOD Inspect the data to see if there is a simple pattern, such as the rate doubling when the concentration doubles. In this Example the second and third data points show that doubling the concentration of A increases the rate by a factor of $6.9 \times 10^{-4}/1.7 \times 10^{-4} \approx 4$, which suggests a second-order reaction. Since the rate law of a second-order reaction is $r = k[A]^2$, check this conclusion by evaluating $r/[A]^2$, which should be a constant. If it is, then its value is the second-order rate constant k.

ANSWER The following table is drawn up from the data provided:

$[A]/mol\,dm^{-3}$	0.090	0.15	0.30	0.54
$r/[A]^2$	7.65×10^{-3}	7.56×10^{-3}	7.67×10^{-3}	7.54×10^{-3}

The ratio is constant within experimental error, so that the reaction is second-order in A and its rate constant (the mean of the values) is $7.6 \times 10^{-3}\,dm^3\,mol^{-1}\,s^{-1}$.

COMMENT The third significant figure in the ratio in the table is not justified, but was retained to avoid rounding errors during the calculation.

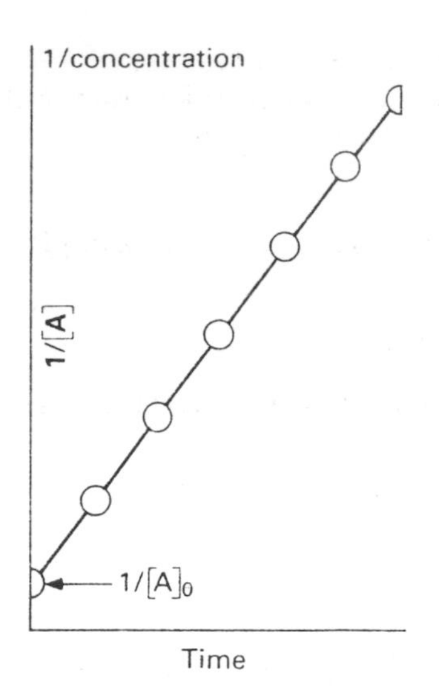

Figure 14.9 Investigating a second-order reaction

A much better method is to solve the rate law and plot the appropriate function of the concentration against the time, looking for a straight line. The solution of equation 14.2.4 is described in Appendix 14.1; the result is

$$\frac{1}{[A]} = \frac{1}{[A]_0} + kt \qquad (14.2.5)$$

Hence, *if* the reaction is second-order in A, a plot of $1/[A]$ against t gives a straight line and its slope is k (see Figure 14.9).

If the data confirm that the reaction is second-order, equation 14.2.5 can be used to predict the concentration of A at any stage by rearranging it into

$$[A] = \frac{[A]_0}{1 + k[A]_0 t} \qquad (14.2.6)$$

This expression is plotted in Figure 14.10 and compared with the form of the first-order decay: the curve shows that at the start a second-order reaction slows down more rapidly than a first-order reaction with the same initial rate.

The time for the concentration of A to fall to half its initial value is obtained easily from equation 14.2.5 by setting $t = t_{\frac{1}{2}}$ and $[A] = \frac{1}{2}[A]_0$:

$$kt_{\frac{1}{2}} = \frac{2}{[A]_0} - \frac{1}{[A]_0}$$

which rearranges to

$$t_{\frac{1}{2}} = \frac{1}{k[A]_0} \qquad (14.2.7)$$

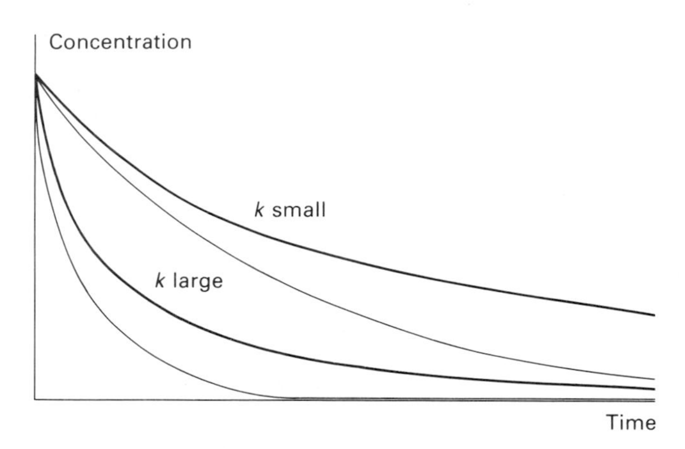

Figure 14.10 The time dependence of the concentration of a reactant in a second-order reaction (compared with first-order behaviour)

Unlike that of a first-order reaction, the half-life of a second-order reaction depends on the initial concentration, and the higher that concentration is, the more rapidly does [A] decrease to half its value (see Figure 14.11). Suppose that [A] falls to one-half its initial value in 10 s; it will take a *further* 20 s to fall to one-quarter its initial value. Had the reaction been first-order, only another 10 s would have been needed.

14.3 Reaction order and reaction mechanism

A reaction might be found to have a rate law that is more complicated than any of those shown so far. This can be taken into account by generalizing what has been said, and leads to additional useful information.

Reaction order

If experiment shows that the rate law for a certain reaction is

$$rate = k[A]^x[B]^y \qquad (14.3.1)$$

then the reaction is said to be of order x in A, of order y in B, and of order

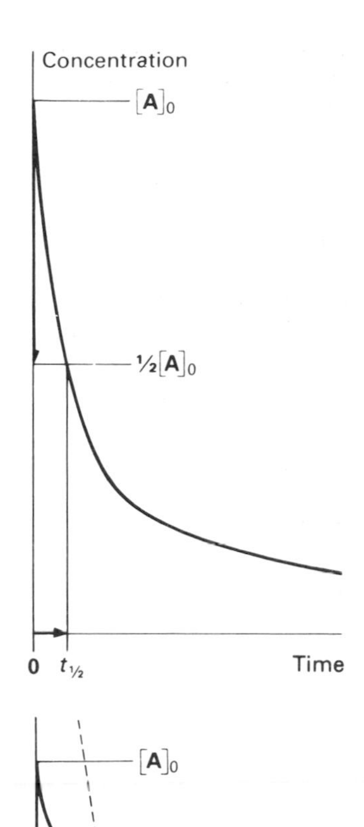

Figure 14.11 The half-life of a second-order reaction increases as the initial concentration of reactant is decreased

$x + y$ overall. In other words, the **order** is the power to which the concentration of a substance (either a reactant or a product) is raised in the rate law. In particular, reactions having rate laws of the form

$$rate = k[A][B]$$

are *first-order in* A, *first-order in* B, and *second-order overall.* An example is the synthesis of hydrogen iodide:

$$H_2(g) + I_2(g) \rightarrow 2HI(g) \qquad rate = k[H_2][I_2]$$

EXAMPLE

Account for the following observations on the iodination of propanone in acidified aqueous solution:
(a) When the propanone concentration is doubled, the rate doubles.
(b) When the iodine concentration is doubled, the rate remains unchanged.
(c) When the pH of the solution is reduced from 1.0 to 0.70, the rate doubles.

METHOD Find the order with respect to each component by deciding the power to which the concentration must be raised to account for the data.

ANSWER (a) The first observation shows that the rate is proportional to the concentration of propanone; therefore, since rate $\propto$ [propanone], the reaction is *first-order in propanone.*
(b) The rate is independent of the iodine concentration, so that the reaction is *zeroth-order in iodine.*
(c) pH = 1.0 corresponds to $[H^+(aq)] = 0.10\,mol\,dm^{-3}$; pH = 0.70 corresponds to $[H^+(aq)] = 0.20\,mol\,dm^{-3}$. Doubling the hydrogen ion concentration therefore doubles the rate, implying that the reaction is *first-order in hydrogen ions.*

COMMENT It follows that the rate law is

$$rate = k[\text{propanone}][H^+(aq)]$$

and that it is second-order overall. The occurrence of the hydrogen ion concentration in the rate law but not in the overall chemical equation indicates an **acid-catalysed reaction**, a term which is explained in Section 14.5.

A reaction may be found to be **zeroth-order** in a reactant A, in which case the reaction rate is independent of the concentration of A (because $[A]^0 = 1$). An example is the decomposition of ammonia on heated tungsten:

$$2NH_3(g) \xrightarrow{\text{W, heat}} N_2(g) + 3H_2(g) \qquad rate = k$$

The concentration of the ammonia decreases at a steady rate until it reaches zero.

EXAMPLE

Which of the graphs **1** to **5** shows the correct variation of initial rate with initial concentration for (a) first-order, (b) second-order reactions?

METHOD Use the information that in a first-order reaction the rate is directly proportional to the concentration (and hence the initial rate is also directly proportional to the initial concentration in a series of runs), and that in a second-order reaction the rate is proportional to the square of the concentration.

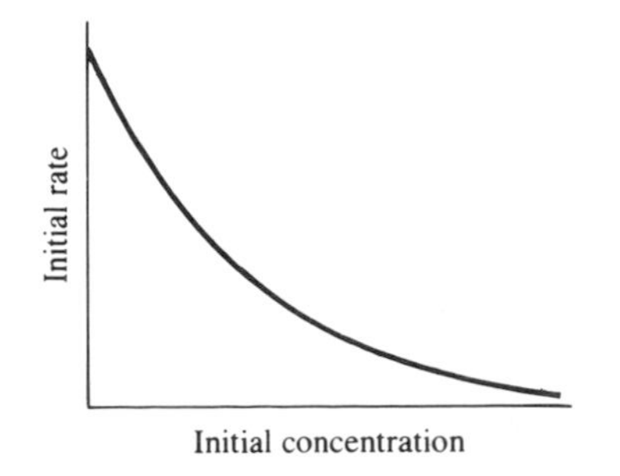

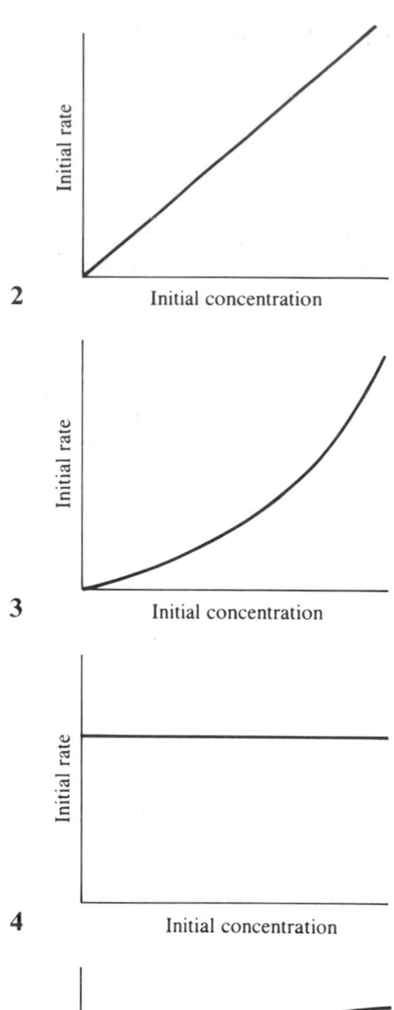

ANSWER (a) The plot of the initial rate of a first-order reaction against initial concentration is a straight line; hence graph **2** corresponds to a first-order reaction.
(b) The plot of the initial rate of a second-order reaction against initial concentration is a parabola ($y = ax^2$); hence graph **3** corresponds to a second-order reaction.

COMMENT The remaining simple graph, **4**, is typical of a zeroth-order reaction.

It sometimes happens that a reactant is in such great excess that its concentration remains almost unchanged throughout the reaction. In this case a true second-order rate law of the form $rate = k'[A][B]$ behaves as though $k'[A]$ were another constant k, and hence becomes a rate law of the form $rate = k[B]$. Such a reaction is called a **pseudofirst-order reaction**, and in it the concentration of B decays exponentially towards zero. Pseudofirst-order reactions often occur in solution, especially if the solvent – often water – takes part in the reaction, for its concentration barely changes.

Not all reactions can be ascribed an order. For instance, the apparently simple reaction

$$H_2(g) + Br_2(g) \rightarrow 2HBr(g)$$

is found to have the following rate law:

$$rate = \frac{k'[H_2][Br_2]^{\frac{1}{2}}}{1 + k''[HBr]/[Br_2]}$$

Although the concept of order breaks down in such cases, the rate law remains perfectly well defined: in order to calculate the rate under given conditions, simply substitute the measured values of the rate constants and the concentrations of the substances involved in the reaction.

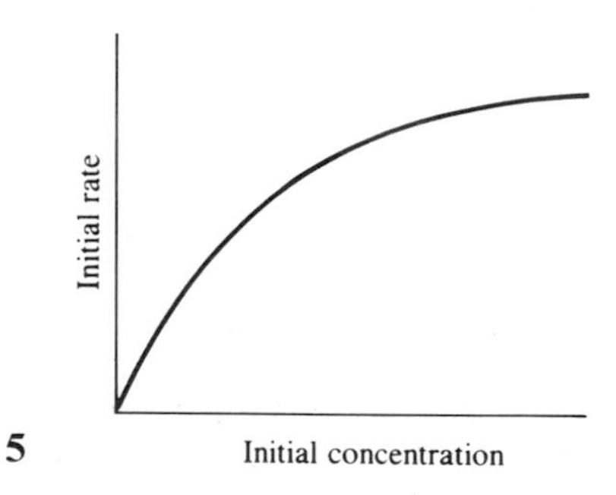

Chain reactions

A complicated rate law implies that the **reaction mechanism**, the individual steps proposed to account for the conversion of reactants into products, is also complicated. For example, the $H_2 + Br_2$ rate law can be explained in terms of a mechanism involving **radicals** – atoms or molecules with unpaired electrons. Since a product of one step acts as a reactant in another, the mechanism is an example of a **chain reaction** in which radicals are the **chain carriers** (another example is described in Section 27.3). All reaction steps are in the gas phase, and the suffix (g) will not be written explicitly in this description.

Step 1: Initiation The Br_2 dissociates as a result of energetic collisions or the absorption of light, forming two atoms (which are radicals, as each has an unpaired electron, denoted by a dot):

$$Br_2 \rightarrow 2Br^{\cdot}$$

Step 2: Propagation The bromine atoms, the chain carriers, attack hydrogen molecules:

$$Br^{\cdot} + H_2 \rightarrow HBr + H^{\cdot}$$

and the hydrogen atoms, the new chain carriers, attack bromine molecules:

$$H^{\cdot} + Br_2 \rightarrow HBr + Br^{\cdot}$$

The bromine atoms produced can in turn attack other hydrogen molecules, and so the chain continues and more hydrogen bromide is formed.

Step 3: Inhibition Some of the HBr molecules already formed may be destroyed by hydrogen atom attack,

$$H^{\cdot} + HBr \rightarrow H_2 + Br^{\cdot}$$

with the result that some of the product is destroyed and the net rate of its production is reduced.

Step 4: Termination The chain of reactions depends on the presence of the radical chain carriers. In particular, the propagation steps cannot occur if the reaction

$$Br^{\cdot} + Br^{\cdot} \rightarrow Br_2$$

takes place, and so it stops the chain.

The overall rate of product formation can be calculated by combining the rates of the individual steps. When this is done, the experimental rate law is obtained, which supports the proposed mechanism.

Explosions

A chain reaction may also lead to an explosion. This may occur if one of the steps leads to **chain branching**, in which one radical attacks a molecule and produces two or more radicals. For example, in the reaction between hydrogen and oxygen, the step

$$H_2 + {}^{\cdot}O^{\cdot} \rightarrow H^{\cdot} + HO^{\cdot}$$

is a chain-branching step. The two radicals so produced can in turn attack other molecules, and so more and more radicals are produced in subsequent reactions. If termination reactions cannot compete, the rate of the overall reaction may increase extremely rapidly, causing an explosion.

A similar type of chain branching occurs in nuclear fission, when capture of a single neutron by a ^{235}U nucleus leads to the production of several neutrons as the nucleus disintegrates. The chain carriers are the neutrons in this **nuclear chain reaction**.

14.4 Rates and conditions

Reaction rates depend not only on *concentrations* (or on *partial pressures* of gases), but also on the *temperature*, the presence of *catalysts*, the *state of subdivision* of solids, and, in some cases, the *intensity of light* incident on the system (see Section 27.3, for example). The effect of the state of subdivision, for example, can be demonstrated by comparing the rates of production of carbon dioxide when dilute hydrochloric acid reacts with either marble chips or powdered calcium carbonate; the powder reacts much more rapidly than the chips do, because it has a much greater surface area.

The concentration dependence

A rate law summarizes the experimentally observed concentration dependence; in this section we explain how it arises.

In order for molecules to react they must meet. If we imagine molecules moving at random, as they do in a gas or in a solution, then the greater their concentrations the more frequently they meet. Hence, the higher the concentration, the greater the rate of the reaction.

This explanation accounts very neatly for the rate law of simple *one-step* gas-phase reactions, such as the attack of a hydrogen atom on a bromine molecule,

$$H^{\cdot} + Br_2 \rightarrow HBr + Br^{\cdot}$$

which experiment shows to be overall second-order:

$$rate = k[H^{\cdot}][Br_2]$$

If we take the view that the reaction occurs as a result of the collision between a hydrogen atom and a bromine molecule, then second-order kinetics is expected because the rate increases as the concentrations of both reactants increase.

Only for elementary, single-step reactions can the rate law be written directly from the chemical equation, as in the example above. In a complex reaction (such as the $H_2 + Br_2$ reaction) the observed rate may be the outcome of several different steps, each having a different rate: the *overall* rate law cannot be predicted by inspection of the *overall* equation.

Consider, for example, the iodination of propanone (CH_3COCH_3) in acidic solution. The reaction is described by the equation:

$$CH_3COCH_3(aq) + I_2(aq) \rightarrow CH_3COCH_2I(aq) + HI(aq)$$

It would be quite wrong to assert that the overall rate law is $rate = k[CH_3COCH_3][I_2]$ unless such a statement was supported by experimental information, *or* unless the reaction was known to occur in a single step. In fact, when the rate law is investigated experimentally (see the first Example in Section 14.3), it turns out to be

$$rate = k[CH_3COCH_3][H^+(aq)]$$

So, although the reaction is second-order overall, it is zeroth-order (not first-order) in iodine, and first-order in hydrogen ions even though they do not appear in the overall reaction.

The form of the iodination rate law can be explained as follows. Suppose that the reaction proceeds in two steps: in the first step the propanone is prepared for attack and in the second iodine attacks. Suppose too that the first step is slow and the second fast. Then the overall rate is limited by the rate of the first step, because the second takes place rapidly once the first has occurred. This is like two towns being joined by a good road crossing two bridges: if the first bridge is narrow the traffic will queue to pass over it, but once they are across, the second broader bridge does not impede them (see Figure 14.12).

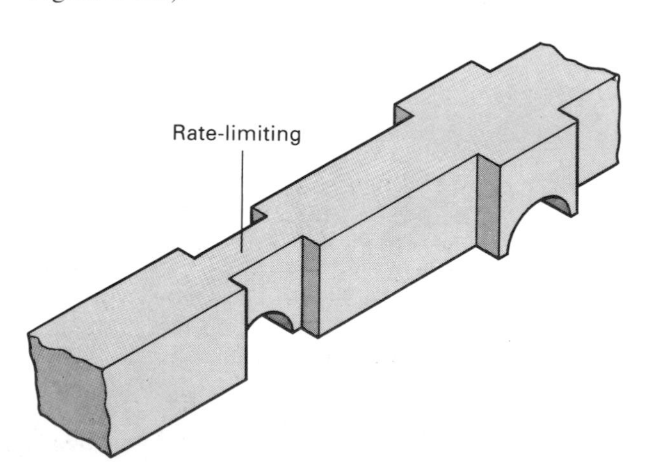

Figure 14.12 An analogy for the rate-limiting step of a reaction

When a single step limits the overall rate of a multi-step reaction, it is called the **rate-limiting step** (or **rate-determining step**) of the reaction. In the iodination example, the rate law is consistent with the rate-limiting step being the attack of a hydrogen ion on a propanone molecule: the rate law for that single-step process is the same as the experimental rate law. Some reaction steps are so fast that the overall rate is limited by the rate at which the reactants can be brought into contact: in one part of the glycolysis reaction in a biochemical cell the rate-limiting step is the rate of supply of materials to triose-phosphate isomerase, a very efficient enzyme.

The rate-limiting step in the propanone reaction involves two substances, a propanone molecule and a hydrogen ion. It is therefore called a **bimolecular step**. In some cases a reaction step consists of a single molecule shaking itself apart, and is called a **unimolecular step**. The terms 'second-order' and 'bimolecular' (and, similarly, 'first-order' and 'unimolecular') must not be confused. The *order* of the reaction is something that must be determined through experiment and inspection of the rate law; the *molecularity* is the number of substances taking part in a specified step in the reaction mechanism. The observed rate law, even if it is second-order, may be the outcome of a complex series of unimolecular and bimolecular processes.

The temperature dependence

Almost all reactions go faster at higher temperatures. When Arrhenius (see Box 8.1) studied the effect of temperature on reaction rates (in 1889), he found that the temperature dependence of the rate constant could usually be expressed as follows:

Arrhenius equation: $k = A\mathrm{e}^{-E_a/RT}$ (14.4.1)

The **Arrhenius parameters** A and E_a depend on the reaction. A is called the **Arrhenius factor** (or the **pre-exponential factor**) and E_a is the **activation energy**. The Arrhenius equation applies to all kinds of reactions; even luminescent tropical fireflies have been found to flash more rapidly on warmer nights in accord with its requirements. Values of the activation energy (the more important parameter in the equation) for some reactions are given in Table 14.1.

Table 14.1 Activation energies of some reactions, E_a/kJ mol^{-1}

Reaction	Conditions	E_a/kJ mol^{-1}
$C_2H_4 + H_2 \rightarrow C_2H_6$	gas phase	180
$2HI \rightarrow H_2 + I_2$	gas phase	183
	Pt catalyst	58
$H_2 + Cl_2 \rightarrow 2HCl$	photochemical	25
$CH_3Cl + CH_3O^- \rightarrow CH_3OCH_3 + Cl^-$	methanol solvent	100

EXAMPLE

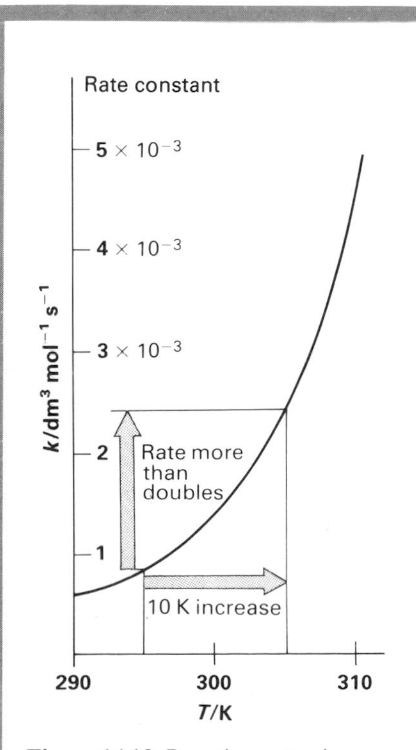

Figure 14.13 Reaction rates increase sharply with temperature

Calculate the ratio of the rate constants at 30 °C and 20 °C for a reaction with activation energy 50 kJ mol^{-1}.

METHOD Use the Arrhenius equation (equation 14.4.1), with the temperature in kelvins. The A-factors cancel when ratios are taken, and so can be ignored. Begin by evaluating E_a/RT at each temperature.

ANSWER At 30 °C,

$$E_a/RT = \frac{50 \times 10^3\ \mathrm{J\ mol^{-1}}}{8.314\ \mathrm{J\ K^{-1}\ mol^{-1}} \times 303.15\ \mathrm{K}} = 19.84$$

At 20 °C,

$$E_a/RT = \frac{50 \times 10^3\ \mathrm{J\ mol^{-1}}}{8.314\ \mathrm{J\ K^{-1}\ mol^{-1}} \times 293.15\ \mathrm{K}} = 20.51$$

The ratio of rate constants is therefore

$$\frac{k(30\,^\circ\mathrm{C})}{k(20\,^\circ\mathrm{C})} = \frac{\mathrm{e}^{-19.84}}{\mathrm{e}^{-20.51}} = \frac{2.4 \times 10^{-9}}{1.2 \times 10^{-9}} = 2.0$$

COMMENT This calculation is the origin of the commonly stated rule that a 10 K increase of temperature doubles the rate of a reaction (see Figure 14.13). Many reactions have activation energies of around 50 kJ mol^{-1}, and for them this prediction is valid. On the other hand many do not, and for them the effect of temperature is quite different.

Activation energies are calculated by measuring rate constants at different temperatures and making an **Arrhenius plot**. First, the Arrhenius equation is rearranged by taking natural logarithms, giving

$$\ln k = \ln A - E_a/RT \qquad (14.4.2)$$

Then $\ln k$ is plotted against $1/T$ to give a straight line of slope $-E_a/R$, as shown in Figure 14.14. (The intercept of the plot with the $\ln k$ axis at $1/T = 0$ is equal to $\ln A$, so that A may be obtained by extrapolating the line.)

EXAMPLE

The decomposition of dinitrogen pentoxide, N_2O_5, was followed at different temperatures and the rate constants for the reaction were found to vary as follows:

T/K	300	310	320	330
k/s^{-1}	3.38×10^{-5}	1.35×10^{-4}	4.98×10^{-4}	1.50×10^{-3}

Calculate the activation energy of the reaction.

METHOD Make an Arrhenius plot of $\ln k$ against $1/T$. The slope of the graph is $-E_a/R$.

ANSWER Draw up the following table:

$1/(T/\text{K})$	3.33×10^{-3}	3.23×10^{-3}	3.13×10^{-3}	3.03×10^{-3}
$\ln (k/\text{s}^{-1})$	-10.3	-8.91	-7.60	-6.50

The points are plotted in Figure 14.14. The slope of the line is

$$\text{slope} = \frac{8.12 - 9.38}{0.100 \times 10^{-3}} = -1.26 \times 10^4$$

It follows that

$$-E_a/R = -1.26 \times 10^4 \text{ K}$$

so that

$$E_a = 1.26 \times 10^4 \text{ K} \times R$$
$$= 1.26 \times 10^4 \text{ K} \times 8.314 \text{ J K}^{-1} \text{ mol}^{-1} = 105 \text{ kJ mol}^{-1}$$

COMMENT Pay special attention to the way the slope of a graph, which is dimensionless (see page 519), is related to a physical observable, which has dimensions.

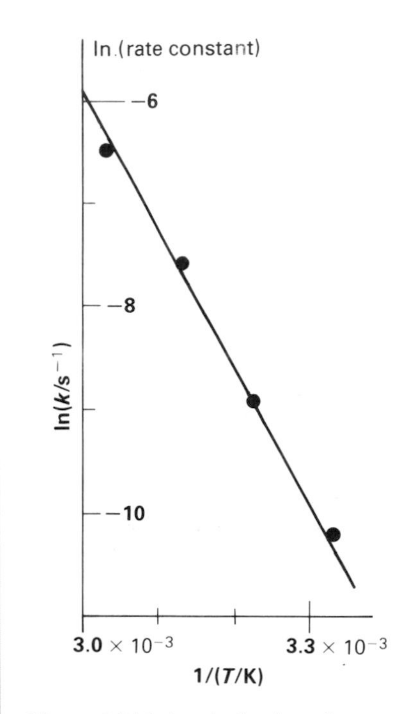

Figure 14.14 An Arrhenius plot

The form of the Arrhenius equation is quite simple to explain. Not only must molecules meet in order to react, but they must also have enough energy to reorganize their bonds. For a simple bimolecular reaction step (like the reaction of H˙ and Br_2), not only must the particles collide, but they must smash together with enough kinetic energy – the activation energy – to make the atomic rearrangement possible. The activation energy is the height of the **reaction barrier**, the hump in Figure 14.15: if the reactants have enough kinetic energy to reach the top of the reaction barrier they merge into a single **activated complex**, and then go on to form products.

The number of molecules having *at least* the kinetic energy E_a at the temperature T is proportional to the shaded area under the Maxwell–Boltzmann curve of kinetic energies (see Figure 14.16, and also Section 4.2). As the temperature is increased, the area of the shaded region increases and more molecules have kinetic energies greater than E_a; hence the reaction rate increases.

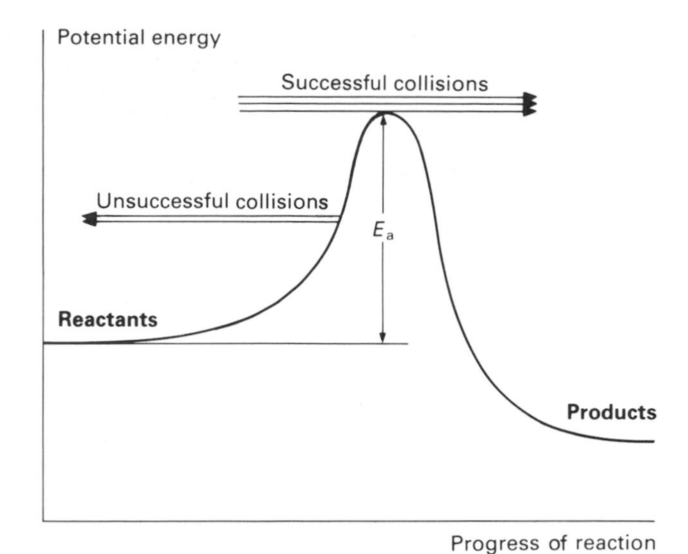

Figure 14.15 The activation energy barrier: the activated complex corresponds to the maximum potential energy

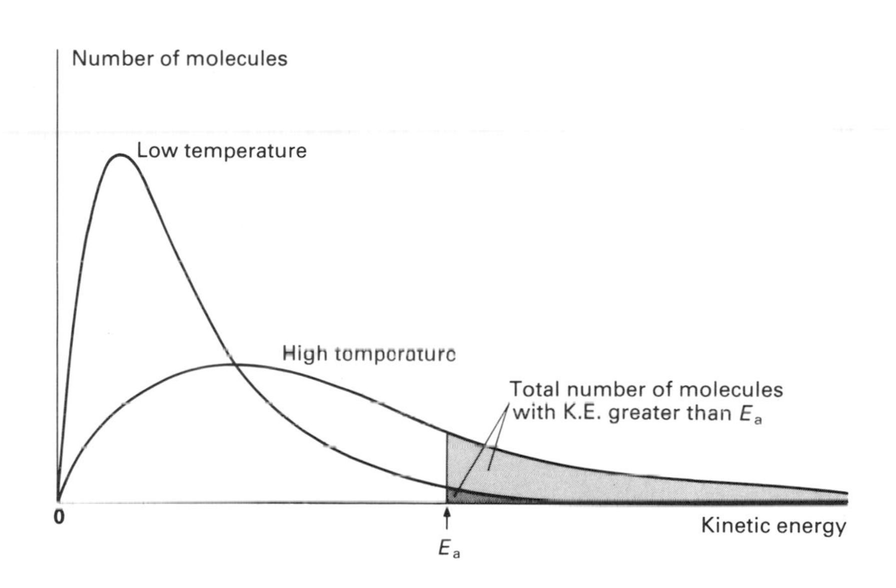

Figure 14.16 The proportions of molecules having at least the activation energy (shaded regions) at two temperatures

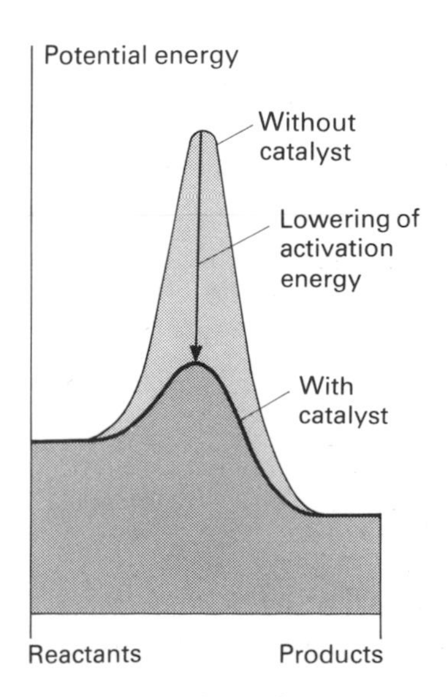

Figure 14.17 The effect of a catalyst on the activation energy

14.5 Catalysis

A **catalyst** is a substance that increases the rate of a chemical reaction without itself being consumed, so that it is recoverable unchanged in mass or chemical composition. Its presence has no effect on the equilibrium composition of the reaction mixture, but it does affect the rate at which that equilibrium is reached. An **enzyme** is a naturally occurring biologically active catalyst.

The action of catalysts

A catalyst is more an active bystander than a mere onlooker. Although it does not appear in the reaction equation, the fact that it affects the reaction rate must mean that it is involved in intermediate steps. A catalyst acts by providing an alternative reaction pathway with a lower activation energy (see Figure 14.17). At a given temperature, a higher proportion of reactant molecules have enough energy to undergo reaction when the catalyst is present (E_a is small) than when it is absent (E_a is large).

The details of the action of a catalyst depend on its identity and that of the reaction, but catalysts have several features in common.

A catalyst increases the rates of the forward *and* reverse reactions. For example, concentrated sulphuric acid is a catalyst for both the hydration of ethene and the dehydration of ethanol (see Sections 29.3 and 29.5). Because the forward and reverse reactions are affected equally, the equilibrium composition is left unchanged: the equilibrium composition in the Haber–Bosch ammonia synthesis, described in Section 21.7, is the same whether or not a catalyst is used, but is reached much more quickly if iron is present.

While very small amounts of catalyst are often effective, sometimes large quantities must be present, as in the Friedel–Crafts reaction (see Section 34.3). Sometimes a **promoter** is necessary: this is a substance that does not itself catalyse the reaction but improves the performance of the actual catalyst. An example is the addition of small quantities of potassium, calcium and aluminium oxides to the finely divided iron catalyst used in the Haber–Bosch process.

Although a catalyst does not undergo a net chemical change, its participation in a reaction may result in it changing its physical form and in some instances crumbling. In the Ostwald process for the production of nitric acid from ammonia, described in Section 21.7, the platinum–rhodium gauze catalyst must be changed periodically because the reaction roughens the surface (see Figure 14.18).

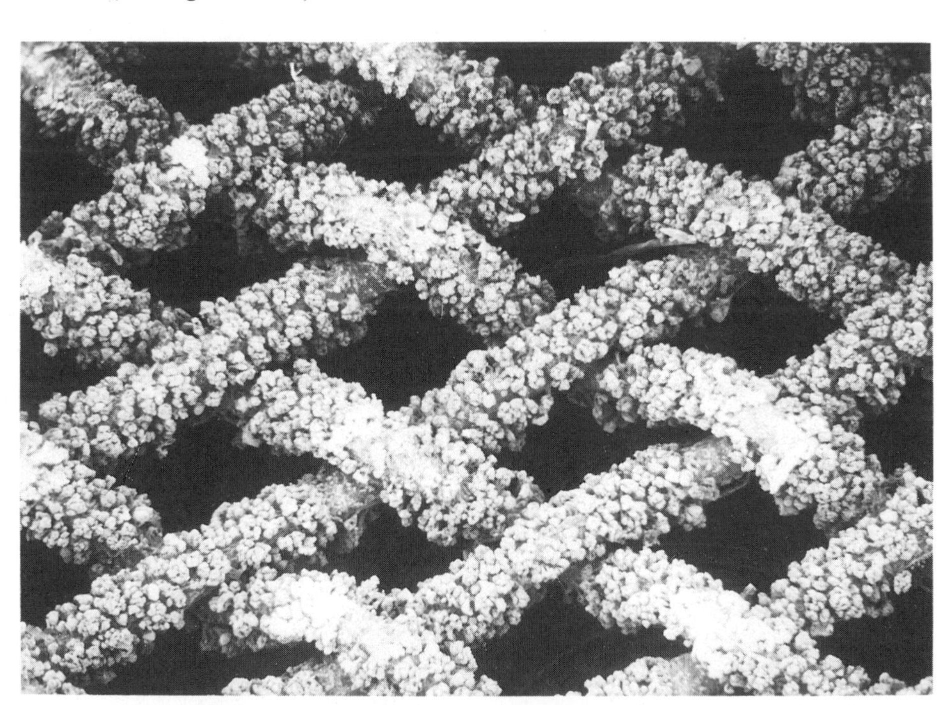

Figure 14.18 The roughened surface of the platinum–rhodium gauze catalyst for the Ostwald process after use. (The gauze is magnified roughly 100 times.)

Catalysts often have to undergo violent chemical change, and sometimes break up under the stress; hence some need to be **supported** on an inert material. Supporting a catalyst often has the additional important advantage of increasing its surface area. An example is phosphoric acid, which is a catalyst for the direct hydration of ethene to ethanol (see Section 29.5), and is supported on a material based on silica.

Catalysts can also be **poisoned** (made ineffective) by foreign material. For instance, carbon monoxide poisons the iron catalyst used in the Haber–Bosch synthesis, and lead poisons the catalytic converters used to catalyse the complete combustion of unburnt hydrocarbons and carbon monoxide in the exhaust systems of cars.

In an **autocatalytic reaction** the product is a catalyst for its own formation. An example is the iodination of propanone discussed in Section 14.4 above. If the solution is not buffered to constant pH, the HI(aq) provides hydrogen ions that catalyse the formation of more HI(aq). Autocatalysis is a little like feedback in an electronic circuit and can give rise to very strange phenomena.

Indeed, like electronic circuits with feedback, some chemical reactions can even break into oscillation, the chemical analogue of the howl heard when a microphone is in range of its loudspeaker; Box 23.1 describes some examples.

The classification of catalysts

A **homogeneous catalyst** is one that is in the same phase as the reactants; for example, the gas nitrogen monoxide, NO, is a homogeneous catalyst for the gas-phase oxidation of sulphur dioxide to sulphur trioxide. A **heterogeneous catalyst** is one in a different phase from the reactants, and is normally a solid for a gas- or liquid-phase reaction: vanadium(V) oxide, V_2O_5, is a heterogeneous catalyst for the same reaction. Some examples of industrially important heterogeneous catalysts are given in Table 14.2.

Table 14.2 Heterogeneous catalysts

Class	Function	Examples
metals	hydrogenation, dehydrogenation	Fe (Haber), Ni, Pd, Pt
semiconducting oxides	oxidation, dehydrogenation	V_2O_5 (contact process)
insulating oxides	dehydration	Al_2O_3, SiO_2
acids	polymerization, isomerization, cracking	$(HO)_2SO_2$, SiO_2

Homogeneous catalysis and enzyme function

Homogeneous catalysis normally operates through the formation of an intermediate compound that is more easily attacked than the original reactant. In the reaction between A and B, for instance, a catalyst C may first combine with A to form AC which is attacked by B, forming products and regenerating C. Important subclasses of homogeneous catalysis are **acid catalysis** and **base catalysis**, for many organic reactions proceed by one or other mechanism and sometimes by both (see, for instance, the hydrolysis of nitriles described in Section 31.4). An example already encountered in this chapter is the iodination of propanone, which may be either acid- or base-catalysed.

Enzymes are homogeneous catalysts that act on other substances, their **substrates**, involved in metabolic processes. An enzyme, which is always a protein molecule of some kind, is generally very specific in its action, and even a small change in its substrate may destroy its activity. Urease, for example, catalyses the hydrolysis of urea, $CO(NH_2)_2$, yet has no effect on that of ethanamide, CH_3CONH_2. Many enzymes require the presence of other non-protein molecules called **coenzymes** before they can function; most water-soluble vitamins are components of coenzymes.

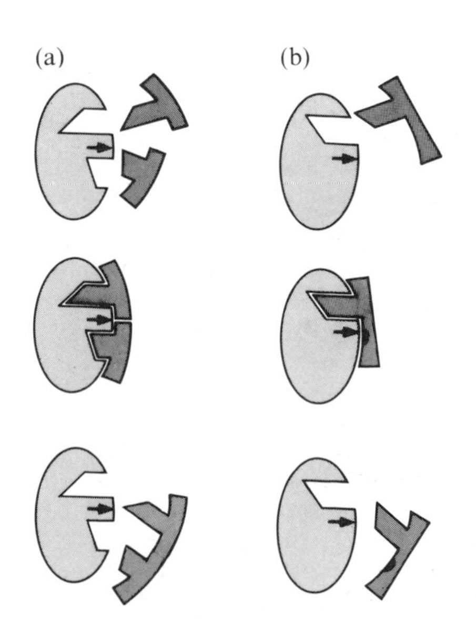

Figure 14.19 An enzyme can (a) join two molecules or (b) alter a molecule

The simplest picture of enzyme action is the **lock-and-key model**. The **active site** (the business end) of an enzyme has a well-defined, rigid structure, and only substrates of a definite, matching shape can fit into it, like a key fitting into a lock. This is illustrated in Figure 14.19, which shows (a) how a pair of molecules may be joined together by an enzyme if they both fit (and hence bond to) the active site and (b) how a single molecule may be doctored if it fits the active site of its enzyme.

The lock-and-key model is consistent with the highly specific character of enzyme action, and is supported by X-ray diffraction and n.m.r. structural studies (see Sections 3.3 and 3.4). Electron microscopy has also been used to obtain pictures of a substrate attached to its enzyme. The model is also consistent with the loss of an enzyme's activity on heating, for the elaborate folding of the protein molecule is disrupted by even quite gentle heating, and as its precise structure disappears so too does the enzyme's ability to function.

Heterogeneous catalysis and adsorption

A heterogeneous catalyst is active on account of the processes that occur at its surface. The first step is the **adsorption** of a reactant on to the surface of the catalyst. Adsorption, which must be distinguished from **absorption** (penetration into the bulk), can take two forms. **Physisorption** is the attachment of a substance to a surface by weak, van der Waals forces (see Section 5.1). **Chemisorption** is more exothermic: it takes place when a substance forms a chemical bond with the surface, and often involves dissociation (see Figure 14.20(a)) – the dissociative adsorption of nitrogen, for instance, is probably the rate-limiting step in the Haber–Bosch process. Chemisorption thus makes the reactant open to attack by other molecules already adsorbed on the surface yet free enough to migrate over it, or by collisions with molecules in the gas or liquid above the surface (Figure 14.20(b)).

As an example, the key steps in the oxidation of propene are illustrated in Figure 14.21. The reaction proceeds at a bismuth molybdate surface, and involves some very complicated chemistry. The first step (Figure 14.21(a)) is the adsorption of the propene molecule, $CH_2{=}CHCH_3$, by loss of a hydrogen atom and the formation of a carbon–surface bond. The hydrogen atom escapes with a surface oxygen atom (Figure 14.21(b)) and goes on to form water. The hydrocarbon molecule loses another hydrogen atom (Figure 14.21(c)), drags an oxygen atom out of the surface (Figure 14.21(d)) and escapes as propenal, $CH_2{=}CHCHO$, which is the starting point for the manufacture of other materials. The removal of oxygen atoms leaves gaps in the surface which are filled by attack by molecular oxygen (Figure 14.21(e)).

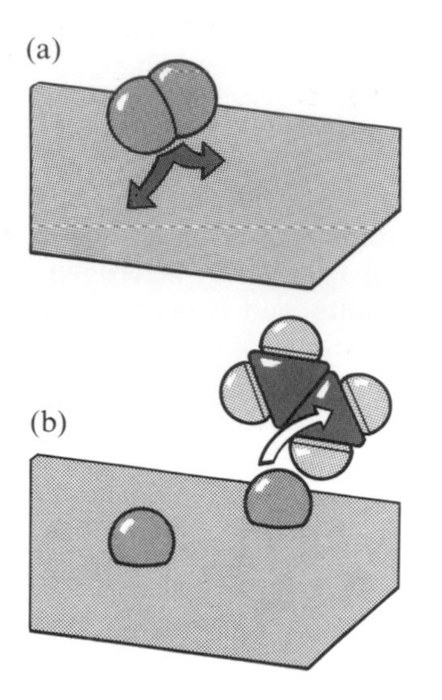

Figure 14.20 Heterogeneous catalysis: (a) adsorption and dissociation, (b) attack

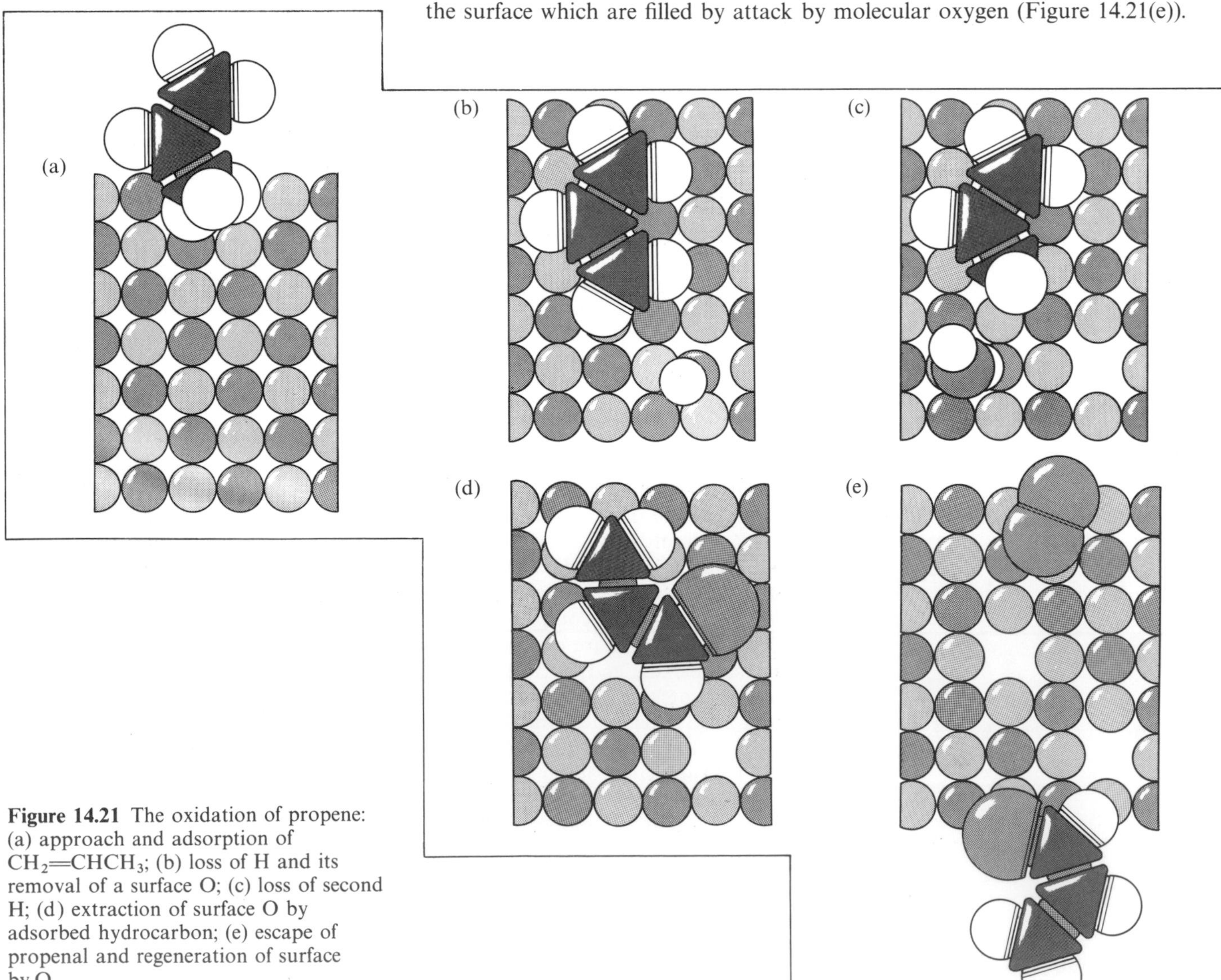

Figure 14.21 The oxidation of propene: (a) approach and adsorption of $CH_2{=}CHCH_3$; (b) loss of H and its removal of a surface O; (c) loss of second H; (d) extraction of surface O by adsorbed hydrocarbon; (e) escape of propenal and regeneration of surface by O_2

Appendix 14.1 Rates; rate laws and their solutions

The precise definition of the **instantaneous rate** of a reaction

$$aA + bB \rightarrow cC + dD$$

may be made in terms of reactants:

$$rate = -\frac{1}{a} \times \frac{d[A]}{dt}$$

$$rate = -\frac{1}{b} \times \frac{d[B]}{dt}$$

or in terms of products:

$$rate = \frac{1}{c} \times \frac{d[C]}{dt}$$

$$rate = \frac{1}{d} \times \frac{d[D]}{dt}$$

and any of these definitions is valid.

The negative signs in the first pair of definitions ensure that the rate is positive (because d[A] and d[B], the changes in [A] and [B], are negative). The precise form of the rate law (equation 14.2.1) is therefore (after minor rearrangement)

$$\frac{d[A]}{dt} = -ak[A]$$

This is a standard type of differential equation, and given that $[A] = [A]_0$ at $t = 0$, it has the solution

$$[A] = [A]_0 e^{-akt}$$

as may be checked by differentiating:

$$\frac{d[A]}{dt} = [A]_0 \frac{de^{-akt}}{dt} = -ak[A]_0 e^{-akt} = -ak[A]$$

This solution, with $a = 1$, was used in the text.

The second-order rate law (equation 14.2.4) is

$$\frac{d[A]}{dt} = -ak[A]^2$$

It is integrated by separating variables:

$$\frac{d[A]}{[A]^2} = -ak\,dt$$

Then, given that $[A] = [A]_0$ at $t = 0$, the solution is

$$\frac{1}{[A]} = \frac{1}{[A]_0} + akt$$

which rearranges into

$$[A] = \frac{[A]_0}{1 + ak[A]_0 t}$$

This solution, with $a = 1$, was used in the text.

Summary

- ☐ The rate of a reaction varies during the course of the reaction.
- ☐ The rate is the change of the concentration of a substance divided by the time taken for the change to occur.
- ☐ Rates are measured by following the time dependence of the concentration by spectroscopy, titration, conductivity, pH and pressure or volume measurements.
- ☐ In order to measure a concentration the reaction must be quenched (slowed to an imperceptible rate).
- ☐ A first-order rate law is $rate = k[A]$. k is the rate constant. The time dependence of the concentration is given by equation 14.2.2.
- ☐ The half-life of a reaction is the time taken for the concentration to decrease to one-half its initial value.
- ☐ For a first-order reaction the half-life is independent of the initial concentration and is related to the rate constant by $t_{\frac{1}{2}} = (\ln 2)/k$.
- ☐ A second-order rate law is $rate = k[A]^2$. k is the rate constant. The time dependence of the concentration is given by equation 14.2.5.
- ☐ For a second-order reaction the half-life decreases with increasing initial concentration and is given by $t_{\frac{1}{2}} = 1/k[A]_0$.
- ☐ If the rate of a reaction between A and B has the form $rate = k[A]^x[B]^y$ then the order in A is x, the order in B is y, and the overall order is $x + y$.
- ☐ The rate law cannot be predicted from the chemical equation. It must be determined experimentally.
- ☐ Reactions often occur in several steps. If one step is much slower than the others it is termed the rate-limiting step.
- ☐ A rate law is explained by proposing a mechanism for the reaction, that is, a sequence of individual reaction steps.
- ☐ The molecularity of a reaction is the number of substances involved in a given step in the reaction mechanism.
- ☐ The temperature dependence of rate constants usually follows the Arrhenius equation: $k = A\,e^{-E_a/RT}$. E_a is the activation energy, A is the Arrhenius factor.
- ☐ An Arrhenius plot of $\ln k$ against $1/T$ has a slope $-E_a/R$, and is used to measure the activation energy.
- ☐ The rates of many reactions typically double for a temperature rise of 10 °C near room temperature.
- ☐ The Arrhenius equation can be explained on the basis that molecules (a) must meet and (b) must have at least the energy E_a in order to react.
- ☐ A catalyst increases the rate of a chemical reaction without itself being consumed.
- ☐ A catalyst (a) does not affect the position of equilibrium, (b) is effective in small quantities, (c) may be physically altered, (d) is often specific, (e) may be promoted or poisoned, (f) may require a support.
- ☐ Catalysts are either in the same phase as the reactants (homogeneous) or in a different phase (heterogeneous).
- ☐ Important examples of homogeneous catalysis are acid catalysis and base catalysis. Frequently an intermediate compound is formed.
- ☐ The simplest picture of enzyme action is the lock-and-key model.
- ☐ Heterogeneous catalysis occurs by adsorption of one or both reactants on to the catalyst surface followed by reaction and then desorption.

PROBLEMS

1 Define the terms *rate*, *rate constant*, *order of reaction* and *half-life*.

2 What is the difference between the *order* and the *molecularity* of a reaction?

3 Explain how a reaction may be tested for (a) first-order kinetics, (b) second-order kinetics.

4 The rate constant of a first-order reaction was measured as $1.11 \times 10^{-3}\,s^{-1}$. What is the half-life of the reaction? What time is needed for the initial concentration to fall to (a) one-eighth of its initial value, (b) three-quarters of its initial value?

5 What fraction of the initial concentration in the reaction in the last question will remain after (a) 15 s, (b) 1 h?

6 The rate of decomposition of hydrogen peroxide in aqueous solution was measured by titration against aqueous potassium manganate(VII) solution. Equal volumes of solution were withdrawn at various times, and titrated:

t/min	0	10	20
$V(KMnO_4)/cm^3$	45.6	27.6	16.5

Confirm that the reaction is first-order in the peroxide and find the half-life and rate constant.

7 In an experiment to determine the rate of inversion of sugar (that is, the rate of its conversion into glucose and fructose) the following data were obtained using a polarimeter:

t/min	0	10	20	30	40	80	∞
Angle/deg	64.8	57.6	51.0	44.8	39.2	20.6	−28.2

Plot the data and determine the order of the reaction with respect to sucrose. Find the rate constant and calculate the half-life of the reaction.

8 Explain why rates of reaction usually increase with temperature, and define and explain the term *activation energy*.

9 (a) State *five* factors which may affect the rate of a chemical reaction.
(b) The reaction between iodide ions and persulphate (peroxodisulphate(VI)) ions in aqueous solution may be represented by the equation:

$$2I^-(aq) + S_2O_8^{2-}(aq) \rightarrow I_2(aq) + 2SO_4^{2-}(aq)$$

Outline a method of investigating the rate of this reaction so that the order of the reaction with respect to persulphate ions may be determined.
(c) The hydrolysis of methyl ethanoate may be represented by the equation:

$$CH_3COOCH_3(l) + H_2O(l) \rightleftharpoons CH_3COOH(aq) + CH_3OH(aq)$$

When the hydrolysis was carried out in the presence of aqueous hydrochloric acid in a constant-temperature bath, the following results were obtained:

Time/s × 10^4	Concentration of ester/mol dm^{-3}
0	0.24
0.36	0.156
0.72	0.104
1.08	0.068
1.44	0.045

(i) State the reasons for using the hydrochloric acid and the constant-temperature bath.
(ii) Plot a graph of ln[ester] against time.
(iii) By reference to the graph, show that the reaction is first-order with respect to the ester.
(iv) Given the equation

$$\frac{d[\text{ester}]}{dt} = -k[\text{ester}]$$

use your graph to determine a value for k and state its units. *(AEB 1985)*

10 (a) The decomposition of dinitrogen oxide, N_2O, to nitrogen, N_2, and oxygen, O_2, is a first-order reaction.
Use this example to answer the following:
(i) Write a rate equation for the decomposition and explain what is meant by the rate of the reaction, the order of the reaction and the rate constant.
(ii) Given that the half-life of the reaction at 1000 K is 60 hours, calculate (1) the rate constant, giving the units, and (2) the total gas pressure after 60 hours at 1000 K, the initial pressure being 2×10^5 Pa, assuming the volume remains constant.
(b) State, and explain, how each of the following affects the rate of a chemical reaction: activation energy, temperature, pressure and catalyst.
(c) Given the following information about dinitrogen oxide:

$\Delta H_f^\ominus$/kJ mol^{-1}	+81.6
$T\Delta S^\ominus$/kJ mol^{-1}	−22.4

give an explanation of why the nitrogen and oxygen of the air would not be expected to combine under standard conditions to give dinitrogen oxide. *(AEB 1984)*

11 What do you understand by the term *order* when applied to a chemical reaction?
Given data on the variation of concentration with time, give two methods which may be used to determine the order of that reaction.
Propanone (acetone) reacts with iodine in the presence of acid according to the equation

$$CH_3COCH_3 + I_2 \rightarrow CH_3COCH_2I + H^+ + I^-$$

In an experiment, 5 cm^3 of propanone, 10 cm^3 of sulphuric acid of concentration 1.0 mol dm^{-3} and 10 cm^3 of a solution of iodine of concentration 1.0×10^{-3} mol dm^{-3} were mixed, made up to 100 cm^3 with distilled water and placed in a thermostat. Every five minutes samples were removed and titrated against a solution of sodium thiosulphate. The experiment was repeated but using 20 cm^3 of sulphuric acid in the reaction mixture. The following results were obtained.

Time/min		5	10	15	20	25
Titre/cm^3	for 10 cm^3 acid	18.5	17.0	15.5	14.0	12.5
	for 20 cm^3 acid	17.0	14.0	11.0	8.0	5.0

From plots of titre against time, determine the order of reaction with respect to iodine and with respect to $[H^+]$.
How would you determine the order with respect to propanone? *(Oxford and Cambridge)*

12 The rate of a reaction between A and B varies as follows:

$$rate = k[A]^2[B]$$

(a) What is the overall order of the reaction?
(b) If the concentration of only one of the reactants can be doubled, which would give the greater increase in the overall rate?
(c) Give the reasons for your answer to (b).
(d) Suggest two other ways by which the rate of the reaction might be increased. *(Based on JMB)*

13 (a) Explain what is meant by the terms (i) *velocity constant* (or *rate constant*), (ii) *activation energy*.
(b) Nitrogen dioxide decomposes in the gaseous phase into nitrogen oxide and oxygen. Suggest an experimental method for determining the rate of this reaction.

(c) The following results were obtained for the velocity constant k of this reaction:

$k/\mathrm{dm^3\,mol^{-1}\,s^{-1}}$	*Temperature*/K
3.16	650
28.2	730
158	800
1120	900
5010	1000

Use this information to find the activation energy for this reaction. *(Oxford S)*

14 When ethanal is heated to about 800 K, it decomposes by a *second-order reaction* according to the equation

$$CH_3CHO(g) \rightarrow CH_4(g) + CO(g)$$

If, however, the ethanal is heated to the same temperature with iodine vapour, the iodine *catalyses* the reaction; this latter reaction is *first-order with respect to both iodine and ethanal* and has a lower *activation energy* than the reaction with ethanal alone.
(a) Explain the meaning of the terms in italics.
(b) Suggest a set of experiments by which you could confirm that the uncatalysed reaction is second-order.
(c) Indicate briefly what experiments would be necessary to show that the reaction with iodine present has a lower activation energy than the decomposition of ethanal itself. *(London)*

15 Define the order and rate constant for a chemical reaction. The hydrolysis of sucrose is found to be first-order with respect to sucrose and first-order with respect to hydrogen ion concentration, with a rate constant $k = 2.3\ \mathrm{dm^3\,mol^{-1}\,s^{-1}}$. Calculate the time for the sucrose concentration to fall to $0.05\ \mathrm{mol\,dm^{-3}}$ in solutions with the following initial concentrations: (a) $0.1\ \mathrm{mol\,dm^{-3}}$ sucrose, $0.1\ \mathrm{mol\,dm^{-3}}$ HCl; (b) $0.1\ \mathrm{mol\,dm^{-3}}$ sucrose in a buffer solution with pH = 4. *(Oxford Entrance)*

16 What do you understand by the terms *rate expression* and *reaction order*? Suggest one method each for measuring the rates of the following reactions:
(a) the hydrolysis of ethyl ethanoate (ethyl acetate),
(b) the oxidation of hydriodic acid by hydrogen peroxide.

The equilibrium between two isomers A and B can be represented:

$$A \underset{k_2}{\overset{k_1}{\rightleftharpoons}} B$$

where k_1 and k_2 are rate constants for the forward and reverse reactions, respectively. Starting with a non-equilibrium mixture of concentrations $[A]_0 = a$ and $[B]_0 = b$ it was found that x moles of A had reacted after time t. Give an expression for the rate, dx/dt, and hence show that the integrated rate expression is

$$\ln\left(\frac{p}{p-x}\right) = (k_1 + k_2)t$$

where

$$p = \frac{k_1 a - k_2 b}{k_1 + k_2}$$

After 69.3 minutes $x = p/2$; calculate k_1 and k_2 if the equilibrium constant $K = 4$. *(Cambridge Entrance)*

17 What do you understand by the *rate constant* of a chemical reaction? Explain the concept of *activation energy* and show how it may be used to explain why some reaction rates increase dramatically for only a modest increase in temperature.

A reaction was found to have the following temperature dependence of the rate constant:

T/K	500	555	625	714
$k/\mathrm{dm^3\,mol^{-1}\,s^{-1}}$	0.025	0.36	5.5	79

Derive the activation energy for this reaction by plotting a suitable graph. Does this graph provide any other information about the reaction? *(Oxford Entrance)*

The figure on the opposite page shows a computer graphic of copper aluminium borate.

PART TWO

INORGANIC

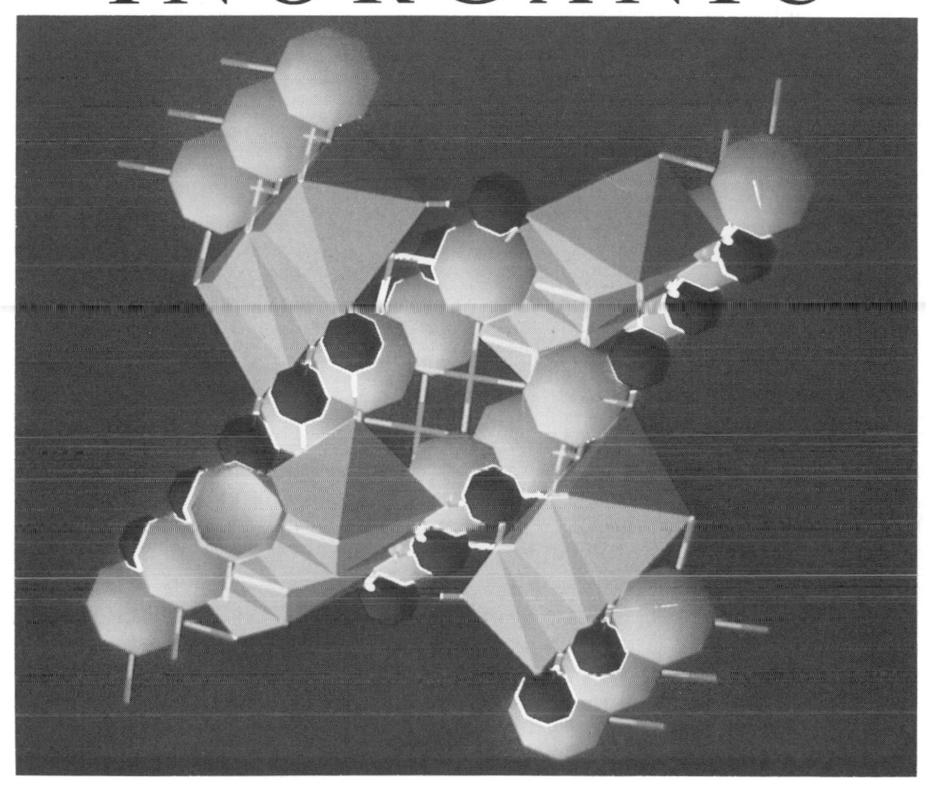

CHEMISTRY

15 Introduction to inorganic chemistry

In the next ten chapters we deal with the structures and properties of individual elements and their compounds. In previous chapters we have described the principles that lie behind their properties; we therefore begin with a survey of the main points that should be borne in mind as we meet the individual elements. Throughout the following chapters we will emphasize the periodicity of the elements.

Chemists have to deal with 109 elements. Ninety of these elements occur naturally; the remaining nineteen (technetium, promethium and those following uranium) occur in such minute quantities that in order to study them they have to be made synthetically by nuclear reactions, as described in Section 1.4. All 109 elements except helium, neon and argon form compounds with other elements, and so chemists have to cope with an enormous number of substances. About a million **inorganic** compounds are known, and so it is essential that their properties are approached systematically. A further five million **organic** compounds (that is, compounds of carbon) are known, and so organic chemistry needs its own special treatment and is dealt with in Chapters 26 to 35 of this book. The modern trend of chemistry is to blur the distinction between organic chemistry and inorganic chemistry; however, initially it is very helpful to treat them separately.

The next ten chapters deal with the following questions:

1. What are the electronic structures of the elements and their compounds?
2. Which reactions are *thermodynamically* feasible?
3. Which reactions are *fast enough* to be chemically significant, and which are too slow to be worth considering?
4. Where, in what quantities and in what forms are the elements found geographically, and how are they extracted?
5. How are the elements and compounds used, and *why* are they used like that?
6. Are there any specially interesting features of the elements, perhaps on account of their involvement in biological processes or in new 'high-technology' materials, or, more negatively, as pollutants?

Some of these questions can be answered only by drawing on information more advanced than is needed here, but we shall see that the principles introduced in the preceding fourteen chapters can be applied successfully to many of them.

Above all, it is important to develop a sense of *pattern* in inorganic chemistry, for otherwise the facts merge into a swamp of boring detail. In order for the pattern to become apparent, we shall always relate the properties of elements to their position in the Periodic Table, and adopt the following structure for each chapter that introduces a Group of elements:

An introduction describes the position of the elements in the Periodic Table. It is important to remember, however, that the properties of an element should also be seen in the context of the members of neighbouring Groups and not solely as a characteristic of the Group under discussion.

1 Group systematics This important section puts the element into the cross-hairs of a chemist's sights, so that it can be seen in the context of its neighbours. We consider trends in the following three properties:

Oxidation states and bonding characteristics. Oxidation numbers, discussed in Section 15.2, indicate the types of compound that an element can form, and whether the compounds are formed by oxidation or reduction. The concept of oxidation number helps to rationalize many of an element's properties.

Periodic trends. This summarizes the trends in properties of the elements and their compounds within a Group.

Shapes and sizes. The periodicity of the elements is shown very clearly by the atomic and ionic radii of the elements. These are important, because the *geometry* of a compound, its shape and dimensions, is an important feature of its chemical behaviour.

2 The elements This section describes the elements themselves, and summarizes the forms in which they are found, their principal physical properties, and how they react with other elements and with acids and bases.

After these sections we describe three important classes of compounds, which demonstrate the main features of the chemical properties of the elements:

3 Oxides, and either hydroxides or oxoacids These compounds reflect the metallic or non-metallic character of the element (as we shall explain later).

4 Halides The halogens, and especially fluorine, bring out the highest oxidation numbers of the elements, so that by considering them we see, in effect, how far the elements can be made to surrender control of their valence electrons. The halides are also very important compounds in their own right.

5 Compounds with hydrogen The hydrogen atom is so small that it often brings out a peculiarity in the chemical properties of an element. The physical properties of compounds containing hydrogen are often highly important, and reflect the ability of hydrogen to participate in hydrogen bonding (see Section 5.1).

Although the oxides, halides and hydrogen compounds are the most revealing and characteristic compounds of the elements, each element forms compounds that are important on account of their industrial or biological significance, or because they illustrate an interesting feature of chemistry. Therefore, we also include a section called

6 Other compounds This is a catch-all category for important compounds, including aqua cations, that do not have a place elsewhere.

7 Industrial chemistry Now that we have seen the range of compounds, and some of their applications that enrich our world, we consider the occurrence in nature and the manufacture of the elements and some of their important compounds. The term 'manufacture' implies the extraction of the element either from a mixture (as for oxygen from air) or from its ores, and the production of compounds of fundamental economic importance in chemical industry.

Although we present a systematic account of chemistry, it must never be forgotten that chemistry is a network of interrelated properties, and the subject is more like a richly patterned tapestry than a simple list. The pleasure of chemistry lies in discovering these patterns and, with luck and some hard work, by the time you reach the end of this book you will have shared something of that pleasure.

15.1 Electronic structures

The Periodic Table is a list of the elements in order of atomic number in an arrangement that shows the periodicity of their properties. This periodicity reflects the periodic repetition of similar electron configurations (see Section 1.2). The Table is divided into **blocks**, according to the subshell being filled using the building-up principle. We shall discuss the elements in blocks, beginning with the s-block elements, with valence configurations s^1 or s^2, followed by the p-block elements, with valence configurations s^2p^1 to s^2p^6, and then finally the d-block elements, in which the d-orbitals are being filled.

The s- and p-blocks are divided vertically into **Groups**, in which all the members of a Group have the same valence configuration, apart from differences in principal quantum number. The number of valence electrons is equal to the Group number. The d-block elements are best treated together, because distinctions between Groups are much less important within that block.

The horizontal rows of the Periodic Table are called **Periods**, and the Period number is equal to the principal quantum number of the valence shell (see Figure 15.1).

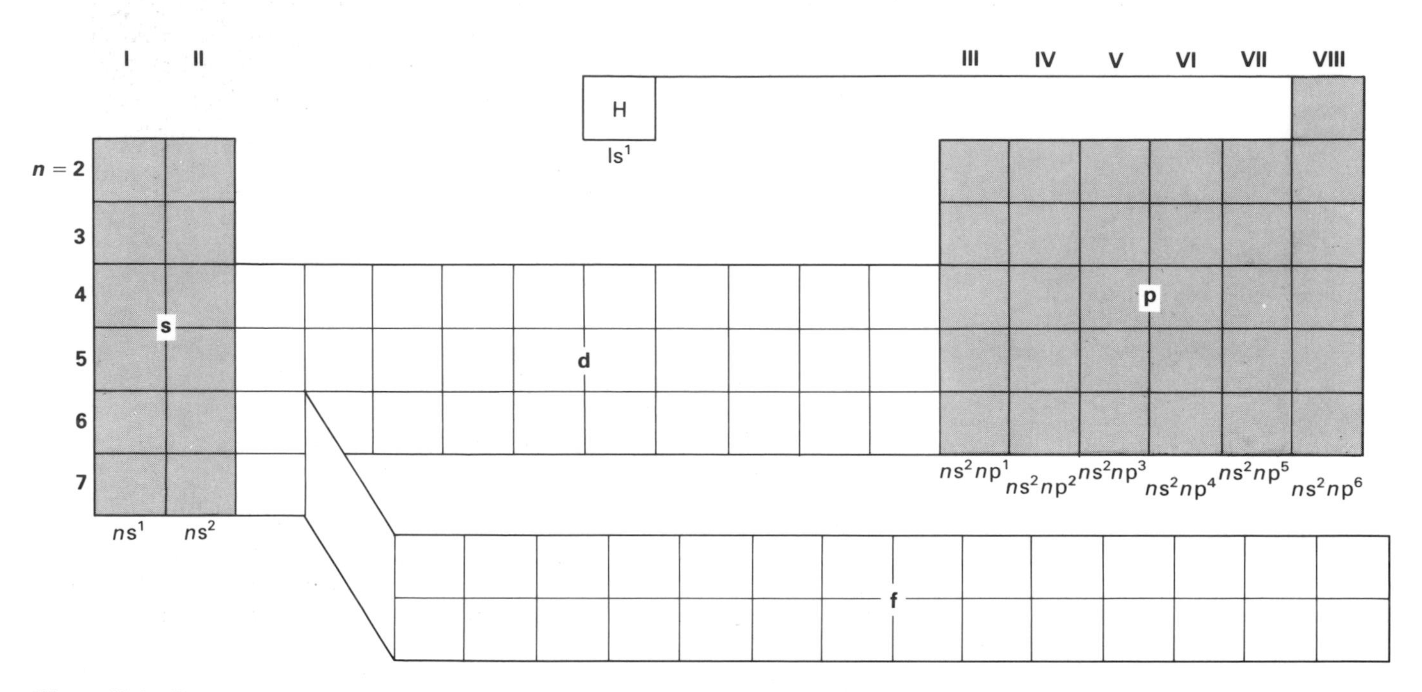

Figure 15.1 The s-, p-, d- and f-blocks in the Periodic Table

A general feature is that the members of the s-block are electropositive metals, since they can readily lose their valence electrons. The elements in the p-block are more varied. On the left of the block (particularly in Group III) are metals such as aluminium, while on the right are the electronegative non-metals, such as oxygen and chlorine, and the noble gases. The **metalloids** (or **semi-metals**), which are intermediate in character, lie roughly on a diagonal line starting at silicon and running downwards and to the right. The d-block elements are all metals, and show a transition in properties between the electropositive s-block metals and the less electropositive metals on the left of the p-block. They are often therefore called the **transition metals**, but as there is some controversy as to whether the metals on the far right of the block, such as zinc, are properly regarded as transition metals, we shall use only the more neutral term 'd-block elements'. These trends are summarized in Figure 15.2.

15.2 Oxidation numbers

It proves to be very convenient to assess the extent to which an atom gains or loses electrons when it forms a compound. While detailed calculations can be made and the change in electron density mapped, it is often sufficient to report the redistribution of electrons by exaggerating it and giving the **oxidation number** (symbol: Ox) of the element (which is defined below). Thus, if an electron is captured by the more electronegative atom in a bond, the more electronegative element has been reduced (in the technical sense – see Section 13.2): its oxidation number is reduced by one. Likewise, the less electronegative element has been oxidized: its oxidation number is increased by one. (Even if the electron is only partially captured – or lost – the oxidation

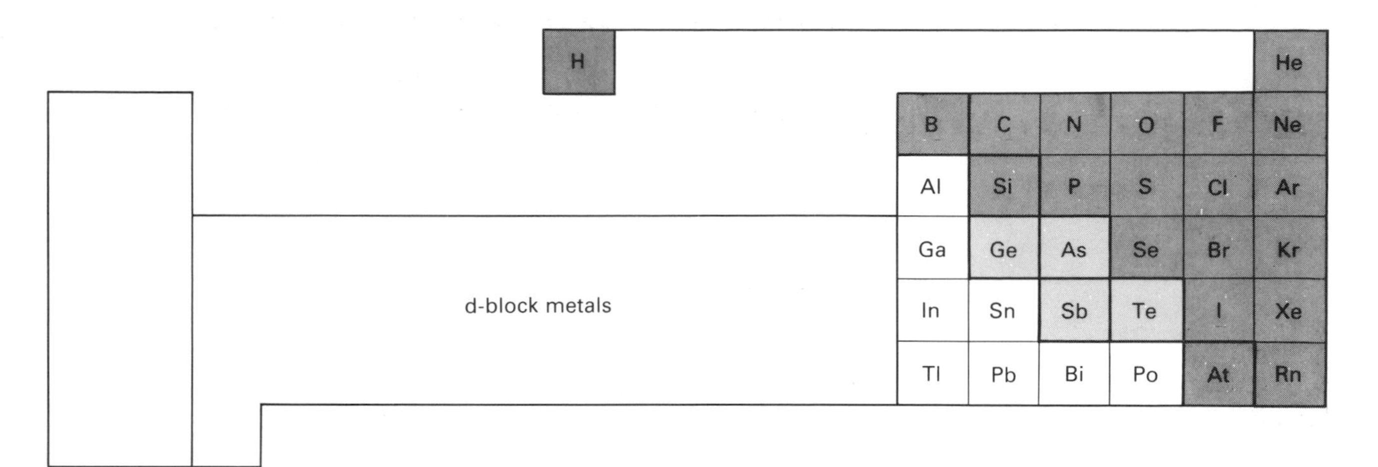

Figure 15.2 The Periodic Table: the heavy tint indicates the non-metals in the p-block, and the lighter tint indicates metalloids

numbers are calculated by supposing that the capture or loss is complete.) Knowing whether an atom's oxidation number has increased, decreased or remained the same during a reaction is a valuable guide to the type of reaction that has occurred and, in practical terms, to the kind of reagent that should be used to bring about the change. More generally speaking, keeping track of oxidation numbers lets us discern patterns of chemical reactivity.

The oxidation number of an element is defined by a set of rules that reflect the tendency of electrons to move towards the more electronegative atom in a bond. Specifically, *the oxidation number is equal to the number of electrons that must be added to the atom in its combined state to form the neutral atom*. For example, one electron must be added to the Na^+ ion in sodium chloride to form the neutral atom; hence the oxidation number of sodium is +1, and this is written as $Ox(Na) = +1$. Similarly, the Cl^- ion must lose one electron; hence its oxidation number is −1, or $Ox(Cl) = -1$.

The oxidation of sodium to sodium cations

$$Na \rightarrow Na^+ + e^-$$

corresponds to an increase of $Ox(Na)$ from 0 to +1, and is denoted

$$Na(0) \rightarrow Na(+1)$$

Likewise, the reduction of chlorine to chloride ion corresponds to a decrease of $Ox(Cl)$ from 0 to −1 and in terms of oxidation numbers is denoted

$$Cl(0) \rightarrow Cl(-1)$$

When naming a compound of an element that has more than one common oxidation number, the number is given as a roman numeral, as in iron(II) and iron(III) for $Ox(Fe) = +2$ and $Ox(Fe) = +3$ respectively.

Oxidation numbers apply to covalent molecules as well as to ions, because any drift of electrons is exaggerated so that the compound is treated as *formally* ionic. Thus, in HCl the molecule is treated *formally* as H^+Cl^-, so that $Ox(H) = +1$ and $Ox(Cl) = -1$.

When several different atoms are present, keeping track of the electron drifts is more difficult and a set of rules is very helpful. The rules give the same values for simple compounds that would be calculated by exaggerating the drifts of electrons, but are easier to apply in more complicated cases. They should be worked through in order until the oxidation number is found.

1 *For an uncombined element* E, $Ox(E) = 0$. For example, $Ox(F) = 0$ in F_2, $Ox(O) = 0$ in O_2 and O_3, and $Ox(Na) = 0$ in solid sodium metal.
2 *The sum of the oxidation numbers of all the atoms in a molecule or ion is equal to the total charge on the entity*. For example, $Ox(Cl) = -1$ for Cl^-, $Ox(O) = -1$ for the peroxide ion, O_2^{2-}, and $Ox(O) = -\frac{1}{2}$ for the superoxide ion, O_2^-.

3 $Ox(F) = -1$ *in all compounds containing fluorine.* For example, $Ox(F) = -1$ in OF_2, and since the sum of the oxidation numbers is 0 (because the overall charge is zero), $Ox(O) = +2$ in OF_2.

4 $Ox(O) = -2$, *unless fluorine is also present or peroxides and superoxides* (see Section 17.3) *are being considered.* For example, $Ox(O) = -2$ in H_2O, so that $Ox(H) = +1$ in H_2O.

5 *Each shared electron pair is assigned to the more electronegative atom, and the oxidation number is identified with the formal charge on the atom:*

$$Ox(E) = \text{number of valence electrons on the uncombined atom E} - \text{number of valence electrons on E after assigning the electrons}$$

For example, in $AlCl_3$ each Al—Cl bonding electron pair is assigned to chlorine, so that $Ox(Cl) = 7 - 8 = -1$; hence by rule 2, $Ox(Al) = +3$.

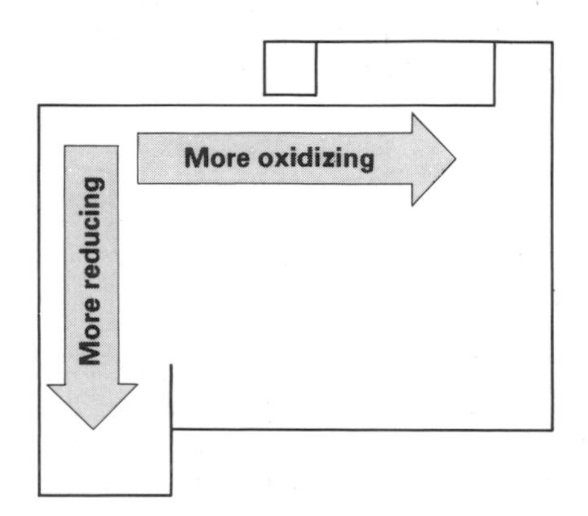

Figure 15.3 Elements on the upper right of the Periodic Table tend to be oxidizing agents and those on the lower left tend to be reducing agents

In this book, numerous examples of the use of oxidation numbers are given in Chapters 16 to 25, and a redox reaction is almost always signalled by a note alongside in smaller type, giving the changes in oxidation numbers of the elements involved, for example:

$$2Na(s) + Cl_2(g) \rightarrow 2NaCl(s) \qquad Cl(0) \rightarrow Cl(-1), \quad Na(0) \rightarrow Na(+1)$$

Keeping track of oxidation numbers helps to rationalize the properties of elements, especially when, as for nitrogen, they can take a wide range of values.

15.3 Periodic trends

In this subsection of each chapter we describe how the properties of both the elements and their compounds vary within a Group. The principal properties of the *elements* considered are their ionization energies, their standard electrode potentials and their electronegativities.

The ionization energy indicates the ease with which an electron may be removed from the atom in the gas phase. In general, elements on the lower left of the Periodic Table have low first ionization energies and those on the upper right have high values (see Section 1.2).

The standard electrode potential indicates the ease with which electrons are lost in aqueous solution, and a large negative value indicates a strong reducing power (see Section 13.5). Trends are more complex than for ionization energies, but elements on the lower left of the Periodic Table tend to have negative values (are reducing agents), and those on the upper right have positive values (are oxidizing agents) (see Figure 15.3).

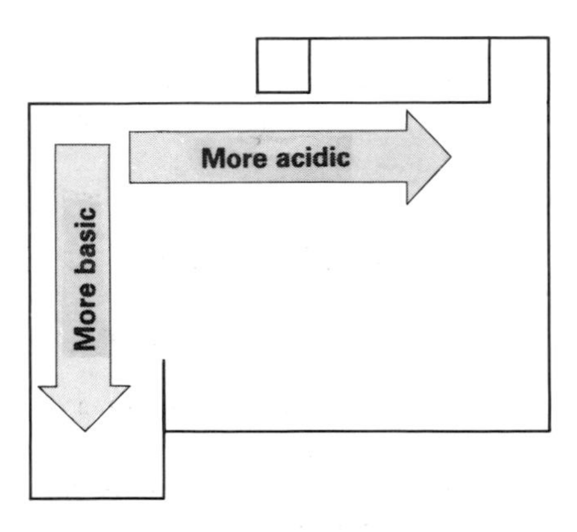

Figure 15.4 Trends in acid–base properties for the oxides in the Periodic Table

Electronegativity indicates the power of an atom to attract shared electron pairs in a compound (see Section 2.6). Large electronegativity differences between atoms imply predominantly ionic bonding in their compounds; similar values imply predominantly covalent bonding. Electronegativities are largest towards the upper right of the Periodic Table: they increase from left to right across a Period and decrease from top to bottom down a Group.

Some typical periodic trends are shown by *compounds*:

The oxides of the elements become more acidic from left to right across a Period and more basic from top to bottom down a Group (see Figure 15.4).

For instance, in Period 3, sodium oxide is basic, aluminium oxide is amphoteric, and the sulphur oxides are acidic. The oxide ion, O^{2-}, is a strong Brønsted base in water, so that the highly ionic oxides (such as Na_2O) are basic. The covalent oxides (such as SO_2) are Lewis acids and react with water to give oxoacids. Therefore, elements with basic oxides usually form hydroxides whereas those with acidic oxides usually form oxoacids.

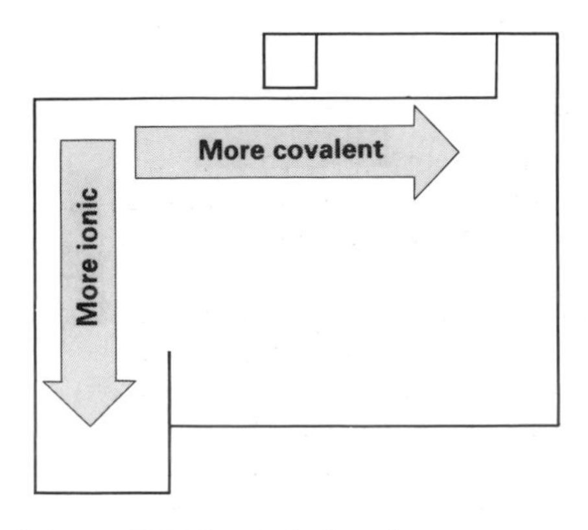

Figure 15.5 Trends in bonding between elements in the Periodic Table

The oxides, chlorides and hydrides all become more covalent from left to right across a Period and more ionic from top to bottom down a Group (see Figure 15.5).

For example, sodium chloride is a typical ionic solid whereas phosphorus trichloride is a typical covalent liquid. Ionic chlorides tend simply to dissolve in water; covalent chlorides are Lewis acids and are often hydrolysed by water.

15.4 Shapes and sizes

In this subsection of each chapter we consider the three-dimensional forms of atoms, ions and molecules.

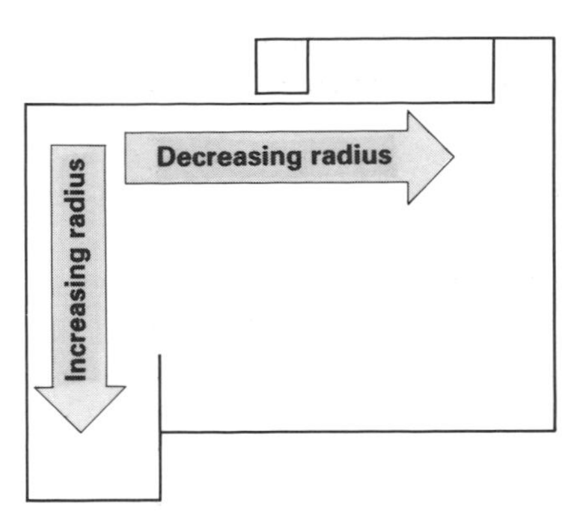

Figure 15.6 Trends in atomic radii in the Periodic Table

Periodic trends in the sizes of atoms and ions

The measure of size is the **atomic** or **ionic radius**; neither is a rigidly definite property, however, and in general radii depend on the identity and number of the surrounding atoms and ions. Atomic radii determine the geometry of compounds, such as whether six ions can attach to a central ion (and hence become its **ligands**) without repelling each other significantly. Ionic radii determine the lattice enthalpies of ionic compounds.

The radii depend on location in the Periodic Table (see Figure 15.6):

Radii increase down a Group.

This increase results from electrons occupying shells of greater principal quantum number in successive Periods.

Radii decrease from left to right across a Period.

Within a Period, the outermost electrons are poorly shielded by the other electrons in the valence shell; hence the increasing effective nuclear charge of successive elements means that the atoms and ions of each element are smaller than those of the previous one.

Cations are always smaller than their parent atoms.

Electrons have been removed from the atom, often as far as an inner closed shell.

Anions are always larger than their parent atoms.

Electrons have been added without any increase in nuclear charge, and the repulsions between them weaken the hold maintained on them by the nucleus.

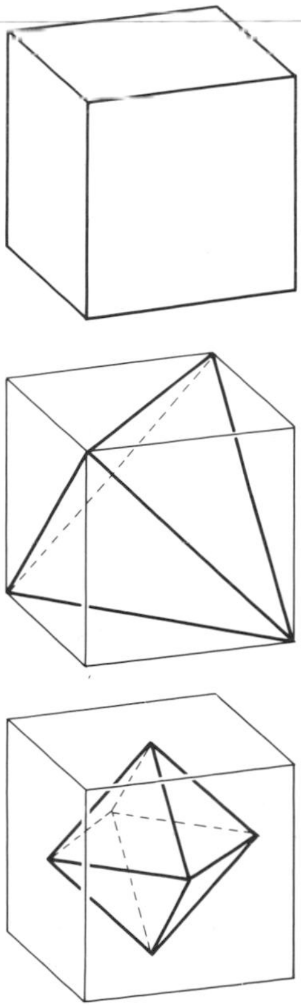

The shapes of covalent molecules

Knowing the shapes of molecules helps us to rationalize their physical properties: for example, if the water molecule were linear it would not be polar, and the properties of liquid water would be profoundly different from those we know.

The valence shell electron pair repulsion theory (VSEPR theory – see Section 2.3) proves to be a reliable way of predicting the shapes of simple molecules. The theory focuses on the electron pairs on the central atom, and predicts the shapes according to the number of bonding pairs and lone pairs.

By far the most important shapes in inorganic chemistry are the tetrahedron and the octahedron. It is helpful to bear in mind the relation of these two shapes to each other and to the cube (see Figure 15.7). These shapes are found both among individual molecules (CH_4, for instance, is tetrahedral) and in the local details of the structures of inorganic solids.

Figure 15.7 The geometrical relationship between the cube, tetrahedron and octahedron

The structures of inorganic solids

While many inorganic compounds are molecular liquids and gases held together weakly by van der Waals forces, most are solids, often consisting of extended arrays of ions interacting strongly with each other electrostatically (see Section 5.1).

At first glance, solid structures may not seem to be based on tetrahedral and octahedral shapes. However, inspection of close-packing arrangements (see Section 5.2) shows that the holes between the ions are either tetrahedral or octahedral. In many solids these holes are filled by other atoms or ions. Sometimes all the holes are filled, and the solid has a well-defined formula. One example is NaCl, where each Na^+ ion occupies an octahedral hole formed by six Cl^- ions. Part of the richness of inorganic chemistry stems from the existence of compounds in which the holes are only incompletely filled,

(a)

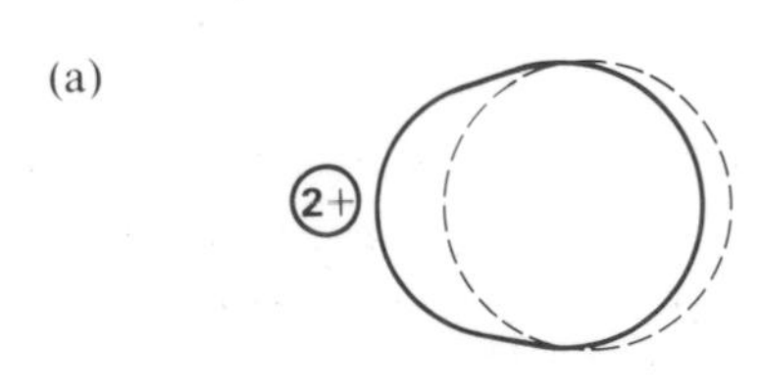

(b)

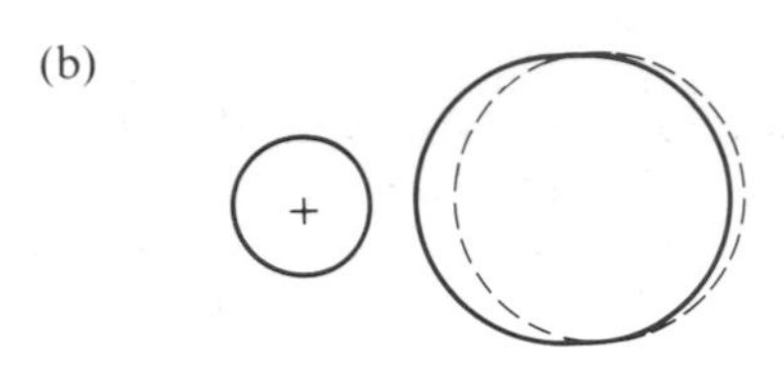

Figure 15.8 The greater polarizing power of an M^{2+} ion (a) as compared with an M^{+} ion (b)

often randomly. These compounds are frequently **non-stoichiometric**, that is, their formulae cannot be expressed using whole-number proportions of atoms; their odd properties have brought them to the centre of much modern research. Despite their non-stoichiometry, however, the *local* neighbourhood of an atom or ion in such a compound is often tetrahedral or octahedral.

Polarization

One simple way of rationalizing trends in the properties of compounds is through the concept of **polarization**.

Polarization is the distortion of an ion by the electric field due to a neighbouring ion, as illustrated in Figure 15.8. A cation attracts the electrons of a neighbouring anion to an extent that depends on the cation's **polarizing power** and the anion's **polarizability**, both of which vary periodically.

The polarizing power of a cation depends on its **charge density**, not on the charge alone. That is, a small cation exerts a stronger distorting field than a less compact ion of the same charge because its charge density is greater. One feature of the Periodic Table is the similarity of the polarizing powers of diagonal neighbours, such as Li^{+} and Mg^{2+}, for the increase in radius on going from Period 2 to Period 3 roughly balances the increase in charge on going from Group I to Group II; in practice, this means that diagonal neighbours are similar. This **diagonal relationship** is particularly important for the early members of Periods 2 and 3 of the Periodic Table.

The polarizability of an anion increases with ionic radius (because the central nucleus exerts a weaker control over the valence electrons of a large ion) and with ionic charge (because the greater electron repulsions in a highly charged ion result in greater ease of distortion). The general trend of polarizability across Period 2 is

$$N^{3-} > O^{2-} > F^{-}$$

which reflects the effect of charge, and the trends within Groups VI and VII are:

$$S^{2-} > O^{2-}; \qquad I^{-} > Br^{-} > Cl^{-} > F^{-}$$

which reflect the effect of radius.

An example of the effect of polarization is shown by the series of compounds

NaF MgO AlN SiC

which have the same numbers of electrons but from left to right have cations with increasing polarizing power and anions of increasing polarizability: NaF is a typical ionic solid whereas SiC is a typical covalent solid.

15.5 The classification of reactions

The pattern of inorganic chemistry becomes much clearer once it is realized that most reactions fall into one of three classes: redox (reduction–oxidation) reactions, Lewis acid–base reactions and precipitation reactions. Identifying the type of reaction taking place is an important part of appreciating that inorganic chemistry is a logical subject and not just a collection of unrelated facts.

Redox reactions: electron-transfer reactions

In a redox reaction one or more electrons are transferred from one substance to another, resulting in changes of oxidation number.

Redox reactions are often best regarded as a combination of two half-reactions (see Section 13.2), one an oxidation (a source of electrons) and the other a reduction (a sink for electrons). For example, in the half-reaction

$$2I^{-}(aq) \rightarrow I_2(aq) + 2e^{-} \qquad I(-1) \rightarrow I(0)$$

Ox(I) has increased from -1 to 0, so that iodine has been oxidized. In the half-reaction

$$Fe^{3+}(aq) + e^- \rightarrow Fe^{2+}(aq) \qquad Fe(+3) \rightarrow Fe(+2)$$

$Ox(Fe)$ has decreased from +3 to +2, so that iron has been reduced. The overall reaction is the sum of these two half-reactions (with coefficients chosen to match the numbers of electrons),

$$2Fe^{3+}(aq) + 2I^-(aq) \rightarrow 2Fe^{2+}(aq) + I_2(aq) \qquad Fe(+3) \rightarrow Fe(+2), \quad I(-1) \rightarrow I(0)$$

a redox reaction in which iodide ions reduce iron(III) ions, themselves being oxidized and producing the characteristic brown colour of iodine.

Redox reactions include the reaction of sodium with water, in which one hydrogen atom in each water molecule is reduced from $Ox(H) = +1$ to 0 and sodium is oxidized from $Ox(Na) = 0$ to +1:

$$\begin{aligned} Na(s) &\rightarrow Na^+(aq) + e^- \\ 2H_2O(l) + 2e^- &\rightarrow 2OH^-(aq) + H_2(g) \\ 2Na(s) + 2H_2O(l) &\rightarrow 2NaOH(aq) + H_2(g) \end{aligned}$$

The reaction of metals with acids is also a redox reaction, *not* an acid–base reaction: electrons are transferred from the metal, which is oxidized, to the hydrogen ions ($Ox(H) = +1$), which are reduced to molecular hydrogen ($Ox(H) = 0$). For example, when magnesium reacts with acid:

$$\begin{aligned} Mg(s) &\rightarrow Mg^{2+}(aq) + 2e^- \\ 2H^+(aq) + 2e^- &\rightarrow H_2(g) \\ Mg(s) + 2H^+(aq) &\rightarrow Mg^{2+}(aq) + H_2(g) \end{aligned}$$

and the magnesium is oxidized from $Ox(Mg) = 0$ to +2.

Lewis acid–base reactions: coordination reactions

In a Lewis acid–base reaction a coordinate bond is formed between an electron pair donor, the Lewis base, and an electron pair acceptor, the Lewis acid. The electron pair donor is the more electronegative atom, so that by the oxidation number rules it is assigned the electrons of the new bond even though they are actually now shared. Therefore neither oxidation number is changed, and a Lewis acid–base reaction is not a redox reaction.

Lewis acid–base reactions include **complexation reactions**, such as when ammonia coordinates to copper(II) ions, which play the role of the Lewis acid:

$$[Cu(H_2O)_6]^{2+}(aq) + 4NH_3(aq) \rightarrow [Cu(NH_3)_4(H_2O)_2]^{2+}(aq) + 4H_2O(l)$$

Other Lewis acid–base reactions include **hydrolysis reactions**, of which an example is

$$PCl_5(s) + 4H_2O(l) \rightarrow (HO)_3PO(aq) + 5HCl(aq)$$

in which H_2O is the Lewis base and PCl_5 is the Lewis acid.

A very important class of Lewis acid–base reactions are those in which the hydrogen ion acts as the Lewis acid: such Brønsted acid–base reactions are usually called simply **neutralization reactions** (see Section 12.2). In all such reactions there is an equilibrium between the acid, HA (such as HCl), and the base B:

$$HA + B \rightleftharpoons HB^+ + A^-$$

In the reverse reaction, A^- plays the role of the base, and is called the **conjugate base** of the acid HA. Likewise, HB^+ is the **conjugate acid** of the base B.

Precipitation reactions

In precipitation reactions a solid precipitates from solution upon exceeding its solubility product (see Section 12.1):

$$Ag^+(aq) + Cl^-(aq) \rightarrow AgCl(s)$$

This reaction differs from redox and Lewis acid–base reactions in the sense that all the ions originally present are still present, but two of them (Ag^+ and Cl^-) are now to be found in the solid while the other two remain in solution.

(No solid is completely insoluble, however, and so there are always a few hydrated ions – $Ag^+(aq)$ and $Cl^-(aq)$ – present in the solution after precipitation.)

15.6 How far and how fast?

As well as classifying reactions, it is also important to know whether a reaction has a thermodynamic tendency to occur and, if so, how quickly it will form products.

The direction of a chemical reaction is governed by the sign of the standard reaction Gibbs energy, $\Delta_r G^\ominus$ (discussed in Section 11.3). If $\Delta_r G^\ominus$ is negative, then the equilibrium favours products; if it is positive, then the equilibrium favours reactants:

$$\Delta_r G^\ominus = -RT \ln K$$

where K is the equilibrium constant for the reaction. If $\Delta_r G^\ominus$ is large and negative, the equilibrium constant is very large, and at equilibrium the reaction mixture will consist almost entirely of products. If $\Delta_r G^\ominus$ is small (in the range -10 to $+10\,\text{kJ mol}^{-1}$) at equilibrium the products and the reactants will be present in the mixture in comparable proportions. While it is important to realize that the equilibrium constant is related to $\Delta_r G^\ominus$, in many reactions the magnitude of $\Delta_r H^\ominus$ so exceeds that of $T\Delta_r S^\ominus$ that *as a first approximation* it is possible to judge the likely direction of a reaction from the value of $\Delta_r H^\ominus$ alone. This is only an approximation, however, and there are some reactions for which, especially at high temperatures, a misleading conclusion might be drawn.

Instead of expressing equilibria in terms of $\Delta_r G^\ominus$ values, when dealing with Brønsted acids it is common to report the value of K_a for the proton-transfer reaction. For historical reasons the actual value reported is $pK_a = -\lg K_a$; a *small* value of pK_a indicates a *strong* tendency to donate a proton (see Section 12.2).

Directly related to $\Delta_r G^\ominus$ for redox reactions is the standard electrode potential, $E^\ominus$ (see Section 13.5): if one half-reaction has a more negative $E^\ominus$ than another, then it has a thermodynamic tendency to reduce the latter. That is, while the concept of oxidation number may be used to *identify* the reducing and oxidizing agents in a reaction, *quantitative* conclusions depend on knowing $E^\ominus$ values. Standard electrode potentials are therefore very valuable for discussing redox reactions, and we will see many examples in the following chapters.

A reaction may be thermodynamically feasible but occur so slowly that in practice it may be ignored (this is the case with the conversion of diamonds to graphite at room temperature). In other words, although the reaction tends to occur, its activation energy (see Section 14.4) is so high that it is extremely slow. After judging the thermodynamic feasibility of a reaction, therefore, we should always consider its rate.

It is here that inorganic chemistry differs so strikingly from organic chemistry. Most simple inorganic reactions do not involve bond breaking and so proceed rapidly, whereas organic reactions are often slow because covalent bonds must be broken. A useful generalization is therefore that most inorganic reactions are **thermodynamically controlled** whereas most organic reactions are **kinetically controlled**.

15.7 Inorganic nomenclature

The rules for naming complex inorganic compounds can be daunting; in this book, fortunately, we need to name only the simplest substances, and for them the rules are straightforward.

Binary compounds

Binary compounds consist of only two elements. The less electronegative element is written first, with its oxidation number in roman numerals, followed by the name of the second element with its ending changed to *-ide*:

$CuCl_2$	copper(II) chloride
FeS	iron(II) sulphide
MnO_2	manganese(IV) oxide

The oxidation number is omitted where no ambiguity can arise:

$MgCl_2$ magnesium chloride

Some compounds between two non-metals are named by omitting the oxidation number but including the relative numbers of atoms:

NO_2 nitrogen dioxide

although the name nitrogen(IV) oxide is also acceptable. A few very common substances have non-systematic names, such as:

NH_3	ammonia
H_2O	water
SiO_2	silica

Oxoanions

The names of oxoanions usually have the ending -*ate* followed by the oxidation number of the non-oxygen element:

MnO_4^- manganate(VII)

(The more precise name for this anion is tetraoxomanganate(VII), but the prefix is omitted when there is no ambiguity.) When there is no ambiguity, the oxidation number is not specified:

CO_3^{2-} carbonate

Although the systematic name for SO_3^{2-} is sulphate(IV), to distinguish it from SO_4^{2-}, sulphate(VI), most chemists still use the non-systematic *sulphite* for the former and *sulphate* for the latter. In general, the ending *-ite* implies a lower oxidation number than -ate does. Some systematic names and the corresponding traditional names are given in Table 15.1. Note that the prefix *hypo-* indicates an oxidation number even lower than -ite (as in hypochlorite, ClO^-, chlorate(I)) and *per-* an oxidation number even higher than -ate (as in perchlorate, ClO_4^-, chlorate(VII)).

Table 15.1 Some systematic and traditional names

Systematic name	Traditional name	Formula
chlorate(VII)	perchlorate	ClO_4^-
chlorate(V)	chlorate	ClO_3^-
chlorate(III)	chlorite	ClO_2^-
chlorate(I)	hypochlorite	ClO^-
manganate(VII)	permanganate	MnO_4^-
iron(II)	ferrous	Fe^{2+}
iron(III)	ferric	Fe^{3+}
hexacyanoferrate(II)	ferrocyanide	$[Fe(CN)_6]^{4-}$
hexacyanoferrate(III)	ferricyanide	$[Fe(CN)_6]^{3-}$

Writing the formulae of oxoacids and their salts requires some care. In order to stress that the acidic hydrogen atoms of sulphuric acid are attached to oxygen atoms the formula $(HO)_2SO_2$ will be adopted, rather than the traditional H_2SO_4 (other oxoacids are formulated on the same principle). The anion is written SO_4^{2-}, however, because in the ion all four oxygen atoms are equivalent as the negative charge is delocalized (see Section 2.5) over all of them equally:

$(HO)_2SO_2$ sulphuric acid
$HOSO_3^-$ hydrogensulphate ion
SO_4^{2-} sulphate ion

Some more examples are given in Table 15.2.

Table 15.2 Formulae of some oxoacids and their salts

Compounds of Period 2 elements		Compounds of Period 3 elements	
$(HO)_3B$	boric acid	$(HO)_3PO$	phosphoric acid
$(HO)_2CO$	carbonic acid	$(HO)_2SO$	sulphurous acid
$HOCO_2^-$	hydrogencarbonate ion	$HOSO_2^-$	hydrogensulphite ion
CO_3^{2-}	carbonate ion	SO_3^{2-}	sulphite ion
$HONO_2$	nitric acid	$(HO)_2SO_2$	sulphuric acid
NO_3^-	nitrate ion	$HOSO_3^-$	hydrogensulphate ion
HONO	nitrous acid	SO_4^{2-}	sulphate ion
NO_2^-	nitrite ion	$HOClO_3$	chloric(VII) acid

The same convention will be adopted in organic chemistry. For example, ethanoic acid is written CH_3COOH, because the hydrogen atom is attached to one of the oxygen atoms, but the ethanoate ion is written $CH_3CO_2^-$ because the charge is delocalized and the two oxygen atoms are equivalent.

Summary

- ☐ The Periodic Table is an arrangement of the elements according to their electron configurations. In the s-block the valence electrons have the configuration ns^1 or ns^2; in the p-block the configuration is ns^2np^x (x has a value from 1 to 6). The Period number is n.
- ☐ The oxidation number is equal to the number of electrons that must be added to the atom in its combined state to form the neutral atom.
- ☐ Elements with low first ionization energies, large negative standard electrode potentials and low electronegativities are found in the bottom left-hand corner of the Periodic Table. Elements in the top right-hand corner have the opposite properties.
- ☐ The oxides, chlorides and hydrides each become more covalent with increasing atomic number across a Period, and more ionic with increasing atomic number down a Group.
- ☐ Atomic size decreases from left to right across a Period and increases down a Group.
- ☐ Cations are always smaller than their parent atoms; anions are always larger.
- ☐ Most inorganic compounds are solids, having either ionic or covalent extended network structures.
- ☐ There are three classes of reaction: redox (or electron-transfer) reactions, Lewis acid–base (or coordination) reactions and precipitation reactions.
- ☐ Most inorganic reactions are thermodynamically controlled, whereas most organic reactions are kinetically controlled.

PROBLEMS

1 Give the oxidation number of the element printed in italics for each of the following:

Fe^{3+}, Cl^-, Cl_2, $\mathit{Al}F_3$, Al_2Cl_6, $\mathit{C}S_2$, $\mathit{Si}F_6^{2-}$, $\mathit{P}Cl_4^+$, $\mathit{N}F_3$, $\mathit{N}O_3^-$, $N\mathit{O}_3^-$, $\mathit{S}O_2$, $\mathit{S}O_3^{2-}$, $\mathit{Cl}F_3$, I^+, $\mathit{Xe}F_4$, $\mathit{Cu}Cl$, $Cu\mathit{Cl}$, Fe_2O_3, $\mathit{Mn}O_4^-$

2 Classify the following reactions as (a) redox, (b) Lewis acid–base, or (c) precipitation reactions:

$2Rb(s) + H_2(g) \rightarrow 2RbH(s)$
$Ba^{2+}(aq) + SO_4^{2-}(aq) \rightarrow BaSO_4(s)$
$Fe(s) + 2HCl(aq) \rightarrow FeCl_2(aq) + H_2(g)$
$SO_2(g) + 2H_2S(g) \rightarrow 3S(s) + 2H_2O(l)$
$H_3O^+(aq) + OH^-(aq) \rightarrow 2H_2O(l)$
$SiF_4(g) + 2F^-(aq) \rightarrow SiF_6^{2-}(aq)$
$Cl_2(g) + 2Br^-(aq) \rightarrow 2Cl^-(aq) + Br_2(l)$
$AgNO_3(aq) + KCl(aq) \rightarrow AgCl(s) + KNO_3(aq)$
$Ni(CN)_2(s) + 2KCN(aq) \rightarrow K_2Ni(CN)_4(aq)$
$MnO_4^-(aq) + 8H^+(aq) + 5e^- \rightarrow Mn^{2+}(aq) + 4H_2O(l)$

3 Name the following compounds:

MnO_2, $KMnO_4$, $PbCl_2$, $PbCl_4$, $[Co(CN)_6]^{3-}$, Cl_2O_7, ClO_4^-, $CuSO_4.5H_2O$, SnS, $FeCl_3$

4 Give the formulae of the following compounds:

chromium(III) oxide	potassium chlorate(III)
strontium sulphate	oxygen difluoride
chromium(VI) oxide	ammonium dihydrogenphosphate
potassium dichromate(VI)	disodium hydrogenphosphate
ammonium hydrogencarbonate	
carbon disulphide	

5 Following his experiments on burning elements in oxygen, Lavoisier concluded in 1777 that all acids contain oxygen. Comment on this theory and write an account of how the understanding of acidity developed from that time, through the ideas of nineteenth-century chemists such as Davy and Arrhenius, to the Brønsted–Lowry theory in the present century.

The most widely applicable theory of acidity in use today is the Lewis theory. This defines an acid as an electron pair acceptor and a base as an electron pair donor. Discuss the following reactions in terms of the Lewis theory, comparing with the Brønsted–Lowry theory in each case:
(a) $HCl(g) + H_2O(l) \rightleftharpoons H_3O^+(aq) + Cl^-(aq)$
(b) $NH_3(aq) + H_2O(l) \rightleftharpoons NH_4^+(aq) + OH^-(aq)$
(c) $NH_3(g) + BF_3(g) \rightarrow BF_3.NH_3(s)$ *(London S)*

6 What is meant by the term 'oxidation number' ('oxidation state')? Discuss the rules for assigning oxidation number, and the applications of the concept in organic and inorganic chemistry.

Place a suitable selection of the compounds of *either* nitrogen *or* sulphur on a scale of oxidation number.

Comment on the oxidation number of *either* nitrogen in HN_3 (hydrazoic acid) *or* sulphur in $Na_2S_2O_3$, in relation to the structures of the compounds. *(London S)*

7 Write an account of the more important oxides of the elements from sodium (Na; $Z = 11$) to chlorine (Cl; $Z = 17$), remembering that some of the elements have more than one well-characterized oxide. You may like to consider their composition, structure, thermal stability and behaviour with water and/or acids and alkalis, but your account need not be concerned with all these points nor be confined to them. *(London)*

8 Giving reasons for your choices, decide which of the following first ionization energies and ionic radii correspond to sodium, magnesium and aluminium.

First ionization energies ($kJ\,mol^{-1}$) 494, 577, 736
Ionic radii (nm) 0.050, 0.068, 0.100

Compare (a) the base strengths and solubilities of the hydroxides of these metals and (b) the bonding of the chlorides of sodium and aluminium. *(Welsh part question)*

9 Write an account of the physical and chemical properties of the chlorides (where formed) of the elements of the third Period (sodium to argon), showing how these properties depend on the position of the element in the Period.

How do the chlorides of the elements of the second Period (lithium to neon) differ from those of the elements of the third Period in each Group? *(Nuffield)*

10 Give an account of the chlorides of the elements from sodium to chlorine in the Periodic Table, paying particular attention to:
(a) methods of preparation, including brief experimental details;
(b) the nature of the bonding;
(c) reactions with water. *(London part question)*

11 Metallic character increases from right to left and from top to bottom in the Periodic Table.

Select any *two* properties of elements which determine metallic character and discuss the trends which occur in (a) the halogens, and (b) the Period from sodium to chlorine.

Illustrate your answer with suitable examples.

How far do you agree with the view that a classification of elements into metals and non-metals is unrealistic? *(London)*

12 The enthalpy changes of vaporization $\Delta_v H$ for the elements in the third Period of the Periodic Table are shown in the table below.

Element	Na	Mg	Al	Si	P	S	Cl	Ar
$\Delta_v H/kJ\,mol^{-1}$	101	132	284	300	52	63	10	6

Plot a graph of $\Delta_v H$ (y-axis) against atomic number (x-axis) for the elements sodium to argon inclusive and explain the variation in these data in terms of the structure and bonding of the elements. *(Cambridge part question)*

13 The first member of a Group in the Periodic Table is often found to have properties which are not typical of the Group.

Discuss the reasons for this, illustrating your answer by comparing the chemistry of *four* of the elements in the series lithium to fluorine, and their compounds, with the corresponding elements in the second short Period. *(London S)*

14 The species

O^{2-} F^- Ne Na^+ Mg^{2+}

all have the same number of electrons. Using information from reference books, describe the way in which the first ionization energy and the radius of each of the particles depends on its atomic number.

Hence make an estimate of
(a) the radius of the ion $Al^{3+}(g)$,
(b) the electron affinity of $N^{2-}(g)$, and
(c) the enthalpy change of the reaction $N(g) \rightarrow N^{3-}(g)$. *(Nuffield S)*

15 To what extent are there grounds for believing that all the molecules of a compound contain the same numbers of the different kinds of atoms of which it is composed, in fixed mean relative positions? *(Cambridge Entrance)*

16 *Either*: 'The chemistry of the elements in the first short Period (Li–Ne) differs markedly from that of corresponding elements in the second short Period (Na–Ar)'. Discuss this statement with reference to the chemistry of one or more of the following pairs of elements: lithium and sodium, oxygen and sulphur, fluorine and chlorine.

Or: Discuss, with examples, the changes in chemistry which take place across the first transition series. *(Oxford Entrance)*

16 Hydrogen

Hydrogen forms more compounds than any other element. The great majority of them are covalent, but the hydrogen cation is also very important on account of its role in Brønsted acid–base reactions. Hydrogen is a powerful reducing agent, and is widely used in industry, for example, in the reduction of nitrogen to ammonia.

Hydrogen is the oldest element in the universe, and was formed during the first few minutes following the Big Bang. An atom of hydrogen is very simple, consisting of a single proton and a single electron with configuration $1s^1$. Although it would therefore seem to be a member of Group I and the s-block of the Periodic Table, the properties of hydrogen are quite unlike those of the members of Group I (for example, it is a gas, not a reactive metal), and so it is unhelpful to include it with them. The first element of a Group often has properties that are out of line with those of the other members; with hydrogen we are dealing with the first member of the entire Table, and it is strikingly different from every other element. For this reason it is sensible to place hydrogen in a special place at the head of the Table, and not to include it in any Group.

There are three isotopes of hydrogen, and they are so important that they have their own names, listed in Table 16.1. Naturally occurring hydrogen consists of 99.98 per cent of ^1H, the remainder being deuterium except for a trace of tritium formed by cosmic rays in the upper atmosphere. The important property of deuterium is that since it has an additional neutron its mass is twice that of protium, ^1H, yet because it has the same electron configuration its chemical properties are almost identical. As a result, the effect of an atom's mass on the rates of chemical reactions can be studied by substituting ^2H for ^1H. Moreover, since the presence of deuterium can be detected quite simply, particularly by nuclear magnetic resonance spectroscopy or mass spectrometry, it is easy to follow the pathway of an organic reaction. Tritium can be used similarly, and its presence detected by its radioactivity: it emits β-particles. For example, the reactions of metabolism can be followed by monitoring the radioactivity and tracing what happens to compounds formed at each stage. One very important deuterium compound is **heavy water**, D_2O (its density is $1.1\,g\,cm^{-3}$, and so it is 10 per cent heavier than ordinary water), which is used in some nuclear reactors as a coolant and also as a **moderator**, that is, a substance that slows the neutrons to speeds at which they can be captured more effectively by fissionable nuclei.

16.1 The principal features of hydrogen's chemistry

Oxidation states and bonding characteristics

The most common oxidation number of hydrogen in its compounds is $+1$, as in hydrogen chloride, HCl, and water, H_2O, but in some compounds – particularly those containing the **hydride ion**, H^- – its oxidation number is -1.

Hydrogen is covalently bonded in almost all its compounds. This is partly due to its high ionization energy ($1312\,kJ\,mol^{-1}$), which makes it energetically difficult to form a cation. A second reason is the extremely small size of the hydrogen cation, H^+, which is a bare proton. A proton is so highly polarizing that it acts as an extremely strong Lewis acid (see Section 12.4), attaching to the electron pairs of any neighbouring species. In water it attaches strongly to a lone pair of a water molecule forming H_3O^+ (**1a**, **1b**), the ion central to

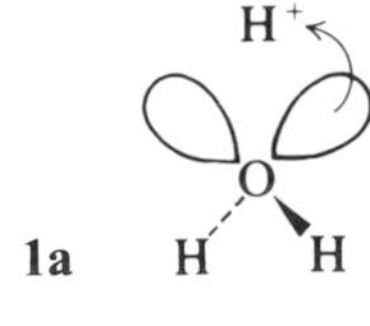

1a

1b

Table 16.1 Isotopes of hydrogen

Name	Mass number	Number of neutrons	Symbol	Alternative symbol
protium	1	0	${}^{1}_{1}H$	–
deuterium	2	1	${}^{2}_{1}H$	D
tritium	3	2	${}^{3}_{1}H$	T

discussions of Brønsted acid–base reactions (see Section 12.2). The uniqueness of hydrogen's chemical properties owes less to its possession of only one electron than to the very small size – even on an atomic scale – of the ion remaining after the loss of that electron, and its exceptionally strong polarizing power.

Electrode potential and hydride formation

By definition, the standard electrode potential of the H^+, H_2 couple is zero (see Section 13.5). Consequently, any couple with a negative standard electrode potential is thermodynamically capable of reducing the hydrogen ion. For example, zinc metal ($E^{\ominus}(Zn^{2+}, Zn) = -0.76$ V) reduces the hydrogen ions in hydrochloric acid to elemental hydrogen:

$$Zn(s) + 2H^+(aq) \rightarrow Zn^{2+}(aq) + H_2(g) \qquad H(+1) \rightarrow H(0),\ Zn(0) \rightarrow Zn(+2)$$

Very strong reducing agents, such as the alkali metals, can reduce elemental hydrogen to a state with oxidation number −1, forming a hydride:

$$2Na(s) + H_2(g) \xrightarrow{\text{heat}} 2NaH(s) \qquad H(0) \rightarrow H(-1),\quad Na(0) \rightarrow Na(+1)$$

The thermodynamic feasibility of the reduction of elemental hydrogen can be demonstrated using a Born–Haber cycle.

EXAMPLE

Confirm that sodium hydride is an exothermic compound.

METHOD Construct a Born–Haber cycle, as described in Section 9.4, taking the thermochemical data from the tables in Chapters 1 and 9.

ANSWER The cycle required is shown in Figure 16.1.

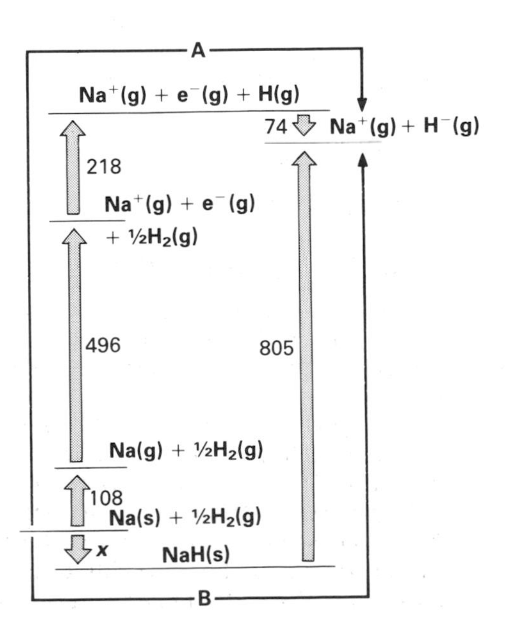

Figure 16.1 The Born–Haber cycle for the formation of sodium hydride

The data are as follows:

Step	Process	$\Delta_r H^\ominus$/kJ mol^{-1}
1	Formation of NaH(s)	$-x$
2	Atomization of Na(s)	+108
3	Ionization of Na(g)	+496
4	Dissociation of $\frac{1}{2}H_2$(g)	+218
5	Electron attachment to H(g)	−74
6	Lattice enthalpy of NaH(s)	+805

The standard formation enthalpy, $-x$, is such that route A is equal to route B. From Figure 16.1 it is clear that

$$\text{route A} = (108 + 496 + 218 - 74)\,\text{kJ mol}^{-1} = 748\,\text{kJ mol}^{-1}$$
$$\text{route B} = (805 - x)\,\text{kJ mol}^{-1}$$

Therefore, $x = (805 - 748)\,\text{kJ mol}^{-1} = 57\,\text{kJ mol}^{-1}$

The standard formation enthalpy is therefore $-57\,\text{kJ mol}^{-1}$, an exothermic quantity.

COMMENT The thermodynamic feasibility of the formation of NaH(s) depends on its standard formation Gibbs energy, and this calculation deals only with the enthalpy; however, in this instance the enthalpy contribution dominates the entropy contribution ($T\Delta_f S^\ominus = -23\,\text{kJ mol}^{-1}$) and so the enthalpy is a good guide.

Atomic and ionic sizes

The relative sizes of the hydrogen atom and the hydride ion are shown in Figure 16.2. The hydride ion is larger than the hydrogen atom because in it two electrons are under the control of a single proton, and their mutual repulsion expands the ion nearly to the size of a chloride ion.

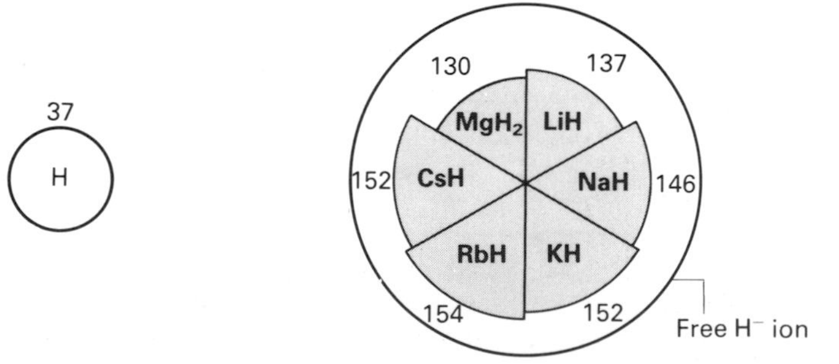

Figure 16.2 The radii (in picometres) of the hydrogen atom and hydride anions; the atom is much smaller than the anion, and the size of the anion depends on the associated cation

The radius of H^+, about 10^{-15} m, is so small that it cannot be represented in the same diagram by a printable dot: on the scale used such a dot would have to be approximately one-thousandth of a millimetre in diameter.

The small size of the hydrogen atom, and the weak shielding of its nucleus by the single electron, accounts for hydrogen's ability to link two highly electronegative atoms by a hydrogen bond, described in Section 5.1. Hydrogen's uniqueness as an element is thus mirrored by its compounds, including the anomalously high boiling points and vaporization enthalpies of water, ammonia and hydrogen fluoride (see Figure 5.8 and Section 9.2), the high density of water as compared with that of ice (Section 5.1), the existence of carboxylic acid molecules as dimers (Section 31.2) and the low acidity of hydrofluoric acid (Section 23.5). Hydrogen bonding also controls the shapes of biologically important molecules, such as starch, proteins and DNA (see Sections 30.5, 32.5 and 5.1 respectively).

16.2 The element

Hydrogen is a colourless, odourless, tasteless gas consisting of H_2 molecules (their electronic structure is discussed in Section 2.5). It is also the gas of lowest density at room temperature and pressure, a property that led to its use in balloons and airships until the *Hindenburg* disaster (Figure 16.3) resulted in its replacement by helium, which is chemically inert.

Figure 16.3 The airship *Hindenburg* was filled with hydrogen. It exploded and caught fire as it came in to land at Lakehurst, New Jersey, in May 1937

The dominant chemical role of hydrogen is as a *reducing agent*: it reduces non-metals with varying degrees of violence. With fluorine it reacts explosively, but with chlorine and oxygen under normal conditions an explosion occurs only when the reaction is initiated with light or a spark. The explosive reaction with oxygen is the basis of the simple test for hydrogen in which it gives a mild explosion (a 'pop') with a lighted splint, and the controlled reaction is used to power space vehicles (see Figure 16.4). The reduction of nitrogen

Figure 16.4 The controlled reaction between hydrogen and oxygen is used to power space vehicles

$$N_2(g) + 3H_2(g) \xrightleftharpoons{450\,^\circ C,\ 250\,atm,\ Fe} 2NH_3(g) \qquad N(0) \rightarrow N(-3),\quad H(0) \rightarrow H(+1)$$

needs a catalyst and high temperatures for an acceptable rate (see Section 21.6). Apart from natural processes, this reaction – the Haber–Bosch synthesis of ammonia – is the principal method by which atmospheric nitrogen is brought into the food chain.

Hydrogen also reduces many metal oxides to the metal. For example, when hydrogen is passed over heated copper(II) oxide, copper is produced, the black oxide turning to shiny metallic copper:

$$CuO(s) + H_2(g) \xrightarrow{heat} Cu(s) + H_2O(g) \qquad Cu(+2) \rightarrow Cu(0),\quad H(0) \rightarrow H(+1)$$

Some of the metals of modern technology are produced by reduction with hydrogen, partly because it is less likely to introduce impurities than reduction with coke. For example, the tungsten used for light bulb filaments is obtained by the reduction of its oxide:

$$WO_3(s) + 3H_2(g) \xrightarrow{850\,^\circ C} W(s) + 3H_2O(g) \qquad W(+6) \rightarrow W(0),\quad H(0) \rightarrow H(+1)$$

Hydrogen is also used as a reducing agent in certain organic reactions, an important example of **hydrogenation** (the addition of hydrogen to a compound) being the conversion of carbon–carbon double bonds to single bonds (see Section 33.3):

$$\mathrm{>C{=}C<} + H_2 \xrightarrow{Pt} \mathrm{-\overset{\displaystyle H}{\underset{|}{\overset{|}{C}}}-\overset{\displaystyle H}{\underset{|}{\overset{|}{C}}}-}$$

This reaction is used to convert vegetable oils to solid fats in the manufacture of margarine, a reaction discussed in Box 31.2.

Very strong reducing agents, such as the alkali metals, can reduce hydrogen, with the formation of hydrides.

16.3 Compounds of hydrogen

Most compounds of hydrogen are discussed under the other elements concerned (see Section 5 of Chapters 17 to 23) or, in the case of its compounds with carbon, as part of the study of organic chemistry. However, the **binary hydrides** are best treated as characteristic compounds of hydrogen, and are described here. The other element in the hydride is not necessarily less electronegative than hydrogen, and so 'binary hydride' is used loosely to mean any binary compound, including ammonia, water and the hydrogen halides such as HCl.

The binary hydrides may be divided into three overlapping classes.

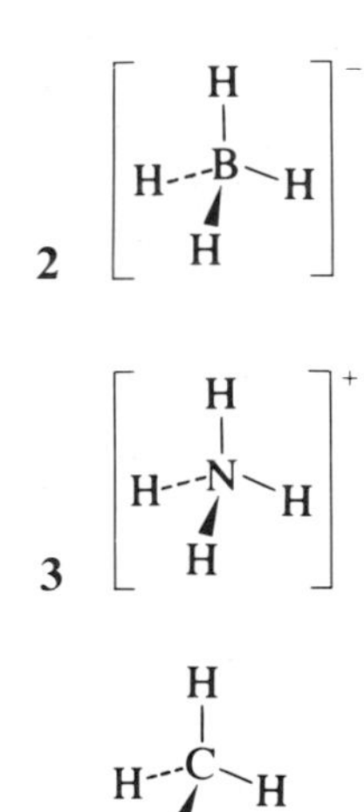

1 Covalent hydrides are compounds in which hydrogen is bonded covalently to another element. This class is by far the largest of the three and includes the ions H_3O^+ (**1b**), BH_4^- (**2**) and NH_4^+ (**3**), for in them the hydrogen is bonded covalently even though the ion is part of an ionic compound. BH_4^- and NH_4^+ have the same shape as CH_4 (**4**).

2 Anionic hydrides are largely ionic compounds containing the hydride anion, H^-. Anionic hydrides are sometimes called **saline hydrides**, because they are salt-like, ionic crystalline solids. Electrolysis of molten lithium hydride produces hydrogen at the *anode*, confirming the presence of H^- anions. Anionic hydrides are prepared by the direct reaction of an electropositive element, such as sodium, heated in hydrogen. They are very rapidly hydrolysed by water:

$$H^-(s) + H_2O(l) \rightarrow H_2(g) + OH^-(aq) \qquad H(-1) + 2H(+1) \rightarrow 2H(0) + H(+1)$$

and so provide a convenient, transportable source of hydrogen.

3 Interstitial hydrides are non-stoichiometric compounds of some of the d-block elements. **Non-stoichiometric** means that the compounds do not have simple formulae such as TiH_2, but are better (but still only approximately) described by formulae such as $TiH_{1.7}$, with the precise proportion of elements dependent on the temperature and pressure. They are called 'interstitial' because the hydrogen atoms occupy the holes (**interstices**) between the cations of the solid metal. For example, titanium forms a grey metallic solid when heated in hydrogen, in which the hydrogen atoms are distributed randomly among its interstices. The interest in interstitial hydrides lies in their ability to release their hydrogen when heated, and so they act as a transportable source of hydrogen in a much more compact form than even liquid hydrogen (see Box 16.1).

EXAMPLE

Predict the chemical properties of the hydrides of rubidium, zirconium and tellurium.

METHOD Locate the elements in the Periodic Table. Decide the class of hydride in each case, and hence suggest their properties.

ANSWER Rubidium is in Group I, and so can be expected to form an anionic hydride; such a compound would be a white crystalline solid that would liberate hydrogen on contact with water. Zirconium is a member of the d-block, and might therefore have a non-stoichiometric interstitial hydride which will probably have a metallic appearance and release hydrogen when heated. Tellurium is in Group VI, and is a member of the p-block; it is likely to form a low-melting, covalent, acidic hydride, H_2Te.

COMMENT Hydrogen telluride, H_2Te, is also a toxic gas with a foul smell; it is absorbed by the skin, and the skunk-like odour persists for a long time.

16.4 The occurrence and manufacture of hydrogen

Occurrence

Most of the universe is made up of hydrogen, which occurs in great interstellar clouds, while the stars themselves are largely hydrogen. Hydrogen is very abundant on Earth, but here it is found almost entirely in combination in the great oceans of water and all the vegetation of the planet, for elemental hydrogen can escape into outer space (see Section 4.2). Hydrogen occurs in combination in many minerals, often as water of crystallization, and it is present in combination with carbon in fossil fuels.

Manufacture

Most hydrogen is obtained from the methane of natural gas by mixing it with steam and passing it over a nickel catalyst:

$$CH_4(g) + H_2O(g) \xrightleftharpoons{900\,^\circ C,\ 30\,atm,\ Ni} CO(g) + 3H_2(g) \qquad H(+1) \rightarrow H(0), \quad C(-4) \rightarrow C(+2)$$

The carbon monoxide is subsequently converted to carbon dioxide in the **shift reaction**:

$$CO(g) + H_2O(g) \xrightleftharpoons{400\,^\circ C,\ Fe_2O_3} CO_2(g) + H_2(g) \qquad H(+1) \rightarrow H(0), \quad C(+2) \rightarrow C(+4)$$

and the carbon dioxide is removed by 'scrubbing' the gas stream with alkali. The last traces of carbon monoxide are removed by passing the scrubbed gas over a nickel catalyst at 350 °C, under which conditions the first equilibrium lies to the left-hand side.

There are various alternative supplies of hydrogen, and economic considerations determine which one an industry adopts. The classic manufacturing processes include the reduction of steam over white-hot coke:

$$C(s) + H_2O(g) \xrightarrow{1000\,^\circ C} CO(g) + H_2(g) \qquad H(+1) \rightarrow H(0), \quad C(0) \rightarrow C(+2)$$

The mixture of gases thus formed is called **water gas**. The reaction is endothermic, and the energy required is obtained by alternating blasts of steam and air, the latter burning the coke. Supplies of hydrogen are also available from the catalytic reforming of hydrocarbons, described in Section 27.4. Some hydrogen is produced electrolytically:

$$2H_2O(l) \xrightarrow{electrolysis} 2H_2(g) + O_2(g) \qquad H(+1) \rightarrow H(0), \quad O(-2) \rightarrow O(0)$$

but as this process is very energy-intensive, requiring abundant supplies of cheap electricity, it is economical only close to hydroelectric plants.

BOX 16.1

Hydrogen as a fuel

In some respects hydrogen can be regarded as a perfect fuel, for on combustion it forms only water and causes no pollution. Some people have therefore speculated about a 'hydrogen economy' in which the predominant fuel, particularly for mobile engines, would be hydrogen. They suppose that the hydrogen could be produced by the electrolysis of water using electricity generated by nuclear fusion. Another important source may turn out to be the decomposition of water by sunlight in the presence of a catalyst:

$$2H_2O(l) \xrightarrow{\text{sunlight, catalyst}} 2H_2(g) + O_2(g)$$

This reaction has already been demonstrated on a laboratory scale, but the catalyst soon loses its activity, and the process is not yet commercially economical.

But would the hydrogen economy become a reality if plentiful hydrogen were available? Many problems would still remain, because the enthalpy of combustion of liquid hydrogen per litre, $11\,000\,kJ\,dm^{-3}$, is low compared with that of liquid hydrocarbon fuels, typically $33\,000\,kJ\,dm^{-3}$, and so large volumes of hydrogen would have to be transported. For example, a car that has a 15 gallon petrol tank would need one holding 45 gallons of liquid hydrogen in order to have the same range. Furthermore, complex, expensive and heavy refrigeration units would have to be fitted so that the fuel remained liquid. One possible chemical solution is to form a metal hydride, and to heat it to release the hydrogen when it is required. The metal hydride with approximate formula $FeTiH_2$ can carry hydrogen in a smaller volume than liquid hydrogen and decomposes when heated. There is a difficulty, however: the hydride is very heavy, and adds considerably to the weight of the vehicle. This is an area where the combined efforts of chemists, physicists and engineers are needed to find a solution.

Summary

- ☐ Hydrogen has the simplest possible electron configuration, $1s^1$.
- ☐ The element is placed alone at the head of the Periodic Table.
- ☐ The three isotopes of hydrogen are protium, deuterium and tritium.
- ☐ The most common oxidation number is +1, and in most compounds hydrogen is covalently bonded. There are some compounds with hydrogen in the −1 oxidation state, however; these contain the anion H^-.
- ☐ Hydrogen is a colourless low-density gas; it consists of H_2 molecules.
- ☐ Hydrogen is flammable in air and is a powerful reducing agent.
- ☐ Hydrogen is manufactured from steam and either coke or natural gas in the presence of a catalyst. In some countries hydrogen is obtained electrolytically.
- ☐ A major use of hydrogen is in the manufacture of ammonia. It is also used in large quantities for the manufacture of margarine, hydrogen chloride, methanol and many other organic compounds.

PROBLEMS

1 (a) List the various methods for the manufacture of hydrogen.
(b) Hydrogen is used as soon as it is made and very little is stored. Why is this?
(c) In the UK, hydrogen is produced by the reaction of natural gas (methane) and steam. What is the other product of this reaction? Explain how the mixture of gases is treated depending on whether the hydrogen is intended for the synthesis of ammonia, methanol or higher alcohols.
(d) In the UK, about $10^9\,m^3$ of hydrogen are produced annually. What mass of methane is required to produce this volume of hydrogen at 20 °C and 1 atm?
(e) List the industrial uses of hydrogen.

2 Hydrogen is sometimes placed at the head of Group I (alkali metals) in the Periodic Table. What are the arguments for and against placing hydrogen in this position?

3 Hydrogen exhibits oxidation numbers of −1, 0 and +1. Elemental hydrogen can therefore act as an oxidizing agent and a reducing agent. Give balanced equations to illustrate the redox reactions of hydrogen.

4 How can a sample of heavy water, D_2O, be converted as efficiently as possible without waste into (a) ND_3, (b) D_2O_2 and (c) C_2D_2?

5 Write an essay on hydrogen bonding.

6 Predict the properties of the binary compounds (if any) formed between hydrogen and (a) francium, (b) radon, (c) antimony, (d) astatine, (e) iron.

7 Hydrogen has a much lower tendency to form the hydride ion, $H^-(g)$, than do the halogens to form halide ions of the type $X^-(g)$. Using appropriate bond energies and electron affinities [negative of electron-gain energies], calculate the enthalpy of formation of the gaseous hydride and four gaseous halide ions and comment on the results. *(Nuffield S)*

8 Compare and contrast the physical and chemical properties of the simple hydrides of each of the following elements:

sodium, calcium, carbon, nitrogen, oxygen and fluorine.

In your answer, attempt to relate these properties to the type of bonding encountered in these hydrides. *(London)*

9 List the formulae of the simplest hydrides of the elements lithium to fluorine, using a book of data if necessary. Classify the hydrides as ionic or covalent, justifying your classification by reference to appropriate data, and draw dot-and-cross diagrams to illustrate their electronic structures.

Are the physical properties of the hydrides of lithium to fluorine consistent with their positions in the Periodic Table? Explain any pattern of inconsistency that you notice. *(Nuffield)*

10 'Hydrogen is an unique element.' What features in the chemistry of hydrogen support this claim?

An element, **R**, reacts with hydrogen at 250 °C to form a black compound, **Q**. On heating to a higher temperature **Q** decomposes to the element and hydrogen in the ratio of 141.2 cm^3 of hydrogen (measured at s.t.p.) for each gram of **R**. **Q** reacts with chlorine at 200 °C to form a green chloride of **R** containing 37.3 per cent by mass of chlorine.

Deduce what you can from the above information. (The molar volume of a gas at s.t.p. is 22.4 $dm^3\,mol^{-1}$.) *(Oxford Entrance)*

17 The s-block elements: Group I

The elements of Group I are called the alkali metals. The most important members of the Group are sodium and potassium, and we concentrate on these. All their compounds are built from the M^+ ion. There is a closer similarity between the elements of this Group than in any other Group of the Periodic Table. All the metals are strong reducing agents, and have alkaline hydroxides. Sodium hydroxide, chloride and carbonate are among the most important industrial chemicals.

	I	II	
2	3 Li	4 Be	
3	11 Na	12 Mg	
4	19 K	20 Ca	21 Sc
5	37 Rb	38 Sr	39 Y
6	55 Cs	56 Ba	57 La
7	87 Fr	88 Ra	89 Ac

Figure 17.1 Group I in the Periodic Table

The elements in Group I, known as the **alkali metals**, are as follows:

lithium	Li	$[He]2s^1$
sodium	Na	$[Ne]3s^1$
potassium	K	$[Ar]4s^1$
rubidium	Rb	$[Kr]5s^1$
caesium	Cs	$[Xe]6s^1$
francium	Fr	$[Rn]7s^1$

Their positions in the Periodic Table are shown in Figure 17.1. In each element the characteristic valence electron configuration is ns^1, where n is the Period number. Because the s-subshell is in the process of being filled, the elements of Groups I and II are members of the s-block (see Section 1.2).

All the isotopes of francium are radioactive, the longest-lived (^{223}Fr) having a half-life of only 22 minutes. It is very rare, and only a few grams are present at any moment in the whole of the Earth's crust. One isotope of caesium (^{133}Cs) has achieved importance because the frequency of one of its spectral transitions is used as the primary time-keeping standard.

17.1 Group systematics

Oxidation states and bonding characteristics

In this Group the only oxidation number of any importance, other than 0, is +1, and all the common compounds of these elements are based on the M^+ ion. The main reason for the dominance of M^+ in their compounds is that the elements have low first ionization energies. These energies decrease down the Group (see Figure 17.2) because the outermost electron is progressively further from the nucleus and hence easier to remove.

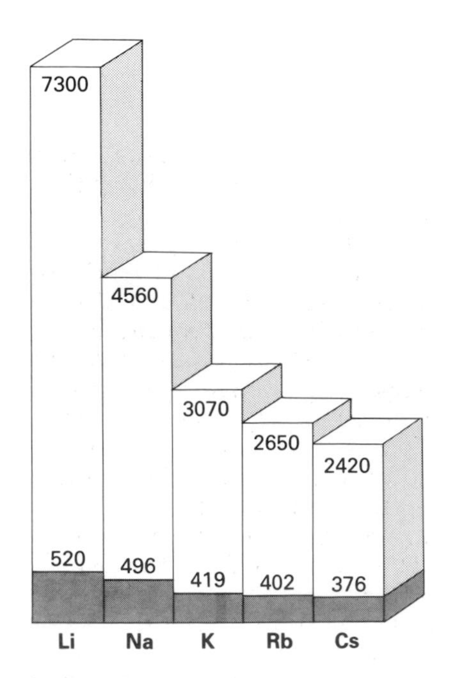

Figure 17.2 The first and second ionization energies of the Group I elements

Figure 17.2 also shows that their second ionization energies are all very much larger than the first (since the electron has to be removed from an inner shell), and therefore M^{2+} ionic compounds are ruled out. This can be demonstrated by setting up a Born–Haber cycle (see Section 9.4) for the standard formation enthalpy of NaCl, as shown in Figure 17.3(a), and comparing it with the cycle for the hypothetical compound $NaCl_2$, shown in Figure 17.3(b). For the latter, the lattice enthalpy has been assumed equal to that of the analogous but real compound $MgCl_2$. The cycles show that whereas NaCl is an exothermic compound, $NaCl_2$ would be strongly endothermic on account of the large second ionization energy of sodium. (As was stressed in Section 11.3, when assessing thermodynamic stability it is necessary to consider the Gibbs energy, but for ionic solids the entropy terms are overshadowed by the large formation enthalpies, and so an enthalpy argument is not misleading.)

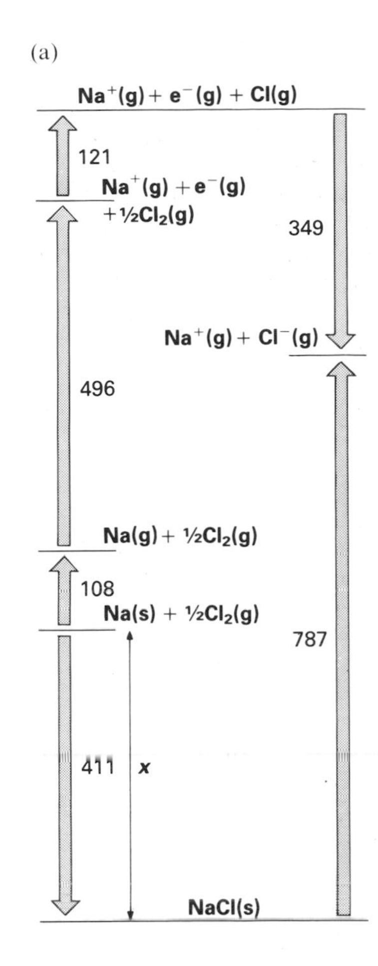

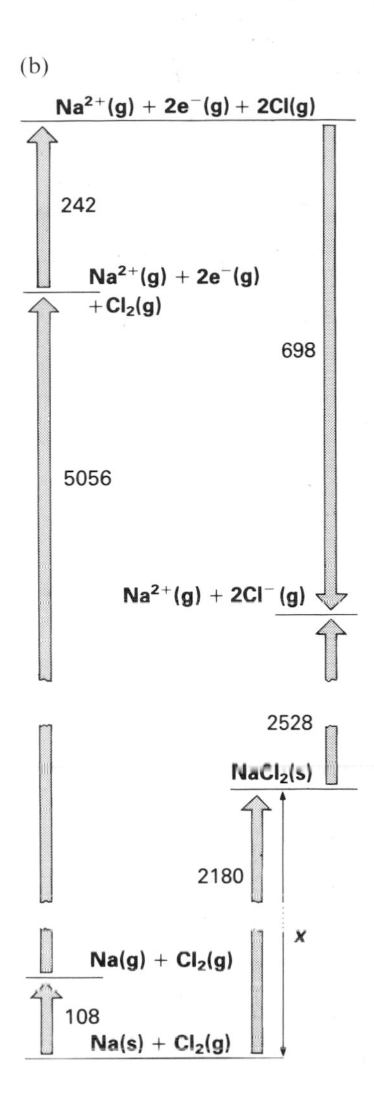

Figure 17.3 The Born–Haber cycle for (a) NaCl and (b) the hypothetical $NaCl_2$

Periodic trends

The alkali metals have chemical and physical properties that are more uniform than in any other Group in the Periodic Table. Lithium, however, being the first member of the Group, does show some anomalies.

All the elements are powerful reducing agents. This is shown by their standard electrode potentials, all of which lie between -2.7 V and -3.0 V (see Figure 17.4), indicating a strong tendency to form cations in solution. The standard electrode potentials do not vary smoothly because they depend on the Gibbs energy for the process $M(s) \rightarrow M^+(aq) + e^-$, which in turn depends on both enthalpy and entropy changes accompanying individual processes, including the atomization and ionization of the metal and the hydration of the ions. No general trend can be predicted without a detailed analysis of each step. A notable feature, however, is that the large hydration enthalpy of the small Li^+ ion contributes to the strongly negative value of $E^\ominus(Li^+, Li)$.

The small size of the lithium cation explains many of lithium's anomalies. It accounts both for its diagonal relationship within the Periodic Table with magnesium, and also for the large lattice enthalpy of Li_2O, which in turn is the reason why lithium oxosalts are decomposed to the oxide on heating more readily than are the corresponding sodium oxosalts. Thus lithium nitrate (like magnesium nitrate) decomposes to the oxide. Lithium (like magnesium) reacts with nitrogen to form a nitride, the reaction being favoured by the high lattice enthalpy of lithium nitride.

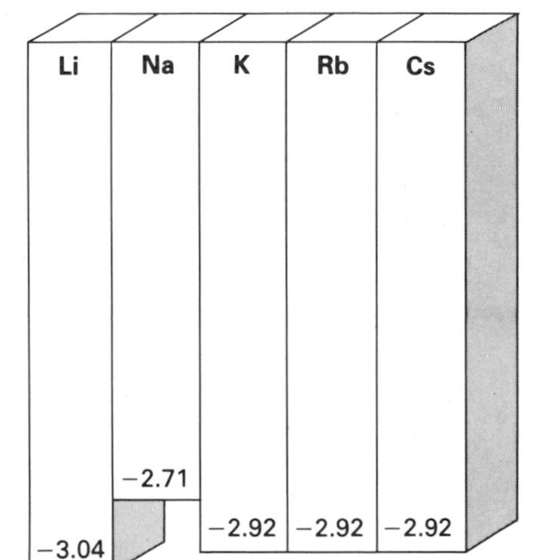

Figure 17.4 The standard electrode potentials of the Group I elements

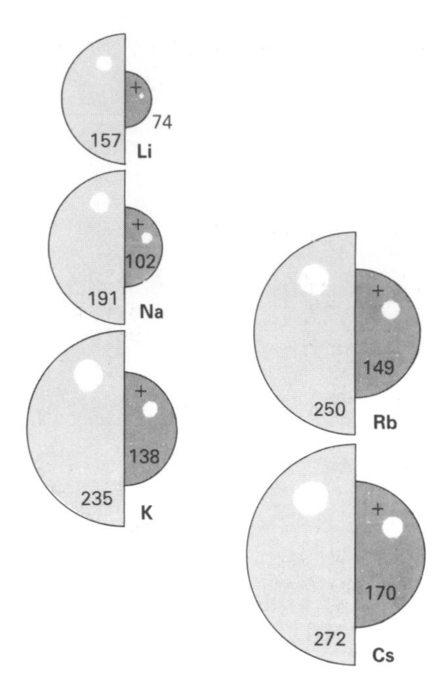

Figure 17.5 The radii (in picometres) of the atoms and M^+ ions of the Group I elements

Shapes and sizes

The sizes of the atoms and cations increase with increasing atomic number (see Figure 17.5). This is partly because more electrons are under the control of the central nucleus and also because they occupy shells with higher principal quantum numbers. The marked reduction in size when the cation is formed from the parent atom is due to the stripping of the single outermost electron from the atom, leaving the inner core.

Ionic radii affect crystal structures through the radius-ratio rules (see Section 5.2). The radius ratio of Na^+ to Cl^- leads to NaCl having the rock-salt structure shown in Figure 17.6(a), with 6:6-coordination, whereas the radius ratio of Cs^+ to Cl^- leads to 8:8-coordination, shown in Figure 17.6(b). The name 'rock-salt structure' is applied to any compound with the structure shown in Figure 17.6(a) even though other elements are involved: the same is true of the caesium chloride structure in Figure 17.6(b).

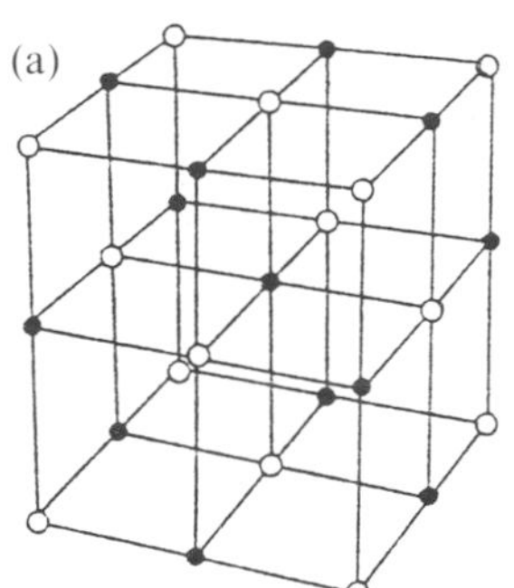

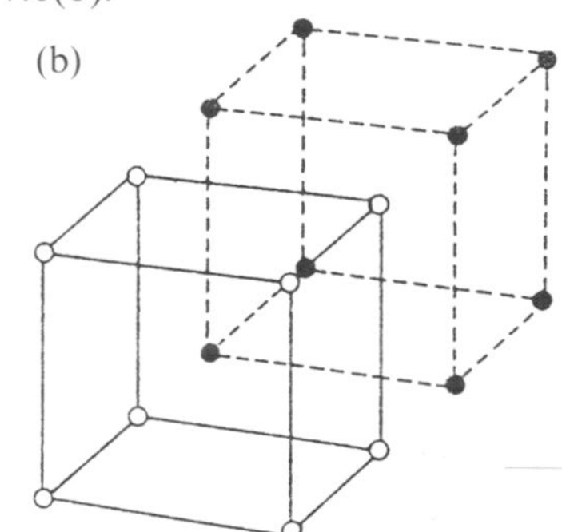

Figure 17.6 (a) The rock-salt structure and (b) the caesium chloride structure

17.2 The elements

The Group I elements are all metals with low melting points, and are soft enough to be cut with a knife. Immediately after cutting the surface is lustrous, but it tarnishes rapidly. The metallic bonding is weak because only one electron is available from each atom. The melting point of sodium (98 °C) is lowered still further by alloying it with potassium, and this liquid alloy is used as a coolant in some nuclear reactors. (From the chemical properties described below, it will become clear that significant engineering problems have to be overcome!) All the metals are less dense than aluminium. This is partly because their crystal structures are body-centred cubic (see Section 5.2), which is not a close-packed structure.

The dominant chemical property of the alkali metals is that they are *strong reducing agents*. This is reflected both in their strongly negative standard electrode potentials and by their ability to reduce oxygen and chlorine:

$$4Li(s) + O_2(g) \rightarrow 2Li_2O(s) \qquad O(0) \rightarrow O(-2), \quad Li(0) \rightarrow Li(+1)$$

$$2Na(s) + Cl_2(g) \rightarrow 2NaCl(s) \qquad Cl(0) \rightarrow Cl(-1), \quad Na(0) \rightarrow Na(+1)$$

The first reaction tarnishes the metals in air, and so they are stored under oil. They cannot be stored under water because they react with it (reduce it) violently, with the evolution of hydrogen:

$$2K(s) + 2H_2O(l) \rightarrow 2KOH(aq) + H_2(g) \qquad H(+1) \rightarrow H(0), \quad K(0) \rightarrow K(+1)$$

The heat of potassium's reaction is enough to ignite the hydrogen. Caesium's reaction is more spectacular: being denser than water it sinks, and the rapid generation of hydrogen under water creates a shock wave that will shatter a glass container (see Figure 17.7). Oxidation of the metal is even more rapid in acid:

$$2Na(s) + 2H^+(aq) \rightarrow 2Na^+(aq) + H_2(g) \qquad H(+1) \rightarrow H(0), \quad Na(0) \rightarrow Na(+1)$$

and even with sodium the energy produced can ignite the hydrogen.

Figure 17.7 Caesium reacts spectacularly with water, and the vessel shatters

The reducing power of the alkali metals is strong enough for them to reduce ammonia and hydrogen:

$$2Na(s) + 2NH_3(g) \xrightarrow{\text{heat}} 2NaNH_2(s) + H_2(g) \qquad H(+1) \rightarrow H(0), \quad Na(0) \rightarrow Na(+1)$$

$$2K(s) + H_2(g) \xrightarrow{\text{heat}} 2KH(s) \qquad H(0) \rightarrow H(-1), \quad K(0) \rightarrow K(+1)$$

$NaNH_2$ is called **sodamide** and KH is **potassium hydride** (see Section 16.3).

The reducing power of sodium is harnessed in industry for the preparation of tetraethyllead(IV), $Pb(CH_2CH_3)_4$, the anti-knock additive for petrol:

$$4Na(s) + Pb(s) + 4CH_3CH_2Cl(g) \xrightarrow{\text{heat}} Pb(CH_2CH_3)_4(l) + 4NaCl(s)$$

Sodium dissolves in liquid ammonia to give a deep-blue solution of sodium cations and **solvated electrons**, e^-(am), where 'am' signifies a solution in liquid ammonia. At higher concentrations the colour of the solution changes to bronze, and it begins to conduct electricity like a metal. The blue solution, which can be thought of as a solution of electrons, is used as a reducing agent.

17.3 Oxides and hydroxides

Oxides

All the alkali metals form ionic solids of composition M_2O, but other compounds with oxygen are also formed when the metals burn in air. The oxides are *basic*, and react with water to form hydroxides:

$$Na_2O(s) + H_2O(l) \rightarrow 2NaOH(aq)$$

Their crystal structures (except that of Cs_2O) resemble the fluorite (CaF_2) structure shown in Figure 5.25, but as the M^+ ions occupy the F^- positions and the O^{2-} ions the Ca^{2+} positions, they are said to have the **antifluorite** structure.

When lithium burns in air it forms mainly Li_2O. When the other metals burn in excess air they form a variety of compounds. Sodium forms mainly the **peroxide**, Na_2O_2, and potassium forms mainly the **superoxide**, KO_2, which is an orange solid. The anions are $O_2{}^{2-}$ and $O_2{}^-$ respectively; both compounds are readily hydrolysed by water:

$$Na_2O_2(s) + 2H_2O(l) \rightarrow 2NaOH(aq) + H_2O_2(aq)$$

$$2KO_2(s) + 2H_2O(l) \rightarrow 2KOH(aq) + H_2O_2(aq) + O_2(g)$$

The products of both reactions include hydrogen peroxide. Potassium superoxide is used in reasonably large quantities for purifying the air in submarines, because it combines with carbon dioxide and simultaneously releases oxygen:

$$4KO_2(s) + 2CO_2(g) \rightarrow 2K_2CO_3(s) + 3O_2(g)$$

The manufacture of potassium superoxide for this purpose is now one of the main uses of potassium.

Hydroxides

The Group I hydroxides are white ionic crystalline solids of formula MOH, and are soluble in water. Except for lithium hydroxide, the solids are **deliquescent**, dissolving in water absorbed from the atmosphere. The aqueous solutions are all strongly alkaline (hence the name 'alkali metals'), and dangerous to handle. For example, the solutions rapidly hydrolyse the protein in the cornea of the eye, and as this tissue does not grow again once damaged a tiny splash can cause permanent injury. Sodium hydroxide ('caustic soda') is used domestically to clear blocked drains, but it should always be used with great caution.

Solutions of alkali metal hydroxides are *strong alkalis*, and neutralize acids to form salts:

$$NaOH(aq) + HCl(aq) \rightarrow NaCl(aq) + H_2O(l)$$

They are also used in the laboratory as a source of hydroxide ions for precipitating other metal hydroxides, for preparing hydroxo-complexes (metal complexes in which OH^- is the ligand), and for hydrolysis reactions in organic chemistry, discussed in Section 31.4:

Precipitation: $Cu^{2+}(aq) + 2OH^-(aq) \rightarrow Cu(OH)_2(s)$

Complexation: $Al^{3+}(aq) + 4OH^-(aq) \rightarrow [Al(OH)_4]^-(aq)$

Hydrolysis: $CH_3COOC_2H_5(aq) + OH^-(aq) \rightarrow CH_3CO_2^-(aq) + C_2H_5OH(aq)$

In each of these reactions *the hydroxide ion is acting as a Lewis base* (see Section 12.4). The hydrolysis of fats to form soap, described in Section 31.6, uses thousands of tonnes of sodium hydroxide each year. Potassium hydroxide is used for long-lasting alkaline batteries because its conductivity is higher than that of the cheaper sodium hydroxide (see Table 8.1); it is also used as the electrolyte in the nickel–iron batteries being developed for electric cars.

17.4 Halides

The alkali metal halides, MHal, are white ionic crystalline solids. All are soluble in water, except lithium fluoride: this exception is largely due to the high lattice enthalpy arising from the strong electrostatic interaction of the small Li^+ and F^- ions. With the exception of the lithium halides, the alkali metal halides are insoluble in organic solvents. Lithium chloride is soluble in ethoxyethane, probably because the high polarizing power of the lithium ion (see Section 15.4) results in some covalent character in the Li—Cl bond. The same high polarizing power also accounts for the formation of a crystalline dihydrate, $LiCl.2H_2O$. The polarizing power of the cations decreases with increasing ionic radius: this reduces their ability to attract water molecules, and neither sodium chloride nor potassium chloride is hydrated.

Caesium chloride, bromide and iodide have the caesium chloride structure (Figure 17.6(b)), and all the other halides have the rock-salt structure (Figure 17.6(a)).

Common salt, sodium chloride, is an essential component of our diet because nerve impulses depend on the transport of sodium ions. As a result it has often been brought under government control: Roman soldiers were given an allowance to buy salt (hence 'salary'), and in some countries the sale of salt is still a government monopoly. Lithium iodide is used in the minute 'lithium batteries' in calculators and heart pace-makers. Lithium batteries are increasingly used for photographic flashes because they can recharge the flash gun more quickly than conventional batteries can.

17.5 Hydrides

The alkali metals are such strong reducing agents that when heated they can reduce hydrogen:

$$2Na(s) + H_2(g) \xrightarrow{\text{heat}} 2NaH(s) \qquad H(0) \rightarrow H(-1),\ Na(0) \rightarrow Na(+1)$$

Sodium hydride is a typical anionic hydride (see Section 16.3). The alkali metal hydrides are themselves strong reducing agents. They are used to prepare the compounds lithium tetrahydridoaluminate, $LiAlH_4$, and sodium tetrahydridoborate, $NaBH_4$, both of which are widely used as reducing agents for industrial and laboratory syntheses (see, for example, Section 30.3):

$$4LiH + AlCl_3 \rightarrow LiAlH_4 + 3LiCl$$

17.6 Other compounds

Aqua cations and oxosalts

Sodium and potassium cations are important to living cells, and compounds of potassium, because of its relative scarcity in freely available form, are added to the soil as 'NPK fertilizer'. In animals, the kidney has the responsibility for regulating the Na^+/K^+ concentration balance, and that in turn controls nerve impulses (see Section 13.4).

Very many oxosalts of the alkali metals are known. They are all essentially ionic, and almost all of them are soluble in water. They are reasonably thermally stable (more so than the corresponding Group II oxosalts, discussed in Section 18.6), but lithium oxosalts are out of line and decompose at lower temperatures than the rest. This is mainly because the small lithium ion stabilizes the oxide.

The alkali metal **carbonates** dissolve in water to give an alkaline solution as a result of salt hydrolysis (see Section 12.3):

$$CO_3{}^{2-}(aq) + H_2O(l) \rightleftharpoons OH^-(aq) + HOCO_2{}^-(aq)$$

On account of this reaction sodium carbonate is often used as an alkali in titrations, in the paper and textile industries, and in the home as a detergent (washing soda is $Na_2CO_3.10H_2O$). Huge quantities of sodium carbonate are used in the manufacture of glass (see Section 20.7). Lithium carbonate is used in the treatment of manic depression: the mode of action is not certain, but it may be due to the regulatory effect of the lithium ion on the Na^+/K^+ concentration balance in nerve cells.

Solid **hydrogencarbonates** are known only for the Group I elements, though not for lithium. Like the carbonates, they are hydrolysed in solution:

$$HOCO_2{}^-(aq) + H_2O(l) \rightleftharpoons OH^-(aq) + (HO)_2CO(aq)$$

Sodium hydrogencarbonate ('sodium bicarbonate', or 'baking soda') decomposes on heating:

$$2NaHOCO_2(s) \xrightarrow{\text{heat}} Na_2CO_3(s) + H_2O(g) + CO_2(g)$$

The release of carbon dioxide is responsible for its use as 'baking soda'. Baking powder contains sodium hydrogencarbonate together with a weak acid, such as 2,3-dihydroxybutanedioic (tartaric) acid; the compounds react to release carbon dioxide, which causes cake or scone mixtures to rise.

The **nitrates** of the alkali metals (except that of lithium) decompose on heating, with the formation of the nitrite and oxygen:

$$2NaNO_3(s) \rightarrow 2NaNO_2(s) + O_2(g) \qquad N(+5) \rightarrow N(+3), \quad O(-2) \rightarrow O(0)$$

This is in contrast to the behaviour of the Group II nitrates (and lithium nitrate), which decompose to the oxide.

Detection

Almost all the compounds of sodium and potassium are soluble in water, and so there are no simple precipitation tests for these elements (but some complex salts are insoluble and can be used). The most common qualitative test relies on the characteristic colour they impart to a flame (see Section 1.3):

lithium: crimson sodium: yellow potassium: lilac

Rubidium (red) and caesium (blue) were first identified in this way (by Bunsen and Kirchhoff) and are named after their flame colours. The flame test for potassium requires care, because the delicate lilac colour is often masked by the brilliant yellow of any sodium impurity. That is why the flame is viewed through cobalt blue glass, which masks (by absorbing) the yellow light but not the lilac. As little as 10^{-9} g of a sodium compound may be detected by the flame test.

17.7 Industrial chemistry

Occurrence

Sodium is the sixth most abundant element in the Earth's crust, and potassium is the seventh. Sodium occurs mainly in sea water and as solid sodium chloride in dried-up sea beds, such as those in Cheshire and Utah (see Figure 17.8). Potassium is more widely distributed, the main commercial sources being **sylvite**, KCl, and **carnallite**, $KCl.MgCl_2.6H_2O$, but it is also extracted from sea water. Sodium is some thirty times more abundant than potassium in sea water, even though the two elements occur in similar abundances in the Earth's crust, partly because the form in which potassium is combined in rocks is less soluble than that of sodium.

Figure 17.8 The Great Salt Desert, Utah, USA

Figure 17.9 The universal time-keeping standard is now the caesium clock

The production of sodium

Because the alkali metals have such negative standard electrode potentials, they can be isolated only by electrolysis of their molten salts. Both sodium and potassium were first isolated in this way by Davy (see Box 17.1). Sodium chloride melts at 800 °C, an uneconomically high temperature, but a eutectic mixture (see Section 7.1) with 60 per cent calcium chloride melts at under 600 °C and is used commercially. The apparatus used, a **Downs cell**, is shown in Figure 17.10: the anode is graphite, which resists attack by chlorine, and the cathode is steel, the cell being designed to keep the chlorine and sodium apart. When a current of about 25 000 A is passed through the cell, the following reactions occur:

At the cathode: $Na^+(l) + e^- \rightarrow Na(l)$ $\quad$ $Na(+1) \rightarrow Na(0)$

At the anode: $2Cl^-(l) \rightarrow Cl_2(g) + 2e^-$ $\quad$ $Cl(-1) \rightarrow Cl(0)$

Notice that sodium is discharged in preference to calcium and that, as in all the best industrial processes, the by-product – chlorine – is also valuable (see Section 23.7).

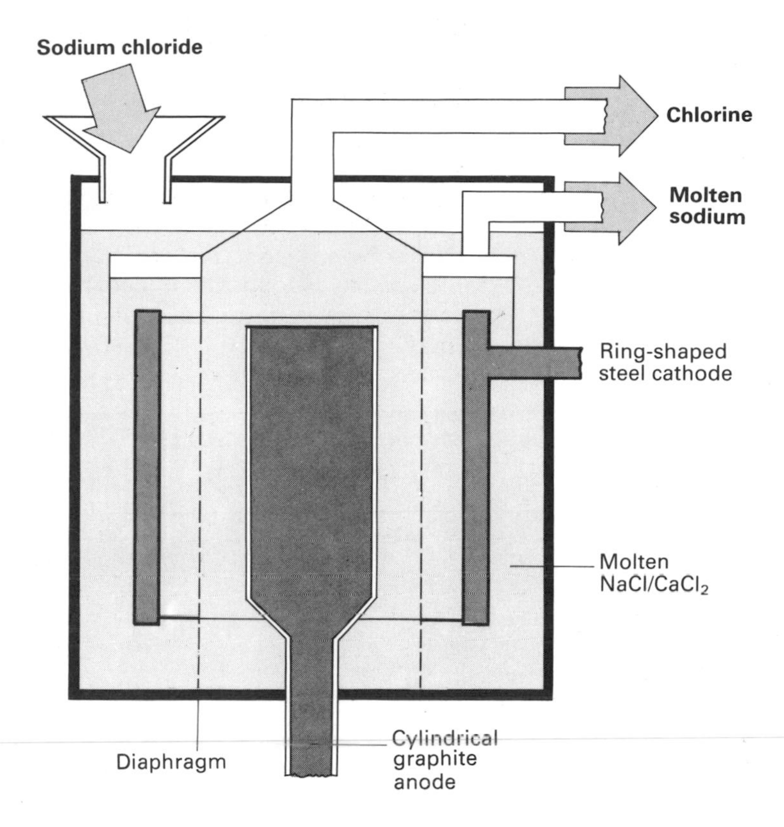

Figure 17.10 A Downs cell for the manufacture of sodium

BOX 17.1

Davy and the discovery of the alkali metals

Within a period of a few days in 1807, Humphry Davy isolated two new elements, potassium and sodium, by passing an electric current through molten caustic potash (KOH) and caustic soda (NaOH) respectively. This was a remarkable experimental achievement because he needed to develop techniques to handle the corrosive melts and the reactive elements they produced; moreover, the idea of using electricity to produce chemical change was new.

It has been said of Davy, perhaps unkindly, that his life consisted of brilliant fragments. His numerous inventions included the miners' safety lamp, and his discoveries included the demonstration that chlorine is an element and the discovery of seven elements (more than anyone else), all by electrolysis. He is also responsible for recognizing the genius of Michael Faraday, whom he appointed his assistant. Davy was no theorist, however, and was reluctant to accept Dalton's atomic theory.

The production of sodium hydroxide

Sodium hydroxide is produced by the electrolysis of saturated **brine** (aqueous sodium chloride) in a cell with steel cathodes and titanium anodes. The reactions are

At the cathode: $2H_2O(l) + 2e^- \rightarrow 2OH^-(aq) + H_2(g)$ $\quad$ $H(+1) \rightarrow H(0)$

At the anode: $2Cl^-(aq) \rightarrow Cl_2(g) + 2e^-$ $\quad$ $Cl(-1) \rightarrow Cl(0)$

The **diaphragm cell** is shown in Figure 17.11; it is designed with a porous asbestos diaphragm, so that the chlorine does not mix with the hydrogen or the sodium hydroxide. The electrolyte is drawn off the cell, and on evaporation the less soluble unchanged sodium chloride crystallizes, leaving a solution of sodium hydroxide.

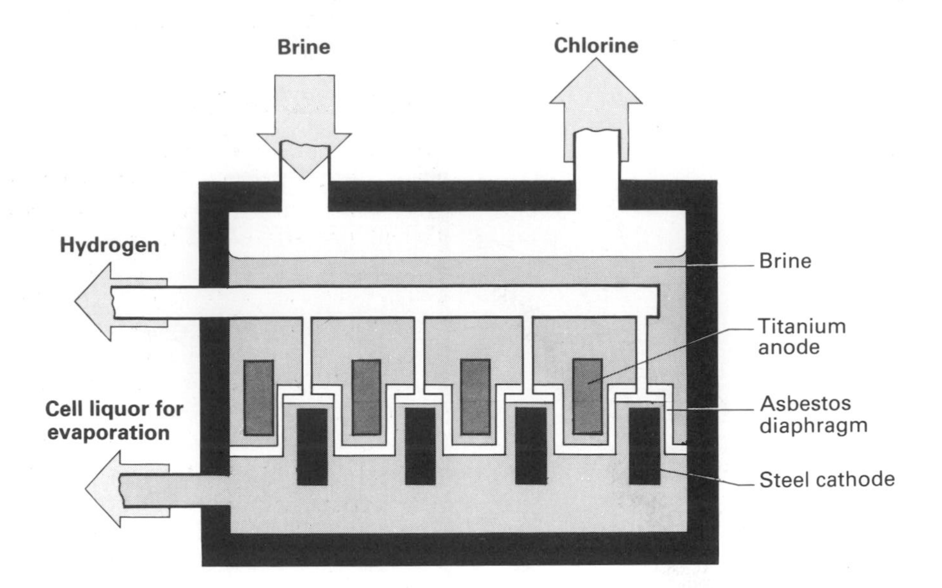

Figure 17.11 The electrolysis of brine in a diaphragm cell for the manufacture of chlorine and sodium hydroxide

An alternative process uses a **Castner–Kellner cell**, shown diagrammatically in Figure 17.12, in which the cathode is a stream of mercury. For reasons connected with the mechanism of the process (as distinct from its thermodynamics) sodium metal is formed at the cathode more rapidly than hydrogen is released, and dissolves in the mercury to form an amalgam. Sodium hydroxide is formed when this amalgam is treated with water over graphite blocks in another unit, the **decomposer**, and the mercury is recycled. Many Castner–Kellner cells are still in operation, but their numbers are declining on account of the environmental hazards of mercury (see Box 17.2), for some always manages to escape during the recycling operations.

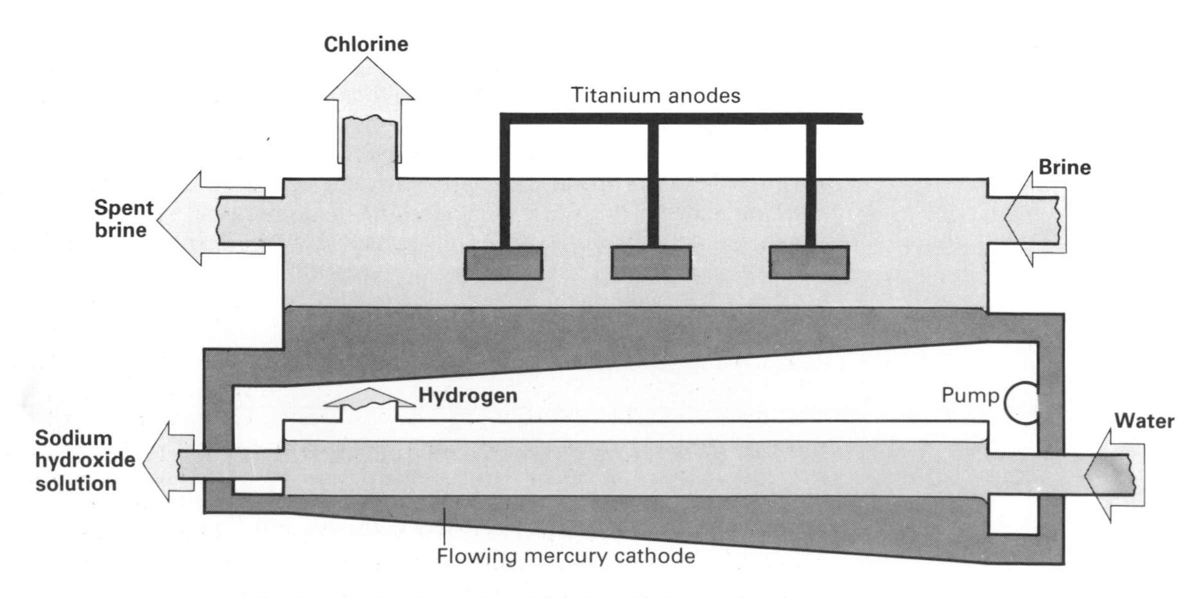

Figure 17.12 The manufacture of sodium hydroxide using a Castner–Kellner cell

BOX 17.2

The Minamata tragedy

The factory town of Minamata in Japan made poly(chloroethene) (PVC) and other chemicals. Chlorine for PVC was made by the electrolysis of brine using a flowing mercury cathode (the Castner–Kellner process). The effluent was discharged into the sea.

The first sign (in the 1950s) that something was wrong was the appearance of a strange illness affecting cats. Their muscles weakened, they became unable to walk and finally they became paralysed, went into a coma, and died. This should have alerted the local health authorities, but nothing was done until the illness began to affect humans, and people started to die and babies were born with deformities.

An investigation showed that mercury compounds were in the effluent from the factory and that these concentrated in fish, the staple diet of the local people (and the cats). As a result of this tragedy, there are now very strict rules worldwide about discharging effluent into natural waters, particularly if there is any chance that heavy metal ions might be present in the effluent.

The production of sodium carbonate

Sodium carbonate, **soda ash**, is made by the **Solvay process** (which is also known as the **ammonia–soda process**). This scheme makes cunning use of the different solubilities of the various substances involved.

There are two linked sequences of reactions. In the **sodium stream** carbon dioxide is passed into ammoniacal brine:

$$CO_2(g) + NH_3(aq) + NaCl(aq) + H_2O(l) \rightarrow NaHOCO_2(s) + NH_4Cl(aq)$$

The sodium hydrogencarbonate thus formed is insoluble under these conditions; it is filtered off, and then heated in rotating **calciners**:

$$2NaHOCO_2(s) \xrightarrow{\text{heat}} Na_2CO_3(s) + H_2O(g) + CO_2(g)$$

This gives the desired product, sodium carbonate, and the carbon dioxide is recycled. However, as 2 mol $NaHOCO_2$ generates only 1 mol CO_2, more carbon dioxide must be supplied in order for the first reaction to proceed. That carbon dioxide comes from the **calcium stream**, where calcium carbonate is thermally decomposed into calcium oxide and carbon dioxide. The calcium oxide is slaked with water to form calcium hydroxide, and that alkali is used to regenerate ammonia from the ammonium chloride produced in the first reaction:

$$Ca(OH)_2(aq) + 2NH_4Cl(aq) \rightarrow 2NH_3(g) + CaCl_2(aq) + 2H_2O(l)$$

The importance of the product and the ingenuity of the process made Ernest Solvay rich enough to found a series of scientific conferences which bear his name. The cost of energy for this energy-intensive process has risen, however, as has the cost of ammonia (which has to be supplied to make good the losses that occur during the cycle); moreover, there are few uses for the principal by-product, calcium chloride, the dumping of which into rivers causes environmental concern. As a result the Solvay process is being displaced, and sodium carbonate obtained by heating the mineral **trona**, $Na_2CO_3.NaHOCO_2.2H_2O$, is becoming increasingly important commercially.

Summary

- The alkali metals have the electron configuration [Noble gas]ns^1.
- Their only oxidation number other than zero is +1.
- All common compounds of the alkali metals contain either M^+ or $M^+(aq)$ ions, and are water-soluble.
- The ionic radii increase with increasing atomic number.
- The metals are strong reducing agents, reducing water to hydrogen, and hydrogen to the hydride ion.
- The oxygen compounds are the oxides (M_2O), peroxides (M_2O_2) and superoxides (MO_2): the oxides are basic.
- The alkali metal hydroxides are alkaline.
- Depending on the cation:anion radius ratio, the halides adopt either the rock-salt or the caesium chloride structure.
- Lithium has a diagonal relationship with magnesium: for example, their carbonates decompose relatively easily.
- Sodium metal is obtained by electrolysis of molten sodium chloride in a Downs cell; sodium hydroxide is obtained by electrolysis of brine in a Castner–Kellner cell or a diaphragm cell.
- Sodium carbonate is produced by the Solvay process or by heating trona.

PROBLEMS

1 Predict the following properties of francium ($Z = 87$) and its compounds:
(a) the first ionization energy (to the nearest 25 kJ mol^{-1});
(b) the reaction of the metal with water;
(c) the melting point of FrCl (to the nearest 50 °C);
(d) the lattice enthalpy of FrF (to the nearest 25 kJ mol^{-1});
(e) the reaction of the metal with hydrogen;
(f) the products of thermal decomposition of $FrNO_3$.

2 Look up data to complete the table below of lattice enthalpies (in kJ mol^{-1}) of the alkali metal halides.

Cation \ Anion	F^-	Cl^-	Br^-	I^-
Li^+	1031 6:6			
Na^+				
K^+				
Rb^+				
Cs^+				613 8:8

Look up the crystal structures of these compounds which will either have the rock-salt structure (6:6 coordination) or the caesium chloride structure (8:8 coordination). Enter the structures in the table using the symbols 6:6 or 8:8.
(a) What trends do you notice in the values of the lattice enthalpies? Explain them.
(b) Explain the changes in crystal structure.

3 Give an account of the thermal stability of the carbonates and nitrates of the elements lithium, sodium and potassium.

4 Write an essay to illustrate how ionization energies help to explain the chemistry of the Group I elements.

5 Hydrogen is evolved when water reacts with a white solid **A** to produce a colourless solution having a pH 14. A flame test on the solid **A** gives a lilac flame. When **A** is allowed to stand in air for some hours a damp impure solid **B** is produced. When **B** is treated with dilute hydrochloric acid a gas **C** is evolved.
(a) What is **A**?
(b) What volume of hydrogen at 20 °C and 1 atm will be produced when 0.200 g of **A** are treated with water?
(c) What compounds will be in the impure solid **B**?
(d) What is the gas **C**?
(e) What gas would have been evolved when water was added if compound **A** had been (i) KO_2, and (ii) KNH_2?

6 Values of the lattice enthalpy can be determined either by using the Born–Haber cycle with experimental data or by theoretical calculation using ionic radii and charges. For sodium chloride the values (in kJ mol^{-1}) are 787 and 769 respectively and are in close agreement; but for silver chloride the values are 905 and 833. Explain.

7 (a) Describe the preparation of sodium hydroxide by electrolysis of brine.
(b) A plant produces 310 tonnes of sodium hydroxide per day.
(i) What mass of chlorine does it produce each day?
(ii) What is the minimum quantity of electricity consumed each day?
(Faraday's constant is 96 500 C mol^{-1}.)

8 Complete the following table to illustrate the diagonal relationship between lithium and magnesium.

	Lithium	Sodium	Magnesium
Reaction of metal with oxygen			
Solubility of fluoride			
Solubility of salts in organic solvents			
Action of heat on the carbonate			
Action of heat on the nitrate			

9 The s-block of the Periodic Table contains Group IA, the alkali metals, and Group IIA, the alkaline earths. Give an account of these two Groups of elements and their compounds, paying particular attention to similarities and differences within the Groups. *(London)*

10 Discuss the structure of sodium chloride. Your answer should consider each of the following:
(a) how the structure may be determined,
(b) a description of the crystal structure,
(c) energetic aspects, and
(d) the relation between structure and properties.

The value of the Avogadro constant is $6.022 \times 10^{23}\,mol^{-1}$. Use this, taking any other data you require from a book of data, to calculate the internuclear separation of adjacent ions in sodium chloride. Compare this value with that obtained using ionic radii ($Na^+ = 0.102$ nm, $Cl^- = 0.180$ nm) and comment on any discrepancy you may find. *(Nuffield S)*

11 Suggest explanations for each of the following statements which refer to Group IA, the alkali metals.
(a) Only lithium reacts with nitrogen to form a nitride.
(b) Although the standard electrode potentials become more negative from sodium to caesium, the standard electrode potential of lithium is the most negative in the Group.
(c) On heating, lithium nitrate forms an oxide, whereas the nitrates of the other metals in the Group form nitrites.
(d) Lithium forms only one oxide, whereas the other alkali metals each form at least two oxides.
(e) Some of the compounds of lithium have a partially covalent character.
(f) Although the ions Na^+, Mg^{2+} and Al^{3+} have the same electronic configuration, they have different ionic radii. *(London)*

12 (a) Give an account of the industrial preparation of sodium carbonate by the Solvay (ammonia–soda) process. Your answer should concentrate on the chemical and physico-chemical principles involved. Details of industrial plant are not required.
(b) Give two industrial uses of sodium carbonate.
(c) What is the action of heat on (i) sodium hydrogensulphate, (ii) potassium nitrate?
(d) Explain what happens when hydrogen chloride gas is passed into a concentrated aqueous solution of sodium chloride. *(Oxford)*

13 What patterns exist in the solubilities in water of the compounds of the metals in the s-block of the Periodic Table?

To what extent is it possible to explain these patterns in terms of ionic size? Suitable data may be found in books of data. *(Nuffield)*

14 Compare and contrast the properties of the salts of lithium with those of the salts of potassium. What explanations can be given for the observed differences? *(Oxford part question)*

15 (a) Give an account of a preparation of (i) sodium hydroxide and (ii) sodium carbonate from sodium chloride.
(b) Suggest methods for the detection of (i) sodium chloride and sodium carbonate as impurities in a sample of sodium hydroxide, (ii) sodium hydroxide as impurity in a sample of sodium carbonate.
(c) How do you account for the fact that many metal hydroxides are much more soluble in aqueous hydroxide than in aqueous sodium carbonate? *(Cambridge Entrance)*

16 Explain what is meant by the *lattice energy* of an alkali metal halide MX, and show by means of a thermochemical cycle how it is related to the enthalpy of formation of MX and appropriate properties of M, M^+, X_2, X and X^-.

The lattice energies of sodium fluoride and sodium chloride are 910 and 770 $kJ\,mol^{-1}$ respectively, and the solubilities of both salts in water are almost independent of temperature. What can you deduce about the relative magnitudes of the enthalpy changes for the process

$$X^-(g) + aq \rightarrow X^-(aq)$$

for X = F and X = Cl? *(Cambridge Entrance)*

18

The s-block elements: Group II

The members of Group II are called the alkaline earth metals. In all their compounds they have the oxidation number +2 and, with a few exceptions (among the compounds of beryllium and magnesium), their compounds are principally ionic. Apart from beryllium, the elements are strong reducing agents and are typical metals, with basic oxides. The solubilities and thermal stabilities of the oxosalts depend on the size of the cation. Calcium carbonate is widespread in nature, and widely used in the manufacture of cement, glass and iron.

	I	II	
2	3 Li	4 Be	
3	11 Na	12 Mg	
4	19 K	20 Ca	21 Sc
5	37 Rb	38 Sr	39 Y
6	55 Cs	56 Ba	57 La
7	87 Fr	88 Ra	89 Ac

Figure 18.1 Group II in the Periodic Table

The elements in Group II are as follows:

beryllium	Be	$[He]2s^2$
magnesium	Mg	$[Ne]3s^2$
calcium	Ca	$[Ar]4s^2$
strontium	Sr	$[Kr]5s^2$
barium	Ba	$[Xe]6s^2$
radium	Ra	$[Rn]7s^2$

Their positions in the Periodic Table are shown in Figure 18.1. Their characteristic valence electron configuration is ns^2, where n is the Period number. The elements calcium, strontium and barium are called the **alkaline earth metals** because their hydroxides are alkaline; however, the name is often applied loosely to all the members of the Group.

All the isotopes of radium are radioactive, and the discovery and eventual isolation of the element by Marie Curie is an excellent example of scientific detective work, patience and determination (see Box 18.1).

BOX 18.1

Figure 18.2 Marie Curie in her laboratory in Paris

Marie Curie and radium

Marie Sklodowska, who was born in Poland, worked for several years as a governess in order to send her elder sister to medical school in France; after qualifying, her sister invited Marie to join her in Paris to study. There she obtained the highest marks in physics at the Sorbonne, and married Pierre Curie, who was already established in research.

Shortly after Henri Becquerel had discovered radioactivity in 1896, it was realized that one uranium ore, **pitchblende**, had a higher activity than pure uranium compounds. This suggested to Marie Curie that there must be a substance in the ore more radioactive than uranium itself, and she set herself the task of identifying and isolating it. With help from Pierre, she treated tonnes of pitchblende, all the while following ('tracing') the substance they were looking for by its intense radioactivity. Eventually, by painstaking and repeated fractional crystallization, they obtained minute amounts of the pure chloride of a new element which they named **radium**. This was a remarkable achievement, for the mass ratio of radium to uranium in pitchblende is only $1:3 \times 10^6$; hence it was necessary to process about ten tonnes of ore to obtain one gram of radium. The Curies shared the 1903 Nobel prize for physics with Becquerel for their work on radioactivity.

Pierre was killed in a road accident in 1906 (he was run over by a horse-drawn carriage) and Marie took over his post, becoming the first woman to be professor of physics at the Sorbonne. She later won the 1911 Nobel prize for chemistry for her discovery of radium and polonium (which she named after her native country). Her daughter Irène added to the family honours by sharing the 1935 Nobel prize for chemistry with her husband Frédéric Joliot.

18.1 Group systematics

Oxidation states and bonding characteristics

In all their compounds the elements of Group II have the oxidation number +2. Nearly all the compounds are based on M^{2+} or $M^{2+}(aq)$ ions, the only exceptions being a few compounds of beryllium and magnesium.

The widespread occurrence of compounds containing M^{2+} ions, and the absence of compounds containing M^{+} ions, can be rationalized in terms of the formation enthalpies of each type of compound since, as in most arguments relating to ionic lattice stability, the enthalpy changes concerned are so large that they dominate any entropy differences (see Section 17.1).

A Born–Haber cycle (see Section 9.4) for the standard formation enthalpy of $CaCl_2(s)$ is drawn in Figure 18.3(a). It shows that the considerable endothermic contribution from the double ionization of Ca to Ca^{2+} is more than compensated by the very large lattice enthalpy (which is calculated using the known crystal structure of the solid). The Born–Haber cycle for the standard formation enthalpy of the hypothetical compound CaCl(s) is drawn in Figure 18.3(b). It shows that although the energy investment for ionization is much less (because only one electron must be removed), the lattice enthalpy – calculated on the assumption that CaCl would have a rock-salt structure (see Section 17.1) – is very much lower: this is because Ca^{+} is larger than Ca^{2+} and has a smaller charge. Because $CaCl_2(s)$ has a much more exothermic formation enthalpy it is by far the more stable species, and in fact is the only known chloride of calcium.

The third ionization energies of magnesium and calcium, corresponding to the removal of an electron from an inner shell, are so large (several thousand kilojoules per mole) that the energy investment cannot be recovered. Hence these elements do not occur with oxidation number +3.

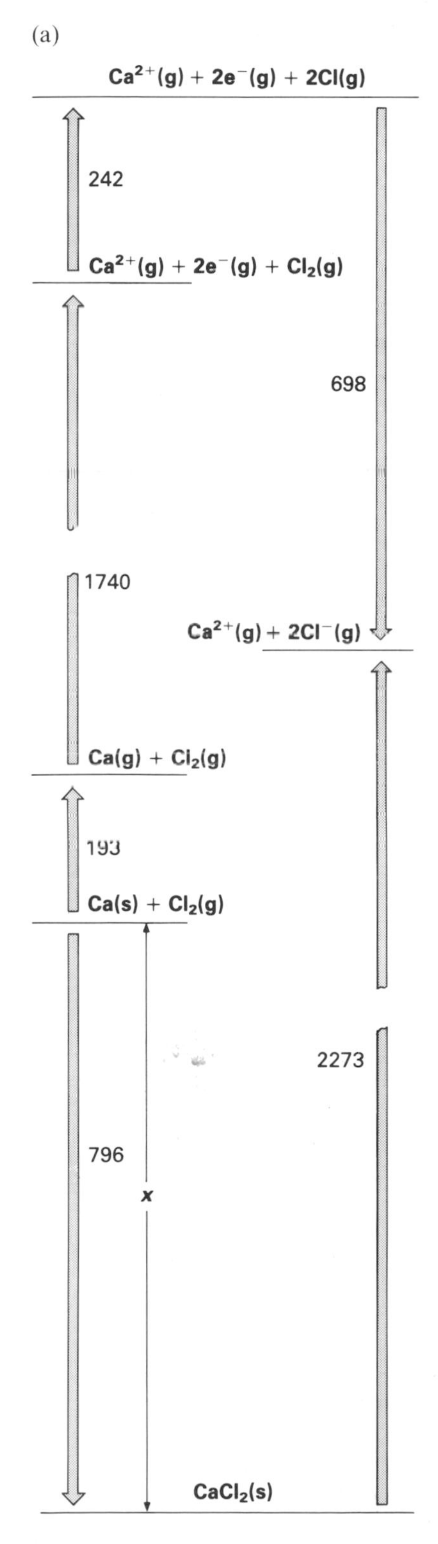

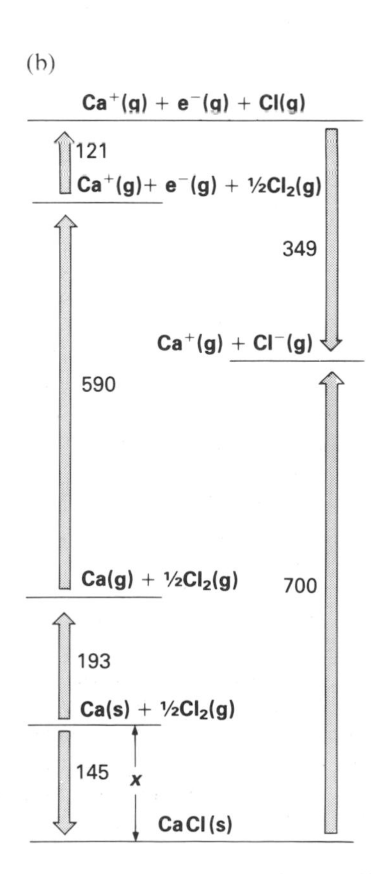

Figure 18.3 The Born–Haber cycle for (a) $CaCl_2$ and (b) the hypothetical CaCl

The sum of the first and second ionization energies *decreases* with increasing atomic number (see Figure 18.4), because the nucleus has less control over the valence electrons in the heavier elements. The high values for beryllium and magnesium account for their tendency to form compounds with significant covalent character; lower down the Group, however, the sum of the ionization energies is small enough for the elements to form predominantly ionic compounds.

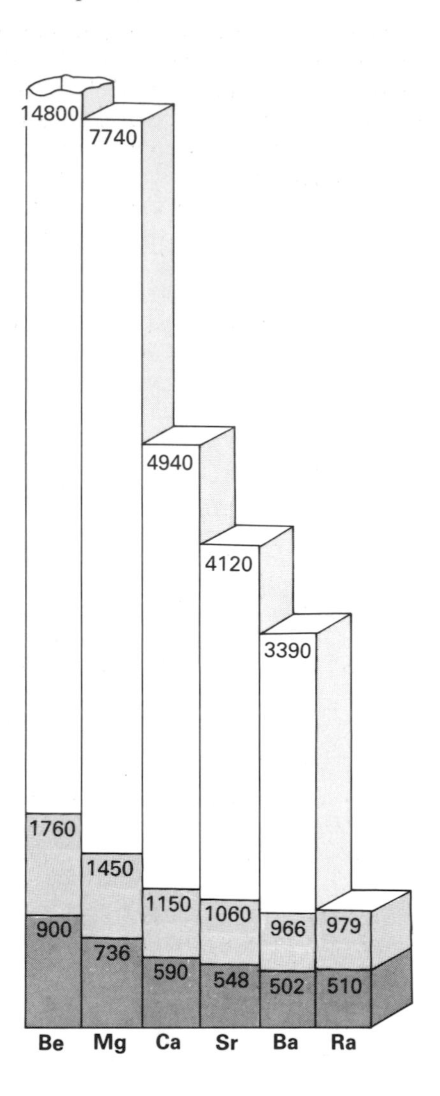

Figure 18.4 The first, second and third ionization energies of the Group II elements

Periodic trends

The ability of the metals to act as reducing agents dominates their chemical properties. This ability is reflected clearly in the standard electrode potentials, which show that the elements become increasingly electropositive ($E^\ominus$ more negative) on descending the Group (see Figure 18.5). Only beryllium ($E^\ominus = -1.85$ V) is not strongly electropositive.

Although we shall not deal in detail with the chemistry of beryllium, it is worth noting that it is markedly different from the other members of the Group, having a strong diagonal relationship within the Periodic Table with aluminium. These differences stem from the small size, and hence the high polarizing power, of the Be^{2+} ion. This strong polarizing power accounts for beryllium's ability to reduce alkaline solutions, such as aqueous sodium hydroxide:

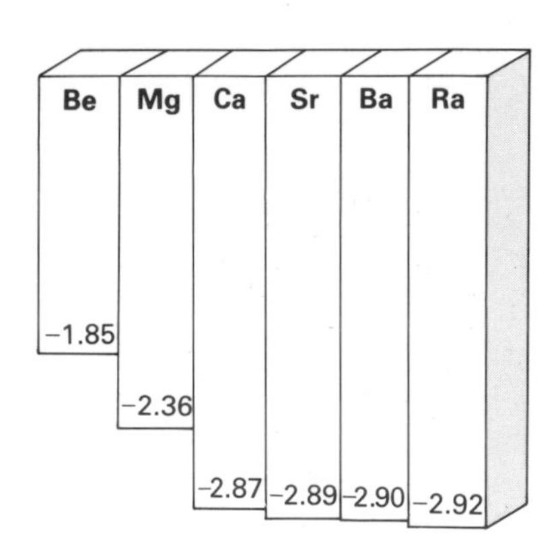

Figure 18.5 The standard electrode potentials of the Group II elements

$$Be(s) + 2OH^-(aq) + 2H_2O(l) \rightarrow [Be(OH)_4]^{2-}(aq) + H_2(g)$$

$$H(+1) \rightarrow H(0), \quad Be(0) \rightarrow Be(+2)$$

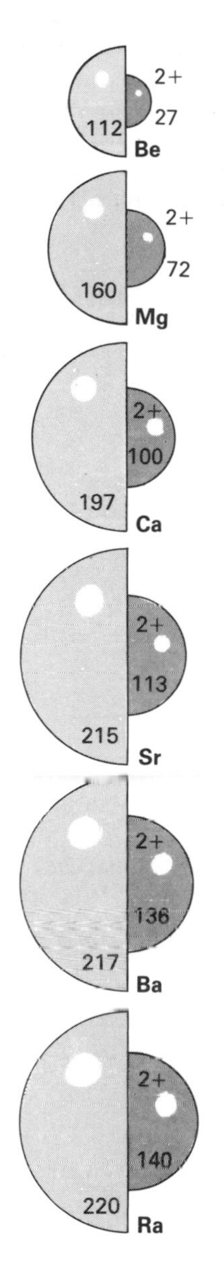

Figure 18.6 The radii (in picometres) of the atoms and M^{2+} ions of the Group II elements

A driving force for this reaction, which produces the **tetrahydroxoberyllate ion**, is the great strength of the bonds between the Be^{2+} ion and the surrounding OH^- ions. Beryllium oxide and hydroxide are amphoteric, for similar reasons. Beryllium is highly toxic, probably because it can replace Mg^{2+} ions in some enzymes.

Shapes and sizes

The atomic and ionic radii of the elements increase smoothly down the Group, as shown in Figure 18.6. The increase is due to the increasing number of electrons under the control of the central nucleus, and the need for them to occupy shells of higher principal quantum number. Notice that the cations are significantly smaller than the parent atoms, principally because the outermost shell of electrons has been lost.

The sizes of the cations affect the crystal structures of their compounds. The radius ratio (see Section 5.2) of Mg^{2+} to F^- is 0.54, whereas the ratio for Ca^{2+} to F^- is 0.75. Magnesium fluoride has the **rutile structure** (the same structure as rutile, TiO_2, itself – see Figure 18.7(a)) with 6:3-coordination; calcium fluoride gives its name to the fluorite structure, shown in Figure 18.7(b), with 8:4-coordination.

18.2 The elements

The elements of Group II are all metals. They are harder and denser than sodium and potassium, and they have higher melting points. All these properties are due largely to the presence of two valence electrons on each atom instead of one, which leads to stronger metallic bonding.

Like those of the members of Group I, the chemical properties of the Group II elements are dominated by the *strong reducing power* of the metals. Once started, the reactions with oxygen and chlorine are vigorous:

$$2Mg(s) + O_2(g) \rightarrow 2MgO(s) \qquad O(0) \rightarrow O(-2),\quad Mg(0) \rightarrow Mg(+2)$$

$$Ca(s) + Cl_2(g) \rightarrow CaCl_2(s) \qquad Cl(0) \rightarrow Cl(-1),\quad Ca(0) \rightarrow Ca(+2)$$

Magnesium burns with a bright white light, and was once widely used for photographic flash bulbs; it is still used in flares, incendiary bombs and tracer bullets. All the metals except beryllium form oxides in air at room temperature. The oxide merely dulls the surface of magnesium and calcium, but the reactivity of barium is so great that it must be stored under oil.

All the metals except beryllium reduce water and dilute acids to hydrogen:

$$Mg(s) + H_2O(g) \xrightarrow{\text{steam}} MgO(s) + H_2(g) \qquad H(+1) \rightarrow H(0),\quad Mg(0) \rightarrow Mg(+2)$$

$$Mg(s) + 2H^+(aq) \rightarrow Mg^{2+}(aq) + H_2(g) \qquad H(+1) \rightarrow H(0),\quad Mg(0) \rightarrow Mg(+2)$$

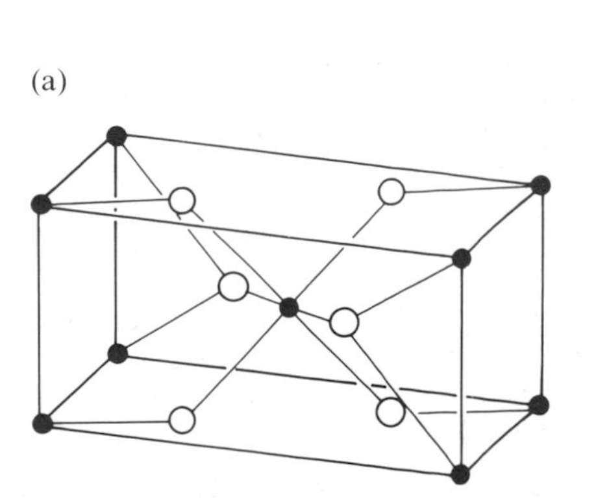

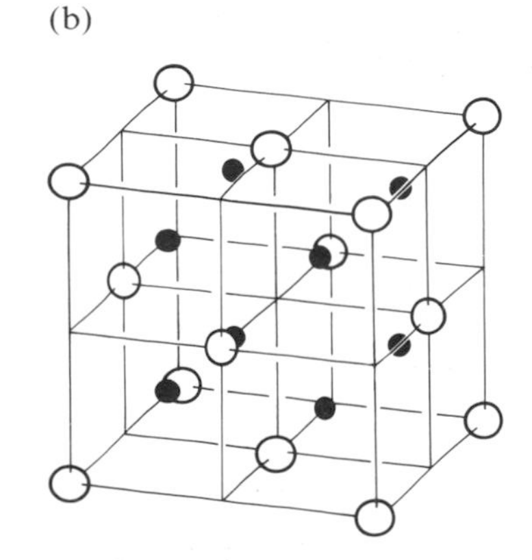

Figure 18.7 (a) The rutile structure and (b) the fluorite structure

Magnesium reacts only slowly with water unless the water is boiling, but calcium reacts readily, even at room temperature, and forms a cloudy white suspension of the sparingly soluble hydroxide:

$$Ca(s) + 2H_2O(l) \rightarrow Ca(OH)_2(s) + H_2(g) \qquad H(+1) \rightarrow H(0), \quad Ca(0) \rightarrow Ca(+2)$$

Calcium (and strontium and barium) can reduce hydrogen gas when heated, forming the hydride:

$$Ca(s) + H_2(g) \xrightarrow{\text{heat}} CaH_2(s) \qquad H(0) \rightarrow H(-1), \quad Ca(0) \rightarrow Ca(+2)$$

The hot metals are also strong enough reducing agents to reduce nitrogen gas and form nitrides:

$$3Mg(s) + N_2(g) \xrightarrow{\text{heat}} Mg_3N_2(s) \qquad N(0) \rightarrow N(-3), \quad Mg(0) \rightarrow Mg(+2)$$

This is an exothermic reaction because, although a large energy investment is required in order to make Mg^{2+} and N^{3-} ions, their electrostatic interaction in the solid is so strong that the lattice enthalpy is very large. The reaction of magnesium with nitrogen is one reason why the metal burns so vigorously in air – it combines with the nitrogen as well as the oxygen. (This is not a practicable way of trapping atmospheric nitrogen to produce fertilizer, however, because the magnesium is so expensive, even if it is recycled.)

Magnesium can reduce, and burn in, carbon dioxide:

$$2Mg(s) + CO_2(g) \xrightarrow{\text{heat}} 2MgO(s) + C(s) \qquad C(+4) \rightarrow C(0), \quad Mg(0) \rightarrow Mg(+2)$$

This means that magnesium fires cannot be put out using carbon dioxide fire extinguishers!

Magnesium is the only element of this Group that is widely used in industry: the other metals are too soft, too reactive or toxic. It is often alloyed with aluminium where low-density, high-strength construction is needed, as in airframes. Magnesium's strong reducing power is used for the cathodic protection of steel objects (see Section 13.5) such as bridges, hulls and outboard motors, and for the extraction of less electropositive metals, as in the **Kroll process** for titanium, which takes place in an argon atmosphere:

$$2Mg(l) + TiCl_4(g) \xrightarrow{1000\,^{\circ}C} Ti(s) + 2MgCl_2(l) \qquad Ti(+4) \rightarrow Ti(0), \quad Mg(0) \rightarrow Mg(+2)$$

18.3 Oxides and hydroxides

Oxides

There are two important series of oxygen compounds of the Group II elements: one contains the oxide ion, O^{2-}, and the other the peroxide ion, $O_2{}^{2-}$ (see Section 22.5).

Barium peroxide, BaO_2, is formed when barium burns in excess air. Peroxides, MO_2, are also known for all the other elements except beryllium. The non-existence of beryllium peroxide can be traced to the small size of Be^{2+} (compare the way in which the low thermal stability of lithium carbonate can be ascribed to the small size of Li^+, see Section 17.1): the lattice enthalpy of BeO is particularly high.

The oxides of the metals, MO, are all white solids. They are normally prepared by heating the carbonates or hydroxides:

$$MgCO_3(s) \xrightarrow{\text{heat}} MgO(s) + CO_2(g)$$

$$Ca(OH)_2(s) \xrightarrow{\text{heat}} CaO(s) + H_2O(g)$$

Apart from beryllium oxide, the oxides have the rock-salt structure shown in Figure 17.6(a), and the high charge of the metal cation results in a high lattice enthalpy and a high melting point. Magnesium oxide has such a high melting point (over 2800 °C) that it is used as a **refractory material** (one that is resistant to heat) in furnace linings. Beryllium oxide is used in rocket nozzles.

Calcium oxide (**quicklime**) glows with a bright white light when strongly heated and was once used for stage lighting: this is the origin of the term 'limelight'.

With the exception of BeO, which is amphoteric, the oxides are basic. This has an important application in iron-making, described in Section 25.9, because calcium oxide, which is formed by heating calcium carbonate (**limestone**), reacts with acidic impurities in the ore, such as silicates, and forms a molten slag:

$$CaO(s) + SiO_2(s) \xrightarrow{900\,^{\circ}C} CaSiO_3(l)$$

Hydroxides

Calcium, strontium and barium oxides react with water to form hydroxides:

$$CaO(s) + H_2O(l) \rightarrow Ca(OH)_2(s)$$

A lump of calcium oxide swells, steams and disintegrates as water is added to it, and forms a white powder, calcium hydroxide (**slaked lime**). Slaked lime is used in agriculture to increase the pH of acidic soils and in the construction industry as a component of lime mortar, which is a mixture of sand and slaked lime in water.

Calcium hydroxide is sparingly soluble in water (to the extent of about $2\,g\,dm^{-3}$), and the mildly alkaline solution is called **lime water**. It is used to test for the acidic gas carbon dioxide, which produces a milky-white suspension of the even less soluble calcium carbonate:

$$Ca(OH)_2(aq) + CO_2(g) \rightarrow CaCO_3(s) + H_2O(l)$$

If more carbon dioxide is bubbled through the solution, the soluble hydrogencarbonate eventually forms:

$$CaCO_3(s) + H_2O(l) + CO_2(g) \rightarrow Ca(HOCO_2)_2(aq)$$

and, as a result, the milky suspension clears again.

Magnesium hydroxide is also a mild alkali and is used as a milky-white suspension in water, known as milk of magnesia, to treat acid indigestion.

EXAMPLE

Explain why magnesium hydroxide is precipitated by the addition of aqueous ammonia but calcium hydroxide is not.

METHOD Precipitation occurs when the product of the ion concentrations exceeds the solubility product (see Section 12.1). Take the solubility products for the two metal hydroxides from Table 12.1 and judge whether the ammonia solution (a weak alkali) is likely to result in an ion product that exceeds them.

ANSWER For magnesium hydroxide, $K_{sp} = 1.1 \times 10^{-11}\,mol^3\,dm^{-9}$, a value easily exceeded for a typical magnesium ion concentration ($0.1\,mol\,dm^{-3}$) and hydroxide ion concentration in ammonia (typically $1.3 \times 10^{-3}\,mol\,dm^{-3}$ for $0.1\,mol\,dm^{-3}$ $NH_3(aq)$). For calcium hydroxide, $K_{sp} = 5.5 \times 10^{-6}\,mol^3\,dm^{-9}$, a figure five orders of magnitude larger, and one less likely to be reached.

COMMENT Both hydroxides precipitate with aqueous sodium hydroxide because then $[OH^-(aq)] = 0.1\,mol\,dm^{-3}$ (for $0.1\,mol\,dm^{-3}$ $NaOH(aq)$) and the solubility products are both readily exceeded.

18.4 Halides

The anhydrous halides are prepared by direct reaction of the elements:

$$Mg(s) + Cl_2(g) \xrightarrow{\text{heat}} MgCl_2(s) \qquad Cl(0) \rightarrow Cl(-1),\ \ Mg(0) \rightarrow Mg(+2)$$

Unlike the Group I halides, all the Group II halides are usually found hydrated: anhydrous calcium chloride has a strong affinity for water, and is used as a drying agent. This property reflects the higher charge on the metal cations. The fluorides are less soluble than the other halides, a characteristic that can be traced to the very high lattice enthalpies that result from the small F^- ion. Magnesium fluoride is used to coat the lenses of cameras, in order to reduce the amount of light reflected from their surfaces (it gives rise to the violet colour on the surface of the lens).

The halides of all the Group II elements except beryllium are predominantly ionic; beryllium halides are predominantly covalent. For example, beryllium chloride vapour consists of linear Cl—Be—Cl molecules, and in the solid it has an extended chain-like structure:

```
    Cl     Cl     Cl     Cl     Cl
  Be     Be     Be     Be     Be
    Cl     Cl     Cl     Cl     Cl
```

In this structure, each beryllium atom is surrounded by four chlorine atoms at the corners of a tetrahedron, and each chlorine atom is shared by two beryllium atoms. The overall Be:Cl atomic ratio is therefore 1:2, corresponding to the formula $BeCl_2$. A coordination number of 4 is typical for beryllium.

18.5 Hydrides

The hydrides of calcium, strontium and barium are prepared by direct reaction of the metals with hydrogen:

$$Ba(s) + H_2(g) \xrightarrow{\text{heat}} BaH_2(s) \qquad H(0) \rightarrow H(-1),\ \ Ba(0) \rightarrow Ba(+2)$$

They are largely ionic, but are hydrolysed by water with the formation of hydrogen:

$$CaH_2(s) + 2H_2O(l) \rightarrow Ca(OH)_2(aq) + 2H_2(g)$$

$$2H(-1) + 4H(+1) \rightarrow 2H(+1) + 4H(0)$$

18.6 Other compounds

Aqua cations and oxosalts

The numerous oxosalts of the Group II metals are predominantly ionic white solids with varying degrees of solubility in water and thermal stability.

The solubility of compounds with doubly charged oxoanions (such as sulphate) generally decreases down the Group. This is mainly because the lower hydration enthalpy of the larger cations cannot overcome the lattice enthalpy, which is fairly constant. The insolubility of barium sulphate is used in the test for sulphates, which form a white precipitate with an acidified solution of barium ions:

$$Ba^{2+}(aq) + SO_4^{2-}(aq) \rightarrow BaSO_4(s)$$

Soluble barium compounds are toxic, but because the sulphate is so insoluble ($K_{sp} = 1.3 \times 10^{-10}\ mol^2\ dm^{-6}$) it is safe to use a suspension of barium sulphate as a 'barium meal' to give to a patient before an X-ray examination of the digestive tract: the barium ions, which are opaque to X-rays because they have so many electrons to scatter the radiation, coat the walls of the stomach and make any imperfections visible (see Figure 18.8). **Plaster of Paris** is an

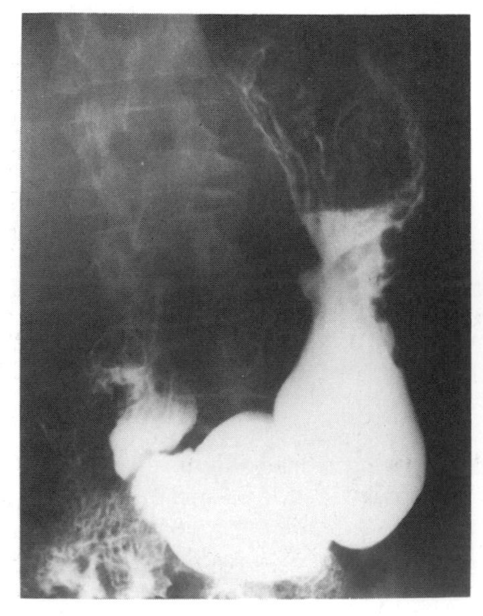

Figure 18.8 This radiograph, taken after a 'barium meal', reveals a benign gastric ulcer near the top of the stomach

insoluble form of calcium sulphate, $2CaSO_4.H_2O$, which hydrates to $CaSO_4.2H_2O$ when mixed with water, and in the process sets hard and expands slightly. Plaster of Paris is obtained from gypsum, $CaSO_4.2H_2O$, the original source of which was Montmartre in Paris.

The thermal stability of the oxosalts increases down the Group. For example, at atmospheric pressure magnesium carbonate begins to decompose rapidly at 500 °C, but calcium carbonate must be heated to about 900 °C:

$$CaCO_3(s) \xrightarrow{\text{heat}} CaO(s) + CO_2(g)$$

EXAMPLE

Suggest a reason why calcium carbonate decomposes at a lower temperature than barium carbonate does.

METHOD Consider the reaction

$$MCO_3(s) \rightarrow MO(s) + CO_2(g)$$

where M = Ca or Ba. Decomposition is thermodynamically favourable when the standard reaction Gibbs energy is negative. Both reactions have a positive standard Gibbs energy at room temperature; therefore, estimate the temperature at which it changes sign. Do this by writing

$$\Delta_r G^{\ominus} = \Delta_r H^{\ominus} - T\Delta_r S^{\ominus}$$

setting $\Delta_r G^{\ominus} = 0$, and hence solving

$$\Delta_r H^{\ominus} = T\Delta_r S^{\ominus}$$

for T, assuming that the standard reaction enthalpy and entropy do not depend on temperature (a reasonable approximation).

ANSWER From the standard formation enthalpies in Table 9.3:

$$\Delta_r H^{\ominus} = \Delta_f H^{\ominus}(MO, s) + \Delta_f H^{\ominus}(CO_2, g) - \Delta_f H^{\ominus}(MCO_3, s)$$

and similarly for the standard reaction entropy.

For $CaCO_3(s)$:

$$\Delta_r H^{\ominus} = \{-635 + (-394) - (-1207)\}\ \text{kJ mol}^{-1} = +178\ \text{kJ mol}^{-1}$$

$$\Delta_r S^{\ominus} = \{40 + 214 - 93\}\ \text{J K}^{-1}\ \text{mol}^{-1} = +161\ \text{J K}^{-1}\ \text{mol}^{-1}$$

For $BaCO_3(s)$:

$$\Delta_r H^{\ominus} = \{-553 + (-394) - (-1219)\}\ \text{kJ mol}^{-1} = +272\ \text{kJ mol}^{-1}$$

$$\Delta_r S^{\ominus} = \{70 + 214 - 112\}\ \text{J K}^{-1}\ \text{mol}^{-1} = +172\ \text{J K}^{-1}\ \text{mol}^{-1}$$

The decomposition temperature (the temperature at which $\Delta_r G^{\ominus}$ becomes zero) is therefore:

$$\text{for } CaCO_3(s)\text{: } T = \frac{178 \times 10^3\ \text{J mol}^{-1}}{161\ \text{J K}^{-1}\ \text{mol}^{-1}} = 1110\ \text{K}$$

$$\text{for } BaCO_3(s)\text{: } T = \frac{272 \times 10^3\ \text{J mol}^{-1}}{172\ \text{J K}^{-1}\ \text{mol}^{-1}} = 1580\ \text{K}$$

The difference is due principally to the less endothermic reaction enthalpy for the decomposition of calcium carbonate. This arises largely because the lattice enthalpy of CaO(s) is greater than that of BaO(s), and that in turn is largely due to the smaller size of the Ca^{2+} ion.

COMMENT The actual decomposition temperature of calcium carbonate is about 1170 K (900 °C), and that of barium carbonate is about 1630 K (1360 °C).

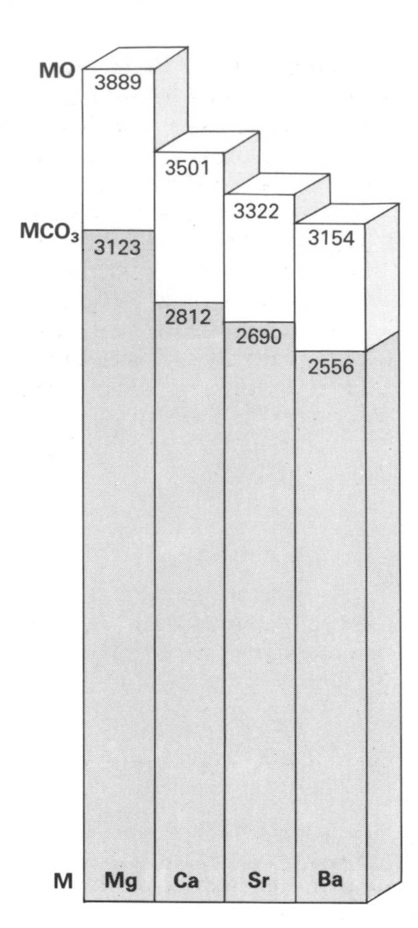

Figure 18.9 The lattice enthalpies of the Group II oxides, MO, are higher than those of the corresponding carbonates, MCO_3: the difference in lattice enthalpy gets smaller from magnesium to barium

The Group II metal nitrates also decompose on heating to give the oxide:

$$2Mg(NO_3)_2(s) \xrightarrow{\text{heat}} 2MgO(s) + 4NO_2(g) + O_2(g)$$

$$N(+5) \rightarrow N(+4), \quad O(-2) \rightarrow O(0)$$

The Group I nitrates decompose only as far as the nitrites, however (see Section 17.6). These observations suggest that the high lattice enthalpies of the Group II oxides are important.

The thermal stabilities can be rationalized in terms of the dependence of the lattice enthalpies of the oxosalt and the oxide on the size of the cation. The packing of the ions is energetically more favourable when both the anion and the cation are small, which is the case for the oxide but not the carbonate. Hence, while the oxosalt lattice enthalpies decrease down the Group as the cation gets larger, the oxide lattice enthalpies decrease even more, and the oxosalt becomes relatively more stable (see Figure 18.9).

Carbonates

The carbonates of the Group II metals are enormously important and occur everywhere around us. Calcium carbonate (see Figure 18.10) occurs in vast quantities in sedimentary rocks, particularly as chalk, limestone and (with magnesium carbonate) dolomite, and together with rocks based on the silicates is responsible for the appearance of much of our landscape. It is used in cement and glass manufacture (see Section 20.7) and in iron-making (Section 25.9), and so it contributes to our townscapes too. Calcium carbonate can also be decorative: marble is one crystalline form, in which the delicate colours are due to impurities, particularly iron oxides.

The hydrogencarbonates of magnesium and calcium are a major cause of the **hardness of water**, although some other soluble salts, including chlorides and sulphates, also contribute. Hardness arises as rain water trickles through carbonate rocks on its way to streams and reservoirs. The rain water contains dissolved carbon dioxide, and reacts with the rock to form the more soluble hydrogencarbonates:

$$CaCO_3(s) + H_2O(l) + CO_2(g) \rightarrow Ca^{2+}(aq) + 2HOCO_2^{-}(aq)$$

As a result, Ca^{2+} and Mg^{2+} ions are carried along with the water. When the water is used with soap, which is typically the sodium salt of a long-chain carboxylic acid such as stearic acid (see Section 31.6), the less soluble calcium and magnesium stearates precipitate (because their solubility products are exceeded), and form a scum.

Hydrogencarbonate hardness is called **temporary hardness** because it is removed by boiling the water, when the carbonate is reformed:

$$Ca^{2+}(aq) + 2HOCO_2^{-}(aq) \rightarrow CaCO_3(s) + H_2O(l) + CO_2(g)$$

This reaction and the preceding one are jointly responsible for the slow transport of our hills and mountains into our boilers and kettles, where they are least wanted because the carbonate deposit impairs their thermal efficiency. The hardness due to the other calcium and magnesium salts is

(a)

(b)

(c)

Figure 18.10 Three of the forms of calcium carbonate: (a) chalk, (b) calcite, (c) marble

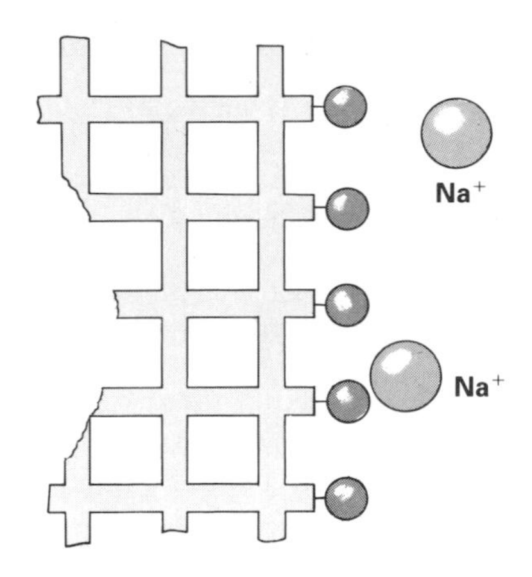

Figure 18.11 Part of a cation-exchange resin

called **permanent hardness**, because the anions do not undergo the change characteristic of hydrogencarbonates, and so are not removed by boiling.

Thermal inefficiency cannot be tolerated in industry, and so water that is to be used as boiler feed must have any calcium and magnesium ions removed, usually by **ion exchange**. In this process, the water is passed through an **ion-exchange resin** which in its active form contains sodium ions electrostatically bound to the polar groups of an insoluble polymer (see Figure 18.11). As a result of their higher charge density, the M^{2+} ions in the water exchange places with the Na^+ ions: they become firmly bound to the resin, while the sodium ions pass into the water. When all the sodium ions of the resin have been exchanged, it must be regenerated by washing through with concentrated aqueous sodium chloride (brine). Another method for ridding water of calcium ions is to add the reagent **edta** (see Section 32.3), often represented by the formula H_4Y, as shown in Figure 18.12. The edta ion envelops the Ca^{2+} and forms a very stable complex (see Section 25.7):

$$Ca^{2+}(aq) + H_4Y(aq) \rightarrow CaH_2Y(aq) + 2H^+(aq)$$

Carbides

When calcium oxide is heated to about 2200 °C with coke, the industrially important compound **calcium dicarbide**, CaC_2, is formed.

$$CaO(s) + 3C(s) \xrightarrow{2200\,^\circ C} CaC_2(s) + CO(g)$$

The anion is the molecular ion $[:C{\equiv}C:]^{2-}$. When treated with water, calcium dicarbide forms ethyne (acetylene, HC≡CH):

$$CaC_2(s) + 2H_2O(l) \rightarrow Ca(OH)_2(s) + C_2H_2(g)$$

This is a potentially valuable route from an abundant mineral, limestone, to an industrially important organic chemical, ethyne, and its importance will increase as the world's oil reserves become exhausted.

Detection

Three of the elements give characteristic colours when their compounds are heated in a flame: calcium gives a brick-red flame, strontium crimson, and barium green. (Fireworks use their salts to bring about the same effects.) The quantitative determination of magnesium is often carried out by titration with edta (see Section 32.3).

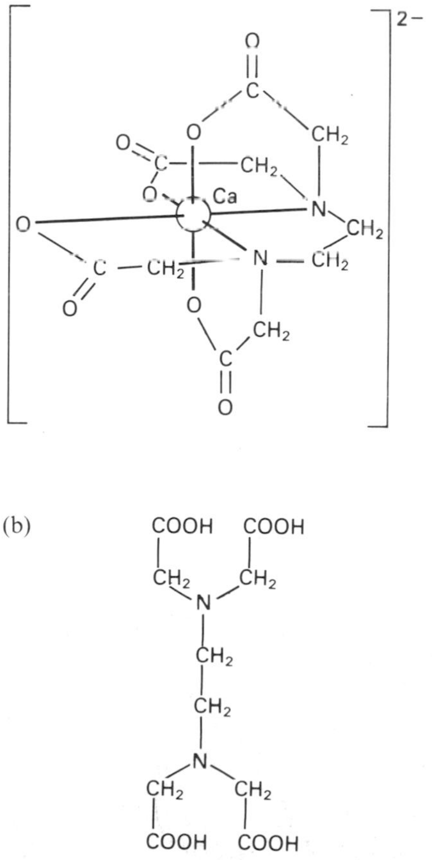

Figure 18.12 (a) The 6-coordinate complex of Ca^{2+} with the anion of edta (CaY^{2-}); (b) edta, H_4Y, can lose four protons to form the anion Y^{4-}

18.7 Industrial chemistry

Occurrence

Magnesium is the eighth most abundant element in the Earth's crust, and calcium is the fifth. The main minerals are **carnallite** ($MgCl_2.KCl.6H_2O$), **magnesite** ($MgCO_3$), **dolomite** ($MgCO_3.CaCO_3$), **chalk** and **limestone** (both $CaCO_3$), **gypsum** ($CaSO_4.2H_2O$) and **anhydrite** ($CaSO_4$). Other important sources include **Epsom salts** ($MgSO_4.7H_2O$), sea shells and coral ($CaCO_3$), and animal teeth and bones (approximately $Ca_3(PO_4)_2$). Strontium and barium occur in small quantities as minerals of composition $SrSO_4$ and $BaSO_4$. A radioactive isotope of strontium, ^{90}Sr, is a dangerous component of nuclear fallout, since Sr^{2+} can replace Ca^{2+} in bones and lead to cancer, especially leukaemia.

One of the most important substances in the world is a compound of magnesium. The molecule of **chlorophyll**, which captures sunlight and makes life possible through photosynthesis, consists of a modified porphyrin ring with a magnesium ion at its centre (see Box 3.1).

The extraction of magnesium

Of the Group II metals, only magnesium is produced on a large scale. Although dolomite is an important mineral source, magnesium is also extracted from sea water, which contains about 1.3 g of Mg^{2+} per kilogram. In the extraction process, calcium hydroxide (from roasted oyster shells) is added to the sea water and the less soluble magnesium hydroxide is precipitated. This hydroxide is then converted to the chloride by reaction with hydrochloric acid followed by evaporation in a current of hydrogen chloride:

$$Mg(OH)_2(s) + 2HCl(aq) \rightarrow MgCl_2(aq) + 2H_2O(l)$$

The metal itself is then obtained by electrolysis of the fused anhydrous chloride in a Downs cell of the kind that is used for sodium extraction and described in Section 17.7.

Summary

- ☐ The Group II metals, which include the alkaline earth metals calcium, strontium and barium, have the electron configurations [Noble gas]ns^2.
- ☐ In their compounds the only oxidation number is +2 and the elements occur as M^{2+} and $M^{2+}(aq)$.
- ☐ The metals (except beryllium) are strong reducing agents, and reduce water to hydrogen.
- ☐ The oxides are basic (except BeO, which is amphoteric) and ionic.
- ☐ The thermal stability of the oxosalts increases with the size of the cation.
- ☐ The most common forms of magnesium and calcium are their carbonates; Mg^{2+} and Ca^{2+} are responsible for the hardness of water.
- ☐ Calcium carbonate is used in cement, glass and iron manufacture, and in the Solvay process.
- ☐ The elements are detected by their flame colorations: calcium gives a brick-red flame, strontium a crimson one and barium a green one.
- ☐ Magnesium is extracted from sea water by electrolysis of the molten chloride, and is alloyed with aluminium to produce a low-density constructional metal.

PROBLEMS

1 Describe the properties of the oxosalts of the elements magnesium, calcium, strontium and barium.

2 Using appropriate data explain why the compounds MgCl and $MgCl_3$ do not exist.

3 Write an essay on the importance of calcium carbonate in industry and in the environment.

4 (a) Illustrate the structures of magnesium fluoride (rutile structure) and calcium fluoride (fluorite structure) by drawing the two unit cells.
(b) What are the coordination numbers of the Mg^{2+}, Ca^{2+} and F^- ions in (i) magnesium fluoride and (ii) calcium fluoride? Account for the different structures of these two fluorides.
(c) The fluorite structure can be described as a cubic close-packed arrangement of Ca^{2+} ions with F^- ions occupying the holes. What type of hole, and what proportion of this type of hole, do the F^- ions occupy?

5 2.14 g of a white solid **A** were treated with dilute hydrochloric acid and carbon dioxide was evolved leaving a colourless solution, which was then divided accurately into two equal parts. One part was evaporated to dryness, and the white solid residue gave a crimson colour in the flame test. The other part was treated with dilute sulphuric acid until there was no further precipitation. The precipitate was filtered off, washed, dried and was found to have a mass of 1.33 g.

Guess the formula of **A** and use the quantitative data to confirm your guess.

6 Describe the manufacture of magnesium metal from sea water. Norway produces approximately 45×10^3 tonnes of magnesium per year. 1 m^3 of sea water contains 1.3 kg of Mg^{2+} ions.
(a) Why is Norway one of the world's major producers of magnesium metal?
(b) What volume of sea water is needed to produce one tonne of magnesium?
(c) What mass of chlorine is produced during the manufacture of one tonne of magnesium?
(d) What are the main uses of magnesium and its compounds?

7 What pattern can be seen from Table 13.1 of standard electrode potentials $E^{\ominus}$ at 25 °C, for lithium, sodium, potassium, beryllium, magnesium and calcium? How can you account for this pattern?

8 Comment on the following statements:
(a) A convenient way of fixing nitrogen would be to burn magnesium in air, add water to the product, and recycle the magnesium.

(b) Barium compounds are toxic, and yet patients are sometimes fed a 'barium meal'.
(c) One of the most dangerous fission products is strontium-90.
(d) The reaction of calcium oxide (readily available from limestone) and coke will become of immense importance in the future.
(e) Metallic calcium bobs up and down when added to water.

9 Barium sulphide can be formed by strongly heating barium sulphate with carbon:

$$BaSO_4(s) + 4C(s) \rightarrow BaS(s) + 4CO(g) \qquad (i)$$

The sulphide can then react with sodium hydroxide according to the equation:

$$BaS(s) + 2NaOH(aq) \rightarrow Ba(OH)_2(s) + Na_2S(aq) \qquad (ii)$$

The sodium sulphide can be removed and barium oxide obtained by thermal decomposition of barium hydroxide.

This series of reactions can be used to make barium oxide from the sulphate.
(a) Suggest a safety precaution which should be used if barium sulphate was heated with carbon in the laboratory. Give a reason.
(b) Outline a simple practical procedure you would use to remove the sodium sulphide in reaction (ii) and then obtain pure, dry barium oxide.
(c) (i) Write an equation for the thermal decomposition of barium hydroxide.
(ii) Calculate the percentage yield if 1.53 g of barium oxide is obtained from 4.66 g of barium sulphate. (A_r: O = 16; S = 32; Ba = 137.)
(d) Write out the electronic configuration of barium (atomic number = 56).
(e) Explain what type of chemical bonding you would expect in barium oxide.
(f) Draw a diagram to show the electronic structure of barium oxide but showing the outer shells (orbitals) of electrons only.
(g) Barium oxide is often used because of its refractory properties. Explain what this means and how this property is related to your answer in (e) above.
(h) What is the flame colour of barium?
(i) Using a sample of barium hydroxide, what practical procedure would you adopt to obtain the flame colour of barium?
(j) Explain, in terms of electron transitions, why the flame colour of barium is obtained by using the procedure above. *(SUJB)*

10 What trends in physical properties and chemical behaviour can be observed in the Group II elements and their compounds? Use a reference book to select appropriate data.

What are the main differences in chemical behaviour between the elements in Groups I and II of the Periodic Table? Explain the differences in thermal stability between the carbonates of Group I and Group II of the Periodic Table. *(Nuffield S)*

11 One of the hazards of nuclear explosions is the generation of a radioactive isotope of strontium which can replace calcium in bones. This radioactive isotope of strontium has a half-life of 28 years. If 1×10^{-6} g of this isotope were absorbed by a newly born child, what mass would remain after 28 years, and after 84 years? *(JMB part question)*

12 The following is a simple account of the production of magnesium from sea water:

'Sea water is concentrated and calcium removed by the controlled addition of carbonate ions (*A*). The mixture is filtered and the clear solution treated by controlled addition of hydroxide ions (*B*). The precipitate is thermally decomposed (*C*) and the residue converted into anhydrous magnesium chloride (*D*). Magnesium is then obtained by electrolysis (*E*).'
(a) Write equations, as simply as possible, for reactions *A*, *B*, *C* and *D*.
(b) What is meant by 'controlled addition'?
(c) Outline the process *E*, emphasizing what you consider to be the important points (details of plant are *not* required).
(d) Suggest why it is difficult to produce magnesium by heating the residue from the thermal decomposition with carbon. Mention a metallic ore reducible by carbon.
(e) Write *two* commercial uses for magnesium or its compounds.
(f) Group II metals can only form one oxidation state. Using a suitable electronic configuration diagram (outer shells only), show how they achieve this.
(g) Arrange the atoms of the Group II metals in order of decreasing atomic size (i.e. put the *smallest* last). Comment on the stability of the metal cation as the Group is ascended.
(h) Suggest why beryllium chloride hydrolyses completely, magnesium chloride hydrolyses partially, but barium chloride does not hydrolyse at all.
(i) State the 'flame' colours for calcium, strontium and barium. Suggest why magnesium compounds fail to produce a visible 'flame' colour. *(SUJB)*

13 Outline the main factors responsible for the observed trends and gradations in physical and chemical properties of elements, and their compounds, in the same Group of the Periodic Table.

The elements in Group II of the Periodic Table are, in order of increasing atomic number, beryllium, magnesium, calcium, strontium, barium and radium (chemical symbol: Ra). In the light of your knowledge of the chemistry of the first five of these elements, make predictions concerning the following points:
(a) the reaction of radium with water,
(b) the solubility of radium salts in water,
(c) the acid/base behaviour of the oxide and hydroxide of radium,
(d) the type of bonding present in radium compounds and the physical properties of these compounds,
(e) methods that could be used to obtain pure radium from its compounds. *(London)*

14 (a) Calcium oxide is an ionic compound having the formula $Ca^{2+}O^{2-}$.
(i) Construct a Born–Haber cycle for the formation of calcium oxide, identifying each energy value involved.
(ii) Explain why the formula of calcium oxide is $Ca^{2+}O^{2-}$ rather than $Ca^{+}O^{-}$.
(b) Give and explain the factors that determine the nature of the bonding in a binary compound. *(London S)*

15 (a) The ionic radius of each of the Group II metals is given below:

Ion	Be^{2+}	Mg^{2+}	Ca^{2+}	Sr^{2+}	Ba^{2+}
Atomic number	4	12	20	38	56
Ionic radius/nm	0.027	0.072	0.100	0.113	0.136

(i) How can you account for the variation in the radius of the Group II ions?
(ii) How can you account for the fact that the ionic radius of the aluminium ion Al^{3+}, atomic number 13, is smaller than that of the magnesium ion?

(b) The thermal stability of the carbonates of the Group II elements increases with the atomic number of the element.
(i) Write an equation for the thermal decomposition of magnesium carbonate, including state symbols.
(ii) Suggest an explanation for the effect of the size of the cation on the thermal stability of the Group II carbonates. *(AEB 1986)*

16 (a) How is calcium metal obtained? (Technical details are not required.) Why is this the preferred method?
(b) Describe the chemical properties of (i) calcium oxide, (ii) calcium hydride, in each case giving reasons for the stated properties.
(c) Some calcium compounds are used industrially on a large scale. Select any *one* such compound and describe in outline how and why it is used in industry.
(d) Calcium does not form any stable compounds in which it has an oxidation state of +3. Briefly discuss the reasons for this. *(Oxford and Cambridge)*

17 Use any book of data and your knowledge of the compounds, the elements and their reactions to show that chemists are justified in considering magnesium, calcium, strontium and barium to be a Group of elements with characteristically similar properties. *(Nuffield)*

18 Describe how you would prepare good specimens of (a) magnesium sulphate, $MgSO_4 . 7H_2O$, from magnesite (i.e., native magnesium carbonate which may contain some calcium carbonate impurity), (b) barium sulphate, $BaSO_4$, from a sample of barium carbonate. *(London part question)*

19 This question concerns the chemistry of strontium (atomic number 38) and its compounds.
(a) Write down the ground-state electron configuration for an isolated strontium atom.
(b) (i) Indicate, using a copy of the diagram below, the pattern of the first five ionization energies for strontium.
(ii) Explain the shape of your graph.

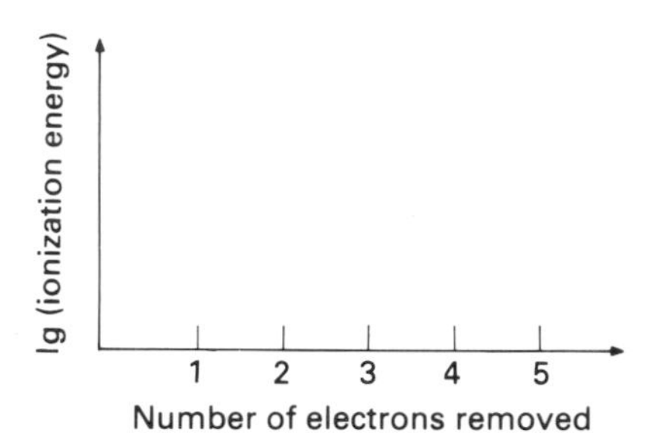

(Nuffield part question)

19

The p-block elements: Group III

In this chapter and the next five, we describe the chemistry of the p-block elements. We start with Group III and concentrate on its first two members, boron and aluminium. In their compounds boron and aluminium occur only with oxidation number +3, and with a few exceptions their compounds are best described as covalent. The chemical properties of these compounds are dominated by their Lewis acid character, and they form a wide range of complexes with Lewis bases. There is an increase in metallic character on going from boron to aluminium, and whereas boron oxide is acidic, aluminium oxide is amphoteric. Aluminium is a major component of many rocks and clays. On account of its low density and the protection of its surface by an oxide layer, aluminium is widely used as a constructional metal.

		III	IV
2		5 B	6 C
3		13 Al	14 Si
4	30 Zn	31 Ga	32 Ge
5	48 Cd	49 In	50 Sn
6	80 Hg	81 Tl	82 Pb

Figure 19.1 Group III in the Periodic Table

The p-block of the Periodic Table begins at Group III and extends as far as Group VIII. The elements of Group III are as follows:

boron	B	$[He]2s^2 2p^1$
aluminium	Al	$[Ne]3s^2 3p^1$
gallium	Ga	$[Ar]3d^{10}4s^2 4p^1$
indium	In	$[Kr]4d^{10}5s^2 5p^1$
thallium	Tl	$[Xe]4f^{14}5d^{10}6s^2 6p^1$

Their positions in the Periodic Table are shown in Figure 19.1. The characteristic valence electron configuration is ns^2np^1, where n is the Period number. We shall consider gallium, indium and thallium only briefly, for they do not occur very widely and as yet have few uses, although gallium arsenide, GaAs, is of increasing importance as a semiconductor (see Box 20.1), it finds an application in the semiconductor lasers in compact disc players.

19.1 Group systematics

Oxidation states and bonding characteristics

The only oxidation number of boron and aluminium in any of their compounds is +3. This does not necessarily mean that the compounds are built from E^{3+} ions, however; it must be remembered that an oxidation number is an artificial measure of the distribution of electrons between atoms and emphasizes the changes that occur when compounds are formed (see Section 15.2).

The heaviest member of the Group, thallium, can occur with another stable oxidation number, +1 (see the discussion of the inert-pair effect in Section 20.1), and thallium(III) chloride decomposes just above room temperature:

$$TlCl_3(s) \rightarrow TlCl(s) + Cl_2(g) \qquad Tl(+3) \rightarrow Tl(+1), \quad 3Cl(-1) \rightarrow Cl(-1) + 2Cl(0)$$

The compound TlI_3, which appears from its formula to be thallium(III) iodide, is in fact thallium(I) triiodide, and contains the triiodide ion, I_3^- (see Section 23.4). Thallium(I) solutions, which contain $Tl^+(aq)$, are poisonous.

Periodic trends

The general trend down Group III is *from non-metallic to metallic character*. This is shown in the structures of the elements, and in their chemical properties. For example, whereas elemental boron shows its non-metallic character in its covalent network structure, aluminium has a close-packed metallic structure. Similarly, the oxide of boron is acidic whereas the oxide of aluminium is amphoteric. This trend continues down the Group, and thallium(III) oxide is basic.

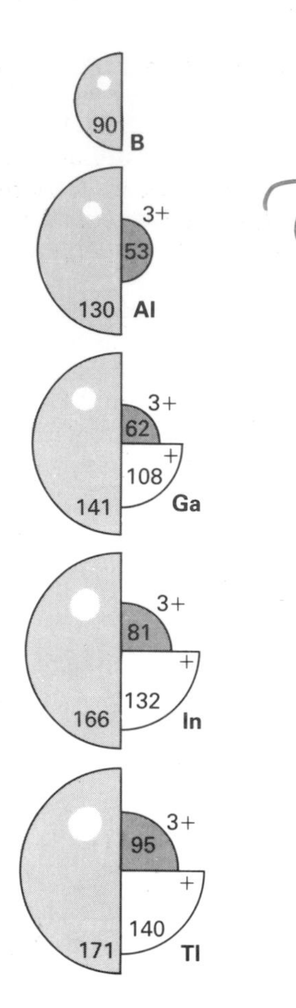

Figure 19.2 The radii (in picometres) of the atoms and of the M^+ and M^{3+} ions of the Group III elements

1 $AlCl_3$ (Cl—Al(Cl)—Cl)

2 $[AlCl_4]^-$

The chemical properties of boron and aluminium reflect the behaviour of their compounds as Lewis acids, but here too there are marked differences between them. For instance, while both boron and aluminium form the complex ions $[E(OH)_4]^-$ with the strong Lewis base OH^-, a hydrated cation $[E(H_2O)_n]^{3+}$ is formed only by aluminium. However, due to the polarizing power of Al^{3+} the cation $[Al(H_2O)_6]^{3+}$ is as acidic as ethanoic acid, and easily loses a proton from a coordinated water molecule to the solvent. The very small, highly polarizing boron ion would encourage proton loss so much that a hydrated boron cation is unknown.

Shapes and sizes

The atomic radii of the elements are shown in Figure 19.2: they increase down the Group as the principal quantum number of the valence shell increases. The marked difference in size between boron and the other elements is another factor responsible for the differences between boron and the other elements in the Group.

The individual molecules of the tricovalent compounds of boron and aluminium, including $BCl_3(g)$ and $AlCl_3(g)$ (**1**), are trigonal planar, as expected from the VSEPR rules (see Section 2.3) because each molecule has three electron pairs. The fourth bonding position is occupied when a complex such as $AlCl_4^-$ (**2**) is formed, and the molecules or ions are then tetrahedral, like the isoelectronic compound $SiCl_4$.

Aluminium is in Period 3 and so its atom, as well as being larger than that of boron, can expand its valence shell by making use of its 3d-orbitals. One of the major differences between boron and aluminium is that whereas the maximum coordination number of boron is four, aluminium forms complexes in which it is attached to up to six groups. For example, both elements form fluoro-complexes; boron forms the BF_4^- ion, but the corresponding aluminium complex is AlF_6^{3-}.

19.2 The elements

Boron

Boron is a non-metal with an unusual structure. It has several allotropes, and even the simplest allotrope consists of groups of twelve boron atoms arranged at the corners of a regular icosahedron (a regular twenty-faced figure) to give structures like tiny geodesic domes (see Figure 19.3). These B_{12} icosahedra

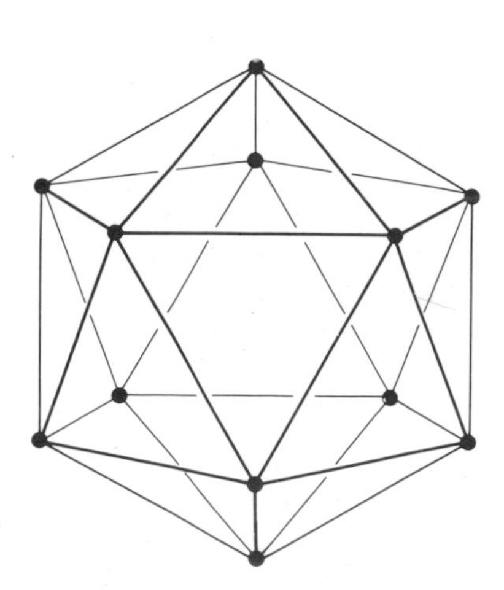

Figure 19.3 An icosahedron, as in B_{12}, has twelve identical corners and twenty equilateral triangular faces

Figure 19.4 Because boron nuclei absorb neutrons so well, it was sprayed over the Chernobyl nuclear plant after it had exploded in an attempt to quench the nuclear chain reaction

then stack together into a network structure; as a result, boron has a very high melting point (2300 °C) and is chemically unreactive except at high temperatures – for example, it is not attacked by molten sodium hydroxide even at 500 °C. Little was known about its properties until recently, when interest was stimulated by its important *nuclear* properties, particularly the ability of ^{10}B to capture neutrons (see Figure 19.4). There is some electron delocalization within the icosahedra, but none between them, and so boron does not conduct electricity.

Aluminium

Aluminium is a typical metal, with a cubic close-packed structure (see Section 5.2). It is a good conductor of electricity and heat, and its low density ($2.7\,g\,cm^{-3}$), widespread availability and ability to accept many surface finishes make it commercially and industrially very important (see Figure 19.5).

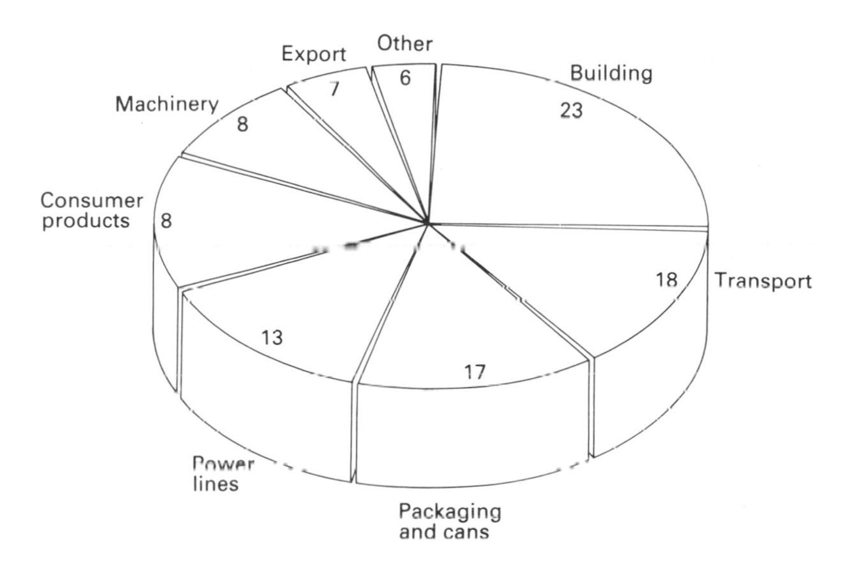

Figure 19.5 The uses of aluminium

At first glance aluminium might be expected to be a highly reactive metal, and not the sort of material from which to build aeroplanes. This is because $E^{\ominus}(Al^{3+}, Al) = -1.66\,V$, a value so negative that aluminium would be expected to be oxidized readily in air and to be susceptible to attack by acids. In fact, it is both; but the attack proceeds only as far as the outer surface layer because the oxide layer produced in air, even though it is only about 10 nm thick, is so tenacious that it seals the metal below from further attack. This is quite different from the reaction of iron, where the oxide layer flakes off the surface, exposing more metal and thus allowing the attack to continue indefinitely.

This protective oxide layer forms on every piece of aluminium that is exposed to the air. Its presence allows aluminium utensils to be used in the kitchen, even for boiling such acidic foods as rhubarb, and it allows aeroplanes to survive flights through damp air for decades. It protects power lines, which are constructed from aluminium because its conductivity per unit mass is higher than that of copper, so that pylons (which are also built from aluminium) can be less numerous and less intrusive on the landscape.

The protective layer is so important that it is often enhanced by **anodizing** the aluminium. In this process an aluminium article, such as a window frame, is made the anode in an electrolytic bath containing sulphuric acid. When a current is passed the oxygen released at the anode combines with the aluminium and the surface oxide layer is artificially thickened to about 20 μm. An advantage of this technique is that dyes can be added to the bath and are incorporated into the oxide, giving surfaces with different colours. This is the technique used for producing the coloured anodized aluminium which is a feature of much modern architecture.

Figure 19.6 The intense heat given out by the thermit process is used to weld railway lines

Naked aluminium, with its surface layer damaged (for example, by violent reaction), shows its strongly electropositive nature to the full. The oxidation

$$4Al(s) + 3O_2(g) \xrightarrow{\text{heat}} 2Al_2O_3(s) \qquad O(0) \rightarrow O(-2), \quad Al(0) \rightarrow Al(+3)$$

is strongly favoured thermodynamically ($\Delta_r G^\ominus(298\,K) = -3165\,kJ\,mol^{-1}$). It can therefore be used to reduce many metal oxides:

$$2Al(s) + Cr_2O_3(s) \xrightarrow{\text{heat}} 2Cr(s) + Al_2O_3(s) \qquad Cr(+3) \rightarrow Cr(0), \quad Al(0) \rightarrow Al(+3)$$

The same reaction occurs with iron(III) oxide, and although there are cheaper ways of making iron, it is used as a source of intense heat in the spectacular **thermit process** (see Figure 19.6):

$$2Al(s) + Fe_2O_3(s) \xrightarrow{\text{heat}} 2Fe(l) + Al_2O_3(s) \qquad Fe(+3) \rightarrow Fe(0), \quad Al(0) \rightarrow Al(+3)$$

Heating is needed to start the reaction: once in progress its exothermicity sustains it. Aluminium powder is used in bombs, and as a fuel in solid rocket boosters, together with ammonium chlorate(VII) (see Section 23.3) as oxidizer.

EXAMPLE

Calculate the standard reaction enthalpy of the thermit process at 298 K.

METHOD Write the reaction

$$2Al(s) + Fe_2O_3(s) \rightarrow 2Fe(s) + Al_2O_3(s)$$

and then use the tables of standard formation enthalpies (Table 9.3) to calculate the standard reaction enthalpy as explained in Section 9.4. The standard formation enthalpies of the elements are zero, and so need not be written.

ANSWER

$$\begin{aligned}\Delta_r H^\ominus &= \Delta_f H^\ominus(Al_2O_3, s) - \Delta_f H^\ominus(Fe_2O_3, s)\\ &= \{(-1675.7) - (-824.2)\}\,kJ\,mol^{-1}\\ &= -851.5\,kJ\,mol^{-1}\end{aligned}$$

COMMENT The actual reaction usually takes place at much higher temperatures than 298 K, but the reaction enthalpy at these high temperatures is not greatly different from the value calculated here. In practice, the reaction produces enough energy to melt iron (m.p. 1535 °C).

Aluminium is sufficiently electropositive to reduce non-metals when heated with them:

$$2Al(s) + N_2(g) \xrightarrow{\text{heat}} 2AlN(s) \qquad N(0) \rightarrow N(-3), \quad Al(0) \rightarrow Al(+3)$$

$$2Al(s) + 3Cl_2(g) \xrightarrow{\text{heat}} 2AlCl_3(s) \qquad Cl(0) \rightarrow Cl(-1), \quad Al(0) \rightarrow Al(+3)$$

The protective aluminium oxide layer is resistant to dilute acids, but is moderately soluble in alkalis. That is, aluminium reduces strong alkalis, a product being the **tetrahydroxoaluminate ion**:

$$2Al(s) + 2OH^-(aq) + 6H_2O(l) \rightarrow 2[Al(OH)_4]^-(aq) + 3H_2(g)$$
$$H(+1) \rightarrow H(0), \quad Al(0) \rightarrow Al(+3)$$

A relevant feature of aluminium's chemistry is the exceptional stability of aluminium oxide, which is largely due to the high lattice enthalpy of $Al_2O_3(s)$.

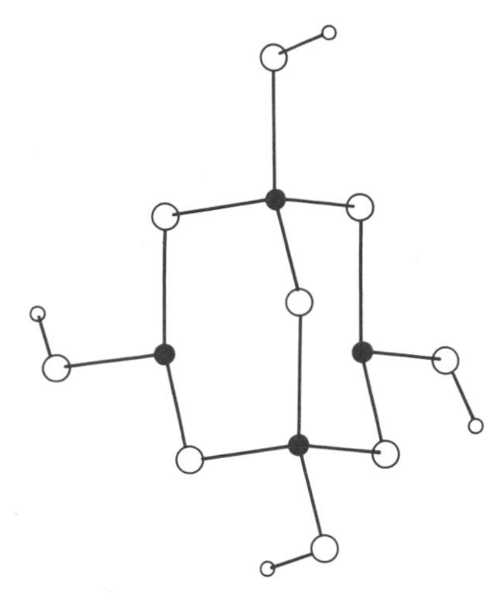

Figure 19.7 The structure of the anion $[B_4O_5(OH)_4]^{2-}$ present in borax

19.3 Oxides

Boron oxide and the borates

Boron oxide, B_2O_3, is an insoluble white solid which melts at 450 °C but does not boil until the temperature has been raised to over 2000 °C. This is explained by its structure, which is an extended, covalently bonded network. This network-making ability is maintained when boron oxide dissolves in silicates (see Section 20.3) and leads to the formation of **borosilicate glasses**, such as 'Pyrex', which expand little when heated and are easily worked.

Boric acid, $(HO)_3B$, is the end-product of the hydrolysis of many boron compounds. It is a white crystalline solid which is used to some extent as a mild antiseptic, and in much greater quantities for the manufacture of borosilicate glass. The solid is soluble in hot water because the molecules are linked only by hydrogen bonds, and not by the strong B—O covalent bonds characteristic of the oxide. It is a very weak acid by virtue of its ability to accept an oxygen lone pair from an H_2O molecule:

$$(HO)_3B(aq) + 2H_2O(l) \rightleftharpoons H_3O^+(aq) + [B(OH)_4]^-(aq)$$

Borates have complicated extended-network structures; in this respect boron resembles its diagonal neighbour in the Periodic Table, silicon. The most important borate is **borax**, $Na_2[B_4O_5(OH)_4].8H_2O$ (see Figure 19.7), which occurs in large quantities in California, Tibet and Turkey. It is the principal source of boric acid, which is produced by the action of acid:

$$[B_4O_5(OH)_4]^{2-}(aq) + H_2O(l) + 2H_3O^+(aq) \rightarrow 4(HO)_3B(aq)$$

Borax is mined in megatonne quantities to supply the glass industry.

Aluminium oxide

Aluminium oxide, Al_2O_3, which is also called **alumina**, is a white, polymorphic solid. It occurs naturally as **bauxite** and as **corundum** (see Box 19.1). Some of the impure forms of corundum are prized more highly than the pure substance (see Colour 4), for they include **ruby** (in which some Al^{3+} ions are replaced by Cr^{3+} and **sapphire** (where two Al^{3+} ions are replaced by one Ti^{4+} ion and one Co^{2+} or Fe^{2+} ion at some locations). Since aluminium oxide is refractory it is used as a support for the catalysts used in cracking and reforming hydrocarbons (see Sections 27.3 and 27.4).

Aluminium oxide is *amphoteric*. Because the compact oxide is inert, this property is best demonstrated using the freshly precipitated hydroxide, which

BOX 19.1

Figure 19.8 Engraving glass using a diamond-tipped 'pencil'

Some very hard substances

The hardness of a substance is reported on **Mohs' scale**, in which any substance can scratch any other substance that is placed lower down the scale. **Diamond**, which can scratch all others (see Figure 19.8), is the hardest known substance, and is given the maximum value of 10 on the scale.

Some of the compounds of Group III elements are very hard. **Corundum**, a mineral form of alumina, Al_2O_3, has hardness 9, and is used as an abrasive (and in toothpaste). A microcrystalline form of corundum contaminated with the iron oxides magnetite and haematite is sold as the abrasive **emery**. Boron nitride, BN, is isoelectronic with carbon (having on average four valence electrons per atom) and although one form is a white refractory (heat-resistant) solid with a slippery feel like that of graphite, under high pressures it converts to **borazon**, with a diamond-like structure and a hardness to match. Boron carbide, $B_{13}C_2$, consists of groups of boron atoms pinned together by carbon atoms: it is very hard but also very light, and used to make bullet-proof vests and in other forms of armour-plating.

is also amphoteric. The precipitate shows its amphoteric nature by reacting in acids to give $[Al(H_2O)_6]^{3+}$ ions, and also reacting in alkalis to give the tetrahydroxoaluminate ion:

$$[Al(H_2O)_6]^{3+}(aq) \underset{3H^+}{\overset{3OH^-}{\rightleftharpoons}} [Al(H_2O)_3(OH)_3](s) \underset{H^+}{\overset{OH^-}{\rightleftharpoons}} [Al(H_2O)_2(OH)_4]^-(aq)$$

In these formulae the aluminium atom is shown with six groups attached; the evidence for that is very sparse, however, and the hydroxide and the anion are often denoted $Al(OH)_3$ and $[Al(OH)_4]^-$.

When metal oxides are fused with alumina, as in the reaction

$$MgO(s) + Al_2O_3(s) \xrightarrow{\text{heat}} MgAl_2O_4(s)$$

the product is a **mixed oxide** (in this case a **spinel**). Mixed oxides are *not* simply mixtures of oxides, but a single phase having two or more types of cation in the holes in an oxide ion array.

19.4 Halides

The halides of aluminium can be prepared by the direct reaction of heated aluminium with the dry halogen or the dry hydrogen halide:

$$2Al(s) + 3Cl_2(g) \xrightarrow{\text{heat}} 2AlCl_3(s) \qquad Cl(0) \rightarrow Cl(-1), \quad Al(0) \rightarrow Al(+3)$$

$$2Al(s) + 6HBr(g) \xrightarrow{\text{heat}} 2AlBr_3(s) + 3H_2(g) \qquad H(+1) \rightarrow H(0), \quad Al(0) \rightarrow Al(+3)$$

Other methods are available. Boron trifluoride, the most important halide of boron, is produced by heating the oxide with calcium fluoride and concentrated sulphuric acid.

Boron trifluoride is a gas consisting of trigonal planar molecules, but aluminium fluoride is an insoluble, high-melting solid (m.p. 1290 °C). This difference reflects the larger size of the Al^{3+} ion and its lower ability to polarize the F^- ion (see Section 15.4), so that aluminium fluoride is best regarded as ionic. The high lattice enthalpy of aluminium fluoride contributes to its low solubility in water.

Aluminium chloride is a volatile solid which sublimes at 180 °C. It is predominantly covalent, largely because the Cl^- ion is more polarizable than F^-. The vapour formed when aluminium chloride sublimes consists of an equilibrium mixture of monomers ($AlCl_3$) and dimers (Al_2Cl_6); the dissociation of the dimers is endothermic, so that the monomers are favoured by high temperatures and above 200 °C they predominate. The formation of the dimer is an example of a molecule behaving simultaneously as a Lewis acid and a Lewis base: two of the chlorine atoms form bonds by donating a lone pair to vacant orbitals of the aluminium atom (see Figure 19.9).

Boron trihalides and aluminium halides act as Lewis acids to a wide range of electron-pair donors. The products are generally addition compounds or **adducts**, and include the white solid H_3NBF_3 and the anion BF_4^-.

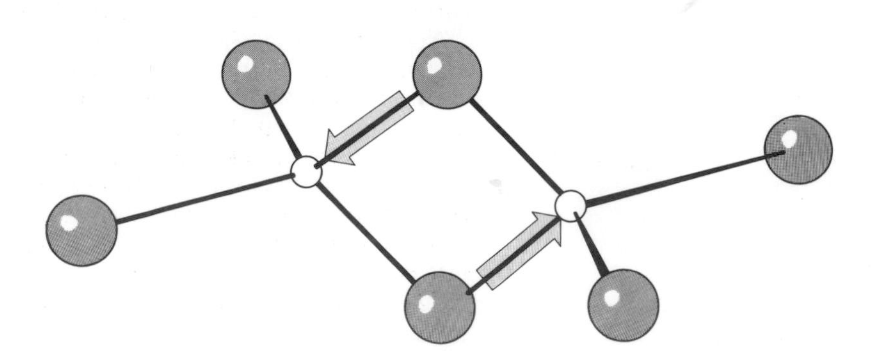

Figure 19.9 The chlorine bridges in Al_2Cl_6

EXAMPLE

Calculate the oxidation number of boron in potassium tetrafluoroborate, KBF_4.

METHOD Potassium tetrafluoroborate is an ionic compound; concentrate therefore on the oxidation number (Ox) of boron in the anion, BF_4^-. Apply the rules given in Section 15.2 (remember that the sum of the oxidation numbers of all the atoms must equal the charge on the anion):

$$Ox(\mathrm{B}) + 4Ox(\mathrm{F}) = -1$$

ANSWER According to the rules, $Ox(\mathrm{F}) = -1$ in all its compounds. Therefore

$$Ox(\mathrm{B}) + 4 \times (-1) = -1$$

so that $Ox(\mathrm{B}) = +3$.

COMMENT When evaluating oxidation numbers, always work systematically through the rules in Section 15.2 in the order given.

Another example of an adduct is the compound formed when aluminium chloride dissolves in ethoxyethane and excess solvent is evaporated:

$$(C_2H_5)_2O(l) + AlCl_3(s) \rightarrow (C_2H_5)_2OAlCl_3(s)$$

in which a lone pair on the oxygen atom spearheads the attack and forms a coordinate bond with the aluminium (see Section 2.6).

The widespread use of aluminium chloride and boron trifluoride as catalysts is due to their ability to act as Lewis acids. Aluminium chloride is used in the important **Friedel–Crafts reactions**, discussed in Section 34.3. Lewis acid–base coordinate bond formation also has a more destructive aspect, for it can be the first step in the hydrolysis of the compound. Solid anhydrous aluminium chloride, for example, fumes in moist air as a result of the reaction

$$AlCl_3(s) + 3H_2O(l) \rightarrow Al(OH)_3(s) + 3HCl(g)$$

When plenty of water is available, as when aluminium chloride dissolves in water, the exothermic reaction

$$AlCl_3(s) + 6H_2O(l) \rightarrow [Al(H_2O)_6]^{3+}(aq) + 3Cl^-(aq)$$

occurs, and the acidic $[Al(H_2O)_6]^{3+}$ ion is formed.

19.5 Hydrides

Boron hydrides

Boron forms an extensive and interesting series of hydrides, the **boranes**, and in this it resembles carbon (although the range of compounds is very much more limited) and silicon. More than two dozen boron hydrides are known.

The simplest hydride is not BH_3, as might be expected, but its dimer **diborane**, B_2H_6. Diborane has a remarkable structure (see Figure 19.10), which is the pattern for all the other boranes. Each boron atom is surrounded tetrahedrally by four hydrogen atoms; two are **terminal** atoms, the other two are **bridging** atoms. The bonding in diborane was originally a puzzle, but we now regard the bonding as a straightforward extension of the molecular orbital description of a chemical bond between two atoms given in Section 2.5. In diborane each bridging hydrogen atom contributes to a molecular orbital that spreads over *three* atoms (B—H—B), and when two electrons occupy this orbital they form a **three-centre bond**, one pair of electrons bonding three atoms together.

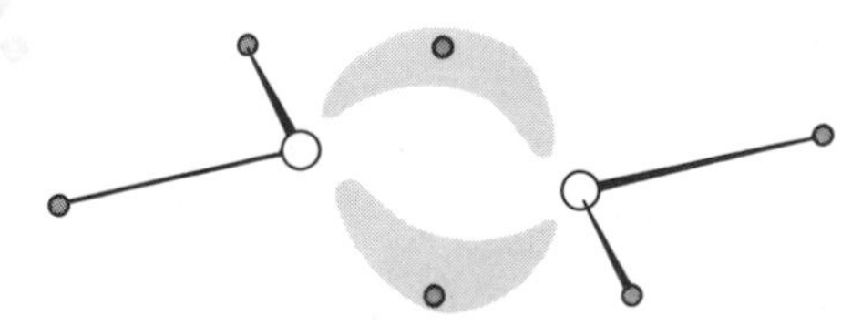

Figure 19.10 Each bridging B—H—B bond contains an electron pair delocalized over three atoms

Diborane is a very reactive gas (b.p. −93 °C). It is spontaneously flammable in air, forming the oxide as it burns:

$$B_2H_6(g) + 3O_2(g) \rightarrow B_2O_3(s) + 3H_2O(l) \qquad O(0) \rightarrow O(-2),\ H(-1) \rightarrow H(+1)$$

The higher boranes, which are networks of boron and hydrogen atoms linked together by two- and three-centre bonds, are formed when diborane is heated in sealed containers above 100 °C. They have complex structures based on clusters of boron atoms in arrangements related to the B_{12} icosahedron, and with shapes that range in appearance from delicate spiders' webs to untidy birds' nests.

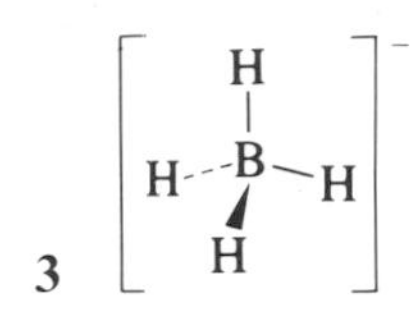

The most important derivatives of the boranes are the **tetrahydridoborates**, compounds containing the tetrahedral ion BH_4^- (**3**). About 2000 tonnes of the sodium salt are produced each year because it is a useful reducing agent. One example of its industrial use is in its reduction of nickel(II) chloride solutions to produce a nickel surface on a plastic, which can then be electroplated to give a metallic finish and used in the production of high-quality record masters. Conventional electrolytic reduction is not possible in this case because the plastic does not conduct.

Aluminium hydride

Aluminium does not form such an interesting range of hydrides. The only one known is AlH_3, which is a complex solid extensively polymerized by Al—H—Al bridges. Its most important derivative is the white crystalline solid **lithium tetrahydridoaluminate**, $LiAlH_4$, which is also often called 'lithium aluminium hydride'. It is a powerful and versatile reducing agent, and is widely used in organic chemistry (for examples of its use see Sections 30.3 and 31.4).

19.6 Other compounds

Aqua cations and oxosalts

Boron is a non-metal and does not form cations. Solutions containing the hydrated aluminium cation ($[Al(H_2O)_6]^{3+}$ – more formally the **hexaaquaaluminium ion**) and aluminium oxosalts are readily prepared by reacting freshly precipitated aluminium hydroxide with the appropriate oxoacid. It is difficult, and sometimes impossible, to obtain the solid oxosalt by evaporation, because of hydrolysis of the product.

The reasons for the hydrolysis of aluminium salts can be traced back to the small size and high charge of the aluminium ion. The high polarizing power of Al^{3+} withdraws electrons from the O—H bonds of the water molecules that are directly linked to it, making the aqua cation a Brønsted acid (see Section 12.2). A solvent water molecule can act as a base and accept a proton:

$$[Al(H_2O)_6]^{3+}(aq) + H_2O(l) \rightleftharpoons [Al(H_2O)_5OH]^{2+}(aq) + H_3O^+(aq)$$

This hydrolysis results in solutions of aluminium salts being significantly acidic (about as acidic as vinegar). The proton donation can continue:

$$[Al(H_2O)_5OH]^{2+}(aq) + H_2O(l) \rightleftharpoons [Al(H_2O)_4(OH)_2]^+(aq) + H_3O^+(aq)$$

$$[Al(H_2O)_4(OH)_2]^+(aq) + H_2O(l) \rightleftharpoons [Al(H_2O)_3(OH)_3](s) + H_3O^+(aq)$$

but the exact identity of the precipitate is complicated. These equilibria shift in favour of products (aluminium hydroxide) if the hydrogen ion is removed by reaction with the conjugate base of a weak acid. For example, with carbonate ions:

$$2[Al(H_2O)_6]^{3+}(aq) + 3CO_3^{2-}(aq) \rightarrow 2[Al(H_2O)_3(OH)_3](s) + 3CO_2(g) + 3H_2O(l)$$

Hence, aluminium carbonate cannot be isolated by evaporating a solution containing hydrated aluminium cations and carbonate ions. Sulphide ions have a similar effect. Aluminium salts of strong acids can be isolated, however. They usually crystallize with several molecules of water per formula unit (a striking example is aluminium sulphate, $Al_2(SO_4)_3 . 18H_2O$), which is another consequence of the high polarizing power of Al^{3+}.

The compound $KAl(SO_4)_2 . 12H_2O$ gives its name to an important class of salts, the **alums**. These are **double salts** of general formula $M(I)M'(III)(SO_4)_2 . 12H_2O$, where M(I) is a singly charged cation, such as NH_4^+ or K^+, and M′(III) is a triply charged cation, such as Al^{3+}, Cr^{3+} or Fe^{3+}. All alums contain the octahedral $[M'(H_2O)_6]^{3+}$ ion.

Alums are easy to crystallize from aqueous solutions, and form attractive crystals which are **isomorphous** (that is, they have the same shape). Their usefulness can once again be traced to the high charge of the Al^{3+} ion. They are added to effluents to neutralize the negative charge on the colloidal particles of sewage (see Box 7.1), which then aggregate sufficiently to be removed by filtration.

19.7 Industrial chemistry

Occurrence

Boron occurs principally as **borax**, $Na_2[B_4O_5(OH)_4] . 8H_2O$, and related minerals. Aluminium is the third most abundant element in the Earth's crust, of which it forms 8 per cent by mass, and the most abundant metal. It occurs widely in the aluminosilicates, such as **feldspars** (the principal component of many igneous rocks, such as granite), **micas** and, in a weathered form, **clays**. Commercially useful minerals include **bauxite,** $Al_2O_3 . xH_2O$, where x lies between 1 and 3 (it is so called because deposits were first discovered near the town of Les Baux in France), **corundum** (Al_2O_3) and **cryolite** (sodium hexafluoroaluminate, Na_3AlF_6, from Greenland; *cryos* is Greek for 'frost').

The manufacture of aluminium oxide from bauxite

The principal impurities in bauxite are silicates and iron(III) oxide (which is responsible for its red discoloration). The amphoteric aluminium oxide and the acidic silicates dissolve in aqueous sodium hydroxide, but the iron(III) oxide does not. After filtration, the aluminium hydroxide is precipitated by treatment with a weak acid, typically carbon dioxide:

$$[Al(OH)_4]^-(aq) + CO_2(g) \rightarrow Al(OH)_3(s) + HOCO_2^-(aq)$$

The oxide is then obtained by heating the filtered solid:

$$2Al(OH)_3(s) \xrightarrow{1200\,^\circ C} Al_2O_3(s) + 3H_2O(g)$$

The manufacture of aluminium

Aluminium is obtained by the electrolysis of aluminium oxide. The oxide's melting point (over 2000 °C) is too high for electrolysis of the melt to be economical, however, and it is dissolved in molten cryolite instead. This is called the **Hall–Héroult process**, after its American and French inventors Charles Hall and Paul Héroult, who invented the process simultaneously in 1886. The conversion of aluminium from a prized, rare metal to one of unparalleled technological usefulness is due to these two young chemists, who were both twenty-three at the time.

In the Hall–Héroult process the temperature is maintained at about 950 °C, and electrolysis using a graphite anode is carried out using very high currents (around 100 000 A) to produce molten aluminium (see Figure 19.11). Aluminium is obtained at the cathode ($Al^{3+} + 3e^- \rightarrow Al$), and oxygen is discharged at the anode ($2O^{2-} \rightarrow O_2 + 4e^-$).

The oxygen attacks the graphite, forming carbon dioxide, so that the electrodes need to be renewed frequently (about every other day). Aluminium manufacture makes heavy demands on electrical power supplies, and is usually carried out where electricity is relatively cheap. A more recent and less energy-intensive process is the **Toth process**, in which bauxite reacts with chlorine to form aluminium chloride, which is then electrolysed; the chlorine is recycled.

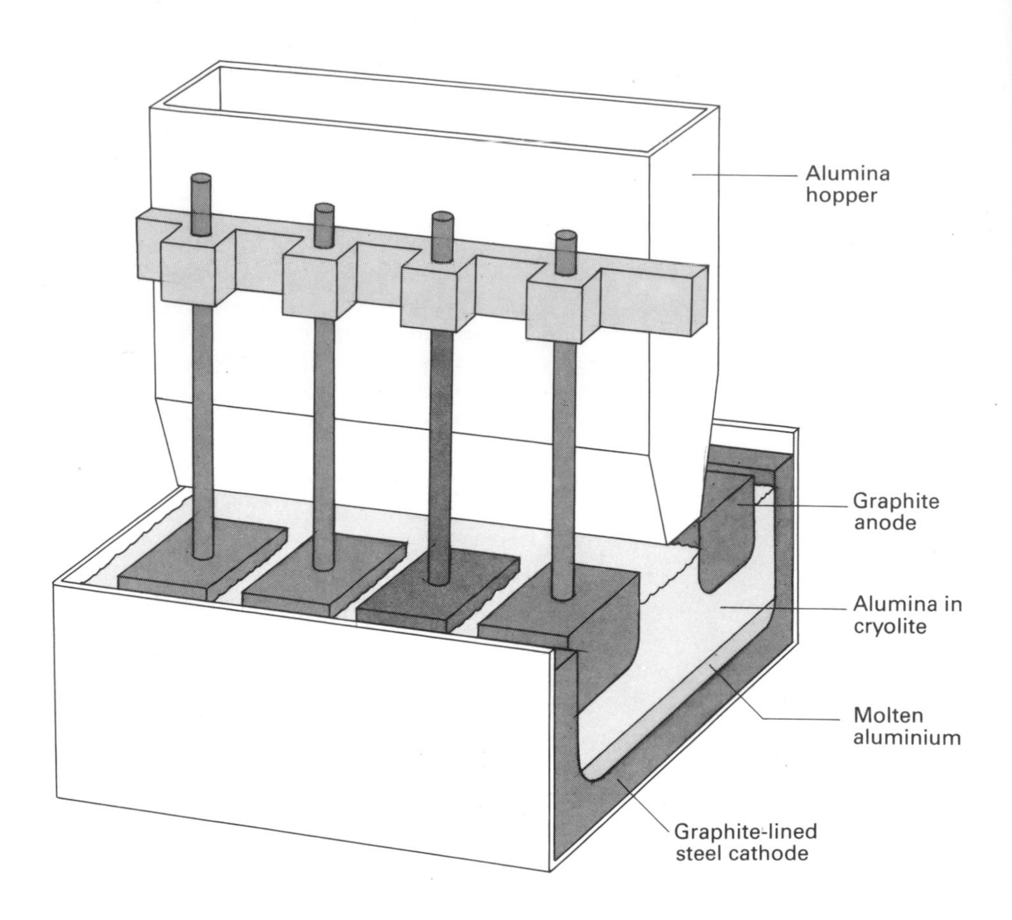

Figure 19.11 The Hall–Héroult process for the manufacture of aluminium

Summary

- Boron and aluminium have the electron configuration [Noble gas]ns^2np^1.
- The only oxidation number for boron and aluminium (apart from 0) is +3.
- Boron is covalently bonded in all its compounds.
- Aluminium forms several compounds in which it is essentially ionic, such as AlF_3 and Al_2O_3, and the ion $[Al(H_2O)_6]^{3+}$.
- Although aluminium has a highly negative standard electrode potential ($E^{\ominus} = -1.66$ V), its surface is protected by a layer of oxide. Aluminium is very widely used because of its low density and resistance to corrosion.
- Aluminium is the most abundant metal in the Earth's crust.
- Boron oxide is acidic whereas aluminium oxide is amphoteric.
- Because of the high polarizing power of Al^{3+}, aqueous solutions of the hydrated cation, $[Al(H_2O)_6]^{3+}$, are acidic.
- The halides of boron and aluminium are strong Lewis acids. They form 4-coordinate adducts with Lewis bases.
- A distinguishing feature of boron is that it forms only 4-coordinate (tetrahedral) complexes, whereas aluminium forms both 4-coordinate and 6-coordinate (octahedral) complexes.
- The hydrido-anions BH_4^- and AlH_4^- have wide uses as reducing agents.
- Alums have the general formula M(I)M′(III)$(SO_4)_2 . 12H_2O$.
- Aluminium is manufactured by the electrolysis of a fused mixture of aluminium oxide and cryolite.

PROBLEMS

1 Complete the following table of melting and boiling points (in °C) of some halides of boron and aluminium.

		F	Cl	Br
B	m.p.			
	b.p.			
Al	m.p.			
	b.p.			

Comment on the trends.

2 In solid compounds and in solution the ion $[Al(H_2O)_6]^{3+}$ is common, but there are no compounds containing the hydrated B^{3+} ion. Explain.

3 A white solid **A** gives an intense yellow colour in the flame test. 0.445 g of the solid were dissolved in water and ammonia was added until there was no further precipitate of hydrated aluminium oxide. The precipitate was filtered off, washed and ignited. The mass of Al_2O_3 so obtained was 0.118 g. The filtrate was acidified with dilute nitric acid to make a volume of exactly 100 cm³. Then a 10 cm³ portion of this solution was treated with aqueous silver nitrate. The precipitate of silver chloride was washed and dried, and weighed 0.133 g.

Calculate the percentages of aluminium and chlorine in **A** and hence deduce its formula.

4 What are the shapes of the following molecules or ions: BCl_3, BF_4^-, Al_2Cl_6, AlF_6^{3-}? Rationalize the shapes using VSEPR theory.

5 Write an essay on the uses of aluminium and its compounds.

6 Write equations for the reactions to convert aluminium into (a) Al_2O_3, (b) $AlCl_3$, (c) $LiAlH_4$, (d) $KAl(SO_4)_2.12H_2O$, (e) an aqueous solution containing $[Al(OH)_4]^-$. Write separate equations for each step of a multi-step process. Label each equation as 'redox' or 'Lewis acid–base'. (Some reactions may be of both kinds.) For redox reactions indicate the changes in oxidation number of all the elements present; and for Lewis acid–base reactions indicate which is the Lewis acid and which is the Lewis base.

7 Explain the following:
(a) Aluminium metal is electropositive ($E^{\ominus}(Al^{3+}, Al) = -1.66$ V), and yet aluminium metal is used as a constructional material.
(b) Solutions of aluminium salts are acidic.
(c) Boron trichloride is always monomeric, but aluminium chloride vapour between 150 and 200 °C consists mainly of Al_2Cl_6 molecules.
(d) Boron oxide is acidic, but aluminium oxide is amphoteric.
(e) Boron nitride exists in two forms. One has a structure similar to graphite and one a structure similar to diamond.
(f) Aluminium metal cannot be obtained by reducing aluminium oxide with carbon.

8 What is the minimum quantity of electricity required to produce 1 tonne of aluminium? In the actual process much more electricity is used. Why?

9 (a) Give an account of the manufacture of aluminium, explaining carefully the principles involved and the reasons for the conditions used.
(b) How and under what conditions does aluminium react with (i) chlorine, (ii) hydrochloric acid, (iii) aqueous potassium hydroxide?
(c) The boiling points of aluminium fluoride and aluminium bromide are 1270 °C and 265 °C respectively. What inferences may be drawn from these boiling points about the structures of these two halides? *(Oxford)*

10 Explain the corrosion resistance of aluminium in atmospheric conditions and how this resistance may be increased commercially.

Mention *three* uses of aluminium, stating clearly in each case the property of the metal which makes it a particularly appropriate material. *(London part question)*

11 Explain what constitutes an alum and outline a laboratory preparation of a pure crystalline sample of an alum starting from aluminium.

4.74 g of an alum which were found to contain 0.39 g of potassium and 0.27 g of aluminium were dissolved in 250 cm³ of water. 20 cm³ of the solution produced 0.373 g of $BaSO_4$ on quantitative precipitation. Calculate the number of molecules of water of crystallization in the alum.

(A_r: H = 1.0; O = 16; Al = 27; S = 32; K = 39; Ba = 137.) *(Welsh part question)*

12 (a) Write an equation, including state symbols, which represents the first ionization energy of an element *M*.
(b) Describe briefly, with the aid of a diagram, how the first ionization energy of an element can be experimentally determined using a simple electron source, electron accelerator and current detector assembled inside a glass bulb filled with sodium vapour at low pressure.
(c) Typical results for the ionization energies of boron are:

Ionization energy	1st	2nd	3rd	4th	5th
kJ mol⁻¹	+800	+2400	+3700	+25 000	+32 800

(i) On a piece of graph paper, plot $\log_{10}$ [ionization energy/kJ mol⁻¹] against the number of electrons removed from boron.
(ii) Label the graph at each point with the electronic configuration (s, p, d or f) of the corresponding electron removed.
(iii) Deduce from the graph, with some explanation, the most likely formula of boron oxide.
(d) Write an equation, including state symbols, for the standard enthalpy change which is known as the lattice energy of boron oxide.
(e) The experimental value of the lattice energy of boron oxide is significantly different from that derived theoretically.
What assumption in the theory must be invalid to account for the discrepancy?
(f) For boron fluoride gas:
(i) draw a bond diagram;
(ii) explain why the structure is electron-deficient;
(iii) explain, using a bond diagram, how it forms an addition compound with ammonia. *(SUJB)*

13 (a) A hydrated aluminium sulphate, $Al_2(SO_4)_3.xH_2O$, contains 8.1% of aluminium by mass.
(i) Find the value of *x*.

(ii) Comment on the colour of an aqueous solution of the salt.

(b) Explain why an aqueous solution of aluminium sulphate has a pH less than 7.

(c) The standard electrode potential for the couple $Al^{3+}(aq)|Al(s)$ is -1.66 V. Why does aluminium not dissolve readily in dilute acids? (*London part question*)

14 (a) Discuss briefly whether the properties of the boron halides suggest that boron is a metal or a non-metal.

(b) Explain what is meant by a *Lewis acid* and give an example of a boron halide acting as a Lewis acid.

(c) The peaks with the highest mass-to-charge ratio in the mass spectrum of boron trifluoride occur at *m/e* 67 and 68. At what mass-to-charge ratios above *m/e* 110 will there be peaks in the mass spectrum of boron trichloride? If boron trifluoride and boron trichloride are mixed, chemical reactions occur and when the mass spectrum is measured some time later several new peaks with, for example, mass-to-charge ratios of 86 and 104 appear. What species could be responsible for the peaks at *m/e* 86 and 104? Indicate, with your reasons, which, if any, of these species would possess a dipole moment.

(*Oxford S*)

15 Boron trichloride reacts with ammonia at room temperature to form a compound **P** of formula BCl_3NH_3. It also reacts with ammonium chloride at 170 °C to form a compound **Q** of formula $B_3Cl_3N_3H_3$. **Q** undergoes reduction to give a volatile liquid **R** containing the elements: boron (40.3 per cent by mass), nitrogen (52.2 per cent by mass) and hydrogen only. 100 cm^3 of the vapour of **R**, measured at 373 K and 1 atmosphere pressure, was found to have a mass of 0.263 g. Electron diffraction studies showed that **R** has a cyclic structure with equal boron–nitrogen bond lengths and bond angles of 120°. Hydrogen chloride reacts with **R** to form substance **S**. At 50 °C, 1 mole of **S** loses 3 moles of hydrogen gas to form **Q**. **R** forms trimethyl derivatives of formula $B_3N_3C_3H_{12}$. On hydrolysis, one of these derivatives **T** yields methylamine and boric acid, $B(OH)_3$; another, **U**, yields ammonia and methylboronic acid, $CH_3B(OH)_2$.

Suggest structures for substances **P** to **U**. Comment on any particular points of interest in their structures. Show clearly how you arrive at your conclusions. Give balanced equations where possible for the reactions involved.

(*Nuffield S*)

16 'Copper was used by man before iron. Aluminium has only recently been used.' Discuss chemical factors which you think would be relevant to this observation.

(*Oxford Entrance*)

20 The p-block elements: Group IV

The elements in Group IV differ considerably from each other. Carbon is a typical non-metal, and in nearly all its compounds it is tetracovalent. Tin and lead are typically metallic, and each forms two series of compounds with oxidation numbers +2 and +4. Carbon and silicon are among the most important of the elements: all living things are made up of molecules containing carbon, and silicon is a major component of the rocks that make up the Earth's crust.

	III	IV	V
2	5 B	6 C	7 N
3	13 Al	14 Si	15 P
4	31 Ga	32 Ge	33 As
5	49 In	50 Sn	51 Sb
6	81 Tl	82 Pb	83 Bi

Figure 20.1 Group IV in the Periodic Table

The elements in Group IV are as follows:

carbon	C	$[He]2s^2 2p^2$
silicon	Si	$[Ne]3s^2 3p^2$
germanium	Ge	$[Ar]3d^{10}4s^2 4p^2$
tin	Sn	$[Kr]4d^{10}5s^2 5p^2$
lead	Pb	$[Xe]4f^{14}5d^{10}6s^2 6p^2$

Their positions in the Periodic Table are shown in Figure 20.1. The characteristic valence electron configuration of the atoms is ns^2np^2, where n is the Period number. We shall not consider germanium any further although, as it has excellent semiconducting properties, it may become very important in a few years' time.

20.1 Group systematics

Oxidation states and bonding characteristics

1a L:E:L (with L above and below, lone-pair dots)

1b tetrahedral E with four L

All the elements form compounds in which they have oxidation number +4. These include tetrachloromethane (CCl_4) and silica (SiO_2) and in general EL_4 (**1**) and EO_2, where E denotes the element. In these compounds all four valence electrons take part in bonding.

The lower members of the Group, particularly tin and lead, also occur in compounds with oxidation number +2; examples include tin(II) chloride, $SnCl_2$, and lead(II) sulphate, $PbSO_4$. The emergence of this oxidation number is an example of the trend in Groups III, IV and V for the oxidation number $N-2$ (where N is the Group number) to become increasingly stable with increasing atomic number. In this Group $N = 4$, and so oxidation number +2 can be expected to be more stable in lead than in tin. This gradual removal of the effective participation of two of the four valence electrons is called the **inert-pair effect**, and it is related in a complicated way to the variation of ionization energies, lattice enthalpies and covalent bond enthalpies. In fact, the variation is so subtle that even today no satisfactory *simple* explanation of the effect is available. Therefore, we shall regard the inert-pair effect as a useful rule of thumb, and not attempt to give any explanation.

The reducing powers of tin and lead can be assessed by referring to their standard electrode potentials (Table 13.1). The values for Sn^{2+},Sn and Pb^{2+},Pb are −0.14 V and −0.13 V respectively: they are both negative, and therefore both metals have a thermodynamic tendency to displace hydrogen from acids. There may be *kinetic* reasons which result in the displacement being unimportant in practice, however. This is often so with lead, because many lead(II) salts are insoluble and provide a protective layer on the metal. This is one of the reasons why lead is widely used as a durable building material.

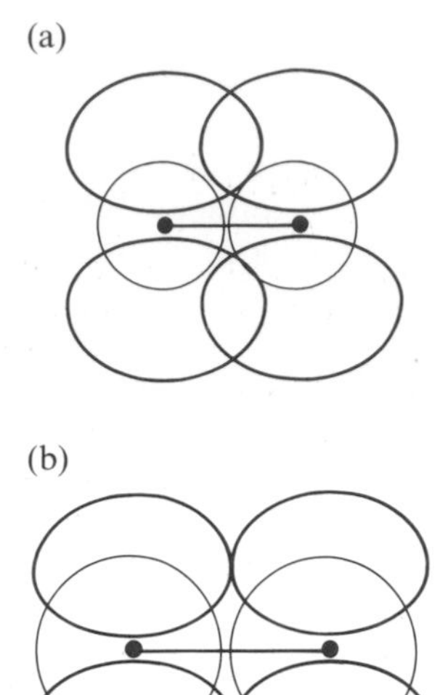

Figure 20.2 p_π–p_π bonding is (a) important when the atoms are close together (as in the C—C bond) but (b) insignificant when the atoms are further apart (as in the Si—Si bond)

Periodic trends

The general trend in the Group is from non-metallic to metallic character with increasing atomic number. This shows itself in several ways. The elements change from the extended covalent structure typical of carbon (the diamond and graphite structures, shown in Figures 5.2 and 5.3) to the typically metallic close-packed, 12-coordinate structure of lead, shown in Figure 5.20. Whereas the dioxides of carbon and silicon are acidic (a typical feature of non-metals), tin(IV) and lead(IV) oxides are amphoteric. Finally, whereas neither carbon nor silicon exists as a simple cation, and their compounds are more covalent than ionic, aqueous solutions containing Sn^{2+}(aq) and Pb^{2+}(aq) are well known and tin and lead compounds include ionic salts.

As usual, the first member of the Group – carbon – is strikingly different from the others. This is due to the small size of its atom and the unavailability of d-orbitals. Its small size means that carbon can form multiple bonds, because two atoms can come so close that broadside overlap of p-orbitals is feasible (see Figure 20.2). The atoms of the other members of the Group are too bulky for this to happen, so that whereas many carbon compounds contain double bonds, the compounds of other elements in the Group have predominantly single bonds. This leads to the striking differences between CO_2 and SiO_2: the former is built from individual O═C═O molecules, whereas the latter is an extended array of SiO_4 units linked by sharing oxygen atoms so that each silicon atom has a half-share of four oxygen atoms. The absence of d-orbitals shows itself in the high *kinetic* stability of carbon compounds: whereas silicon and the elements lower in the Group can accommodate an electron pair from an attacking Lewis base, such as water, carbon cannot expand its octet and its compounds are kinetically more stable (see Section 26.1).

Shapes and sizes

The size of an atom depends on its environment and on the type of bonding; we must distinguish between ionic and covalent compounds.

For tin and lead, which form ionic compounds with oxidation number +2, it is instructive to compare their *ionic radii* (see Figure 20.3): Pb^{2+} is bigger than Sn^{2+} because its electrons are more numerous and its valence shell has a larger principal quantum number.

All the elements of Group IV form covalent compounds of formula EL_4, and so for them it is also instructive to compare the *covalent radii*: these, like the ionic radii and for the same reasons, increase down the Group.

It is interesting to examine the shapes of the molecules. These are all based on the tetrahedron, as predicted by VSEPR theory (see Section 2.3), except that carbon's ability to form double and triple covalent bonds results in molecules with a richer variety of shapes as shown in Section 2.4.

The elements other than carbon can form compounds using their d-orbitals. Silicon has 3d-orbitals that are sufficiently low in energy to take part in bond formation. This shows itself in the existence of octahedral silicon complexes and in the reactions of silicon tetrafluoride, SiF_4. This compound acts as a Lewis acid because silicon's d-orbitals can accept lone pairs, and in excess hydrofluoric acid it forms the octahedral $SiF_6{}^{2-}$ ion (**2**):

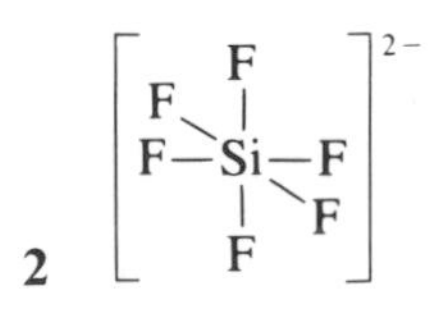

$$SiF_4(g) + 2F^-(aq) \rightarrow SiF_6{}^{2-}(aq)$$

The covalent compounds of the Group IV elements with oxidation number +2 show yet another variation in shape. For these compounds we have to take into account the lone pair of electrons, and then the VSEPR rules predict distorted trigonal planar molecules. This is observed for tin(II) chloride, for example, which in the vapour consists of angular $SnCl_2$ molecules, the lone pair lying in the same plane as the two Sn—Cl bonds (**3**).

3 Sn, Cl, Cl

20.2 The elements

Carbon

Carbon exists in two important solid allotropic forms: **diamond** and **graphite**. Diamond has an extended covalently bonded structure in which each carbon atom is bonded to four others at the corners of a tetrahedron (see Figure 5.2). This compact, rigid arrangement explains why diamond is both extremely hard and chemically inert. Its high refractive index (the reason for its sparkle), its transparency and its rarity have resulted in diamonds being coveted as jewels (see Figure 20.4). Diamonds are also important industrially; because they are excellent and unusually hard-wearing abrasives, they are used to tip the bits used for oil-drilling, as bearings in precision instruments and for record styli. A lesser known property of diamond is its extremely high thermal conductivity (another consequence of the rigidity of the lattice, because thermal motion is readily transmitted): as a result, diamonds are sometimes used for the bases of integrated circuits that need to be kept cool. Natural diamonds are found embedded in the rock **kimberlite**, often in very large 'pipes' (vertical veins of ore); synthetic diamonds are also widely available, and are made by heating graphite to about 2000 °C under extremely high pressure in the presence of molten chromium or nickel, either of which catalyses the conversion.

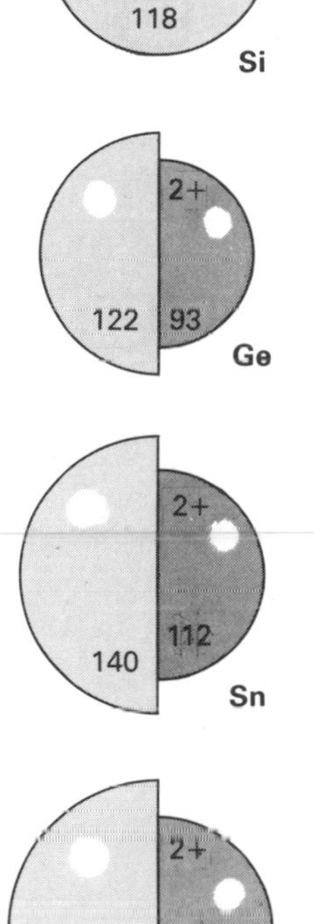

Figure 20.3 The radii (in picometres) of the atoms and M^{2+} ions of the Group IV elements

(a)

(b)

Figure 20.4 The Cullinan diamond is the largest ever found: (a) shows a model of the original gem, from which was cut the 'First Star of Africa' (b), which is in the head of the Royal Sceptre

Graphite has a layer structure, shown in Figure 5.3. The planes of hexagonal rings of covalently bonded carbon atoms are held together by weak van der Waals forces, and slide over each other easily. As a result, graphite is used as a lubricant. The 'lead' in pencils is a crude form of graphite; as the pencil writes, the layers of graphite are rubbed off and deposited on the paper. Graphite also has interesting electrical properties, because the electrons are delocalized within each layer. The electrical conductivity is therefore high in the direction parallel to the layers, but much lower (about 5000 times less) perpendicular to them. The high electrical conductivity and low reactivity of graphite account for its use as electrodes in electrolytic cells, such as in the Downs cell for extracting sodium (see Figure 17.10).

A modern application of graphite is in the form of **carbon fibres** incorporated as strengthening material in polymers of various kinds, for example, in lightweight turbine blades and as brakes (for some Formula 1 racing cars). The fibres are formed by heating acrylic fibres in air to about 300 °C, and then in an inert atmosphere to 1500 °C to produce the carbon fibre itself.

Other forms of graphite include **charcoal**, which is made by heating wood or bones in the absence of air, **soot**, **coke** and **carbon black**, the last being used for reinforcing rubber tyres. These four are all impure, microcrystalline forms of graphite.

All forms of carbon are thermodynamically unstable with respect to oxidation. (Put less formally: charcoal is an excellent fuel for barbecues.) At temperatures below about 1000 K combustion of carbon results in carbon dioxide (see Section 11.3):

$$C(s) + O_2(g) \xrightarrow{<1000\,K} CO_2(g) \qquad O(0) \rightarrow O(-2),\quad C(0) \rightarrow C(+4)$$

Above 1000 K the principal product is carbon monoxide:

$$2C(s) + O_2(g) \xrightarrow{>1000\,K} 2CO(g) \qquad O(0) \rightarrow O(-2),\quad C(0) \rightarrow C(+2)$$

At high temperatures carbon also reduces other non-metals and steam, typical reactions being

$$4C(s) + S_8(s) \xrightarrow{1000\,^\circ C} 4CS_2(g) \qquad S(0) \rightarrow S(-2),\quad C(0) \rightarrow C(+4)$$

$$C(s) + H_2O(g) \xrightarrow{1000\,^\circ C} CO(g) + H_2(g) \qquad H(+1) \rightarrow H(0),\quad C(0) \rightarrow C(+2)$$

The product of this last reaction, which is carried out by passing steam over white-hot coke, is called **water gas**, and until the development of the petrochemical industry was a major source of hydrogen.

Silicon

The structure of silicon, a lustrous blue-grey solid, is like that of diamond. Its most important physical property is that it is a semiconductor (see Box 20.1). Silicon is chemically unreactive on account of an oxide layer that seals the surface of the bulk solid against attack, and high temperatures are usually needed before extensive oxidation occurs:

$$Si(s) + O_2(g) \xrightarrow{1000\,^\circ C} SiO_2(s) \qquad O(0) \rightarrow O(-2),\quad Si(0) \rightarrow Si(+4)$$

$$Si(s) + 2Cl_2(g) \xrightarrow{300\,^\circ C} SiCl_4(g) \qquad Cl(0) \rightarrow Cl(-1),\quad Si(0) \rightarrow Si(+4)$$

An exception is its reaction with fluorine, which takes place at room temperature:

$$Si(s) + 2F_2(g) \xrightarrow{20\,^\circ C} SiF_4(g) \qquad F(0) \rightarrow F(-1),\quad Si(0) \rightarrow Si(+4)$$

Silicon is not attacked by aqueous acids, but it does react with concentrated alkalis. This reaction is an oxidation, and hydrogen is evolved (compare the behaviour of aluminium – see Section 19.2):

$$Si(s) + 4OH^-(aq) \xrightarrow{heat} SiO_4^{4-}(aq) + 2H_2(g) \qquad H(+1) \rightarrow H(0), Si(0) \rightarrow Si(+4)$$

BOX 20.1

Figure 20.5 A 32-bit 'transputer': this small silicon chip contains a quarter of a million transistors

Semiconductors

The Group IV elements silicon and germanium have revolutionized whole areas of technology, just as carbon revolutionized the natural world. Whereas the first computers were huge, energy-intensive structures with wires linking each memory location, the whole of a modern microprocessor can be fitted on to a chip a few millimetres square (see Figure 20.5). This change is a result of the discovery and development of **semiconductors**.

Metals conduct electricity because the valence electrons of their atoms occupy orbitals that spread throughout the solid (see Section 2.5). Electrons in such orbitals, which are called **conduction bands**, can move in response to an applied potential difference (p.d.). In an electrical insulator, the electrons are localized on individual atoms or in bonds between them, and are not free to wander through the solid. To put it another way, there is a large energy gap, a **band gap**, between the orbitals occupied by these localized electrons and the next available empty conduction band.

In a semiconductor, a small additional supply of electrons, or their removal, can result in the solid becoming conducting. Thus, pure germanium can be **doped** with a minute quantity of either gallium or arsenic, its neighbours in Period 4. The atoms of both elements are similar in size to a germanium atom, and so do not greatly distort the structure. An arsenic atom (from Group V) has one more electron than a germanium atom, however, and it occupies the conduction band. When a p.d. is applied to the doped germanium, the current is carried by the negatively charged electrons supplied by the arsenic atoms. This substance is called an ***n*-type semiconductor** (*n* denoting *n*egative electrons). Alternatively, if germanium is doped with gallium, the solid has fewer electrons overall (since each gallium atom has one less electron than a germanium atom). This is expressed by saying that the gallium introduces **positive holes** into the solid. The presence of the positive holes (that is, the absence of electrons) enables the remaining electrons to move through the solid more freely, and the gallium-doped solid is called a ***p*-type semiconductor** (*p* for the *p*ositive holes).

Tin and lead

Tin has two common allotropes. At room temperature the stable form is the metallic allotrope **white tin** (denoted β-tin). The crackle heard when sheets of pure tin are bent is due to the crystal faces rubbing against each other. Below 13.2 °C the thermodynamically stable form is the allotrope **grey tin** (α-tin), which has a diamond-like structure. The transition between the allotropes is slow near the transition temperature, and α-tin is formed only after long exposure of β-tin to low temperatures (the change is known as 'tin plague' – see Section 6.2). The only form of lead is metallic, and both β-tin and lead readily form alloys with each other and with other metals (Table 20.1). The discovery of alloys of this kind was a major step forward for humanity, for it led to the Bronze Age (approximately 3500–1000 B.C.), which was

Table 20.1 Alloys of tin

Name	Mass % Sn	Other metal(s)
pewter	≈90	Sb
solder	≈33	Pb
bronze	5–10	Cu
type metal	≈5	Pb, Sb

Note: Most alloys contain several other metals. The metals shown in the last column are major (>5 per cent) constituents.

characterized by a surge in the range of artefacts that the new materials made available. Much the same kind of progress continues to this day, with chemists extending the range of materials and hence the range of objects that can be made.

The main use of tin is as a surface layer to protect iron and steel objects. Since it is resistant to weak acids, such as the citric acid of citrus fruit, it has been widely used for protecting cans containing food and drink. Nowadays food and drink cans are increasingly made from aluminium rather than tin-plated steel. Tin-plate is produced either by electrolysis or by dipping the object into molten tin. Tin is also used in the **float-glass process** for the production of plate glass, in which molten glass is poured on to the flat surface of a slowly moving stream of molten tin.

Lead is a dense, malleable, soft metal with a low melting point (327 °C). It is very easy to work, and from ancient times until recently it was used for water pipes; since its Latin name is *plumbum*, water supply workers became known as plumbers. As we now know, lead is a cumulative, heavy-metal poison – it has been speculated that one reason for the fall of the Roman Empire was the mental decline brought about by the ingestion of lead from the water supply and from goblets – and it is no longer used in this way. The current principal use of lead, on a megatonne scale each year, is as the electrodes in car batteries (see Box 13.1). Another use is as a screen against radiation.

Tin and lead are quite easily oxidized, tin usually to Sn(IV) and lead to Pb(II). When heated they react with the oxygen in air to form the oxides:

$$Sn(s) + O_2(g) \xrightarrow{\text{heat}} SnO_2(s) \qquad O(0) \rightarrow O(-2), \quad Sn(0) \rightarrow Sn(+4)$$

$$2Pb(l) + O_2(g) \xrightarrow{\text{heat}} 2PbO(s) \qquad O(0) \rightarrow O(-2), \quad Pb(0) \rightarrow Pb(+2)$$

Bulk lead is normally protected by an oxide layer, but unprotected and powdered lead (made by heating lead(II) 2,3-dihydroxybutanedioate (tartrate) in the absence of air) is **pyrophoric**, that is, it catches fire on exposure to air.

Both tin and lead reduce the halogens, and undergo the same changes of oxidation number as with oxygen:

$$Sn(s) + 2Cl_2(g) \longrightarrow SnCl_4(l) \qquad Cl(0) \rightarrow Cl(-1), \quad Sn(0) \rightarrow Sn(+4)$$

$$Pb(s) + Cl_2(g) \xrightarrow{\text{heat}} PbCl_2(s) \qquad Cl(0) \rightarrow Cl(-1), \quad Pb(0) \rightarrow Pb(+2)$$

The lower oxidation number is produced in *both* cases when hydrogen halides are used in place of the halogens, and both tin and lead react as follows (E = Sn, Pb):

$$E(s) + 2HCl(g) \xrightarrow{\text{heat}} ECl_2(s) + H_2(g) \qquad H(+1) \rightarrow H(0), \quad E(0) \rightarrow E(+2)$$

EXAMPLE

Will lead displace tin from aqueous tin(II) chloride?

METHOD The answer to this type of question depends on knowing the standard electrode potentials of the redox couples concerned, and using the information given in Section 13.5. Refer to Table 13.1 for standard electrode potentials; remember that a metal M will displace another M′ only if it has a couple with a more negative value.

ANSWER From Table 13.1,

$$E^{\ominus}(Sn^{2+}, Sn) = -0.14\,V, \quad E^{\ominus}(Pb^{2+}, Pb) = -0.13\,V$$

Since the Sn^{2+}, Sn couple is thus slightly more strongly reducing than the Pb^{2+}, Pb couple in aqueous solution, under standard conditions, *tin has a tendency to displace lead*, not vice versa.

COMMENT The conclusion that lead cannot displace tin might be thought surprising in view of the relative locations of the elements in the Group, with tin above the 'more metallic' lead. In chemistry small shifts of values may upset general tendencies, however; for reliable conclusions quantitative data must always be used.

Lead resists attack by sulphuric acid, because a protective layer of insoluble lead(II) sulphate is formed, and lead vessels are sometimes used for transporting the acid. However, lead does react with acids, such as dilute nitric acid and ethanoic acid, that form soluble lead(II) salts.

20.3 Oxides

Survey and structures

The formulae and principal characteristics of the major oxides of Group IV are given in Table 20.2. The dioxides clearly differ widely in their structures, and hence in their properties: we breathe carbon dioxide (a molecular gas) but we stand on silicon dioxide (silica, a solid with an extended covalently bonded network structure). Tin and lead dioxides (i.e., tin(IV) oxide and lead(IV) oxide) are best thought of as ionic compounds; each metal ion is at the centre of an octahedron of six oxide ions, and each oxide ion is surrounded by three metal ions.

Table 20.2 The oxides of Group IV elements

Formula	Characteristics
CO	gas; molecular compound; neutral
CO_2	gas; molecular compound; acidic
SiO_2	solid; extended covalent network; acidic
SnO	solid; ionic compound; amphoteric
SnO_2	solid; ionic compound; amphoteric
PbO	solid; ionic compound; amphoteric, but mainly basic
Pb_3O_4	solid; a mixed oxide containing Pb^{2+} and Pb^{4+} ions
PbO_2	solid; strong oxidizing agent

In agreement with the Group trends, oxides with lower oxidation numbers become more stable with increasing atomic number. Lead(IV) oxide decomposes when heated in air, forming lead(II) oxide:

$$2PbO_2(s) \xrightarrow{600\,^{\circ}C} 2PbO(s) + O_2(g) \qquad Pb(+4) \rightarrow Pb(+2),\quad O(-2) \rightarrow O(0)$$

This is easily detected, because the dioxide is brown and the monoxide is yellow. Because of the relative instability of the compounds in which lead has oxidation number +4, lead(IV) oxide is a strong oxidizing agent, and can be used to oxidize concentrated hydrochloric acid to chlorine:

$$PbO_2(s) + 4HCl(aq, conc.) \xrightarrow{heat} PbCl_2(s) + Cl_2(g) + 2H_2O(l) \qquad Pb(+4) \rightarrow Pb(+2),\quad Cl(-1) \rightarrow Cl(0)$$

Brown lead(IV) oxide oxidizes sulphur dioxide to white lead(II) sulphate:

$$PbO_2(s) + SO_2(g) \rightarrow PbSO_4(s) \qquad Pb(+4) \rightarrow Pb(+2),\quad S(+4) \rightarrow S(+6)$$

In contrast, tin(II) sulphate decomposes above 380 °C:

$$SnSO_4(s) \xrightarrow{>380\,^{\circ}C} SnO_2(s) + SO_2(g) \qquad S(+6) \rightarrow S(+4),\quad Sn(+2) \rightarrow Sn(+4)$$

Carbon monoxide

Carbon monoxide, CO, is prepared in the laboratory by warming methanoic or ethanedioic acid (or their sodium salts) with concentrated sulphuric acid:

$$HCOOH(l) \xrightarrow{\text{heat, sulphuric acid}} CO(g) + H_2O(l)$$

$$(COOH)_2(s) \xrightarrow{\text{heat, sulphuric acid}} CO(g) + CO_2(g) + H_2O(l)$$

$$2C(+3) \rightarrow C(+2) + C(+4)$$

(The former reaction is a dehydration, an acid–base reaction; the second is a disproportionation as well as a dehydration.) Carbon monoxide is also formed when carbon-containing materials burn in a limited supply of oxygen. It is a colourless gas, which burns with a blue flame to form carbon dioxide:

$$2CO(g) + O_2(g) \rightarrow 2CO_2(g) \qquad O(0) \rightarrow O(-2), \quad C(+2) \rightarrow C(+4)$$

Carbon monoxide is dangerously toxic because it binds much more strongly than oxygen with the iron in haemoglobin, forming carboxyhaemoglobin. The formation of this relatively stable compound blocks the oxygen transport system of the blood, and the victim suffocates. Since the gas is odourless, its presence cannot easily be detected, and this adds to its danger.

The most important chemical property of carbon monoxide is its reducing power. It reduces iron(III) oxide to iron:

$$Fe_2O_3(s) + 3CO(g) \xrightarrow{\text{heat}} 2Fe(l) + 3CO_2(g) \qquad Fe(+3) \rightarrow Fe(0), \quad C(+2) \rightarrow C(+4)$$

This is the main reaction responsible for the production of iron from its ores in blast furnaces, described in Section 25.9.

Although carbon monoxide is formally the anhydride of methanoic acid (a relationship used in its laboratory preparation, described above), the rate of formation of the acid when the gas dissolves in water is so slow that in practice carbon monoxide is a neutral oxide. At 200 °C and 8 atm it does react with sodium hydroxide, with the formation of sodium methanoate, $Na(HCO_2)$.

When heated in the presence of a catalyst (a mixture of zinc oxide and chromium(III) oxide), carbon monoxide combines with hydrogen, forming methanol:

$$CO(g) + 2H_2(g) \xrightarrow{300\,^{\circ}C,\ 300\,\text{atm, catalyst}} CH_3OH(g)$$

This is an important industrial method for the synthesis of methanol, and a bridge between inorganic and organic chemicals.

Carbon dioxide and the carbonates

Carbon dioxide is essential to life since, being a gas, it is a readily available source of carbon for growing plants. They incorporate it by the process of **photosynthesis**, in which it is combined with water in the presence of sunlight to form carbohydrates (see Section 30.5). This extremely complex reaction can be summarized as follows:

$$6CO_2(g) + 6H_2O(l) \xrightarrow{\text{light}} C_6H_{12}O_6(aq) + 6O_2(g)$$

This is the head of the food chain, for by it the energy of the Sun is trapped and made available for other kinds of life.

Carbon dioxide is a gas under normal conditions, but freezes at −78 °C to a white solid. The liquid forms only under a pressure of at least 5.1 atm (see Section 6.2). At 1 atm the solid sublimes directly to the gas, and so solid carbon dioxide ('dry ice') is widely used as a cheap, convenient refrigerant (for example, for storing ice cream).

The gas dissolves in water to give an acidic solution:

$$CO_2(g) + H_2O(l) \rightleftharpoons (HO)_2CO(aq) \rightleftharpoons H^+(aq) + HOCO_2^-(aq)$$
$$\rightleftharpoons 2H^+(aq) + CO_3^{2-}(aq)$$

A simple consequence of Le Chatelier's principle (see Section 10.2), which is also the principle behind the production of soda water, is that if the pressure increases, more gas dissolves. Evaporating the solution releases carbon dioxide, and pure carbonic acid cannot be obtained. The addition of acid to the solution shifts the equilibria to the left, and carbon dioxide is evolved. Conversely, alkali shifts the equilibria to the right. An example is the reaction with lime water (see Section 18.3) which is turned milky by carbon dioxide:

4

$$Ca(OH)_2(aq) + CO_2(g) \rightarrow CaCO_3(s) + H_2O(l)$$

The carbonate ion (CO_3^{2-}, **4**) is trigonal planar and the double negative charge is delocalized over all three oxygen atoms (see Section 2.5). It is extensively hydrolysed in solution, and solutions of soluble carbonates, such as sodium carbonate, are alkaline (see Section 12.3). Most carbonates are insoluble (with the usual exception of the alkali metal carbonates), partly on account of the high lattice enthalpy arising from the doubly charged anion. Hydrogencarbonates, which contain the $HOCO_2^-$ ion, are generally more soluble (see Section 18.6).

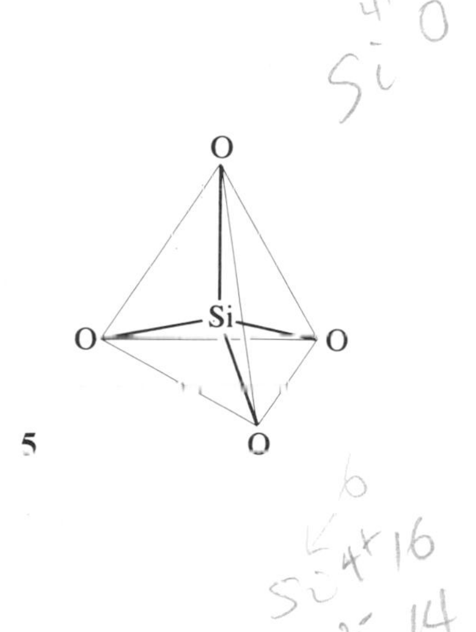

5

Silica and silicates

One of silicon's most important features is the strong bond that it forms with oxygen. Silica, SiO_2, is polymorphic, and an important form is **quartz**, which is used to control the frequency of oscillators (in clocks and watches, for example). **Sand** is an impure form of quartz; its pale brown colour is due to iron impurities.

Since all its polymorphs have an extended network structure of covalently linked SiO_4 tetrahedra (**5**), silica is insoluble in water. In a blast furnace it reacts with the base calcium oxide to form a slag:

$$CaO(s) + SiO_2(s) \xrightarrow{900\,^\circ C} CaSiO_3(l)$$

Oxoanions derived from silica are called **silicates.** They are complex and varied, but extremely important in everyday life – one of the best known, indeed notorious, is **asbestos** (see Box 20.2) – and account for about 95 per cent of the Earth's crustal rocks. All silicates have structures based on SiO_4 tetrahedra, with each tetrahedron sharing some of the oxygen atoms at its corners with its neighbours. In silica itself, each SiO_4 tetrahedron shares its corners with four other SiO_4 tetrahedra. Each silicon atom therefore has a half-share in four oxygen atoms, and so the average formula is SiO_2 (see Figure 20.6). If *none* of the oxygen atoms is shared, then the **orthosilicate** SiO_4^{4-} anion results. The richness of the structures of the silicates then arises because the sharing may be intermediate between these two extremes, and various ring, chain and sheet structures are possible (see Figure 20.7).

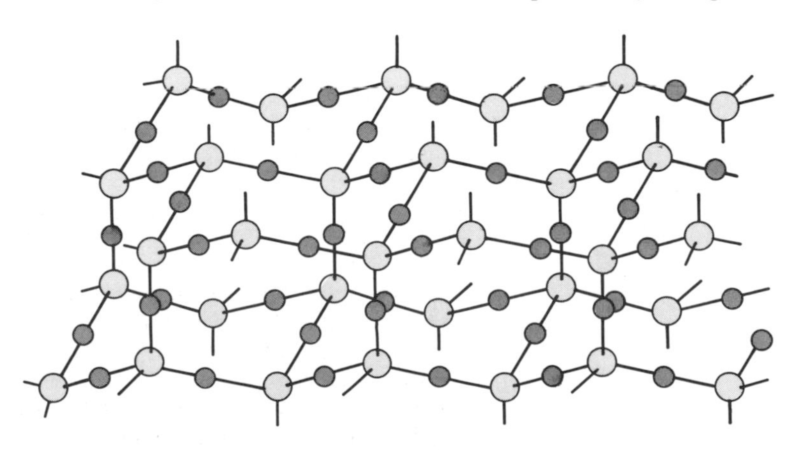

Figure 20.6 Cristobalite, a form of silica

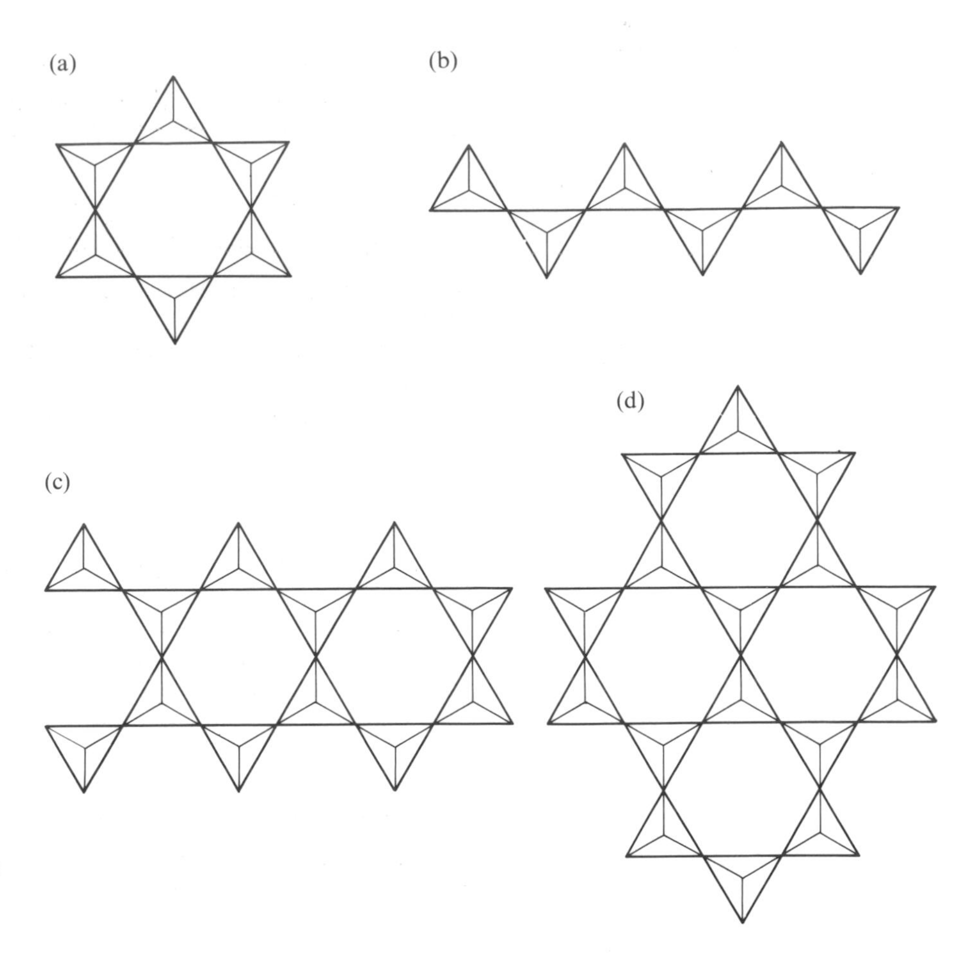

Figure 20.7 SiO_4 tetrahedra join to form (a) rings, (b) chains, (c) double chains and (d) sheets

BOX 20.2

Figure 20.8 Fibres of asbestos

Asbestos

Asbestos is a silicate that has been known since antiquity for its ability to resist heat. Recently, however, it has become clear that it is also a dangerous health hazard. This is because when asbestos is powdered or subjected to stress, microscopic particles of its fibres can become airborne and eventually lodge in people's lungs, resulting in the disease of **asbestosis** and causing **mesothelioma** (a type of cancer). Sufferers from asbestosis experience great difficulty in breathing, and for some even walking becomes a problem. The form known as **blue asbestos** is the most dangerous and its use is prohibited by many countries, and the major form, **white asbestos**, is now carefully controlled.

The complexity of the silicates themselves is exceeded by that of the **aluminosilicates** (see Section 19.7), in which some silicon atoms are replaced by aluminium atoms. Industrially important aluminosilicates include the **zeolites**, which have a very open structure, resembling tunnels and cavities built on a molecular scale. The sizes of these openings depend on the composition of the zeolite, and allow them to act like sieves, admitting molecules only if they are small enough. Zeolites have important applications as 'molecular sieves' and as ion-exchange materials (see Section 18.6). The name 'zeolite' comes from the Greek for 'boiling stone', because water trapped in the structure boils off when the stone is heated.

Tin and lead oxides

The general tendency towards metallic character on descending the Group is shown by the oxides. The metal(IV) oxides become more basic, and tin(IV) oxide is amphoteric:

$$SnO_2(s) + 2OH^-(aq,conc.) + 2H_2O(l) \rightarrow [Sn(OH)_6]^{2-}(aq)$$

$$SnO_2(s) + 6HCl(aq,conc.) \rightarrow SnCl_6{}^{2-}(aq) + 2H_3O^+(aq)$$

$[Sn(OH)_6]^{2-}$ is called the **stannate(IV)** ion, and the corresponding lead ion is called the **plumbate(IV)** ion.

The tin(II) and lead(II) oxides are important. Both are amphoteric, and react with nitric acid and also with alkali:

$$PbO(s) + 2HONO_2(aq) \rightarrow Pb(NO_3)_2(aq) + H_2O(l)$$

$$PbO(s) + 4OH^-(aq) + H_2O(l) \rightarrow [Pb(OH)_6]^{4-}(aq)$$

$[Pb(OH)_6]^{4-}$ is the principal complex ion in solutions containing the **plumbate(II)** ion.

Lead(II) oxide is polymorphic, one form being red and the other yellow. It is used for making lead glass, which is denser and tougher than ordinary glass, and has a higher refractive index. Lead glass is used for making prisms and 'crystal' glassware, the sparkle being due to the high refractive index, which in turn is due to the large number of electrons in the lead ion and its resulting high polarizability. If lead(II) oxide is heated in air near 450 °C, the mixed oxide Pb_3O_4, **red lead**, is formed:

$$6PbO(s) + O_2(g) \xrightarrow{450\,^{\circ}C} 2Pb_3O_4(s) \qquad O(0) \rightarrow O(-2),\ 6Pb(+2) \rightarrow 4Pb(+2) + 2Pb(+4)$$

Red lead is used in paint to protect steel from corrosion by sea water, and in primers (for example, for the corroded areas of cars). The 'unexpected' formula Pb_3O_4 can be rationalized by its structure. It is a mixed oxide, like spinel (see Section 19.3), derived from $2Pb^{2+}$, Pb^{4+} and $4O^{2-}$. This mixed oxide structure also explains the behaviour of red lead when it is treated with nitric acid: it turns brown, the colour of PbO_2, because the Pb^{2+} ions form lead(II) nitrate:

$$Pb_3O_4(s) + 4HONO_2(aq) \rightarrow PbO_2(s) + 2Pb(NO_3)_2(aq) + 2H_2O(l)$$

20.4 Halides

All the elements of Group IV form tetrahalides, $E(Hal)_4$, but only tin and lead form dihalides. The thermal stability of the tetrahalides decreases with increasing atomic number as compounds with oxidation number +4 become less stable; lead(IV) chloride decomposes near room temperature:

$$PbCl_4(l) \rightarrow PbCl_2(s) + Cl_2(g) \qquad Pb(+4) \rightarrow Pb(+2),\ Cl(-1) \rightarrow Cl(0)$$

and its bromide and iodide are even less stable.

The tetrahalides are covalent tetrahedral molecules. **Tetrachloromethane**, CCl_4, also commonly called **carbon tetrachloride**, is a colourless, toxic liquid with a pungent smell, and is best prepared by the reaction of carbon disulphide and chlorine:

$$CS_2(l) + 3Cl_2(g) \rightarrow CCl_4(l) + S_2Cl_2(l) \qquad Cl(0) \rightarrow Cl(-1),\ S(-2) \rightarrow S(+1)$$

Silicon tetrachloride and tin(IV) chloride are prepared by direct reaction between the elements. There are marked differences between the tetrachlorides of carbon and silicon: whereas carbon tetrachloride is unaffected by water, silicon tetrachloride is rapidly hydrolysed to a hydrated form of silica and hydrogen chloride:

$$SiCl_4(l) + 2H_2O(l) \rightarrow SiO_2(s) + 4HCl(aq)$$

A practical consequence of this hydrolysis is that the glass stoppers of bottles containing silicon tetrachloride are often tightly stuck because the silica produced seals the stopper to the bottle. Part of the explanation for this difference in behaviour has already been indicated: silicon can use its 3d-orbitals to expand its valence shell, so becoming a Lewis acid and able to accommodate the electron pair of the attacking Lewis base, H_2O. Tin(IV) chloride behaves in the same way, and is also hydrolysed; for the same reason it is also an effective Friedel–Crafts catalyst (see Section 34.3).

The dihalides, $M(Hal)_2$, where M is either tin or lead, are best regarded as ionic. Lead(II) chloride, while soluble in hot water, is insoluble in cold water and is precipitated as a white solid when a solution of chloride ions is added to a lead(II) solution:

$$Pb^{2+}(aq) + 2Cl^-(aq) \rightarrow PbCl_2(s)$$

The precipitate dissolves in excess concentrated hydrochloric acid by the formation of complex ions, including tetrachloroplumbate(II):

$$PbCl_2(s) + 2Cl^-(aq) \rightarrow PbCl_4{}^{2-}(aq)$$

The reaction between lead(II) nitrate and potassium iodide is unusual in that it occurs at a noticeable rate in the solid state (as well as rapidly in solution); yellow lead(II) iodide is produced:

$$Pb^{2+}(s) + 2I^-(s) \rightarrow PbI_2(s)$$

20.5 Hydrides

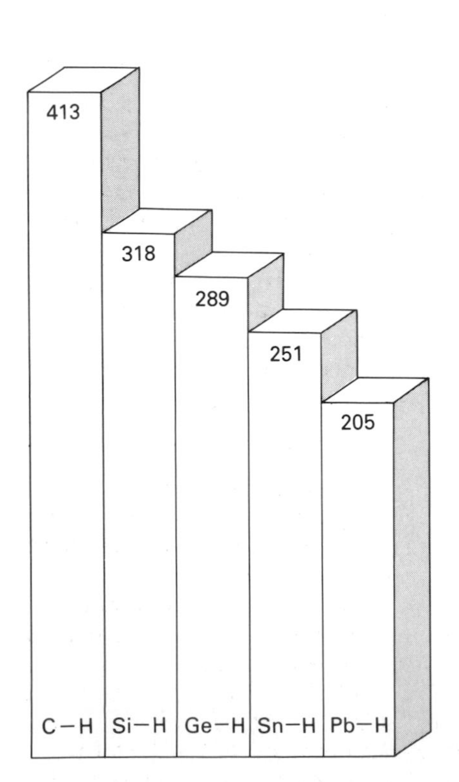

Figure 20.9 The E—H average bond enthalpies

The ability of carbon atoms to **catenate** – to link together into a chain – is exceptional and the hydrides of carbon, the **hydrocarbons**, are so important that they are given separate treatment, in Chapters 27, 33 and 34. Silicon forms a series of hydrides of general formula Si_nH_{2n+2}, where n has a value from 1 to 8; these are known as the **silanes.** Tin forms only the thermally unstable stannanes SnH_4 and Sn_2H_6, and PbH_4, plumbane, has been formed only in trace amounts. This decreased stability correlates well with the decrease in element–hydrogen bond strength down the Group (see Figure 20.9).

The silanes are either gases or volatile liquids. **Monosilane**, SiH_4, and **stannane**, SnH_4, are prepared by the reaction of the appropriate tetrachloride with lithium tetrahydridoaluminate in dry ethoxyethane as solvent:

$$SiCl_4 + LiAlH_4 \rightarrow SiH_4 + LiCl + AlCl_3$$

Silanes are much more reactive than are the alkanes, such as methane, CH_4. The principal reason is once again that silanes are Lewis acids on account of the availability of d-orbitals on silicon, and are susceptible to attack by Lewis bases. Thus, silanes are readily hydrolysed by water in the presence of even a trace of alkali, and form hydrated silica:

$$SiH_4(g) + 2H_2O(l) \xrightarrow{pH > 7} SiO_2(s) + 4H_2(g) \qquad 4H(-1) + 4H(+1) \rightarrow 8H(0)$$

Like the alkanes, the silanes are flammable:

$$SiH_4(g) + 2O_2(g) \xrightarrow{heat} SiO_2(s) + 2H_2O(l) \qquad O(0) \rightarrow O(-2),\quad H(-1) \rightarrow H(+1)$$

and for $n > 1$ they ignite spontaneously in air at room temperature.

20.6 Other compounds

Aqua cations and oxosalts

Tin(II) chloride is soluble in water, producing the $Sn^{2+}(aq)$ ion, which is a strong reducing agent. For example, it reduces mercury(II) chloride to mercury(I) chloride, Hg_2Cl_2, and then to mercury; the initial precipitate is white, becoming dirty grey as the mercury is formed:

$$SnCl_2(aq) + 2HgCl_2(aq) \rightarrow Hg_2Cl_2(s) + SnCl_4(aq)$$

$Hg(+2) \rightarrow Hg(+1)$, $Sn(+2) \rightarrow Sn(+4)$

$$SnCl_2(aq) + Hg_2Cl_2(s) \rightarrow 2Hg(l) + SnCl_4(aq)$$

$Hg(+1) \rightarrow Hg(0)$, $Sn(+2) \rightarrow Sn(+4)$

On account of the greater stability of the +2 oxidation number for lead, lead(II) is not a reducing agent. Whereas $Sn^{2+}(aq)$ reduces acidified solutions of dichromate(VI), which thereon turn green due to the appearance of Cr^{3+}, $Pb^{2+}(aq)$ precipitates yellow lead(II) chromate, $PbCrO_4$, the pigment **chrome yellow**, which is widely used for road markings.

Both $Sn^{2+}(aq)$ and $Pb^{2+}(aq)$ precipitate amphoteric hydroxides with alkali:

$$Pb^{2+}(aq) + 2OH^-(aq) \rightarrow Pb(OH)_2(s)$$

and the addition of more alkali results in their dissolution by complex formation. Most lead(II) salts are insoluble (the nitrate and ethanoate being exceptions) and some are highly coloured; lead(II) iodide, for example, is bright yellow and lead(II) sulphide is black. The latter is formed when **white lead**, once used as the pigment in white paint, is exposed to polluted air; this is the reason for the discoloration of paint that was once often seen in industrial areas. White lead incorporates 'basic lead carbonate', which has the approximate formula $2PbCO_3 . Pb(OH)_2$. It has been replaced as a pigment by titanium(IV) oxide (see Section 25.3). Old oil paintings continue to be marred in this way, however, and restorers use hydrogen peroxide to oxidize the sulphide to white lead(II) sulphate.

Organic compounds of silicon, tin and lead

The C—C and C—Si bond enthalpies (348 and 305 $kJ\,mol^{-1}$ respectively) are similar, and so it should not be surprising that many thousands of organosilicon compounds have been prepared. However, whereas carbon–carbon multiple bonds are common, silicon does not form multiple bonds, and so much of the variety that characterizes carbon's chemistry is absent from silicon's.

One important series of organosilicon compounds is the **siloxanes** (which are also commonly called the **silicones**). Just as silicon tetrachloride is readily hydrolysed to the characteristic network structure of silica, so dichlorodimethylsilane, $(CH_3)_2SiCl_2$, is hydrolysed to a linear polymeric chainlike structure with repeating $—(CH_3)_2SiO—$ units:

$$n(CH_3)_2SiCl_2 + nH_2O \rightarrow [—(CH_3)_2SiO—]_n + 2nHCl$$

Siloxanes are examples of synthetic polymers, but instead of the more usual carbon chain the 'backbone' consists of alternating silicon and oxygen atoms. They are thermally stable and water-repellent oils, waxes that are widely used for polishing surfaces and where resilience is required (silicone wax) and elastomers (rubbers). One of their advantages over natural rubber is that their viscosities and elasticities vary only slightly with temperature. A siloxane fluid is used in the viscous coupling systems of some car transmissions.

Organotin compounds include R_2SnX_2 and R_3SnX, where R is an alkyl group and X an anion such as ethanoate or hydroxide. The R_2SnX_2 compounds are used as stabilizers to prevent discoloration of PVC, and the R_3SnX series as pesticides, for it is believed that although they kill insects they have no effect on mammals. An important organolead compound is **tetraethyllead**(IV), $Pb(C_2H_5)_4$ (see Box 27.3).

20.7 Industrial chemistry

Occurrence

Carbon occurs as the free element as **diamond** (in central and southern Africa and in the USSR) and as **graphite** (in Korea, Mexico and the USSR). In combined form, carbon is widespread in fossil fuels (coal, oil, shale and natural gas), as calcium carbonate (see Section 18.6), in all plants and animals, and as the carbon dioxide of the air (see Box 20.3).

Silicon is the second most abundant element (after oxygen) in the Earth's crust, of which it accounts for 27 per cent of the mass. It never occurs free but always combined with oxygen in **quartz**, **sandstone** and the silicates. Thousands of silicate minerals have been identified. Tin and lead are comparatively rare. The principal ore of tin is **cassiterite**, SnO_2, and that of lead is **galena**, PbS (see Colour 5).

BOX 20.3

Carbon dioxide in the atmosphere

The proportion of carbon dioxide in the atmosphere is increasing, not only because greater amounts of fossil fuels are being burnt but also because, as civilization develops, more cement is made. The destruction of the rain forests may also be reducing the rate at which carbon dioxide is withdrawn from the air by photosynthesis.

The environmental consequences of the rise in carbon dioxide levels are uncertain, but potentially very serious. One is that more carbon dioxide will dissolve in sea and lake water and decrease its pH. A second is that more energy will be trapped in the atmosphere by the **greenhouse effect** (that is, the way in which carbon dioxide allows visible radiation from the Sun to reach the Earth, but traps infrared radiation from the Earth instead of allowing its free radiation back into space); there is a consequent warming of the atmosphere. It is not yet possible to make reliable predictions, but some experts suggest that the mean temperature of the surface of the Earth may have increased by 0.1 °C by the year 2020. Although small, that rise of temperature may be enough to melt a part of the polar ice caps, to the extent that worldwide flooding might be caused.

The manufacture of silicon

Very pure silicon for semiconductors is obtained from its chloride, which is first purified by fractional distillation and then reduced with very pure magnesium or zinc:

$$SiCl_4(l) + 2Mg(s) \rightarrow Si(s) + 2MgCl_2(s) \qquad Si(+4) \rightarrow Si(0), \quad Mg(0) \rightarrow Mg(+2)$$

The element is then further purified by **zone refining**, in which a molten zone is moved along a silicon rod several times; during each passage it collects the impurities and carries them to one end, which is then removed (see Figure 20.10).

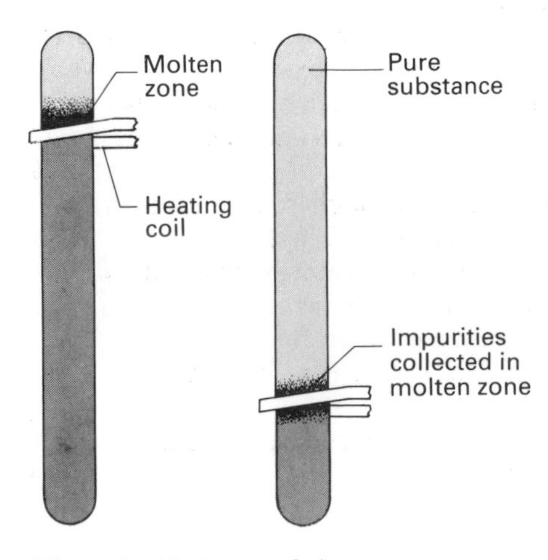

Figure 20.10 Zone refining

The manufacture of cement

Portland cement, which is so called because its inventor thought it resembled Portland stone, is made by heating a finely ground mixture of limestone with carefully controlled amounts of clay (aluminosilicate) in a rotary kiln at 1500 °C. The resultant **clinker** is reground, and mixed with about 3 per cent of gypsum, $CaSO_4 . 2H_2O$. As cement, like the clay that is used to make it, is a complex mixture of silicates and aluminosilicates, any equation that tries to represent its manufacture must be a simplification. An approximation, which conveys the kind of reactions that take place, is

$$CaAl_2Si_2O_8(s) + 8CaCO_3(s) \xrightarrow{1500\,^\circ C} \underbrace{\{2Ca_3SiO_5(s) + Ca_3Al_2O_6(s)\}}_{\text{cement}} + 8CO_2(g)$$

The setting of cement is also a complex process, in which the cement reacts with water together with carbon dioxide from the air. A cement kiln produces a few thousand tonnes a day, and the worldwide annual production is nearly 1 billion tonnes.

The manufacture of glass

In the manufacture of glass a mixture containing sandstone (SiO_2), limestone ($CaCO_3$) and sodium carbonate (Na_2CO_3) is fused in a furnace at about 1500 °C, and the components react approximately as follows:

$$CaCO_3(s) + SiO_2(s) \rightarrow CaSiO_3(l) + CO_2(g)$$

$$Na_2CO_3(s) + SiO_2(s) \rightarrow Na_2SiO_3(l) + CO_2(g)$$

The resultant transparent mixture of silicates sets to a glassy solid, which is better regarded as an extremely viscous liquid. Other oxides, such as B_2O_3 for borosilicate glasses ('Pyrex'), or oxosalts (including K_2CO_3) are added to modify the properties of the resulting glass. Coloured glass is obtained by adding various materials to the mixture, such as cadmium selenide or colloidal gold for red glass and cobalt(II) oxide for blue glass.

Summary

- ☐ The elements have the characteristic valence electron configuration ns^2np^2.
- ☐ There are considerable differences between the elements in this Group: carbon and silicon are typical non-metals; tin and lead are typical metals.
- ☐ Features of carbon chemistry are the strength of the carbon–carbon covalent bond, the formation of chains and rings of carbon atoms, and the ability of carbon to form multiple bonds.
- ☐ Unlike carbon, silicon does not form multiple bonds.
- ☐ The +4 oxidation number is exhibited by all the elements. Tin and lead also form many compounds with the element having oxidation number +2.
- ☐ Silicon can form compounds with coordination numbers greater than 4 (SiF_6^{2-}, for example) and this is consistent with the greater reactivity of many silicon compounds compared with those of carbon (for instance, silicon tetrachloride is readily hydrolysed, tetrachloromethane is not).
- ☐ Very pure silicon, 'doped' with other elements, is a semiconductor and is of prime importance in making solid-state electronic devices.
- ☐ Many important alloys (bronze, solder, pewter) contain tin and/or lead.
- ☐ Carbon dioxide is a simple covalent molecule; silica has a three-dimensional network structure.
- ☐ Silicates have ring, chain, sheet and network structures.
- ☐ The oxides SnO_2, SnO and PbO are amphoteric.
- ☐ Silicon tetrachloride is rapidly hydrolysed because it is a Lewis acid as a result of the availability of the 3d-orbitals.
- ☐ Tin and lead form dihalides as well as tetrahalides.
- ☐ With the exception of lead(II) nitrate and lead(II) ethanoate, lead(II) salts are insoluble.
- ☐ Cement is a complex aluminosilicate formed by heating clay and limestone. Glass is formed by fusing sandstone, limestone and sodium carbonate.

PROBLEMS

1 The mass percentage composition of two white powders **A** and **B** are as follows:

	C	Sn	F
A	24	–	76
B	–	76	24

Calculate the empirical formulae of **A** and **B**. Compare and explain the structures of **A** and **B**.

2 Using appropriate data, calculate the standard combustion enthalpies of methane (CH_4) and silane (SiH_4). List the reasons why silane is not used as a fuel.

3 Use $E^\ominus$ values to calculate the equilibrium constant of the reaction $Sn^{4+}(aq) + Sn(s) \rightarrow 2Sn^{2+}(aq)$.

4 Complete the table below to show the main reactions of the elements carbon, silicon, tin and lead. Use the symbol E for the elements when writing the equation for the reaction in column 2, as in the example on the first line.

Reagent	Equation	Elements which undergo reaction	Exceptions
Cl_2 and heat	$E + 2Cl_2 \rightarrow ECl_4$	E = C, Si, Sn	Pb gives $PbCl_2$
air and heat			
hot concentrated HCl			
hot concentrated $HONO_2$			
aqueous NaOH			
molten NaOH			

Explain the trends apparent in the table.

5 (a) The following molecules or ions contain tin and chlorine: $SnCl_2$, $SnCl_3^-$, $SnCl_4$ and $SnCl_6^{2-}$. Construct a table showing for each of these four species (i) the coordination number of tin, (ii) the oxidation number of tin and (iii) its shape.
(b) Write equations to illustrate (i) a redox reaction and (ii) a Lewis acid–base reaction for $SnCl_2$.
(c) Write equations to illustrate (i) a redox reaction and (ii) a Lewis acid–base reaction for $SnCl_4$.

6 Write an essay on the chemistry and industrial importance of carbon monoxide.
What do the species CO, N_2, CN^- and NO^+ have in common?

7 (a) How is the element silicon manufactured?
(b) How is very pure silicon made?
(c) Describe the manufacture of (i) cement and (ii) glass.
(d) List the uses of lead and its compounds.

8 The most widespread compounds in the Earth's crust are silicates. Their structures are based on the SiO_4^{4-} tetrahedral unit. These units are joined by sharing oxygen atoms at the corners of the tetrahedra to produce discrete ions, chains, sheets or three-dimensional networks. Fill in the gaps in the table below.

Number of shared corners	Si:O ratio	Formula	Type of structure	Example of a compound or mineral containing this species	Electric charge on the species per Si atom
0	1:4	SiO_4^{4-}	discrete ion	$ZrSiO_4$	4
1	1:3.5				3
2		SiO_3^{2-}			2
3			sheets		1
4				SiO_2	0

9 'Whereas the study of the alkali metals reveals the similarities that exist between elements in a particular Group of the Periodic Table, the study of the elements C to Pb brings out the differences and dissimilarities between them.'
Discuss and justify this statement. *(London)*

10 Give a concise account of the chemistry of the elements in Group IV of the Periodic Table (carbon, silicon, germanium, tin and lead), referring in your answer to *all* of the following aspects:
(a) the variation in first ionization energy,
(b) the hydrolysis of the tetrachlorides,
(c) the acid–base character of the oxides of the elements in oxidation states II and IV,
(d) the relative stability of the II and IV oxidation states of the elements in the oxides and chlorides.
Show how the electron pair repulsion theory can be used to predict the shape of the silicon tetrachloride molecule. *(Cambridge)*

11 The elements carbon to lead in Group IV of the Periodic Table illustrate differences and similarities between elements in a particular Group.
By reference to carbon and lead *only*, illustrate this statement by considering each of the following and explaining any differences:
(a) methods of formation and structures of the dioxides;
(b) the type of bonds that the elements form;
(c) the relative stabilities of the lower and higher oxides;
(d) the relative stabilities of the tetrachlorides;
(e) the formation of a large number of hydrides of carbon.
In each case make a prediction for germanium, remembering that germanium is between carbon and lead in the Periodic Table. *(London)*

12 (a) Describe, with essential practical details, how tin(IV) chloride can be prepared from metallic tin. Draw diagrams of the apparatus you would use and show how to obtain a pure product.
(b) Explain what happens when
(i) aqueous solutions of tin(II) chloride and mercury(II) chloride are mixed;
(ii) hydrogen sulphide is passed through an aqueous solution of tin(II) chloride;

(iii) sodium hydroxide solution is added to an aqueous solution of tin(II) chloride until the alkali is in excess.

(c) Arrange the +2 oxidation states of germanium, lead and tin in order of *increasing* ease of oxidation. Explain your answer. (*SUJB*)

13 (a) Tin(II) compounds show reducing properties. Give *one* example of this behaviour, and write an equation for the reaction.

(b) Lead(IV) oxide has oxidizing properties. Describe *one* example of this behaviour, and write an equation for the reaction.

When dilute hydrochloric acid is added to a solution of lead(II) nitrate, a white precipitate is formed; but, if concentrated acid is used, much less precipitate is formed.

(i) What is the white solid formed? Write an ionic equation for the reaction.

(ii) Explain this difference in behaviour.

(*London part question*)

14 On heating the soft silvery metal, **A**, in air it first melted and then gave a yellow solid, **B**. Further heating of **B** in oxygen yielded the bright red solid, **C**. Treatment of **C** with nitric acid gave a dark brown solid, **D**. The solid **D** dissolved in an acidified nitrite solution. Addition of potassium iodide to this solution precipitated a bright yellow solid, **E**. The solid **E** dissolved in hot water to give a colourless solution, but yellow crystalline **E** separated on cooling. Another yellow solid **F** was obtained by the addition of potassium chromate solution to the solution obtained from **D**. On passing chlorine gas over **E**, contained in a small boat, iodine was displaced and eventually nothing was left in the boat.

Account for these observations and identify **A** to **F**.

(*Cambridge Entrance*)

15 Radiocarbon, ^{14}C, half-life 5600 years, is produced in small amount in the upper atmosphere by the cosmic ray bombardment. The carbon of living matter is continuously being replaced by atmospheric carbon by natural processes.

Suggest how these observations can be made the basis for a method of dating organic remains, drawing attention to the assumptions necessary to your argument.

(*Cambridge Entrance*)

21 The p-block elements: Group V

The two most important elements in Group V are nitrogen and phosphorus, and we concentrate on these. They are important in nature because they are both essential to plant growth, and hence are major components of fertilizers. Both elements are typical non-metals, forming acidic oxides and volatile, covalent halides and hydrides. Nitrogen is remarkable for its wide range of oxidation numbers, from −3 to +5.

	IV	V	VI
2	6 C	7 N	8 O
3	14 Si	15 P	16 S
4	32 Ge	33 As	34 Se
5	50 Sn	51 Sb	52 Te
6	82 Pb	83 Bi	84 Po

Figure 21.1 Group V in the Periodic Table

The elements in Group V are as follows:

nitrogen	N	$[He]2s^2 2p^3$
phosphorus	P	$[Ne]3s^2 3p^3$
arsenic	As	$[Ar]3d^{10}4s^2 4p^3$
antimony	Sb	$[Kr]4d^{10}5s^2 5p^3$
bismuth	Bi	$[Xe]4f^{14}5d^{10}6s^2 6p^3$

Their positions in the Periodic Table are shown in Figure 21.1. The characteristic valence electron configuration is ns^2np^3, where n is the Period number. We shall consider arsenic, antimony and bismuth only where they illustrate the increasing metallic character of the elements with increasing atomic number: nitrogen is a non-metal, bismuth is a metal. Arsenic (principally as arsenic(III) oxide) is a notorious poison, yet organic compounds of arsenic were used in pharmaceuticals (see Box 21.1); antimony is best known in its alloy with tin, namely **pewter**. Its symbol comes from the name *stibium*.

BOX 21.1

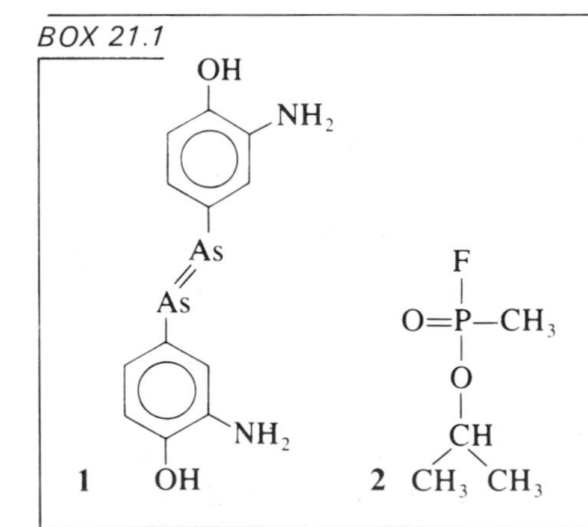

Organophosphorus and organoarsenic compounds

Numerous organic chemicals contain phosphorus or arsenic, and many of these compounds are physiologically active. One of the first to be recognized as a treatment for a specific disease was **Salvarsan (1)**, which was used to treat syphilis. Some organophosphorus and organoarsenic compounds are used as pesticides and herbicides, but there is some concern over their use because others, including **Sarin (2)**, have been developed for use as nerve gases. Nerve gases act by blocking the function of the enzyme acetylcholinesterase, so preventing the action of acetylcholine (which is the chemical responsible for transmitting impulses from one nerve cell to another) and hence are known as **anticholinesterase poisons**. Even minute quantities can cause paralysis.

21.1 Group systematics

Oxidation states and bonding characteristics

The sums of the first three ionization energies for nitrogen and phosphorus are 8850 kJ mol^{-1} and 5880 kJ mol^{-1} respectively, and these enormous values suggest that neither element is likely to form a triply charged cation. There is only partial loss of electrons when forming covalent bonds, however, and both elements can form covalent compounds in which they have a positive oxidation number; in each case the maximum oxidation number is +5.

The lowest oxidation number for both elements is −3. The formation of N^{3-}(g) is strongly endothermic, and for nitrides to form there must be a very large compensating lattice enthalpy. This is so only for very small cations, such as Li^+ and Mg^{2+}, and even then the nitrides Li_3N and Mg_3N_2 have

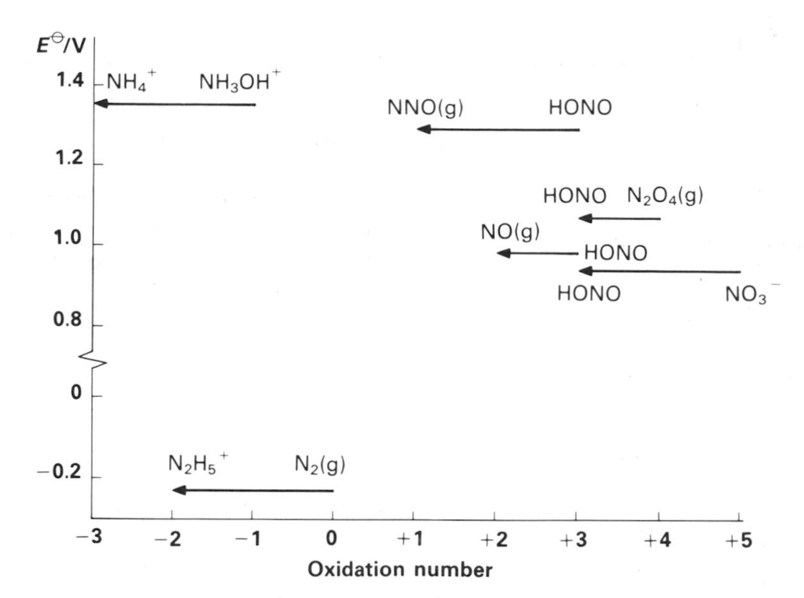

Figure 21.2 The standard electrode potentials for some common redox reactions of nitrogen compounds

significant covalent character. The structures of the other nitrides, and also of the phosphides, are best regarded as covalent. The important hydride NH_3 is an example of a covalent compound in which nitrogen has oxidation number −3.

Nitrogen occurs with a wide variety of oxidation numbers (all integer values from −3 to +5) and a systematic approach to its chemistry is therefore imperative. When elements possess a range of oxidation numbers it is convenient to show their standard electrode potentials on a diagram such as Figure 21.2. The illustration can be used in conjunction with Table 13.1 to rationalize the element's redox behaviour and to suggest how one compound can be converted to another by the appropriate choice of oxidizing or reducing agent. For example, consider the values for nitrous acid, HONO:

$$NO_3^-(aq) + 3H^+(aq) + 2e^- \rightarrow HONO(aq) + H_2O(l), \quad E^{\ominus} = +0.94\ V$$
$$N(+5) \rightarrow N(+3)$$

$$HONO(aq) + H^+(aq) + e^- \rightarrow NO(g) + H_2O(l), \quad E^{\ominus} = +0.99\ V$$
$$N(+3) \rightarrow N(+2)$$

These show that it can reduce (and hence decolorize) acidified potassium manganate(VII), for which $E^{\ominus} = +1.52$ V, and bromine ($E^{\ominus} = +1.07$ V), but oxidize (and hence turn brown) a solution of iodide ions ($E^{\ominus} = +0.54$ V).

Periodic trends

The elements in Group V show a marked trend towards metallic character on descending the Group. This is shown most clearly in the structures of the elements: nitrogen is a gas, phosphorus a non-metallic solid, and bismuth a lustrous metal. The trend is also shown in their chemical properties. For example, the oxides become increasingly basic: phosphorus(III) oxide is acidic, arsenic(III) and antimony(III) oxides are amphoteric and bismuth(III) oxide is basic. Bismuth's most stable oxidation number is +3 (see Section 20.1), and bismuth(V) is so unstable relative to bismuth(III) that sodium bismuthate(V), $NaBiO_3$, even oxidizes colourless Mn^{2+}(aq) to purple MnO_4^-(aq).

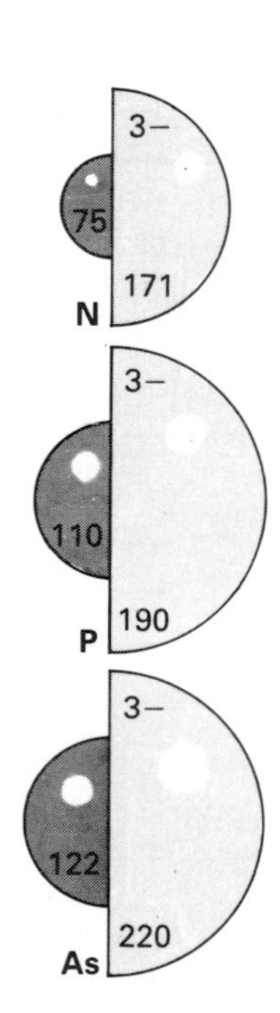

Figure 21.3 The radii (in picometres) of the atoms and M^{3-} anions of the Group V elements

Shapes and sizes

The covalent radii of the atoms increase with increasing atomic number (see Figure 21.3). The nitrogen atom is anomalously small; this enables it to participate in multiple bonding with itself and with its neighbours carbon and oxygen, because broadside overlap of p-orbitals can significantly lower the energy.

VSEPR theory (see Section 2.3) readily explains the shapes of molecules containing nitrogen, and also of those containing phosphorus so long as its d-orbitals are taken into account.

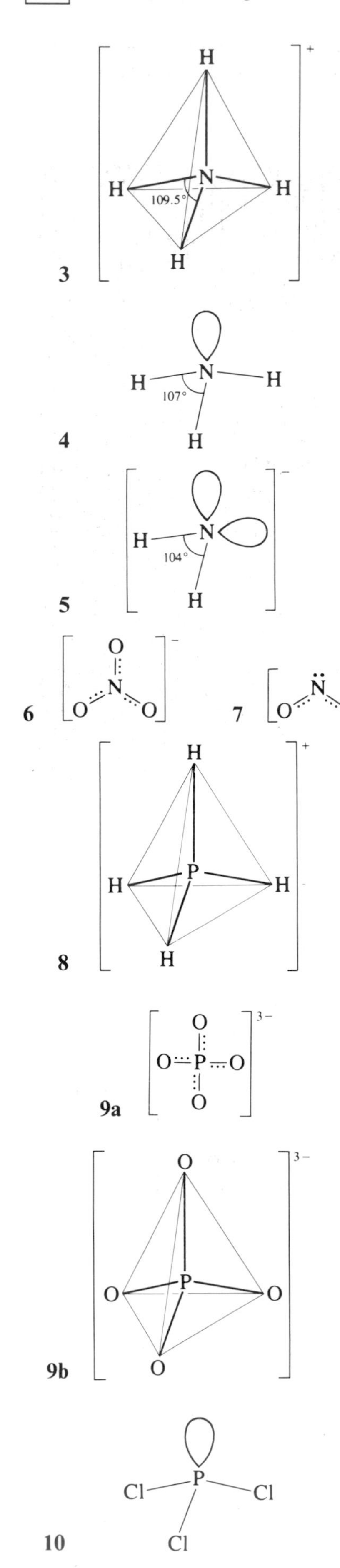

If there are four electron pairs around the atom, then VSEPR theory leads us to expect a tetrahedral arrangement. This is the case with the ammonium ion (NH_4^+, **3**), in which all four hydrogen atoms are equivalent; it is also the basic shape of the pyramidal ammonia molecule, NH_3, in which one electron pair is a lone pair (**4**). Similarly, the electron pairs of the amide ion (NH_2^-, **5**) are arranged tetrahedrally, but the distortion from the two lone pairs results in an angular ion with bond angle 104°.

The VSEPR rules require that in entities which have three electron pairs, or which contain a multiple bond (which is to be treated as a single electron pair – see Section 2.4), the basic shape is trigonal planar. Both the nitrate ion (NO_3^-, **6**) and the nitrite ion (NO_2^-, **7**) are examples, although the latter has a single lone pair in one of the trigonal positions. The nitrate ion is isoelectronic with the carbonate ion, and has the same shape but one unit less charge; in both NO_3^- and CO_3^{2-} the charge is delocalized over all three oxygen atoms (see Section 2.5).

Phosphorus is unable to form multiple bonds using its p-orbitals, and so it forms no planar ions; its tetrahedral ions include the phosphonium ion (PH_4^+, **8**) and the phosphate ion (PO_4^{3-}, **9**). When only three other atoms are present, as in phosphine (PH_3) and phosphorus trichloride (PCl_3, **10**), the shape is pyramidal, with a lone pair occupying the remaining tetrahedral site (as in NH_3).

A greater variety of shapes is possible when phosphorus uses its d-orbitals. In the vapour of phosphorus pentachloride (PCl_5) in which there are five electron pairs around the central phosphorus atom (so that one d-orbital is being used to expand the valence shell from eight to ten electrons), the molecule is a trigonal bipyramid (**11**). Similarly, the complex anion PCl_6^- has six electron pairs, two d-orbitals being involved, and is octahedral (**12**); this ion occurs in solid phosphorus pentachloride, which has the structure $PCl_4^+PCl_6^-$. Solid phosphorus pentabromide is $PBr_4^+Br^-$, this structure reflecting the difficulty of packing more than four large bromine atoms round the central phosphorus atom.

21.2 The elements

Nitrogen

Nitrogen exists as the diatomic molecule N_2. Its formal name is **dinitrogen**, but this is normally used only when ambiguity must be avoided, and 'nitrogen' is normally sufficient. It is a colourless, odourless gas, which condenses to a colourless liquid at −196 °C (77 K). Nitrogen is very unreactive, largely because its bond enthalpy is so high (944 kJ mol^{-1} – compare O_2, 496 kJ mol^{-1}, and F_2, 158 kJ mol^{-1}). The strength of the bond and the short bond length (110 pm compared with 146 pm for the N—N single bond) provide excellent experimental evidence for the theoretical description of the molecule as having a triple bond, N≡N.

The only element to react with nitrogen at room temperature is lithium, which forms the nitride, Li_3N:

$$6Li(s) + N_2(g) \rightarrow 2Li_3N(s) \qquad N(0) \rightarrow N(-3), \quad Li(0) \rightarrow Li(+1)$$

Magnesium reacts directly, but only when ignited (see Section 18.2). Some micro-organisms, however, have evolved a method for bringing about direct reaction with atmospheric nitrogen and building it into protein; this **biological nitrogen fixation** is characteristic of the bacteria in the roots of leguminous plants such as beans, peas and clovers (see Figure 21.4). This is an important early step in the food chain. Subsequent steps, which are part of the **nitrogen cycle**, are shown in a simplified form in Figure 21.5.

Some atmospheric nitrogen is also converted into useable form by lightning:

$$N_2(g) + O_2(g) \xrightarrow{\text{electric discharge, lightning}} 2NO(g) \qquad O(0) \rightarrow O(-2), \quad N(0) \rightarrow N(+2)$$

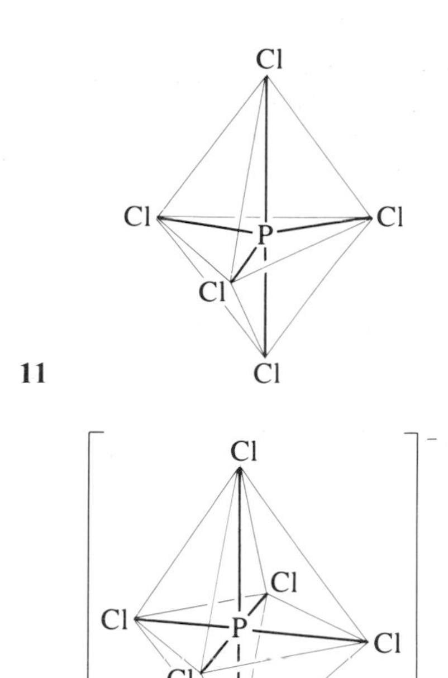

This is called **atmospheric nitrogen fixation**, and the product, nitrogen monoxide, is oxidized further to several other oxides, such as nitrogen dioxide, NO_2:

$$2NO(g) + O_2(g) \rightarrow 2NO_2(g) \qquad O(0) \rightarrow O(-2),\quad N(+2) \rightarrow N(+4)$$

Figure 21.4 Nodules containing nitrogen-fixing bacteria on the roots of a pea plant

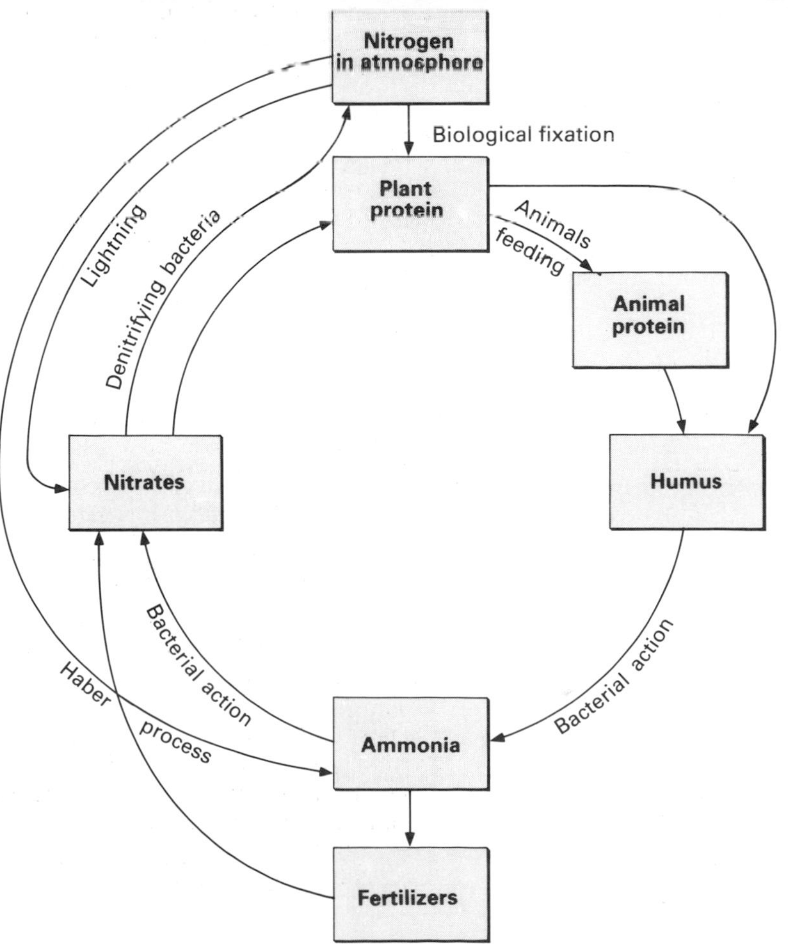

Figure 21.5 The nitrogen cycle

The various oxides formed in this way and by similar reactions in internal combustion engines are collectively denoted NO_x, and are one of the causes of acid rain (see Figure 12.8 and Box 22.2). They also contribute to the nitrogen cycle, and transfer about 10 megatonnes of nitrogen to the land each year.

The need to grow more food and to use land more efficiently has stimulated the development of methods for intervening in the nitrogen cycle by fixing the nitrogen industrially and adding it to the soil as fertilizers. The Haber–Bosch process, described in Section 21.7, is an example of this **industrial nitrogen fixation**.

The low reactivity of dinitrogen has led to its widespread use as an inert atmosphere for chemical and metallurgical processes where the presence of oxygen may be a hazard. For example, it is used to prevent the oxidation of tin used in the manufacture of float-glass (see Section 20.2) and to inhibit oxidation when packaging air-sensitive pharmaceuticals. About one-third of all elemental nitrogen produced is supplied as **liquid nitrogen**, which acts as an inert, safe refrigerant and maintains a constant temperature of around $-196\,^{\circ}C$.

P P P P

13

Phosphorus

Phosphorus has at least two allotropes, **red phosphorus** and **white phosphorus**. Their properties are compared in Table 21.1. White phosphorus is a solid containing covalent tetrahedral P_4 molecules (**13**) and red phosphorus is an amorphous solid which is known to have an extended covalent structure, although its exact form is uncertain.

Table 21.1 White and red phosphorus

	White phosphorus	Red phosphorus
Appearance	yellow waxy solid	red powder
Structure	P_4	three-dimensional network of atoms
t_m	$44\,^{\circ}C$	sublimes at $400\,^{\circ}C$
t_b	$280\,^{\circ}C$	
Solubility in CS_2	soluble	insoluble
Toxicity	toxic	non-toxic
Reaction with air	ignites spontaneously above $35\,^{\circ}C$	starts to burn at about $250\,^{\circ}C$
Reaction with hot NaOH(aq)	PH_3 produced	no reaction
Preparation	condensing P_4 vapour	heating white P in the absence of air

EXAMPLE

Examine the reason why nitrogen exists as dinitrogen rather than, by analogy with white phosphorus, as N_4 by calculating the standard formation enthalpy of N_4.

METHOD Estimate the standard formation enthalpy (using the values of bond enthalpies given in Table 9.4) for the process $2N_2(g) \rightarrow N_4(g)$.

ANSWER The reaction requires two triple bonds to be broken and six single bonds to be formed. The enthalpy change is thus *approximately*

$$\Delta_r H^{\ominus} = 2 \times 944\,\text{kJ mol}^{-1} - 6 \times 163\,\text{kJ mol}^{-1}$$
$$= +910\,\text{kJ mol}^{-1}$$

COMMENT The N_4 molecule is strongly endothermic compared with two N_2 molecules; moreover, the standard formation entropy is negative (because two molecules join to form one). Hence $N_4(g)$ is thermodynamically very unstable with respect to $2N_2(g)$. In a diphosphorus molecule the bonding would be weak (because multiple bonding between adjacent phosphorus atoms is unlikely) and so P_4 is more stable than P_2.

Figure 21.6 Chemiluminescence of P_4: the water contains a small piece of white phosphorus. As the water boils, some phosphorus vaporizes, and the light given off during its chemiluminescence is just sufficient for photography

White phosphorus is much more reactive than red phosphorus, largely because the small bond angle in the P_4 molecules is a source of strain. It glows in air as a result of its oxidation, which is the origin of its name (from the Greek for 'light bearer'); this is an example of a **chemiluminescent reaction**, that is, one that generates light (see Figure 21.6). At about 35 °C it spontaneously and spectacularly ignites, producing white fumes of the oxides. In a limited supply of air, phosphorus(III) oxide is formed:

$$P_4(s) + 3O_2(g) \rightarrow P_4O_6(s) \qquad O(0) \rightarrow O(-2),\quad P(0) \rightarrow P(+3)$$

but when the oxygen supply is plentiful oxidation proceeds as far as phosphorus(V) oxide:

$$P_4(s) + 5O_2(g) \rightarrow P_4O_{10}(s) \qquad O(0) \rightarrow O(-2),\quad P(0) \rightarrow P(+5)$$

In view of this reactivity and its low ignition temperature and also because it is toxic, white phosphorus is usually stored under water, and great care should be taken when using it. On account of its extended covalent structure, red phosphorus is much less reactive than white; nevertheless it undergoes the same reactions, but at higher temperatures.

Over a million tonnes of phosphorus are produced annually, largely for conversion to phosphoric acid, phosphorus halides and sulphides, and organophosphorus compounds (organic compounds that contain phosphorus). Small amounts of elemental phosphorus are used in weaponry; it was once used in matches but has now been replaced by its sulphide, P_4S_3.

21.3 Oxides and oxoacids

Nitrogen oxides

The oxides of nitrogen, together with some of their properties, are listed in Table 21.2. We consider here only the three most important nitrogen oxides: N_2O, NO and NO_2. N_2O_3 and N_2O_5, which are the anhydrides of nitrous and nitric acids, are also known.

Dinitrogen oxide, which can be written as N_2O but more explicitly as NNO, is also called **nitrogen(I) oxide** (and sometimes **nitrous oxide**). It was also once called 'laughing gas' on account of its physiological effect. It was among the

Table 21.2 The oxides of nitrogen

Formula	Oxidation number	Name (traditional name)	Physical state at 25 °C and 1 atm	Effect of water
N_2O	+1	dinitrogen oxide (nitrous oxide)	gas	insoluble
NO	+2	nitrogen monoxide (nitric oxide)	gas	insoluble
N_2O_3	+3	dinitrogen trioxide	dissociates to NO + NO_2	acidic – HONO
NO_2	+4	nitrogen dioxide	gas: $2NO_2 \rightleftharpoons N_2O_4$	acidic – HONO and $HONO_2$
N_2O_5	+5	dinitrogen pentoxide	solid	acidic – $HONO_2$

first anaesthetics introduced (see Box 28.1) and is still used, mixed with oxygen, to relieve pain during childbirth.

Dinitrogen oxide is prepared by heating a mixture of ammonium sulphate and sodium nitrate:

$$(NH_4)_2SO_4(s) + 2NaNO_3(s) \rightarrow 2NNO(g) + Na_2SO_4(s) + 4H_2O(g)$$

$$2N(-3) + 2N(+5) \rightarrow 4N(+1)$$

This method is much safer than the alternative of heating ammonium nitrate, the hazards of which were demonstrated on a large scale in 1947 when a cargo ship carrying ammonium nitrate exploded in a harbour in Texas, doing $67 million worth of damage. (Ammonium nitrate and fuel oil mixtures, **ANFO explosives**, are common commercial explosives.) The reaction is an oxidation of the ammonium ion ($Ox(N) = -3$) by the nitrate ion ($Ox(N) = +5$) to give the oxide, in which $Ox(N)$ has the intermediate value of $+1$.

Dinitrogen oxide is a moderately unreactive gas, but it will relight a glowing splint because heating decomposes it to the elements. On account of its unreactivity, and because it is colourless, odourless and tasteless, it is used as a propellant for some types of canned food, including whipped cream, and for making whipped ice cream.

Nitrogen monoxide, NO, is also known as nitrogen(II) oxide (and still occasionally as **nitric oxide**). It is prepared industrially by the oxidation of ammonia:

$$4NH_3(g) + 5O_2(g) \xrightarrow{\text{Pt/Rh catalyst, 850 °C}} 4NO(g) + 6H_2O(g)$$

$$O(0) \rightarrow O(-2), \quad N(-3) \rightarrow N(+2)$$

In the laboratory it is most conveniently prepared by the reduction of a nitrite with potassium iodide:

$$2NO_2^-(aq) + 2I^-(aq) + 4H^+(aq) \rightarrow 2NO(g) + 2H_2O(l) + I_2(s)$$

$$N(+3) \rightarrow N(+2), \quad I(-1) \rightarrow I(0)$$

or of 50 per cent nitric acid with copper, as described in the next section.

Nitrogen monoxide is a colourless, sparingly soluble gas. It is an example of an **odd-electron molecule**, since it has eleven valence electrons. It is highly reactive, and combines rapidly with oxygen to give brown fumes of nitrogen dioxide:

$$2NO(g) + O_2(g) \rightarrow 2NO_2(g) \qquad O(0) \rightarrow O(-2), \quad N(+2) \rightarrow N(+4)$$

Nitrogen dioxide, NO_2, is also called **nitrogen(IV) oxide**. It is prepared in the laboratory by heating solid lead(II) nitrate:

$$2Pb(NO_3)_2(s) \xrightarrow{400\,°C} 2PbO(s) + 4NO_2(g) + O_2(g)$$

$$N(+5) \rightarrow N(+4), \quad O(-2) \rightarrow O(0)$$

or by reducing concentrated nitric acid with copper (see the next section).

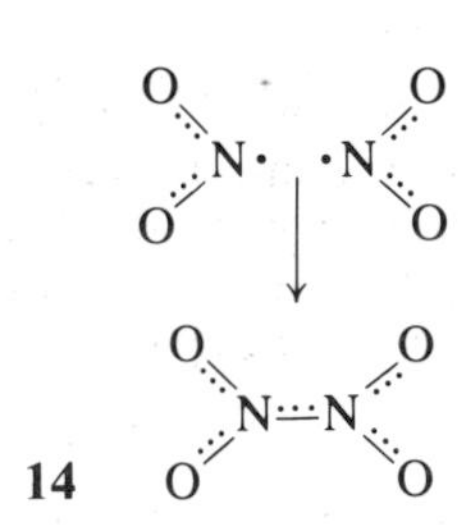

Nitrogen dioxide is a dark brown, toxic gas and is responsible for the brown colour of smog (see Figure 21.7). Like nitrogen monoxide, it is an odd-electron molecule, and one consequence is that it dimerizes (that is, two molecules combine to form a single molecule) to produce the colourless compound N_2O_4 (**14** – see Section 11.2):

$$2NO_2 \rightleftharpoons N_2O_4$$

Below its melting point ($-11\,°C$) it is a colourless solid because the equilibrium lies almost completely in favour of the dimer. Above 140 °C the equilibrium lies almost completely in favour of the monomer (NO_2), and so a sample is a dark brown gas. The response of the equilibrium to the temperature thus follows Le Chatelier's principle (see Section 10.2), because the dimerization is exothermic, and therefore favoured by lowering the temperature.

Nitrogen dioxide reacts with water to give a mixture of **nitrous acid**, HONO, and **nitric acid**, $HONO_2$:

$$2NO_2(g) + H_2O(l) \rightarrow HONO(aq) + HONO_2(aq) \qquad 2N(+4) \rightarrow N(+3) + N(+5)$$

Figure 21.7 A sequence of photographs, taken in Boston, USA, in 1970, showing the onset of photochemical smog during a day

The oxoacids of nitrogen, like those of phosphorus, are so important that they deserve a separate subsection to themselves.

Oxoacids of nitrogen

There are two important oxoacids of nitrogen: **nitric(III) acid**, HONO, more commonly known as **nitrous acid**, and **nitric(V) acid**, or simply **nitric acid**, $HONO_2$, often denoted HNO_3.

An aqueous solution of nitrous acid is obtained when cold aqueous solutions of nitrites are acidified with hydrochloric acid:

$$NO_2^-(aq) + HCl(aq) \rightarrow HONO(aq) + Cl^-(aq)$$

The solution, which is pale blue due to the presence of the unstable oxide N_2O_3, cannot be concentrated because nitrous acid is thermally unstable:

$$3HONO(aq) \rightarrow HONO_2(aq) + H_2O(l) + 2NO(g) \quad 3N(+3) \rightarrow N(+5) + 2N(+2)$$

Nitrous acid acts both as an oxidizing agent and as a reducing agent as described in Section 21.1.

Nitric acid is made industrially in huge quantities by the Ostwald process (see Section 21.7). In the laboratory it can be prepared by the action of concentrated sulphuric acid on potassium nitrate:

$$NO_3^-(s) + (HO)_2SO_2(aq,conc.) \rightarrow HONO_2(aq) + HOSO_3^-(aq)$$

Distillation of the product results in an azeotrope (see Section 7.1), containing 68 per cent $HONO_2$, which can be purified by vacuum distillation over phosphorus(V) oxide. Pure nitric acid is a colourless liquid boiling at 83 °C; the nitric acid used as a laboratory reagent is usually yellow as a result of its partial decomposition to nitrogen dioxide in the presence of sunlight:

$$4HONO_2(aq) \rightarrow 4NO_2(aq) + 2H_2O(l) + O_2(g) \quad N(+5) \rightarrow N(+4), \quad O(-2) \rightarrow O(0)$$

Nitric acid is highly reactive, and behaves as an acid, an oxidizing agent and as a nitrating agent (see Section 34.3). Its acidic reactions arise from the highly electronegative oxygen atoms, which stabilize the anion by delocalizing the charge:

$$HONO_2(aq) + H_2O(l) \rightarrow H_3O^+(aq) + NO_3^-(aq)$$

The acidic reactions include the formation of salts with bases and the release of carbon dioxide from carbonates:

$$CO_3^{2-}(s) + 2HONO_2(aq) \rightarrow CO_2(g) + H_2O(l) + 2NO_3^-(aq)$$

All nitrates are soluble in water, partly on account of the low lattice enthalpies arising from the nitrate anion with its single charge spread over the three oxygen atoms.

Very dilute nitric acid oxidizes magnesium to Mg^{2+} with the evolution of hydrogen; in this reaction the *hydrogen* is reduced and the oxidation number of nitrogen remains unchanged:

$$Mg(s) + 2HONO_2(aq,dil.) \rightarrow Mg(NO_3)_2(aq) + H_2(g)$$

$$H(+1) \rightarrow H(0), \quad Mg(0) \rightarrow Mg(+2)$$

The oxidizing reactions of nitric acid reflect the presence of nitrogen with its highest oxidation number (+5). These reactions include the oxidation of both metals and non-metals, and in the process of oxidation the acid is reduced to a variety of products. For example, nitric acid oxidizes copper with the evolution of oxides of nitrogen:

$$3Cu(s) + 8HONO_2(aq,50\%) \rightarrow 3Cu(NO_3)_2(aq) + 4H_2O(l) + 2NO(g)$$

$$N(+5) \rightarrow N(+2), \quad Cu(0) \rightarrow Cu(+2)$$

$$Cu(s) + 4HONO_2(aq, conc.) \rightarrow Cu(NO_3)_2(aq) + 2H_2O(l) + 2NO_2(g)$$

$$N(+5) \rightarrow N(+4), \quad Cu(0) \rightarrow Cu(+2)$$

In alkaline solution **Devarda's alloy** (an alloy of aluminium, zinc and copper) reduces the nitrogen in nitric acid to its lowest oxidation number, −3, and ammonia is evolved. Some metals, including iron and aluminium, are apparently unreactive towards the acid, but that is often due to the formation of a strongly bound, protective layer of oxide that makes the metal **passive**, or kinetically unreactive. Gold is unreactive, but reacts with **aqua regia** (a 1:3 mixture of concentrated nitric and hydrochloric acids) because as soon as the gold ions are formed, they are complexed by chloride ions to $AuCl_4^-$.

Non-metals, such as sulphur and phosphorus, are oxidized to their oxoacids, the nitric acid being reduced to a lower oxide:

$$S(s) + 6HONO_2(aq,conc.) \xrightarrow{heat} (HO)_2SO_2(aq) + 2H_2O(l) + 6NO_2(g)$$

$$N(+5) \rightarrow N(+4), \quad S(0) \rightarrow S(+6)$$

$$P(s) + 5HONO_2(aq,conc.) \xrightarrow{heat} (HO)_3PO(aq) + H_2O(l) + 5NO_2(g)$$

$$N(+5) \rightarrow N(+4), \quad P(0) \rightarrow P(+5)$$

(In the former reaction sulphur is present as S_8; so, to be precise, the equation should be adjusted accordingly.)

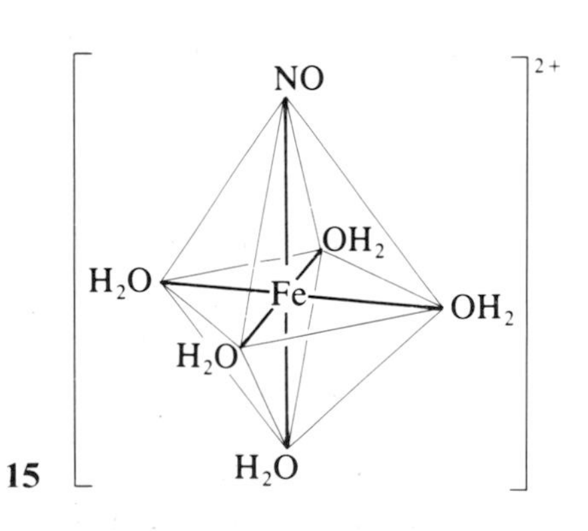

15

As all common nitrates are soluble, there is no simple precipitation test for them. One test for nitrates therefore uses their ability to react with Devarda's alloy in alkali, evolving ammonia. A traditional test (the **brown ring test**) is to add aqueous iron(II) sulphate to the solution; then, when concentrated sulphuric acid is slowly added to the mixture, a brown ring forms at the junction of the liquids as a result of the formation of $[Fe(H_2O)_5NO]^{2+}$ (**15**). (Both these tests are inconclusive if nitrite ions are present.)

Nitric acid is produced on a megatonne scale. About 80 per cent of production is converted to ammonium nitrate fertilizer (by direct reaction with ammonia), and some of the rest is used for the production of explosives, including nitroglycerine and trinitrotoluene or TNT (see Sections 29.4 and 35.1 respectively). Why these act so effectively can be seen from the reaction that occurs when nitroglycerine is detonated by a shock:

$$4\,\begin{matrix} CH_2ONO_2 \\ | \\ CHONO_2 \\ | \\ CH_2ONO_2 \end{matrix}(l) \rightarrow 6N_2(g) + 12CO_2(g) + 10H_2O(g) + O_2(g)$$

In theory, four molecules of reactant in the liquid phase give rise to 29 molecules of product in the gaseous phase, and the sudden increase of volume creates the shock wave of the blast. **Dynamite** was made by absorbing nitroglycerine in a clay (**kieselguhr**), which makes it less sensitive to shock and hence safer to handle: its invention made Alfred Nobel the fortune now used to fund the prizes named after him.

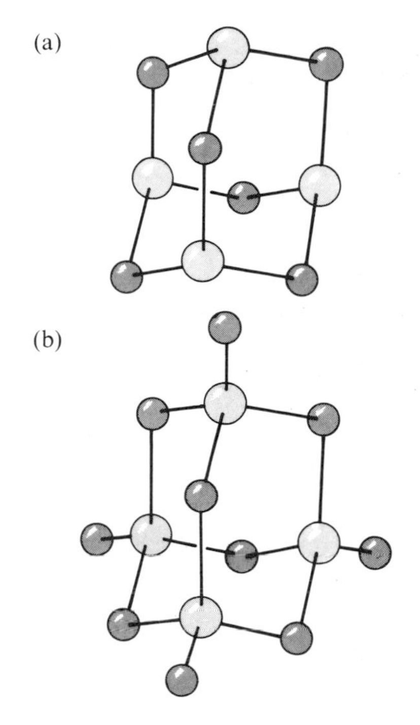

Figure 21.8 The structures of (a) P_4O_6 and (b) P_4O_{10} in the gas phase

Phosphorus oxides

The two oxides of phosphorus, phosphorus(III) oxide and phosphorus(V) oxide (which used to be called 'phosphorus pentoxide' because its empirical formula is P_2O_5), are formed by burning phosphorus in a limited or an excess supply of air respectively (see Section 21.2). The molecules of both oxides consist of tetrahedra of phosphorus atoms. In phosphorus(III) oxide, P_4O_6, a phosphorus atom lies at each corner of a regular tetrahedron, with an oxygen atom acting as a bridge along each edge (see Figure 21.8(a)). A phosphorus(V) oxide molecule, P_4O_{10}, which exists in the vapour, has in addition another oxygen atom attached to each corner of the tetrahedron (see Figure 21.8(b)). Solid phosphorus(V) oxide has a polymeric structure.

Both oxides are white solids. Phosphorus(III) oxide is the anhydride of **phosphonic acid**, $(HO)_2HPO$, and phosphorus(V) oxide is the anhydride of **phosphoric acid**, $(HO)_3PO$:

$$P_4O_6(s) + 6H_2O(l) \rightarrow 4(HO)_2HPO(aq)$$

$$P_4O_{10}(s) + 6H_2O(l) \rightarrow 4(HO)_3PO(aq)$$

Phosphorus(V) oxide reacts so readily with water that it is a powerful dehydrating agent (see, for example, Section 31.4); it will even dehydrate concentrated sulphuric acid. It is therefore used to produce the anhydrides of other acids, and sometimes to keep an atmosphere very dry, although in this its action is impaired by the skin of metaphosphoric acid, $(HOPO_2)_n$, that forms on its surface.

Oxoacids of phosphorus

O
HO:P:OH
OH

16a

O
P
HO OH
OH

16b

There are many oxoacids of phosphorus, but we shall concentrate on the most important, **phosphoric acid** ($(HO)_3PO$, **16**), which is also called **orthophosphoric acid**. The existence of **phosphonic acid** ($(HO)_2HPO$, **17**) should also be mentioned: one of its hydrogen atoms is attached directly to the phosphorus atom, and so the acid is only diprotic.

Phosphoric acid is prepared either by adding water to phosphorus(V) oxide (its anhydride) or by the hydrolysis of phosphorus pentachloride:

$$PCl_5(s) + 4H_2O(l) \rightarrow (HO)_3PO(aq) + 5HCl(aq)$$

It is a colourless, syrupy liquid, its high viscosity being due to extensive hydrogen bonding between the molecules in the liquid. Chemically it is a triprotic acid:

$$NaOH(aq) + (HO)_3PO(aq) \rightarrow Na^+(aq) + (HO)_2PO_2^-(aq) + H_2O(l)$$

$$2NaOH(aq) + (HO)_3PO(aq) \rightarrow 2Na^+(aq) + HOPO_3^{2-}(aq) + 2H_2O(l)$$

$$3NaOH(aq) + (HO)_3PO(aq) \rightarrow 3Na^+(aq) + PO_4^{3-}(aq) + 3H_2O(l)$$

Whereas nitric acid is a powerful oxidizing agent, phosphoric acid has almost no oxidizing action.

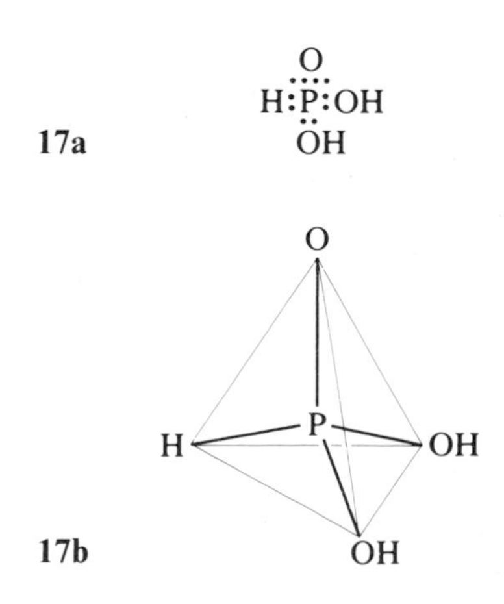

Phosphoric acid is produced on a large scale industrially. One of its principal uses is for the manufacture of fertilizers: since most phosphates are insoluble it is even more important that their journey through the ecosystem should be speeded up than in the case of the soluble nitrates, for otherwise their circulation is so slow that land becomes exhausted. Their presence in high concentration can cause problems, however (see Box 21.2). Another main use of phosphoric acid is for rust-proofing metal surfaces. It is extensively used for treating car bodies before they are painted: a strongly bonded protective layer is formed. Phosphoric acid is also added to cola drinks to produce a sharp taste.

Like silicates (see Section 20.3), phosphorus oxoacids and their salts condense to form complex structures consisting of chains and rings. The chain polymers formed by PO_4 tetrahedra sharing corners can be denoted $(PO_3^-)_n$; these are isoelectronic with those formed by phosphorus's two neighbours, silicon and sulphur, which form complex silicates based on $(SiO_3^{2-})_n$ chains, and polymeric sulphur trioxide, $(SO_3)_n$, respectively. The tripolyphosphates

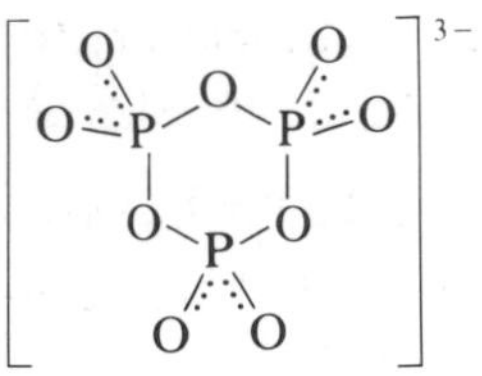

18

(**18**) are used in detergents because they bind to ions such as Ca^{2+} to produce a soluble complex, thus eliminating the formation of scum in hard water by the precipitation of insoluble calcium salts (see Section 18.6).

BOX 21.2

Eutrophication

The complex web of beneficial and damaging effects that accompany any technological advance is well illustrated by the example of fertilizers. Their use has the obvious advantage of improving the yield of crops to feed the hungry, but as they become washed out of the soil by rain and accumulate in rivers and lakes they can cause **eutrophication** (from the Greek for 'well nourished'). In this condition there is excessive growth of algae and other plants, and consequently a heavy demand on the oxygen dissolved in the water; its concentration may fall to a point at which other aquatic plants and fish die. This results in dead, putrid lakes, and in extreme cases to swamps and the eventual disappearance of the lake. Eutrophication is encouraged by the discharge of hot water from power stations, because oxygen is less soluble in hot water than in cold, so that the falling oxygen concentration reaches a critically low level more quickly. A striking example of the early stages of eutrophication was observed in Lake Balaton in Hungary, the largest lake in central Europe, which suddenly turned green in 1982 when there was a surge in the growth of green and blue-green algae.

The main causes of chemically induced eutrophication are nitrates and phosphates used as fertilizers, phosphates from detergents (which eventually find their way into the effluent of sewage plants) and untreated sewage (which acts as a natural, organic fertilizer). One chemical treatment of small eutrophic lakes is to add soluble aluminium salts, which precipitate the phosphates. The nitrate load cannot be removed in this way, however, since all common nitrates are soluble, and their control at source is essential.

21.4 Halides

Nitrogen halides

The nitrogen halides, $N(Hal)_3$, all have covalent, pyramidal structures. Nitrogen trichloride, NCl_3, is a yellow oil, which is dangerously explosive and difficult to handle. It is hydrolysed by water to ammonia and chloric(I) acid (HOCl). It is used industrially (for example, to bleach flour) as a gas, because in that form it is less dangerous.

Phosphorus halides

Much more important than the nitrogen halides are the two series of phosphorus halides, $P(Hal)_3$ and $P(Hal)_5$. The phosphorus chlorides are prepared by direct reaction between white phosphorus and either a limited supply (for the trihalide) or an excess (for the pentahalide) of the halogen:

$$P_4(s) + 6Cl_2(g) \rightarrow 4PCl_3(l) \qquad Cl(0) \rightarrow Cl(-1),\ P(0) \rightarrow P(+3)$$

$$P_4(s) + 10Cl_2(g) \rightarrow 4PCl_5(s) \qquad Cl(0) \rightarrow Cl(-1),\ P(0) \rightarrow P(+5)$$

Iodine, however, is too weak an oxidizing agent to form PI_5.

The structures of the phosphorus chlorides were described in Section 21.1. They are typical non-metal halides, and are hydrolysed by water to phosphonic acid and phosphoric acid respectively:

$$PCl_3(l) + 3H_2O(l) \rightarrow (HO)_2HPO(aq) + 3HCl(aq)$$

$$PCl_5(s) + 4H_2O(l) \rightarrow (HO)_3PO(aq) + 5HCl(aq)$$

Phosphorus pentachloride is used extensively in organic chemistry to chlorinate hydroxyl groups (see, for example, Sections 29.3 and 31.4).

EXAMPLE

Solid phosphorus pentachloride is an ionic solid in which the cation is PCl_4^+ (see Section 21.1). Predict the shape of the cation.

METHOD Write a Lewis structure, count the number of electron pairs, and then use VSEPR theory as described in Section 2.3.

ANSWER Phosphorus supplies five electrons and each chlorine supplies seven. The single charge on the cation signifies that one of these electrons has been lost. A Lewis structure is

$$\left[\begin{array}{ccc} & :\ddot{Cl}: & \\ :\ddot{Cl}: & \ddot{P} & :\ddot{Cl}: \\ & :\ddot{Cl}: & \end{array}\right]^+$$

There are four electron pairs on phosphorus, and so the basic shape is tetrahedral. None of the pairs is a lone pair, and so the ion is predicted to be a regular tetrahedron.

COMMENT The three neighbouring elements aluminium, silicon and phosphorus in Period 3 form the isoelectronic entities $AlCl_4^-$, $SiCl_4$ and PCl_4^+; the shapes of all these are regular tetrahedra.

21.5 Compounds with hydrogen

Compounds of nitrogen and hydrogen

The most important compounds of nitrogen with hydrogen are **ammonia**, NH_3, and **hydrazine**, N_2H_4, and we shall concentrate on these. Ammonia is by far the more important, for in its industrial synthesis (described in Section 21.7) the inert nitrogen of the air is made into a reactive compound, thus making the abundant supply of atmospheric nitrogen available to industry – because the ammonia can be oxidized to nitric acid – and to agriculture. In the laboratory preparation of ammonia, an ammonium salt is heated gently with an alkali, since the higher temperature favours the evolution of gas:

$$NH_4^+(aq) + OH^-(aq) \xrightarrow{\text{heat}} NH_3(g) + H_2O(l)$$

Ammonia is a gas at room temperature; it condenses to a colourless liquid at $-33\,°C$. It has a sharp and characteristic smell, and is highly soluble in water owing to its ability to form hydrogen bonds with the surrounding H_2O molecules. The nitrogen has its lowest oxidation number (-3), and in many reactions this shows itself by ammonia's role as a reducing agent:

$$4NH_3(g) + 3O_2(g) \xrightarrow{\text{ignite}} 2N_2(g) + 6H_2O(l) \qquad O(0) \rightarrow O(-2),\ \ N(-3) \rightarrow N(0)$$

$$8NH_3(g) + 3Cl_2(g) \rightarrow N_2(g) + 6NH_4Cl(s) \qquad Cl(0) \rightarrow Cl(-1),\ \ N(-3) \rightarrow N(0)$$

$$2NH_3(g) + 3ClO^-(aq) \rightarrow N_2(g) + 3Cl^-(aq) + 3H_2O(l) \qquad Cl(+1) \rightarrow Cl(-1),\ \ N(-3) \rightarrow N(0)$$

$$2NH_3(g) + 3CuO(s) \xrightarrow{\text{heat}} N_2(g) + 3Cu(s) + 3H_2O(g) \qquad Cu(+2) \rightarrow Cu(0),\ \ N(-3) \rightarrow N(0)$$

A yellow-green flame is sometimes observed, as in the reaction with oxygen. All these equations show the product as dinitrogen, but the dominant product actually depends on the exact conditions. For instance, in the reaction with chlorate(I), hydrazine is formed if gelatine is present, and nitrogen trichloride is produced with excess chlorine. The reaction with oxygen produces nitrogen with an oxidation number higher than zero if a catalyst is used to speed up an otherwise slow reaction:

$$4NH_3(g) + 5O_2(g) \xrightarrow{\text{Pt/Rh, 850 °C}} 4NO(g) + 6H_2O(g)$$

$$O(0) \rightarrow O(-2), \quad N(-3) \rightarrow N(+2)$$

We see more of this important reaction in Section 21.7.

Although ammonia is normally a reducing agent, it can be reduced by even stronger reducing agents, such as sodium or potassium metal. In these reactions the oxidation number of nitrogen remains -3 and the hydrogen is reduced:

$$2Na(s) + 2NH_3(g) \xrightarrow{\text{heat}} 2NaNH_2(s) + H_2(g) \quad H(+1) \rightarrow H(0), \quad Na(0) \rightarrow Na(+1)$$

$NaNH_2$ is sodium amide ('sodamide').

Ammonia is basic, and neutralizes acids:

$$NH_3(g) + HCl(aq) \rightarrow NH_4Cl(aq)$$

Its solution in water is weakly alkaline ($pK_b = 4.75$) as a result of the equilibrium

$$NH_3(aq) + H_2O(l) \rightleftharpoons NH_4^+(aq) + OH^-(aq)$$

A typical reaction of aqueous ammonia is the precipitation of metal hydroxides. For instance, cobalt(II) hydroxide is precipitated when aqueous ammonia is added to a solution containing $Co^{2+}(aq)$ because it disturbs the equilibrium

$$[Co(H_2O)_6]^{2+}(aq) \rightleftharpoons [Co(OH)_2(H_2O)_4](s) + 2H^+(aq)$$

in favour of the right-hand side by removing the hydrogen ions.

A more general description of ammonia is that it is a Lewis base, on account of its lone pair of electrons. This explains its action as a Brønsted base, its ability to form complexes by coordinate bonding and its behaviour as a nucleophile (see Sections 28.3 and 32.3) in organic chemistry. An example of its Lewis base behaviour is its ability to replace water molecules ('ligands' – Section 25.7) in d-block aqua ions:

$$[Cu(H_2O)_6]^{2+}(aq) + 4NH_3(aq) \rightleftharpoons [Cu(NH_3)_4(H_2O)_2]^{2+}(aq) + 4H_2O(l)$$

As the complex ion with water ligands is light blue and that with ammonia ligands is dark blue, this reaction is a test for the presence of copper(II) ions.

In summary, ammonia is (a) *a reducing agent* (on account of the low oxidation number of nitrogen in the molecule) and (b) *a Lewis base* (on account of its lone pair of electrons).

EXAMPLE

Suppose that the labels on three laboratory cylinders containing boron trifluoride, tetrafluoromethane and ammonia had become illegible. How could they be identified without using any other chemicals?

METHOD Identify the significant chemical features of the three compounds, namely that ammonia is a Lewis base, boron trifluoride a Lewis acid and tetrafluoromethane is neither. Identify any other characteristic physical properties of the gases, such as the pungent odour of ammonia.

ANSWER First identify ammonia by its smell; the Lewis acid boron trifluoride can then be recognized by its reaction with the Lewis base ammonia to form a white smoke. Tetrafluoromethane reacts with neither of the other gases at room temperature.

> COMMENT Detecting unknown gases by their smell must be done **with the utmost caution**. If moist litmus paper had been available, its blue colour would have been a much safer way to identify the cylinder containing ammonia.

Ammonia has numerous uses, and about 100 megatonnes are produced worldwide each year (see Figure 21.9). It is used as a fertilizer both directly (applied from pipes in liquid form) and indirectly after conversion to other nitrogenous fertilizers, such as ammonium sulphate, ammonium nitrate and urea, $CO(NH_2)_2$. Ammonia is also a raw material for nitric acid manufacture and for the production of nylon (see Box 32.3). It was once heavily used in the Solvay process for the production of sodium carbonate, described in Section 17.7.

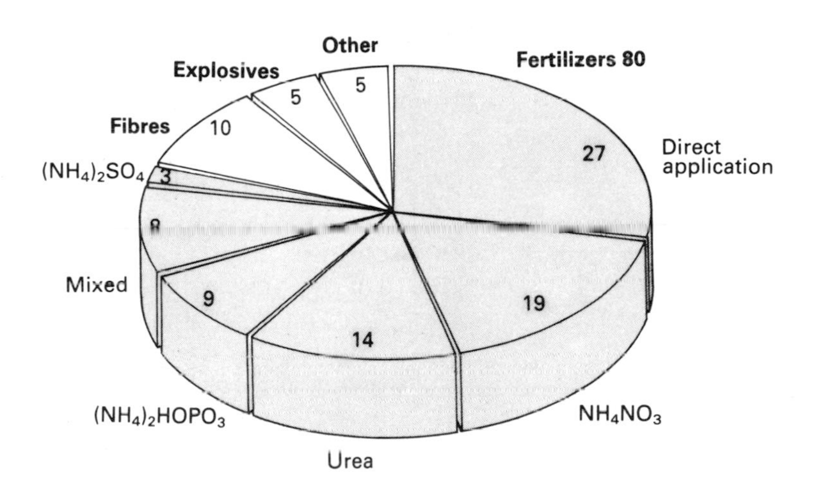

Figure 21.9 The uses of ammonia

The other important compound of nitrogen and hydrogen is **hydrazine**, N_2H_4 (the nitrogen analogue of hydrogen peroxide), which is more precisely represented by the formula H_2NNH_2. Like ammonia, it is a base and a reducing agent. For example, the traditional technique for making silver mirrors uses aqueous hydrazine to reduce Ag^+ to Ag (ammonia, a weaker reducing agent, merely forms a complex with the silver ions). A more modern application has been as the reducing agent in rocket fuels: to leave the surface of the Moon the Apollo lunar module used methyl-substituted hydrazine as fuel with the oxidizing agent liquid dinitrogen tetroxide, N_2O_4. This mixture is **hypergolic**, that is, the reactants burst into flame on mixing.

Phosphine

The only stable compound of phosphorus and hydrogen is the gas **phosphine**, PH_3, the analogue of ammonia. Like ammonia, phosphine is a Lewis base, but it is a more powerful reducing agent and reduces both copper(II) and silver(I) ions to the metals:

$$4Cu^{2+}(aq) + PH_3(g) + 4H_2O(l) \rightarrow 4Cu(s) + (HO)_3PO(aq) + 8H^+(aq)$$

$$Cu(+2) \rightarrow Cu(0), \quad P(-3) \rightarrow P(+5)$$

(Ammonia merely forms a complex.) With oxygen phosphine burns to form phosphoric acid:

$$PH_3(g) + 2O_2(g) \rightarrow (HO)_3PO(l) \qquad O(0) \rightarrow O(-2), \quad P(-3) \rightarrow P(+5)$$

Pure phosphine is not spontaneously flammable in air, but the gas normally ignites as a result of the presence of trace amounts of P_2H_4, which is very unstable.

Phosphine is much less soluble in water than ammonia because it does not form hydrogen bonds. It is also a weaker Brønsted base than ammonia ($pK_b = 27$), and forms very few salts. For example, whereas ammonia and hydrogen iodide form the reasonably stable salt ammonium iodide, the analogous formation of phosphonium iodide

$$PH_3(g) + HI(g) \rightarrow PH_4I(s)$$

is reversed at temperatures above 50 °C.

21.6 Other compounds

There are no simple aqua cations of the Group V elements except those of bismuth. The ion Bi^{3+} exists in oxosalts such as $Bi_2(SO_4)_3$ and $Bi(NO_3)_3$, but in aqueous solution they are hydrolysed to compounds containing the **bismuthyl ion**, BiO^+:

$$BiCl_3(aq) + H_2O(l) \rightleftharpoons BiOCl(s) + 2HCl(aq)$$

In practice, solid bismuth(III) chloride dissolves in concentrated hydrochloric acid to give a clear solution; adding water displaces the equilibrium to the right, and the solution turns opalescent. The addition of concentrated hydrochloric acid restores the clear solution.

21.7 Industrial chemistry

Occurrence

Dinitrogen makes up 78 per cent by volume of the atmosphere, and nitrogen occurs in combination in the protein and DNA of living things, and in the minerals **saltpetre**, KNO_3, and **Chile saltpetre**, $NaNO_3$.

Phosphorus is the eleventh most abundant element in the Earth's crustal rocks. By far the most important sources are the **apatites**, $Ca_5(PO_4)_3X$ where X = F, Cl or OH, the decayed and compressed remains of bones. Tooth enamel is mainly hydroxyapatite (X = OH), and its fluoridation converts it to the less soluble, harder fluorapatite (X = F). There are vast deposits of apatite in North Africa, and phosphates are widespread in the biosphere.

The production of ammonia

The Haber–Bosch process, already mentioned in Section 10.1 and Box 10.1, is the catalysed synthesis of ammonia (see Figure 21.10):

$$N_2(g) + 3H_2(g) \rightleftharpoons 2NH_3(g) \qquad N(0) \rightarrow N(-3), \quad H(0) \rightarrow H(+1)$$

Figure 21.10 An industrial plant at Billingham, UK, using the Haber–Bosch process for the synthesis of ammonia

The nitrogen is obtained from air, and the hydrogen from natural gas (see Section 16.4) with care being taken to remove all traces of carbon monoxide, which poisons the catalyst. The mixture is compressed to about 250 atm and passed over a finely divided iron catalyst at about 450 °C; the catalyst contains the oxides of potassium, calcium and aluminium as promoters to enhance its activity. The yield of ammonia is increased by working at high pressure, but this adds substantially to the cost of the plant, and a compromise between cost and yield is needed (see Section 10.2). A similar compromise governs the temperature selected, for although at low temperatures the yield is favoured (the synthesis is exothermic) equilibrium is reached too slowly to be economical. At the temperatures and pressures used in practice, about 15 per cent conversion is attained. The ammonia is condensed, and the unchanged gases are recirculated.

The production of nitric acid

Nitric acid is manufactured by the **Ostwald process**. The first stage is the catalytic oxidation of ammonia in which air containing about 10 per cent of ammonia is passed over platinum/rhodium alloy gauzes (see Figure 21.11) at about 850 °C and 5 atm:

$$4NH_3(g) + 5O_2(g) \xrightarrow{\text{Pt/Rh, 850 °C, 5 atm}} 4NO(g) + 6H_2O(g) \qquad O(0) \rightarrow O(-2), \quad N(-3) \rightarrow N(+2)$$

This produces about 96 per cent conversion. The nitrogen monoxide is cooled, mixed with more air, and then passed through a water absorption column:

$$2NO(g) + O_2(g) \rightarrow 2NO_2(g) \qquad O(0) \rightarrow O(-2), \quad N(+2) \rightarrow N(+4)$$

$$4NO_2(g) + 2H_2O(l) + O_2(g) \rightarrow 4HONO_2(aq) \qquad O(0) \rightarrow O(-2), \quad N(+4) \rightarrow N(+5)$$

This combination of oxidation and dissolution produces reasonably concentrated (60 per cent) nitric acid. Further concentration, to the azeotropic concentration of 68 per cent (see Section 7.1), is achieved by distillation.

Figure 21.11 The platinum–rhodium gauze catalyst at the heart of the nitric acid plant has to be changed periodically (see also Figure 14.18)

Summary

- The elements have the valence electron configuration ns^2np^3.
- Nitrogen and phosphorus arc typical non-metals, arsenic and antimony are metalloids and bismuth is a metal.
- Nitrogen is restricted to a coordination number of four, but phosphorus forms compounds with ten and twelve electrons (for example, PO_4^{3-} and PF_6^- respectively) in its valence shell.
- The main oxidation numbers of phosphorus are +5 and +3, whereas nitrogen has a range of oxidation numbers from +5 to −3.
- Delocalized double bonds are a feature of the structures of nitrogen oxides such as N_2O_4, and of the nitrite and nitrate anions.
- Dinitrogen is converted into ammonia or nitrates in nature by biological and atmospheric nitrogen fixation.
- Elemental phosphorus is polymorphic. White phosphorus consists of tetrahedral P_4 molecules. Red phosphorus has an extended covalent structure and is less reactive.
- Nitrogen dioxide and the oxides of phosphorus (P_4O_6 and P_4O_{10}) are acidic.
- Nitric acid, $HONO_2$, behaves as an acid, an oxidizing agent and a nitrating agent.
- Phosphoric acid, $(HO)_3PO$, forms condensed polyphosphate anions of formula $(PO_3^-)_n$ which consist of PO_4 tetrahedra joined by sharing corners to form either rings or chains.
- Phosphorus chlorides, PCl_3 and PCl_5, are covalent and readily hydrolysed.
- Ammonia has both Lewis base and reducing properties.
- Ammonia is manufactured by the high-pressure reaction of nitrogen and hydrogen (Haber–Bosch process) over an iron catalyst; nitric acid is manufactured by the catalytic oxidation of ammonia (Ostwald process).

PROBLEMS

1 (a) State the total number of outer shell (valence) electrons in the species NO_2^+, NO_2 and NO_2^-.
(b) Describe and explain the shapes of NO_2^+, NO_2 and NO_2^-.
(c) Explain why NO_2 dimerizes but NO_2^+ and NO_2^- do not.
(d) NO_2^+ is a Lewis acid whereas NO_2^- is a Lewis base. Illustrate and explain this statement.

2 (a) State the coordination number and oxidation number of phosphorus in each of the species PCl_3, PCl_4^+, PCl_5 and PCl_6^-.
(b) Describe, and rationalize using VSEPR theory, the shape of these four species.
(c) What is the structure of solid phosphorus pentachloride?
(d) Why is it unlikely that a compound NCl_5 will ever be prepared?

3 (a) Compare the physical and chemical properties of ammonia and phosphine.
(b) Why is liquid ammonia sometimes used as a solvent for chemical reactions?
(c) Give examples, with balanced equations, of ammonia acting as (i) a Brønsted base, (ii) a Lewis base and (iii) a reducing agent.

4 Draw up a table to illustrate the gradation of properties of the elements nitrogen, phosphorus, arsenic, antimony and bismuth, using the chemical and physical properties of the compounds E_2O_3 and ECl_3 as examples.

5 What is the structural similarity between P_4, P_4O_6 and P_4O_{10}? Draw the three structures. Mark on the drawing of P_4O_{10} the short bonds between phosphorus and oxygen atoms.

6 (a) Draw the structure of hydrazine.
(b) The standard electrode potential of hydrazine in acidic solution at 25 °C for the reaction:

$$N_2(g) + 5H^+(aq) + 4e^- \rightarrow N_2H_5^+(aq)$$

is $E^\ominus = -0.23$ V. Write a balanced equation for the reaction of hydrazine with chlorine in acidic solution.
(c) Aqueous solutions of hydrazine are used to remove, by reduction, the dissolved oxygen in the water in boilers of power stations. Write an equation for the reaction. What mass of oxygen can be removed by 1 kg of hydrazine? Assuming that the dissolved oxygen is present at 0.01 p.p.m., what mass of water can be treated by 1 kg of hydrazine? What are the advantages of using hydrazine over using sodium sulphite solution?

7 (a) Write an essay on the fixation of nitrogen.
(b) Write balanced equations to represent the conversion of atmospheric nitrogen to ammonium nitrate for use in fertilizers. What mass of ammonium nitrate can be obtained from one tonne of dinitrogen?

8 Give the oxidation number of nitrogen in each of the following species:
(a) NH_2^-, (b) N_2, (c) NO_2^+, (d) $N_2H_5^+$, (e) NF_3, (f) N_2F_4, (g) NH_4^+, (h) N_2O, (i) NO^+.

9 Comment on the following:
(a) Nitric acid is a stronger acid than nitrous acid.
(b) There is no nitrogen analogue of phosphoric acid, $(HO)_3PO$.
(c) Ammonia is a stronger base than nitrogen trifluoride.
(d) There are at least five oxides of nitrogen but only two oxides of phosphorus.

10 Give an account of the oxidizing properties of nitric acid.

11 Describe the Haber process for the manufacture of ammonia, paying particular attention to the sources of raw materials and to the physico-chemical principles involved.
Mention briefly *three* important uses of nitrogen-containing compounds, indicating in each case why they are important. (*London part question*)

12 (a) Compare the elements nitrogen and phosphorus by referring to
(i) the bonding, structure and basic nature of the trihydrides,
(ii) the bonding, structure and reaction with water of the chlorides,
(iii) the bonding and structure of the oxides, M_2O_3 and M_2O_5 (where M represents either nitrogen or phosphorus).
(b) 6.00 g of a sample of ammonium chloride known to be contaminated with sodium chloride were added to $100\,cm^3$ of aqueous sodium hydroxide of concentration $2.00\,mol\,dm^{-3}$ in a suitable apparatus. The mixture was boiled until all the ammonia had been expelled and then titrated with aqueous sulphuric acid of concentration $1.00\,mol\,dm^{-3}$. $47.0\,cm^3$ of the acid were required for neutralization.
Calculate the percentage by mass of ammonium chloride in the sample. (*AEB 1986*)

13 (a) Describe the manufacture of nitric acid from nitrogen and hydrogen, with special reference to the physico-chemical principles and economic factors involved. (Technical details of the plant are *not* required.)
(b) (i) How and under what conditions does nitric acid react with iodine?
(ii) How may NO be prepared from nitric acid?
(c) How and under what conditions does ammonia react with chlorine? (*Oxford*)

14 This question concerns the chemistry of *either* sulphur *or* nitrogen. Write your answer in terms of *one* of these elements only.
(a) Review the principal oxidation states of the element, giving an example of a compound for each oxidation number. Describe and explain *one* 'redox' reaction, showing how the element changes its oxidation number in the course of the reaction described.
(b) Give the name of a hydride of the element and comment on the shape of the molecule. Describe the action of the hydride with water, commenting on points of physico-chemical interest. How does an aqueous solution of the hydride react with a solution of a named metal ion?
(c) Give the names *or* formulae of (i) a simple salt, (ii) a 'large' organic molecule, containing the element, which are manufactured in the chemical industry. Comment on the properties of each compound, and the reasons for its use; give an outline of its manufacture. (*London*)

15 What oxidation numbers are shown by nitrogen in its compounds? Give *one* example of a compound or an ion for each different oxidation number.
Under suitable conditions the hydrazinium ion, $N_2H_5^+$, reacts quantitatively with bromate(V) ions, BrO_3^-, to form nitrogen gas and a colourless solution containing bromide ions. $25.00\,cm^3$ of 0.0200 M hydrazinium hydrogensulphate solution on titration required $22.34\,cm^3$ of 0.0150 M potassium bromate(V) solution. Calculate the stoichiometry of the reaction and hence deduce a balanced equation for the reaction.
The reaction mixture requires careful warming to ensure completion of the reaction during the titration. If the mixture is overheated, rather more potassium bromate(V) solution is required and the final solution has a pale yellow colour. What explanation can you suggest for such results? Support your suggestions with appropriate redox potential values obtained from reference books. (*Nuffield*)

16 Group V of the Periodic Table contains the elements N, P, As, Sb, Bi, in increasing order of atomic number. The most Common oxidation states of this Group are -3, 0, $+3$, $+5$.
(a) Write the formula of (i) the hydride of nitrogen in the -3 oxidation state, (ii) an oxide of arsenic in the $+3$ oxidation state, (iii) an oxoacid of phosphorus in the $+5$ oxidation state.
(b) Write an equation showing the thermal dissociation of the compound in (a)(i), naming the required catalyst.
(c) Draw a bond diagram to show the molecular structure of the compound in (a)(i), commenting on the bond angles and the geometrical shape of the structure.
(d) Put the hydrides of this Group in order of *increasing* stability to heat (put the *most stable last*).
(e) Red phosphorus and concentrated nitric acid react according to the equation:

$$P_4(s) + 20HNO_3(l) \rightarrow P_4O_{10} + 20NO_2(g) + 10H_2O(l)$$

(i) Name the *type* of reaction involved.
(ii) 1.42 g P_4O_{10}, in water, produces an acidic solution which neutralizes $20.0\,cm^3$ 1.00M sodium hydroxide solution. Deduce the formula of the phosphorus acid formed. (A_r: P = 31, O = 16.)
(iii) Say whether the N-containing acid of comparable formula will be more (or less) acidic than the acid in (ii). Explain your answer.
(f) Describe the action of heat on Group I and II nitrates, emphasizing patterns of behaviour. Indicate, very briefly, how ionic radii can be used to interpret these. (*SUJB*)

17 (a) Give *one* example of a difference, apart from colour, between the physical properties and *one* example of a difference between the chemical properties of the red and white allotropes of phosphorus. What is the reason for these differences?
(b) How may red phosphorus be converted into white phosphorus?
(c) Why does nitrogen not form allotropes similar to those of phosphorus?
(d) A compound PBr_x contains 88.6 per cent by mass of bromine. Deduce the molecular formula for PBr_x and sketch the shape of the molecule, giving your reasons.
(e) Under certain conditions PBr_x reacts with bromide ions to form the ion $[PBr_{x+1}]^-$. Predict the shape of this ion, giving your reasons. (*Oxford and Cambridge*)

18 White phosphorus, on exposure to a plentiful supply of air, reacts slowly at first and finally vigorously as it is oxidized. Write an equation and draw an energy diagram for this reaction. Comment on the molecular structure of the product.

A piece of phosphorus was burned in a flask of excess oxygen and, following the reaction, the flask was rinsed with cold water to dissolve the product, the solution being transferred and made up to $100\,cm^3$.

$50\,cm^3$ of this solution required $10\,cm^3$ of a solution of sodium hydroxide for neutralization. The remaining $50\,cm^3$ of solution was boiled for ten minutes and subsequently found to require almost $30\,cm^3$ of the sodium hydroxide solution for neutralization. How do you account for this difference?

Describe briefly the use of phosphorus pentachloride and phosphorus(V) oxide in the preparation of acid chlorides and nitriles in organic chemistry.

State briefly the differences between the simple hydrides of phosphorus and nitrogen, PH_3 and NH_3. *(London)*

19 Some reactions of nitrogen and its compounds are given below.

Stating clearly your reasons, identify the compounds **A** to **G** and describe the stereochemistry and the electronic structures of the *nitrogen*-containing ions or molecules in these compounds.

When magnesium is heated in nitrogen, a pale grey compound, **A**, is produced. When **A** is hydrolysed, a colourless gas **B** is produced which dissolves in water to give an alkaline solution. The reaction of **B** with sodium chlorate(I) leads to the formation of a colourless liquid **C** with empirical formula NH_2. The reaction of **C** with sulphuric acid in a 1:1 molecular ratio produces a salt **D**, $N_2H_6SO_4$, which contains one cation and one anion per formula unit. An aqueous solution of **D** reacts with nitrous acid to give a solution which, when neutralized with ammonia, produces a salt **E**, with empirical formula NH, which contains one anion and one cation per 'molecular' unit.

The gas **B** reacts with heated sodium to give a solid **F** and hydrogen. The reaction of **F** with N_2O in a 1:1 molecular ratio gives a solid **G**, which contains the same anion as **E**, and water. **G** decomposes when heated to give sodium and nitrogen. *(Cambridge S)*

20 Give an account of the chemistry of ammonia and phosphine. How would you attempt to show that:
(a) ammonia forms a stable complex with calcium chloride?
(b) phosphine in aqueous solution is present almost entirely as PH_3 molecules? *(Cambridge Entrance)*

21 Discuss the physico-chemical principles underlying the preparation of ammonia from nitrogen and hydrogen.

How do you account for the high solubility of ammonia in water and the observed shape of the ammonia molecule?

Describe briefly how you would attempt to find whether ammonia reacts with chlorate(I) ion (hypochlorite) in aqueous alkaline solution according to the equation

$$2NH_3 + 3ClO^- \rightarrow N_2 + 3Cl^- + 3H_2O$$

(Cambridge Entrance)

22 Give an account of the Haber process for the manufacture of ammonia, discussing any principles of physical chemistry involved. How is nitric acid obtained from ammonia?

Explain the observation: When sulphur reacts with hot concentrated nitric acid, a red-brown gas is evolved, but when hydrogen sulphide reacts with more dilute nitric acid, a colourless gas is evolved which turns red on exposure to the atmosphere. *(Cambridge Entrance)*

22 The p-block elements: Group VI

The two most important elements in Group VI are oxygen and sulphur. Oxygen is the most abundant element in the Earth's crust, and is vital to life. It forms compounds with almost all the other elements, and its high electronegativity brings out their high oxidation numbers. Sulphur is of immense industrial importance, largely in the form of sulphuric acid, and a nation's annual consumption of the latter is often taken as an index of its economic prosperity. The oxidation number of oxygen in its compounds is almost always -2; for sulphur and the heavier elements oxidation numbers range from -2 to $+6$. All the elements in the Group, except polonium, are either non-metals or metalloids.

	V	VI	VII
2	7 N	8 O	9 F
3	15 P	16 S	17 Cl
4	33 As	34 Se	35 Br
5	51 Sb	52 Te	53 I
6	83 Bi	84 Po	85 At

Figure 22.1 Group VI in the Periodic Table

The elements in Group VI are as follows:

oxygen	O	$[He]2s^2 2p^4$
sulphur	S	$[Ne]3s^2 3p^4$
selenium	Se	$[Ar]3d^{10} 4s^2 4p^4$
tellurium	Te	$[Kr]4d^{10} 5s^2 5p^4$
polonium	Po	$[Xe]4f^{14} 5d^{10} 6s^2 6p^4$

Their positions in the Periodic Table are shown in Figure 22.1. The characteristic valence electron configuration is ns^2np^4, where n is the Period number.

Selenium and tellurium are rare elements. The electrical conductivity of selenium changes when light shines on it, and so it is used in photoelectric cells (in some cameras, for example) and in photocopying machines (the Xerox process, for example, depends on the electrostatic attraction of selenium for carbon powder). Polonium is radioactive and occurs only in trace amounts among radioactive decay products. It was isolated by Marie Curie, who named it after her native Poland (see Box 18.1). These rarer elements will be considered only to illustrate Group trends.

22.1 Group systematics

Oxidation states and bonding characteristics

The electronegativity of oxygen ($\chi = 3.4$) is second only to that of fluorine. By the rules for assigning oxidation numbers given in Section 15.2, $Ox(O)$ is therefore always negative except in the compounds of oxygen with fluorine. In the combined state, $Ox(O)$ is almost always -2, although exceptions include the peroxide ion, O_2^{2-}, in which $Ox(O) = -1$, and the superoxide ion, O_2^-, in which $Ox(O) = -\frac{1}{2}$.

The oxidation numbers of sulphur range from -2 (in H_2S, for example) to $+6$ (in SO_4^{2-}, for example), but the most common are -2, $+4$ and $+6$. This wide range is partly due to sulphur's ability to accommodate extra electrons in its valence shell by making use of its d-orbitals: since the other atoms concerned are usually more electronegative than sulphur, all the electrons involved in bonding are formally assigned to the other atoms leaving the sulphur with a high oxidation number ($+6$ for sulphur hexafluoride, SF_6, for example).

The dominant characteristic of the highly electronegative members of the Group is their ability to *gain* electrons. Although the *first* electron gain is exothermic, however, the second is strongly endothermic and overall the formation of O^{2-} is markedly endothermic:

$$O(g) + 2e^- \rightarrow O^{2-}(g) \qquad \Delta_r H^\ominus = +703\,\text{kJ mol}^{-1}$$

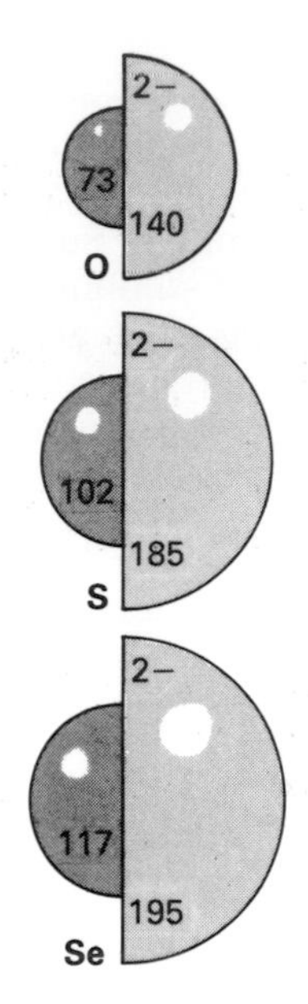

Figure 22.2 The radii (in picometres) of the atoms and M^{2-} anions of oxygen, sulphur and selenium

The formation of O^{2-} therefore requires a substantial enthalpy investment; but because the anion is both doubly charged and small the investment is likely to be recovered from a high lattice enthalpy, especially if the cations are also small and highly charged. This balance of enthalpy changes results in many metal oxides having exothermic standard formation enthalpies; examples of predominantly ionic oxides include Na_2O, MgO and Al_2O_3. When the ionization energies of the elements are high the lattice enthalpy cannot compensate for the energy investment, and oxides such as SiO_2, P_4O_{10} and SO_3 are predominantly covalent.

Periodic trends

As usual, the metallic character of the elements increases down the Group, but only polonium has the characteristics of a metal. Oxygen, sulphur and selenium are non-metals and tellurium is a metalloid.

The marked differences between oxygen and the other members of the Group stem from:

1. the small size of the oxygen atom, which enables it to form multiple bonds (especially with carbon and nitrogen) using its p-orbitals;
2. the absence of accessible d-orbitals, which results in the atom being unable to expand its valence shell;
3. oxygen's high electronegativity; this enables it to take part in hydrogen bonding, which is not possible for the less electronegative members of the Group. This feature has profound consequences: for example, the existence of water as a liquid rather than a gas is due to oxygen's ability to participate in hydrogen bonding (see Section 5.1).

Shapes and sizes

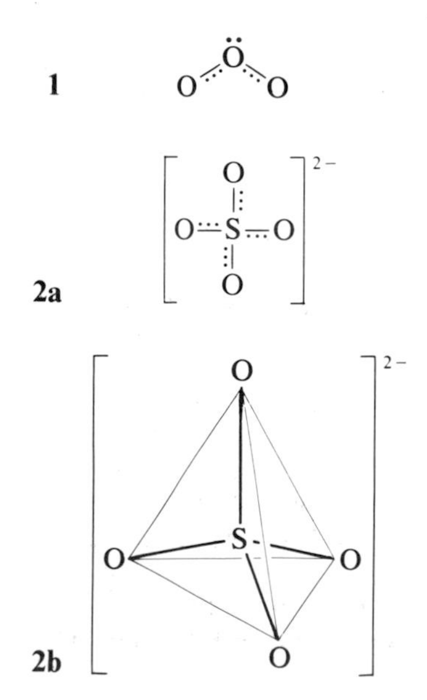

The covalent and ionic radii of some of the Group VI elements are shown in Figure 22.2. As expected, these increase down the Group, as electrons occupy shells with higher principal quantum numbers.

The shapes of the covalent molecules are readily explained in terms of VSEPR theory (see Sections 2.3 and 2.4). Four electron pairs can be accommodated in the valence shells of oxygen, and up to six in those of sulphur and of the heavier members of the Group. When predicting the shapes of these molecules, the rule should be remembered that multiple bonds are treated like single electron pairs. Thus trioxygen, which is much more commonly called ozone, is an angular molecule (**1**) because it has *effectively* three electron pairs on the central atom, and its bond angle of 117° is slightly less than the 120° expected for three equivalent pairs. SO_4^{2-} is tetrahedral (**2**), for even though there are six electron pairs around the sulphur atom, it is treated as having *effectively* four pairs. The shapes that result are summarized in Table 22.1.

Table 22.1 Shapes of molecules or ions containing oxygen or sulphur

Number of lone pairs	Number of neighbouring atoms	Shape	Example
0	3	trigonal planar	SO_3
1	2	angular	SO_2
0	4	tetrahedral	SO_4^{2-}
1	3	pyramidal	H_3O^+
2	2	angular	H_2O
0	6	octahedral	SF_6

EXAMPLE

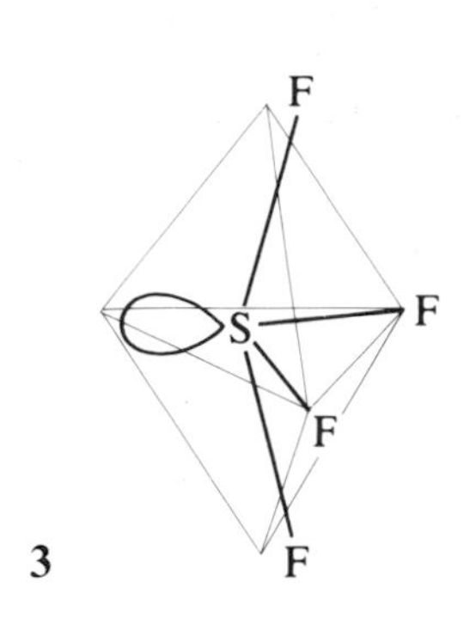

3

Predict the shape of the SF_4 molecule.

METHOD Write the Lewis structure, count the number of bonding pairs and lone pairs on the central sulphur atom, and then use the VSEPR rules in Section 2.3.

ANSWER Sulphur supplies six valence electrons, and each fluorine supplies one to the bond it forms: hence in SF_4 sulphur has ten electrons in five pairs. The basic shape is therefore a trigonal bipyramid. The electrostatic repulsions between the pairs are least if the single lone pair occupies an equatorial position and the two axial fluorine atoms bend away from it, resulting in the structure shown in **3**.

COMMENT It is found experimentally that the equatorial F—S—F angle is 102° and the axial S—F bonds are 3° away from vertical.

22.2 The elements

Oxygen

Oxygen occurs as two gaseous allotropes, O_2 and O_3.

Dioxygen, O_2, is usually called simply 'oxygen', and is by far the more common allotrope. It is an odourless, tasteless gas which condenses to a pale blue liquid at −183 °C. Oxygen is also **paramagnetic** (see Box 22.1) and the liquid is attracted to the poles of a magnet.

BOX 22.1

The structure of dioxygen

Michael Faraday was the first to notice that oxygen is paramagnetic. At the time the phenomenon of molecular magnetism was mysterious, but we now know that paramagnetism is due to the presence of one or more *unpaired* electrons: because each electron has spin (as described in Section 1.2) it behaves as a tiny bar magnet.

The paramagnetism of odd-electron molecules – such as nitrogen monoxide, NO, which has eleven valence electrons (five from nitrogen and six from oxygen) – is not surprising because each molecule must have at least one unpaired electron. Dioxygen is an *even-electron molecule*, however, with twelve valence electrons, and it is not at all obvious that it should have unpaired electrons.

The paramagnetism of dioxygen was puzzling for many years. The Lewis theory could not account for it, for it leads to all twelve electrons being paired:

$$\ddot{\underset{\cdot\cdot}{O}}::\ddot{\underset{\cdot\cdot}{O}}$$

An explanation was obtained only when molecular orbital theory was developed. Then it was realized that the orbitals of an O_2 molecule are arranged in such a way that a lower energy is obtained if two of the twelve electrons occupy different orbitals of the same energy, rather than crowding into a single orbital. Since they are in different orbitals, the electrons are not paired (see Section 1.2).

Oxygen is very reactive as an oxidizing agent, its two most important reactions being **combustion**, in which compounds containing carbon, hydrogen and oxygen, for example, are oxidized to carbon dioxide and water, and **respiration**, in which the overall effect is the same, but which occurs in a much more subtle way. Typical combustion reactions (which are vital to the economic prosperity of the world) are the burning of carbohydrates, such as those of wood, and of hydrocarbons such as those of petroleum and natural gas.

Oxygen is a major industrial chemical; it is used principally in the steel industry, as described in Section 25.9, and on a smaller scale for reviving polluted rivers and treating patients with breathing difficulties. A bulk-tonnage use of more recent origin is that of liquid oxygen as oxidant in space rockets (the fuel being liquid hydrogen).

Since oxygen is readily available in cylinders (which are filled by compressing the gas obtained by fractional distillation of liquid air), there is rarely a need to prepare it in the laboratory. However, it can be prepared by the electrolysis of dilute aqueous electrolytes, by the catalytic decomposition of hydrogen peroxide or by the thermal decomposition of some oxosalts – for example, by heating potassium chlorate(V) in the presence of manganese(IV) oxide (which acts as a catalyst):

$$2KClO_3(s) \xrightarrow{100\,^\circ C,\ MnO_2} 2KCl(s) + 3O_2(g) \qquad Cl(+5) \rightarrow Cl(-1),\quad O(-2) \rightarrow O(0)$$

Trioxygen, O_3, the other allotrope of oxygen, is much better known as **ozone**. Up to ten per cent ozone is prepared by passing oxygen gas through a silent electric discharge: this forms oxygen atoms which attack dioxygen molecules (noisy discharges, such as lightning, signify that ionization has occurred). Ozone (from the Greek for 'to smell') is a pungent, pale blue gas which condenses to an inky-blue liquid at $-112\,^\circ C$. Its presence in the upper atmosphere is an important shield against harmful ultraviolet radiation from the Sun (but see Colour 8).

Ozone is thermally unstable; it is highly reactive, and a powerful oxidizing agent. For example, it cleaves carbon–carbon double bonds (see Section 33.3), and oxidizes iodide ions to iodine quantitatively in alkaline solution:

$$O_3(g) + 2I^-(aq) + H_2O(l) \rightarrow O_2(g) + I_2(s) + 2OH^-(aq)$$
$$O(0) \rightarrow O(-2),\quad I(-1) \rightarrow I(0)$$

Ozone is used in place of chlorine in some water purification plants, one advantage being that it avoids the formation of possibly harmful chlorinated organic molecules, and another that it acts more rapidly.

Sulphur

Sulphur has more allotropes than oxygen, and its ability to form chains and rings is a striking feature of its chemistry.

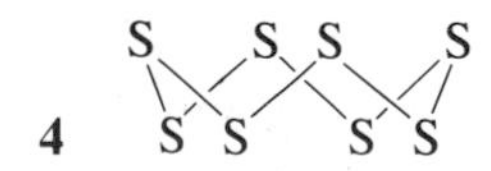

At room temperature the thermodynamically most stable form is **rhombic sulphur** (or **α-sulphur**). This yellow allotrope occurs as short, squat crystals (Figure 22.3(a)) and consists of puckered, crown-like S_8 rings (**4**). When the temperature is raised to $96\,^\circ C$ **monoclinic sulphur** (or **β-sulphur**), which consists of needle-like crystals (Figure 22.3(b)), becomes the most stable allotrope (see Section 6.2) but the change is slow. Monoclinic sulphur is also built from S_8 rings, but they are packed together with less order than in rhombic sulphur. At about $120\,^\circ C$ monoclinic sulphur melts to an amber liquid, and the colour darkens as the temperature is raised. On further heating the liquid becomes much more viscous as the S_8 rings break, with the formation of long entangled chains of sulphur atoms, up to a quarter of a million atoms long. At still higher temperatures, but below the boiling point of sulphur at $445\,^\circ C$, the viscosity decreases again because the long chains break into much smaller fragments (about a thousand atoms long at $400\,^\circ C$). If this liquid is cooled very quickly by pouring it into water the chains are preserved, and a rubbery mass called **plastic sulphur** is formed. Plastic sulphur gradually crumbles as the S_8 rings form again and it slowly reverts to the thermodynamically more stable rhombic sulphur.

(a)

(b)

Figure 22.3 Crystals of (a) rhombic, and (b) monoclinic sulphur

Sulphur is reactive in all its forms. It is oxidized by elements of higher electronegativity, and burns in oxygen with a blue flame to form the chokingly pungent dioxide:

$$S_8(s) + 8O_2(g) \xrightarrow{\text{heat}} 8SO_2(g) \qquad O(0) \rightarrow O(-2), \quad S(0) \rightarrow S(+4)$$

With elements of lower electronegativity, such as metals, sulphur acts as an oxidizing agent and forms **sulphides**. The reaction with some metals is vigorous, especially if the metal is finely divided and has a very large surface area:

$$8Zn(s) + S_8(s) \xrightarrow{\text{heat}} 8ZnS(s) \qquad S(0) \rightarrow S(-2), \quad Zn(0) \rightarrow Zn(+2)$$

Sulphur is not as strong an oxidizing agent as oxygen; for example, when iron is heated with sulphur it forms iron(II) sulphide but with oxygen it forms a mixture of oxides containing both iron(II) and iron(III). The only elements that do not combine directly with sulphur are the noble gases, nitrogen, iodine and some noble metals, such as platinum and gold.

Huge quantities of sulphur are used for the manufacture of sulphuric acid, described in Section 22.6. Another major use is in the **vulcanization** of rubber, in which sulphur is used to form cross-links between the rubber molecules (see Box 33.3).

22.3 Oxides and oxoacids

In this section we need consider only the oxides and oxoacids of sulphur. The most important oxides are **sulphur dioxide**, SO_2, and **sulphur trioxide**, SO_3. They are both acidic oxides, and are the anhydrides of **sulphurous acid**, $(HO)_2SO$, and **sulphuric acid**, $(HO)_2SO_2$, respectively.

Sulphur dioxide

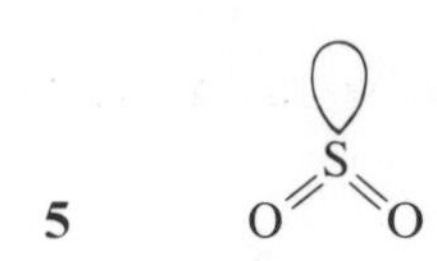

5

Sulphur dioxide (**5**) is formed when sulphur burns in air or oxygen. Since fossil fuels all contain sulphur, it is also formed when they burn, and contributes to the problem of acid rain (see Box 22.2). It is prepared in the laboratory by heating copper with concentrated sulphuric acid:

$$Cu(s) + 2(HO)_2SO_2(aq, conc.) \xrightarrow{heat} SO_2(g) + CuSO_4(aq) + 2H_2O(l)$$

$$S(+6) \rightarrow S(+4), \quad Cu(0) \rightarrow Cu(+2)$$

Sulphur dioxide is a colourless, toxic gas with a choking smell. It is an acidic oxide, and dissolves readily in water to give a complicated mixture generally referred to as 'sulphurous acid' and denoted either $(HO)_2SO$ or H_2SO_3:

$$SO_2(g) + H_2O(l) \rightarrow (HO)_2SO(aq)$$

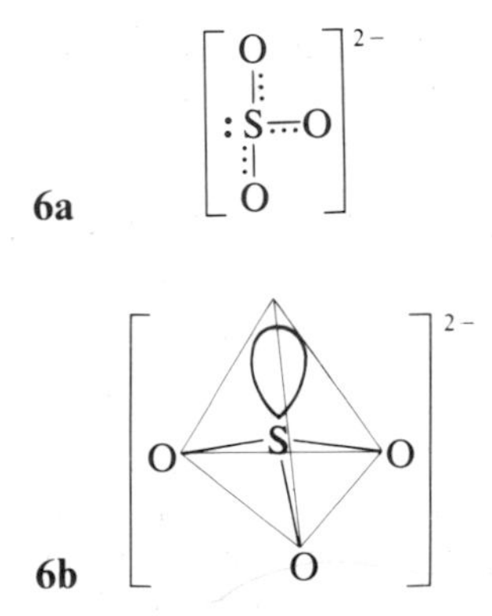

6a

6b

Pure sulphurous acid itself has never been isolated because as the solution is made more concentrated by heating, the equilibrium of this exothermic reaction shifts to the left and sulphur dioxide is evolved. Its salts are well known, however, and contain the **sulphite ion** (SO_3^{2-}, **6**).

The oxidation number of sulphur in sulphur dioxide is +4, which is intermediate in sulphur's range (from −2 to +6). Consequently, sulphur dioxide can be expected to act both as an oxidizing agent and as a reducing agent. This is the case, but its most important reactions are as a reducing agent. For example, it reduces acidified potassium dichromate(VI), turning it from orange to green. It reacts with oxygen to form sulphur trioxide, and with chlorine to form the colourless liquid sulphur dichloride dioxide, SCl_2O_2 (although the rates are appreciable only in the presence of a catalyst). Because of its anti-microbial properties, sulphur dioxide is used as a preservative in foods. Canned orange juice and poor-quality wines sometimes smell of the gas.

Sulphur dioxide behaves as an oxidizing agent when the other substance is even more strongly reducing; it is then reduced to elemental sulphur. For

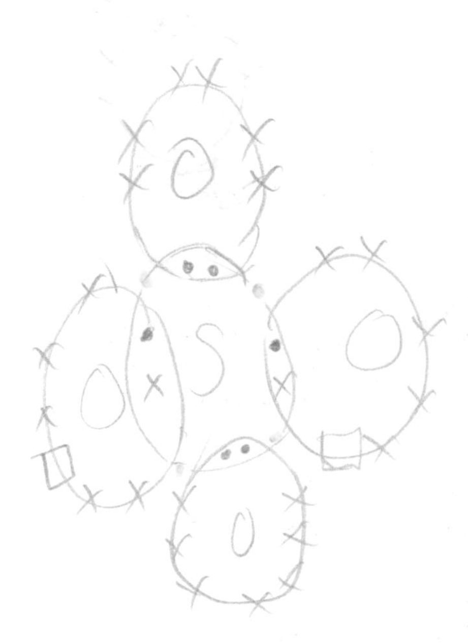

BOX 22.2

Acid rain

Oxides of nitrogen – collectively denoted NO_x – from car exhausts combine with sulphur dioxide (from burning fossil fuels in power stations and from natural sources such as volcanoes), ozone and water vapour in the atmosphere to turn rain, which is already slightly acidic due to the presence of dissolved carbon dioxide, into **acid rain**, a more corrosive, dilute solution of nitrogen and sulphur oxoacids.

Acid rain attacks the limestone and marble of buildings, washes nutrients from the soil, and lowers the pH of rivers and lakes. By leaching aluminium and other metal cations from clays and minerals it increases the metal concentrations in lakes and rivers. The effect of reduced pH is made more serious by the presence of aluminium ions because although fish can survive in moderately acidic waters, they die when $Al^{3+}(aq)$ ions are present too: there is a **synergistic** effect, the presence of one pollutant increasing the effect of the other.

Since the late 1970s concern has been growing that many forests in Europe have been severely affected by acid rain (see Figure 12.8), either by the direct consequences of lowered pH or indirectly through the viral infections that it may encourage. The economic cost of eliminating the problem (for example, by using $Ca(OH)_2$ to remove the sulphur dioxide from the flue gases of power stations) is enormous: but so is the political cost, because one country's emissions can damage another country's environment.

example, sulphur dioxide oxidizes hydrogen sulphide in the presence of water as a catalyst, forming yellow sulphur:

$$SO_2(g) + 2H_2S(g) \rightarrow 3S(s) + 2H_2O(l) \qquad S(+4) + 2S(-2) \rightarrow 3S(0)$$

Sulphurous acid, sulphites and thiosulphates

In aqueous solution sulphurous acid, the hydrogensulphite ion (once called the bisulphite ion) and the sulphite ion are interrelated by the following equilibria:

$$(HO)_2SO(aq) \underset{H^+(aq)}{\overset{OH^-(aq)}{\rightleftharpoons}} HOSO_2^-(aq) \underset{H^+(aq)}{\overset{OH^-(aq)}{\rightleftharpoons}} SO_3^{2-}(aq)$$

The redox properties of the ions follow the pattern shown by sulphur dioxide. The Campden tablets used as preservatives in fruit-bottling, brewing and home wine-making are sulphites of various kinds; however, sulphite food additives such as sodium sulphite (the additive coded as E221) should be used with care because they are potentially hazardous to asthmatics as they deoxygenate blood.

There is one important additional oxidizing reaction of SO_3^{2-}: its reaction with sulphur itself to form the **thiosulphate ion**, SSO_3^{2-} (also written $S_2O_3^{2-}$, **7**):

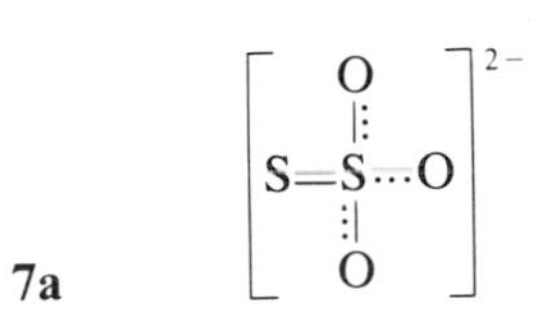

7a

$$8SO_3^{2-}(aq) + S_8(s) \rightarrow 8SSO_3^{2-}(aq) \qquad S(+4) + S(0) \rightarrow 2S(+2)$$

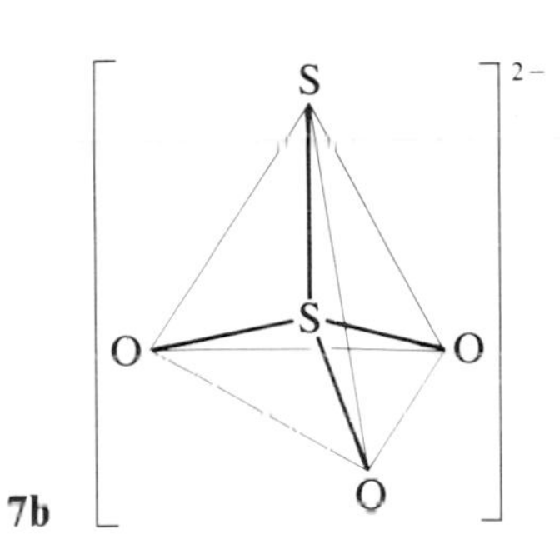

7b

'*Thio-*' (from the Greek word for 'sulphur') is used to signify that an oxygen atom has been replaced by a sulphur atom. The reaction is easy to carry out: flowers of sulphur are added to a boiling aqueous solution of sodium sulphite, and the product is obtained as large, transparent crystals of the pentahydrate, $Na_2SSO_3.5H_2O$. The addition of acid reverses this reaction and amorphous sulphur is formed; because of the microscopic size of the particles the reaction mixture turns white initially, and then the characteristic yellow appears:

$$2H^+(aq) + SSO_3^{2-}(aq) \rightarrow SO_2(g) + S(s) + H_2O(l) \qquad 2S(+2) \rightarrow S(+4) + S(0)$$

Sodium thiosulphate is photographer's 'hypo' (see Box 25.2). It is also used in the laboratory to reduce iodine quantitatively, which makes it useful for the titrimetric determination of iodine (see Section 23.2):

$$I_2(aq) + 2SSO_3^{2-}(aq) \rightarrow 2I^-(aq) + (SSO_3^-)_2(aq)$$

or, more simply:

$$I_2(aq) + 2S_2O_3^{2-}(aq) \rightarrow 2I^-(aq) + S_4O_6^{2-}(aq) \qquad I(0) \rightarrow I(-1),\ S(+2) \rightarrow S(+2\tfrac{1}{2})$$

8

$S_4O_6^{2-}$ is called the **tetrathionate ion** (**8**).

Sulphur trioxide

Sulphur trioxide is formed by the oxidation of the dioxide in the presence of a catalyst, such as platinum or vanadium(V) oxide. It is a volatile white solid (b.p. 45 °C) that reacts violently with water:

$$SO_3(s) + H_2O(l) \rightarrow (HO)_2SO_2(aq)$$

that is, it is the anhydride of sulphuric acid.

In the gas phase sulphur trioxide is mainly monomeric, and the SO_3 molecules are trigonal planar (**9**). However, in the solid polymerization occurs with the formation of $(SO_3)_3$ rings and long polymeric chains (see Figure 22.4).

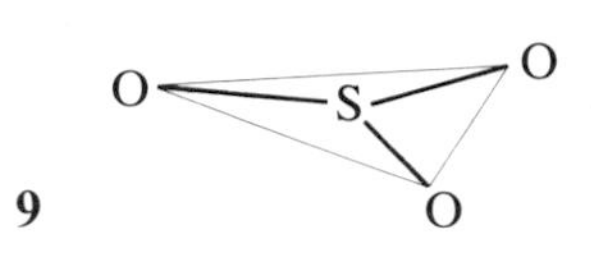

9

Sulphuric acid and the sulphates

In aqueous solution sulphuric acid, the hydrogensulphate ion (once called the bisulphate ion) and the sulphate ion all coexist in equilibrium:

$$(HO)_2SO_2(aq) \underset{H^+(aq)}{\overset{OH^-(aq)}{\rightleftharpoons}} HOSO_3^-(aq) \underset{H^+(aq)}{\overset{OH^-(aq)}{\rightleftharpoons}} SO_4^{2-}(aq)$$

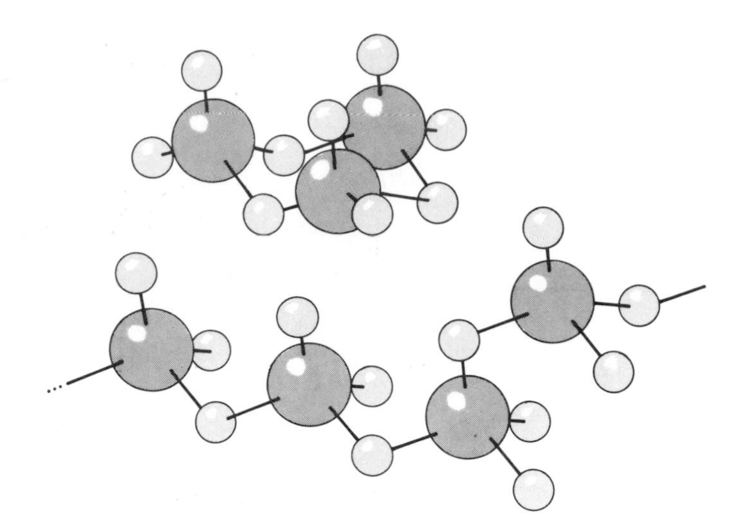

Figure 22.4 The ring and chain forms of sulphur trioxide

The first ionization of sulphuric acid is virtually complete and hence it is a strong acid. The second ionization is very much less complete ($pK_{a2} = 1.92$), so that $HOSO_3^-$ is only a weak acid. This marked difference is due to the difficulty of removing a second hydrogen ion from the negatively charged hydrogensulphate ion.

Pure sulphuric acid is a colourless, viscous liquid that boils at 338 °C. Its chemical importance stems from its ability to act as three kinds of reagent: as an acid, an oxidizing agent and a dehydrating agent.

Sulphuric acid is an important acid partly because, having such a low volatility, it can be used to displace more volatile acids from their salts, as in the laboratory preparation of nitric acid:

$$KNO_3(s) + (HO)_2SO_2(aq,conc.) \xrightarrow{\text{heat}} KHOSO_3(s) + HONO_2(l)$$

Care must be taken when the displaced acid is a substance that can be oxidized, for then the sulphuric acid in its second role, that of an oxidizing agent, can destroy the product.

The role of *hot concentrated* sulphuric acid as an oxidizing agent stems from the stability of gaseous sulphur dioxide, which is formed when the acid is heated, as in the reaction with copper (see the preparation of sulphur dioxide described above) and in

$$2HBr(aq) + (HO)_2SO_2(aq,conc.) \rightarrow Br_2(aq) + 2H_2O(l) + SO_2(g)$$

$$S(+6) \rightarrow S(+4), \quad Br(-1) \rightarrow Br(0)$$

The second reaction explains why hydrogen bromide cannot be prepared by the action of sulphuric acid on bromides (or hydrogen iodide from iodides); similarly, hydrogen sulphide cannot be dried by bubbling the gas through sulphuric acid.

The dehydrating power of concentrated sulphuric acid is a sign of its high affinity for water. Not only does it dry compounds containing water; it also extracts the elements of water from compounds. For example, when sulphuric acid acts on carbohydrates such as sucrose (sugar) or paper, a black frothy mass of carbon is left. Its action on sucrose can be expressed as follows:

$$C_{12}H_{22}O_{11}(s) \xrightarrow{(HO)_2SO_2(aq,\,conc.)} 12C(s) + 11H_2O(l)$$

The severe burns it causes when it comes in contact with the skin is due to the exothermicity of this dehydration.

Because of its three-fold role in chemistry, and its cheapness, sulphuric acid is widely used in industry. It is said that almost every manufactured product has come into contact with sulphuric acid at some stage of its manufacture, and therefore a nation's consumption of the acid is an indicator of its economic prosperity. The annual UK consumption of sulphuric acid for various purposes is shown in Figure 22.5.

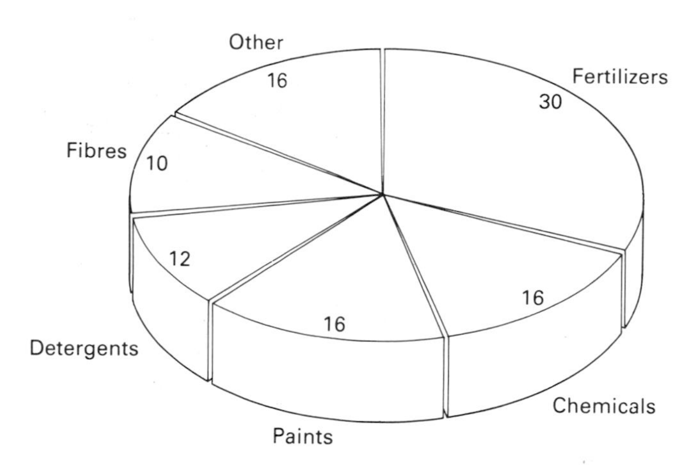

Figure 22.5 The uses of sulphuric acid in the UK

Most sulphates are soluble, partly because the double charge is delocalized over all four oxygen atoms. The principal exceptions are the sulphates of calcium (which is sparingly soluble), strontium, barium and lead(II). The formation of a white precipitate when barium ions are added to a solution is the usual test for sulphates:

$$Ba^{2+}(aq) + SO_4{}^{2-}(aq) \rightarrow BaSO_4(s)$$

The precipitate is distinguished from barium sulphite, $BaSO_3$, by not reacting with dilute acid.

22.4 Halides

Halides of oxygen

Only fluorine is more electronegative than oxygen, and so the fluorides are the only true halides of oxygen. All the other compounds of the halogens with oxygen are better regarded as oxides of the halogens, and are discussed in Section 23.3.

Oxygen difluoride, OF_2, which is a colourless, toxic gas, is prepared by passing fluorine through dilute aqueous sodium hydroxide:

$$2F_2(g) + 2NaOH(aq) \rightarrow OF_2(g) + 2NaF(aq) + H_2O(l)$$

$$F(0) \rightarrow F(-1), \quad O(-2) \rightarrow O(+2)$$

In this reaction, fluorine is acting as an oxidizing agent; in OF_2, $Ox(O)$ is $+2$, which is the highest value it ever reaches.

Halides of sulphur

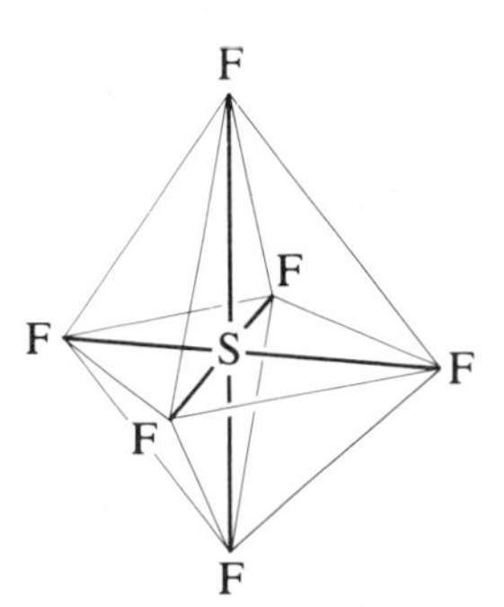

10

The most important of the numerous sulphur halides are **sulphur hexafluoride**, SF_6, and **disulphur dichloride**, S_2Cl_2.

Sulphur hexafluoride is prepared by direct reaction of the elements, and is a gas of exceptionally high chemical and thermal stability. For instance, it is not attacked by molten potassium hydroxide even at 500 °C. This stability can be pictured as arising from the sheath of six fluorine atoms packed round the sulphur atom (**10**). Their tightly held electrons repel the electrons of any attacking Lewis base. Because of this unreactivity, and particularly the difficulty with which the molecule can be ionized by electric fields, sulphur hexafluoride is widely used as a gaseous insulator in transformers and electrical switch gear.

EXAMPLE

A colourless, volatile liquid is found to contain 25.2 per cent by mass of sulphur and 74.8 per cent by mass of fluorine. At 293 K and 0.10 atm, 0.254 g of its vapour occupies 240 cm^3. Find its formula and suggest a structure for its molecules.

METHOD First determine the empirical formula; then use the gas density information to deduce the molecular formula by treating it as a perfect gas. Finally, suggest a structural formula by analogy with other sulphur–fluorine binary compounds, allowing for the possibility that an S—S link may be present.

ANSWER The ratio of S:F atoms is

$$\text{S:F} = \frac{25.2}{32.06}:\frac{74.8}{19.00}$$
$$= 0.786:3.94$$
$$= 1:5$$

Hence the empirical formula is SF_5.

The molar volume at 293 K and 0.10 atm of a perfect gas is $240 \times 10^3\ \text{cm}^3\ \text{mol}^{-1}$; hence the amount of molecules n in the vapour is

$$n = \frac{240\ \text{cm}^3}{240 \times 10^3\ \text{cm}^3\ \text{mol}^{-1}} = 1.00 \times 10^{-3}\ \text{mol}$$

The mass of the vapour is 0.254 g, so that the molar mass is

$$\frac{0.254\ \text{g}}{1.00 \times 10^{-3}\ \text{mol}} = 254\ \text{g mol}^{-1}$$

Since the molar mass of SF_5 is 127.1 g mol^{-1}, the molecular formula is S_2F_{10}.

The compound is molecular (this is suggested by its volatility), and a possible structure is one in which each sulphur atom is surrounded by five fluorine atoms and the other sulphur atom at the corners of an octahedron (**11**).

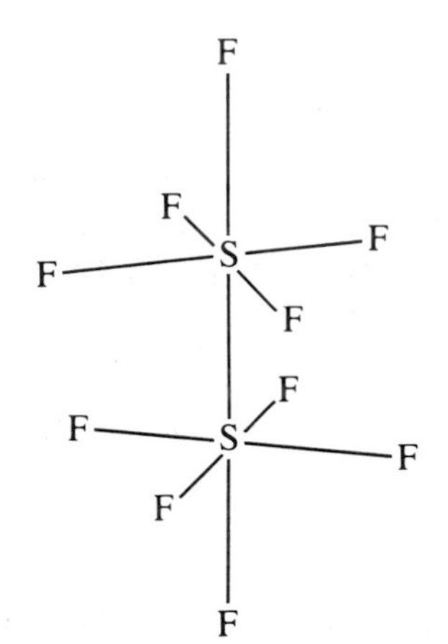

11

COMMENT The suggested structure has been confirmed spectroscopically.

Disulphur dichloride is the best known and most stable of the chlorides of sulphur; it is prepared by direct reaction of the elements, and purified by fractional distillation. It is a toxic golden-yellow liquid with a particularly revolting smell. Nevertheless it is used in large quantities in industry, principally in rubber vulcanization (see Box 33.3).

EXAMPLE

Two colourless, volatile liquids have formulae (a) SCl_2O and (b) SCl_2O_2; both are hydrolysed by water. Which of the following reagents might you use to treat the hydrolysis products in order to distinguish between (a) and (b): acidified aqueous silver nitrate, aqueous barium chloride and acidified aqueous potassium dichromate(VI)?

METHOD Write the equations for the hydrolysis reactions of each substance, and consider the reactions of these products with the stated reagents.

ANSWER The hydrolysis reactions are

$$SCl_2O(l) + 2H_2O(l) \rightarrow (HO)_2SO(aq) + 2HCl(aq) \quad \mathbf{1}$$

$$SCl_2O_2(l) + 2H_2O(l) \rightarrow (HO)_2SO_2(aq) + 2HCl(aq) \quad \mathbf{2}$$

Acidified silver nitrate solution will give a white precipitate of silver chloride with the hydrolysis products of both **1** and **2**, and so will not distinguish them.

Barium chloride will give a white precipitate of barium sulphite with the hydrolysis products of **1** and of barium sulphate with those of **2**. Barium sulphite reacts with the hydrochloric acid formed in **1**, but the sulphate does not; the precipitate will thus be less copious from **1** than from **2**.

Acidified potassium dichromate(VI) will oxidize sulphurous acid formed from **1** but not sulphuric acid formed from **2**, because in the latter the sulphur already has its highest oxidation number (+6). This reagent is therefore best for distinguishing **1** from **2**.

COMMENT A clue to the solution is that the principal difference between (a) and (b) is the oxidation number of sulphur, and so the best distinguishing reagent is likely to be the oxidizing agent.

22.5 Compounds with hydrogen

Some of the compounds to be considered here are already familiar in other contexts, namely **water** (H_2O), **hydrogen peroxide** (H_2O_2) and **hydrogen sulphide** (H_2S), but we also mention a less familiar class, the **polysulphanes** (H_2S_m, $m = 2$ to 8). The last show the much greater tendency of sulphur as compared with oxygen to **catenate**, that is, to form chains of atoms. The corresponding ions are also familiar: they are the oxide and hydroxide ions (O^{2-}, OH^-), the peroxide ion (O_2^{2-}), the **sulphide** and **hydrogensulphide ions** (S^{2-} and HS^-) and the **polysulphide ions** (S_m^{2-}).

Water

Water is important not merely because it is so widespread but also because it has unique properties, which stem largely from oxygen's ability to participate in hydrogen bonding. The effects on its physical properties include its high melting and boiling points and the occurrence of its maximum density at 4 °C (see Section 5.1), as well as its ability to attach through hydrogen bonds to many non-ionic compounds, such as sugar. Its large dipole moment also enables it to attach to ions electrostatically, and therefore to dissolve many ionic substances.

An important chemical property of water is its self-ionization, mentioned already in Section 12.2:

$$2H_2O(l) \rightleftharpoons H_3O^+(aq) + OH^-(aq)$$

The **oxonium ion**, $H_3O^+(aq)$, is a hydrated proton; the less specific formula $H^+(aq)$ is often used instead.

Water is one of the most versatile of chemicals. It can act as a Brønsted acid, a Brønsted base, a Lewis base, an oxidizing agent and a reducing agent.

An example of water's action as a Brønsted acid is its reaction with ammonia:

$$NH_3(aq) + H_2O(l) \rightleftharpoons NH_4^+(aq) + OH^-(aq)$$

while it acts as a Brønsted base in its reaction with ethanoic acid:

$$CH_3COOH(aq) + H_2O(l) \rightleftharpoons H_3O^+(aq) + CH_3CO_2^-(aq)$$

Its Lewis basicity is due to the lone pairs of electrons on the oxygen atom, which enable it to form a coordinate bond with a Lewis acid, such as a proton (this accounts for its Brønsted basicity) or a metal cation (this explains the hydration of such ions in aqueous solution). An actual chemical reaction occurs with some Lewis acids, causing **hydrolysis**:

$$PCl_5(s) + 4H_2O(l) \rightarrow (HO)_3PO(aq) + 5HCl(aq)$$

A very important class of reactions of this kind occurs in organic chemistry where Lewis base attack is called **nucleophilic attack** (see, for example, Sections 28.3 and 31.4).

The nature of the redox reactions of water depends on the other reactant. With strong reducing agents, such as electropositive metals, water is reduced to hydrogen:

$$2Na(s) + 2H_2O(l) \rightarrow 2NaOH(aq) + H_2(g) \qquad H(+1) \rightarrow H(0), \quad Na(0) \rightarrow Na(+1)$$

With strong oxidizing agents, including the anode of an electrolytic cell, water is oxidized to oxygen. For example, with fluorine:

$$2F_2(g) + 2H_2O(l) \rightarrow 4HF(aq) + O_2(g) \qquad F(0) \rightarrow F(-1), \quad O(-2) \rightarrow O(0)$$

Hydrogen peroxide

Hydrogen peroxide, H_2O_2, is formed industrially by the catalytic oxidation of hydrogen, and in the laboratory by the action of dilute sulphuric acid on barium peroxide:

$$BaO_2(s) + (HO)_2SO_2(aq,dil.) \rightarrow H_2O_2(aq) + BaSO_4(s)$$

Barium peroxide is chosen because its sulphate is insoluble, and can be removed by filtration.

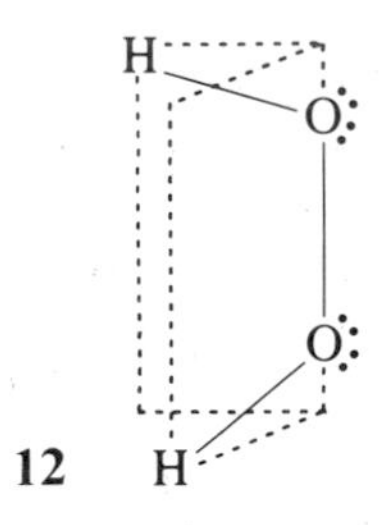

12

Pure hydrogen peroxide is a pale blue liquid with physical properties rather similar to those of water. For example, it freezes at $-0.4\,^\circ C$ and boils at $150\,^\circ C$. This similarity is due to the extensive hydrogen bonding that affects the structures of both liquids. The shape of the molecule is shown in **12**, the O—O bond acting as a kind of hinge; the precise angle that one O—H bond makes to the other depends on the state.

Hydrogen peroxide can be identified by the formation of chromium peroxide, $CrO(O_2)_2$, with acidified potassium dichromate(VI): this forms a deep blue solution in ethoxyethane.

Hydrogen peroxide is easily decomposed if catalysts such as manganese(IV) oxide are present:

$$2H_2O_2(aq) \rightarrow 2H_2O(l) + O_2(g) \qquad 4O(-1) \rightarrow 2O(-2) + 2O(0)$$

EXAMPLE

Calculate the standard reaction Gibbs energy for the decomposition of pure hydrogen peroxide into liquid water and oxygen at 298 K.

METHOD Write the reaction, with the states specified; then use standard formation Gibbs energies (Table 11.2) as explained in Section 11.3.

ANSWER The standard reaction Gibbs energy for the reaction

$$2H_2O_2(l) \rightarrow 2H_2O(l) + O_2(g)$$

is

$$\begin{aligned}\Delta_r G^\ominus &= \{2\Delta_f G^\ominus(H_2O,l) + \Delta_f G^\ominus(O_2,g)\} - 2\Delta_f G^\ominus(H_2O_2,l)\\ &= \{2(-237.2\,kJ\,mol^{-1}) + 0\} - 2(-120.4\,kJ\,mol^{-1})\\ &= -233.6\,kJ\,mol^{-1}\end{aligned}$$

COMMENT The decomposition is thermodynamically favoured, but in practice the reaction is very slow at room temperature unless a catalyst is present.

If a reducing agent is present, then that is oxidized in preference to this disproportionation. Because of its strong oxidizing ability, hydrogen peroxide is used domestically for bleaching hair and as a mild disinfectant. The precise products depend on the pH. For example, in acidic solution water is formed (see Section 14.1):

$$H_2O_2(aq) + 2I^-(aq) + 2H^+(aq) \rightarrow 2H_2O(l) + I_2(s)$$

$$O(-1) \rightarrow O(-2), \quad I(-1) \rightarrow I(0)$$

Hydrogen peroxide can act as a reducing agent with a strong oxidizing agent. One example is its ability to decolorize acidified potassium manganate(VII):

$$2MnO_4^-(aq) + 5H_2O_2(aq) + 6H^+(aq) \rightarrow 2Mn^{2+}(aq) + 8H_2O(l) + 5O_2(g)$$

$$Mn(+7) \rightarrow Mn(+2), \quad O(-1) \rightarrow O(0)$$

The industrial importance of hydrogen peroxide stems from its oxidizing role. It is used to bleach textiles and in pollution control – for example, in the conversion of foul-smelling sulphides to innocuous sulphates.

Hydrogen sulphide

Hydrogen sulphide, H_2S, is prepared by the action of dilute acids on sulphides:

$$FeS(s) + 2H^+(aq) \rightarrow H_2S(g) + Fe^{2+}(aq)$$

It is a toxic gas with a disgusting smell reminiscent of bad eggs (that of hydrogen telluride is even worse). One danger of hydrogen sulphide is its ability to deaden the sense of smell, however, and its presence cannot always be detected. It dissolves readily in water to form a weakly acidic solution:

$$H_2S(aq) + H_2O(l) \rightleftharpoons H_3O^+(aq) + HS^-(aq) \qquad pK_{a1} = 6.9$$

$$HS^-(aq) + H_2O(l) \rightleftharpoons H_3O^+(aq) + S^{2-}(aq) \qquad pK_{a2} = 14.1$$

Consequently there are two series of salts, the hydrogensulphides, containing the HS^- ion, and the sulphides, containing the S^{2-} ion.

Sulphides are known for most metals, and are formed either by direct reaction of the elements:

$$8Fe(s) + S_8(s) \xrightarrow{\text{heat}} 8FeS(s) \qquad S(0) \rightarrow S(-2), \quad Fe(0) \rightarrow Fe(+2)$$

or, more commonly, by passing hydrogen sulphide gas through an aqueous solution of a metal salt:

$$\underset{\text{blue}}{CuSO_4(aq)} + H_2S(g) \rightarrow \underset{\text{black}}{CuS(s)} + (HO)_2SO_2(aq)$$

The latter procedure is effective because most sulphides are insoluble, and precipitate. Hydrogen sulphide can be distinguished from sulphur dioxide (both turn acidified potassium dichromate(VI) from orange to green) by the formation of black lead(II) sulphide on paper that has been soaked in lead(II) nitrate or ethanoate, both of which are colourless. Because the sulphide ion is so polarizable, sulphides have considerable covalent character, in contrast to the more ionic oxides.

Hydrogen sulphide is also a strong reducing agent: the oxidation number of sulphur is -2, and it can be readily oxidized to higher states. In most of these reactions (with potassium manganate(VII) or dichromate(VI), for example) the oxidation number rises to 0, and elemental sulphur is produced. This is the case even when the gas burns in a restricted supply of air:

$$8H_2S(g) + 4O_2(g) \rightarrow S_8(s) + 8H_2O(l) \qquad O(0) \rightarrow O(-2), \quad S(-2) \rightarrow S(0)$$

When excess air is present, however, oxidation produces sulphur dioxide:

$$2H_2S(g) + 3O_2(g) \rightarrow 2SO_2(g) + 2H_2O(l) \qquad O(0) \rightarrow O(-2), \quad S(-2) \rightarrow S(+4)$$

Polysulphanes

The polysulphanes, H_2S_m, consist of a chain of sulphur atoms with a hydrogen atom at each end. They are thermally unstable yellow oils (except for H_2S_2, which is colourless), which disproportionate into sulphur and hydrogen sulphide. They can be regarded as the parent acids of the **polysulphides**, which contain S_m^{2-} ions.

Polysulphides are sometimes found as minerals. For example, **iron pyrites**, FeS_2, is a polysulphide with $m = 2$; it has a rock-salt structure, with Fe^{2+} cations and S_2^{2-} anions. The deep blues and greens of **lapis lazuli** are due to polysulphide *radical* anions: the mineral is an aluminosilicate in which S_3^- ions are responsible for its blue colour and S_2^- ions for its green.

22.6 Industrial chemistry

Occurrence

Despite its reactivity, oxygen occurs extremely widely as the free element, principally as O_2 (comprising 21 per cent by volume of air) but also as O_3 in moderate concentrations at high altitudes in the **ozone layer**. In combined form it is also a major component of many minerals and of the oceans in the form of water.

Sulphur is also quite abundant, being the sixteenth most abundant element in the Earth's crust. It too occurs as the free element, particularly in the USA and Mexico. Many minerals are either sulphides – including **zinc blende**, ZnS, and **cinnabar**, HgS – or sulphates, such as **anhydrite**, $CaSO_4$.

The extraction of sulphur

Native (uncombined) sulphur is extracted from the ground by the **Frasch process**, illustrated diagrammatically in Figure 22.6. Superheated water is pumped down into the deposit, and melts the sulphur; compressed air is pumped down a concentric pipe at the same time, and a frothy mixture of sulphur and water is pushed to the surface. This is an energy-intensive process, however, and confined to regions rich in native sulphur, such as Louisiana and Texas. Most sulphur is now obtained from fossil fuels, a change of emphasis to which environmental pressures have also contributed; an advantage of cutting down the sulphur load of the effluent from a furnace is that it helps to reduce the formation of acid rain (see Box 22.2).

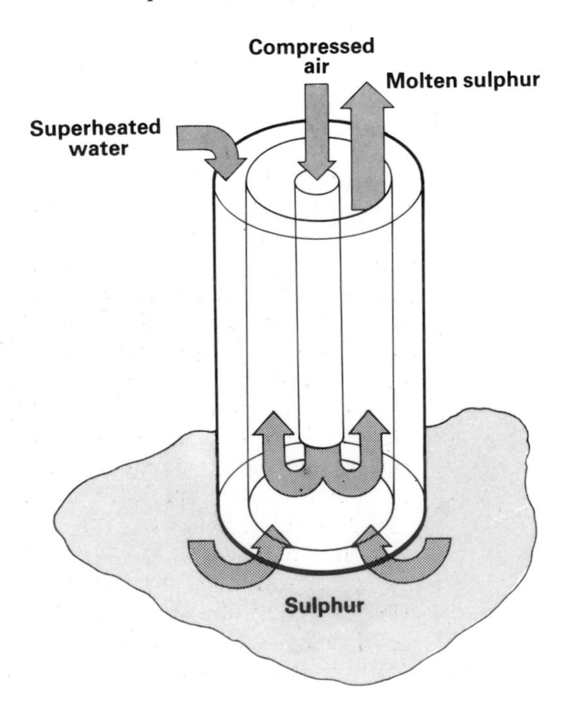

Figure 22.6 The Frasch process for extracting sulphur

Sulphuric acid: the contact process

Sulphuric acid manufacture begins with the production of sulphur dioxide, either by burning molten sulphur in air or as a by-product of the roasting (that is, heating in air) of sulphide minerals such as Cu_2S, ZnS and FeS_2 in the course of the extraction of the metals.

Sulphur trioxide is then produced by mixing the dioxide with excess air at about 450 °C and passing the purified gases over a vanadium(v) oxide (V_2O_5) catalyst, promoted by potassium sulphate, on a silica support:

$$2SO_2(g) + O_2(g) \rightarrow 2SO_3(g) \qquad O(0) \rightarrow O(-2), \quad S(+4) \rightarrow S(+6)$$

The oxidation is exothermic, and the temperature rises to about 600 °C. The hot gases are cooled again to 450 °C, and passed over a second catalyst; 99.5 per cent conversion to the trioxide is achieved after two more repetitions of the entire procedure. Notice how the temperature is a compromise: the reaction is exothermic, and the equilibrium yield is lower at high temperatures (see Section 10.2), but the rate of the reaction is greater.

Next, the sulphur trioxide is dissolved in 98 per cent concentrated sulphuric acid, leading to the formation of **oleum**, a complex substance represented as $(HOSO_2)_2O$ (**13**), or $H_2S_2O_7$ for simplicity:

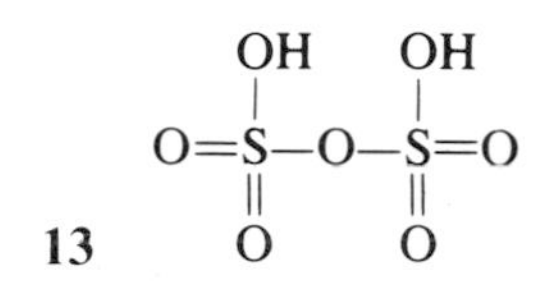

$$SO_3(g) + (HO)_2SO_2(l) \rightarrow H_2S_2O_7(l)$$

Dilution of the oleum then results in sulphuric acid:

$$H_2S_2O_7(l) + H_2O(l) \rightarrow 2(HO)_2SO_2(aq)$$

Direct addition of sulphur trioxide to water is not possible because the reaction is so exothermic that a fine mist forms which does not condense.

Summary

- ☐ The elements have the valence electron configuration ns^2np^4.
- ☐ The most common oxidation number for oxygen is −2, but for sulphur a range of oxidation numbers from −2 to +6 is known.
- ☐ Oxygen is the second most electronegative element.
- ☐ Oxygen does not have energetically low-lying d-orbitals accessible and so is never surrounded by more than eight electrons; whereas sulphur, by employing its low-energy 3d-orbitals, can form compounds with ten or twelve electrons surrounding the sulphur.
- ☐ Oxygen and sulphur are non-metals. Polonium is the only element in the Group to show metallic character.
- ☐ There are two allotropes of oxygen: dioxygen, O_2, and ozone, O_3.
- ☐ Two important and related reactions of oxygen are combustion and respiration.
- ☐ Sulphur has two major allotropes: rhombic and monoclinic sulphur have S_8 rings packed in different ways, whereas plastic sulphur consists of long chains of sulphur atoms.
- ☐ Sulphur can act both as an oxidizing agent (e.g. it reacts with Zn to form ZnS) and as a reducing agent (e.g. it reacts with O_2 to form SO_2).
- ☐ The important oxides of sulphur are SO_2 and SO_3, and the important oxoacids are $(HO)_2SO$ and $(HO)_2SO_2$.
- ☐ Sulphur dioxide can act both as an oxidizing agent (e.g. it reacts with H_2S to form S) or as a reducing agent (e.g. it reacts with Cl_2 to form SCl_2O_2).
- ☐ Elemental sulphur reacts with sodium sulphite solutions to produce the important reducing agent sodium thiosulphate, $Na_2SSO_3 . 5H_2O$.
- ☐ Sulphur trioxide is made by the catalytic oxidation of sulphur dioxide with vanadium(v) oxide as catalyst.
- ☐ Sulphuric acid is a valuable acid in industry. It is also an oxidizing agent (when hot and concentrated) and a dehydrating agent (when concentrated).
- ☐ The formation of a white precipitate of barium sulphate, $BaSO_4$, insoluble in dilute hydrochloric acid, is a test for the presence of sulphate ions.
- ☐ Sulphur hexafluoride is a gas of exceptionally high stability.

- ☐ The hydrides of oxygen are water and hydrogen peroxide, H_2O_2. Water has anomalous physical properties arising from hydrogen bonding. It is an excellent solvent. Water can behave as a Brønsted acid, a Brønsted base, a Lewis base, an oxidizing agent and a reducing agent.
- ☐ Hydrogen sulphide is a diprotic acid and a reducing agent.

PROBLEMS

1 How is oxygen obtained on an industrial scale? Make a list of the industrial uses of oxygen.

2 (a) Give the oxidation numbers of oxygen in the following compounds or ions:

$O^{2-}, O_2^{2-}, O_2^{-}, OF_2, Cl_2O, H_2O, H_2O_2$

(b) Give the oxidation numbers of sulphur in the following compounds or ions:

$S^{2-}, SCl_2, S_2Cl_2, S_2O_3^{2-}, SO_3^{2-}, S_4O_6^{2-}, S_2F_{10}$

(c) Give two examples of disproportionation reactions of compounds containing sulphur. Show the oxidation number of sulphur in all the reactants and products.

3 Describe the preparation, properties and uses of hydrogen peroxide.

4 Explain the following:
(a) CO_2 is linear, but SO_2 and O_3 are angular;
(b) the two O—H bonds in H_2O_2 do not lie in the same plane;
(c) SF_4 is not tetrahedral;
(d) the six-membered ring in S_3O_9 is not planar;
(e) the eight-membered ring in S_8 is not planar.
Your answers should show the structures of the compounds and describe the nature of the bonding.

5 Make a list of the types of reaction water undergoes by writing balanced equations. Reactions should be classified as follows:
(a) water acting as an oxidizing agent (note the changes in oxidation number);
(b) water acting as a reducing agent (note the changes in oxidation number);
(c) water acting as a Brønsted acid;
(d) water acting as a Brønsted base;
(e) water acting as a Lewis base.

6 A colourless liquid **A** can be prepared by the reaction of hydrogen chloride and sulphur trioxide. The liquid reacts violently with water. In one experiment 0.117 g of **A** were completely hydrolysed by water and then excess aqueous barium chloride was added. The barium sulphate precipitate was filtered, washed and dried; it weighed 0.233 g. In another experiment 0.203 g of **A** were completely hydrolysed by water and then excess silver nitrate was added. The silver chloride precipitate was filtered, washed and dried; it weighed 0.250 g.
(a) What is the formula of **A**?
(b) Write balanced equations to represent (i) the preparation of **A** from HCl and SO_3 and (ii) its reaction with water.

7 Compare the properties, reactions and structures of oxides and sulphides. Use H_2O, CaO and CuO as examples of oxides and H_2S, CaS and CuS as sulphides.

8 Write balanced equations for the following reactions:
(a) SO_3^{2-} with (i) Cl_2, (ii) Fe^{3+} and (iii) H_2S;
(b) H_2O_2 with (i) I^-, (ii) Fe^{2+} and (iii) $K_2Cr_2O_7$.
Write the half-reactions, look up the standard electrode potentials and predict the products of the reactions.

9 Compare the properties of sulphur dioxide and sulphur trioxide, including physical properties, structure, reaction with water, acid–base reactions and redox reactions.

10 From your knowledge of trends in the Periodic Table, make predictions about the following:
(a) the boiling point and acid strength of H_2Te,
(b) the existence of Po^{2+}(aq) ions,
(c) the reactivity of TeF_6 compared with SF_6.

11 How would you distinguish between (a) sodium sulphite and sodium thiosulphate, (b) sodium sulphate and sodium hydrogensulphate?

12 (a) Describe the contact process for the manufacture of sulphuric acid from sulphur, paying particular attention to the reasons for the choice of operating conditions.
(b) When dry hydrogen chloride is passed into fuming sulphuric acid (oleum) one of the products is a colourless fuming liquid **L**, which reacts violently with water to give a mixture of sulphuric and hydrochloric acids. In each of two separate experiments 0.466 g of **L** was hydrolysed completely. In the first experiment an excess of aqueous silver nitrate was added to the resulting solution, and 0.574 g of a white precipitate was obtained. In the second experiment an excess of aqueous barium chloride was added to the resulting solution, and 0.932 g of a white precipitate was obtained. Use these data to determine the empirical formula of **L** and write a balanced equation for its reaction with water.
(Cambridge)

13 (a) Describe in outline how oxygen (dioxygen) may be converted into (i) a sample of oxygen containing trioxygen (ozone), and (ii) an aqueous solution of hydrogen peroxide.
(b) Suggest how the concentrations of trioxygen in sample (i) and of hydrogen peroxide in sample (ii) could be determined.
(c) Trioxygen is said to react readily with hydrogen peroxide according to the equation

$$H_2O_2 + O_3 \rightarrow H_2O + 2O_2$$

Suggest a method that would enable you to show that this equation is correct, assuming that an aqueous solution of hydrogen peroxide of known concentration and a sample of oxygen containing a known amount of trioxygen are available. *(Oxford and Cambridge)*

14 List *one* example of a compound for each positive oxidation number of sulphur. Draw diagrams of the electronic structure of the compounds you list and describe some of their redox reactions, using reference books where appropriate.

10.0 cm^3 of 0.02M potassium manganate(VII) solution was decolorized by 12.1 cm^3 of 0.05M sodium sulphite solution which had been acidified by 2M sulphuric acid. Calculate the oxidation number change of the sulphur in this reaction and suggest an explanation of your result. *(Nuffield)*

15 Write an essay on the properties and uses of sulphuric acid. Include in your answer examples, where possible, from both inorganic and organic chemistry, of H_2SO_4 acting as (a) an acid, (b) a catalyst, (c) an oxidizing agent, (d) a dehydrating agent, (e) an electrophile and (f) a sulphonating agent. *(JMB)*

16 Use the following experimental data to deduce an equation for the reaction between chlorine and sodium thiosulphate in aqueous solution, and explain your reasoning.

3.10 g of $Na_2S_2O_3.5H_2O$ were dissolved in water to give 250 cm^3 of solution, and chlorine was passed through until reaction was complete. Excess chlorine was then swept from the solution using gaseous nitrogen.

25.0 cm^3 samples of the resulting solution were found

(a) to require 12.5 cm^3 of 1.00M KOH for neutralization,
(b) to require 10.0 cm^3 of 1.00M $AgNO_3$ to complete the precipitation of chloride ions, and
(c) to give 0.583 g of white precipitate when treated with an excess of aqueous $BaCl_2$.

Iodine reacts with sodium thiosulphate according to the equation

$$I_2 + 2Na_2S_2O_3 \rightarrow 2NaI + Na_2S_4O_6$$

Compare the changes in oxidation number of the chlorine and sulphur in the first reaction with those of iodine and sulphur in the second. What explanation can you offer for this difference between the two halogen elements?
(Relative atomic masses: H = 1, O = 16, Na = 23, S = 32, Ba = 137) *(London)*

17 Suppose you were provided with a sample of a sulphur chloride which you believed to have the molecular formula SCl_2. It is a liquid at room temperature.

Describe experiments which you could yourself perform in order to confirm this formula. Give actual experimental details, including some idea of quantities, concentrations of solutions, etc, and show how the molecular formula could be calculated from the results. *(Nuffield)*

18 A sample of α-sulphur (rhombic) was placed inside a closed container from which all the air had been removed. The α-sulphur was heated *slowly* and the vapour pressure measured. At 96 °C the α-sulphur was transformed into β-sulphur (monoclinic) and at 119 °C the β-sulphur melted. In a second experiment the α-sulphur was heated *rapidly* and at 115 °C it melted. In a third experiment a triple point was found at 151 °C and 1290 atm pressure at which α-sulphur, β-sulphur and liquid sulphur were in equilibrium.

Using the above information, draw a large (half-page) schematic phase equilibrium diagram for the sulphur system, including any metastable states.

(a) On the diagram, label each separate area and each triple point.
(b) Describe what each line on the diagram represents.

The molecular formula of both α-sulphur and β-sulphur is S_8. Suggest how this could be verified in a school laboratory. *(Oxford and Cambridge)*

19 (a) Draw a diagram of the molecular structure of solid sulphur.
(b) Describe what happens when solid sulphur is gently heated until it reaches boiling point, explaining the observations in terms of structural changes.
(c) When the mineral anhydrite is heated to high temperature, the reactions which occur are:

$$\underset{\textbf{(A)}}{CaSO_4} \rightarrow \underset{\textbf{(B)}}{CaS} + 2O_2$$

$$CaSO_4 \rightarrow CaO + SO_2 + \tfrac{1}{2}O_2$$

$$CaSO_4 \rightarrow \underset{\textbf{(C)}}{CaSO_3} + \tfrac{1}{2}O_2$$

Give the *chemical* names for **A**, **B**, **C**, quoting the oxidation number of sulphur in each.
(d) Suggest how the gaseous products of this thermal decomposition could be used to produce sulphuric acid on a large scale. Mention reaction conditions, write equations, and name catalysts if necessary.
(e) Write equations for the action of dilute sulphuric acid on **B** and **C** above, giving tests for any gases evolved.
(f) The sulphur trioxide structure has trigonal planar geometry. Explain this using a bond diagram.
(g) $3SO_2 + 2H_2O \rightarrow S(s) + 2H_2SO_4(aq)$

This equation represents the thermal decomposition of sulphur dioxide solution. Explain the *type* of chemical reaction involved. *(SUJB)*

20 Water plays a vital role in chemistry.

Consider the structure of this compound in the three physical states, and give reasons for any physical behaviour you believe to be anomalous.

Survey the action of water as a solvating agent, complexing agent and agent of hydrolysis. *(London S)*

21 Give a brief account of the physical and chemical properties of either sulphuric acid or nitric acid.

The action of phosphorus pentachloride on sulphuric acid yields a compound **A**, $ClHO_3S$, which reacts with anhydrous hydrogen peroxide to yield **B**, H_2O_5S, and **C**, $H_2O_8S_2$. When **A** is heated in a sealed tube it forms equimolar amounts of sulphuric acid and **D**. **D** reacts with excess of ammonia to form **E**, $H_4N_2O_2S$. Thermal decomposition of **E** yields ammonia and **F**, HNO_2S. Suggest a structural formula for each of the compounds **A** to **F** inclusive. *(Cambridge Entrance)*

22 A sample of 1.44 g of sulphur compound was dissolved in water. On evaporating the solution at room temperature, when only water vapour was lost, the solid residue weighed 1.98 g. The solid residue was treated with sodium hydroxide solution and the ammonia evolved absorbed in an excess of 0.50 M hydrochloric acid. Back titration of the excess acid showed that the ammonia had neutralized 60 cm^3 of the acid. On acidifying the sodium hydroxide solution with an excess of hydrochloric acid and adding barium chloride solution a dense white precipitate was obtained. The supernatant solution now contained only sodium and barium chlorides in addition to hydrochloric acid. The weight of the dried precipitate was 3.51 g.

Suggest a formula for the original compound and give equations for the reactions taking place. *(Cambridge Entrance)*

23 Compare the physical and chemical properties of the oxides of the elements of the first two short Periods (lithium to chlorine). *(Oxford Entrance)*

23 The p-block elements: Group VII

The elements of Group VII, the halogens, are very similar to each other. They are all non-metals, and oxidize metals to form their halides. The halogen oxides are acidic, and their hydrides are covalent. Except for fluorine, the most electronegative element of all, the range of oxidation numbers of the halogens is wide. Fluorine shows special characteristics as its high electronegativity and small size enable it to bring out the highest oxidation numbers of elements. It forms compounds with all elements other than helium, neon and argon.

The elements of Group VII – the **halogens** – are as follows:

fluorine	F	$[He]2s^2 2p^5$
chlorine	Cl	$[Ne]3s^2 3p^5$
bromine	Br	$[Ar]3d^{10}4s^2 4p^5$
iodine	I	$[Kr]4d^{10}5s^2 5p^5$
astatine	At	$[Xe]4f^{14}5d^{10}6s^2 6p^5$

	VI	VII	VIII
1			2 He
2	8 O	9 F	10 Ne
3	16 S	17 Cl	18 Ar
4	34 Se	35 Br	36 Kr
5	52 Te	53 I	54 Xe
6	84 Po	85 At	86 Rn

Figure 23.1 Group VII in the Periodic Table

Their positions in the Periodic Table are shown in Figure 23.1; the characteristic valence electron configuration is ns^2np^5, where n is the Period number. All the isotopes of astatine (the name comes from the Greek for 'unstable') are radioactive, the longest-lived nuclide, ^{210}At, having a half-life of only 8.3 hours. As a result, weighable amounts of the element cannot be obtained, and its chemical properties have been inferred from **tracer** experiments; in these the radioactivity of analogous compounds, such as those of iodine contaminated with a little astatine, are monitored.

The name *halogen*, which is derived from the Greek for 'salt producer', reflects the ability of the members of the Group to react with metals to produce salts. Chlorine is so called because it is a yellow-green gas (from *chloros*, the Greek for 'yellowish green'), bromine because of its pungent smell (*bromos*, stench) and iodine because it is a purple-black solid with a violet vapour (*iodes*, violet) – see Colour 6. Fluorine derives its name from *fluo*, the Latin for 'I flow', because calcium fluoride was used as a flux for dissolving metal oxides.

23.1 Group systematics

Oxidation states and bonding characteristics

Fluorine is the most electronegative of all the elements and so, by definition, in all its compounds it is assigned oxidation number -1.

Chlorine is more electronegative than all the elements except fluorine and oxygen, and so its oxidation number is positive only in its fluorides, oxides and oxoacids: in all other compounds its oxidation number is -1. In common with the other elements of Period 3, chlorine has 3d-orbitals close in energy to its 3s- and 3p-orbitals, and can make use of them to expand its octet. Since there are seven electrons in the valence shell of the free atom, the octet can be expanded as far as the formation of seven covalent bonds, as in dichlorine heptoxide, Cl_2O_7 (**1**), in which $Ox(Cl) = +7$. If fewer than seven valence electrons are used in covalent bonding, the atom can have one, two or three lone pairs, and the oxidation numbers will be $+5$, $+3$ and $+1$ respectively. Similar considerations apply to the heavier halogens. The oxidation numbers most commonly shown are usually odd; the few compounds with even oxidation numbers (such as chlorine dioxide, ClO_2, $Ox(Cl) = +4$) tend to be thermally unstable.

1 $O_3Cl{-}O{-}ClO_3$ (each Cl bonded by double bonds to three O atoms: O=Cl(=O)(=O)–O–Cl(=O)(=O)=O)

The wide range of oxidation numbers of the halogens means that the elements have a rich redox chemistry. Some of the important redox couples are listed in Table 23.1: remember that the more positive the value of $E^\ominus$, the greater the oxidizing ability of the couple (see Section 13.5). The very high value for the F_2, F^- couple emphasizes fluorine's exceptional strength as an oxidizing agent.

Table 23.1 Some standard electrode potentials at 298 K, $E^\ominus/V$

$0 \rightarrow -1$
$Hal_2 + 2e^- \rightarrow 2Hal^-(aq)$

$+1 \rightarrow -1$
$HOHal(aq) + H^+(aq) + 2e^- \rightarrow Hal^-(aq) + H_2O(l)$

$+5 \rightarrow 0$
$2HalO_3^-(aq) + 12H^+(aq) + 10e^- \rightarrow Hal_2 + 6H_2O(l)$

Hal_2	$F_2(g)$	$Cl_2(g)$	$Br_2(l)$	$I_2(s)$
$0 \rightarrow -1$	+2.87	+1.36	+1.07	+0.54
$+1 \rightarrow -1$	–	+1.49	+1.33	+0.99
$+5 \rightarrow 0$	–	+1.47	+1.52	+1.19

EXAMPLE

Explain what happens when chlorine gas is bubbled through aqueous potassium iodide and the mixture is then shaken with the organic solvent 1,1,1-trichloroethane (TCE).

METHOD Judge the direction of the redox reaction by referring to the standard electrode potentials of the relevant couples in Table 23.1. The organic solvent dissolves the more electron-rich substance preferentially.

ANSWER $E^\ominus(Cl_2, Cl^-) = +1.36$ V is more positive than $E^\ominus(I_2, I^-) = +0.54$ V, so that chlorine tends to oxidize iodide ions to iodine. Iodine is much more soluble in TCE than in water, and so it can be predicted that a violet solution of iodine in TCE will be formed.

COMMENT In solvents that do not form specific bonds with iodine molecules, the solute particles are I_2 molecules, which give the same colour to the solution as to the vapour. In solvents that do form bonds, such as ethanol, the colour of the solution is brown.

Periodic trends

The properties of the halogens and their compounds generally vary smoothly with increasing atomic number, and cover a narrower range than in other p-block Groups; for example, the differences between fluorine and iodine are less pronounced than those between carbon and tin. Moreover, two main tendencies help to systematize a great deal of halogen chemistry:

1 electronegativity decreases with increasing atomic number;
2 oxidizing ability decreases with increasing atomic number.

The result of decreased electronegativity is increased covalent character in compounds; for example, AlF_3 is a largely ionic solid that melts at 1290 °C, while $AlCl_3$ is a largely covalent solid that sublimes at 180 °C. The oxidizing strength of the elements is in the order

$$F_2 > Cl_2 > Br_2 > I_2$$

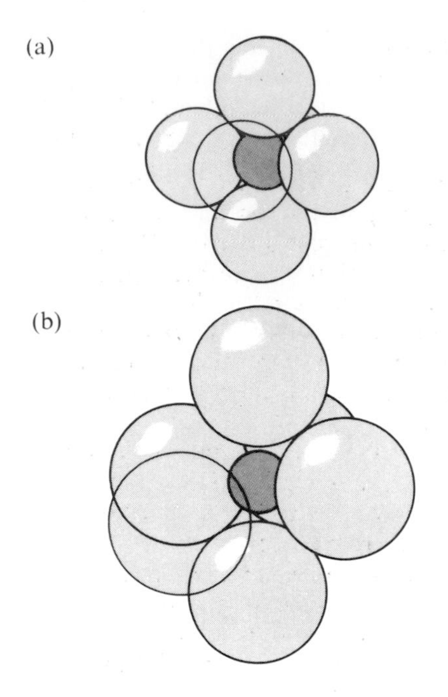

Figure 23.2 An aluminium ion (a) can accommodate six fluoride ions, but (b) cannot accommodate six chloride ions

and if either chlorine water or bromine water (their dilute aqueous solutions) is added to aqueous iodide ions, a dark brown solution of iodine is produced by oxidation of the iodide ions:

$$Cl_2(aq) + 2I^-(aq) \rightarrow I_2(aq) + 2Cl^-(aq) \qquad Cl(0) \rightarrow Cl(-1),\ I(-1) \rightarrow I(0)$$

The manufacture of bromine from solutions of bromides depends on chlorine's ability to oxidize bromides:

$$Cl_2(g) + 2Br^-(aq) \rightarrow Br_2(aq) + 2Cl^-(aq) \qquad Cl(0) \rightarrow Cl(-1),\ Br(-1) \rightarrow Br(0)$$

All the elements are non-metals, but iodine shows some slight metallic character: it has a lustrous appearance, and in a few compounds it exists as a complex cation. The bulk of its properties suggest classification as a non-metal, however, and it is just more electronegative than sulphur (see Table 2.1).

In common with other Period 2 elements, fluorine shows some anomalies. It occurs with only one oxidation number (−1). The small size of the fluorine atom (and of the fluoride ion) allows several to pack around a central atom; this is *one* factor in its ability to induce atoms to show their highest oxidation states and highest coordination numbers, as in the existence of AlF_6^{3-} but only $AlCl_4^-$ (see Figure 23.2). Other factors that affect the formation of these compounds include the oxidizing power of fluorine and the strengths of the bonds that it forms.

The small size of a fluorine atom has two further effects. First, the F—F bond dissociation enthalpy is unusually low, because of the strong electrostatic repulsion between the lone pairs on the two atoms (see Figure 23.3). Second, the electron-gain energy of fluorine is also anomalously low (see Table 1.2), because the additional electron enters a compact valence shell and so experiences strong repulsion from the seven electrons already present (the average separation between the electrons is smaller than in chlorine).

Shapes and sizes

As expected, the covalent and ionic radii of the halogen atoms and ions increase down the Group (see Figure 23.4).

The shapes of the covalent molecules and ions containing halogen atoms are readily rationalized by VSEPR theory, and depend in the usual way on the numbers of lone pairs and bonding pairs, as described in Section 2.3.

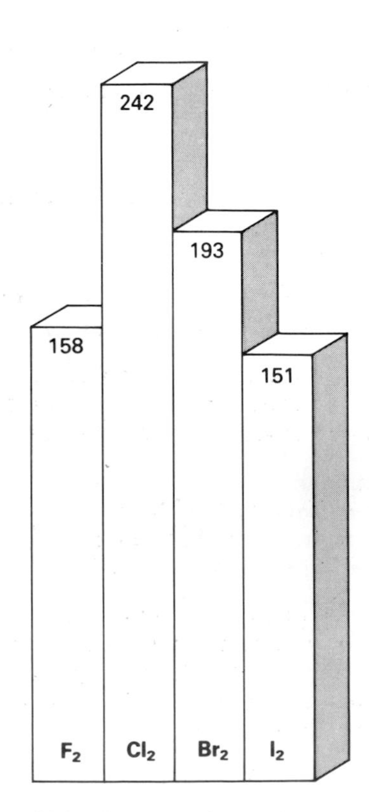

Figure 23.3 The bond dissociation enthalpies for the halogens

23.2 The elements

All the halogens exist as diatomic molecules at room temperature. Their melting and boiling points increase with increasing atomic number, because the dispersion forces between the molecules become stronger as the number of electrons increases, and whereas fluorine and chlorine are normally gases, bromine is a liquid and iodine is a solid. The colour of the elements also deepens from almost colourless fluorine (where the electrons are so tightly held that only ultraviolet light can excite them) through yellow-green chlorine and red-brown bromine to the deep, lustrous purple-black of iodine (where the electrons are so loosely held that most of the visible frequencies are absorbed).

Their most characteristic chemical property is their ability to oxidize: this will be the recurring feature of this chapter.

Fluorine

Fluorine is too strong an oxidizing agent to be prepared by any process other than electrolysis, in which the anode can be made so positive that it can strip an electron even from a fluoride ion. Henri Moissan succeeded where Davy and Faraday had failed, and the process he developed in 1886 is still essentially the one used today. In it, potassium fluoride is dissolved in anhydrous liquid hydrogen fluoride, and the solution is electrolysed at 100 °C

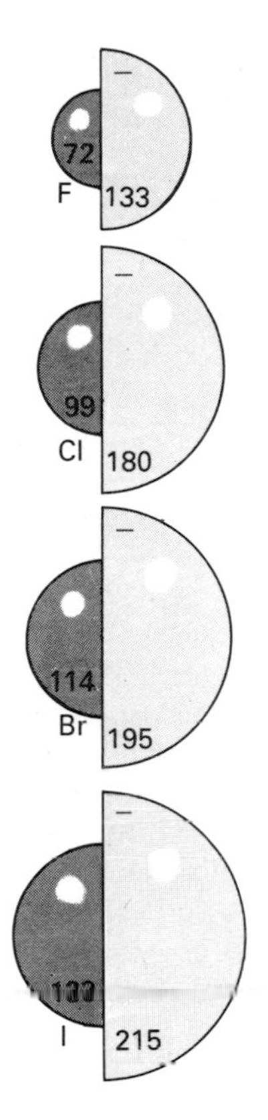

Figure 23.4 The radii (in picometres) of the atoms and M^- anions of the halogens

in a cell with a carbon anode and a steel cathode:

At the anode: $2F^- \rightarrow F_2(g) + 2e^-$ $\quad$ $F(-1) \rightarrow F(0)$

At the cathode: $2H^+ + 2e^- \rightarrow H_2(g)$ $\quad$ $H(+1) \rightarrow H(0)$

The practical difficulties of preparing fluorine are formidable. As a result of its thermodynamic instability and kinetic reactivity (due to its high standard electrode potential and the weak F—F bond respectively) it attacks all organic matter vigorously, and forms a violently explosive mixture with hydrogen, the other product of the electrolysis. The hydrogen fluoride used in the preparation is also highly corrosive, causing severe burns when it touches the skin, and etching glass. Nevertheless, fluorine is such an important chemical that the difficulties have been overcome, and industrial plants now exist that produce around 10 tonnes of the element a day.

The vigour of fluorine as an oxidizing agent is shown by its ability to form compounds with all elements except helium, neon and argon. Many of the direct reactions are spectacular; for example, finely powdered iron wool burns in fluorine (forming FeF_3) like cotton wool burns in air (see Figure 23.5).

All the reactions of fluorine are highly exothermic. This is mainly due to the low bond dissociation enthalpy of the fluorine molecule, as well as to the strength of any covalent X—F bonds that are formed or, if the product is an ionic compound, to its high lattice enthalpy or hydration enthalpy. All these features reflect the small sizes of the fluorine atom and fluoride ion.

On account of the strong oxidizing ability of fluorine, the products of its direct reactions are often compounds in which the other elements have their highest oxidation numbers:

$$S_8(s) + 24F_2(g) \rightarrow 8SF_6(g) \qquad F(0) \rightarrow F(-1),\quad S(0) \rightarrow S(+6)$$

(Oxygen oxidizes sulphur only to oxidation number +4, to sulphur dioxide.) Even chlorine is oxidized by fluorine:

$$Cl_2(g) + 3F_2(g) \xrightarrow{\text{excess } F_2,\ 250\ C} 2ClF_3(g) \qquad F(0) \rightarrow F(-1),\quad Cl(0) \rightarrow Cl(+3)$$

Water is oxidized to oxygen:

$$2H_2O(l) + 2F_2(g) \rightarrow 4HF(aq) + O_2(g) \qquad F(0) \rightarrow F(-1),\quad O(-2) \rightarrow O(0)$$

and dilute alkali is oxidized to oxygen difluoride, OF_2 (see Section 22.4).

Chlorine, bromine and iodine

Chlorine is less strongly oxidizing than fluorine, and can be made by chemical oxidation. In the laboratory it is prepared by oxidizing concentrated hydrochloric acid with potassium manganate(VII):

$$2KMnO_4(s) + 16HCl(aq) \rightarrow 5Cl_2(g) + 2MnCl_2(aq) + 2KCl(aq) + 8H_2O(l)$$
$$Mn(+7) \rightarrow Mn(+2),\quad Cl(-1) \rightarrow Cl(0)$$

It may also be prepared by the oxidation of sodium chloride in concentrated sulphuric acid using manganese(IV) oxide and heating. This is the preferred method for preparing bromine and iodine, and the general reaction (using the potassium halide, KHal) is:

$$2KHal(s) + MnO_2(s) + 2(HO)_2SO_2(l) \rightarrow Hal_2(g) + MnSO_4(aq) + K_2SO_4(aq) + 2H_2O(l)$$
$$Mn(+4) \rightarrow Mn(+2),\quad Hal(-1) \rightarrow Hal(0)$$

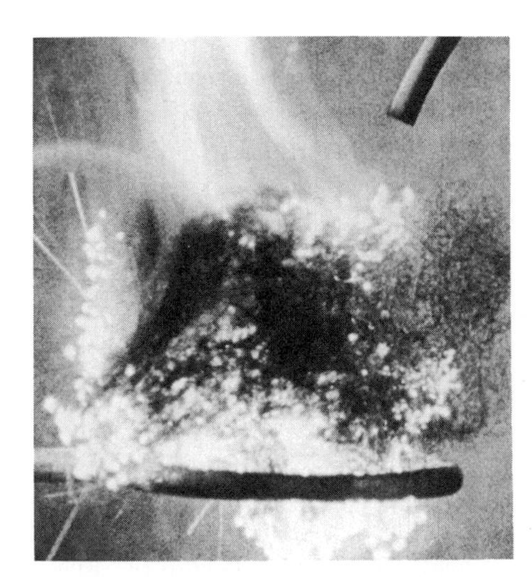

Figure 23.5 Iron burns as vigorously in fluorine as cotton wool burns in oxygen. The fluorine is delivered through the pipe at top right

The bromine is collected by distillation and the iodine by sublimation.

Most elements are oxidized directly by chlorine, bromine and iodine, but often the reaction must be activated by ultraviolet light or by heating. For example, a mixture of chlorine and hydrogen does not react in the dark at room temperature, but a flash of light initiates an explosion:

$$H_2(g) + Cl_2(g) \rightarrow 2HCl(g) \qquad Cl(0) \rightarrow Cl(-1),\quad H(0) \rightarrow H(+1)$$

In general, the reactivity of the halogens with other elements decreases down the Group. With iron, chlorine produces a red-brown sublimate of iron(III) chloride:

$$2Fe(s) + 3Cl_2(g) \xrightarrow{\text{heat}} 2FeCl_3(s) \qquad Cl(0) \rightarrow Cl(-1), \quad Fe(0) \rightarrow Fe(+3)$$

and bromine produces a mixture of $FeBr_3$ and $FeBr_2$, while iodine is too weak an oxidizing agent to oxidize iron beyond iron(II) iodide:

$$Fe(s) + I_2(s) \xrightarrow{\text{heat}} FeI_2(s) \qquad I(0) \rightarrow I(-1), \quad Fe(0) \rightarrow Fe(+2)$$

The halogens can oxidize many compounds. For example, hydrogen sulphide is oxidized to sulphur:

$$8H_2S(g) + 8Cl_2(g) \rightarrow S_8(s) + 16HCl(g) \qquad Cl(0) \rightarrow Cl(-1), \quad S(-2) \rightarrow S(0)$$

All the halogens oxidize thiosulphate ions, normally to sulphate ions, but the less strongly oxidizing iodine produces tetrathionate ions:

$$I_2(aq) + 2S_2O_3^{2-}(aq) \rightarrow 2I^-(aq) + S_4O_6^{2-}(aq) \qquad I(0) \rightarrow I(-1), \quad S(+2) \rightarrow S(+2\tfrac{1}{2})$$

This reaction is very important in redox chemistry since it is quantitative. Oxidizing agents can therefore be estimated by adding excess iodide ions to the acidified solution and titrating the liberated iodine with standard aqueous sodium thiosulphate. The equivalence-point is often detected with freshly prepared starch solution, which forms a blue complex with iodine.

The disproportionation reactions of chlorine, bromine and iodine with water and alkalis are important both in the laboratory and in everyday life. Chlorine reacts with *cold, dilute* aqueous sodium hydroxide to give chloride and chlorate(I) (ClO^-) ions:

$$Cl_2(g) + 2OH^-(aq,dil.) \rightarrow Cl^-(aq) + ClO^-(aq) + H_2O(l)$$
$$2Cl(0) \rightarrow Cl(-1) + Cl(+1)$$

and similarly with water, with which dissolved chlorine reacts to produce a mixture of hydrochloric and chloric(I) acids:

$$Cl_2(g) + H_2O(l) \rightarrow HCl(aq) + HOCl(aq) \qquad 2Cl(0) \rightarrow Cl(-1) + Cl(+1)$$

The equilibrium does not lie strongly to the right, and so some dissolved chlorine remains. The corresponding equilibrium for bromine lies further to the left, and it is even less favourable for iodine.

Since in the halate(I) ions, such as ClO^- and BrO^-, the halogens have oxidation numbers intermediate in their range, further disproportionation is possible. When sodium chlorate(I) is heated gently it forms sodium chloride and sodium chlorate(V):

$$3ClO^-(aq) \xrightarrow{75\,^\circ C} 2Cl^-(aq) + ClO_3^-(aq) \qquad 3Cl(+1) \rightarrow 2Cl(-1) + Cl(+5)$$

The significance of this reaction is that when chlorine is passed into *hot, concentrated* aqueous sodium hydroxide, disproportionation occurs to chloride and chlorate(V) ions:

$$3Cl_2(g) + 6OH^-(aq,conc.) \xrightarrow{75\,^\circ C} 5Cl^-(aq) + ClO_3^-(aq) + 3H_2O(l)$$
$$6Cl(0) \rightarrow 5Cl(-1) + Cl(+5)$$

The BrO^- and IO^- ions are less stable and the corresponding reactions of bromine and iodine occur at lower temperatures:

$$3Hal_2(aq) + 6OH^-(aq) \rightarrow 5Hal^-(aq) + HalO_3^-(aq) + 3H_2O(l)$$
$$6Hal(0) \rightarrow 5Hal(-1) + Hal(+5)$$

The disproportionation can be shown by heating iodine in aqueous sodium hydroxide: on adding aqueous barium nitrate an immediate white precipitate of barium iodate(V) is obtained. The precipitation is slower at room temperature. These reactions are reversed in acidic solution, and $HalO_3^-$ ions oxidize Hal^- ions to the element:

$$5Hal^-(aq) + HalO_3^-(aq) + 6H^+(aq) \rightarrow 3Hal_2 + 3H_2O(l)$$
$$5Hal(-1) + Hal(+5) \rightarrow 6Hal(0)$$

Most of the uses of the halogens depend on their oxidizing properties. About 70 per cent of all manufactured chlorine is converted to chlorinated organic compounds, such as chloroethene for the manufacture of PVC, and chloroalkanes for use as solvents. About 20 per cent of the chlorine produced is used to treat water supplies, and to bleach paper and fabrics (both of which depend on its oxidizing power, either directly or through chlorate(I) ions); the 'chlorine' smell so characteristic of swimming pools is in fact the odour of chlorinated amines. The principal use of bromine is in the preparation of 1,2-dibromoethane, which is added to petrol (see Box 27.3). There are no major industrial uses of elemental iodine, but small quantities are used domestically in ethanol solution, which is called **tincture of iodine**, as a mild antiseptic (the element's oxidizing power again). Iodine plays an important role in the body as it regulates the activity of the thyroid gland, and the addition of trace quantities of iodide or iodate(V) ions to table salt has helped to eliminate **goitre**, an iodine deficiency disease of the thyroid. 'Anti-radiation tablets', which are effective if taken within about two hours of mild exposure to radiation, contain potassium iodate(V), which saturates the thyroid gland with iodine and prevents the uptake of radioactive iodine-131.

23.3 Oxides and oxoacids

Since fluorine is more electronegative than oxygen there are no fluorine oxides, only oxygen fluorides (discussed in Section 22.4). The only fluorine oxoacid, HOF, is unstable at room temperature.

Oxides

Chlorine, bromine and iodine each form several oxides, which are thermally – and sometimes explosively – unstable. As oxides of typical non-metals they are covalent, volatile molecular compounds which dissolve in water to give acidic solutions. The bromine oxides are much less well known than those of chlorine and iodine, and will not be considered further here.

The lowest oxidation number of chlorine in an oxide, +1, is found in **dichlorine oxide**, Cl_2O, a yellow-brown gas prepared on a large scale in industry by the reaction of chlorine and moist sodium carbonate:

$$2Cl_2(g) + 2Na_2CO_3(aq) + H_2O(l) \rightarrow 2NaHOCO_2(aq) + 2NaCl(aq) + Cl_2O(g)$$

$$2Cl(0) \rightarrow Cl(-1) + Cl(+1)$$

This disproportionation reaction is similar to that between chlorine and cold dilute sodium hydroxide, described in the last section.

The next oxide of importance is **chlorine dioxide**, ClO_2, in which $Ox(Cl) = +4$. Over a hundred thousand tonnes of this highly explosive yellow gas are made each year, mainly by reducing sodium chlorate(V) in acidic solution with sulphur dioxide:

$$2ClO_3^-(aq) + SO_2(g) \rightarrow 2ClO_2(g) + SO_4^{2-}(aq)$$

$$Cl(+5) \rightarrow Cl(+4), \quad S(+4) \rightarrow S(+6)$$

It is also formed when chloric(V) acid, $HOClO_2$, decomposes by disproportionation:

$$3HOClO_2(aq) \rightarrow 2ClO_2(g) + HOClO_3(aq) + H_2O(l)$$

$$3Cl(+5) \rightarrow 2Cl(+4) + Cl(+7)$$

This is why concentrated sulphuric acid should never be added to a chlorate(V) salt: the chloric(V) acid then formed immediately decomposes to chlorine dioxide, which explodes so violently that it can shatter the vessel.

Chlorine dioxide (and dichlorine oxide to a smaller extent) is used in the papermaking industry to bleach wood pulp, where it gives a good whiteness without degrading the fibre.

The chlorine oxide with the highest oxidation number, +7, is dichlorine heptoxide, $O(ClO_3)_2$ (**1** above) or simply Cl_2O_7, a shock-sensitive colourless oil which is the anhydride of chloric(VII) acid, $HOClO_3$.

In contrast to the explosive character of the chlorine oxides, iodine's principal oxide, iodine(V) oxide, $O(IO_2)_2$ or simply I_2O_5, is a thermally stable, white, crystalline solid. It is notable because it rapidly and quantitatively oxidizes carbon monoxide:

$$5CO(g) + I_2O_5(s) \rightarrow I_2(s) + 5CO_2(g) \qquad I(+5) \rightarrow I(0), \quad C(+2) \rightarrow C(+4)$$

and hence gives a simple way of determining the concentration of carbon monoxide in air, as the iodine liberated can be titrated with standard aqueous sodium thiosulphate.

Oxoacids and oxosalts

Only the most important of the many halogen oxoacids will be described here (Table 23.2). Two general rules help to rationalize their acid–base properties:

1. *The greater the electronegativity of the halogen, the stronger the acid.*
2. *The greater the number of oxygen atoms present (or, equivalently, the higher the halogen's oxidation number), the stronger the acid.*

Both factors contribute to the dispersal of the negative charge of the anion: the first by allowing the halogen atom to share some of it, and the second through delocalization.

Table 23.2 The four chlorate anions

Oxidation number	Formula	Number of lone pairs	Shape	Name	Traditional name
+7	ClO_4^-	0	tetrahedral	chlorate(VII)	perchlorate
+5	ClO_3^-	1	pyramidal	chlorate(V)	chlorate
+3	ClO_2^-	2	angular	chlorate(III)	chlorite
+1	ClO^-	3	—	chlorate(I)	hypochlorite

Halate(I) ions, such as ClO^-, are strong oxidizing agents. Iodide ions are oxidized to iodine in acidic solution:

$$2I^-(aq) + 2H^+(aq) + ClO^-(aq) \rightarrow I_2(s) + H_2O(l) + Cl^-(aq) \qquad Cl(+1) \rightarrow Cl(-1), \quad I(-1) \rightarrow I(0)$$

Salts containing chlorate(I) ions, which are still sometimes called **hypochlorites**, are used in domestic liquid bleaches. 'Bleaching powder' is a mixture of calcium chlorate(I) ($Ca(ClO)_2$), calcium chloride ($CaCl_2$) and calcium hydroxide ($Ca(OH)_2$).

The halate(III) salts are unstable, and chloric(III) acid cannot be isolated. The best known is sodium chlorate(III), previously called **sodium chlorite**, which disproportionates on heating:

$$3NaClO_2(s) \xrightarrow{\text{heat}} NaCl(s) + 2NaClO_3(s) \qquad 3Cl(+3) \rightarrow Cl(-1) + 2Cl(+5)$$

EXAMPLE

Write balanced equations showing how iodide ions are oxidized by chlorate(V) ions.

METHOD Write balanced equations for the oxidation and reduction half-reactions, multiply them by factors that ensure that the number of electrons supplied by the oxidation matches the number used in the reduction, then add them together.

ANSWER The balanced half-equations are

Reduction: $ClO_3^-(aq) + 6H^+(aq) + 6e^- \rightarrow Cl^-(aq) + 3H_2O(l)$

Oxidation: $2I^-(aq) \rightarrow I_2(s) + 2e^-$

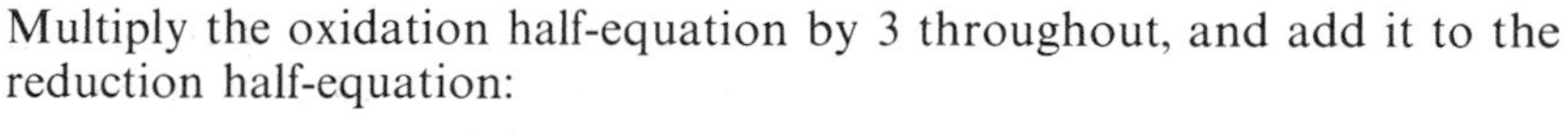

Multiply the oxidation half-equation by 3 throughout, and add it to the reduction half-equation:

$$6I^-(aq) + ClO_3^-(aq) + 6H^+(aq) \rightarrow 3I_2(s) + Cl^-(aq) + 3H_2O(l)$$

COMMENT It is instructive to compare this equation with that for the same oxidation using chlorate(I) ions. Here 6 mol I^- are needed for 1 mol ClO_3^-, but only 2 mol I^- are needed for 1 mol ClO^-. The difference reflects the much larger change of chlorine's oxidation number (from +5 to −1) in the chlorate(V) reaction as compared with that of chlorate(I) (from +1 to −1).

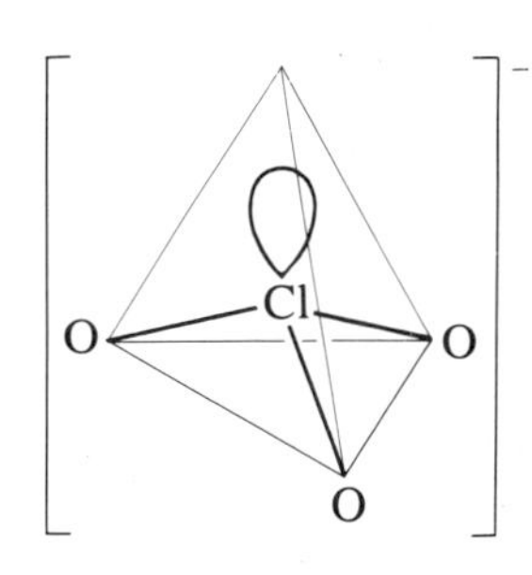

2

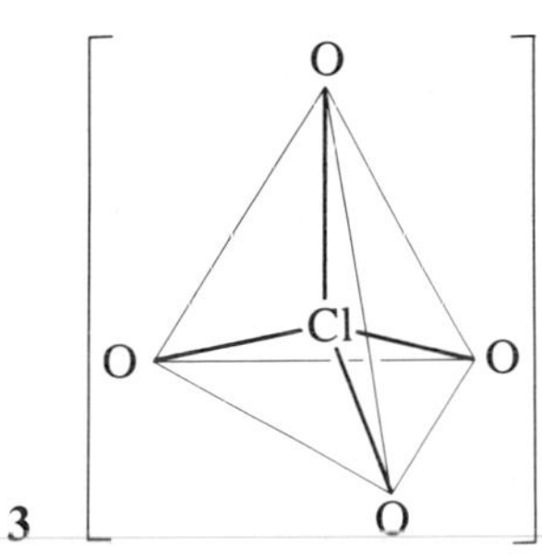

3

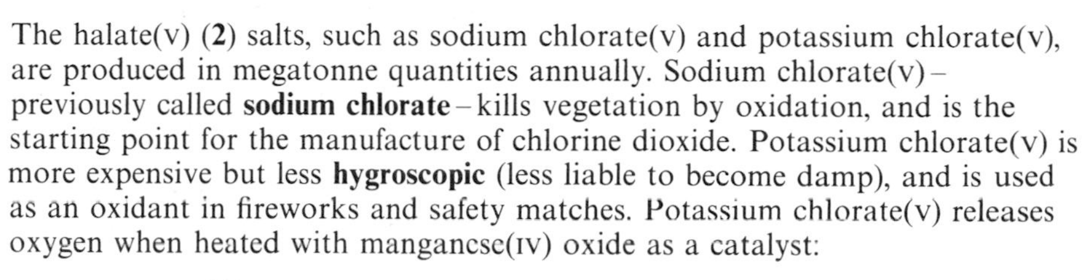

The halate(V) (**2**) salts, such as sodium chlorate(V) and potassium chlorate(V), are produced in megatonne quantities annually. Sodium chlorate(V) – previously called **sodium chlorate** – kills vegetation by oxidation, and is the starting point for the manufacture of chlorine dioxide. Potassium chlorate(V) is more expensive but less **hygroscopic** (less liable to become damp), and is used as an oxidant in fireworks and safety matches. Potassium chlorate(V) releases oxygen when heated with manganese(IV) oxide as a catalyst:

$$2KClO_3(s) \xrightarrow{100\,°C,\ MnO_2} 2KCl(s) + 3O_2(g) \qquad Cl(+5) \rightarrow Cl(-1),\quad O(-2) \rightarrow O(0)$$

The oxoanions in which the halogens have their highest oxidation numbers are the halate(VII) ions (**3**), which occur in magnesium chlorate(VII), or **magnesium perchlorate**, $Mg(ClO_4)_2$. They are prepared by the electrolytic oxidation of halate(V) salts, or in disproportionation reactions such as

$$4KClO_3(s) \xrightarrow[\text{below } 400\,°C]{\text{controlled heating}} KCl(s) + 3KClO_4(s) \qquad 4Cl(+5) \rightarrow Cl(-1) + 3Cl(+7)$$

Halate(VII) salts are very powerful oxidizing agents. Although they are moderately stable in the absence of organic material at room temperature, they are dangerously explosive when in contact with even a trace of oxidizable material. They also oxidize inorganic material, as in some space rockets in which the fuel is the potent combination of ammonium chlorate(VII) and powdered aluminium.

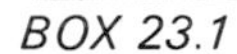

BOX 23.1

Clock reactions and oscillating reactions

Some reactions do not glide smoothly towards products, but suddenly produce a shower of products; in others, the concentrations of reactants and products oscillate. For example, if solutions of sodium sulphite, excess sodium iodate(V) and starch are mixed with hydrochloric acid nothing seems to happen, but after about thirty seconds the solution suddenly turns blue.

The explanation is as follows. The iodate(V) ions are reduced by sulphite ions with the formation of iodide ions:

$$IO_3^-(aq) + 3SO_3^{2-}(aq) \rightarrow I^-(aq) + 3SO_4^{2-}(aq)$$

and the iodide ions are oxidized by the iodate(V) ions, with the formation of iodine:

$$IO_3^-(aq) + 5I^-(aq) + 6H^+(aq) \rightarrow 3I_2(aq) + 3H_2O(l)$$

(The iodine actually remains in solution as I_3^-, the triiodide ion, but that is a detail.) Although iodine normally forms a blue complex with the starch, in the presence of sulphite ions it is immediately reduced to iodide ions:

$$3I_2(aq) + 3SO_3^{2-}(aq) + 3H_2O(l) \rightarrow 6I^-(aq) + 6H^+(aq) + 3SO_4^{2-}(aq)$$

and the solution remains colourless. Once all the sulphite ions have been removed in the first of these three reactions, the iodine can then survive, and the blue complex is formed. Hence, there is a sudden surge of blue.

By a suitable choice of conditions it is possible to arrange for even more striking effects. Iodine and iodic(V) acid can compete for hydrogen peroxide:

$$I_2(aq) + 5H_2O_2(aq) \rightarrow 2HOIO_2(aq) + 4H_2O(l)$$

$$2HOIO_2(aq) + 5H_2O_2(aq) \rightarrow I_2(aq) + 5O_2(g) + 6H_2O(l)$$

and one of the products in each of these reactions is a reactant in the other. This is like the condition of 'positive feedback' in an oscillating electric circuit, such as the howl that develops when a microphone picks up the sound from its own loudspeaker; the result is that the reaction breaks into oscillation. The iodine concentration increases and decays periodically, and the colour of the solution oscillates from blue to colourless until all the hydrogen peroxide has been used.

23.4 Halides

As the halogens combine with almost every other element, it is no surprise that they also combine among themselves to form the **interhalogens** and the related **polyhalide anions**.

Interhalogens

All the interhalogens have the formula XY_n, where Y is the more electronegative halogen, and $n = 1, 3, 5$ or 7; they are listed in Table 23.3. All possible combinations of halogens are found only for $n = 1$. The shapes of the molecules can be predicted using VSEPR theory; two examples are ClF_3 (**4**) and IF_5 (**5**).

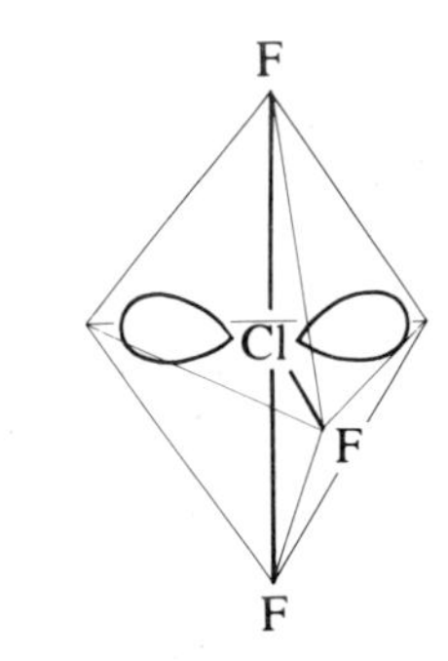

4

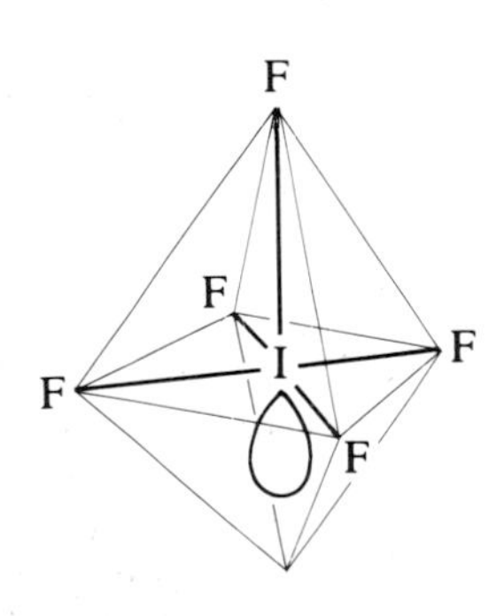

5

Table 23.3 The interhalogens

XY	XY_3	XY_5	XY_7
ClF	ClF_3	ClF_5	
BrF	BrF_3	BrF_5	
IF	IF_3	IF_5	IF_7
BrCl			
ICl	I_2Cl_6		
IBr			

The binary XY interhalogens are prepared by direct reaction of the elements:

$$I_2(s) + Cl_2(g) \rightarrow 2ICl(l) \qquad Cl(0) \rightarrow Cl(-1),\ I(0) \rightarrow I(+1)$$

Their physical properties are intermediate between those of the parent halogens; for example, ICl is a red liquid. When excess Y_2 is used, reaction occurs as far as the higher interhalogen:

$$2ICl(l) + 2Cl_2(g) \rightleftharpoons I_2Cl_6(s) \qquad Cl(0) \rightarrow Cl(-1),\ I(+1) \rightarrow I(+3)$$

Excess fluorine reacts with chlorine at 250 °C to form ClF_3, a colourless gas which is a violent fluorinating agent – in fact, it reacts with almost every substance, including asbestos, platinum and xenon. It can be stored in nickel vessels, however, as a protective layer of nickel(II) fluoride forms on nickel surfaces. Chlorine trifluoride is used in nuclear fuel reprocessing, because the fluorides of plutonium and most other fission products are non-volatile whereas UF_6 is volatile and can be separated from them.

Polyhalide anions

The interhalogens act as Lewis acids towards halide ions, such as F^-:

$$ClF(g) + CsF(s) \rightarrow CsClF_2(s)$$

where the anion formed is ClF_2^-.

The most common polyhalide anions are those of structure XY_2^-, which are linear, $[Y{-}X{-}Y]^-$. Fluorine cannot occupy the central position, X, because it cannot expand its octet. Of the XY_2^- polyhalide anions, the most familiar is the **triiodide ion**, I_3^-:

$$I_2(s) + I^-(aq) \rightleftharpoons I_3^-(aq)$$

This is important for iodometric titrations, because the covalent molecules of elemental iodine are much more soluble in aqueous potassium iodide solution than in water. The brown colour of these solutions is in fact due to the triiodide ion, not to iodine itself.

EXAMPLE

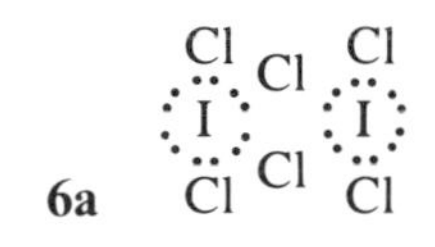

6a

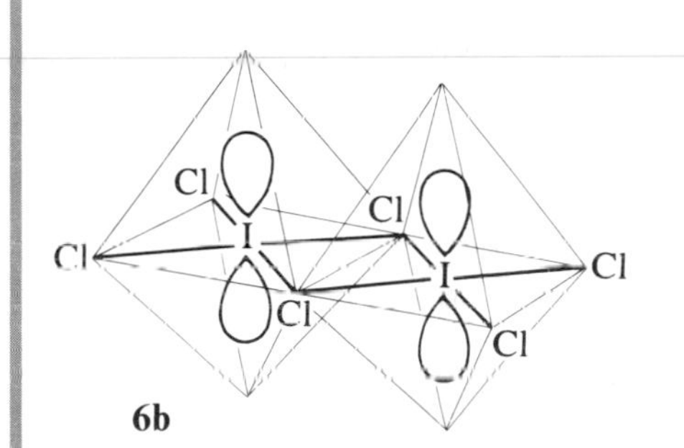

6b

Predict the shape around each iodine atom in I_2Cl_6, given that there are two bridging chlorine atoms (as in Al_2Cl_6, discussed in Section 19.4).

METHOD Draw a Lewis structure for the molecule given that each iodine atom supplies seven electrons and can expand its octet. Count the number of electron pairs and use the VSEPR rules in Section 2.3 to predict the shape around each iodine atom.

ANSWER A Lewis structure is shown in **6a**. Each iodine atom has four bonding pairs and two lone pairs, giving six electron pairs in all. Hence the basic shape is octahedral. The two lone pairs will lie on opposite sides of the plane formed by the four bonding pairs.

COMMENT The structure of the molecule is shown in **6b**: all eight atoms lie in a plane, and although the molecule has a superficial resemblance to Al_2Cl_6, the presence of the iodine lone pairs results in a different arrangement around the central atoms.

23.5 Compounds with hydrogen

The compounds of formula HHal are called **hydrogen halides**, and include the very important hydrogen chloride, HCl, and hydrogen fluoride, HF. All can be formed by direct reaction of the elements, but the vigour of the reaction decreases with increasing atomic number. Fluorine and hydrogen react explosively at room temperature. Chlorine and hydrogen also explode, but only after initiation of the reaction by a spark or a flash of light. The reaction between bromine and hydrogen is slower, and for commercially acceptable rates requires heating and a catalyst:

$$H_2(g) + Br_2(g) \xrightarrow{300\,^\circ C,\,Pt} 2HBr(g) \qquad Br(0) \rightarrow Br(-1), \quad H(0) \rightarrow H(+1)$$

The formation of hydrogen iodide is less favoured thermodynamically, and a red-hot wire placed in hydrogen iodide produces copious violet clouds of iodine.

Hydrogen fluoride and hydrogen chloride are much more conveniently prepared by heating a halide with concentrated sulphuric acid:

$$NaCl(s) + (HO)_2SO_2(aq,conc.) \xrightarrow{heat} NaHOSO_3(s) + HCl(g)$$

This method is unsuitable for the preparation of hydrogen bromide and especially hydrogen iodide, because these compounds are oxidized by the

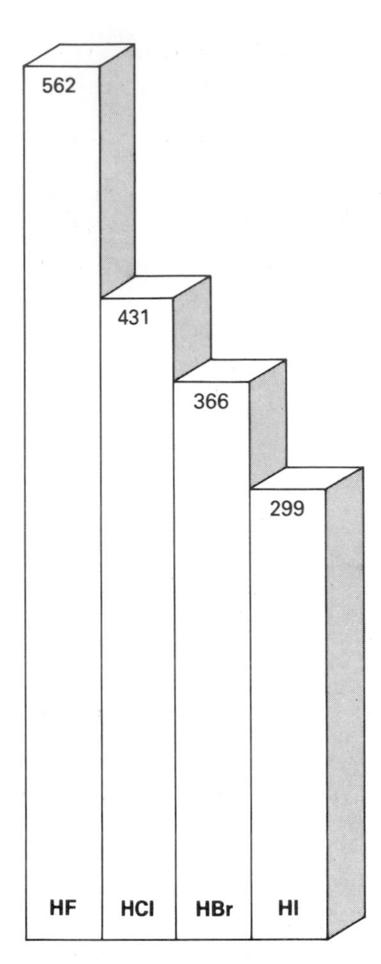

Figure 23.6 The H—Hal bond dissociation enthalpies decrease with increasing atomic number of the halogen

sulphuric acid; a less oxidizing acid, such as concentrated phosphoric acid, must be used instead. Hydrogen bromide and hydrogen iodide are also formed by the reaction between red phosphorus and bromine or iodine in water, which can be thought of as the formation of the phosphorus trihalide, immediately followed by its hydrolysis.

Hydrogen fluoride is a colourless liquid which boils at 19.5 °C; all the other hydrogen halides are colourless gases. This difference is due to the hydrogen bonding between HF molecules in the liquid, like that between H_2O molecules in water (see Section 5.1).

All the hydrogen halides are very soluble in water and give acidic solutions, such as the familiar **hydrochloric acid**, HCl(aq). Hydrofluoric acid is a weak acid ($pK_a = 3.25$), not only because of the strong H—F bond (see Figure 23.6), but also because hydrogen bonding occurs between F^- and HF in solution, so that the proton is less readily lost. The linear $[FHF]^-$ ion survives crystallization – in the compound KHF_2, for example.

A special reaction of hydrofluoric acid is its ability to attack glass (see Figure 23.7) which is the basis of etching, including the frosting of the interiors of light bulbs. This is a two-stage process: first, silicon tetrafluoride is formed:

$$SiO_2(s) + 4HF(aq) \rightarrow SiF_4(g) + 2H_2O(l)$$

which then reacts with excess hydrofluoric acid:

$$SiF_4(g) + 2HF(aq) + 2H_2O(l) \rightarrow 2H_3O^+(aq) + SiF_6^{2-}(aq)$$

The reactions of the other hydrohalic acids are those of typical acids: they form salts with basic oxides, and liberate carbon dioxide from carbonates. Their salts are the metal halides, such as sodium chloride, which are discussed in Sections 17.4, 18.4 and 19.4. Methods for distinguishing between the halide ions are summarized in Table 23.4.

Halide ions are Lewis bases, forming complex ions (such as AlF_6^{3-}) with

Figure 23.7 Decorative windows, such as this one, are made by etching the glass with hydrofluoric acid

Table 23.4 Tests to distinguish between F^-, Cl^-, Br^- and I^- in aqueous solution

	F^-	Cl^-	Br^-	I^-
Precipitation tests:				
1 $AgNO_3(aq)$ and dilute $HONO_2(aq)$	no ppt	white ppt (AgCl)	cream ppt (AgBr)	yellow ppt (AgI)
Solubility of AgHal in aqueous ammonia	—	soluble[a]	partially soluble	insoluble
2 $CaCl_2(aq)$	white ppt of CaF_2	—	—	—
Redox tests:				
1 Concentrated $(HO)_2SO_2$ and MnO_2; heat	**CARE!** HF etches glass	Cl_2[b]	Br_2[c]	I_2[d]
2 $Cl_2(g)$	—	—	Br_2	I_2
3 $Br_2(aq)$	—	—	—	I_2

[a] Silver chloride dissolves in aqueous ammonia by forming the complex ion $[Ag(NH_3)_2]^+$.
[b] Chlorine can be detected by the faint green colour of the gas, and by the bleaching of moist litmus paper.
[c] Bromine can be detected by the red colour and characteristic smell.
[d] Iodine can be detected by the black precipitate formed and by the blue colour a dilute aqueous solution forms with freshly prepared starch solution.

strong Lewis acids. With organic compounds they act as nucleophiles (see, for instance, Sections 29.3 and 35.4).

23.6 Other compounds

The most important 'other' compounds of the halogens are their organic compounds. Familiar examples include PVC, DDT and TCP (see Section 33.3, Box 34.2 and Section 35.3 respectively).

23.7 Industrial chemistry

Fluorine is the thirteenth most abundant element in the Earth's crust, and its most important minerals are **fluorite** (or fluorspar, CaF_2, see Colour 5), **cryolite** (Na_3AlF_6) and **fluorapatite** ($Ca_5(PO_4)_3F$). Chlorine is a constituent of certain minerals, such as **rock-salt** (NaCl) and **carnallite** ($KCl.MgCl_2.6H_2O$), but huge quantities of chloride ions occur in sea water, inland lakes and subterranean brine wells. Bromine and iodine are much less abundant than chlorine; bromide ions occur in sea water, and iodine occurs as sodium iodate(v) impurity in Chile saltpetre and in brine wells.

Chlorine is manufactured by the electrolysis of brine (as described in Section 17.7), either in a Castner-Kellner cell or, because of the risks of using mercury, increasingly in a diaphragm cell.

Summary

- ☐ The elements have the characteristic valence electron configuration ns^2np^5.
- ☐ As fluorine is the most electronegative element, by definition it can only have compounds with the oxidation number -1.
- ☐ Chlorine, bromine and iodine exhibit a range of oxidation numbers from -1 to $+7$, with odd numbers being the most important.
- ☐ Fluorine is never surrounded by more than eight electrons, whereas up to fourteen electrons may be involved in the bonding of chlorine, bromine and iodine.
- ☐ All the elements are non-metals: the oxides are acidic, the hydrides and interhalogen compounds are covalent and there are no aqua cations.

- ☐ As a result of electron pair repulsions, the electron-gain energy and bond dissociation enthalpy of fluorine are unexpectedly low.
- ☐ The standard electrode potentials for the halogens show that the order of oxidizing power is $F_2 > Cl_2 > Br_2 > I_2$.
- ☐ There is a series of halate anions – for example, ClO^-, ClO_2^-, ClO_3^- and ClO_4^-. Compounds containing these anions are powerful oxidizing agents. The related acids have an acid strength in the order HOCl < HOClO < $HOClO_2$ < $HOClO_3$.
- ☐ Chlorine reacts with sodium hydroxide to give Cl^- and either ClO^- (cold, dilute NaOH) or ClO_3^- (hot, concentrated NaOH).
- ☐ Compounds in which halogens have oxidation numbers intermediate between −1 and +7 tend to disproportionate.
- ☐ The interhalogens XY_n, with $n = 1, 3, 5, 7$, are covalent. The only compound with $n = 7$ is IF_7 which has a coordination number of 7.
- ☐ Iodine dissolves in aqueous potassium iodide to form the linear triiodide ion, I_3^-.
- ☐ The hydrogen halides have both acidic and reducing properties. The reducing power increases with increasing atomic number.
- ☐ Hydrogen fluoride has an abnormally high boiling point, and is a weak acid.
- ☐ The silver halides are insoluble, except for silver fluoride. Silver iodide is yellow and insoluble in aqueous ammonia.
- ☐ Chlorine is manufactured by the electrolysis of brine.

PROBLEMS

1 From your knowledge of the halogens and of trends in the Periodic Table predict the physical and chemical properties of astatine and its major compounds.

2 How is chlorine manufactured in industry? How is it made in the laboratory? Make a list of the major uses of chlorine.

3 What are the shapes of the following molecules or ions: IF_2^-, ICl_4^-, BrF_3, BrF_2^+, BrF_4^-, BrF_5, BrF_6^+, IF_7?

Construct a table showing for each of these entities the oxidation number of the heavier halogen, its coordination number and the number of lone pairs. (Use VSEPR theory to predict the shapes.)

4 'Hydrogen bonding is very important in compounds containing hydrogen and fluorine.' Illustrate this statement by reference to (a) the physical and chemical properties of hydrogen fluoride, (b) the structure of ammonium fluoride, and (c) the hydrogen difluoride anion, HF_2^-.

5 Use a data book to construct two tables to show the difference between fluorides and chlorides. The first table should contain the values of the lattice enthalpies of fluorides and the corresponding chlorides of about ten metals. The second table should contain the values of the E—F and E—Cl bond energies, where E = B, C, Si, N, P, O, S, F and Cl. Comment on the patterns that emerge.

6 Chlorine exhibits a range of oxidation numbers from −1 to +7. By writing balanced equations, give examples of disproportionation reactions of compounds in which chlorine has intermediate oxidation numbers. Indicate the changes in oxidation number.

7 Hydrogen chloride is formed when concentrated sulphuric acid is heated with sodium chloride. Why cannot hydrogen bromide and hydrogen iodide be prepared in this way? How are small samples of each prepared in the laboratory? Write equations throughout.

8 Use standard electrode potentials to predict the results of mixing (a) NaCl and $NaBrO_3$ in acidic solution and (b) Cl_2 and IO_3^- in alkaline solution.

9 Iodine(V) oxide, I_2O_5, reacts quantitatively at room temperature with carbon monoxide:

$$5CO(g) + I_2O_5(s) \rightarrow I_2(s) + 5CO_2(g)$$

A 2.00 dm^3 sample of a gas at 20 °C and 1 atm containing some carbon monoxide was passed over I_2O_5. The iodine liberated was separated from unreacted I_2O_5, dissolved in aqueous potassium iodide and titrated against 0.0500 mol dm^{-3} sodium thiosulphate solution. 20.5 cm^3 of thiosulphate solution were required. What is the percentage by volume of carbon monoxide in the gas?

10 Explain the following observations:
(a) A gas is evolved when dilute hydrochloric acid is added to calcium chlorate(I).
(b) Chloric(VII) acid is a stronger acid than chloric(I) acid.
(c) Fluorine reacts more vigorously than chlorine.
(d) Iodine monochloride is a liquid whereas diiodine hexachloride is a solid.

11 (a) Fluorine and its compounds often have properties noticeably different from those of chlorine and other halogens. Discuss the following, relating the differences where possible to the data given at the end of this section.
(i) Chlorine gas can be prepared from chlorides by 'chemical' methods but fluorine gas has to be prepared by electrolysis.
(ii) The vapour pressures of aluminium fluoride, chloride and bromide reach a value of one atmosphere at 1564 K, 696 K and 530 K respectively.
(iii) The pH of an aqueous solution of sodium fluoride is greater than 7 while that of a similar solution of sodium chloride is 7.

Size of ions/nm
F^- 0.133; Cl^- 0.181; Br^- 0.196

Standard electrode potentials/V
$F_2(g), 2F^-(aq)$ + 2.87
$Cl_2(aq), 2Cl^-(aq)$ + 1.36
$Br_2(aq), 2Br^-(aq)$ + 1.09

The dissociation constant of hydrofluoric acid is 5.6×10^{-4} mol dm^{-3}.

(b) Aqueous copper(II) sulphate solution is blue in colour; when aqueous potassium fluoride is added, a green *precipitate* is formed but, when aqueous potassium chloride is added instead, a bright green *solution* is formed. What do you think is happening in the two cases? (*London*)

12 (a) By reference to the following standard electrode potentials, explain what happens when chlorine is bubbled into aqueous potassium bromide.

$\frac{1}{2}Cl_2 + e^- \rightleftharpoons Cl^-$ $E^\ominus = +1.36$ V
$\frac{1}{2}Br_2 + e^- \rightleftharpoons Br^-$ $E^\ominus = +1.09$ V

(b) Given the standard electrode potentials

$$MnO_4^- + 8H^+ + 5e^- \rightleftharpoons Mn^{2+} + 4H_2O \quad E^\ominus = +1.52\ V$$

$$MnO_2 + 4H^+ + 2e^- \rightleftharpoons Mn^{2+} + 2H_2O \quad E^\ominus = +1.23\ V$$

how do you account for the fact that chlorine may be prepared from hydrochloric acid by the action of either potassium manganate(VII) or manganese(IV) oxide? (*London part question*)

13 (a) Chlorine is regarded as an oxidizing agent. Briefly explain this statement in terms of electron transfer.
(b) Illustrate the answer to (a) by discussing the reaction of chlorine with
(i) $Br^-(aq)$;
(ii) $Sn^{2+}(aq)$;
(iii) $H_2S(aq)$.
(c) Describe the observations which can be noted in each of the following experiments, and write equations:
(i) potassium manganate(VII) (permanganate) crystals are mixed with concentrated hydrochloric acid;
(ii) aqueous potassium chloride and aqueous silver nitrate are mixed and concentrated aqueous ammonia is then added;
(iii) aqueous bromine (i.e. a dilute solution) is mixed with an excess of potassium iodide solution, and aqueous sodium thiosulphate, $Na_2S_2O_3$, is then added.
(d) List the hydrogen halides in order of *decreasing* ease of oxidation by concentrated sulphuric acid. Explain the sequence in bond energy terms. (*SUJB*)

14 (a) If potassium chlorate(V) is heated above its melting point, oxygen is evolved as the only gaseous product and the residue contains no oxygen.
(i) Write an equation for the reaction.
(ii) Write the change in oxidation number, if any, for (1) potassium, (2) chlorine, and (3) oxygen.
(iii) Describe briefly a simple test tube experiment by which you could confirm the nature of the anion present in the residue.
(b) If potassium chlorate(V) is heated to its melting point, a reaction in which no oxygen is evolved occurs according to the equation:

$$4K^+ClO_3^-(s) \rightarrow 3K^+ClO_4^-(s) + K^+Cl^-(s)$$

(i) What is the oxidation number of chlorine in ClO_4^-?
(ii) What is the oxidation number of chlorine in Cl^-?
(iii) What type of reaction does this illustrate? Give a reason for your answer.
(iv) Name the oxosalt in the product and write an equation for its thermal decomposition.
(v) If the potassium oxosalt is the less soluble of the two products, describe briefly how you might isolate it from the residue. (*SUJB part question*)

15 Discuss the trends which occur in Group VII of the Periodic Table, the halogens. Illustrate your answer with suitable examples, paying attention to the abnormal properties of fluorine. (*London*)

16 Give an account of the manufacture of *two* of the halogens fluorine, chlorine, bromine and iodine, and briefly describe *two* important uses for each of the four halogens. (*Nuffield*)

17 Discuss the comparative chemistry of the Group VII elements, fluorine to iodine, emphasizing the trends in descending the Group.
Give brief answers to the following questions:
(a) The name of the halogen astatine is derived from a Greek word meaning unstable – how do you think astatine might decompose?
(b) In what physical state would you expect astatine to exist at room temperature – solid, liquid or gas?
(c) Would you expect AgAt to be soluble in (i) water, (ii) concentrated aqueous ammonia?
(d) Would you expect HAt to be a strong or a weak acid in aqueous solution?
(e) Place the series of ionic halides MF, MCl, MBr, MI, MAt in order of expected decreasing melting point and state why you have chosen this order. (*Cambridge Entrance*)

18 Show how the physical and chemical properties of the chlorides of *five* of the following elements are related to the position of the element in the Periodic Table: hydrogen, sodium, aluminium, silicon, nitrogen, phosphorus, iron, tin. (*Oxford Entrance*)

19 The element francium (Fr) is the last of the alkali metals and the element astatine (At) is the last of the halogens. Both are radioactive and available only in *very* small quantities. Outline how you might investigate the chemistry of *one* of these elements, and suggest what you would expect to find. (*Oxford Entrance*)

24 The noble gases

The noble gases belong to Group VIII of the Periodic Table. They have closed-shell configurations, and are generally unreactive. Xenon, and to a lesser extent krypton, can be oxidized by sufficiently powerful oxidizing agents. Several compounds of xenon are known, but only with the most electronegative elements, fluorine and oxygen. All the gases are monatomic, and boil at low temperatures.

	VI	VII	VIII
1			2 He
2	8 O	9 F	10 Ne
3	16 S	17 Cl	18 Ar
4	34 Se	35 Br	36 Kr
5	52 Te	53 I	54 Xe
6	84 Po	85 At	86 Rn

Figure 24.1 Group VIII in the Periodic Table

The elements of Group VIII, which is also sometimes called Group 0, are as follows:

helium	He	$1s^2$
neon	Ne	$[He]2s^22p^6$
argon	Ar	$[Ne]3s^23p^6$
krypton	Kr	$[Ar]3d^{10}4s^24p^6$
xenon	Xe	$[Kr]4d^{10}5s^25p^6$
radon	Rn	$[Xe]4f^{14}5d^{10}6s^26p^6$

The members of the Group are known as the **noble gases**: their positions in the Periodic Table are shown in Figure 24.1. For helium the first shell is full; for the other elements the characteristic valence electron configuration is ns^2np^6, where n is the Period number.

Helium was detected spectroscopically on the Sun (hence its name, derived from the Greek, *helios*, for 'sun') nearly thirty years before it was isolated on Earth by William Ramsay. The discovery of argon (from the Greek, *argos*, for 'the idle one') was a result of Lord Rayleigh's careful measurements of the density of nitrogen samples from different sources: he noticed that nitrogen obtained from air was always slightly denser than that obtained from ammonia. When Ramsay burnt magnesium in nitrogen from air he found that a small amount of gas remained, and identified it as a new element. The other noble gases were isolated by careful fractional distillation of liquid air; the names of three reflect their unexpectedness (neon, the new one; krypton, the hidden one; xenon, the stranger), and that of radon comes from its radioactivity. All the gases are now obtained commercially by the fractional distillation of liquid air, except for helium, which is obtained from natural gas wells where it has accumulated as a result of the α-decay of heavy elements (see Section 1.4).

24.1 Group systematics

After the hectic chemistry of the halogens, the noble gases display an almost arrogant reluctance to react with anything (hence their name). Some of them do form some compounds, but the discovery of their chemical properties dates only from 1962, when the first of these were prepared (see Box 24.1).

Oxidation states and bonding characteristics

Helium, neon and argon form no compounds (it is still believed). Krypton combines with fluorine to form the colourless solid KrF_2. Xenon forms a wider range of compounds, with oxidation numbers +2, +4, +6 and +8. Only the most important will be dealt with here. Radon is intensely radioactive, and is therefore regarded as dangerous to study; its compounds are largely unknown.

BOX 24.1

Figure 24.2 Crystals of xenon tetrafluoride, XeF_4

From inertness to nobility: the discovery of the first xenon compound

Work in one field can often lead to a discovery in another. In 1962 the British-born Canadian chemist Neil Bartlett was working on platinum fluorides. By accident he exposed the vapour of platinum(VI) fluoride, PtF_6, to air and found that an orange solid was formed, which he identified as a compound containing a **dioxygenyl cation**:

$$O_2(g) + PtF_6(g) \rightarrow O_2^+PtF_6^-(s)$$

He realized that PtF_6 is an oxidizing agent so powerful that it can even oxidize dioxygen. If PtF_6 can remove an electron from dioxygen (ionization energy 1175 kJ mol^{-1}), he reasoned, then it ought to be able to oxidize xenon (ionization energy 1170 kJ mol^{-1}), and thus open up a route to compounds of the – until then – 'inert' gases.

It is reported that Bartlett's colleagues were more than a little surprised when he asked for some xenon 'so that I can try some reactions'. The experiment of mixing xenon and PtF_6 vapour was successful, and an orange-yellow solid of approximate formula $XePtF_6$ was formed:

$$Xe(g) + PtF_6(g) \rightarrow Xe^+PtF_6^-(s)$$

With the announcement of the preparation of this compound, the inert gases were shown not to be totally inert (as, indeed, Linus Pauling had predicted in 1933). A psychological barrier had been broken, and chemists around the world immediately tried to prepare other compounds of xenon. Within months XeF_4 (see Figure 24.2) and XeF_2 had been prepared: from then on, the gases of Group VIII were no longer inert, but merely noble.

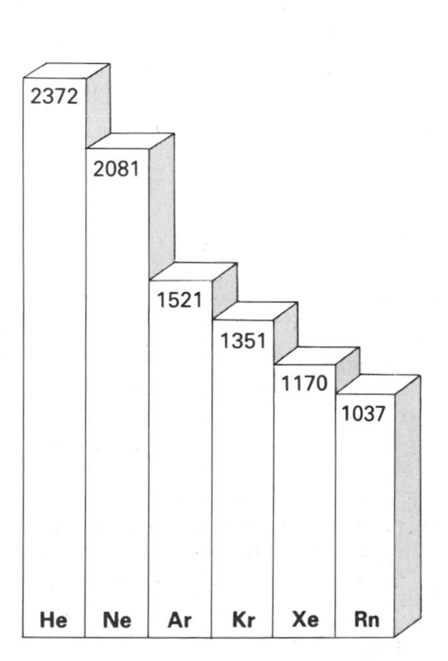

Figure 24.3 The first ionization energies of the noble gases decrease with increasing atomic number

Periodic trends

The crucially important trend is that the first ionization energy decreases with increasing atomic number (see Figure 24.3). The first ionization energy of xenon is comparable to that of bromine, and as bromine combines with more electronegative elements xenon's ability to form compounds with fluorine and oxygen should come as no surprise. Nevertheless chemists were surprised when the first compounds were reported, and quickly replaced the original name for the Group, the **inert gases**, by the present one.

EXAMPLE

Chemists often try to assess whether a compound is likely to be sufficiently stable to be worth attempting to prepare by estimating its standard formation enthalpy. Is it worth attempting to prepare $Xe^+F^-(s)$?

METHOD Construct a Born–Haber cycle (see Section 9.4), using data for xenon and fluorine as far as possible, and estimating unknown values (such as the lattice enthalpy) from data for an analogous known compound: in this calculation CsF(s) is suitable. If the compound is strongly endothermic, it is unlikely to be stable.

ANSWER The Born-Haber cycle is shown in Figure 24.4. Data are as follows.

Step	Process	$\Delta_r H^\ominus$/kJ mol^{-1}
1	Formation of XeF(s)	x
2	Ionization of Xe(g)	+1170
3	Dissociation of $\frac{1}{2}F_2(g)$	+79
4	Electron attachment to F(g)	−333
5	Lattice enthalpy of XeF(s)	≈ +740 (value for CsF(s))

Figure 24.4 A Born–Haber cycle for the formation of the hypothetical XeF

Route A must equal route B; it follows that

$$(x + 740)\,\text{kJ mol}^{-1} = (1170 + 79 - 333)\,\text{kJ mol}^{-1}$$

hence $x = +176$

This indicates that ionic XeF(s) is likely to be a strongly endothermic compound (for comparison, the standard formation enthalpy of CsF(s) is -532 kJ mol^{-1}), and unlikely to be prepared.

COMMENT Once again, we are assuming that entropy terms are negligible. In fact they also work against the stability of XeF(s), because an entropy decrease occurs when the two gases form a solid.

Shapes and sizes

Atomic radii increase from helium to xenon (see Figure 24.5), and xenon is big enough to allow several fluorine or oxygen atoms to cluster around.

All the compounds of xenon involve the expansion of its octet, and the shapes of the compounds can be treated by the usual VSEPR rules (given in Section 2.3). In XeF_4, for instance, there are six pairs of electrons in the valence shell of xenon: four bonding pairs and two lone pairs. The basic shape is therefore octahedral. The two lone pairs take up positions as far apart as possible, on either side of the plane of the four bonding pairs; hence XeF_4 is a square planar molecule (**1**).

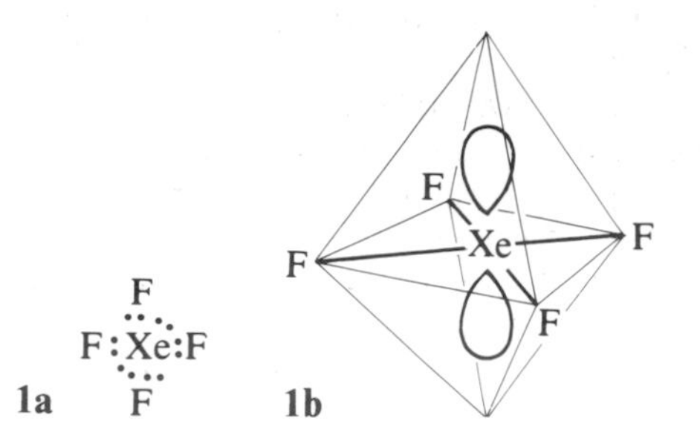

24.2 The elements

The elements exist as monatomic gases. Only dispersion forces act between the atoms, and the lighter gases condense to liquids only at low temperatures. Helium in fact has the lowest boiling point of any substance (4.2 K), and one of its principal uses is in **cryogenics**, the production of low temperatures for superconducting magnets and supercomputers. Helium is the only substance to possess two *liquid* phases: it makes a transition to a very strange phase, called helium-II, when its temperature falls below 2.2 K. This peculiar liquid is **superfluid**, and flows without viscosity: it creeps over the surface of any vessel that contains it – and out of the vessel if it is open (see Figure 24.6).

Helium is used to dilute the oxygen used by divers (see Section 10.3) and as a non-flammable buoyant gas for airships. Argon is widely used in industry as an inert atmosphere in high-temperature metallurgical processes where nitrogen would be too reactive, such as the reduction of titanium(IV) chloride to titanium. Neon and argon are used for filling discharge tubes, neon giving an orange-red glow and argon blue; the wide range of colours of 'neon lights' is obtained by varying the mixture of noble gases used to fill the tubes and by using coloured glass to make them.

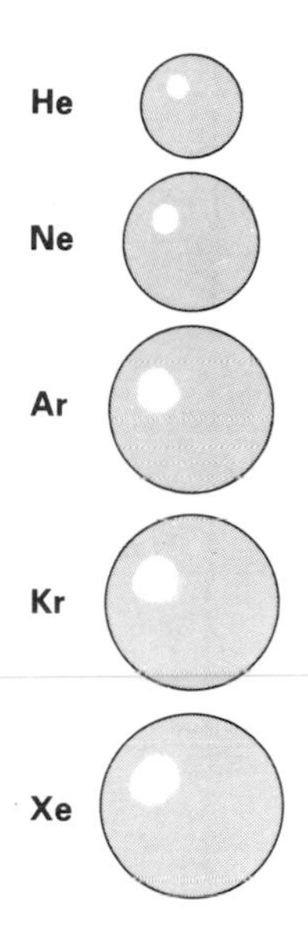

Figure 24.5 Relative atomic radii of the noble gases

24.3 Oxides

Xenon hexafluoride is very rapidly hydrolysed:

$$XeF_6(s) + 3H_2O(l) \rightarrow XeO_3(s) + 6HF(aq)$$

Magnesium oxide is added to the reaction mixture in order to precipitate fluoride ions as MgF_2, and the xenon(VI) oxide is obtained as a colourless deliquescent solid by evaporation of the solvent. The oxide is an explosively violent oxidizing agent, and must be treated with the greatest caution.

24.4 Halides

Only the most powerful oxidizing agents can attack xenon, and the production of all xenon compounds starts with its reaction with fluorine. Depending on the conditions, XeF_2, XeF_4 and XeF_6 may all be produced by heating the elements under pressure:

$$Xe(g) + 2F_2(g) \xrightarrow[400\,^\circ C,\ 6\ atm]{\text{slight excess } F_2} XeF_4(s) \qquad F(0) \rightarrow F(-1),\ \ Xe(0) \rightarrow Xe(+4)$$

Xenon tetrafluoride is a white solid that is useful as a powerful fluorinating agent, even attacking platinum:

$$Pt(s) + XeF_4(s) \xrightarrow{\text{heat}} PtF_4(s) + Xe(g) \qquad Xe(+4) \rightarrow Xe(0),\ \ Pt(0) \rightarrow Pt(+4)$$

Figure 24.6 Superfluid helium spontaneously flows out of any unstoppered container

Summary

- ☐ The noble gases have closed-shell electron configurations. Thus neon and argon have configurations [Noble gas]ns^2np^6, and krypton and xenon have configurations [Noble gas]$(n-1)d^{10}ns^2np^6$.
- ☐ They are monatomic gases, and occur in the atmosphere.
- ☐ Argon is used in industry for creating inert atmospheres; neon and argon are used in discharge tubes; helium, which has the lowest boiling point of any substance, is used for producing low temperatures.
- ☐ Only krypton, xenon and radon form compounds, and even then only with the most electronegative elements (oxygen and fluorine).
- ☐ Xenon has oxidation numbers +2, +4, +6 and +8 in its compounds.

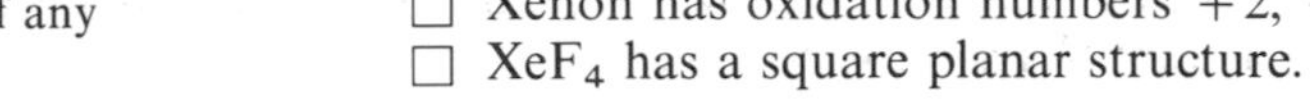

- ☐ XeF_4 has a square planar structure.

PROBLEMS

1 Write an essay on the discovery of the noble gases. Comment on their position in the Periodic Table.

2 Use appropriate data to account for the observation that helium, neon and argon do not form any compounds and that most of the compounds of the noble gases are formed by xenon.

3 How are the noble gases obtained commercially? List the uses of these elements.

4 Consider the following series of molecules or ions:

SbF_5	TeF_6	IF_7	x
SbF_3	TeF_4	IF_5	XeF_6
y	z	IF_4^-	XeF_4
SbF_5	TeF_4	IF_3	XeF_2

(a) Construct a table to show for each entity (i) the oxidation number of the element other than fluorine, (ii) the number of lone pairs of electrons of the element, and (iii) the shape of the molecule or ion.
(b) The spaces marked x, y and z are blank because the molecules or ions which would complete the pattern have not so far been prepared. Write the formulae of these hypothetical entities in spaces x, y and z.
(c) Give the two most likely reasons why the compound to fit in space x has not been prepared.
(d) Which series is isoelectronic?

5 Plot atomic radius versus atomic number for the noble gas elements. Account for the irregularities in the graph.

6 How may XeF_4 be prepared?
What is the spatial arrangement of the atoms in XeF_4 and how may this be accounted for in terms of the Sidgwick–Powell theory of electron pair repulsion?
(Oxford part question)

7 This question concerns compounds of the element xenon.
(a) Xenon and fluorine react to form at least three different fluorides. After separation and purification, a sample of one of these compounds was analysed:

Mass of xenon fluoride $= 0.490\,g$
Volume of xenon obtained on removing the fluorine from the sample (measured at 25 °C and 1 atm) $= 48\,cm^3$

Calculate the empirical formula for this fluoride of xenon.
(Relative atomic masses: Xe = 131, F = 19. 1 mole of gas at 25 °C and 1 atm occupies $24\,dm^3$.)
(b) Calculate the standard [enthalpy] of formation of the fluoride $XeF_4(g)$ from the following bond energy terms, and comment on its likely stability:

$E\,(Xe—F) = 130\,kJ\,mol^{-1}$
$E\,(F—F) = 158\,kJ\,mol^{-1}$

(Note: You may wish to construct an energy cycle diagram and use it to answer this question.)
(c) The oxide XeO_3 is produced by the hydrolysis of the fluoride XeF_6. Write an equation for this reaction.
(Nuffield)

8 This question is concerned mainly with aspects of the chemistry of the fluorides of xenon.
(a) Assuming that the lattice energy of xenon monofluoride, XeF, would be equal to that of caesium fluoride, CsF, calculate the standard enthalpy of formation, $\Delta_f H^\ominus_{298}$, for xenon monofluoride, a hypothetical crystalline solid.
(b) Make a table comparing your calculated value for the standard enthalpy of formation of xenon monofluoride with values listed in reference books for other xenon fluorides. What are the corresponding values per mole of fluorine atoms? Comment on the likely energetic stability of these fluorides in relation to their constituent elements.
(c) The xenon fluorides which exist are, in fact, molecular substances. Suggest the most likely shapes for molecules of XeF_2 and XeF_4.
(d) Given that the standard enthalpy change for

$$XeF_4(s) \rightarrow XeF_4(g)$$

is $+62\,kJ\,mol^{-1}$, and using an appropriate energy cycle, calculate an approximate value for the Xe—F bond energy.
(e) Why do you think compounds of neon and fluorine, and compounds of xenon and iodine, are unknown?
(Nuffield S)

25 The d-block elements

The elements of the d-block include many of the metals used in industry as constructional materials or as catalysts, with iron pre-eminent in importance. Their chemical properties are characterized by a wide range of oxidation numbers, which is important for their function as catalysts. They also form numerous complexes, of which many are coloured, and some are paramagnetic. In their low oxidation numbers their oxides are basic and their halides ionic; in their high oxidation numbers the oxides are acidic and the halides covalent.

The elements of the d-block, in which the d-subshell is being filled, are shown in Figure 25.1. They are sometimes referred to as the 'transition elements', but as there is some disagreement about the precise meaning of this term (not everyone includes copper, and some chemists exclude scandium), we will use only the name 'd-block elements'. Those in the first row of the block, the only ones we shall consider, have the following electron configurations:

scandium	Sc	$[Ar]3d^1 4s^2$
titanium	Ti	$[Ar]3d^2 4s^2$
vanadium	V	$[Ar]3d^3 4s^2$
chromium	Cr	$[Ar]3d^5 4s^1$
manganese	Mn	$[Ar]3d^5 4s^2$
iron	Fe	$[Ar]3d^6 4s^2$
cobalt	Co	$[Ar]3d^7 4s^2$
nickel	Ni	$[Ar]3d^8 4s^2$
copper	Cu	$[Ar]3d^{10} 4s^1$
zinc	Zn	$[Ar]3d^{10} 4s^2$

3	12 Mg											13 Al
4	20 Ca	21 Sc	22 Ti	23 V	24 Cr	25 Mn	26 Fe	27 Co	28 Ni	29 Cu	30 Zn	31 Ga
5	38 Sr	39 Y	40 Zr	41 Nb	42 Mo	43 Tc	44 Ru	45 Rh	46 Pd	47 Ag	48 Cd	49 In
6	56 Ba	57 La	72 Hf	73 Ta	74 W	75 Re	76 Os	77 Ir	78 Pt	79 Au	80 Hg	81 Tl
7	88 Ra	89 Ac	104 Unq	105 Unp	106 Unh	107 Uns	108 Uno	109 Une				

Figure 25.1 The d-block in the Periodic Table

The valence electron configurations of chromium and copper ($3d^5 4s^1$ and $3d^{10} 4s^1$ respectively) are exceptions to the $3d^x 4s^2$ rule that applies to the other elements, because of the lower energy of half-full (d^5) and full (d^{10}) d-subshells.

The form of the Periodic Table shown in Figure 25.1 conceals the existence of the **f-block elements** in Periods 6 and 7, in which an f-subshell is being filled. There are two rows of f-block elements: the fourteen elements from cerium to lutetium, squeezed in between lanthanum and hafnium, and another fourteen elements from thorium to lawrencium, which lie between actinium and unnilquadium (the systematic designation of element 104). The former group is known as the **lanthanoids** (or **lanthanides**) because they resemble lanthanum, and the latter as the **actinoids** (or **actinides**). The lanthanoids were once called the **rare earth metals**, but most are in fact more abundant than iodine. They have very similar chemical properties and are used only for specialized purposes; they will not be discussed further in this book. All the actinoids are radioactive. They include uranium and plutonium, the elements of central importance to the production of nuclear energy.

Group distinctions are less sharp within the d-block than in the s- and p-blocks and the block is best considered as a whole.

25.1 Block systematics

Oxidation states and bonding characteristics

The most characteristic feature of the elements of the d-block is that, except for scandium and zinc, *they each exhibit several oxidation numbers*. The reason is that the energies of all the d-electrons are very similar, so that in general there are only small energy differences between the removal of different numbers of electrons from the 3d- and 4s-subshells. The range of oxidation numbers of the first row of d-block elements is shown in Table 25.1. The table may appear daunting at first sight, but only the numbers in bold type need be remembered. The entries are arranged so that oxidation numbers corresponding to the same number of electrons outside the argon core appear on the same line. For example, both Mn^{2+} and Fe^{3+} are $[Ar]3d^5$ ions.

Table 25.1 Common oxidation numbers of the first-row d-block elements

Number of electrons outside the argon core	Sc	Ti	V	Cr	Mn	Fe	Co	Ni	Cu	Zn
0	**3** only	**4**	**5**	**6**	**7**					
1		3	**4**		6					
2		2	3			6				
3			2	**3**	**4**					
4				2	3					
5			0		**2**	**3**				
6				0		**2**	**3**			
7					0		**2**	3		
8						0		**2**	3	
9							0		**2**	
10								0	**1**	**2** only

Several patterns may be observed in Table 25.1.

1 *The first and last members of the row have only one oxidation number.* This is why scandium and zinc are unlike the other d-block elements, and are often treated separately.
2 *All the members of the row except zinc can have oxidation number +3* (although it is sometimes of little importance). Typical examples include M_2O_3, MF_3 and $M^{3+}(aq)$; Cu^{3+}, however, does not exist.
3 *All the elements except scandium can have oxidation number +2.* Typical examples include MO, MCl_2 and $M^{2+}(aq)$.
4 *The highest oxidation number for the elements from scandium to manganese is equal to the number of 4s- and 3d-electrons in their atoms.*
5 *Low oxidation numbers are common after manganese.* This is largely due to the increase in ionization energy on going to the right along the row, for the atoms become smaller.
6 *Only copper has important compounds with oxidation number +1.*

The high oxidation numbers arise through covalent bonding with more electronegative elements, as in $TiCl_4$ and MnO_4^-. The existence of intermediate oxidation numbers (as, for example, in FeO and Fe_2O_3) is due to a subtle balance between the energy investment required to remove another electron and the greater lattice enthalpy that then arises.

General characteristics and trends

All the d-block elements are metals, and most have high melting and boiling points. Most of them are hard and rigid, and hence are useful constructional materials. They are all good conductors of electricity. Their d-electrons result in some special characteristics that distinguish them from other metals, such as the colours of many of their compounds, and the paramagnetism of some. Moreover, since the polarizing powers of their ions are high, they readily form complexes by acting as Lewis acids.

The ease with which the d-block elements change their oxidation numbers is connected to their usefulness as catalysts. Aqueous cobalt(II) chloride catalyses the reaction between hydrogen peroxide and Rochelle salt (sodium potassium 2,3-dihydroxybutanedioate). The colour of the cobalt changes from pink to green, reverting to pink when the reaction is complete. Other d-block metals also provide examples of useful catalysts: titanium(IV) chloride in the Ziegler–Natta catalysts, vanadium(V) oxide for the contact process, a mixed oxide of copper and chromium for the catalytic converters for car exhaust gases, manganese(IV) oxide for potassium chlorate(V) decomposition, iron for the Haber–Bosch process, nickel for hydrogenation, and copper(I) salts for the Sandmeyer reaction.

The colours of the compounds of the d-block elements arise because they absorb visible light (see Box 3.1). In a **charge-transfer transition** an electron migrates from a ligand to the metal, as in the manganate(VII) ion, MnO_4^-, which absorbs in the green region of the spectrum, and appears purple. The orange of the dichromate(VI) ion also arises from a charge-transfer transition, in this case resulting from the absorption of blue light. A second type of transition is due to the excitation of a d-electron to another d-orbital *within* the metal ion (a **d–d transition**). The wavelength of light absorbed depends on the separation of the energies of the d-orbitals in the compound caused by the ligands (see Figure 25.2); **ligand field theory**, a subject we shall not consider in this book, gives a detailed explanation of the colours.

Neither $Sc^{3+}(aq)$ nor $Zn^{2+}(aq)$ is coloured, the d-subshell in these ions being empty and full respectively – that is, a *partially* filled d-subshell is needed for colour.

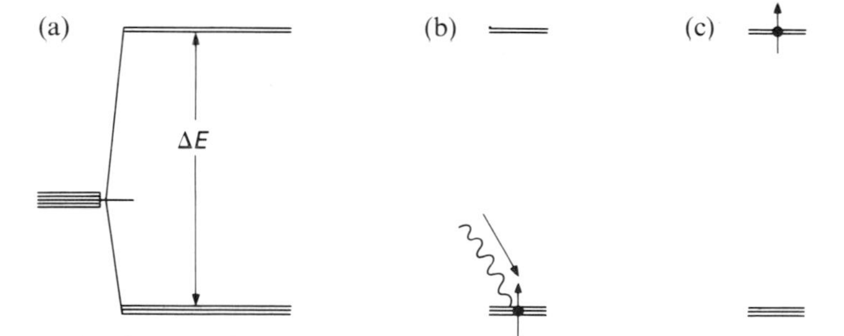

Figure 25.2 (a) The splitting of d-orbitals by an octahedral ligand field: (b) an electron in the lower group of orbitals absorbs radiation and (c) is promoted to the higher group of orbitals

Shapes and sizes

Ionic radii decrease with increasing atomic number (see Figure 25.3), as the d-electrons shield the increased nuclear charge poorly.

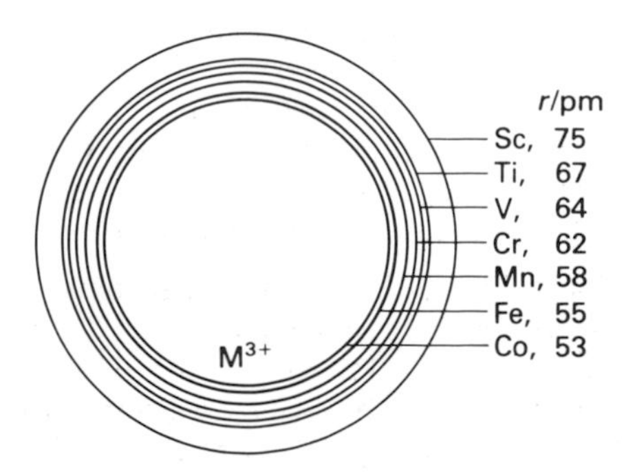

Figure 25.3 The radii of the M^{3+} ions of the d-block elements (in octahedral complexes)

The complexes of the elements show a wide variety of shapes, but two dominate: the **octahedral complex** with six ligands attached to the central ion, and the **tetrahedral complex** with four usually negatively charged ligands (see Figures 25.4(a) and 25.4(b) respectively). Square planar complexes (Figure 25.4(c)), are also reasonably common for ions with a d^8 electron configuration (including Ni^{2+}).

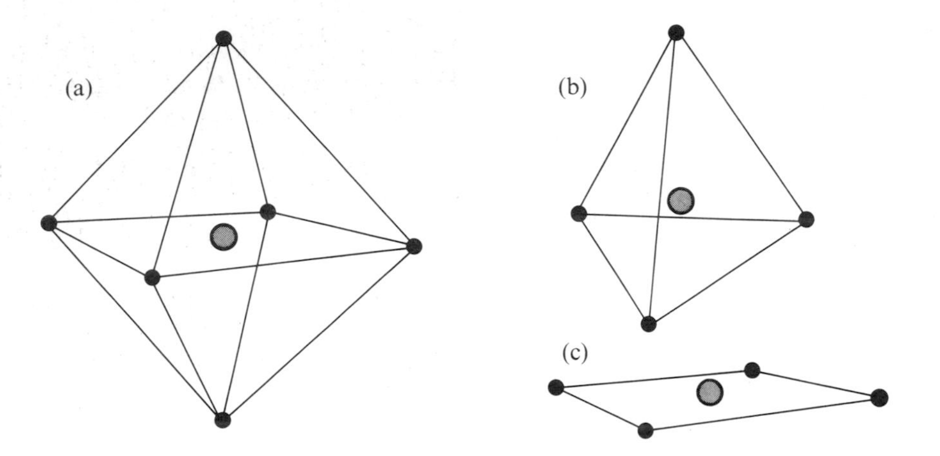

Figure 25.4 (a) An octahedral complex, (b) a tetrahedral complex and (c) a square planar complex

25.2 The elements

Physical properties and uses of the elements

The electron configurations of the atoms of the d-block elements are all similar, and the physical properties of the elements are correspondingly similar. All are malleable, ductile metals, with a lustrous appearance and good electrical and thermal conductivities. Because the atomic radii are small (as a result of the poor shielding abilities of the d-electrons) the solids are quite dense (see Figure 25.5); indeed, the densest elements of all – iridium and osmium (density 22.6 g cm^{-3}) – are d-block elements.

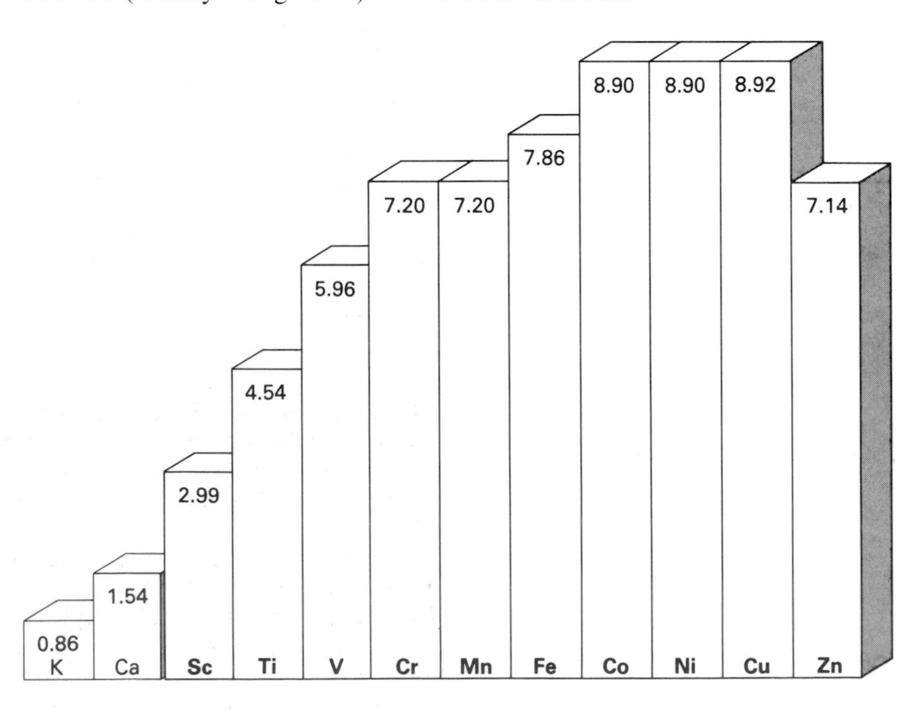

Figure 25.5 The densities (in g cm^{-3}) of the elements from potassium to zinc

Figure 25.6 The framework of the Lockheed *Blackbird* aircraft is made from titanium: the extra cost is only justifiable because aluminium would be less suitable in the conditions of the extreme upper atmosphere

Titanium has a larger atomic radius than iron, and is therefore less dense. Its combination of mechanical strength and low density makes it an attractive material for aircraft components, including both engines and airframes (see Figure 25.6).

Chromium is best known for the lustrous surface it gives when electroplated on to metal objects. Nickel is alloyed with chromium to make **nichrome** (60% Ni, 40% Cr), an alloy whose electrical resistance varies little with temperature.

Iron is the most widely used of all the d-block elements, partly because it is so abundant, but also because it can be converted into **steel**, which has superior resistance to corrosion and is harder. Steels consist mainly of iron, but with the addition of 0.2–1.7 per cent of carbon, which helps to pin the iron atoms together and give a stronger material. Special steels (see Table 25.2) are prepared by adding small amounts of other elements. A typical **stainless steel**, for example, contains 18 per cent of chromium and 8 per cent of nickel. **Hadfield steel**, which contains 13 per cent of manganese, is used where extreme hardness is required, as in excavators and safes. The ability of the d-block metals to be alloyed together in this way reflects their similar atomic radii.

Table 25.2 Alloy steels

Name	Fe/%	C/%	Other metals	Property	Use
mild steel	≈99.8	≈0.2	—	easily shaped	car bodies
high-carbon steel	≈98	≈1.7	Mn	hard	tools
tungsten steel	≈95	≈1.0	W	very hard	armour
Hadfield steel	≈85	≈1.2	Mn	very hard	springs
stainless steel	≈75	≈0.3	Cr, Ni	corrosion-resistant	cutlery
Permalloy	≈20	≈0.2	Ni, Mo	high magnetic permeability	magnets

Iron is **ferromagnetic,** which means that permanent magnets can be made from it (and from its alloys with cobalt and nickel). In ferromagnetic materials large numbers of electrons spin in the same direction throughout a **domain** of the metal, and their combined magnetic moments give rise to a large magnetic field.

Copper is more resistant to oxidation than are the other members of the first row of the d-block (it is the only member with a positive standard electrode potential), and although expensive it is widely used in the form of protective sheeting. A copper object exposed to air turns green as a result of the formation of a patina of basic copper(II) carbonate, sulphate or chloride (unless, as sometimes happens in a city, a black sulphide is formed first). The main use of unalloyed copper is in electrical cables, but aluminium is replacing it (see Section 19.2). Copper forms several important alloys, listed in Table 25.3.

Zinc is widely used, both in alloys and as a protective coating on iron and steel. An object may be coated with zinc, a process called **galvanizing** (see Section 13.5), either by immersing it in molten zinc, or by electroplating, or by spraying it with a zinc powder paint.

Table 25.3 Alloys of copper

Name	Mass % Cu	Other metal
bronze	90–95	Sn
brass	55–90	Zn
'silver' coinage	≈75	Ni
nickel silver	55–65	Ni, Zn
Monel*	≈30	Ni

* Monel metal is used for handling corrosive substances such as fluorine

Chemical properties

Scandium reacts with water about as vigorously as calcium does:

$$2Sc(s) + 6H_2O(l) \rightarrow 2Sc(OH)_3(s) + 3H_2(g) \qquad H(+1) \rightarrow H(0),\quad Sc(0) \rightarrow Sc(+3)$$

With the exception of copper, all the first-row d-block elements have negative standard electrode potentials for the M^{2+}, M couple, which means that there is a thermodynamic tendency for the metals to be oxidized and to go into solution as $M^{2+}(aq)$ ions. The rates of the oxidation vary widely along the row, however, and depend on the amount of surface exposed; iron reacts only slowly. Chromium is resistant to attack by water, but it reduces dilute acid with the evolution of hydrogen, as does iron:

$$Fe(s) + 2HCl(aq) \rightarrow FeCl_2(aq) + H_2(g) \qquad H(+1) \rightarrow H(0),\quad Fe(0) \rightarrow Fe(+2)$$

Chromium and iron are protected (**passivated**) by concentrated nitric acid, which forms an oxide layer on the metal surface (compare the behaviour of aluminium, described in Section 19.2).

The oxidation of iron by water in the presence of oxygen is a reaction of extreme economic importance, absorbing a significant fraction of the gross national product of every industrialized country, for it leads to the formation of **rust**:

$$4Fe(s) + 3O_2(g) + 2H_2O(l) \rightarrow 4FeO(OH)(s) \qquad O(0) \rightarrow O(-2),\quad Fe(0) \rightarrow Fe(+3)$$

The thermodynamics of a model of this process were mentioned briefly in Section 11.2. The overall process is a complex sequence of electrochemical steps, which depends on the existence of different oxygen concentrations in the water near different parts of the iron. Iron can be protected from rusting by covering its surface with an *unbroken* layer of paint or other adherent substance, such as a phosphate layer (see Section 21.3), or by connecting it to a more electropositive metal which can supply its electrons to satisfy the demands of the oxygen reduction reaction. Magnesium is often used for this purpose, in the form of a sacrificial anode to protect ship hulls, bridges and outboard motors, as described in Section 13.5. The zinc of galvanized iron can also be regarded as a sacrificial anode.

25.3 Oxides, hydroxides and oxoacids

At first sight, the d-block elements seem to form a bewildering array of oxides (see Table 25.4). Various patterns emerge when their properties are examined, however, and it is convenient to classify them as follows:

1. highest oxides: V_2O_5, CrO_3, Mn_2O_7
2. +4 oxides: MO_2 (M = Ti, V, Cr, Mn)
3. +3 oxides: M_2O_3 (M = Sc, Ti, V, Cr, Mn, Fe)
4. mixed +3, +2 oxides: M_3O_4 (M = Mn, Fe, Co)
5. +2 oxides: MO (M = all except Sc, Cr)

Table 25.4 The major oxides of the first-row d-block elements

Oxidation number	Sc	Ti	V	Cr	Mn	Fe	Co	Ni	Cu	Zn
+7					Mn_2O_7					
+6				CrO_3						
+5			V_2O_5							
+4		TiO_2	VO_2	CrO_2	MnO_2					
+3	Sc_2O_3	Ti_2O_3	V_2O_3	Cr_2O_3	Mn_2O_3	Fe_2O_3				
mixed (+2, +3)					Mn_3O_4	Fe_3O_4	Co_3O_4			
+2		TiO	VO		MnO	FeO	CoO	NiO	CuO	ZnO
+1									Cu_2O	

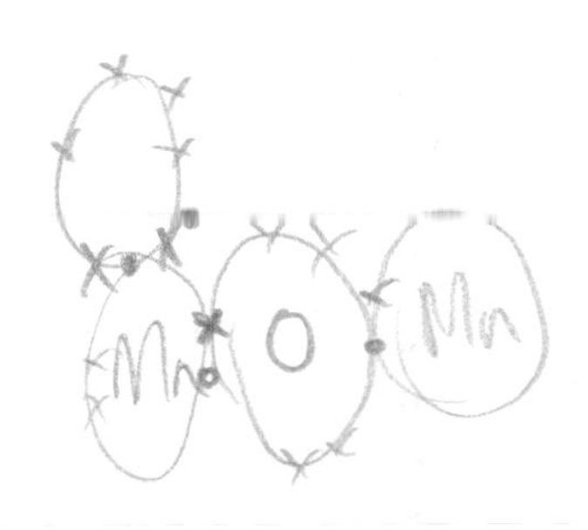

As a result of the high polarizing power of the highly charged metal ion, the three highest oxides are largely covalent. The other oxides are largely ionic, with close-packed oxide ions and metal ions. Many of the ionic oxides are non-stoichiometric compounds (see Section 15.4), and the composition of the substance denoted FeO in Table 25.4 in fact ranges from $Fe_{0.84}O$ to $Fe_{0.95}O$, with some Fe^{3+} ions replacing Fe^{2+} ions.

As in the p-block the oxides with the highest oxidation numbers are acidic (Mn_2O_7 and CrO_3), those with intermediate oxidation numbers are neutral (MnO_2) or amphoteric (Cr_2O_3), and those with the lowest oxidation numbers are basic (MnO).

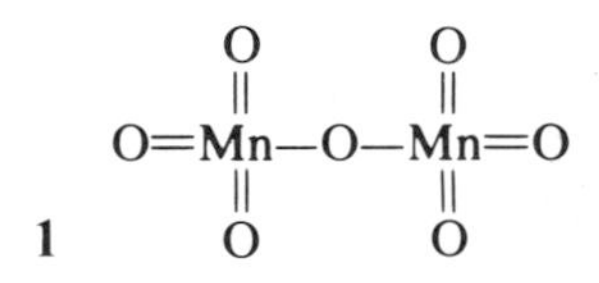

1

The highest oxides

The covalence of the oxides of highest oxidation number is obvious from their physical appearance. Manganese(VII) oxide (Mn_2O_7, **1**) is a green oil that freezes at 6 °C, chromium(VI) oxide, CrO_3, is a deep red solid that melts at 197 °C, and vanadium(V) oxide, V_2O_5, is an orange-yellow solid melting at 690 °C. Their structures account for this increase in melting point: the oily Mn_2O_7 is a simple molecular compound consisting of a pair of tetrahedra sharing one corner, solid CrO_3 has tetrahedra sharing two corners to form a chain-like structure, and the high-melting V_2O_5 consists of zig-zag double chains.

The attractive colours of the oxides are carried over into the oxoanions formed when they react with alkali. The **chromate(VI) anion**, CrO_4^{2-}, is bright yellow, and is in equilibrium with the orange **dichromate(VI) anion** ($(O(CrO_3)_2)^{2-}$ or $Cr_2O_7^{2-}$, **2**):

$$[O_3Cr{-}O{-}CrO_3]^{2-}\quad \mathbf{2}$$

$$2CrO_4^{2-}(aq) + 2H^+(aq) \rightleftharpoons Cr_2O_7^{2-}(aq) + H_2O(l)$$

yellow orange

This is a **condensation reaction** (see Section 30.3); CrO_4^{2-} is favoured in alkaline and $Cr_2O_7^{2-}$ in acidic solution. A similar chameleon-like variety of colours is also shown by manganese(VII) oxide: when this green oil reacts with alkali it forms a purple solution containing the **manganate(VII) ion**, MnO_4^- (which is also commonly called the **permanganate ion**).

As the high oxidation numbers in these compounds lead us to expect, all the oxides and their anions are powerful oxidizing agents, increasingly so as the oxidation number increases (for vanadium(V) oxide, this ability is related to its catalytic action in the oxidation of sulphur dioxide, described in Section 22.6). Acidified dichromate(VI) and manganate(VII) ions are common oxidizing agents in the laboratory both for redox titrations and in organic chemistry. Orange dichromate(VI) solutions are reduced to green chromium(III) ions:

$$Cr_2O_7^{2-}(aq) + 14H^+(aq) + 6e^- \rightarrow 2Cr^{3+}(aq) + 7H_2O(l), \quad E^\ominus = +1.33\,V$$

$$Cr(+6) \rightarrow Cr(+3)$$

The manganate(VII) ion is reduced in alkaline solution only to the brown manganese(IV) oxide (sometimes via the green manganate(VI) ion, MnO_4^{2-}):

$$MnO_4^-(aq) + 2H_2O(l) + 3e^- \rightarrow MnO_2(s) + 4OH^-(aq), \quad E^\ominus = +0.59\,V$$

$$Mn(+7) \rightarrow Mn(+4)$$

but in acidic solution reduction proceeds as far as the almost colourless manganese(II) ions, and the equivalence-point is easier to detect:

$$MnO_4^-(aq) + 8H^+(aq) + 5e^- \rightarrow Mn^{2+}(aq) + 4H_2O(l), \quad E^\ominus = +1.51\,V$$

$$Mn(+7) \rightarrow Mn(+2)$$

Potassium manganate(VII) is used as a disinfectant (in common with similar compounds, it kills bacteria by oxidation) and for the purification of water. Its advantage over chlorine is that it leaves no taste, and the product, MnO_2, causes coagulation of colloidal particles in the water to a product that can be filtered off.

The metal(IV) oxides, MO_2

Titanium(IV) oxide, TiO_2, is white, but the other three oxides we consider have strong colours: VO_2 is blue while CrO_2 and MnO_2 are brownish-black. All four oxides are insoluble solids with high melting points. Titanium(IV) oxide occurs as the mineral **rutile**, and its crystal structure is common to all four oxides and to a number of metal(II) halides, including magnesium fluoride (see Section 18.1, and in particular Figure 18.7(a)).

Titanium(IV) oxide is used on a megatonne scale as the white pigment in paints, since its refractive index is high, so that the numerous small grains of the pigment scatter incident light very strongly. It has a double advantage over the lead salts that were once used: it is non-toxic, and it does not blacken with age (compare Section 20.6). Titanium(IV) oxide may also be fused with other metal oxides, such as BaO, to form mixed oxides, such as $BaTiO_3$. These compounds do not contain TiO_3^{2-} anions. They are ceramics with useful electrical properties: $BaTiO_3$, for example, is **piezoelectric** (that is, a potential difference develops across the crystal when it is mechanically stressed) and is used as a transducer in microphones and record player pick-ups.

Chromium(IV) oxide is of increasing importance in the formulation of 'chrome' audio and video tapes, which have a better high-frequency response than 'ferric' (Fe_2O_3) tapes.

Manganese(IV) oxide is a useful oxidizing agent, and in acidic solution is reduced to manganese(II):

$$MnO_2(s) + 4HCl(aq) \xrightarrow{\text{heat}} MnCl_2(aq) + Cl_2(g) + 2H_2O(l)$$

$$Mn(+4) \rightarrow Mn(+2), \quad Cl(-1) \rightarrow Cl(0)$$

Its principal use (approximately 0.5 megatonne a year) is in dry cell batteries (see Box 25.1), and so it can be regarded as the principal depository of the electrons dumped in countless electrical circuits around the world after they have been used to drive radios, lamps, cassette players and so on.

BOX 25.1

The role of manganese(IV) oxide in a dry cell

The first dry cell was patented by the French engineer and inventor Georges Leclanché in 1866: he set up a factory to manufacture them, and the same basic design is still used today. The aim was to design a cell that could not be spilt and that would produce neither gases nor liquids (which would burst the sealed container).

The negative electrode of a typical dry cell is made of zinc, and usually takes the form of the container for the entire cell; the positive electrode is a carbon rod (see Figure 13.4). The electrolyte is a paste of carbon, ammonium chloride and manganese(IV) oxide, MnO_2.

Zinc is oxidized at the negative electrode (the anode):

$$Zn \rightarrow Zn^{2+} + 2e^-$$

The ammonium chloride is present in order to form a stable complex with the ions produced in the reaction:

$$Zn^{2+} + 2NH_4Cl + 2OH^- \rightarrow [ZnCl_2(NH_3)_2] + 2H_2O$$

Around the carbon electrode the manganese(IV) oxide is reduced and forms a hydrated form of manganese(III) oxide:

$$2MnO_2 + 2H_2O + 2e^- \rightarrow 2MnO(OH) + 2OH^-$$

The overall cell reaction is therefore

$$Zn + 2NH_4Cl + 2MnO_2 \rightarrow [ZnCl_2(NH_3)_2] + 2MnO(OH)$$

and there is no net formation of water: the cell remains 'dry'.

The metal(III) oxides, M_2O_3

We consider only five of these oxides: those of titanium (dark violet), vanadium (black), chromium (green), manganese (black) and iron (red-brown). All of them are high-melting insoluble solids.

The oxides may be prepared either by acid–base neutralization or by redox reaction. In the neutralization method, an alkali is added to a solution of $M^{3+}(aq)$ ions, when the hydrated oxide precipitates:

$$Fe^{3+}(aq) + 3OH^-(aq) \rightarrow FeO(OH).H_2O(s)$$

(The precipitate is often written $Fe(OH)_3$, but there is no evidence that such a species exists.) The precipitate is dehydrated on heating, to give the oxide:

$$2FeO(OH).H_2O(s) \xrightarrow{200\,^{\circ}C} Fe_2O_3(s) + 3H_2O(l)$$

The redox preparations start with compounds of the metal with an oxidation number other than +3. For example, chromium(III) oxide is formed when the metal is heated in air:

$$4Cr(s) + 3O_2(g) \xrightarrow{heat} 2Cr_2O_3(s) \qquad O(0) \rightarrow O(-2),\quad Cr(0) \rightarrow Cr(+3)$$

Chromium(III) oxide is also formed when ammonium dichromate(VI) is heated, because the ammonium cation is oxidized by the dichromate(VI) anion:

$$(NH_4)_2Cr_2O_7(s) \xrightarrow{heat} Cr_2O_3(s) + N_2(g) + 4H_2O(g)$$

$$Cr(+6) \rightarrow Cr(+3),\quad N(-3) \rightarrow N(0)$$

The volcano-like reaction spills a lava of green chromium(III) oxide.

The oxides also form mixed oxides; for example:

$$CaCO_3(s) + Fe_2O_3(s) \xrightarrow{heat} CaFe_2O_4(s) + CO_2(g)$$

(These do not contain 'FeO_2^-' anions.) They are called **ferrites** and are used as radio aerials.

The mixed oxides, M_3O_4

The mixed oxides contain two M^{3+} ions and one M^{2+} ion per formula unit, where M is Mn, Fe or Co, and they are black solids. Their colours arise from a charge-transfer transition in which an electron migrates from an M^{2+} ion to an M^{3+} ion. The most famous example is **magnetite**, Fe_3O_4, the mineral which as **lodestone** first drew people's attention to magnetism and its use to navigators.

The metal(II) oxides

All the first-row d-block elements except scandium and chromium form non-stoichiometric metal(II) oxides of approximate formula MO. They all have the rock-salt structure, and all except ZnO are coloured (usually grey, green or black) and basic. For example, copper(II) oxide, which is black, reacts with dilute sulphuric acid to form blue aqueous copper(II) sulphate:

$$CuO(s) + (HO)_2SO_2(aq) \xrightarrow{\text{heat}} CuSO_4(aq) + H_2O(l)$$

Zinc oxide is amphoteric, and has the unusual property of being white when cold and yellow when hot.

Hydroxides, $M(OH)_2$, are precipitated when sodium hydroxide is added to aqueous solutions of M^{2+} ions:

$$Ni^{2+}(aq) + 2OH^-(aq) \rightarrow Ni(OH)_2(s)$$

If the element is able to form an oxide or hydroxide with a higher oxidation number, air must be excluded; otherwise, oxidation follows precipitation – for example, green iron(II) hydroxide turns brown on standing:

$$Fe^{2+}(aq) + 2OH^-(aq) \xrightarrow{\text{air absent}} Fe(OH)_2(s)$$

$$4Fe^{2+}(aq) + 8OH^-(aq) + O_2(g) \xrightarrow{\text{air present}} 4FeO(OH)(s) + 2H_2O(l)$$

$$O(0) \rightarrow O(-2), \quad Fe(+2) \rightarrow Fe(+3)$$

The colours of these hydroxides are given in Table 25.5. They all react with acids:

$$Ni(OH)_2(s) + 2H^+(aq) \rightarrow Ni^{2+}(aq) + 2H_2O(l)$$

Zinc hydroxide is amphoteric, however, and reacts with alkali as well, giving the **tetrahydroxozincate ion**, $[Zn(OH)_4]^{2-}$:

$$Zn^{2+}(aq) + 2OH^-(aq) \rightarrow Zn(OH)_2(s)$$

$$Zn(OH)_2(s) + 2OH^-(aq) \rightarrow [Zn(OH)_4]^{2-}(aq)$$

The hydroxides of cobalt(II), nickel(II), copper(II) and zinc also redissolve in excess aqueous ammonia as a result of complexation, a topic dealt with in detail in Section 25.7:

$$Ni(OH)_2(s) + 6NH_3(aq) \rightarrow [Ni(NH_3)_6]^{2+}(aq) + 2OH^-(aq)$$

$$Zn(OH)_2(s) + 4NH_3(aq) \rightarrow [Zn(NH_3)_4]^{2+}(aq) + 2OH^-(aq)$$

Table 25.5 The hydroxides of the first-row d-block elements with oxidation number +2

Formula	Colour	Notes
$Mn(OH)_2$	white	
$Fe(OH)_2$	pale green	rapidly turns brown due to oxidation
$Co(OH)_2$	blue/pink	polymorphic
$Ni(OH)_2$	green	
$Cu(OH)_2$	blue	
$Zn(OH)_2$	white	soluble in excess $OH^-(aq)$

$Ti(OH)_2$, $V(OH)_2$ and $Cr(OH)_2$ cannot be isolated: there is immediate oxidation

Copper(I) oxide

The only oxide with oxidation number +1 is the red solid copper(I) oxide, Cu_2O, which is formed together with CuO when copper is heated in air. Copper(I) oxide is insoluble; if forced into solution – by the addition of sulphuric acid, for instance – it rapidly disproportionates:

$$\underset{\text{red}}{Cu_2O(s)} + (HO)_2SO_2(aq) \rightarrow Cu(s) + \underset{\text{blue}}{Cu^{2+}(aq)} + SO_4^{2-}(aq) + H_2O(l)$$

$$2Cu(+1) \rightarrow Cu(0) + Cu(+2)$$

Copper(I) oxide is also formed as a red precipitate when an aldehyde or a reducing sugar is tested with Fehling's solution (see Section 30.3).

25.4 Halides

Among so much chemistry, it may be a relief to find that the halides of the first-row d-block elements have few uses and can be dismissed reasonably briskly. The highest oxidation number found for chlorides is +4, partly because of the steric difficulty of packing more chlorine atoms around the small, central metal atom but more importantly because the elements with the highest oxidation numbers are strongly oxidizing and oxidize the halide ions to halogens. Only the fluoride ion can survive, because it is so difficult to oxidize: both VF_5 and CrF_6 exist, showing yet again fluorine's ability to coax an element into its highest oxidation number.

The halides with higher oxidation numbers are essentially covalent, and are hydrolysed by water:

$$TiCl_4(l) + 2H_2O(l) \rightarrow TiO_2(s) + 4HCl(aq)$$

(The dense white clouds of titanium(IV) oxide formed in this reaction are used as a naval smoke-screen and for smoke trails from aerobatic aircraft.) Halides with lower oxidation numbers are largely ionic and dissolve in water, often giving coloured solutions. Aqueous iron(II) chloride solution is pale green and turns brown slowly unless air is rigorously excluded, while aqueous manganese(II) chloride is very pale pink; the colour of aqueous cobalt(II) chloride depends on its concentration (see Section 25.7).

The higher halides can be prepared by direct synthesis. For instance, iron powder and chlorine react vigorously on heating to produce iron(III) chloride, which sublimes as a brown solid:

$$2Fe(s) + 3Cl_2(g) \xrightarrow{\text{heat}} 2FeCl_3(s) \qquad Cl(0) \rightarrow Cl(-1),\quad Fe(0) \rightarrow Fe(+3)$$

In the vapour iron(III) chloride is dimeric, Fe_2Cl_6 (compare Al_2Cl_6, Section 19.4), and it is an important Lewis acid catalyst in the Friedel–Crafts reaction (see Section 34.3). When hydrogen chloride replaces chlorine in the reaction, the lower chloride, $FeCl_2$, is formed:

$$Fe(s) + 2HCl(g) \xrightarrow{\text{heat}} FeCl_2(s) + H_2(g) \qquad H(+1) \rightarrow H(0),\quad Fe(0) \rightarrow Fe(+2)$$

Copper(I) halides

Of the first-row d-block metals, only copper forms halides with oxidation number +1, although copper(I) fluoride does not exist (on account of fluorine's aggressive oxidizing ability). In addition, iodide ions are such good reducing agents that when they are added to a copper(II) solution, copper(I) iodide is produced, together with iodine:

$$2Cu^{2+}(aq) + 4I^-(aq) \rightarrow 2CuI(s) + I_2(s) \qquad Cu(+2) \rightarrow Cu(+1),\quad I(-1) \rightarrow I(0)$$

The white precipitate of copper(I) iodide can be made visible by reducing the iodine with sodium thiosulphate, but excess thiosulphate forms a complex with copper(I), and dissolves the precipitate. Silver bromide reacts with the thiosulphate ion in the same way (see Box 25.2).

BOX 25.2

Photography

The first photographs were based on the chance observation by the German physician Heinrich Schulze in 1727 that silver nitrate is blackened by sunlight; this was turned into a useful process by Louis Daguerre and William Fox Talbot in the late 1830s. Modern photographic emulsions consist of gelatine containing tiny, uniform crystals of silver bromide. Silver iodide is used for fast film, and dyes are incorporated for colour photography.

Each small crystal or **grain** contains about 10^{12} Ag^+ ions. When the film is exposed, a **latent image** is formed, in which some of the ions are reduced to silver atoms by electrons released from the bromide ions by the incident light:

$$Br^- + h\nu \rightarrow Br + e^-$$

$$e^- + Ag^+ \rightarrow Ag$$

Grains containing as few as four silver atoms are sensitive to further reduction in the **development** stage. The silver atoms act as catalysts for the selective and complete reduction to silver of these sensitized grains by a mild reducing agent, such as benzene-1,4-diol (hydroquinone, **3**). The other (unchanged) grains of silver bromide are removed in the **fixing** stage, by forming a soluble complex with, for monochrome films, sodium thiosulphate ('hypo'):

OH

3 OH

$$AgBr(s) + 2SSO_3^{2-}(aq) \rightarrow [Ag(SSO_3)_2]^{3-}(aq) + Br^-(aq)$$

Both products are soluble, and once they have been washed away the image on the film is permanent (insensitive to further exposure to light).

The image at this stage is a **negative**, because it is dark where light fell (and the grains are reduced to silver) and transparent where it was not exposed. Exposure of a paper coated with silver bromide to light passing through the negative, followed by repetition of the processing stages, leads to the formation of a positive image.

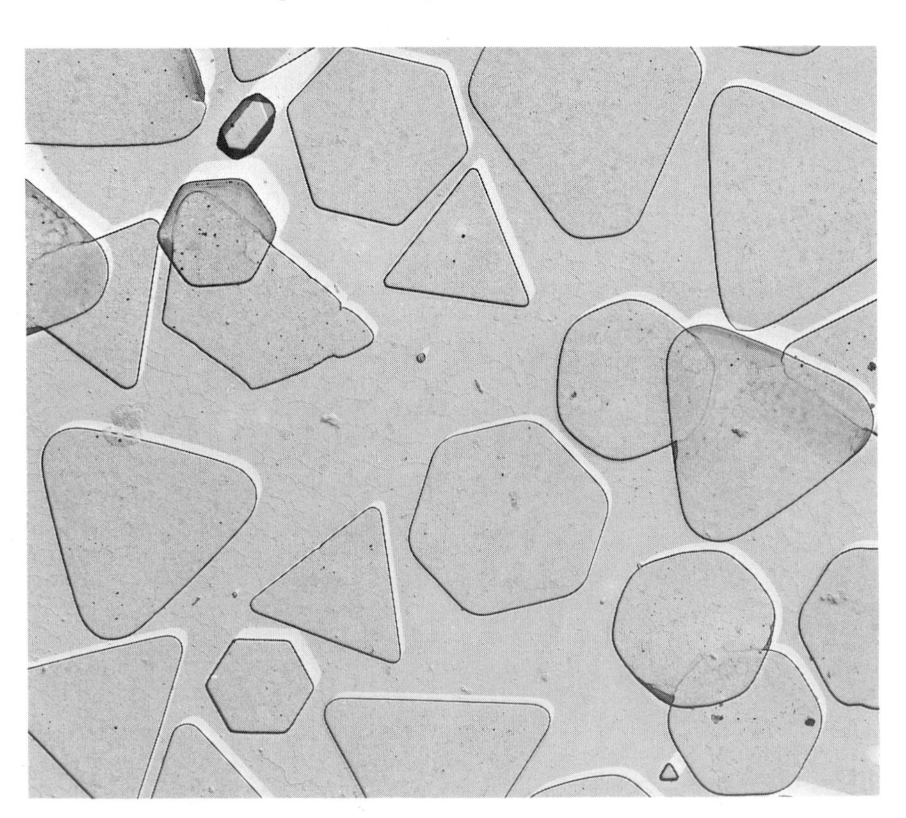

Figure 25.7 Grains of a high-speed photographic film, magnified approximately 4500 times

Copper(I) chloride is prepared from copper(II) chloride (chlorine is too strong an oxidizing agent to be used directly) by heating it with excess copper in concentrated hydrochloric acid:

$$Cu(s) + CuCl_2(aq) + 2Cl^-(aq) \rightarrow 2CuCl_2^-(aq) \qquad Cu(0) + Cu(+2) \rightarrow 2Cu(+1)$$

When the solution containing the complex is diluted, copper(I) chloride precipitates as a white solid. The lack of colour is due to the electron configuration of copper(I), which is d^{10} – the complete d-subshell.

25.5 Hydrides

The d-block elements form an unusual yet important series of compounds with hydrogen. They are unusual because they do not have a definite stoichiometry, and cannot be accurately described as either ionic or covalent. They are important because the reaction between hydrogen and a d-block element is a crucial step in many catalytic processes (see Section 33.3).

The d-block elements are truly transitional with respect to hydrides, for they separate the ionic hydrides of the s-block elements (such as K^+H^-) from the covalent hydrides of the p-block elements (such as H—Br). For example, hydrogen and titanium produce a substance TiH_x of variable composition (with x less than 2) which is metallic in appearance and conducts electricity well. Structurally, the hydrides are best regarded as interstitial compounds (see Section 16.3), where the hydrogen atoms occupy the tetrahedral holes between the metal ions which still retain almost their original positions in the metal crystal. Similar interstitial compounds are formed with carbon and nitrogen, but their larger atoms tend to occupy the larger octahedral holes.

Palladium is the d-block element that reacts most readily with hydrogen; it can absorb nearly a thousand times its own volume of the gas. The hydrogen is released when the compound is heated, so that it is a convenient 'sponge' for storing the gas. Palladium is permeable to hydrogen but not to other gases, and a foil of the metal can act as a kind of sieve to remove hydrogen from a mixture of gases.

25.6 Aqua ions

Aqua ions are metal ions with up to about six water molecules attached by coordinate bonds. These are, in fact, a special case of the very broad class of coordination compounds treated in Section 25.7.

M^{2+} and M^{3+} ions

All the first-row d-block metals except scandium form M^{2+} ions, and all except the last three – nickel, copper and zinc – also form M^{3+} ions. The configurations and (for $M^{2+}(aq)$) the colours of these ions are listed in Table 25.6: note that a d^{10} ion is colourless.

Table 25.6 Properties and colours of d-block ions in water

$M^{2+}(aq)$										
ion		Ti^{2+}	V^{2+}	Cr^{2+}	Mn^{2+}	Fe^{2+}	Co^{2+}	Ni^{2+}	Cu^{2+}	Zn^{2+}
electron configuration outside argon core		d^2	d^3	d^4	d^5	d^6	d^7	d^8	d^9	d^{10}
colour		unstable	violet	blue	pale pink	pale green	pink	green	pale blue	colourless
$M^{3+}(aq)$										
ion	Sc^{3+}	Ti^{3+}	V^{3+}	Cr^{3+}	Mn^{3+}	Fe^{3+}	Co^{3+}			
electron configuration outside argon core	d^0	d^1	d^2	d^3	d^4	d^5	d^6			
$E^\ominus(298\ K)/V$ for $M^{3+}(aq) + e^- \rightarrow M^{2+}(aq)$		−0.37	−0.26	−0.41	+1.51	+0.77	+1.81			

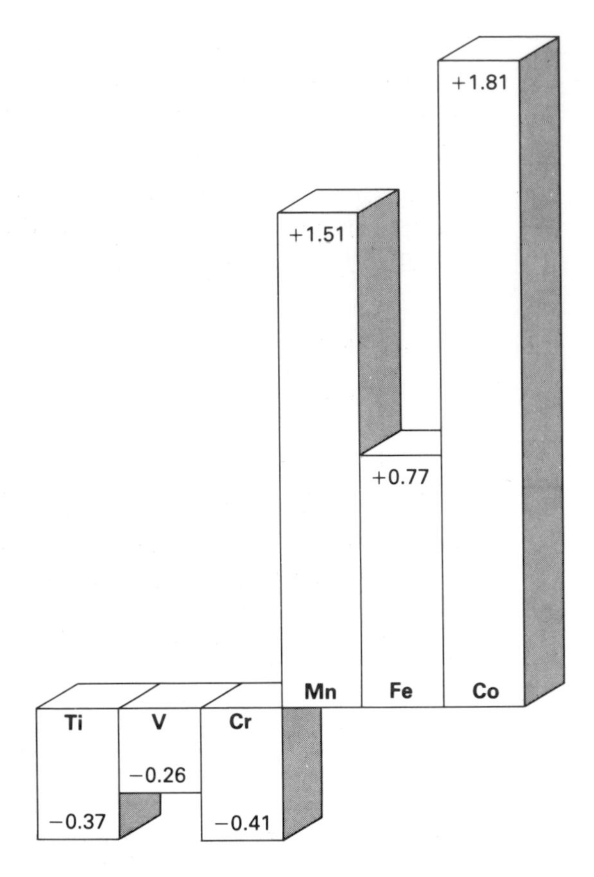

Figure 25.8 The standard electrode potentials for $M^{3+}(aq) + e^- \rightarrow M^{2+}(aq)$

The relative stabilities of the ions to oxidation and reduction can be judged from the standard electrode potentials of the M^{3+}, M^{2+} couples, shown in Figure 25.8. The negative values for the titanium, vanadium and chromium couples indicate that the oxidized form (M^{3+}) is favoured, and indeed the $M^{2+}(aq)$ ions of those three elements are strong reducing agents which survive only in the absence of air. The vanadium(II) ion even reduces water to hydrogen:

$$2V^{2+}(aq) + 2H_2O(l) \rightarrow 2V^{3+}(aq) + 2OH^-(aq) + H_2(g)$$

$H(+1) \rightarrow H(0), \quad V(+2) \rightarrow V(+3)$

The high positive values of the standard electrode potentials of the manganese and cobalt couples indicate that reduction is favoured, and $Mn^{3+}(aq)$ and $Co^{3+}(aq)$ are both strong oxidizing agents. The cobalt(III) ion oxidizes water to oxygen:

$$4Co^{3+}(aq) + 2H_2O(l) \rightarrow 4Co^{2+}(aq) + O_2(g) + 4H^+(aq)$$

$Co(+3) \rightarrow Co(+2), \quad O(-2) \rightarrow O(0)$

The standard electrode potential of the iron couple has an intermediate value (partly because Fe^{3+} has a relatively stable half-filled d^5 configuration), and both Fe^{2+} and Fe^{3+} can exist with several accompanying anions. Nevertheless, iron(II) compounds do tend to be oxidized in air, and green $Fe(OH)_2$ turns brown on standing. One exception is **Mohr's salt** – ammonium iron(II) sulphate, $(NH_4)_2Fe(SO_4)_2 . 6H_2O$ – which is stable in air and is used as a primary standard in manganate(VII) redox titrations. The reaction between iron(III) and copper,

$$2Fe^{3+}(aq) + Cu(s) \rightarrow 2Fe^{2+}(aq) + Cu^{2+}(aq)$$

$Fe(+3) \rightarrow Fe(+2), \quad Cu(0) \rightarrow Cu(+2)$

is used to etch copper for printed-circuit boards.

The aqua ions also undergo characteristic Lewis acid–base reactions. The first is the **ligand replacement reaction**, in which one or more of the hydrating water molecules is replaced by another group. These groups, or ligands, may

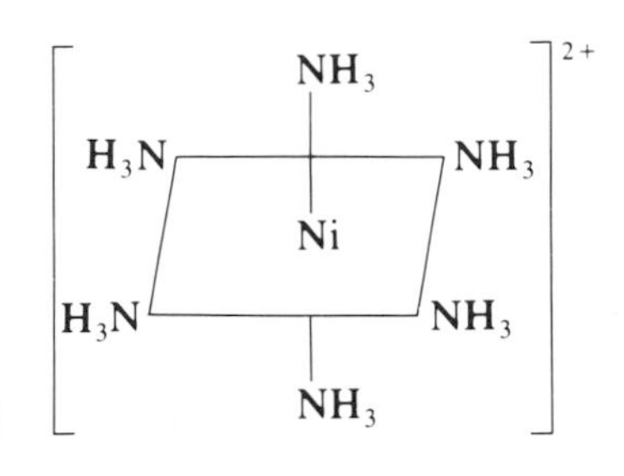

be neutral molecules (such as H_2O and NH_3), anions (such as CN^- and Cl^-) or, exceptionally, cations (such as NO^+). The essential feature of a ligand is that it is a Lewis base, that is, that it has at least one pair of electrons that can be used to form a coordinate bond with the central metal ion, which plays the role of a Lewis acid. For example, when excess concentrated aqueous ammonia is added to nickel(II) sulphate, the green solution becomes blue-violet as a result of the formation of the hexaammine complex (**4**):

$$\underset{\text{green}}{[Ni(H_2O)_6]^{2+}(aq)} + 6NH_3(aq) \rightarrow \underset{\text{blue-violet}}{[Ni(NH_3)_6]^{2+}(aq)} + 6H_2O(l)$$

More examples, and their structures, are discussed in Section 25.7.

A second type of characteristic reaction is proton loss from the hydrating water molecules (as in $[Al(H_2O)_6]^{3+}(aq)$, discussed in Section 19.6); that is, the ions are acidic. Metal(III) aqua ions, such as $Fe^{3+}(aq)$, for example, are strikingly acidic:

$$[Fe(H_2O)_6]^{3+}(aq) + H_2O(l) \rightleftharpoons [Fe(H_2O)_5OH]^{2+}(aq) + H_3O^+(aq)$$

and the brown hydroxo-complex is present in such a high concentration that its colour is characteristic of iron(III) solutions. (Solid ammonium iron(III) alum, which does contain the hexaaqua ion, is pale violet.) Proton loss is easier when the charge density of the central metal ion is high, because the O—H bond is weakened by electron withdrawal; the Fe^{3+} ion is both highly charged and small, and the acidity of $[Fe(H_2O)_6]^{3+}$ is correspondingly high – when sodium carbonate is added to aqueous iron(III) chloride, carbon dioxide is evolved and iron(III) hydroxide precipitates.

$$[Fe(H_2O)_6]^{3+}(aq) \rightleftharpoons FeO(OH)(s) + H_2O(l) + 3H_3O^+(aq)$$

$$CO_3{}^{2-}(aq) + 2H_3O^+(aq) \rightleftharpoons CO_2(g) + 3H_2O(l)$$

Similarly, adding sodium sulphide to aqueous titanium(III) sulphate produces hydrogen sulphide and precipitates titanium(III) hydroxide.

M^+ and M^{4+} ions

Since the M^{4+} ions are such strong oxidizing agents, all – except one – are unstable in aqueous solution. The exception, although it needs special precautions to survive, is the vanadium(IV) ion. This highly charged, small ion binds a water molecule so strongly, and loosens its O—H bonds so effectively, that it exists in solution as the **oxovanadium(IV) ion**, VO^{2+}, commonly known as the **vanadyl ion**. Solutions of the vanadyl ion are blue, and a pleasing demonstration of the origin of vanadium's name (it is called after Vanadis, the Scandinavian goddess of beauty) is the effect of zinc amalgam added to an acidic solution of ammonium vanadate(V):

$$\underset{\substack{\text{yellow}\\+5}}{VO_4{}^{3-}(aq)} \xrightarrow{\text{reduce}} \underset{\substack{\text{blue}\\+4}}{VO^{2+}(aq)} \xrightarrow{\text{reduce}} \underset{\substack{\text{green}\\+3}}{V^{3+}(aq)} \xrightarrow{\text{reduce}} \underset{\substack{\text{violet}\\+2}}{V^{2+}(aq)}$$

Ox(V)

The instability of M^+ ions is shown in a different way: they disproportionate. Whenever an attempt is made to prepare an aqueous solution of Cu^+ ions, for example, metallic copper is deposited (see Section 25.3 above).

EXAMPLE

Predict whether $Cu^+(aq)$ is stable with respect to disproportionation into copper metal and $Cu^{2+}(aq)$.

METHOD Calculate the standard reaction Gibbs energy for the disproportionation

$$2Cu^+(aq) \rightarrow Cu(s) + Cu^{2+}(aq)$$

using the standard electrode potentials (Table 13.1) of the half-reactions at 298 K:

$$Cu^+(aq) + e^- \rightarrow Cu(s), \quad E^\ominus = +0.52\,V \qquad (a)$$

$$Cu^{2+}(aq) + 2e^- \rightarrow Cu(s), \quad E^\ominus = +0.34\,V \qquad (b)$$

Convert these potentials to standard reaction Gibbs energies using equation 13.4.1*b*:

$$\Delta_r G^\ominus = -zFE^\ominus$$

ANSWER Convert to standard reaction Gibbs energies:

$$\Delta_r G^\ominus(a) = -1 \times 9.65 \times 10^4\,C\,mol^{-1} \times 0.52\,V$$
$$= -50\,kJ\,mol^{-1}$$

(since $1\,C\,V = 1\,J$), and

$$\Delta_r G^\ominus(b) = -2 \times 9.65 \times 10^4\,C\,mol^{-1} \times 0.34\,V$$
$$= -66\,kJ\,mol^{-1}$$

The disproportionation reaction is the sum $2(a) - (b)$, so that

$$\Delta_r G^\ominus = 2\Delta_r G^\ominus(a) - \Delta_r G^\ominus(b)$$
$$= 2(-50\,kJ\,mol^{-1}) - (-66\,kJ\,mol^{-1})$$
$$= -34\,kJ\,mol^{-1}$$

The standard reaction Gibbs energy is negative, so that under standard conditions the disproportionation is spontaneous.

COMMENT Copper(I) compounds are known, but are either insoluble (copper(I) oxide and copper(I) iodide, for example) or exist in solution as complexes (such as $CuCl_2^-$).

25.7 Coordination compounds

A striking feature of the d-block elements is their ability to act as Lewis acids and to form **coordination compounds** in which several (often six) ligands act as Lewis bases and are attached to the central metal ion and form its **coordination sphere**. The cluster is usually, though not always, charged, and is then called a **complex ion**. Some typical ligands are listed in Table 25.7.

Types of ligand

Ligands that occupy a single position in the coordination sphere are called **unidentate** ('one-toothed'), those that occupy two are **bidentate** ('two-toothed'), and so on. In a complex formed by a bidentate ligand the metal atom forms part of a five- or six-membered ring, as if it were held in a claw; these complexes are called **chelates** (from the Greek for 'crab's claw').

Stability constants and the chelate effect

Different ligands bond with different strengths, and an important task in organizing the chemistry of these compounds is to find patterns of stability.

In aqueous solution, the stability of a complex can be expressed in terms of

Table 25.7 Some typical ligands

Unidentate: Ligand	General name for complexes	Bidentate: Ligand	General name for complexes	Polydentate:
Neutral ligands				
$:OH_2$	aqua	en		
$:NH_3$	ammine*			
:CO	carbonyl			
Negatively charged ligands				
$:Cl^-$	chloro-	$C_2O_4^{2-}$	oxalato-	Y^{4-}, where H_4Y = edta (Section 18.6)
$:OH^-$	hydroxo-			
$:NO_2^-$	nitro-			
$:ONO^-$	nitrito-			
$:NCS^-$	*N*-thiocyanato-			
$:SCN^-$	*S*-thiocyanato-			
$:CN^-$	cyano-			
$:NC^-$	isocyano-			

The donor atom is shown with a lone pair. Some donor atoms have two ($:\ddot{O}H_2$), three ($:\ddot{O}H^-$) and four ($:\ddot{C}l:^-$) electron pairs. Note the possibility of isomerism with NO_2^-, NCS^- and CN^-

* Compare the spelling for ammonia complexes with the spelling of amines for RNH_2 (Chapter 32)

equilibrium constants. Consider, as an example, the formation of complex ions of copper(II) with ammonia as ligand. The **overall stability constant**, β_4, is the equilibrium constant for the formation of a deep blue **tetraamminecopper(II) ion**, in which four of the water ligands have been replaced by ammonia ligands; the reaction is

$$[Cu(H_2O)_6]^{2+}(aq) + 4NH_3(aq) \rightleftharpoons [Cu(NH_3)_4(H_2O)_2]^{2+}(aq) + 4H_2O(l)$$

and hence

$$\beta_4 = \frac{[[Cu(NH_3)_4(H_2O)_2]^{2+}]}{[[Cu(H_2O)_6]^{2+}][NH_3]^4}$$

The water concentration is virtually constant, and so is not included in the definition of the constants. Similar **stepwise stability constants** can be defined for the stepwise replacement of ligands, as in

$$[Cu(H_2O)_6]^{2+}(aq) + NH_3(aq) \rightleftharpoons [Cu(NH_3)(H_2O)_5]^{2+}(aq) + H_2O(l)$$

$$K_1 = \frac{[[Cu(NH_3)(H_2O)_5]^{2+}]}{[[Cu(H_2O)_6]^{2+}][NH_3]}$$

and

$$[Cu(NH_3)(H_2O)_5]^{2+}(aq) + NH_3(aq) \rightleftharpoons [Cu(NH_3)_2(H_2O)_4]^{2+}(aq) + H_2O(l)$$

$$K_2 = \frac{[[Cu(NH_3)_2(H_2O)_4]^{2+}]}{[[Cu(NH_3)(H_2O)_5]^{2+}][NH_3]}$$

and so on. The overall stability constant is the product of the four (in this case) successive stepwise stability constants:

$$\beta_4 = K_1K_2K_3K_4$$

Some typical values of overall stability constants are given in Table 25.8: the higher the value, the greater the thermodynamic stability of the complex. The relative stabilities of complexes with different ligands can be demonstrated by noting how they can be replaced by successively stronger ligands. For example, silver ions can undergo the following sequence of reactions:

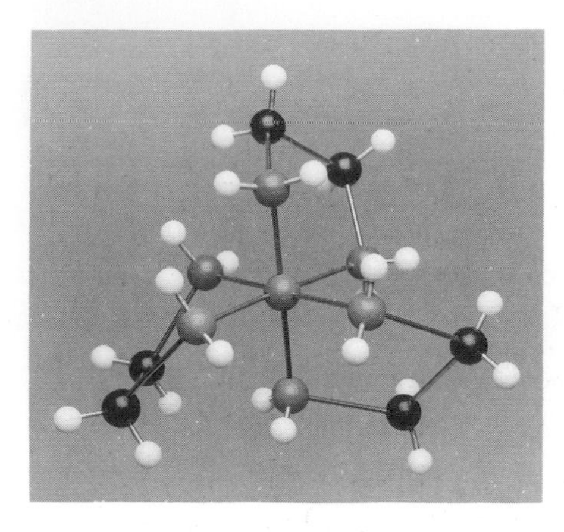

Figure 25.9 Molecular model of the chelate complex $[Ni\,en_3]^{2+}$

$$Ag^+(aq) + Cl^-(aq) \rightarrow \underset{\text{white ppt}}{AgCl(s)} \xrightarrow{NH_3(aq)} \underset{\text{clear}}{[Ag(NH_3)_2]^+(aq)} \xrightarrow{Br^-(aq)} \underset{\text{cream ppt}}{AgBr(s)}$$

$$\xrightarrow{SSO_3^{2-}(aq)} \underset{\text{clear}}{[Ag(SSO_3)_2]^{3-}(aq)} \xrightarrow{I^-(aq)} \underset{\text{yellow ppt}}{AgI(s)}$$

Notice the strikingly large stability constants for the chelating ligands denoted by en (ethylenediamine, ethane-1,2-diamine) and edta (see Figures 25.9 and 18.12). (Remember that the values in Table 25.8 are the logarithms of the stability constants, so that a value of 18.3 represents a stability constant nearly 10^{10} times larger than a value of 8.6.) The explanation of this greater stability, which is called the **chelate effect**, takes us back to the discussion of entropy in Chapter 11. In chelate formation a single ligand molecule liberates at least two ligand molecules from the original coordination sphere of the central ion (six in the case of the sexadentate edta ligand); the reaction thus increases the number of particles present. This contribution to the entropy shifts the equilibrium in favour of products (the enthalpy changes are all very similar as each has six metal—N bonds).

Table 25.8 Stability constants of some complexes

Complex	$\lg \beta_n$	n
$[Cu(NH_3)_4(H_2O)_2]^{2+}$	13.1	4
$CuCl_4^{2-}$	5.6	4
CuY^{2-}*	18.8	1
$[Ni(NH_3)_6]^{2+}$	8.6	6
$[Ni\,en_3]^{2+}$†	18.3	3
$[Fe(SCN)(H_2O)_5]^{2+}$	3.9	1
CaY^{2-}*	10.7	1

* H_4Y = edta
† en = ethane-1,2-diamine

5

The chelate effect is important in analytical chemistry (for instance, nickel(II) forms a characteristic red precipitate with butanedione dioxime, **5**), in water softening (see Section 18.6) and in biology (the iron in haemoglobin, for example, is clasped strongly by the four 'teeth' of the quadridentate porphyrin ring, described in Box 3.1).

Colour changes on complexation

One of the appealing features of chemistry is the opportunity it gives to produce substances having a wide range of striking colours. Indeed, this is the feature that often draws future chemists into the subject. The complex ions of the d-block elements display this feature spectacularly: in fact, many of the colours found on artists' palettes are suspensions of complex ion salts in oil or some synthetic medium.

Striking colour changes often accompany ligand substitution reactions (see Colour 7). For instance, when aqueous potassium thiocyanate is added to a solution of iron(III) ions, a deep blood-red solution is formed:

$$[Fe(H_2O)_6]^{3+}(aq) + SCN^-(aq) \rightarrow [Fe(SCN)(H_2O)_5]^{2+}(aq) + H_2O(l)$$

6

The effect of ligands on the energies of the d-electrons is responsible for some well-known colour changes. For example, the pale blue of copper(II) sulphate, which is due to a d–d transition in the $[Cu(H_2O)_6]^{2+}$ complex ion (**6**), changes to a deep blue when four of the water ligands are replaced by ammonia ligands (**7**). Similarly, cobalt(II) forms a tetrahedral complex anion, $CoCl_4^{2-}$ (**8**), which is deep blue, but when the chloride ligands are replaced by water, the colour changes to the pink of the hydrated Co^{2+} cation:

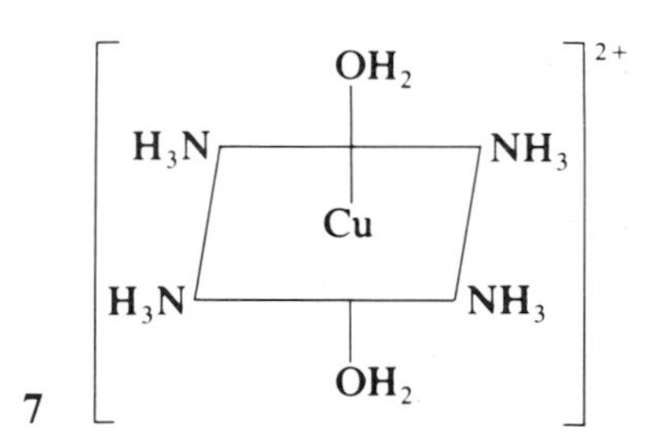

7

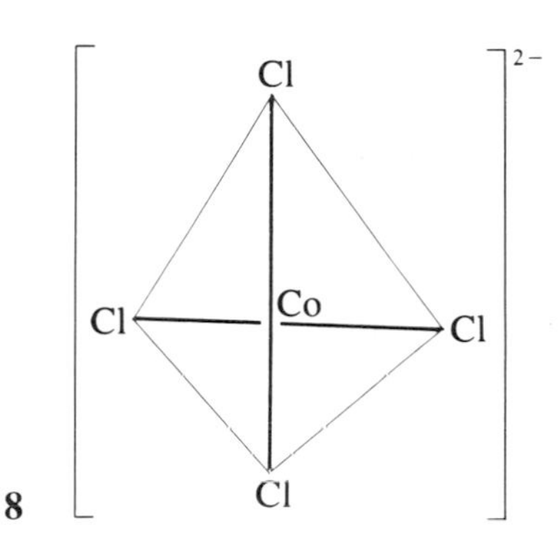

8

$$CoCl_4^{2-}(aq) + 6H_2O(l) \rightarrow [Co(H_2O)_6]^{2+}(aq) + 4Cl^-(aq)$$

blue pink

This colour change is used as a test for the detection of water (in a desiccator, for instance). Another interesting colour change is seen when solutions of iron salts are treated with hexacyanoferrate ions. The reactions of iron(II) with hexacyanoferrate(III) and of iron(III) with hexacyanoferrate(II) lead to an intensely blue precipitate called **Turnbull's blue** and **Prussian blue** respectively.

Modern spectroscopic methods have shown that Prussian blue and Turnbull's blue are identical and have a formula $Fe_4^{III}[Fe^{II}(CN)_6]_3(s)$. The structure is an infinite three-dimensional network structure and this explains its insolubility. Each cyanide ligand donates electron pairs from both the carbon and the nitrogen atoms, and so forms a bridge between two iron ions. The intense colour is due to a charge-transfer transition, in which an electron migrates from an iron(II) ion to an iron(III) ion. As well as being used for Prussian military uniforms, Prussian blue was also used for making the original 'blue prints' prepared by generations of engineering designers.

Standard electrode potentials

The redox properties of the central metal ion of a complex are strongly affected by the ligands. This can be seen very clearly by comparing Co(III), Co(II) couples with water and ammonia as ligands:

$$[Co(H_2O)_6]^{3+}(aq) + e^- \rightarrow [Co(H_2O)_6]^{2+}(aq), \quad E^{\ominus} = +1.81\text{ V}$$

$Co(+3) \rightarrow Co(+2)$

$$[Co(NH_3)_6]^{3+}(aq) + e^- \rightarrow [Co(NH_3)_6]^{2+}(aq), \quad E^{\ominus} = +0.11\text{ V}$$

$Co(+3) \rightarrow Co(+2)$

Hydrated Co^{3+} ions are such powerful oxidizing agents that they oxidize the water they are formed in, as mentioned in Section 25.6, whereas hundreds of ammine complexes, such as $[Co(NH_3)_6]^{3+}(aq)$ itself and $[Co(NH_3)_5Cl]^{2+}(aq)$, are known.

EXAMPLE

Show that iron(III) ions can oxidize iodide ions when in the form of $Fe^{3+}(aq)$, but not when in the form of $[Fe(CN)_6]^{3-}(aq)$.

METHOD Begin by judging the thermodynamic spontaneity of a reaction. In this case, do so by referring to the standard electrode potentials of the couples, to be found in Table 13.1. The couple with the more positive potential can oxidize a couple with a less positive potential.

ANSWER The relevant couples are

$$E^{\ominus}(I_2, I^-) = +0.54\text{ V}$$

$$E^{\ominus}(Fe^{3+}, Fe^{2+}) = +0.77\text{ V}$$

$$E^{\ominus}([Fe(CN)_6]^{3-}, [Fe(CN)_6]^{4-}) = +0.36\text{ V}$$

These are in the order

$$E^{\ominus}(Fe^{3+}, Fe^{2+}) > E^{\ominus}(I_2, I^-) > E^{\ominus}([Fe(CN)_6]^{3-}, [Fe(CN)_6]^{4-})$$

so that only the Fe^{3+}, Fe^{2+} couple can oxidize the I_2, I^- couple.

COMMENT The ligands attached to a metal ion have a considerable effect on its properties, and it is important to refer to the exact complex specified when using thermodynamic arguments.

Isomerism

The common feature of all complexes is that a central metal atom or ion is surrounded by several ligands. When several different ligands are present in the same complex, there are opportunities for isomerism. This topic is discussed fully in Section 26.4; for the present we can regard it simply as the construction of different compounds from the same kit of parts – in these compounds, ligands. Several varieties of isomerism are known, but the three most important are:

1 **Ionization isomerism**, in which the same type of ligand may be a part of the cation or the anion of a compound.
2 **Geometric isomerism**, in which the same ligands have different spatial arrangements around a central ion.
3 **Optical isomerism**, in which one complex ion is the mirror image of another.

As an example of ionization isomerism, consider the compounds of formula $CrCl_3.6H_2O$. Three such compounds are known, all of different colours; their molar conductivities show that the compounds differ in the numbers of Cl^- ions present as ligands and as unattached anions:

$[Cr(H_2O)_6]^{3+}3Cl^-$	$[Cr(H_2O)_5Cl]^{2+}2Cl^-$	$[Cr(H_2O)_4Cl_2]^+Cl^-$
violet	pale green	dark green

Only the chloride ions *outside* the complex will form an immediate white precipitate with silver nitrate, so that these isomers can be distinguished by adding aqueous silver nitrate to a solution of a weighed sample, and weighing the silver chloride formed.

The dark green complex cation $[Cr(H_2O)_4Cl_2]^+$ illustrates geometric isomerism. Each ligand is at the corner of a nearly regular octahedron, but this leaves two possibilities. In one, called the *trans*-isomer (from the Latin for 'across'), the Cl^- ligands occupy opposite corners (**9**) whereas in the second, called the *cis*-isomer (from the Latin for 'on this side of'), the Cl^- ligands are neighbours (**10**). It has been shown experimentally that $[Cr(H_2O)_4Cl_2]^+$ occurs as the *trans*-isomer, presumably on account of repulsion between the chloride ligands. For many compounds, both geometric isomers are known: for instance, the *trans*-isomer of $[Co(NH_3)_4Cl_2]^+$ is green and the *cis*-isomer violet.

The complex ion *cis*-$[Coen_2Cl_2]^+$ illustrates optical isomerism, which is discussed in detail in Section 32.6. The two en ligands can be attached in different ways (**11** and **12**), like two sets of spiral staircases in a building. Each isomer is the mirror image of the other, and neither optical isomer can be superimposed on its own mirror image. A molecule or ion that cannot be superimposed on its own mirror image is said to be **chiral**. Chiral molecules are **optically active**, that is, they rotate the plane of polarization of light. A chiral molecule and its mirror-image isomer, like a pair of hands, are called **enantiomers** and rotate the plane of plane-polarized light by exactly equal amounts in opposite directions.

9

10

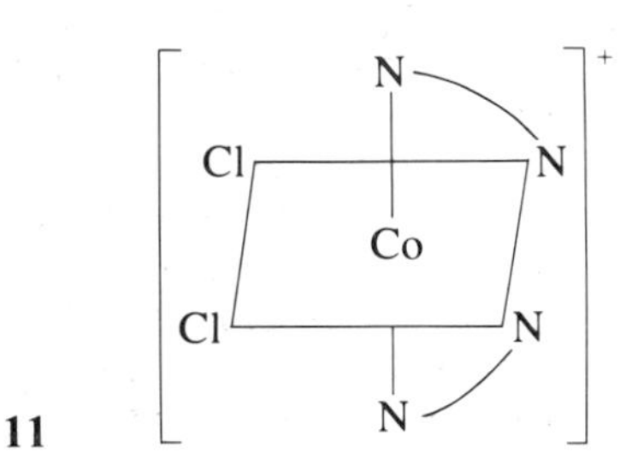

11

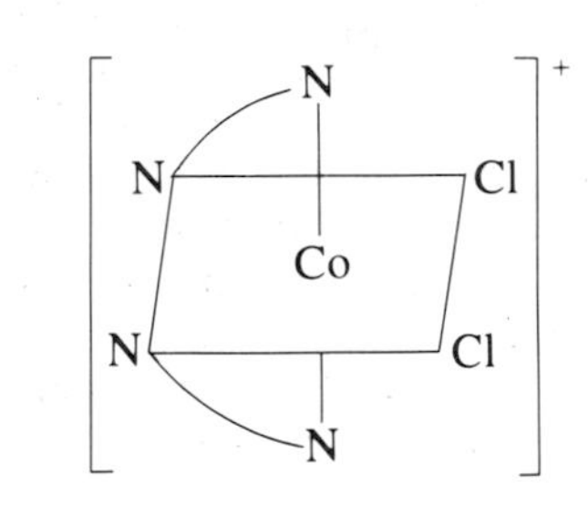

12

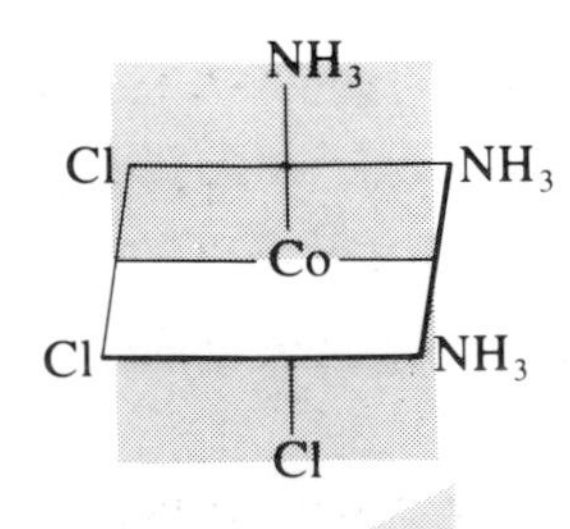

13

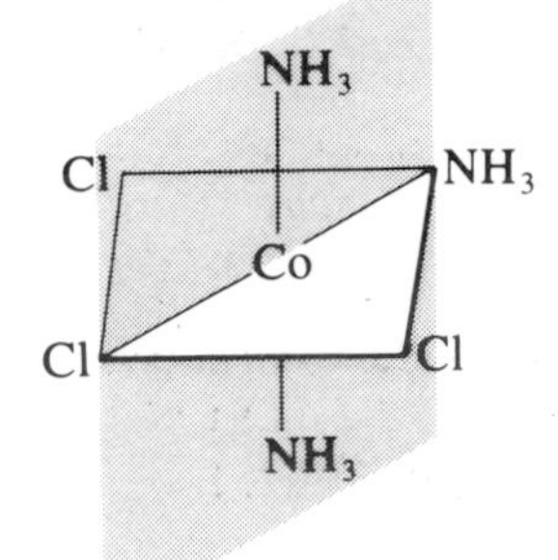

14

EXAMPLE

How many isomers are there of the coordination compound $[Co(NH_3)_3Cl_3]$? Are there any optical isomers?

METHOD Arrange the six ligands at the corners of an octahedron. Identify the different arrangements (that is, arrangements that cannot be interconverted simply by rotating the whole molecule). Judge whether there are any optical isomers by seeing whether any of the isomers lack a plane or centre of symmetry.

ANSWER There are only two isomers, **13** and **14**. Both have planes of symmetry (shaded in the diagrams), and so there are no optical isomers.

COMMENT In modern chemistry the isomers can be distinguished by spectroscopy and X-ray diffraction, although at one time classical chemical techniques had to be used. The Swiss chemist Alfred Werner was the first to propose and confirm the octahedral structure of cobalt(III) complexes, particularly by studying isomerism. He was awarded the Nobel prize for his work (in 1913).

25.8 Low oxidation numbers

In 1889 Ludwig Mond noted that some nickel components in apparatus he used for handling carbon monoxide were attacked. Further investigation led to the discovery of **nickel carbonyl**, $Ni(CO)_4$, a substance that is now used to obtain pure nickel from its ores.

Nickel reacts with carbon monoxide at atmospheric pressure and about 50 °C:

$$Ni(s) + 4CO(g) \xrightarrow{50\,^{\circ}C} Ni(CO)_4(g)$$

Carbon monoxide is a neutral ligand, the molecule has no overall charge, $Ox(Ni) = 0$, and so the formal name of the product is tetracarbonylnickel(0). The compound is a colourless liquid which is both highly toxic and carcinogenic (cancer-producing); it freezes at −25 °C and boils at 43 °C. It can be purified by distillation, but above 200 °C it decomposes to nickel metal and carbon monoxide. This is the cycle used in the **Mond process** for nickel extraction: the carbonyl is prepared, purified and then decomposed.

The existence of nickel carbonyl points to a pattern among d-block elements which is similar to the octet rule of the other blocks: that a common electron configuration is *eighteen* electrons outside the noble gas core. In other words, in many compounds of the first-row elements of the d-block, all nine orbitals of the 4s-, 4p- and 3d-subshells are occupied – the so-called **eighteen-electron rule.**

The eighteen-electron rule has been found particularly appropriate when the oxidation number of the metal is low, such as the zero shown by nickel in nickel carbonyl. In this compound nickel, which has configuration $[Ar]3d^84s^2$, provides ten electrons and each of the four carbonyl ligands donates a lone pair, leading to eighteen electrons in all. The rule suggests that since iron, with configuration $[Ar]3d^64s^2$, has eight electrons outside its argon core, it can accept ten electrons from a neutral Lewis base, and should form a pentacarbonyl, $Fe(CO)_5$; in fact, this compound is well known. Chromium, $[Ar]3d^54s^1$, needs twelve electrons, so that the predicted compound is $Cr(CO)_6$, which has also been prepared.

EXAMPLE

Suggest a reason why manganese carbonyl, $Mn_2(CO)_{10}$, is a dimer, but its chromium and iron analogues, $Cr(CO)_6$ and $Fe(CO)_5$, are not.

METHOD Apply the eighteen-electron rule to the compound: if the 'monomer' does not fulfil the rule, consider whether the dimer does.

ANSWER In $Mn(CO)_5$, manganese ($3d^54s^2$) supplies seven valence electrons, and each ligand supplies two (as a lone pair). Hence the total number of valence electrons around the Mn atom is $7 + (5 \times 2) = 17$, which is less than eighteen. The two $Mn(CO)_5$ units can share an electron, forming a Mn—Mn bond, and each atom can acquire a share in eighteen electrons (**15**).

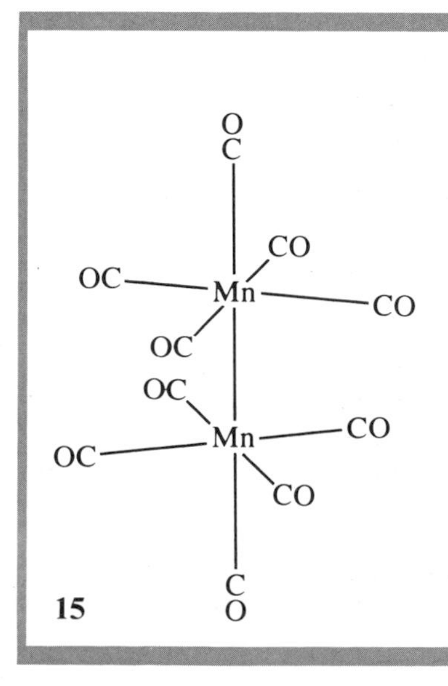

15

COMMENT Another way for a seventeen-electron molecule to reach eighteen electrons is by reduction. Manganese carbonyl is reduced by sodium metal, and forms the eighteen-electron ion $[Mn(CO)_5]^-$:

$$Mn_2(CO)_{10} + 2e^- \rightarrow 2[Mn(CO)_5]^-$$

Carbonyls and related low-oxidation-state compounds are important catalysts. For example, $Co_2(CO)_8$ is used as the catalyst in the **Fischer–Tropsch synthesis** of methanol from synthesis gas (a mixture of hydrogen and carbon monoxide – see Section 16.4), a reaction that might prove to be of overwhelming importance in this world of diminishing hydrocarbon reserves.

25.9 Industrial chemistry

Occurrence

The first six elements of the first-row d-block elements (scandium to iron) occur mainly as oxides, including **rutile** (TiO_2), **ilmenite** ($FeTiO_3$), **chromite** ($FeCr_2O_4$) and **pyrolusite** (MnO_2). Iron is the most abundant d-block element; it is the fourth most abundant element in the Earth's crust, where it is found as **haematite** (Fe_2O_3, see Colour 5) and **magnetite** (Fe_3O_4). The cosmic abundance of iron is also high, because the iron nucleus is very stable. For example, siderite meteorites consist largely of iron, the Earth's core is largely molten iron, and when astronauts first landed on the Moon it was no surprise that they found metallic iron on its surface. Iron oxides occur on Earth on account of the oxygen in its atmosphere, and it is interesting that the deposits of iron oxide ores date from the geological epoch when the early forms of life began to excrete dioxygen, and the primeval reducing atmosphere was converted to one that could oxidize: the surface of the Earth rusted as photosynthesis began. Titanium is the ninth most abundant element in the Earth's crust, being significantly more abundant than the familiar metals chromium, copper and zinc.

The later members of the d-block occur mainly as sulphides, such as **zinc blende** (ZnS, see Colour 5) and **copper pyrites** ($CuFeS_2$). The elements of the second and third rows of the d-block are much rarer. Those close to platinum (particularly iridium, gold and palladium) are so resistant to oxidation (so 'noble') that they are often found native (see Colour 5).

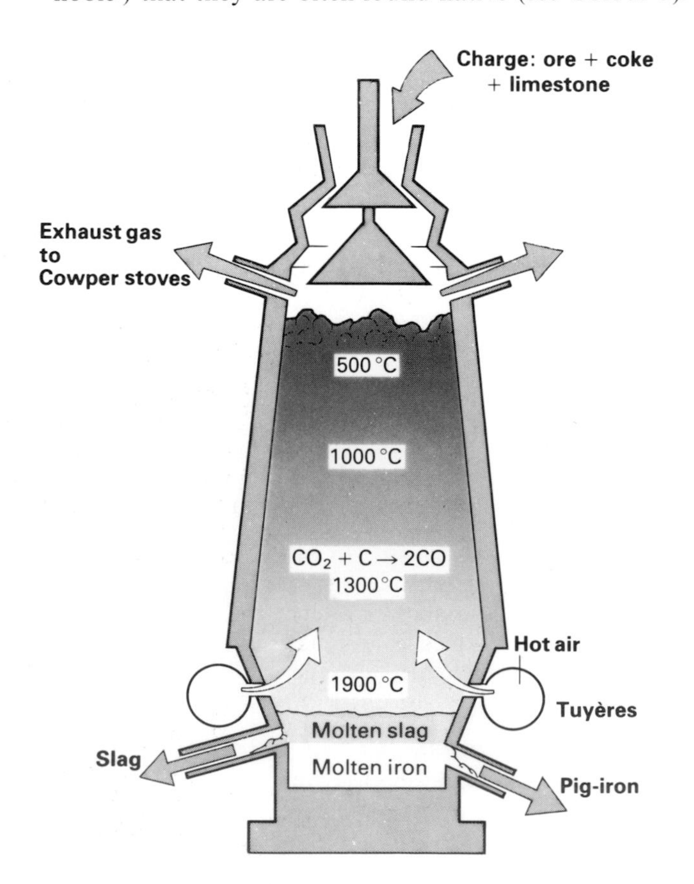

Figure 25.10 The blast furnace can produce molten iron continuously

Figure 25.11 The Queen Victoria blast furnace at Scunthorpe, UK

Manufacture of iron and steel

About 700 megatonnes of steel are produced worldwide each year, the largest amount of any metal. Steel is produced from iron ore in two main stages: first, impure iron is produced from its ore in a **blast furnace**, and then the iron is purified by oxidizing its impurities, and alloyed with other metals.

Figure 25.10 is a diagram of a blast furnace. Iron ore, coke and limestone are fed in through the top, and a blast of hot air at about 900 °C is fed in from the bottom. The temperature is much higher at the bottom of the furnace, where it reaches nearly 2000 °C as the coke burns, than at the top, and the reactions taking place depend on the temperature and therefore the location in the furnace. The *overall* reactions are reasonably simple, however. They can be summarized as follows:

1 Carbon monoxide is produced:

$$C(s) + O_2(g) \rightarrow CO_2(g) \qquad O(0) \rightarrow O(-2),\quad C(0) \rightarrow C(+4)$$

$$CaCO_3(s) \rightarrow CaO(s) + CO_2(g)$$

$$CO_2(g) + C(s) \rightarrow 2CO(g) \qquad C(+4) + C(0) \rightarrow 2C(+2)$$

2 Iron oxides are reduced by the carbon monoxide:

$$Fe_2O_3(s) + 3CO(g) \rightarrow 2Fe(l) + 3CO_2(g) \qquad Fe(+3) \rightarrow Fe(0),\quad C(+2) \rightarrow C(+4)$$

Under the conditions in the furnace, the reducing agent is carbon monoxide, not the coke directly.

Molten iron is run from the bottom of the furnace. Impurities in the ore, largely silica, are removed by reaction with the quicklime produced by the decomposition of the limestone:

$$CaO(s) + SiO_2(s) \rightarrow CaSiO_3(l)$$

At the temperature of the furnace, the calcium silicate is molten: it forms a **slag** that floats on the surface of the molten iron and can be run off and used for making building blocks.

The impure **pig-iron** produced by the furnace contains about 4 per cent of carbon, and smaller amounts of silicon, sulphur and phosphorus. Although it is hard, pig-iron is very brittle, and of little direct use. The non-metallic impurities are removed by oxidation, often in the **basic oxygen process**, shown diagrammatically in Figure 25.12, and also in Colour 9. In this process a

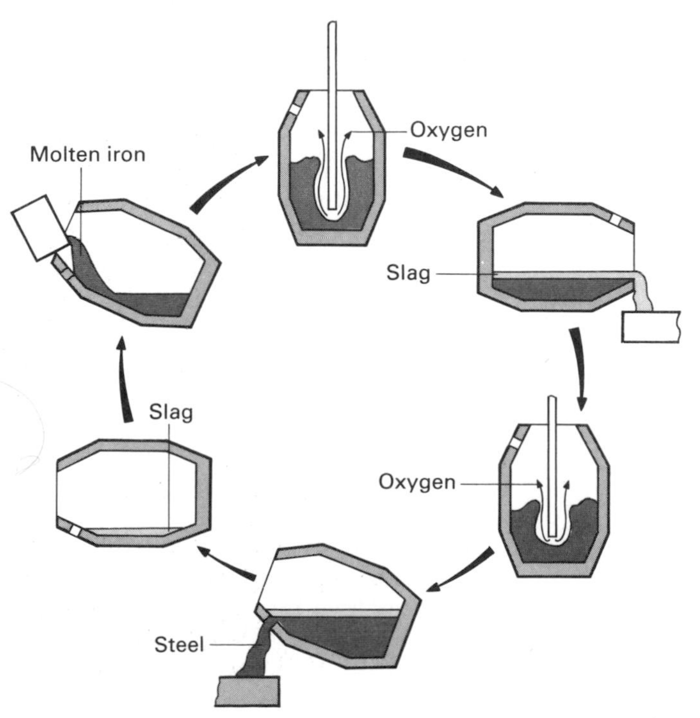

Figure 25.12 The basic oxygen process for manufacturing steel

jet of pure oxygen is blown into a molten mixture of pig-iron and scrap steel through water-cooled 'lances'. The liquid iron is held in a container lined with blocks of calcium oxide or magnesium oxide, and the impurities either combine with the basic oxides to form a slag or are removed as volatile oxides. Conditions are arranged so that between 0.2 and 1.7 per cent of carbon remains in the iron. The product is a crude form of steel: the wide range of modern steels is made by alloying other metals during the basic oxygen process.

Summary

- ☐ The first-row d-block elements have the electron configurations $[Ar]3d^x4s^2$ where $x = 1$ to 10, except for chromium and copper which are $[Ar]3d^5 4s^1$ and $[Ar]3d^{10}4s^1$ respectively.
- ☐ Except for scandium and zinc, the elements exhibit variable oxidation numbers. The highest oxidation numbers for titanium, vanadium, chromium and manganese are +4, +5, +6 and +7 respectively.
- ☐ Scandium has oxidation number +3 in all its compounds, and zinc has oxidation number +2 in all its compounds.
- ☐ For the elements after manganese, the main oxidation numbers are +2 and +3.
- ☐ All the d-block elements are metallic.
- ☐ The standard electrode potential for the couple Cu^{2+}, Cu is positive, but the $E^\ominus$ values for the corresponding couples are negative for all the other first-row d-block elements.
- ☐ The general characteristics of the elements from titanium to copper are their variable oxidation numbers, and their formation of complex ions and coloured ions in solution. Some compounds are paramagnetic. The elements and their compounds are useful catalysts.
- ☐ The shapes of molecules and complex ions of the first-row d-block elements are usually octahedral if the coordination number is 6, or tetrahedral if the coordination number is 4.
- ☐ Steels are alloys of iron and other metals together with 0.2–1.7 per cent of carbon.
- ☐ The rusting of iron, and its prevention, are processes of major economic importance. Rust has the approximate formula FeO(OH) and it is formed by an electrolytic process requiring water, oxygen and an electrolyte.
- ☐ There are five groups of oxides: the highest oxides (V_2O_5, CrO_3, Mn_2O_7), the +4 oxides, the +3 oxides, the mixed oxides (M_3O_4) and the +2 oxides.
- ☐ The +3 oxides are basic or amphoteric; the +2 oxides are basic, except ZnO which is amphoteric.
- ☐ The MnO_4^- and $Cr_2O_7^{2-}$ ions are powerful oxidizing agents.
- ☐ The +4 chlorides e.g. $TiCl_4$ are covalent, but the +3 and +2 chlorides are ionic.
- ☐ The Cu^{2+}(aq) ion oxidizes I^-(aq) to iodine, being reduced to insoluble copper(I) iodide.
- ☐ M^{2+}(aq) ions (M = Ti, V or Cr) are reducing agents, whereas M^{3+}(aq) ions (M = Mn or Co) are oxidizing agents.
- ☐ $[Fe(H_2O)_6]^{3+}$ is acidic.
- ☐ Complexes formed with bidentate ligands are much more thermodynamically stable than the corresponding complexes with unidentate ligands.
- ☐ There are three ionization isomers of $CrCl_3 . 6H_2O$, *cis–trans* isomers of $[Co(NH_3)_4Cl_2]^+$ and optical isomers of *cis*-$[Coen_2Cl_2]^+$.
- ☐ There is an eighteen-electron rule for the later d-block elements similar to the octet rule for Period 2 elements.
- ☐ Steel is prepared by a two-stage process: reduction of iron ore, such as Fe_2O_3, with carbon monoxide in a blast furnace followed by oxidation of the non-metals by the basic oxygen process.

PROBLEMS

1 Construct a table with the columns labelled $d^0, d^1 \ldots d^{10}$ and the rows labelled with the symbols of the d-block elements Sc, Ti...Zn. In each cell of the table write the formula of an important compound or ion of the element with the corresponding oxidation number, for example, in the d^0/Mn cell you might write MnO_4^-, and in the d^{10}/Zn cell you might write $Zn^{2+}(aq)$.

2 Make a list of all the reactions you have studied which are catalysed by a d-block element itself, or by one of its compounds.

3 This problem concerns the reduction of vanadium(V) to vanadium(II). Write equations for the three half-reactions for the reduction by one-electron steps. Write the colours of the ions under the equation. Look up the standard electrode potentials for each half-reaction. From tables of standard electrode potentials select reducing agents which are suitable for each step.

4 Plot graphs of ionic radius and first ionization energy against atomic number for the elements from scandium to zinc. Account for the trends.

5 For each d-block element, and its compounds, make a list of uses.

6 Write balanced equations to show how the following conversions could be made:
(a) $(NH_4)_2Cr_2O_7 \rightarrow CrCl_2$
(b) $MnCl_2 \rightarrow KMnO_4$
(c) $Fe_2O_3 \rightarrow FeCl_3$
(d) $NiCl_2 \rightarrow K_2[Ni(CN)_4]$
(e) $CuSO_4.5H_2O \rightarrow CuI$
(f) $ZnCl_2 \rightarrow K_2[Zn(OH)_4]$

7 The overall stability constant, β_n, of a complex formed between a metal ion $M^{x+}(aq)$ and a ligand L^y is given by the expression:

$$\beta_n = \frac{[ML_n^{(x-ny)+}]}{[M^{x+}][L^{y-}]^n}$$

for the reaction

$$[M(H_2O)_n]^{x+} + nL^{y-} \rightleftharpoons [ML_n]^{(x-ny)+} + nH_2O$$

For $[Ni(NH_3)_6]^{2+}$, $\lg \beta_6 = 8.6$, and for $[Nien_3]^{2+}$, $\lg \beta_3 = 18.3$, where en = ethane-1,2-diamine. Why is the value for the en complex so much greater?

8 State the oxidation number of the chromium atom in the following compounds or ions: (a) K_2CrO_4, (b) $K_2Cr_2O_7$, (c) CrO_2Cl_2, (d) $[CrCl_2(H_2O)_4]^{2+}$, (e) $[Cr(NH_3)_6]^{3+}$, (f) $Cr(CO)_6$, (g) $[Cr(CN)_6]^{3-}$.

9 The standard electrode potentials for the process $M^{2+}(aq) + 2e^- \rightarrow M(s)$ are +0.34 V and −0.76 V for M = Cu and Zn respectively. These elements are neighbours in the Periodic Table. How do you account for such a large difference?

10 Construct a table with column headings as follows:

Compound or ion	Oxidation number of metal	Number of d-electrons on metal	Colour	Paramagnetic (yes/no)

Insert the following formulae in the first column and then complete the table: $[Ti(H_2O)_6]^{3+}$, $Cr_2O_7^{2-}$, MnO_4^-, Fe_3O_4, $KFe_2(CN)_6$, $[Cu(NH_3)_4(H_2O)_2]^{2+}$, $[Zn(H_2O)_6]^{2+}$.

Which of the colours are likely to be due to d–d transitions? How can you explain the colours of the other compounds?

11 Make a list of the known chlorides of the d-block elements. Look up their melting points and boiling points. What pattern emerges? How do you account for it?

12 Write an account of the chemistry of zinc. In what ways do zinc and its compounds differ from calcium and its compounds? How do you account for these differences?

13 Compare and contrast the chemistry of iron with that of aluminium. Give reasons for similarities and differences.

14 (a) State five typical properties of the transition (d-block) elements, illustrating them with suitable specific examples.
(b) Compare the reactions of $Ca^{2+}(aq)$, $Fe^{2+}(aq)$ and $Zn^{2+}(aq)$ with (i) aqueous sodium hydroxide, (ii) aqueous ammonia. Write equations to illustrate your answers.
(c) Outline how $Cr^{3+}(aq)$ ions may be converted to $CrO_4^{2-}(aq)$ ions, and how $Cr_2O_7^{2-}(aq)$ ions may be converted to $Cr^{3+}(aq)$ ions. *(AEB 1986)*

15 Explain the following facts and in each case give *one* appropriate example.
(a) Transition metals are usually hard solids with high melting points.
(b) Transition metals commonly show a number of relatively stable oxidation states.
(c) Transition metals form a large number of coordination compounds.
(d) The aqua-cations of transition metals are often acidic in solution. *(Oxford and Cambridge)*

16 Write an account of the properties of the first-row transition metals by considering with examples wherever possible
(a) the physical properties which are characteristic of transition metals,
(b) the common oxidation states in aqueous solution for each element,
(c) the formation of, and bonding in, complexes,
(d) the reactions of complexes, restricted to substitution reactions of hydrated ions. *(JMB)*

17 Briefly explain why the elements in the first transition series have similar physical and chemical properties.

Comment on the following values of the standard redox potentials for the elements below.

Electrode reaction	$E^{\ominus}$/V
$Cr^{3+}(aq) + e^- \rightleftharpoons Cr^{2+}(aq)$	−0.41
$Mn^{3+}(aq) + e^- \rightleftharpoons Mn^{2+}(aq)$	+1.49
$Fe^{3+}(aq) + e^- \rightleftharpoons Fe^{2+}(aq)$	+0.77

(*Cambridge part question*)

18 (a) Describe the extraction of titanium from its ore rutile (TiO_2).
(b) Give *two* large-scale uses of titanium. Why is titanium preferred to iron for these uses?
(c) Write the full electron configuration of a titanium atom.
(d) On reduction of a colourless aqueous solution containing titanium(IV), a violet solution containing the $[Ti(H_2O)_6]^{3+}$ ion is obtained. The violet solution must be stored in a sealed vessel or handled in an inert atmosphere.
(i) Suggest a possible shape for this ion.
(ii) Account for the observations that the aqueous solution containing titanium(IV) is colourless, while that containing titanium(III) is violet.
(iii) Suggest a reason for the precautions which must be taken in the storage and handling of the aqueous solution containing titanium(III).
[Standard redox potential, Ti(IV)/Ti(III) = +0.10 V]
(*Cambridge*)

19 (a) Describe how, using a simple chemical test in each case, you could distinguish between $FeCl_3(aq)$ and $K_2Cr_2O_7(aq)$. Write chemical equations and give the formulae of the reaction products where appropriate.
(b) A mixture of zinc and iron was dissolved in an excess of dilute sulphuric acid. 0.0672 litre of hydrogen, measured dry at s.t.p., was evolved. When the reaction was complete, the resulting solution required 20.0 cm^3 of 0.02M potassium manganate(VII) for complete oxidation. Determine the mass of zinc in the original mixture.
(*JMB part question*)

20 Give an account of complex ions, paying particular attention to bonding, stereochemistry, stability, stoichiometry and different ligand types. (*London*)

21 Describe carefully *one* of the modern conversion processes for the manufacture of a steel from pig-iron, referring to the physico-chemical aspects of the process. (No account of the blast furnace process for the production of pig-iron is required.)

Refer to *one* particular steel, commenting on its composition and properties in relation to the purpose for which it is made.

Discuss, in electrolytic terms, the corrosion of iron. Describe the prevention of corrosion based on sacrificial processes. (*London part question*)

22 If cobalt(II) carbonate is treated at 90 °C with pentane-2,4-dione, $CH_3COCH_2COCH_3$, and hydrogen peroxide, a vigorous reaction occurs. Carbon dioxide is evolved because the pentane-2,4-dione acts as a weak acid.

When the reaction mixture is cooled, green crystals are obtained, melting point 213 °C. The crystals are insoluble in water but soluble in benzene.

Some crystals prepared as above were analysed as follows:
(a) The percentage of cobalt was determined and was found to be 16.25 per cent by mass.
(b) On ignition, 1.53 g of the green crystals produced 2.81 g of carbon dioxide and 0.81 g of water.
(c) 0.88 g of the green crystals were dissolved in dilute sulphuric acid and treated with excess potassium iodide solution, which reduced the cobalt back to cobalt(II). Iodine was liberated and was found by titration to be equivalent to 24.4 cm^3 of 0.1M sodium thiosulphate solution.

From the data given, calculate
(i) the formula of the green compound,
(ii) the oxidation number of cobalt in the green compound.

Draw a diagram to show a possible structure for the green compound. (*Nuffield*)

23 (a) Explain why the lattice energy of zinc sulphide is different from that based on a purely ionic model.
(b) Give the reactions for the production of zinc from its sulphide and mention *briefly* the reaction conditions. If the enthalpy changes of formation (kJ mol^{-1}) for ZnO(s) and CO(g) are −348.1 and −110.5 respectively, deduce the enthalpy change for (i) the reduction of zinc oxide with carbon and (ii) for the decomposition of zinc oxide to zinc and oxygen. Give the effect of temperature on the reactions and comment on why reaction (i) and not (ii) is used in the extraction of zinc.
(c) Discuss similarities and differences in the chemistry of zinc and copper. Include in your answer the respective electronic configurations, the valencies in their compounds, the characteristics of their ions in solution including reactions of the ions with aqueous ammonia, potassium iodide, hydrochloric acid and potassium cyanide.
(*Welsh*)

24 A complex ethanedioate salt can be prepared by the following procedure, which has to be carried out in a darkened room. An aqueous solution of ethanedioic acid is heated at 75 °C and powdered potassium manganate(VII) added in small portions. When the solution becomes colourless, potassium carbonate is added. The mixture is diluted and cooled to 2 °C before a further portion of potassium manganate(VII) is added. Then the solution is diluted with ethanol and further chilled, when deep reddish purple crystals of the complex form.

The analysis of pure crystals of the complex gave the following results:
(a) 1.0 g of the complex ignited with phosphoric(V) acid left a residue of 0.29 g of manganese(II) diphosphate(V) ($Mn_2P_2O_7$).
(b) 0.10 g of the complex, treated with excess potassium iodide, liberated iodine equivalent to 20.4 cm^3 of 0.10M sodium thiosulphate solution.
(c) 0.10 g of the complex in acidic solution reduced 10.2 cm^3 of 0.02 M potassium manganate(VII) solution.
(d) 1.0 g of the complex ignited wtih sulphuric acid left a mixed residue of 0.84 g of manganese(II) sulphate and potassium sulphate.
(e) When 1.0 g of the complex was heated, 0.11 g of water were collected.

Calculate an empirical formula for the complex salt and write a balanced equation for its formation from ethanedioic acid, potassium manganate(VII) and potassium carbonate. (*Nuffield S*)

25 Discuss the isomerism, if any, exhibited by each of the following compounds:
(a) $Pt(NH_3)_2Cl_2$,
(b) $Co(NH_3)_5ClBr_2$,
(c) $Cr(NH_3)_4Cl_3$,
(d) $Co(en)_3Cl_3$ (en is the bidentate ligand $H_2NCH_2CH_2NH_2$). *(London S)*

26 (a) Chromium occurs as the ore chromite, $FeCr_2O_4$, a mixed oxide of iron(II) and chromium(III). When this is strongly heated with sodium carbonate and excess air, carbon dioxide is evolved and a mixture of sodium chromate(VI) and iron(III) oxide is formed. Write a balanced equation for this reaction. How could this mixture be used to prepare (i) potassium dichromate(VI), (ii) chromium(VI) oxide and (iii) ammonium dichromate(VI)?
(b) What is the structural formula of (i) the chromate(VI) ion and (ii) the dichromate(VI) ion, and how do they differ in colour? Explain what is meant by the oxidation state of an atom and why the oxidation state of chromium is the same in both these ions although the proportion of oxygen is greater in the former.
(c) When potassium dichromate(VI) is heated with sulphur and the residue is subsequently extracted with water a green solid is formed. Treatment of the aqueous extract with aqueous barium chloride gives a white precipitate which is insoluble in dilute hydrochloric acid. The same green solid is also formed by heating ammonium dichromate(VI), but in the absence of sulphur. Discuss the oxidation and reduction processes which occur in these two thermal reactions.
(d) Aqueous potassium dichromate(VI), acidified with sulphuric acid, on treatment with sulphur dioxide at room temperature gives a solution from which violet crystals can be obtained which are isomorphous with those of aluminium potassium sulphate-12-water. Explain how the isomorphism establishes the nature of the violet crystals and give a balanced equation for the reaction leading to their formation.
(e) How is a thin layer of chromium metal normally obtained on a piece of steel, and why is such 'chromium plating' of importance in everyday life? *(Oxford S)*

27 Give concise explanations for the following.
(a) Manganese can exist in oxidation states up to a maximum of +7, whereas scandium has no higher oxidation state than +3.
(b) In aqueous solution, the scandium(III) ion is colourless whereas the iron(II) ion is coloured.
(c) The chromium(II) ion is a better reducing agent than the manganese(II) ion.
(d) The electron configuration of chromium is not $[Ar]3d^44s^2$ as might be expected, but $[Ar]3d^54s^1$.
(e) The compound $Pt(NH_3)_2Cl_2$ and the ion $[Cr(NH_3)_4Cl_2]^+$ each exist in two isomeric forms. *(London)*

28 A solution was prepared by dissolving 8.74 g of hydrated copper(II) sulphate ($CuSO_4.5H_2O$) to give 250 cm^3 of aqueous solution. 25 cm^3 of this solution was treated with an excess of potassium iodide solution. A colourless solid was separated by filtration and washing. The red–brown filtrate and washings were combined and the iodine they contained was titrated with a solution containing 29.76 g sodium thiosulphate ($Na_2S_2O_3.5H_2O$) per dm^3. 29.20 cm^3 of the thiosulphate solution were needed to discharge completely the blue colour, produced by the addition of starch after most of the iodine had already reacted.

The colourless precipitate dissolved in aqueous ammonia in the absence of air and gave a colourless solution. This solution avidly absorbed oxygen giving a deep blue solution. It also absorbs carbon monoxide.

Discuss the reactions taking place in this sequence of operations. *(Cambridge Entrance)*

29 Give an account of the complex ions formed by any *one* transition element with different ligands in aqueous solution, including reference to their stoichiometry and redox properties.

Titration of 25 cm^3 of a hot dilute sulphuric acid solution of a potassium salt of a complex anion, containing iron and $C_2O_4^{2-}$ ligands, required 34.8 cm^3 of a manganate(VII) solution for complete oxidation. The iron in this solution was then reduced to iron(II) by shaking with zinc amalgam. A further titration of the separated aqueous phase then required 5.8 cm^3 of the same manganate(VII) solution. Suggest one, or more, possible compositions for the complex salt. *(Cambridge Entrance)*

30 In what ways are the properties of transition metal ions in aqueous solutions dependent on the nature of the ligands surrounding them?

Suggest explanations of the following observations:
(a) The compound of formula $Cr(H_2O)_6Cl_3$ exists in more than one isomeric form.
(b) Cu_2SO_4 deposits copper metal when treated with water.
(c) $Fe_2(SO_4)_3$ and KI react in aqueous solution to give iodine but this reaction is suppressed when a large excess of fluoride ion is added. *(Cambridge Entrance)*

The figure on the opposite page shows a computer graphic of a molecule of haemoglobin from a human red blood cell.

PART THREE

ORGANIC

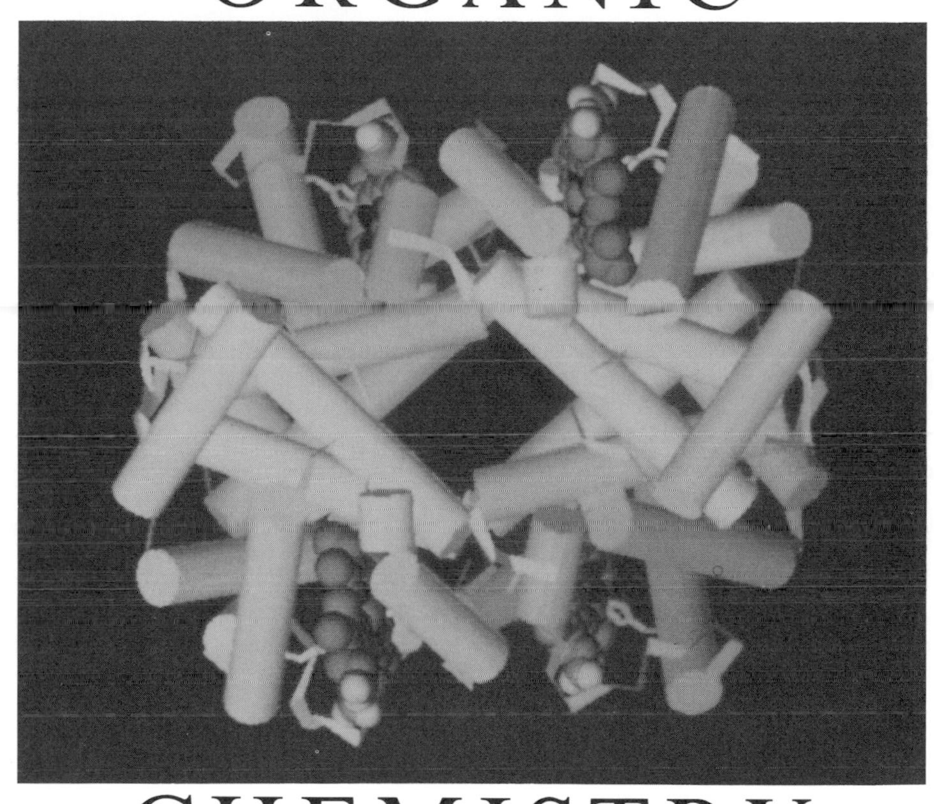

CHEMISTRY

26

Introduction to organic chemistry

In this chapter we look at the special role of carbon as the key element in a vast number of compounds. We consider the ways in which organic compounds can react. We discuss different ways of representing complicated molecular structures on paper. We see how molecules containing the same set of atoms may differ in the ways those atoms are bound together. Finally we consider the techniques available for investigating the structures of organic molecules.

Figure 26.1 Nearly everything in this picture is organic – living things, goods made from animal or vegetable materials, and those made of plastics or synthetic fibres. The only major inorganic constituent is water

Try the following thought-experiment. Look at the things around you in the room and out of doors, and imagine turning the flame of a blow-lamp on them. How many would burn or char? This book would, for one. So would most of the furniture in the room, the flooring and the fabrics, the shoes on your feet (and your feet with them), wood and vegetation in general, the whole range of plastics materials from a ball-point pen to the casing of a television set, and of course the fuel in the blow-lamp. The list is almost endless. The only materials to resist the onslaught would be glass (though it might crack or melt, of course), brick, stone and ceramics, metals and water. The walls might still be there, but the paint would be gone.

The distinguishing characteristic of all the materials that burned or charred is that they consist largely of compounds of carbon. Indeed, the word 'char' means 'turn to charcoal', and charcoal is mainly carbon. The blow-lamp test is not an infallible way of identifying carbon compounds. The metallic headshell of a record player might burn spectacularly: it could be made of magnesium. And you might try to put out the fires using fire extinguishers filled with the carbon compounds CO_2 or CCl_4. But the thought-experiment does illustrate the unique role of one element, carbon, as a component of an extraordinarily wide range of materials, both natural and synthetic. There are around five million known compounds of carbon (nearly all of which also contain hydrogen), far more than the compounds of all the other elements put together.

BOX 26.1

Organic chemistry

Attempts to make a systematic study of chemicals derived from vegetable (and less often animal) sources began towards the end of the eighteenth century, notably by the Swedish chemist Carl Scheele, who isolated such compounds as citric and tartaric acids from fruits, lactic acid from milk, glycerol from olive oil, and uric acid from urine (as well as discovering chlorine and oxygen). The great French scientist and politician Antoine Lavoisier showed that all these naturally occurring compounds contain carbon and hydrogen. Most also contain oxygen and many contain nitrogen.

The name 'organic' was originally used to distinguish compounds obtained from living material ('organisms') from those derived from minerals, but gradually, as synthetic chemicals with similar properties were prepared, the term was extended to cover all carbon compounds except the oxides of carbon and metal carbonates.

26.1 Why carbon?

There are four main reasons for the special properties of carbon.

1 Strong bonds can be formed between two carbon atoms Table 9.4 lists some mean bond enthalpies: carbon–carbon (and carbon–hydrogen) bonds are seen to be strong. This allows molecules to be constructed with 'skeletons' of carbon atoms – sometimes just a few, sometimes hundreds or even thousands – linked together in chains, rings or networks. A nitrogen–nitrogen single bond is much weaker, largely because of the electrostatic repulsion between the lone pairs of electrons on the two atoms (**1**, compare also Section 2.3). A silicon–silicon bond is also relatively weak: silicon atoms are larger than carbon atoms and the bonding electrons in silicon, being further from the nuclei, are less strongly held by them than in carbon atoms.

1

2 Carbon is tetracovalent The hydrogen–hydrogen bond is even stronger than the carbon–carbon bond, but that single bond completes the valence shell of both hydrogen atoms: a chain cannot build up. Bonds between atoms of sulphur are moderately strong, and chains and rings of sulphur atoms do exist: at room temperature and pressure elemental sulphur is largely S_8 (see Section 22.2). However, with only two bonds to each sulphur atom there is nowhere for other atoms or groups to be attached to the sulphur chain. Carbon chains, on the other hand, can branch out into side chains, and other groups can be attached to the carbon atoms to give the observed variety of compounds.

3 Carbon forms strong covalent bonds with other elements Carbon, with its small atom, forms strong bonds to nitrogen, oxygen, the lower halogens and many other elements, enabling a large number of different structures to be built up through bonds between a carbon skeleton and various other groups.

4 Carbon compounds are kinetically stable under normal conditions Science fiction writers sometimes introduce us to life forms based on silicon chemistry rather than carbon. Such an alien biochemistry is not utterly beyond the bounds of possibility even though, as we have seen, silicon–silicon bonds are weaker than carbon–carbon so that silicon compounds are less stable than their carbon analogues. But silicon life forms could not be oxygen-breathers. Disilane, Si_2H_6, and similar compounds with silicon–silicon and silicon–hydrogen bonds spontaneously ignite in air, so that a silicon-based life form would be destroyed before it could build up any complexity. The carbon analogue of disilane is ethane, C_2H_6. It is a minor constituent of natural gas; not surprisingly, therefore, it too burns in air, but only if ignited by a spark or a flame. Although ethane and most other carbon-based compounds are thermodynamically less stable than their combustion products, they are *kinetically* stable: they react only very slowly, if at all, with oxygen at room temperature. The most important reason for the difference is that, by using its 3d-orbitals, silicon can readily accept more than eight electrons into its outer shell: carbon cannot. When silicon compounds react they can do so by forming a new bond from the reagent to the silicon atom. In the reactions of carbon compounds one bond to a carbon atom must be broken to allow another to form: the endothermic bond breaking contributes to the relatively large activation energy (see Section 14.4).

2

3

26.2 The reactivity of organic molecules

The simplest stable organic molecule is methane, CH_4. It contains only carbon–hydrogen bonds, arranged tetrahedrally around the carbon atom (**2**). The giant molecule of diamond (**3**) contains only carbon–carbon bonds, also arranged tetrahedrally, each carbon atom being bonded to four others.

In between these two extremes lie all the so-called **saturated hydrocarbons**. The term *hydrocarbon* indicates that the molecules contain only hydrogen and carbon, and *saturated* means that the molecule contains no double or triple bonds. Each carbon atom is attached to four other atoms, either hydrogen or another carbon. The molecules are based on a skeleton of carbon–carbon bonds each about the same length as those in diamond (154 pm). Except in very strained molecules with rings of three or four carbon atoms, the bond angles are close to the tetrahedral value (109.5°).

The electronegativities of carbon and hydrogen are similar (see Table 2.1), so that carbon–carbon and carbon–hydrogen bonds are of low polarity; neither end of the bond exerts a dominant attraction on the bonding electrons. Consequently most of the reactions of saturated hydrocarbons involve **homolysis** or **homolytic fission** of the bonds ('breaking into similar parts' from the Greek: *homos*, same; *lysis*, loosening). The bonding electrons separate, one going to each end of the breaking bond:

$$-\overset{|}{\underset{|}{C}}:\overset{|}{\underset{|}{C}}- \longrightarrow -\dot{C}\!\lt + \gt\!\dot{C}-$$

Such reactions, which are discussed in more detail in Section 27.3, lead to the formation of **radicals**, fragments with unpaired electrons (the obsolete term *free radical* is still encountered). The reactions are usually initiated by other radicals formed by the homolysis of bonds weaker than the carbon–carbon or the carbon–hydrogen bond. For example, the weak bond in molecular fluorine (see Table 9.4) is easily broken by heat or light to give fluorine atoms that can set off an explosive reaction with saturated hydrocarbons.

When other elements are attached to a hydrocarbon skeleton the pattern of reactivity is changed. Bonds between carbon and more electronegative elements such as oxygen or chlorine are polarized in the sense $C^{\delta+}—Z^{\delta-}$. The bonding electrons are attracted towards the electronegative end of the bond and reactions tend to occur by **heterolysis** or **heterolytic fission** (Greek: *heteros*, different), with the electrons remaining paired. Such reactions involve ions in the reactants or the products or in both. It is often convenient to follow the movement of electron pairs through a reaction. This is done by using 'curly arrows' (↷): the convention is that an arrow represents the transfer of an *electron pair* (not a single electron) *from* the tail *to* the head of the arrow. The breaking of a carbon–chlorine bond can therefore be shown as:

$$-\overset{|}{\underset{|}{C}}:\ddot{\underset{..}{Cl}}: \longrightarrow -C^{+}\!\lt + :\ddot{\underset{..}{Cl}}:^{-}$$

or more simply as:

$$-\overset{|}{\underset{|}{C}}—Cl \longrightarrow -C^{+}\!\lt + Cl^{-}$$

In such compounds, therefore, one part of the molecule, the saturated hydrocarbon skeleton, is virtually inert towards many of the common reagents, notably acids and bases, that attack the other parts. Indeed, the old name for the saturated hydrocarbons was 'paraffins', meaning 'unreactives' (Latin: *parum*, little; *affinitas*, affinity). Thus to a large extent the chemistry of these compounds is confined to the reactions of the 'other parts', the so-called **functional groups** (some of which are listed in Table 26.1), and the carbon skeleton does little more than provide a framework from which the functional groups hang.

Table 26.1 Common functional groups (the symbol R stands for the hydrocarbon skeleton)

R—Cl	chlorides	**R—OH**	alcohols	R—C(=O)—H	R—C(=O)—R′	R—C(=O)—O—H
R—O—R′	ethers	**R—NH$_2$**	amines	aldehydes	ketones	carboxylic acids

On the other hand, in molecules that contain carbon–carbon double or triple bonds there can be an interaction between the functional group and the π-electrons (see Section 2.5) of the multiple bonds, and this interaction can profoundly affect the chemistry of both. In the following chapters we shall therefore need to consider the chemistry of organic compounds under three main headings: the saturated hydrocarbons, the functional groups and the multiply bonded systems.

BOX 26.2

Reaction mechanisms

It is not so long since organic chemistry was a subject to be learned by rote. Textbooks were little more than catalogues of the preparations and reactions of different classes of compounds. It cannot be denied that a good memory for facts remains a valuable attribute of an organic chemist (or any other chemist, for that matter), but we can try to clear a path through what Friedrich Wöhler, one of the first men to establish some sort of order in organic chemistry, called 'a primeval forest'.

During the past fifty years organic chemists have slowly built up an understanding of the 'how' and the 'why' of organic reactions: an understanding of their **mechanisms**. They have learned to recognize patterns of reactivity – similarities in the behaviour of a particular compound towards different reagents and of a series of compounds towards a particular reagent. Recognition led to rationalization and thence to a quantitative description of reactions in terms of their kinetics and thermodynamics. In this book we shall base the discussion of organic reactions on mechanistic principles, especially in rationalizing the patterns of making and breaking bonds. In particular, we shall see that a great deal of organic chemistry depends on just two properties of the reacting molecules: the polar character of bonds and the behaviour of lone pairs of electrons.

26.3 Representing the structures of organic molecules

4 (methane: C bonded to four H)

The methane molecule is tetrahedral. Illustrating three dimensions on a two-dimensional sheet of paper is not easy, and chemists make use of several conventional representations to try to overcome the problem. The diagram of the methane molecule in **4** is more or less self-explanatory. It is shown in perspective, with plain lines standing for bonds in the plane of the paper, a wedge shape for a bond coming up towards the reader, and a dashed line for a bond going down behind the paper. This representation is related to the sort of model that can be made with balls and sticks (Figure 26.2).

A different type of model ('space-filling') illustrates the electron clouds around each atom but cannot conveniently be shown in two dimensions except as a photograph (Figure 26.3).

Figure 26.2 (*left*) Ball-and-stick model of the methane molecule

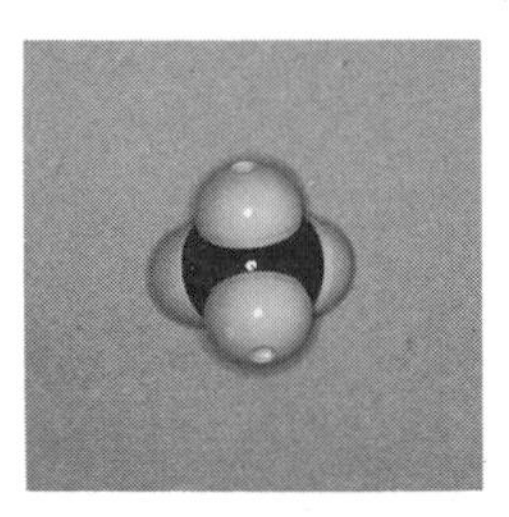

Figure 26.3 (*right*) Space-filling model of the methane molecule

5
```
    H
    |
H—C—H
    |
    H
```

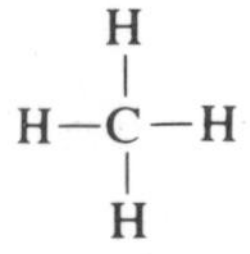

6 CH_4

7 C_4H_{10}

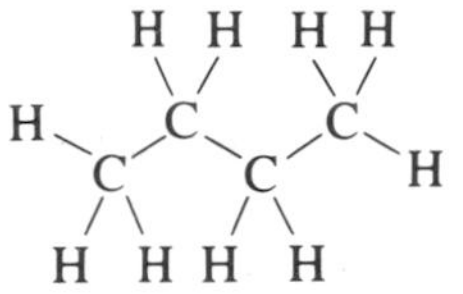

8 (structural diagram of butane)

9 (structural diagram of methylpropane)

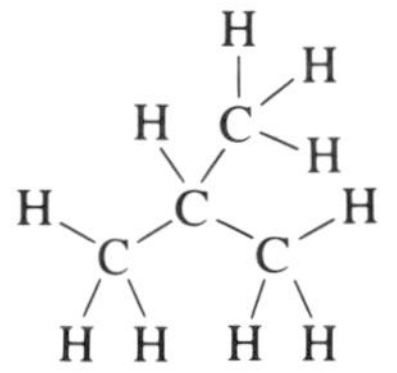

C_4H_{10}

10

11

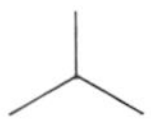

12
```
 |
—C—Cl
 |
```

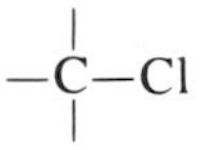

13 $C_4H_{10}O$

14 C_4H_9OH

15

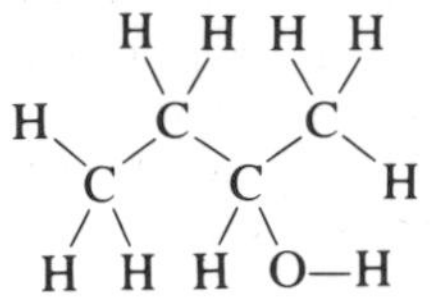

16 $CH_3—CH_2—CH(OH)—CH_3$

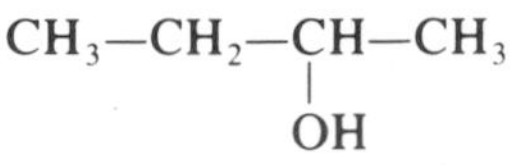

17 $CH_3CH_2CH(OH)CH_3$

If we take the tetrahedral shape of methane for granted we can use simpler representations. A common one is to show the four bonds from carbon at right angles to each other (**5**). It is important to remember that this is purely conventional. *The real molecule is not planar, and the bond angles are 109.5°, not 90°.*

These two-dimensional representations are called **structural formulae**, because they attempt to show the structure of the methane molecule. We could also use simply the **molecular formula** (**6**): given the rules of molecular geometry, this is unambiguous.

Similar representations may be used for larger molecules, such as ethane and the three-carbon molecule, propane:

C_2H_6 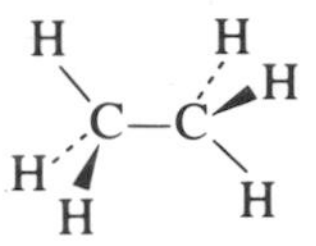 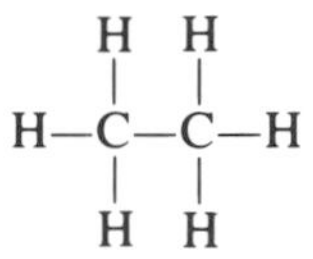 ethane

C_3H_8 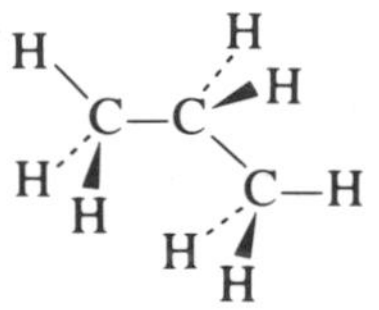 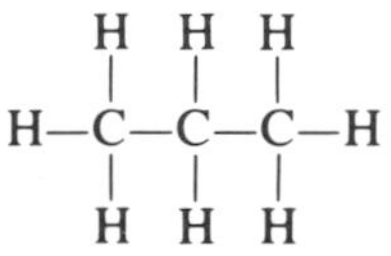 propane

It becomes increasingly difficult to use any of these conventions to represent the structures of larger molecules. With four carbon atoms the molecular formula (**7**) no longer defines the structure unambiguously, as the planar representations show. This is because there are two ways of linking four carbon atoms together – in a continuous chain (**8**) or with chain branching (**9**). Although the term 'straight chain' is commonly used to describe an unbranched chain it is misleading. The bond angles are roughly tetrahedral and the chain follows a zig-zag pattern: the variation of the planar representation shown in **8** and **9** is sometimes used to illustrate this.

The problems of representation get rapidly worse as the molecules we are trying to describe get larger. There are two ways out of the difficulty. One is to break down the molecular formula into parts to show the pattern of the carbon skeleton. There are many ways of writing such extended molecular formulae, sometimes called **condensed structural formulae**; some are illustrated here:

C_4H_{10} $CH_3—CH_2—CH_2—CH_3$ $CH_3CH_2CH_2CH_3$

$CH_3(CH_2)_2CH_3$ $CH_3—CH(CH_3)—CH_3$ $(CH_3)_3CH$

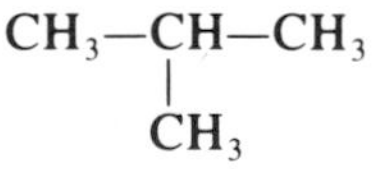

Another way is to simplify the planar or perspective representations so as to make them less cumbersome. A carbon–carbon bond is shown as a straight line, so that the end of a line or the angle between two lines represents a carbon atom. The hydrogen atoms are not shown, but are assumed to be present wherever they are needed to complete the valence shell of carbon. Such representations (**10, 11**) are called **line diagrams**: we shall use line diagrams in this book to indicate rings of carbon atoms. (The lines in these representations should not be confused with those used in the previous section – as in **12** – to mean simply 'a bond to some other atom'.)

Any of these conventions can be extended to represent molecules with functional groups attached to the carbon skeleton. The structures and formulae **13** to **17** all refer to the same alcohol molecule, though not all are unambiguous. It is important to get used to these different ways of writing organic structures: they are part of the language of organic chemistry.

All these representations disguise an important feature of the real molecules: they are not rigid but in constant motion because of rotation about the carbon–carbon bonds – a can of worms rather than a packet of spaghetti.

26.4 Structural isomers

We saw above that the molecular formula C_4H_{10} describes two different molecules:

$$CH_3{-}CH_2{-}CH_2{-}CH_3 \qquad \begin{array}{ccc} CH_3{-} & CH & {-}CH_3 \\ & | & \\ & CH_3 & \end{array}$$

Molecules with the same molecular formula but with different structural formulae are called **structural isomers** (Greek: *isos*, equal; *meros*, part). As the number of atoms in an organic molecule increases the number of possible isomers increases very rapidly: there are two structural isomers of C_4H_{10}, three of C_5H_{12}, five of C_6H_{14}, and nine of C_7H_{16}. For $C_{10}H_{22}$ there are no less than 75 structural isomers.

EXAMPLE

Draw representations of all the structural isomers of C_6H_{14}.

METHOD Start with the isomer that has all six carbon atoms in a continuous chain. Then consider structures with five carbon atoms in a main chain and one in a side chain, and so on.

ANSWER

6 + 0 $\quad CH_3{-}CH_2{-}CH_2{-}CH_2{-}CH_2{-}CH_3$

5 + 1

$$\begin{array}{ccccc} CH_3{-} & CH & {-}CH_2{-} & CH_2{-} & CH_3 \\ & | & & & \\ & CH_3 & & & \end{array} \qquad \begin{array}{ccccc} CH_3{-} & CH_2{-} & CH & {-}CH_2{-} & CH_3 \\ & & | & & \\ & & CH_3 & & \end{array}$$

4 + 2

$$\begin{array}{cccc} & & CH_3 & \\ & & | & \\ CH_3{-} & CH & {-}CH & {-}CH_3 \\ & | & & \\ & CH_3 & & \end{array} \qquad \begin{array}{cccc} & CH_3 & & \\ & | & & \\ CH_3{-} & C & {-}CH_2{-} & CH_3 \\ & | & & \\ & CH_3 & & \end{array}$$

COMMENT Note that there can be no '3 + 3' isomers: three carbon atoms cannot be attached to a three-carbon chain without forming at least a four-carbon chain.

Structural isomers are distinct compounds with distinct chemical and physical properties. Boiling points, for example, depend (among other things) on how compact molecules are: a straight-chain hydrocarbon usually has a higher boiling point than a branched-chain isomer because there are larger areas of contact across which the dispersion forces (see Section 5.1) can operate. Melting points, on the other hand, depend markedly on molecular symmetry. Highly symmetrical molecules can pack together very efficiently in the solid and they need to acquire a relatively high energy before they can break free from the lattice; they therefore have high melting points. The properties of three of the isomers of C_8H_{18} (see Figure 26.4) illustrate this:

	$t_b/°C$	$t_m/°C$
$CH_3CH_2CH_2CH_2CH_2CH_2CH_2CH_3$	126	−57
$CH_3CH_2CH_2CH_2CH_2CH(CH_3)_2$	118	−109
$(CH_3)_3C{-}C(CH_3)_3$	107	+101

Structural isomers can belong to quite different classes of compounds. For instance, the molecular formula C_2H_6O can represent an alcohol, $CH_3{-}CH_2{-}OH$, or an ether, $CH_3{-}O{-}CH_3$.

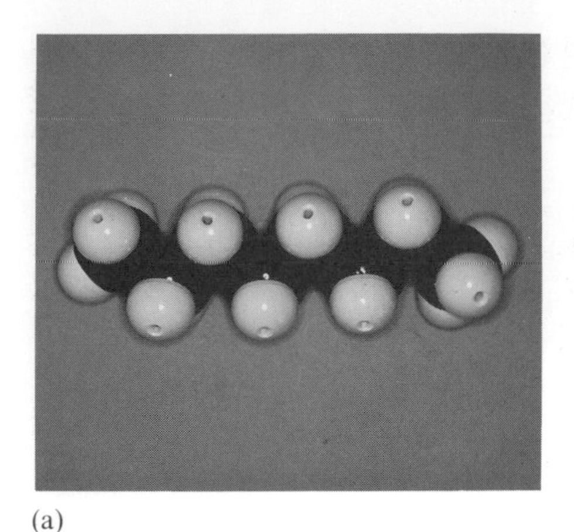
(a)

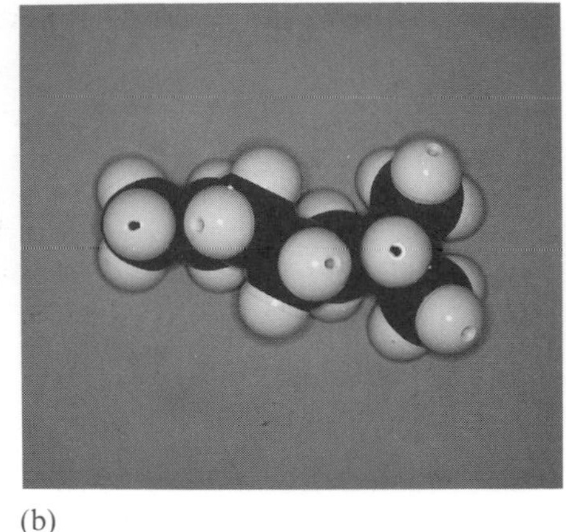
(b)

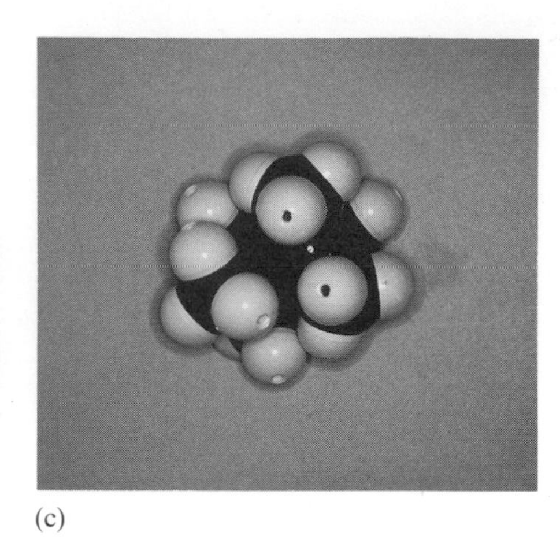
(c)

Figure 26.4 Space-filling models of (a) octane, (b) 2-methylheptane and (c) 2,2,3,3-tetramethylbutane: note that (b) is a little more compact than (a) but is less symmetrical, whereas (c) is both highly compact and highly symmetrical

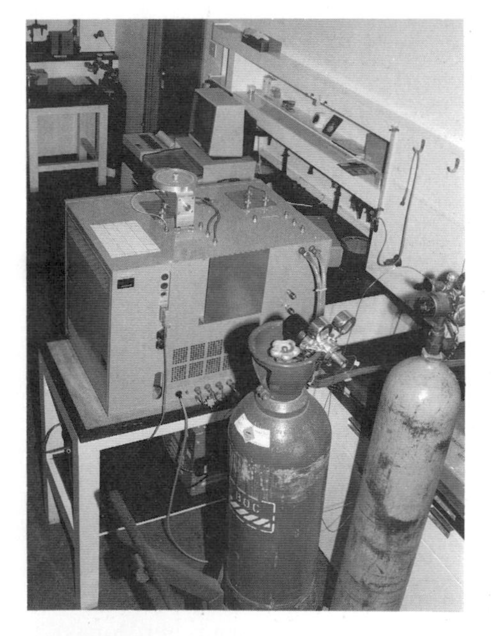

Figure 26.5 Combustion apparatus for carbon, hydrogen and nitrogen analysis. Note the gas cylinders supplying oxygen and helium

26.5 Determining the structures of organic compounds

To analyse an organic compound we need first to establish the molecular formula and then to work out the shape of the carbon skeleton and the natures and positions of the functional groups attached to it.

For crystalline solids, X-ray diffraction can provide almost a photographic picture of the molecular structure (see Section 3.4). A few years ago it took many months to analyse the diffraction patterns from even a fairly simple molecule. Modern computer programs have greatly simplified and speeded up the work, but it is still very expensive and in any case not all organic compounds form suitable crystals. The technique has therefore not displaced older chemical and spectroscopic methods of analysis.

Determining the molecular formula When an organic compound is burned in a stream of oxygen the carbon is converted into carbon dioxide and the hydrogen into water (see Figures 26.5 and 26.6). In the past (going right back to Lavoisier) these were trapped and weighed, but nowadays their masses are determined automatically by instruments. The gas stream, containing helium as a carrier, is passed through a series of chemicals which absorb all volatile combustion products except water, carbon dioxide, and nitrogen or oxides of nitrogen. Fine copper mesh is used to remove all the excess oxygen and to reduce the oxides of nitrogen to nitrogen gas. The mass of water vapour is then determined from measurements of the thermal conductivity of the gas stream before and after the water is absorbed by a bed of hygroscopic magnesium chlorate(VII); carbon dioxide is determined similarly, using the alkali **soda-lime** ($NaOH + Ca(OH)_2$) as the absorbent, and the nitrogen is determined by comparing the thermal conductivity of the nitrogen–helium stream with that of pure helium. If no other elements are present, any oxygen in the original molecule can be calculated by difference.

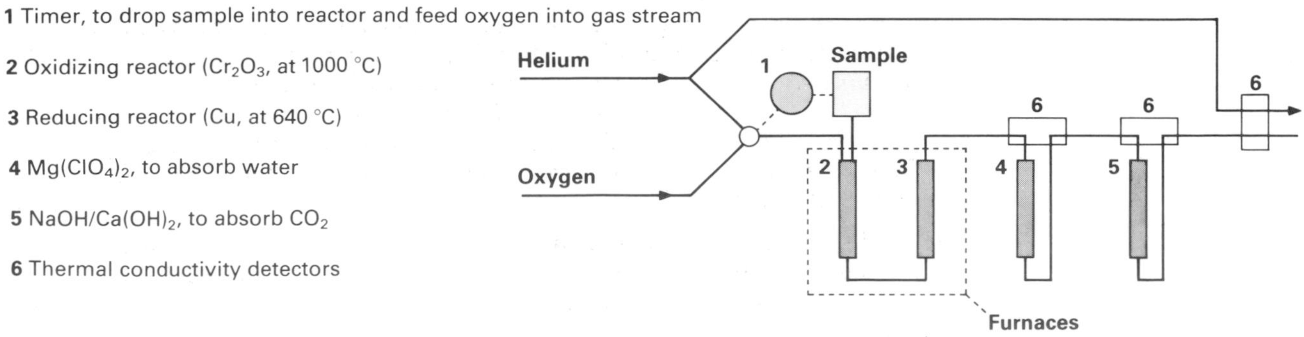

Figure 26.6 Schematic diagram of microanalytical combustion apparatus

EXAMPLE

Compound X contains only C, H and O. 10.3 mg of X is burned to give 20.2 mg of CO_2 and 10.2 mg of H_2O. What is the formula of X?

METHOD Calculate the masses of carbon, hydrogen and oxygen present in the sample of X, and from these find out the relative numbers of atoms of each element. Then apply the normal rules of covalency to suggest the actual numbers of atoms of each element in one molecule of X.

ANSWER 10.3 mg of X contains:

$$20.2\,\text{mg} \times \frac{12.0}{44.0} = 5.51\,\text{mg of carbon}$$

$$10.2\,\text{mg} \times \frac{2.02}{18.0} = 1.14\,\text{mg of hydrogen}$$

and, by difference, 3.65 mg of oxygen

The molar masses of carbon, hydrogen and oxygen are 12.0, 1.01 and 16.0 g mol^{-1} respectively, so that the amounts of atoms of the three elements in the sample of X are:

carbon: $5.51 \times 10^{-3}\,\text{g} \div 12.0\,\text{g mol}^{-1} = 0.459 \times 10^{-3}\,\text{mol}$
hydrogen: $1.14 \times 10^{-3}\,\text{g} \div 1.01\,\text{g mol}^{-1} = 1.13 \times 10^{-3}\,\text{mol}$
oxygen: $3.65 \times 10^{-3}\,\text{g} \div 16.0\,\text{g mol}^{-1} = 0.228 \times 10^{-3}\,\text{mol}$

These amounts must be in the same ratio as the numbers of atoms, but in the molecule elements must be present in whole number ratios. Let us try the assumption that there is a single atom of oxygen. Then there would have to be:

$0.459 \div 0.228 \approx 2$ atoms of carbon, and
$1.13 \div 0.228 \approx 5$ atoms of hydrogen

within experimental error.

This gives the **empirical formula**, C_2H_5O, the simplest whole number ratio of atoms. This cannot be the molecular formula, however, because carbon and oxygen both have even covalencies and a molecule containing only carbon, hydrogen and oxygen must have an even number of hydrogen atoms. The *molecular* formula is therefore $C_4H_{10}O_2$ or some other even multiple of C_2H_5O.

COMMENT To determine the molecular formula unambiguously we need to know the molar mass of X, perhaps from a mass spectrum (see Section 3.1), or at least to have some clue about the size of the molecule. For example, if the boiling point of the sample of X is 83 °C we can be sure its molecular formula is $C_4H_{10}O_2$, because a compound with as many as eight carbon atoms would have a boiling point of at least 100 °C.

Another way of finding the molecular formula is to use high-resolution mass spectrometry, as described in Section 3.1, to measure the mass/charge ratio of the molecular ion (and hence the relative molecular mass) with a precision of around 1 in 10^6. For example, the M_r value of isotopically pure $C_4H_{10}O_2$ (^{12}C, ^{1}H, ^{16}O) is 90.0681, distinguishable from, for example, $C_3H_6O_3$, $C_3H_{10}N_2O$ and C_3H_6OS for which the M_r values are 90.0317, 90.0793 and 90.0139 respectively.

Detailed structural investigations Many functional groups can be detected through their characteristic reactions. Alcohols, for example, react with metallic sodium to give hydrogen gas and sodium alkoxides (see Section 29.4):

$$2CH_3OH + 2Na \rightarrow 2Na^+CH_3O^- + H_2$$

Table 26.2 Characteristic i.r. wavenumbers for stretching vibrations

Group	$\tilde{\nu}$/cm^{-1}
O—H	3700–3400
N—H	3500–3100
C—H	3200–2800
C≡N	2400–2200
C≡C	2300–2100
C=O	1800–1650
C=C	1700–1600
C—O	1250–1000
C—Cl	800–600

Table 26.3 Typical ^{1}H n.m.r. chemical shifts

Group	δ
CH_3—C	0.9–1.1
CH_3—C=C	1.8–2.2
CH_3—C=O	1.8–2.2
CH_3—N	2.5–3.0
CH_3—O	3.3–3.8
C=CH	4.5–6.0
C_6H_5—H	6.5–8.0
O=C—H	9.5–10.0
O=C—O—H	10–14

R—CH_2—X groups usually appear at slightly higher δ values than CH_3—X

However, most structural investigations nowadays are based on instrumental techniques such as those described in Chapter 3, especially infrared (i.r.) and nuclear magnetic resonance (n.m.r.) spectroscopy and mass spectrometry (m.s.).

Infrared spectroscopy is particularly useful for detecting the presence of functional groups from their characteristic vibrational frequencies. The values in Table 26.2 are illustrative, though simplified: i.r. spectroscopy can distinguish, for example, between C—H bonds attached to singly and doubly bonded carbon atoms. The region between 1400 and 1000 cm^{-1} is often very complicated, because most organic molecules contain many single bonds with stretching vibrations in this range and because many bending vibrations also lie in this region. Because this part of the spectrum is characteristic of each individual compound it is called the 'fingerprint region'.

Nuclear magnetic resonance spectroscopy is valuable for three reasons: the chemical shifts characterize differently situated hydrogen atoms (see Table 26.3), the area under each peak is proportional to the number of hydrogen atoms that cause it, and the fine structure shows how many hydrogen atoms there are in neighbouring groups.

EXAMPLE

Figure 26.7 shows the i.r. and n.m.r. spectra of the same compound, which has the molecular formula $C_4H_8O_2$. What is its structure?

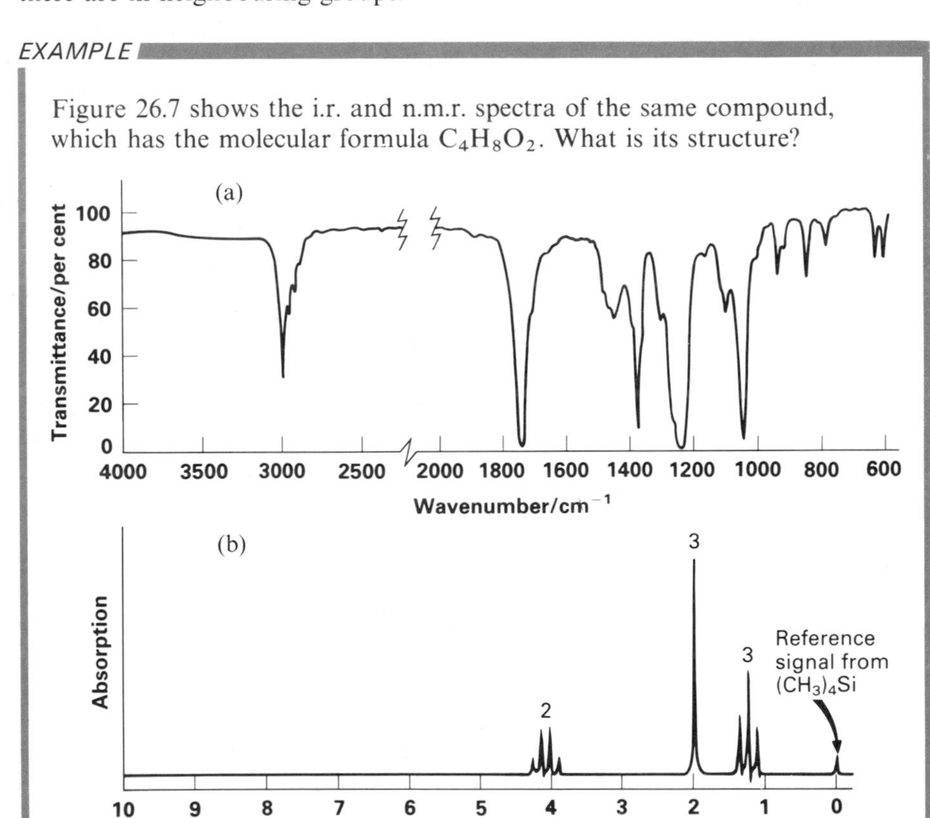

Figure 26.7 (a) I.r. and (b) n.m.r. spectra of $C_4H_8O_2$: the numbers above each group of peaks in (b) give their relative areas

METHOD Use Table 26.2 to identify possible functional groups, and Table 26.3 to suggest possible environments for hydrogen atoms. The areas in the n.m.r. spectrum characterize the relative numbers of hydrogen atoms in each environment. Splitting into n peaks implies $n - 1$ neighbouring hydrogen atoms.

ANSWER Outside the i.r. fingerprint region the two strong peaks at 3000 cm^{-1} and 1740 cm^{-1} correspond to C—H and C=O stretching. The very strong peak at 1240 cm^{-1} is probably a C—O single bond stretch. We need not consider other peaks in the fingerprint region. Note that there is no O—H stretch.

The n.m.r. signal at $\delta = 2$ suggests that the C=O group is attached to a CH_3 group. The other two n.m.r. signals at $\delta = 4.1$ and 1.2 correspond to two and three hydrogen atoms respectively (areas) and the splitting patterns suggest they are neighbours: $—CH_2—CH_3$: the CH_2 signal is split into four by the three CH_3 hydrogens, and the CH_3 signal is split into three by the CH_2 hydrogens. The chemical shift of the CH_2 signal suggests it is bonded to an oxygen atom. Putting all these pieces of information together indicates that the sample is probably $CH_3—CO—O—CH_2—CH_3$, ethyl ethanoate.

COMMENT Note the way the two techniques complement each other: in this Example, the i.r. spectrum gives information about the functional group, and the n.m.r. spectrum about the carbon chains. The spectra contain much more information than we have extracted: their detailed interpretation is a complicated task requiring considerable skill.

Mass spectrometry, unlike the spectroscopic techniques, does not involve the interaction of matter and radiation. We have already referred to the use of high-resolution mass spectrometry to determine molecular formulae by precisely measuring the mass of the molecular ion that is formed when an electron is expelled from the molecule. Low-resolution mass spectrometry can measure relative molecular masses to rather better than ± 1, and can thus be used to investigate molecular structure by recording how the molecular ion breaks down into smaller fragments. As an example of the type of information that can be obtained, refer to the mass spectrum of methanol, CH_3OH, discussed in Section 3.1.

The mass spectrum of a compound containing a chain of CH_2 groups is characterized by a series of peaks separated by 14 m/z units, as seen in Figure 26.8(a), corresponding to fragments formed by breaking different carbon–carbon bonds. In a branched-chain isomer the pattern is less regular, as Figure 26.8(b) shows; fragment ions with the positive charge on the chain-branching carbon atom are, as we shall see in Section 28.3, more stable than those with the charge at other positions, so that the chain preferentially breaks at the branching point.

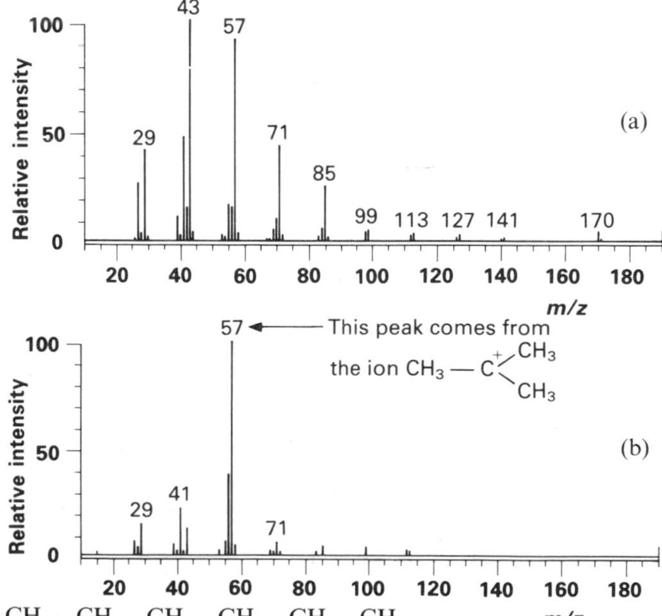

Figure 26.8 Mass spectra of two isomers of $C_{12}H_{26}$:

(a) $CH_3—CH_2—CH_2—CH_2—CH_2—CH_2—CH_2—CH_2—CH_2—CH_2—CH_2—CH_3$

(b) $CH_3—C(CH_3)_2—CH_2—CH(CH_3)—CH_2—C(CH_3)_2—CH_3$

Summary

- ☐ The special role of carbon as the basis of organic chemistry is due to the strength of covalent bonds to carbon atoms, to the tetracovalency of carbon that enables a wide variety of structures to be built up, and to the kinetic stability of carbon compounds under normal conditions on Earth.
- ☐ Saturated hydrocarbons undergo mainly homolytic, radical reactions.
- ☐ Compounds containing carbon bonded to electronegative elements undergo mainly heterolytic, ionic reactions.
- ☐ Those parts of the molecule other than the saturated hydrocarbon skeleton are called functional groups.
- ☐ The three-dimensional structures of small organic molecules can be represented by perspective diagrams. For larger molecules, conventional representations include structural and condensed structural formulae.
- ☐ Molecules with the same molecular formula but with different structural formulae are called structural isomers. They are distinct chemical entities.
- ☐ Empirical formulae can be determined from combustion analysis. They show the simplest integer ratio of the elements present in a molecule.
- ☐ Molecular formulae can be deduced from empirical formulae if the approximate molar mass is known, or they can be determined from high-resolution mass spectrometry.
- ☐ Infrared spectroscopy can reveal the presence of particular functional groups from their characteristic vibrational frequencies.
- ☐ Nuclear magnetic resonance spectroscopy gives information about the type and number of differently situated hydrogen atoms in a molecule, and about the numbers of neighbouring hydrogen atoms.
- ☐ Low-resolution mass spectrometry shows how a molecular ion breaks down into smaller fragments and hence gives information about the structure of the molecule.

PROBLEMS

1 Summarize the reasons for classifying organic chemistry, the chemistry of carbon compounds, separately from that of all the other elements.

2 Write down three possible ways in which one of the C—Cl bonds in CCl_4 could be broken, showing clearly the numbers of valence shell electrons in the products. Classify the processes as homolytic or heterolytic.

3 Draw full structural formulae for the following compounds:

$(CH_3)_2CHC_2H_5$, $CH_3CH(OH)CH_3$, $CH_3(CH_2)_2CHClCH_3$.

4 Explain briefly how i.r. and/or n.m.r. spectroscopy could be used to distinguish between the following pairs of isomers:
(a) CH_3COCH_3 and CH_3CH_2CHO;
(b) $CH_3CH_2CH_2OH$ and $CH_3CH_2OCH_3$;
(c) $CH_3CH_2COOCH_3$ and $CH_3COOCH_2CH_3$.

5 How are the procedures of (a) refluxing, (b) distillation and (c) recrystallization carried out in chemistry laboratories?
Explain the purpose of carrying out the procedures by reference to a *different* chemical reaction in each part of your answer, giving formulae and equations where appropriate.
(*Nuffield*)

6 It may be said that 'organic chemistry is the chemistry of functional groups'. Discuss this statement by comparing and contrasting the reactions of
(a) the OH groups in C_6H_5COOH and C_6H_5OH,
(b) the unsaturated carbon to carbon bonds in $CH_2{=}CHCH_3$ and C_6H_6,
(c) the Cl groups in CH_3COCl and CH_3CH_2Cl. (*JMB*)

7 (a) When a sample of trichloromethane is placed in a magnetic field, the protons in the molecules may absorb energy in the radiofrequency region of the spectrum. Explain why protons behave in this way.
(b) If the sample above is changed to 1,1,2-trichloroethane, the spectrum shown below is obtained at low resolution. Explain why there are now two absorptions, and account for their relative sizes.

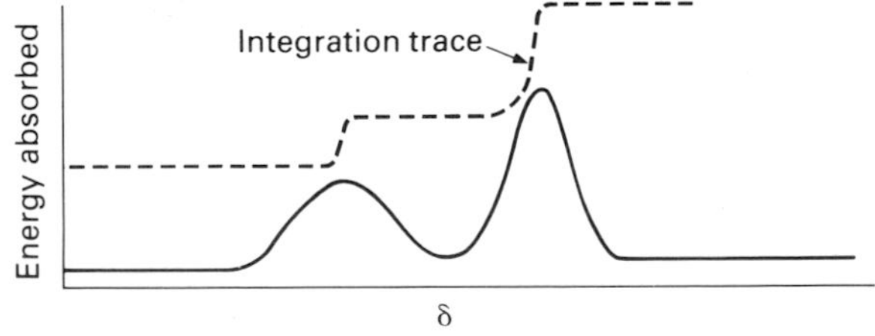

(c) At high resolution, both the peaks shown above split into a number of components. Predict the splitting pattern shown by 1,1,2-trichloroethane and explain how it arises.
(d) The diagram below shows the high-resolution n.m.r. spectrum of the halogen compound C_4H_9Cl. Explain the splitting pattern and hence deduce the structure of the compound.

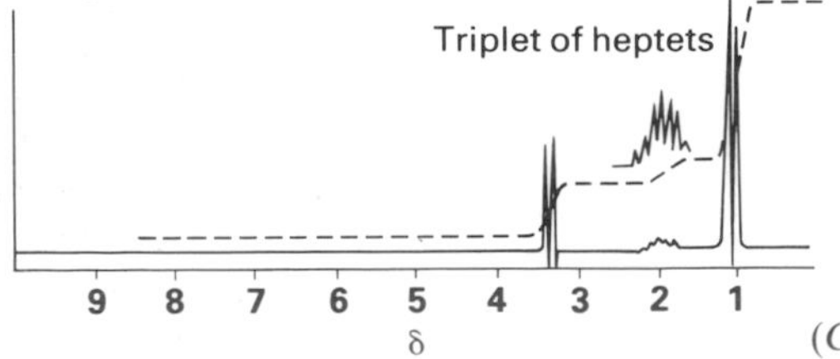

(*Cambridge*)

27 Alkanes: C—H bonds and C—C single bonds

In this chapter we learn the names of the alkanes, which form the basis of the names of most other organic compounds. We look at the physical and chemical properties of alkanes, both of which depend mainly on the low polarity of carbon–carbon and carbon–hydrogen bonds. Finally we consider the vital importance of the alkanes as the major constituents of petroleum.

The alkanes contain only carbon and hydrogen atoms linked by single bonds. We can imagine building them up from methane, CH_4, by inserting $—CH_2—$ groups into carbon–hydrogen bonds: $CH_3—H$, $CH_3—CH_2—H$, $CH_3—CH_2—CH_2—H$ and so on. The general formula for an alkane is therefore C_nH_{2n+2}.

In the previous chapter we mentioned that the alkanes used to be called the paraffins – the 'unreactives'. That description is only partly true: in this chapter we see that some of their reactions are central to the economy and industry of the whole world. First, though, we need to discuss in some detail the naming (**nomenclature**) of alkanes, because this forms the basis for naming most other organic compounds.

27.1 Nomenclature

The general name for a saturated hydrocarbon is **alkane**. The names of all the individual alkanes end in *-ane*. For the straight-chain alkanes the first part of the name describes how many carbon atoms there are (Table 27.1). The prefixes *meth-*, *eth-*, *prop-* and *but-*, signifying from one to four carbon atoms, come from old pre-systematic names; the rest come from Greek or Latin names for the numbers.

Table 27.1 Names of the straight-chain alkanes

Number of carbon atoms	Molecular formula	Name
1	CH_4	methane
2	C_2H_6	ethane
3	C_3H_8	propane
4	C_4H_{10}	butane
5	C_5H_{12}	pentane
6	C_6H_{14}	hexane
7	C_7H_{16}	heptane
8	C_8H_{18}	octane
9	C_9H_{20}	nonane
10	$C_{10}H_{22}$	decane

The groups that are obtained by removing a hydrogen atom from an alkane are called **alkyl** groups, with general formula C_nH_{2n+1}. They are named by replacing the -ane ending with *-yl*: $CH_3—$ is methyl, $C_2H_5—$ ($CH_3—CH_2—$) is ethyl, and so on. The symbol R— is commonly used to denote an unspecified alkyl group.

The chemistry of many compounds is modified by the extent of chain branching in an alkyl group, and special terms are used to describe this, as shown in Table 27.2.

Table 27.2 Classes of alkyl groups

Structure	Class
$R-C(H)(H)-$	**primary** alkyl group
$R-C(R')(H)-$	**secondary** alkyl group
$R-C(R')(R'')-$	**tertiary** alkyl group

In the previous chapter we saw that there are structural isomers of all the alkanes with four or more carbon atoms. Used in a general sense the names in Table 27.1 are often loosely applied to any isomer: we might refer to 'pentanes' as a shorthand for 'isomers with molecular formula C_5H_{12}'. Strictly, however, the names in Table 27.1 should be used only for the straight-chain alkanes: 'pentane' is specifically $CH_3—CH_2—CH_2—CH_2—CH_3$.

BOX 27.1

Naming organic compounds

In the early days of organic chemistry compounds were usually given a name that related to their origin (urea from urine, citric acid from citrus fruits) or to some characteristic physical property (glycerine, from the Greek for 'sweet', *glykeros*). The number of known organic compounds grew rapidly: by 1880 it had reached 20 000 and it has been doubling every twelve or thirteen years ever since.

Each compound needed its own distinctive name, and it soon became obvious that there would have to be some system for naming compounds according to their structure. At first individual chemists put forward their own suggestions but this led to many conflicting usages. In 1892 an international meeting was held in Geneva to devise a uniform system, and the so-called Geneva system is the basis of the names used today. The body now responsible for nomenclature is the International Union of Pure and Applied Chemistry (IUPAC).

Those who try to devise a systematic nomenclature have to face up to the inherent conservatism of human nature, a reluctance to abandon the familiar. For many compounds IUPAC recognizes two different names: one based on the standard system, the other based on older names that are still in common use. In this book we shall, in the main, use systematic IUPAC names, but other common names will be mentioned because professional chemists often use them in preference to systematic names: 'citric acid', for example, is easier to say than '2-hydroxypropane-1,2,3-tricarboxylic acid'.

Branched-chain isomers are named by the following IUPAC rules (see Box 27.1).

1 The longest continuous chain of carbon atoms (the **spine**) is named according to Table 27.1.

2 The carbon atoms of the spine are numbered. The name for the branched-chain alkane is built up from the positional number of the branching point, the name of the alkyl side chain, and the alkane name of the spine. For example:

$$\overset{1}{C}H_3-\overset{2}{C}H_2-\overset{3}{C}H(CH_3)-\overset{4}{C}H_2-\overset{5}{C}H_3$$

3-methyl**pentane**

(3 = point of attachment; methyl = side chain; pentane = spine)

The number may be omitted if there is only one possible position for a side chain:

$$CH_3—CH(CH_3)—CH_3$$

2-methylpropane
or methylpropane

3 If there is a choice, the main chain is numbered from the end that gives the *lower* number at the branching point:

$$CH_3-CH_2-CH_2-\underset{\underset{CH_3}{|}}{CH}-CH_3$$

2-methylpentane
(not 4-methylpentane)

4 Two or more identical side chains are described as in the following example:

$$CH_3-\overset{\overset{CH_3}{|}}{\underset{\underset{CH_3}{|}}{C}}-CH_2-\underset{\underset{CH_3}{|}}{CH}-CH_3$$

2,2,4-trimethylpentane

5 Different side chains are named in *alphabetical* order, ignoring the prefixes such as di- or tri-:

$$CH_3-\underset{\underset{CH_3}{|}}{CH}-\overset{\overset{C_2H_5}{|}}{\underset{\underset{CH_3}{|}}{C}}-CH_2-CH_3$$

3-ethyl-2,3-dimethylpentane

EXAMPLE

We are now equipped to name an alkane even if it has a fairly outlandish-looking structure. Give the IUPAC name for:

$$C_2H_5-\underset{\underset{C(C_2H_5)_3}{|}}{CH}-CH_2-CH_2-CH(CH_3)_2$$

METHOD Locate the longest continuous chain of carbon atoms in the molecule to find the spine. Give names and positional numbers to the side chains.

ANSWER The longest chain contains eight carbon atoms (from one of the CH_3 groups at the top right of the structure as drawn above to one of the C_2H_5 groups at bottom right). For clarity we redraw the structure with the eight-carbon chain horizontal:

$$\overset{8}{C}H_3-\overset{7}{C}H_2-\overset{\overset{C_2H_5}{|}}{\underset{\underset{C_2H_5}{|}}{C}}-\underset{\underset{C_2H_5}{|}}{\overset{5}{C}H}-\overset{4}{C}H_2-\overset{3}{C}H_2-\underset{\underset{CH_3}{|}}{\overset{2}{C}H}-\overset{1}{C}H_3$$

Numbering from the right-hand end and naming the alkyl groups in alphabetical order gives *5,6,6-triethyl-2-methyloctane*. (Remember that the prefix tri- is ignored in the alphabetical sequence.)

COMMENT If we had numbered from the left the positional numbers would have been 3,3,4,7 instead of 2,5,6,6. The correct order is the one that gives the lower number *at the first point of difference*: in this case 2 in preference to 3.

Saturated hydrocarbons that contain a ring of carbon atoms are called **cycloalkanes**. We can imagine forming the ring by removing two hydrogen atoms from an alkane chain and forming a new carbon–carbon bond between the vacant sites. The general formula for a cycloalkane is therefore C_nH_{2n}.

(a)

(b)

Figure 27.1 (a) Ball-and-stick and (b) space-filling models of cyclohexane

The nomenclature of the cycloalkanes is illustrated in the following examples.

```
    CH2                      CH2
   /   \                    /   \
CH2     CH2              CH2     CH—CH3
  \     /                  \     /
  CH2—CH2                  CH2—CH2
```

cyclopentane — methylcyclopentane

(In naming methylcyclopentane a positional number is not needed, because there is no distinction between the carbon atoms in the cyclopentane ring.)

```
     CH2
    /   \
 CH2     CH—CH3
  |      |
 CH2     CH—CH3
    \   /
     CH2
```

1,2-dimethylcyclohexane

(Positional numbers are chosen to be as low as possible.)

27.2 Physical properties

Under normal conditions of temperature and pressure the first four alkanes are gases. The familiar **natural gas** is about 94 per cent methane and 3 per cent ethane; LPG (liquefied petroleum gas) is propane or butane. Campers who use Calor gas (normally butane) have to switch to propane in winter: the boiling point of butane is 0 °C and on a cold day its vapour pressure is too low to run a gas cooker.

From pentane to about pentadecane ($C_{15}H_{32}$) the straight-chain alkanes are liquids at room temperature; alkanes with larger molecules are waxy solids. In the depths of winter diesel-powered trucks can be stranded as their fuel starts to freeze. Boiling points rise steadily with increasing chain length, as shown in Figure 27.2, because the dispersion forces (see Section 5.1) also increase. Branched-chain alkanes have lower boiling points than their straight-chain isomers as discussed in Section 26.4. Figure 27.2 also shows that the melting points of straight-chain alkanes with an odd number of carbon atoms are similar to those of alkanes with one carbon atom fewer; this reflects the different packing in the solid of molecules with odd and even carbon chains.

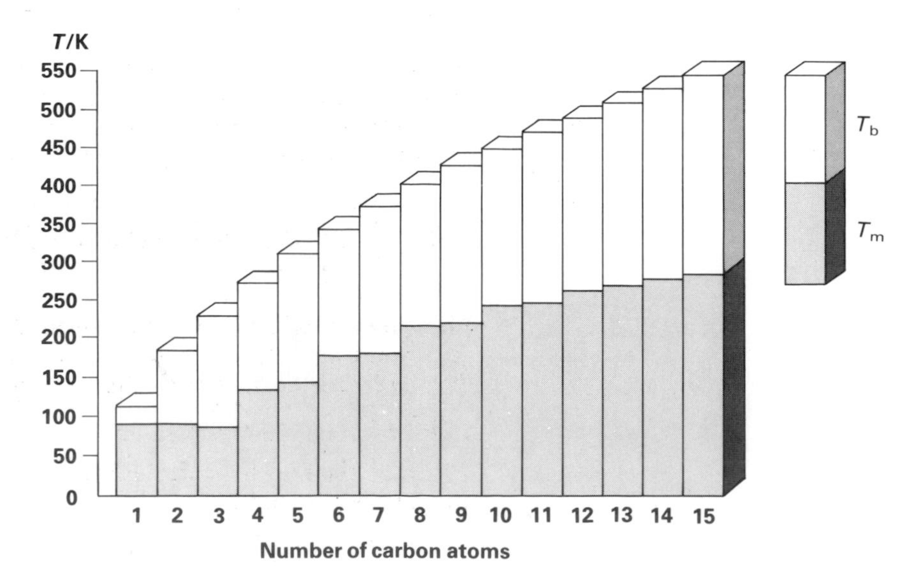

Figure 27.2 Melting and boiling points of straight-chain alkanes

Because of the low polarity of carbon–carbon and carbon–hydrogen bonds the alkanes have zero or very small permanent dipole moments. Consequently, dispersion forces are the only significant interactions between their molecules. They readily dissolve, or dissolve in, other compounds of low polarity: the dispersion forces within a sample of pure hexane, for example, are comparable to those within a mixture of hexane and benzene, and the two liquids mix together in all proportions. On the other hand, the attraction between hexane and water molecules is very much weaker than the hydrogen bonding between one water molecule and another. If hexane and water are shaken together the water molecules all clump together, leaving the hexane as a separate layer: the two liquids are immiscible (see Section 7.1). Such behaviour is often summarized as 'like dissolves like', though this is an over-simplification. Hexane, like all the other liquid alkanes, is less dense than water, so that it floats on top of the water.

27.3 Reactivity patterns and important reactions

Bromination

We saw in Section 26.2 that the low polarity of carbon–carbon and carbon–hydrogen bonds leads to homolytic, *radical* reactions. Examples are the reactions of alkanes with chlorine or bromine, which are important in the manufacture of halogen-containing solvents, refrigerants and anaesthetics (see Box 28.1 and Section 28.5). A few drops of bromine dissolved in hexane give a brown solution. In the dark the colour remains for a long time but in sunlight it slowly fades, showing that there has been a reaction.

A similar reaction, but one that is less easy to demonstrate, occurs in the gas phase between bromine and methane:

$$CH_4(g) + Br_2(g) \xrightarrow{\text{light}} CH_3Br(g) + HBr(g) \qquad \Delta_r H^{\ominus}(298\ K) = -30\ kJ\ mol^{-1}$$

This is an example of a **substitution reaction**: a hydrogen atom of the methane has been substituted by a bromine atom.

The reaction depends on the absorption of photons of light (see Section 3.2). Alkanes are transparent to visible light and to the part of ultraviolet light that penetrates the Earth's atmosphere. It must therefore be the coloured bromine that **initiates** the reaction. The bromine–bromine bond is fairly weak and the bromine molecule can be split into atoms by photons of the right energy:

$$:\ddot{\underset{..}{Br}}:\ddot{\underset{..}{Br}}: \longrightarrow 2:\ddot{\underset{..}{Br}}\cdot \qquad \Delta_r H^{\ominus}(298\ K) = +193\ kJ\ mol^{-1}$$

(Alternatively the bromine atoms can be generated by raising the temperature: at 200 °C bromine reacts with methane even in the dark.)

A bromine atom can now attack a methane molecule, pulling off a hydrogen atom and leaving a **methyl radical**:

$$:\ddot{\underset{..}{Br}}\cdot + H:\overset{H}{\underset{H}{\ddot{\underset{..}{C}}}}:H \longrightarrow :\ddot{\underset{..}{Br}}:H + \cdot\overset{H}{\underset{H}{\ddot{\underset{..}{C}}}}:H$$

The symbols $Br^{\cdot}$ and $CH_3{}^{\cdot}$ may be used for the bromine atom and the methyl radical respectively, to emphasize that each has an unpaired electron. Using these symbols we may show this reaction step more simply as:

$$Br^{\cdot} + CH_4 \rightarrow HBr + CH_3{}^{\cdot}$$

In turn the methyl radical reacts with a bromine molecule to give CH_3Br and another bromine atom:

$$CH_3{}^{\cdot} + Br_2 \rightarrow CH_3Br + Br^{\cdot}$$

The two steps $Br^{\cdot} + CH_4$ and $CH_3^{\cdot} + Br_2$ alternate in a **chain reaction** (see Section 14.3), each step generating a new radical to fuel the next step. Together, these two **propagation** steps consume one CH_4 molecule and one Br_2 molecule and produce one HBr molecule and one CH_3Br molecule, giving the net reaction described above.

From time to time two methyl radicals or two bromine atoms combine, and the chain **terminates**. The products contain small quantities of ethane, formed when two methyl radicals react together, and this provides evidence for the radical mechanism:

$$2CH_3^{\cdot} \rightarrow CH_3\text{—}CH_3$$

As the reaction proceeds a new step becomes possible. A bromine atom might react with a molecule of CH_3Br instead of CH_4, leading to a new chain and a second substitution product, CH_2Br_2:

$$Br^{\cdot} + CH_3Br \rightarrow HBr + CH_2Br^{\cdot}$$

$$CH_2Br^{\cdot} + Br_2 \rightarrow CH_2Br_2 + Br^{\cdot}$$

Indeed, the reaction produces all the possible substitution products: CH_3Br, CH_2Br_2, $CHBr_3$ and CBr_4. The relative amounts depend on the proportions of methane and bromine in the original mixture. We say the reaction is not 'clean', meaning that it does not give a single organic product.

Now let us return to the reaction of bromine with hexane. The pattern is exactly the same, but the reaction is even less clean than that with methane. The first substitution can give one of three different isomers: $CH_3CH_2CH_2CH_2CH_2CH_2Br$, $CH_3CH_2CH_2CH_2CHBrCH_3$ or $CH_3CH_2CH_2CHBrCH_2CH_3$. Further substitutions can give a wide variety of products. This is typical of reactions of alkanes: they tend to be **indiscriminate**, because there is little difference between the strengths of the differently situated carbon–hydrogen bonds.

Halogenation

The reactions of alkanes with the other halogens are largely controlled by the strengths of the carbon–halogen and hydrogen–halogen bonds. Bonds to iodine are weak because of the large size of the iodine atom. The energy released in forming the C—I and H—I bonds is less than that needed to break the I—I and strong C—H bonds. In other words, the gas-phase iodination of methane (and of other alkanes) is endothermic:

$$CH_4(g) + I_2(g) \xrightarrow{\text{light}} CH_3I(g) + HI(g) \qquad \Delta_r H^{\ominus}(298\,K) = +59\,kJ\,mol^{-1}$$

$\Delta_r G^{\ominus}$ is also positive and consequently the equilibrium lies strongly in favour of the reactants.

Chlorination, on the other hand, is more exothermic than bromination:

$$CH_4(g) + Cl_2(g) \xrightarrow{\text{light}} CH_3Cl(g) + HCl(g) \qquad \Delta_r H^{\ominus}(298\,K) = -99\,kJ\,mol^{-1}$$

It is also much *faster*. The rate-limiting step (see Section 14.3) is the attack of $Cl^{\cdot}$ on CH_4. The activation energy for this process is relatively low, because as the new, strong chlorine–hydrogen bond starts to form the energy released contributes to that needed to break the hydrogen–carbon bond.

Fluorine reacts explosively with alkanes and with most other organic compounds. The reaction is strongly exothermic; for example:

$$CH_4(g) + F_2(g) \rightarrow CH_3F(g) + HF(g) \qquad \Delta_r H^{\ominus}(298\,K) = -443\,kJ\,mol^{-1}$$

Each link in the propagating chain liberates enough energy to break more fluorine molecules into atoms: these in turn initiate more chains and the reaction accelerates explosively.

Combustion

The combustion of alkanes in air is also a radical substitution reaction, though the detailed mechanism is much more complicated than that of halogenation. The bonding in molecular oxygen is stronger than that in the halogen molecules (see Table 9.4), so that initiation requires a greater input of energy. A mixture of oxygen and methane is stable unless ignited by a spark (as in lighting a bunsen burner or a gas cooker) or irradiated with ultraviolet light, that is, with energetic photons. As with fluorination, the reaction, once started, accelerates explosively. The strongly exothermic reactions fuel a large part of the world's transport, industry and heating (see Section 27.4):

$$CH_4(g) + 2O_2(g) \rightarrow CO_2(g) + 2H_2O(l) \qquad \Delta_c H^\ominus(298\,K) = -890\,kJ\,mol^{-1}$$

When methane is burned with insufficient oxygen present the products of combustion include carbon monoxide: this is why inadequate ventilation of gas-fired space or water heaters can all too easily prove fatal.

Reactions of cycloalkanes

In general the reactions of cycloalkanes are very similar to those of the alkanes. Cyclopropane and cyclobutane are more reactive, however: in these compounds the angles between the carbon–carbon bonds are much smaller than usual, causing **ring strain**. For example, cyclopropane reacts with bromine in a ring-opening reaction that is much faster than substitution:

$$\text{cyclo-}(CH_2)_3 + Br_2 \longrightarrow Br{-}CH_2{-}CH_2{-}CH_2{-}Br$$

27.4 Petroleum and petrochemicals

The Industrial Revolution was built on 'King Coal', and coal remained the primary source of energy and a major source of industrial organic chemicals until after the Second World War. Its place was usurped, over a very few years, by the cheaper and seemingly limitless flow of oil coming mainly from the Middle East. Coal-fired locomotives gave place to diesel, electricity was increasingly generated in oil-burning power stations, and the massive increase in the numbers of private cars (35 million worldwide in 1938, 200 million in 1971) was, at least in part, a result of the ready availability of petrol. Likewise petroleum rather than coal became the primary source of organic chemicals. So dependent had the industrial world become on this single raw material that the decision by the major oil-producing countries in 1973 to reduce their output and to raise prices threw the whole world economy into disarray.

Petroleum, like coal, is a **fossil fuel** that derives from once-living matter: the decaying remains of small marine organisms which, millions of years ago, were covered by layers of sediment. In the refining of crude oil this mixture is partly separated by fractional distillation (see Section 7.1 and Figure 27.3). The

Figure 27.3 Part of BP's oil refinery at Rotterdam, Holland

(a)

(b)

(c)

Figure 27.4 Variations in the viscosity of oil fractions: (a) petrol, (b) diesel oil, (c) heavy fuel oil

fractions collected vary from one company to another; a typical distribution is described in Table 27.3. Further refining of the residue, using vacuum distillation, gives a mixture of C_{26}–C_{28} alkanes that can be separated by extracting the mixture with lower alkanes into **lubricating oils** and **paraffin wax**. The solid residue from the vacuum distillation is **asphalt** or **bitumen**, used in surfacing roads and covering flat roofs.

About 90 per cent of the world's oil supply is used as fuel. The demand for the lighter fractions – particularly for motor and aviation fuels – is much greater than for the more abundant heavy fractions. Cracking, discussed in Box 27.2, is therefore an essential part of petroleum refining, together with a related process called **catalytic reforming**. This is similar to catalytic cracking except that the conditions are chosen so as to minimize the actual cracking and to encourage the rearrangement to branched-chain alkanes. Reforming also produces some aromatic hydrocarbons (see Chapter 34), such as methylbenzene, by **dehydrogenation**:

$$C_7H_{16} \rightarrow C_6H_5CH_3 + 4H_2$$

Branched-chain alkanes burn more smoothly in petrol engines than their straight-chain isomers, with less tendency to 'knock', that is, to detonate under pressure alone before being ignited by a spark. The aromatic hydrocarbons also improve the anti-knock properties of the fuel. The ability of a petrol to resist knocking is described by its **octane rating**, or octane number. This is measured by comparing its ease of burning with that of mixtures of heptane and 2,2,4-trimethylpentane, which is commonly, but erroneously, called 'iso-octane' (the prefix iso- refers to isomers in which the carbon chain ends in $(CH_3)_2CH—$ but is otherwise straight). Pure heptane is given an octane rating of 0, that of pure iso-octane is 100. Petrol with an octane rating of 97 behaves like a 97:3 mixture by volume of iso-octane and heptane.

The 10 per cent of petroleum that is not used as fuel is the basis of the **petrochemical industry** – the production of synthetic organic chemicals from petroleum. Plastics, paints, drugs, detergents, synthetic fibres and many other products in everyday use are obtained largely from petrochemicals; the fact that all these constitute only 10 per cent of the total petroleum industry emphasizes the extent of the world's dependence on oil as a fuel.

The single most important petrochemical, from which many others are derived, is the unsaturated hydrocarbon **ethene**, $CH_2{=}CH_2$ (Chapter 33). In Europe this is obtained mainly by the thermal cracking of naphtha. In the USA, however, an abundant supply of ethane from natural gas has until recently been a more economic source of ethene. The petrochemical industry is discussed further in Section 33.5.

Table 27.3 Products from the fractional distillation of crude oil (compare Figure 27.4)

Fraction	Percentage by mass of crude oil	Boiling range/°C	Number of carbon atoms in main components	Uses
refinery gas	1–2	<20	1–4	fuel (mainly in the refinery), petrochemicals
gasoline/naphtha	15–25	20–175	5–10	petrol ('gasoline'), petrochemicals via cracking
kerosene	10–15	175–250	11–14	aviation fuel, domestic heating
gas oil	15–25	250–400	15–25	diesel fuel, industrial heating, a source of gasoline via cracking
residue	40–50	>400	>25	power station fuel, asphalt for roads

BOX 27.2

Cracking

Crude oil or **petroleum** (see Section 27.4) contains a wide range of alkanes, from methane to alkanes with fifty carbon atoms or more. The larger molecules can be broken down into smaller ones by **cracking**.

Thermal cracking, or **steam cracking**, involves heating the alkanes with superheated steam, typically to around 800 °C. At this temperature even the strong carbon–carbon bond can be broken in the violent collisions between molecules; for example:

$$R—CH_2—CH_2—CH_3 \rightarrow R—CH_2—CH_2^{\bullet} + CH_3^{\bullet}$$

Again radical chain reactions occur, in particular reactions such as

$$CH_3^{\bullet} + R—H \rightarrow CH_4 + R^{\bullet}$$

$$R—CH_2—CH_2^{\bullet} \rightarrow R—CH{=}CH_2 + H^{\bullet}$$

The final products of the cracking are thus hydrocarbons with fewer carbon atoms than the starting alkanes. Some are lower alkanes; others are **alkenes**, containing carbon–carbon double bonds (these are discussed in Chapter 33). The exact composition of the product mixture depends on the alkanes being cracked, the temperature, the time the reaction is held at high temperature (typically less than half a second) and the ratio of steam to alkane.

Catalytic cracking (colloquially called **cat-cracking**) requires lower temperatures, around 500 °C, and uses oxides of silicon and aluminium as catalysts. Here again, chain reactions occur, but unlike most other alkane reactions they involve *ions* not radicals. Although it is not entirely certain how they are formed, the key species is an alkyl cation (a **carbocation**), for example: $CH_3—CH_2—CH_2—\overset{+}{C}H—CH_3$. In such an ion the positively charged carbon atom is bonded to three other atoms and has only six valence electrons. These ions then react in three main ways:

$$CH_3—CH_2—CH_2—\overset{+}{C}H—CH_3 \rightarrow CH_3—CH_2^{+} + CH_2{=}CH—CH_3$$

$$CH_3—CH_2^{+} + R—H \rightarrow CH_3—CH_3 + R^{+}$$

$$CH_3—CH_2—CH_2—\overset{+}{C}H—CH_3 \rightarrow CH_3—\overset{+}{\underset{\displaystyle CH_3—CH_2}{\underset{|}{C}}}—CH_3$$

The first reaction is the actual cracking, breaking down the original alkane into smaller molecules. The second is the reversion of a carbocation to an alkane molecule, while at the same time a new carbocation is produced to initiate a new reaction chain. The third is an example of a **rearrangement** in which carbon–carbon bonds are made and broken to give a branched carbon skeleton, very important in the petroleum industry. Carbocations rearrange more readily than alkyl radicals do, so that rearrangement is more common in catalytic cracking than in thermal cracking. A second difference arises from the relative instability of the methyl cation, CH_3^{+} (see Section 28.3), which means that catalytic cracking gives very little methane.

BOX 27.3

Figure 27.5 Because of the health hazard from lead from motor car exhausts, engines are now being made that will run on lead-free petrol

Lead in petrol

Knocking reduces the power output of a petrol engine and increases engine wear. The onset of knocking also limits the compression ratio of the engine – the extent to which the fuel–air mixture can be compressed before being ignited (see Section 4.1). High compression ratios are desirable because they give more fuel-efficient engines. In 1921 it was found that the addition of small amounts of tetraethyllead(IV), $(CH_3CH_2)_4Pb$, greatly improved petrol's anti-knock behaviour. The weak carbon–lead bonds are easily broken, giving $CH_3CH_2^{\cdot}$ radicals that can initiate smooth burning. To prevent the lead being deposited in the cylinders, 1,2-dibromoethane, $BrCH_2CH_2Br$, is also added. During the combustion this forms lead(II) bromide, which is swept out with the exhaust gases. Low-grade petrol with an octane rating of 87 can be brought up to 93-octane grade ('regular' or 'two-star') by adding about 0.6 g of tetraethyllead(IV) per litre.

In recent years there has been widespread concern that the lead being discharged into the atmosphere from car exhausts (about 400 000 tonnes worldwide at its peak in 1972) may cause brain damage, especially in young children. In most industrialized countries the addition of lead to petrol is now limited or even prohibited. One consequence of this can be measured: engines use about 3.5 per cent more lead-free fuel than leaded fuel to produce the same power. A second consequence is less clear: lead-free fuels have a higher than normal proportion of aromatic hydrocarbons and there is a possibility that some of the products of their incomplete combustion may be carcinogenic (cancer-producing).

Summary

- ☐ Alkanes are saturated hydrocarbons: they contain only carbon and hydrogen atoms linked by single bonds.
- ☐ The general formula of straight-chain and branched-chain alkanes is C_nH_{2n+2}. The general formula of alkanes containing a ring of carbon atoms (cycloalkanes) is C_nH_{2n}.
- ☐ The names of the straight-chain alkanes (Table 27.1) form the basis of most other organic names.
- ☐ A branched-chain alkane is named using the positional numbers of the branching points, the names of the side chains and the alkane name of the longest continuous chain.
- ☐ The C_1–C_4 straight-chain alkanes are gases; the C_5–C_{15} compounds are liquids.
- ☐ Alkanes are miscible with other compounds of low polarity but not with water.
- ☐ Alkanes undergo radical substitution reactions with bromine and chlorine. These tend to occur indiscriminately at different carbon atoms and multiple substitution is common.
- ☐ Alkanes burn in air or oxygen to form carbon dioxide and water.
- ☐ Cyclopropane and cyclobutane have strained rings and are more reactive than alkanes.
- ☐ Petroleum is a fossil fuel and is separated by fractional distillation.
- ☐ Alkanes can be broken down into smaller molecules (alkanes and alkenes) by cracking. Thermal cracking is a radical reaction, mainly used to manufacture ethene and other petrochemicals from naphtha. Catalytic cracking and the related catalytic reforming are the only important ionic reactions of alkanes. They are used to produce light fuels, especially petrol (gasoline), from the heavier fractions of crude oil.

PROBLEMS

1 Write down structures showing all the isomers with molecular formula C_5H_{12}. Name each isomer.

2 The following representations all describe molecules with the molecular formula C_7H_{16}:

A $CH_3CH_2CH_2CH_2CH_2CH_2CH_3$

B
```
CH3—CH—CH2—CH2—CH3
     |
     CH2—CH3
```

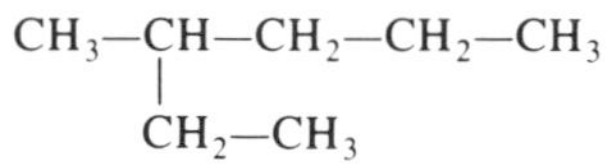

C
```
CH3—CH—CH2—CH3
     |
     CH2—CH2—CH3
```

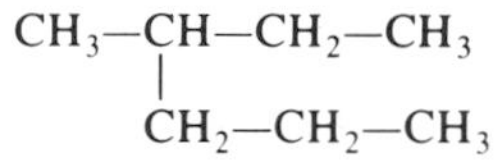

D
```
     H H H H H H
      \/   \/   \/
  H    C    C    C    H
   \  / \  / \  / \  /
    C    C    C    C
   /\   /\   /\   /\
  H  H H  H H  H H  H
```

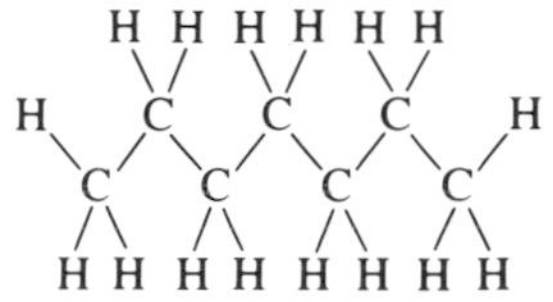

E
```
             CH3
             |
CH3—CH2—C—CH2—CH3
             |
             CH3
```

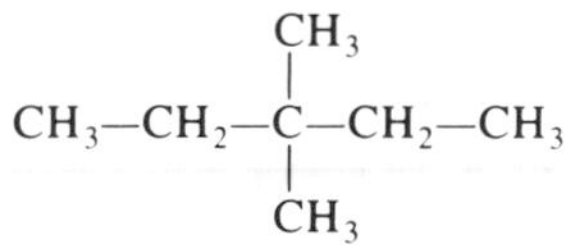

F $(CH_3)_2C(C_2H_5)_2$

G

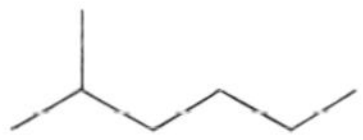

(a) Which representations are of the same molecules?
(b) Name all the isomers shown.

3 A hydrocarbon contains 17.2 per cent of hydrogen by mass. What is its empirical formula? Its boiling point is −12 °C: what is its molecular formula? Write down the isomers with this molecular formula and name them.

4 A mixture of $10\,cm^3$ of a gaseous hydrocarbon and $200\,cm^3$ of oxygen gave on explosion a volume of $175\,cm^3$. When this was shaken with aqueous potassium hydroxide the volume was reduced to $135\,cm^3$. Calculate the molecular formula of the hydrocarbon. (Volumes are measured at 20 °C and 1 atm.)

5 What is *cracking*? Describe the two main methods by which it is carried out and say why they are industrially important.

6 Show the mechanism of the chlorination of methane. Outline the evidence for this mechanism.

7 Write brief notes on the fractional distillation of petroleum, explaining the uses of two of the fractions.

8 (a) For each of parts (i) to (ix) below *only one* of the alternatives **A**, **B**, **C**, **D**, **E** is correct. Answer each part by giving the appropriate letter.
(i) Which structure is under the greatest strain?
(ii) Which structure will exist as optical isomers?
(iii) Which is isomeric with pentane?
(iv) Which has a systematic name ending in -propane?
(v) Which structure could form *two and only two* different monochloro-compounds when hydrogen atoms are replaced by chlorine?
(vi) Which structure has the most carbon atoms in *one* plane?
(vii) Which structure requires the least amount of oxygen per mole for complete combustion?
(viii) Which is the least volatile?
(ix) Which could *not* be produced by addition of hydrogen to an alkene?
(b) Write the systematic names for compounds **A**, **B**, **C**, **D** and **E** in part (a).
(c) Using the structures **A**, **B**, **C**, **D** and **E**, and others as required, explain and illustrate what you understand by the terms (i) structural isomerism, and (ii) optical isomerism.

A
```
              CH3
              |
CH3—CH—CH—CH3
       |
       CH3
```

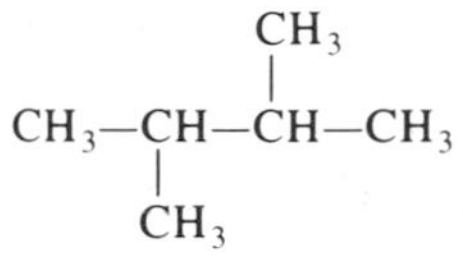

B
```
                 CH3 CH3
                 |    |
CH3—CH2—CH—CH—CH3
```

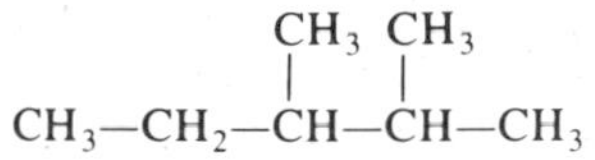

C
```
           H
          /
H2C—C
 |    | \
 |    |  CH3
H2C—CH2
```

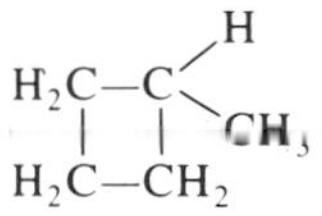

D
```
        CH3
        |
CH3—C—CH3
        |
        CH3
```

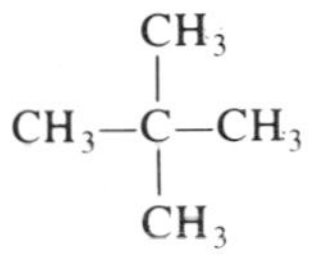

E
```
      CH3
      |
      CH
     /  \
H2C    CH2
 |       |
H2C——CH2
```

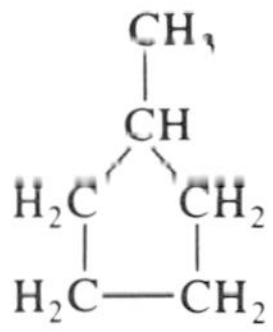

(*SUJB*)

9 (a) Describe the meaning of the term *homolytic fission*, using an organic reaction as your example.
(b) Give an example of an industrial process in which homolytic reaction(s) are involved in making organic compounds, and explain the part played by this type of reaction.
(c) Homolytic bromination of $C_6H_5CH_2CH{=}CHC_6H_5$ gives **K**, $C_6H_5CHBrCH{=}CHC_6H_5$. Draw clear diagrams to show all the stereoisomers of compound **K** and explain the different types of isomerism shown by **K**.
(d) State whether you consider the methyl radical to be neutral or charged and give reasons for your choice.

(*Oxford*)

10 The following substances either mix completely with water or dissolve very readily in it:

ethanol, glucose, propylamine, sodium ethanedioate (sodium oxalate).

On the other hand, the following substances are either sparingly soluble or insoluble in water:

cyclohexane, heptan-1-ol, phenylamine (aniline), calcium ethanedioate (calcium oxalate).

Discuss, with reference to these or other examples, some of the factors which determine whether an organic compound is likely to be soluble or insoluble in water. (*Nuffield*)

28 *Halogenoalkanes*

When a hydrogen atom of an alkane is replaced by a halogen atom, the chemistry of the resulting halogenoalkane is controlled largely by the properties of the carbon–halogen bond. Whereas in the alkanes reactions occur more or less indiscriminately at different parts of the molecule, in the halogenoalkanes they are localized at the halogen functional group.

In the **halogenoalkanes** a hydrogen atom of an alkane is replaced by a halogen atom, giving a compound with the general formula $C_nH_{2n+1}Hal$. The compounds contain a functional group, the halogen atom, attached to an alkane skeleton, and the chemistry of such compounds is largely controlled by the properties of the functional group, with the skeleton acting as a relatively inert support. The chemistry of $CH_3—Cl$ is similar to that of $CH_3CH_2—Cl$ and of $CH_3CH_2CH_2—Cl$, and so on. This series of compounds is an example of a **homologous series** (Greek: *homologos*, agreeing), each compound differing from the previous member of the series by the addition of a CH_2 group. The members of a homologous series all have similar chemical properties, and their physical properties, such as boiling point, show a steady change along the series. The term 'homologous series' should not be used to describe compounds that have the same functional group but with different patterns of chain branching, such as $CH_3CH_2—Cl$, $(CH_3)_2CH—Cl$, $(CH_3)_3C—Cl$ because, as we shall see, they show significant differences in their chemical reactivity. The terms **primary**, **secondary** and **tertiary** (see Section 27.1) that describe the alkyl groups in these three compounds are commonly used, not strictly correctly, to describe the halogenoalkanes themselves: $CH_3CH_2—Cl$, for example, is said to be a primary chloroalkane.

28.1 Nomenclature

The preferred way of naming halogenoalkanes is similar to that used for branched-chain alkanes, described in Section 27.1. The spine is taken as the longest continuous carbon chain bearing the halogeno group, which is named as a substituent just like an alkyl side chain, as in the following examples:

$CH_3—Cl$ chloromethane

$CH_3—CH_2—CH_2—Br$ 1-bromopropane

$CH_3—CH(Br)—CH_3$ 2-bromopropane

(Note that there are two isomers of each monohalogenopropane.) Different substituents are named in alphabetical order; for example:

$CH_3—C(CH_3)(Br)—CI_3$ 2-bromo-1,1,1-triiodo-2-methylpropane

An older convention, still widely used for the simpler molecules, names the halogenoalkanes in inorganic style as **alkyl halides**: $CH_3—Cl$, in this system, is called methyl chloride.

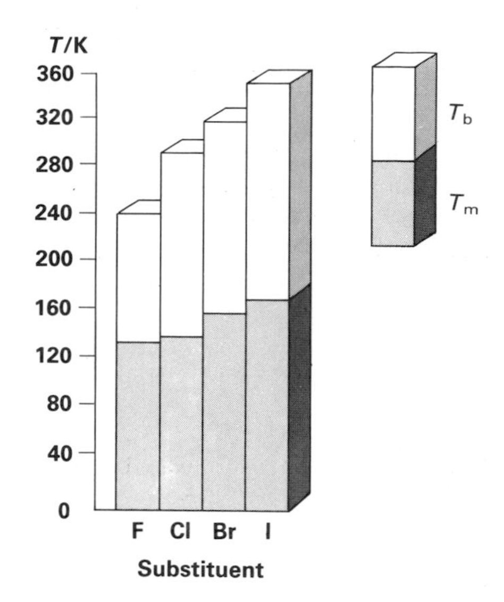

Figure 28.1 Melting and boiling points of halogenoethanes

28.2 Physical properties

When a hydrogen atom of an alkane is replaced by a halogen atom, the melting and boiling points are both raised. The increase in the attractive van der Waals forces between molecules is partly due to electrostatic interactions between their permanent dipoles, which arise from the electronegativity of the halogens. Nevertheless, the melting and boiling points rise in the order $RF < RCl < RBr < RI$, the order of *decreasing* electronegativity (see Figure 28.1): this indicates that the main effect is that the larger, more polarizable halogen atoms increase the dispersion forces between the molecules, as for the hydrogen halides discussed in Section 5.1. Dispersion forces also allow the halogenoalkanes to dissolve in solvents of low polarity and high polarizability, such as carbon disulphide or tetrachloromethane, whereas they are only very slightly soluble in water.

The chloroalkanes are more widely used than the other halogenoalkanes because they are cheaper. Chloromethane and chloroethane are gases at room temperature and pressure, however, and the liquids iodomethane and bromoethane are generally used in the laboratory because they are easier to handle. Chloroethane ($t_b = 12\,°C$) is readily liquefied under moderate pressure. It can be kept in a small spray can and squirted on to the skin, cooling it rapidly as the liquid evaporates and acting as a local anaesthetic; travellers in the American deserts use it for snake bites because the cooling reduces the circulation of the poisoned blood.

28.3 Reactivity patterns and important reactions

The chemistry of the halogenoalkanes is controlled mainly by the properties of the carbon–halogen bonds.

1 The strengths of the bonds The carbon–fluorine bond is very strong (compare Table 9.4), stronger even than carbon–hydrogen, and it is not easily broken. The fluoroalkanes are therefore generally unreactive, a commercially useful property (see Section 28.5). With the larger halogens – chlorine, bromine and iodine – the bonds to carbon get progressively weaker, mainly because of the difference in the sizes of the atoms being bonded together. The reactivity of the halogenoalkanes increases in the order $RF < RCl < RBr < RI$. The pattern is illustrated, for example, by their reactions with aqueous silver nitrate, which can be used to test for the presence of halogen in many organic compounds as well as in inorganic halides. 1-Iodobutane dissolved in warm ethanol rapidly gives a yellow precipitate of silver iodide with aqueous silver nitrate; 1-chlorobutane gives a white precipitate of silver chloride only after several minutes.

2 The polarity of the bonds Carbon–halogen bonds are polar. The electronegative halogen atom attracts the bonding electrons, polarizing the bond in the direction $C^{\delta+}$—$Hal^{\delta-}$. This encourages reactions in which the bond is broken heterolytically (see Section 26.2), with both bonding electrons being taken by the halogen to form the halide anion, Hal^-, which is called the **leaving group**.

A further consideration is the nature of the alkyl group to which the halogen is attached; we shall consider in particular the following examples:

H	H	H	CH_3
H—C—Br	CH_3—C—Br	CH_3—C—Br	CH_3—C—Br
H	H	CH_3	CH_3
bromomethane	bromoethane	2-bromopropane	2-bromo-2-methylpropane
	primary alkyl group	secondary alkyl group	tertiary alkyl group

Nucleophilic substitution

When 2-bromo-2-methylpropane is dissolved in a mixture of water and ethanol an equilibrium is set up between the covalent halogenoalkane and a small proportion of an ionic species in which the carbon–halogen bond is completely broken and the carbon atom is left bearing a positive charge, that is, it forms a **carbocation**. The 'curly' arrow shows the transfer of an electron pair:

$$CH_3-\overset{\overset{CH_3}{|}}{\underset{\underset{CH_3}{|}}{C}}-\ddot{\underset{..}{Br}}: \rightleftharpoons CH_3-\overset{+}{C}\begin{matrix} CH_3 \\ CH_3 \end{matrix} + :\ddot{\underset{..}{Br}}:^-$$

dimethylethyl cation

2-bromo-2-methylpropane dissolved in warm ethanol gives an immediate cream precipitate of silver bromide when aqueous silver nitrate is added to it. Carbocations are highly reactive, because the central carbon atom bears only six valence electrons and can form a new bond to a reagent that can provide an unshared electron pair. One way it can do this is to react with the liberated bromide ion to regenerate the halogenoalkane, but if the reaction is carried out in aqueous sodium hydroxide it can react with the hydroxide ion to give 2-methylpropan-2-ol:

$$H\ddot{\underset{..}{O}}:^- + CH_3-\overset{+}{C}\begin{matrix} CH_3 \\ CH_3 \end{matrix} \longrightarrow CH_3-\overset{\overset{CH_3}{|}}{\underset{\underset{CH_3}{|}}{C}}-\ddot{\underset{..}{O}}H$$

An attacking group that forms a new covalent bond by providing the electron pair is called a **nucleophile** (nucleus-loving, from the Greek: *philos*, loving). A group that forms a new bond by accepting the electron pair, the carbocation in this example, is called an **electrophile** (electron-loving). A nucleophile is thus a Lewis base (see Section 12.4), and an electrophile is a Lewis acid; however, when organic chemists talk about Lewis acids they are usually referring only to inorganic entities, such as $AlCl_3$.

Nucleophiles are often negatively charged, like the hydroxide ion in the reaction above, but they need not be. Water, for example, is a nucleophile because of the lone electron pairs on the oxygen atom, and it can react in a similar way:

$$H_2\ddot{\underset{..}{O}} + CH_3-\overset{+}{C}\begin{matrix} CH_3 \\ CH_3 \end{matrix} \longrightarrow CH_3-\overset{\overset{CH_3}{|}}{\underset{\underset{CH_3}{|}}{C}}-\ddot{O}H_2^+$$

The resulting cation can lose a proton to the solvent, so that the product is again the alcohol.

The complete reaction, starting from 2-bromo-2-methylpropane, is a substitution, in which the bromide ion is replaced by a hydroxide ion. It is therefore called a **nucleophilic substitution** reaction. The rate-limiting step (see Section 14.3) is the unimolecular breaking of the carbon–bromine bond. The reaction is commonly denoted S_N1: substitution, **n**ucleophilic, **uni**molecular. The nucleophile does not enter the reaction till the second, rapid step. Consequently it cannot affect the reaction rate, and the reaction follows first-order kinetics (see Section 14.2):

$$rate = k[(CH_3)_3CBr]$$

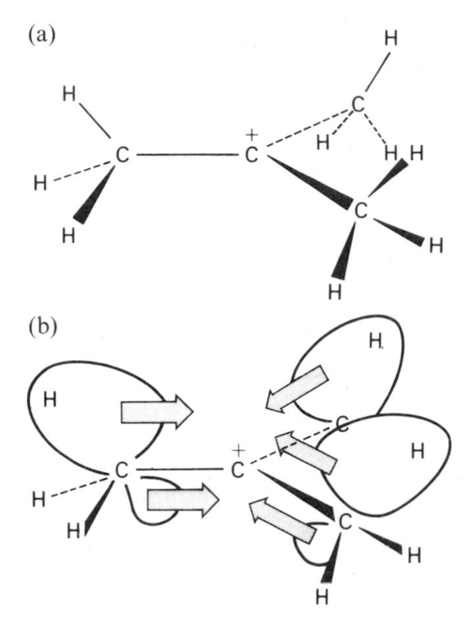

Figure 28.2 (a) Geometry of the tertiary carbocation $(CH_3)_3C^+$; (b) electronic stabilization by attraction of C—H bonding electrons towards positive centre: all nine C—H bonds are distorted in this way (two hydrogen atoms attached to the rear carbon atom are omitted for clarity)

The reaction with aqueous sodium hydroxide goes at the same rate as that with water alone.

In a carbocation the three electron pairs around the positively charged carbon atom are arranged in a plane (see Figure 2.12). For a tertiary alkyl cation, such as dimethylethyl, all three bonds lead to other carbon atoms, which provide a plentiful supply of electrons (Figure 28.2(a)). Electrostatic attraction pulls them towards the positive charge, as illustrated in Figure 28.2(b), distorting the electron distribution in the carbon–hydrogen bonds and leaving the hydrogen atoms bearing a little of the positive charge, which is thereby spread out instead of being fixed entirely on one atom. A dispersed charge is more stable than a concentrated one, so that the effect of the alkyl groups around the cationic centre is to stabilize it. A hydrogen atom bonded to the cationic carbon atom cannot provide a similar source of electrons, and the stability of carbocations therefore increases in the order:

$$\underset{\text{methyl}}{H{-}\overset{+}{C}H_2} < \underset{\text{primary}}{CH_3{-}\overset{+}{C}H_2} < \underset{\text{secondary}}{CH_3{-}\overset{+}{C}H(CH_3)} < \underset{\text{tertiary}}{CH_3{-}\overset{+}{C}(CH_3)_2}$$

For this reason bromomethane, unlike 2-bromo-2-methylpropane, cannot simply dissociate into ions. Instead, the nucleophile approaches the reaction centre from one side *while the bromide ion is leaving from the other*. In this way a full charge does not build up on the carbon atom. The reaction is said to be **concerted**, because bond making and bond breaking take place in concert: the new carbon–nucleophile bond is forming *while* the old carbon–bromine bond is breaking. In this reaction mechanism there is only a single **bi**molecular reaction step and the mechanism is denoted SN2. The kinetics are second-order:

$$rate = k[HO^-][CH_3Br]$$

During the reaction the three bonds that stay attached to the central carbon atom are turned inside out, like an umbrella in a high wind; the process is called **Walden inversion**, after the Latvian chemist Paul Walden who first showed it happens.

$$H\ddot{O}:^- \quad H_3C{-}\ddot{B}r: \longrightarrow \left[H\ddot{O}\cdots CH_3\cdots\ddot{B}r:\right]^- \longrightarrow H\ddot{O}{-}CH_3 \quad :\ddot{B}r:^-$$

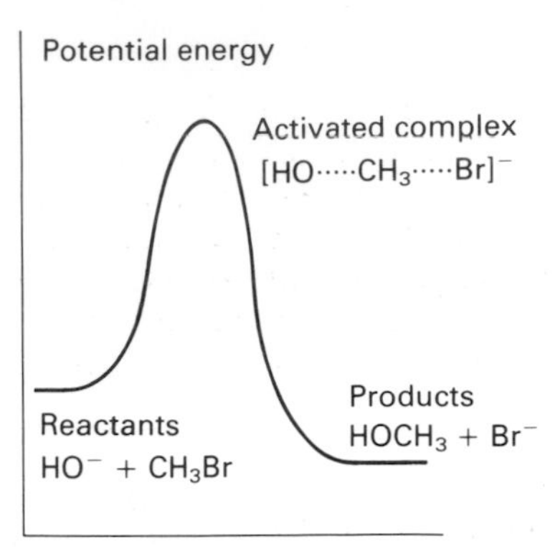

Figure 28.3 Changes in potential energy during an SN2 reaction

While the reaction is proceeding the central carbon atom has five groups attached to it: three are bonded normally whereas the incoming and leaving groups are more weakly bonded by partial bonds, one in the process of forming, the other being broken. This structure is not a normal compound with a definite lifetime: it is called an **activated complex** and represents the arrangement of atoms with maximum potential energy on the path from reactants to products (see Figure 28.3, which should be compared with Figure 14.15).

If the five groups crowded round the central carbon atom of the activated complex are bulky their mutual repulsions raise its energy, increase the activation energy of the reaction, and consequently slow it down. The order of reactivity for the concerted (SN2) mechanism is therefore:

tertiary < secondary < primary < methyl

This is the exact opposite of the order for the carbocation (SN1) mechanism. Consequently, methyl and primary compounds normally react by the concerted mechanism, tertiary compounds by the carbocation mechanism, and secondary compounds by either or both routes. We may picture a gradual merging of one mechanism into the other (see Figure 28.4): starting with the reactions of the halogenomethanes the progressive replacement of hydrogen atoms by methyl groups both hinders the approach of the attacking nucleophile and also helps to expel the halide by stabilizing the carbocation.

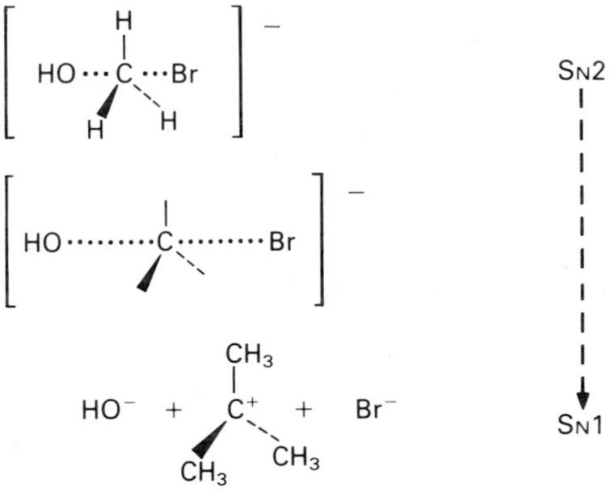

$[HO\cdots CH_3 \cdots Br]^-$ (SN2)

$[HO\cdots\cdots C \cdots\cdots Br]^-$

$HO^- + C^+(CH_3)_3 + Br^-$ (SN1)

Figure 28.4 Gradual progression from SN2 mechanism (illustrated for bromomethane) to SN1 mechanism (illustrated for 2-bromo-2-methylpropane); the midpoint of the reaction is an activated complex in SN2 and a carbocation in SN1

Elimination

We saw above that a nucleophile is a Lewis base. It is therefore also a Brønsted base (see Sections 12.2 and 12.4). The hydroxide ion can therefore also react with a halogenoalkane by pulling a proton from the carbon atom next to the one bearing the halogen. For example, with 2-bromo-2-methylpropane:

$$HO:^- + H_3C\text{–}C(CH_3)_2\text{–}Br: \longrightarrow HOH + H_2C{=}C(CH_3)_2 + :Br:^-$$

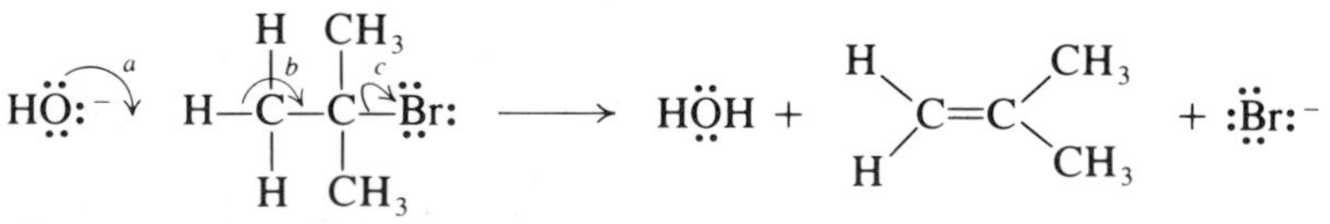

The 'curly' arrows labelled *a*, *b* and *c* show the transfer of electron pairs during the reaction. Arrow *a* shows the basic hydroxide ion providing two electrons to form a new bond to the hydrogen atom, which is effectively lost as a proton, H^+. Arrow *b* shows that the hydrogen–carbon bond is broken heterolytically, the electron pair going to form a second bond between the two carbon atoms. Arrow *c* shows the heterolytic breaking of the carbon–bromine bond with the expulsion of the bromide anion from the organic part of the molecule. The overall reaction is an **elimination**: H^+ and Br^- ions are eliminated from the bromoalkane with the formation of an alkene (alkenes are discussed in detail in Chapter 33).

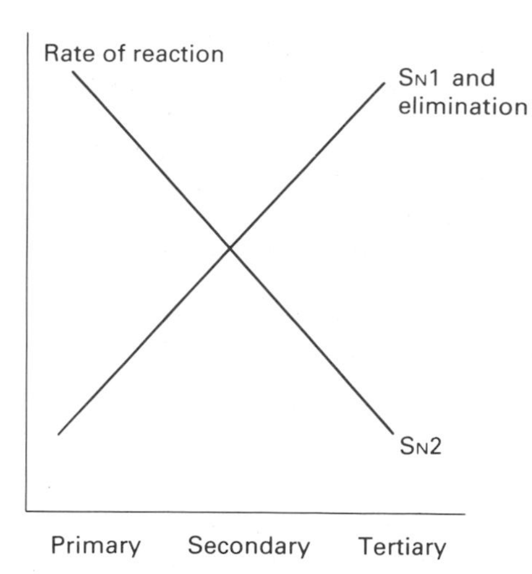

Figure 28.5 Effects of the structure of the alkyl group on the rates of substitution and elimination reactions

Substitution v. elimination

All halogenoalkanes that contain an H—C—C—Hal grouping undergo both substitution and elimination when treated with a Lewis base. The balance between the two depends on two main factors. The first is the base strength of the attacking group: strong bases encourage elimination. The second is the structure of the alkyl group – whether it is primary, secondary or tertiary. We have seen that in substitutions structure affects SN1 and SN2 reactions in opposite ways. In eliminations the order of reactivity is the same as in SN1:

primary < secondary < tertiary

because bulky groups attached to the same carbon atom as the halogen atom help to push it out. Figure 28.5 summarizes the effects of the structure of the alkyl group on the different reactions.

It is clear that for primary halogenoalkanes the normal reaction is substitution, via the SN2 mechanism. Bromoethane reacts with many nucleophiles, as exemplified in the summary that follows, to give substitution products with only a few per cent of elimination. With primary alkyl groups larger than ethyl the elimination reaction can be encouraged by using the strongly basic solution formed by dissolving potassium hydroxide in ethanol.

For secondary halogenoalkanes, substitution and elimination are more finely balanced. Strong bases give mostly elimination, whereas nucleophiles that are only weakly basic give mostly substitution:

$$CH_3\text{—}CH(CH_3)\text{—}Br + KOH \xrightarrow[55^\circ C]{C_2H_5OH} CH_2{=}C(H)CH_3 \qquad 80\%\ \text{elimination}$$

$$CH_3\text{—}CH(CH_3)\text{—}Br + NaI(aq) \longrightarrow I\text{—}CH(CH_3)\text{—}CH_3 \qquad 90\%\ \text{substitution}$$

Except for the SN1 reaction with water, discussed above, tertiary halogenoalkanes give mostly elimination.

Summary of important nucleophilic substitution reactions

A variety of nucleophiles can be used in reactions with halogenomethanes and with primary and secondary halogenoalkanes (except for fluoroalkanes, which are unreactive). As explained above, these reactions are not usually satisfactory with tertiary halogenoalkanes. Nucleophilic substitution offers useful ways of introducing various functional groups into organic molecules. Some important ones are shown below, using the reactions of bromoethane as examples. Typical reaction conditions are quoted: see also Section 30.2 for a discussion of the use of different solvents, and Box 32.1.

1 $$HO^- + C_2H_5\text{—}Br \xrightarrow[\text{reflux}]{NaOH(aq)} C_2H_5\text{—}OH + Br^-$$

ethanol

The reaction is carried out by heating the halogenoalkane under reflux (see Figure 28.6) with aqueous sodium hydroxide to give an alcohol. Tertiary alcohols can be made by treating a tertiary halogenoalkane with water alone. (Alcohols are discussed in detail in Chapter 29.)

2 $$CH_3O^- + C_2H_5\text{—}Br \xrightarrow[\text{reflux}]{CH_3OH} C_2H_5\text{—}OCH_3 + Br^-$$

methoxyethane

This reaction is often called **Williamson's ether synthesis**. The usual reagent is sodium methoxide, $NaOCH_3$, in methanol, CH_3OH, made from the reaction of sodium metal with methanol.

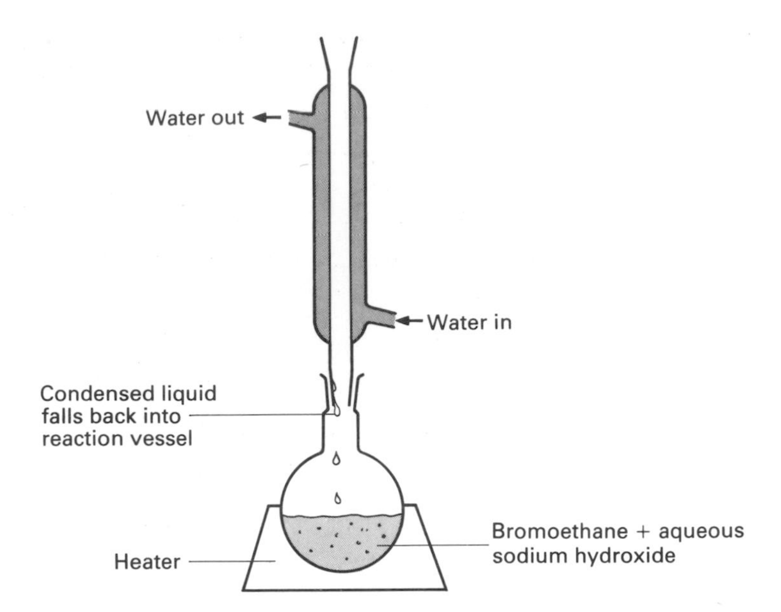

Figure 28.6 Reaction under reflux: volatile compounds are condensed and returned to the reaction mixture

3 $NC^- + C_2H_5{-}Br \xrightarrow[\text{reflux}]{C_2H_5OH} C_2H_5{-}CN + Br^-$

propanenitrile

This reaction, carried out with potassium cyanide in ethanol, is an important way of extending the carbon chain with the formation of a new carbon–carbon bond. Nitriles, and their use in organic syntheses, are discussed further in Section 31.4.

4 $H_3N + C_2H_5{-}Br \longrightarrow C_2H_5{-}NH_3^+ \; Br^-$

ethylammonium bromide

The reaction may be carried out with aqueous ammonia or with ammonia gas in a sealed tube: it is discussed further in Section 32.3.

28.4 Preparative methods for halogenoalkanes

There are two main ways of making halogenoalkanes. They are discussed in detail in other chapters and only summarized here.

1 Alcohols, ROH, can be converted into halogenoalkanes with phosphorus halides (see Section 29.3):

$$ROH \xrightarrow{PCl_5} RCl$$

$$ROH \xrightarrow{PBr_3} RBr$$

$$ROH \xrightarrow{PI_3} RI$$

(The phosphorus tribromide and triiodide used in the last two reactions are sometimes made in the reaction mixture by treating the alcohol with bromine or iodine and red phosphorus.)

The reactions, especially that with phosphorus pentachloride, give a variety of side products, including alkenes and organic phosphorus compounds, as well as the desired halogenoalkane. For chloroalkanes a cleaner reaction is that of alcohols with sulphur dichloride oxide (thionyl chloride) under reflux. This reaction is also experimentally very convenient because the other products are gases that escape from the reaction mixture:

$$ROH + SCl_2O \xrightarrow{\text{reflux}} RCl + HCl + SO_2$$

Bromoalkanes can also be made using concentrated hydrobromic acid, cold for tertiary bromoalkanes, under reflux for primary ones:

$$ROH + HBr \rightarrow RBr + H_2O$$

Hydrogen chloride is less reactive, and although hydrogen iodide does give the iodoalkane it often reacts further: $RI + HI \rightarrow RH + I_2$ (see the discussion of the reaction of iodine with alkanes in Section 27.3).

2 **Alkenes** add hydrogen bromide (from concentrated hydrobromic acid) to give bromoalkanes:

$$CH_2{=}CH_2 + HBr \rightarrow CH_3{-}CH_2{-}Br$$

Concentrated hydrochloric acid similarly gives chloroalkanes. If the double bond has alkyl groups at one end and not at the other, the addition product has the halogen at the alkylated end of the double bond:

$$CH_3{-}CH{=}CH_2 + HBr \rightarrow CH_3{-}CHBr{-}CH_3$$

This is an application of Markovnikov's rule, discussed in Section 33.3.

28.5 The importance of halogenoalkanes

For the laboratory chemist the halogenoalkanes are important for their role as **synthetic intermediates**, that is, as compounds that are easy to make and are in turn easily converted into a wide variety of other compounds via their nucleophilic substitution reactions.

Commercially the only important monohalogenoalkanes are chloroethane and chloromethane. Chloroethane is used for making tetraethyllead(IV) (see Box 27.3):

$$4C_2H_5Cl + \underset{\text{sodium–lead alloy}}{4Na + Pb} \xrightarrow{\text{heat}} (C_2H_5)_4Pb + 4NaCl$$

Chloromethane is used for making $(CH_3)_2SiCl_2$, which is in turn used to make siloxanes, described in Section 20.6.

Compounds with more than one halogen atom are very widely used. Dichloromethane is an important solvent, used in paint strippers and in the spinning of viscose fibres (see Section 30.5). Trichloromethane (**chloroform**) was once widely used as an anaesthetic and tetrachloromethane (carbon tetrachloride) as an industrial and dry-cleaning solvent. These uses ceased when it was found that both compounds are potentially carcinogenic (cancer-inducing). Both are still used, however, in the manufacture of the important chlorofluoromethanes (**Freons**) CCl_2F_2, CCl_3F and $CHClF_2$ using hydrogen fluoride and an antimony(V) halide catalyst; for example:

$$CCl_4 + 2HF \xrightarrow{SbCl_4F} CCl_2F_2 + 2HCl$$

The chlorofluoromethanes are very stable: the strong carbon–fluorine bond is not easily broken, and the electron-attracting fluorine atoms oppose the polarity of the carbon–chlorine bonds so reducing their reactivity. The compounds do not burn, nor do they react in biological systems so that they are non-toxic. These properties, and their low boiling points ($t_b(CCl_2F_2) = -30\,°C$) have led to their widespread use as refrigerants and aerosol propellants. They are also used in sprays for treating muscular sports injuries.

In recent years there has been some anxiety about the use of Freons in aerosols. As they diffuse into the upper atmosphere they can be broken into radicals by the intense ultraviolet radiation there; for example, CCl_2F_2 can give $CClF_2^{\bullet}$ radicals and chlorine atoms. It is thought that these might react with the ozone, itself produced photochemically in the upper atmosphere, and which in turn absorbs a high proportion of the potentially damaging ultraviolet radiation before it reaches the Earth's surface (see Colour 8).

Bromochlorodifluoromethane, $CBrClF_2$, though considerably more expensive than tetrachloromethane, CCl_4, is now widely used instead as a fire extinguisher. Tetrachloromethane works by blanketing the fire in a dense vapour, thus excluding oxygen. Bromochlorodifluoromethane does the same, but also enters into the combustion reaction; bromine atoms react with the chain-propagating radicals of the flame, so helping to quench the fire.

BOX 28.1

Anaesthetics

Most societies, even the most primitive, know some naturally occurring substance that eases pain – opium, for example. In the mid-nineteenth century three chemicals were found to induce sleep during which pain was not felt at all: these were dinitrogen oxide ('laughing gas'), ethoxyethane (ether) and trichloromethane (chloroform). Anaesthetics (the word comes from the Greek for 'insensible') are absorbed into the bloodstream and are then taken up by the fatty tissues because they are more soluble in fat than in water (see Section 28.2). Nerve endings are surrounded by fatty molecules (lipids), and the anaesthetics interfere with the transmission of nerve impulses.

The new anaesthetics (together with antiseptics) transformed surgery, but they all suffer from some disadvantages. Dinitrogen oxide is only weakly anaesthetic, useful for minor operations but not for major surgery. Ether often causes unpleasant side effects such as nausea. Chloroform is potentially dangerous because the lethal dose is only slightly higher than the effective dose. In the 1920s cyclopropane was found to be a powerful and safe anaesthetic without side effects; however, it forms dangerously explosive mixtures with air. In 1956 chemists at ICI found $CF_3CHBrCl$ (**halothane**) to be a near-ideal general anaesthetic: non-toxic, non-flammable and without apparent side effects. Unfortunately there now seems to be a possibility that prolonged exposure – as experienced by workers in the operating theatre – may occasionally cause liver damage.

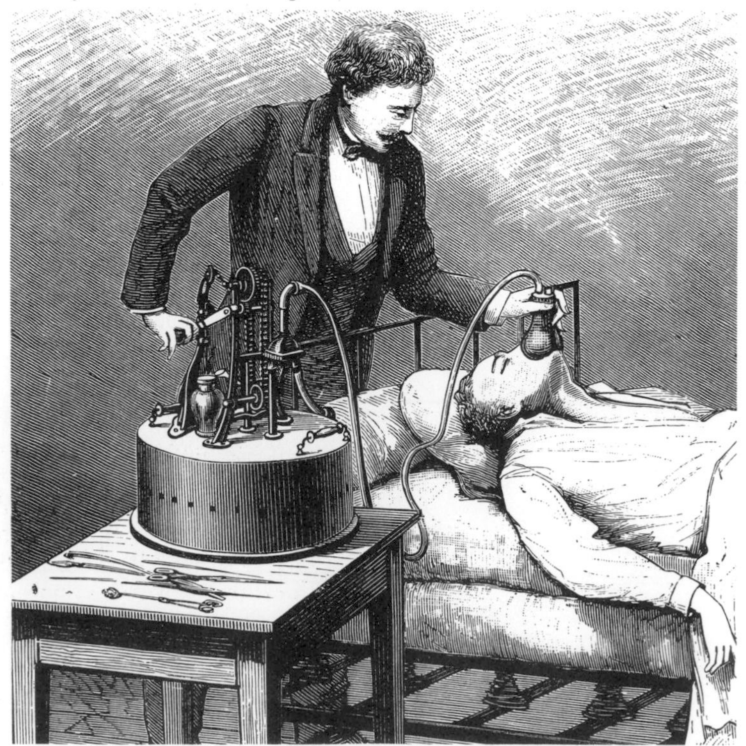

Figure 28.7 Chloroform being administered as an anaesthetic in the 1880s

Summary

- ☐ The halogenoalkanes contain a halogen functional group attached to an alkane skeleton.
- ☐ A series of halogenoalkanes with the general structure $R(CH_2)_nHal$ is an example of a homologous series: the compounds all have similar chemical properties and their physical properties change in a regular fashion.
- ☐ The halogenoalkanes are named by the same rules that are used for branched-chain alkanes, treating the halogen group as a substituent on the main alkane chain. They may also be named as alkyl halides.
- ☐ The polarity of the carbon–halogen bond and, more especially, the polarizability of the heavier halogens increase the intermolecular attractions as compared with those between alkane molecules, and raise the melting and boiling points.
- ☐ The reactivity of the halogenoalkanes increases in the order $RF < RCl < RBr < RI$, the order of decreasing carbon–halogen bond strength.
- ☐ The halogenoalkanes react chiefly by heterolysis of the carbon–halogen bond, with the expulsion of the halide anion.
- ☐ Nucleophiles are Lewis bases, reagents that form new covalent bonds by providing both the bonding electrons. Electrophiles (Lewis acids) form new covalent bonds by accepting an electron pair from a nucleophile.
- ☐ Halogenoalkanes undergo nucleophilic substitution reactions of the type:

$$Nu^- + R{-}Hal \rightarrow R{-}Nu + Hal^-$$

(where Nu^- is an abbreviation for a nucleophile, although not all nucleophiles are negatively charged).

- ☐ Halogenoalkanes undergo elimination reactions of the type:

$$B^- + H{-}\overset{|}{\underset{|}{C}}{-}\overset{|}{\underset{|}{C}}{-}Hal \rightarrow BH + \;>C{=}C<\; + Hal^-$$

(where B^- is an abbreviation for a base: again, not all bases are negatively charged).

- ☐ Primary halogenoalkanes react predominantly by the concerted SN2 mechanism.
- ☐ Tertiary halogenoalkanes react mainly by elimination, except for the SN1 reaction with water.
- ☐ The behaviour of secondary halogenoalkanes is intermediate between that of primary and that of tertiary halogenoalkanes. Substitution predominates with weakly basic nucleophiles. Elimination can be encouraged by using potassium hydroxide in ethanol.
- ☐ Important substitution reactions are summarized below:

$$HO^- + RHal \rightarrow ROH + Hal^-$$
$$R'O^- + RHal \rightarrow ROR' + Hal^-$$
$$NC^- + RHal \rightarrow RCN + Hal^-$$
$$H_3N + RHal \rightarrow RNH_3^+\ Hal^-$$

- ☐ Halogenoalkanes are made from alcohols using phosphorus halides or sulphur dichloride oxide, SCl_2O, and by the addition of hydrogen halides to alkenes.
- ☐ Chloroethane is used in the manufacture of tetraethyllead(IV). Chloromethane is used in making siloxane oils. Dichloromethane is an important solvent. Chlorofluoromethanes (Freons) are refrigerants and aerosol propellants. Bromochlorodifluoromethane, $CBrClF_2$, is a fire extinguisher. 2-Bromo-2-chloro-1,1,1-trifluoroethane (halothane), $CF_3CHBrCl$, is an important anaesthetic.

PROBLEMS

1 Show all the structural isomers of molecular formula C_4H_9F. Classify the alkyl groups in them as primary, secondary or tertiary.

2 State the reagents and reaction conditions for converting C_2H_5Cl into C_2H_5OH, and for the reverse reaction.

3 Outline the mechanisms of the following reactions:

$C_2H_5Br \rightarrow C_2H_5OH$
$(CH_3)_3CBr \rightarrow (CH_3)_3COH$

Briefly account for the differences between them.

4 What are the two main types of reaction that halogenoalkanes undergo? Illustrate them by describing the reactions of 1-chloropropane with (a) NaOH(aq) and (b) KOH in ethanol.

5 Show how the following conversions can be carried out, using only inorganic reagents.

(a) $CH_4 \rightarrow CH_3OH$
(b) $C_2H_5OH \rightarrow C_2H_5CN$
(c) $CH_3I \rightarrow CH_3OCH_3$
(d) $CH_3CH_2CH_2Br \rightarrow CH_3CHBrCH_3$

6 (a) Give a mechanism for the conversion of an alkane into corresponding halogeno-compounds. Further give a mechanism for the conversion of a halogeno-compound into a hydroxy-compound.
(b) Describe how a typical hydroxy-compound can be formed directly from an alkene.
(c) Starting from the halogenoalkane RCH_2X (where R = alkyl and X = halogen) suggest how the amines RCH_2NH_2 and $RCH_2CH_2NH_2$ and the carboxylic acids RCOOH and RCH_2COOH can each be synthesized.
(d) Discuss the relative reactivity of the chlorine atom in 1-chlorobutane, (chloromethyl)benzene and chlorobenzene and describe how you would establish, experimentally, the differences in reactivity. (*Welsh*)

7 (a) Draw a well-labelled diagram of the apparatus you would use to prepare *either* bromoethane (ethyl bromide), b.p. 38 °C, *or* chloroethane (ethyl chloride), b.p. 13 °C, in the laboratory. Your diagram should indicate the reagents and conditions to be used.
(b) Write an equation (or equations) for the reaction involved.
(c) Name *three* impurities which might contaminate the initial sample of bromoethane (or chloroethane).
(d) Describe how you would purify your initial sample of bromoethane (or chloroethane).
(e) How, and under what conditions, does bromoethane (or chloroethane) react with sodium hydroxide? (*SUJB*)

8 (a) Give the reagents, conditions and equations for *two* routes by which bromoethane (ethyl bromide) can be synthesized.
(b) Write down *two* equations to illustrate the different types of reaction shown by bromoethane, and outline the reason for this behaviour in *one* case.
(c) Explain the meaning of the term *homolytic*, using a homolytic chlorination reaction as your example. (*Oxford*)

9 Consider the compounds whose formulae are shown below.

```
   H  H  H  H              H  H  H   O
   |  |  |  |              |  |  |  //
H—C—C—C—C—Br          H—C—C—C—C
   |  |  |  |              |  |  |  \
   H  H  H  H              H  H  H   Br

   H  H  H  H
   |  |  |  |
H—C—C—C—C—H            (benzene ring)—Br
   |  |  |  |
   H  H  Br H

      H
      |
   H—C—H
   H  |  H
   |  |  |
H—C—C—C—H
   |  |  |
   H  Br H
```

Compare the reactions of these bromo-compounds with reagents such as water, hydroxide ions and ammonia, giving reasons for any differences which exist. (*Nuffield*)

10 Write an account of the chemistry of halogenoalkanes.
You should consider the reactions suitable for their preparation and their characteristic reactions. You should also refer to some industrial applications and to the mechanism of at least one of their reactions. (*Nuffield*)

11 Study the reaction scheme below and answer the questions which follow.

C_4H_9Br (**A**) $\xrightarrow{OH^-(aq)}$ C_4H_9OH (**B**) $\xrightarrow{Cr_2O_7^{2-}(aq)/H^+(aq)}$ C_4H_8O (**C**) $\xrightarrow{Ag^+/NH_3(aq)}$ no reaction

C_4H_9Br (**A**) $\xrightarrow{OH^-(ethanolic)}$ C_4H_8 (**D**) $\xrightarrow[(ii)\ H^+(aq)/Zn]{(i)\ O_3}$ $2C_2H_4O$ (**E**) $\xrightarrow{Ag^+/NH_3(aq)}$ silver mirror

(a) Give the structural formula of each of the compounds **A** to **E**.
(b) Which, if any, of the compounds **A** to **E** give(s) a yellow precipitate when warmed with a solution of iodine in aqueous alkali?
(c) Which, if any, of the compounds **A** to **E** exist(s) in optically isomeric forms?
(d) Which, if any, of the compounds **A** to **E** exist(s) in geometrically isomeric forms? (*JMB*)

29 Alcohols and ethers: C—O single bonds and O—H bonds

In the alcohols, R—O—H (**1**), the presence of C—O and O—H bonds leads to a wide range of reactions. Ethers, R—O—R′ (**2**), are much less reactive. This chapter, therefore, deals mainly with the chemistry of the alcohols.

An alcohol can be regarded as having the structure of an alkane (C_nH_{2n+2}) with one of the carbon–hydrogen bonds opened up and an oxygen atom inserted (**3**). An ether has the oxygen atom inserted into a carbon–carbon bond (**4**). The general formula of an ether or an alcohol is therefore $C_nH_{2n+2}O$. We have seen that for alkanes with four or more carbon atoms there can be structural isomers with different patterns of chain branching, and that for halogenoalkanes there are even more varieties of structural isomers because the halogen atom can be attached to different carbon atoms on a chain. Now we see that there can be structural isomers that have different functional groups.

1 R—Ö—H

2 R—Ö—R′

3 $CH_3CH_2CH_2$—O—H (H–C–C–C–O–H, each carbon bearing H atoms)

4 CH_3CH_2—O—CH_3 (H–C–C–O–C–H, each carbon bearing H atoms)

EXAMPLE

How many structural isomers are there with the molecular formula $C_4H_{10}O$?

METHOD First consider the ethers. These must contain the fragment C—O—C: decide how the remaining carbon atoms can be attached to this fragment. The alcohols can all be represented as C_4H_9—OH: decide how many alkyl groups have the formula C_4H_9.

ANSWER There are three ethers, represented here by their frameworks of carbon and oxygen atoms:

C—C—C—O—C   C—C(C)—O—C   C—C—O—C—C

There are four alcohols:

C—C—C—C—OH   C—C—C(C)—OH   C—C(C)—C—OH   C—C(C)(C)—OH

COMMENT See Section 32.6 for further discussion of the isomerism of the second of the alcohols, $CH_3CH_2CH(OH)CH_3$ (butan-2-ol).

29.1 Nomenclature

The ethers are named in a way similar to that used for the halogenoalkanes. The longest carbon chain in the molecule, the spine, is found, and the RO— group attached to it is named as an **alkoxy** substituent: CH_3O— is the methoxy group, C_3H_7O— the propoxy group and so on. In an older form of nomenclature the names of the two alkyl groups attached to the oxygen atom were given, followed by the word 'ether'. Thus we have, for example:

$CH_3OCH_2CH_2CH_3$	1-methoxypropane	(methyl propyl ether)
$CH_3CH_2OCH_2CH_3$	ethoxyethane	(diethyl ether, commonly called just 'ether')

Figure 29.1 A cognac brandy distillery, with small, traditional pot-stills heated by coal fires

BOX 29.1

'Alcohol'

The word 'alcohol' comes from the Arabic *al kuhul*, which referred to a fine powder (actually antimony sulphide) used as a cosmetic eye shadow. Gradually the name came to be used for any powder that could be sublimed, and then for a liquid that could be obtained by distillation. In this sense it was synonymous with the word 'spirit' – as it still is! The liquid obtained by distilling wine was known as 'spirit of wine' or 'alcohol of wine'. 'Spirit' continued to have this general meaning: 'spirit of salt' meant hydrochloric acid, 'spirit of hartshorn' aqueous ammonia. 'Alcohol', however, became a specific name for spirit of wine, and only later was its meaning extended again and applied to the class of chemically related compounds.

A different system of nomenclature is used for the alcohols. The name is based on that of the alkane with the same carbon skeleton: the end of the name is then changed from *-ane* to *-anol*. The position of the hydroxyl (OH) group is given, where necessary, by a number before the 'ol':

$CH_3{-}OH$	methanol	$CH_3{-}CH_2{-}OH$	ethanol
$CH_3{-}CH_2{-}CH_2{-}OH$	propan-1-ol	$CH_3{-}CH(OH){-}CH_3$	propan-2-ol

Older names for alcohols are based on the name of the alkyl group – 'ethyl alcohol' for C_2H_5OH, for example.

Alcohols with two or more hydroxyl groups are said to be **dihydric**, **trihydric** and so on. They are named as in the following examples (traditional names that are still in common use are also given):

$CH_2(OH){-}CH_2(OH)$	ethane-1,2-diol	(ethylene glycol or glycol)
$CH_2(OH){-}CH(OH){-}CH_2(OH)$	propane-1,2,3-triol	(glycerol or, incorrectly, glycerine)

The descriptions **primary**, **secondary** and **tertiary** are used for alcohols (more strictly, for the alkyl groups in alcohols) in the same way as for halogenoalkanes:

$CH_3{-}C(H)(H){-}OH$	$CH_3{-}C(H)(CH_3){-}OH$	$CH_3{-}C(CH_3)(CH_3){-}OH$
primary alkyl group	secondary alkyl group	tertiary alkyl group

29.2 Physical properties

The physical properties of the ethers do not differ greatly from those of halogenoalkanes of similar molar mass. Methoxymethane is a gas at room temperature and pressure, ethoxyethane is a very volatile liquid ($t_b = 34.5\,°C$) that is widely used as a solvent. Ethers are more polar than alkanes but less so than ketones (see Section 30.2) and alcohols. Ethoxyethane needs to be handled with extreme care: its dense vapour can spread as a layer at floor level or along the bench and may be ignited explosively by a flame, a spark or even a hotplate some distance away. Compounds with two or more ether groups, such as $C_2H_5OCH_2CH_2OC_2H_5$ (1,2-diethoxyethane), are widely used as high-boiling solvents.

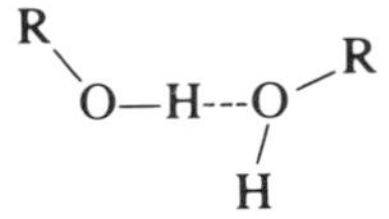

$R-O-H\cdots O(R)-H$ (5)

Alcohol molecules cling to each other more strongly than ether molecules do, because of hydrogen bonding between the hydroxyl hydrogen atom of one molecule and the oxygen atom of another (**5**). The melting points of alcohols, and especially their boiling points, are much higher than those of isomeric ethers:

	CH_3CH_2OH	CH_3OCH_3	$CH_3CH_2CH_2CH_2OH$	$CH_3CH_2OCH_2CH_3$
$t_m/°C$	−117	−138.5	−90	−116
$t_b/°C$	+78.5	−23	+117	+34.5

The lower alcohols mix with water in all proportions because of mutual hydrogen bonding between the water and alcohol molecules. (The *Guinness Book of Records* describes a German beer of the First World War that had an ethanol content of only about 0.1 per cent by volume, and a liquor from Estonia that contained 98 per cent.) Solvation of ions and partially charged groups by the hydroxyl group may overcome lattice enthalpies (see Section 9.4), so that lower alcohols dissolve many ionic and polar solutes. With increasing molar mass, as the hydrocarbon fragment becomes larger, the solubility of the alcohols in water decreases, because the dispersion forces holding the alkyl groups together become more important than the hydrogen bonding between alcohol and water. The higher alcohols are also less good solvents for ionic and polar solutes.

29.3 Reactivity patterns: reactions in which the C—O bond is broken

The properties of the carbon–oxygen single bond are similar to those of the carbon–fluorine bond (see Section 28.3). It is polarized in the sense $\overset{\delta+}{C}—\overset{\delta-}{O}$, but is strong ($360\ kJ\ mol^{-1}$) and not easily broken.

Alcohols (and ethers) are weak Brønsted bases in water. They can become protonated on one of the lone electron pairs on the oxygen atom (as for $H_2O \rightarrow H_3O^+$):

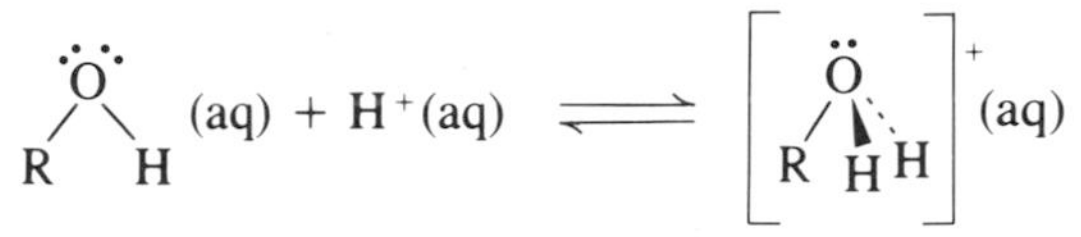

$$ROH(aq) + H^+(aq) \rightleftharpoons [ROH_2]^+(aq)$$

The resulting $C—OH_2^+$ bond is more easily broken than the uncharged C—OH bond, because the bonding electrons are strongly attracted towards the positive charge. Thus in the presence of a strong Brønsted acid (or, indeed, with a Lewis acid), alcohols can undergo nucleophilic substitution and elimination reactions such as those described below.

Reactions with sulphuric acid

Alcohols can react in different ways with sulphuric acid. The predominant reaction can be varied by changing the conditions, but there is always a mixture of products. We shall illustrate the differences by describing the reactions of ethanol (see Figure 29.2).

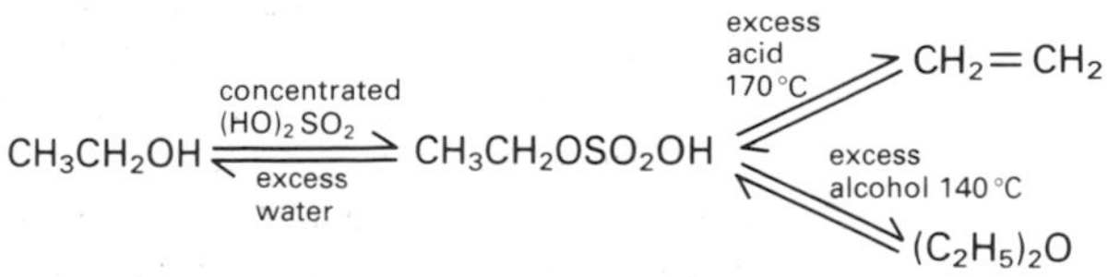

$$CH_3CH_2OH \underset{\text{excess water}}{\overset{\text{concentrated }(HO)_2SO_2}{\rightleftharpoons}} CH_3CH_2OSO_2OH$$

$$CH_3CH_2OSO_2OH \xrightarrow{\text{excess acid, 170 °C}} CH_2{=}CH_2$$

$$CH_3CH_2OSO_2OH \underset{}{\overset{\text{excess alcohol 140 °C}}{\rightleftharpoons}} (C_2H_5)_2O$$

Figure 29.2 Summary of reactions of ethanol with sulphuric acid

$$C_2H_5{-}O{-}\overset{\overset{\displaystyle O}{\|}}{\underset{\underset{\displaystyle O}{\|}}{S}}{-}OH$$

6

1 When ethanol is warmed gently with an *excess of concentrated sulphuric acid* it reacts to form **ethyl hydrogensulphate**, $C_2H_5O{-}SO_2{-}OH$ (**6**), (via attack of the Lewis acid SO_3 on an oxygen lone electron pair). In practice the ethyl hydrogensulphate is not isolated but is allowed to react further as described below.

2 Above about 170 °C ethyl hydrogensulphate undergoes an elimination reaction to give ethene gas:

$$H{-}CH_2{-}CH_2{-}OSO_2OH \rightarrow CH_2{=}CH_2 + (HO)_2SO_2$$

The reaction is similar to the elimination of HHal from halogenoalkanes, described in Section 28.3. Starting from the alcohol the net reaction is a **dehydration**, the removal of H and OH, the components of water:

$$H{-}CH_2{-}CH_2{-}OH \xrightarrow[170\,^\circ C]{\text{excess conc. }(HO)_2SO_2} CH_2{=}CH_2$$

Sulphuric acid is regenerated in the elimination step, so that its role in the dehydration is as a catalyst.

3 As with the halogenoalkanes, elimination is accompanied by substitution. This can be made the predominant reaction by changing the conditions: when an *excess of ethanol* is heated to about 140 °C with concentrated sulphuric acid ethoxyethane is produced, and can be distilled out of the reaction mixture:

$$C_2H_5OH + C_2H_5{-}OSO_2OH \xrightarrow[140\,^\circ C]{\text{excess } C_2H_5OH} C_2H_5{-}O{-}C_2H_5 + (HO)_2SO_2$$

All the reactions are readily reversed, so that, for example, ethene can be converted into ethanol by treating it with concentrated sulphuric acid, diluting with water, and distilling off the ethanol.

Dehydration may also be catalysed by other Lewis acids. Phosphoric acid is commonly used, and alkenes may also be produced in good yield by passing the alcohol vapour over aluminium oxide at about 300 °C (see Figure 29.3 and Table 14.2).

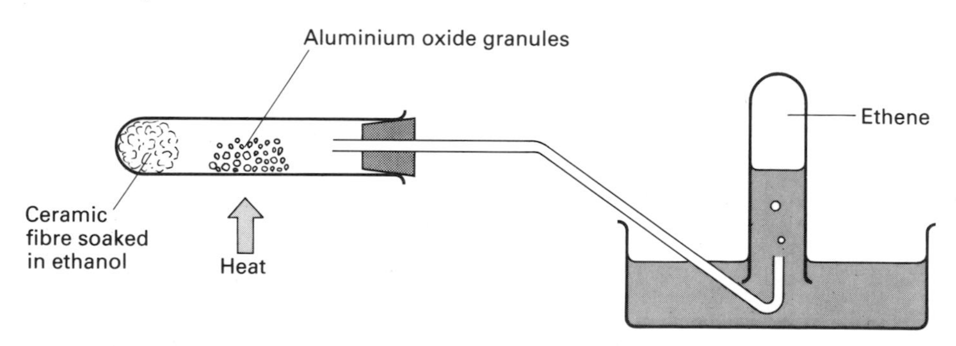

Figure 29.3 Apparatus for the dehydration of ethanol to ethene

Formation of halogenoalkanes

When heated with concentrated aqueous hydrogen bromide or hydrogen iodide, alcohols are converted into the corresponding halogenoalkanes. For example:

$$C_2H_5{-}OH + HBr \rightleftharpoons C_2H_5{-}OH_2^+ + Br^- \rightarrow C_2H_5{-}Br + H_2O$$

Alternatively the reaction may be carried out with concentrated sulphuric acid (to protonate the alcohol) and potassium bromide (or concentrated phosphoric acid and potassium iodide – see Section 23.5).

Ethers react similarly to give one molecule of halogenoalkane and one of alcohol (which, with an excess of acid, will react further):

$$C_2H_5OC_2H_5 + HBr \rightleftharpoons (C_2H_5)_2OH^+ + Br^- \rightarrow C_2H_5Br + C_2H_5OH$$

The reactions with concentrated hydrochloric acid are much slower, for two reasons. First, although it is a strong acid it is less strong than hydrobromic or hydriodic acid, so that there is a lower concentration of the O-protonated cation. Secondly, the chloride anion is a weaker nucleophile than either bromide or iodide, reacting more slowly in most nucleophilic substitutions. However, alcohols can be converted into chloroalkanes by treating them with phosphorus pentachloride or, preferably, sulphur dichloride oxide:

$$ROH + PCl_5 \rightarrow RCl + HCl + POCl_3$$

$$ROH + SCl_2O \rightarrow RCl + HCl + SO_2$$

The reaction with PCl_5 gives a variety of side products, including alkenes and organic phosphorus compounds. Using SCl_2O is cleaner and has the advantage that the non-organic products are gaseous and escape from the reaction mixture. The HCl evolved in these reactions is a useful test for a hydroxyl group.

Phosphorus tribromide and phosphorus triiodide, which may be generated in the reaction mixture from red phosphorus and bromine or iodine, similarly convert alcohols into bromoalkanes and iodoalkanes, respectively.

29.4 Reactivity patterns: reactions in which the C—O bond is not broken

Formation of esters

Esters are compounds of the general form RO—X, where ROH is an alcohol and HOX is an acid. Ethyl hydrogensulphate is an example. Another is the trinitrate ester of propane-1,2,3-triol, commonly called **nitroglycerine**: its use as an explosive is well known, less so its ability to relieve the pain of the heart disorder known as angina pectoris, by dilating the arteries of the heart. The word 'ester' without further description usually means a carboxylic ester, formed from a carboxylic acid, R′—CO—OH (see Section 31.1). Carboxylic esters may be made by refluxing an alcohol with the acid, with concentrated sulphuric acid as a catalyst (the reaction is discussed in detail in Section 31.4):

$$R'COOH + ROH \xrightarrow{\text{conc. } (HO)_2SO_2} R'\text{—CO—}OR + H_2O$$

Formation of alkoxides

The hydroxyl group of alcohols behaves in many ways like that of water. Alcohols are very weak acids with pK_a values around 16. In the presence of sodium or potassium hydroxide they are partly ionized to form **alkoxides**:

$$ROH + HO^- \rightleftharpoons \underset{\text{alkoxide}}{RO^-} + H_2O$$

Alkoxides are also formed when alcohols are reduced by alkali metals:

$$2C_2H_5OH + 2Na \rightarrow \underset{\text{sodium ethoxide}}{2Na^+ C_2H_5O^-} + H_2$$

Evolution of hydrogen gas from a neutral liquid when sodium is added is a common laboratory test for an alcohol.

Oxidation

Primary alcohols are oxidized by warm acidic or alkaline potassium manganate(VII) or warm acidic potassium dichromate(VI) to give **aldehydes**, and these can in turn be oxidized to carboxylic acids. The orange-to-green colour change of the $Cr_2O_7^{2-}$ ion to the Cr^{3+} ion is the basis of the 'breathalyser' test for ethanol.

$$\underset{\text{primary alcohol}}{RCH_2OH} \xrightarrow[\text{H}^+(aq)\ \text{distil}]{Cr_2O_7^{2-}} \underset{\text{aldehyde}}{RCHO} \xrightarrow[\text{H}^+(aq)\ \text{reflux}]{Cr_2O_7^{2-}} \underset{\text{carboxylic acid}}{RCOOH}$$

The aldehyde can be made by distilling a mixture of the alcohol and oxidizing agent. Because it lacks the hydroxyl group and its hydrogen-bonded interactions, the aldehyde is more volatile than the alcohol and distils out of the reaction mixture as it is formed. If the reaction is done under reflux so that the aldehyde falls back into contact with the oxidizing agent, the carboxylic acid is formed (compare Figures 29.4 and 29.5).

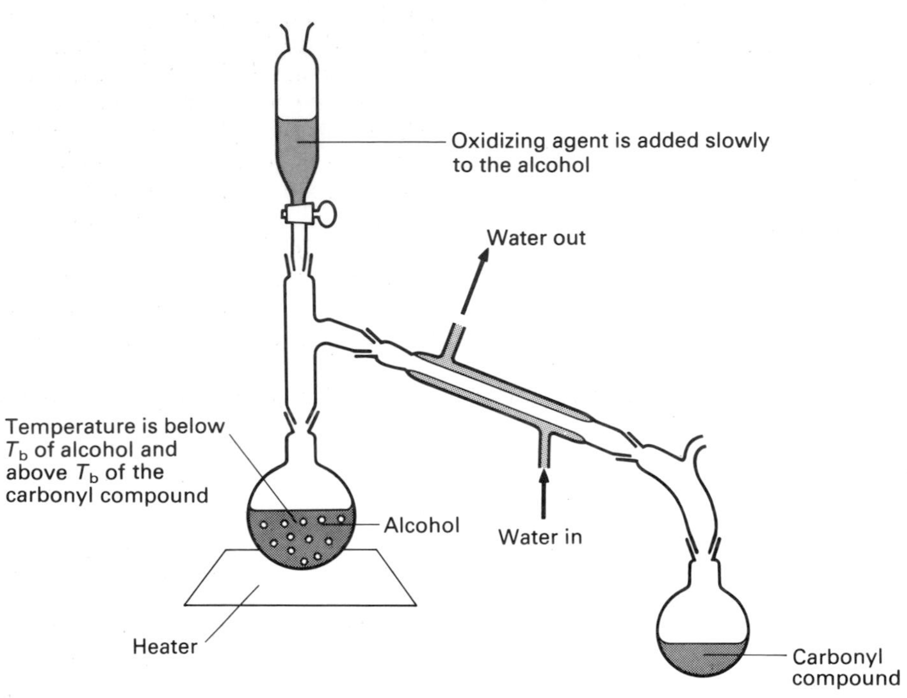

Figure 29.4 Apparatus for the oxidation of a primary alcohol to an aldehyde: the volatile aldehyde is distilled out of the reaction mixture before it can be further oxidized

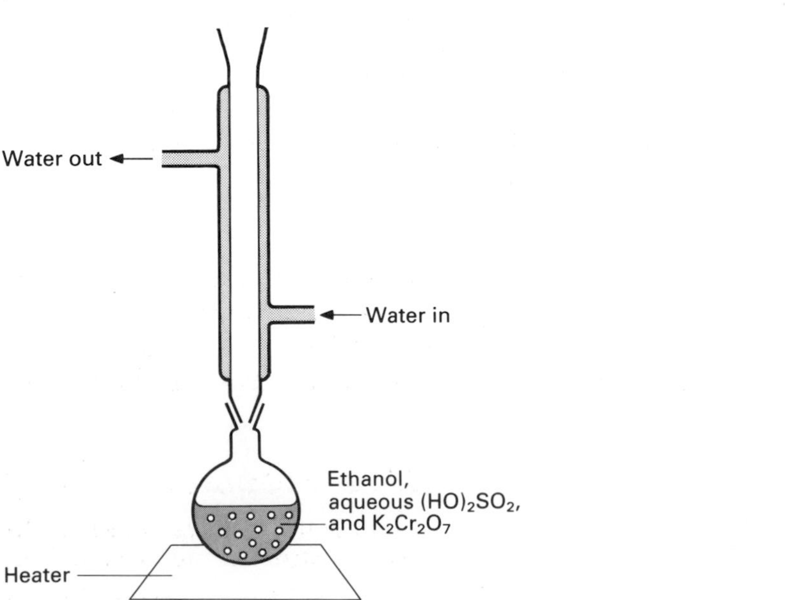

Figure 29.5 Apparatus for the oxidation of a primary alcohol to a carboxylic acid: the volatile aldehyde is returned from the condenser to the reaction mixture where it is further oxidized

Secondary alcohols undergo the first stage of oxidation to give **ketones**; these can be oxidized further only under very vigorous conditions that break carbon–carbon bonds and lead to a variety of products.

$$\underset{\text{secondary alcohol}}{RR'CHOH} \xrightarrow[\text{H}^+\text{(aq)},\ 30\,^\circ\text{C}]{Cr_2O_7^{2-}} \underset{\text{ketone}}{RR'C{=}O}$$

Tertiary alcohols do not have a hydrogen atom on the same carbon as the hydroxyl group, and cannot undergo a similar reaction. Like ketones, they are only oxidized under vigorous conditions with the breaking of carbon–carbon bonds.

The haloform reaction

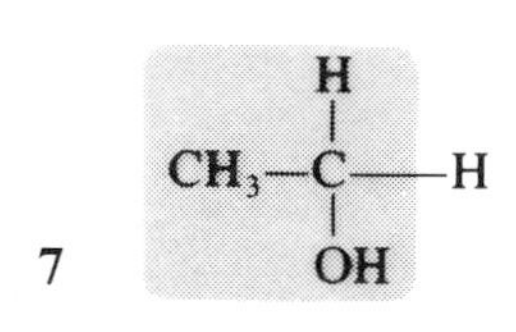

7

H
CH_3—C—CH_3
OH

8

Alcohols containing the group $CH_3CH(OH)$—, when heated with iodine in alkaline solution, give a pale yellow precipitate of triiodomethane (**iodoform**, CHI_3), which has a very characteristic 'hospital' smell. (Only a small quantity of alcohol should be used because an excess dissolves the triiodomethane.) The reaction provides a convenient test for the group: ethanol (**7**) and propan-2-ol (**8**), for example, give iodoform; methanol and propan-1-ol, which do not contain the group, do not. Similar reactions occur with sodium chlorate(I) (hypochlorite) to give trichloromethane, and with sodium bromate(I) (hypobromite) to give tribromomethane (chloroform and bromoform respectively).

The first step in all three reactions is oxidation to the aldehyde or ketone, CH_3CO—. Such compounds therefore also give the iodoform reaction with iodine and alkali; the reaction is discussed further in Section 30.3.

EXAMPLE

One of the products of the bromination of an alkane **A**, C_4H_{10}, has the molecular formula C_4H_9Br(**B**). This is treated with aqueous sodium hydroxide to give an alcohol **C**, C_4H_9OH, that reacts with iodine and alkali to give a yellow precipitate of iodoform. What can you deduce about the structures of the compounds **A**, **B** and **C**?

METHOD First decide which of the isomeric alcohols, C_4H_9OH, can give the iodoform reaction. Note that the carbon skeleton remains intact during the sequence **A** → **B** → **C**, and hence determine the structures of the bromoalkane and alkane.

ANSWER

$$\underset{\mathbf{C}}{CH_3-CH_2-\underset{|\atop OH}{CH}-CH_3} \xleftarrow[H_2O]{NaOH} \underset{\mathbf{B}}{CH_3-CH_2-\underset{|\atop Br}{CH}-CH_3} \xleftarrow[u.v.]{Br_2} \underset{\mathbf{A}}{CH_3-CH_2-CH_2-CH_3}$$

COMMENT The isomeric branched-chain alkane, methylpropane, cannot lead to an alcohol that contains the characteristic $CH_3CH(OH)$— grouping.

29.5 Preparative methods for alcohols

There are three general routes to alcohols. They are discussed in detail in other chapters and only summarized here.

1 Halogenoalkanes undergo nucleophilic substitution with hydroxide ion to give alcohols (see Section 28.3), typically by heating them under reflux with aqueous sodium hydroxide:

$$RHal + HO^- \rightarrow ROH + Hal^-$$

This is a good preparation for primary alcohols, but less good for secondary alcohols because elimination competes with substitution. Tertiary alcohols are formed when the halogenoalkane reacts with water alone.

2 Aldehydes and **carboxylic acids or esters** can be reduced to primary alcohols (see Sections 30.3 and 31.4). The most commonly used reducing agent is lithium tetrahydridoaluminate, $LiAlH_4$, in dry ethoxyethane, followed by water. Sodium tetrahydridoborate, $NaBH_4$, in water reduces aldehydes (and also ketones), but not carboxylic acids or esters.

$$\left.\begin{matrix} RCHO \\ RCOOH \\ RCOOR' \end{matrix}\right\} \xrightarrow[\text{(ii) } H_2O]{\text{(i) } LiAlH_4} RCH_2OH$$

Ketones, under the same conditions, are reduced to secondary alcohols (see Section 30.3):

$$RR'CO \xrightarrow[\text{(ii) } H_2O]{\text{(i) } LiAlH_4} RR'CHOH$$

3 Alkenes react with cold concentrated sulphuric acid to give alkyl hydrogensulphates. When the solution is diluted with water and distilled, these are hydrolysed to the alcohol.

$$CH_2{=}CH_2 + (HO)_2SO_2 \rightarrow CH_3{-}CH_2{-}OSO_2OH \xrightarrow{H_2O} CH_3CH_2OH + (HO)_2SO_2$$

If the alkene has alkyl groups at one end of the double bond and not at the other, addition occurs so as to give the alcohol with the hydroxyl group at the alkylated end of the bond (by Markovnikov's rule, discussed in Section 33.3):

$$CH_3{-}CH{=}CH_2 \xrightarrow[H_2O]{(HO)_2SO_2} CH_3{-}\underset{\displaystyle OH}{\underset{|}{C}H}{-}CH_3 \quad \text{(and not } CH_3CH_2CH_2OH\text{)}$$

The *manufacture* of ethanol is carried out either by the sulphuric acid method described above or, more commonly, by the **direct hydration of ethene**, carried out by passing ethene and water vapour at about 70 atmospheres and 300 °C over a catalyst of phosphoric acid supported on Celite (silica).

29.6 The importance of alcohols

The **fermentation** of sugar or starch by yeast enzymes to give ethanol has been carried out for thousands of years for making intoxicating drinks:

$$\underset{\text{sucrose}}{C_{12}H_{22}O_{11}} + H_2O \rightarrow 4C_2H_5OH + 4CO_2$$

The total annual consumption of alcoholic drinks throughout the world probably approaches 10^{12} litres, or about 10^{11} litres of ethanol – more than 10 litres of ethanol for every man, woman and child.

For many years ethanol produced by fermentation was also a major source of organic chemicals. The ethene from which 'polythene' was first made came from fermented ethanol. Nowadays industrial ethanol comes mainly from petroleum-derived ethene (see Section 29.5, above). When the price of petroleum is high the economics of fermentation become more favourable, however. In Brazil, for example, where there is great potential for agricultural development and little native petroleum, the use of ethanol from fermentation as motor fuel is growing rapidly. In the future we may well see ethanol becoming an increasingly important raw material for industrial organic chemistry. At present the major industrial uses of ethanol are as a solvent and in the manufacture of ethanoic acid.

Figure 29.6 A fermentation vessel in a brewery

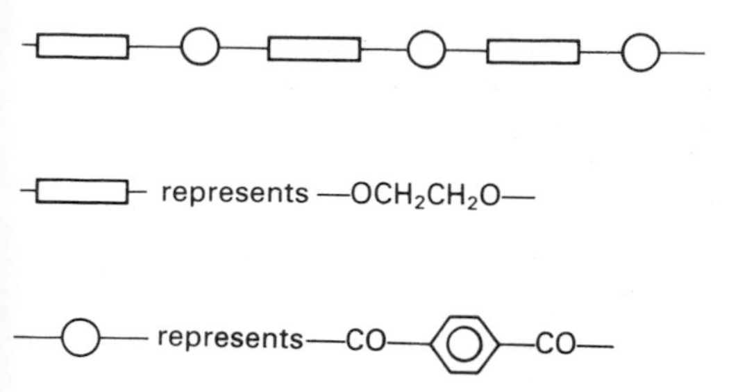

Figure 29.7 Structure of Terylene

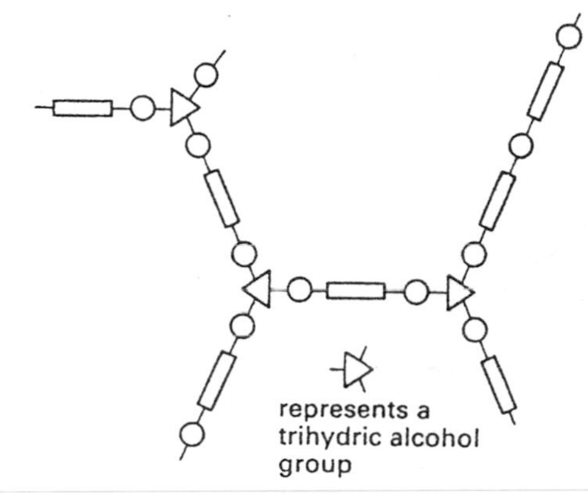

Figure 29.8 Structure of a cross-linked polymer

Higher alcohols are used widely in the manufacture of surfactants, the main ingredients of **detergents**, described in Box 34.3, and of **plasticizers**. Plasticizers are liquids of low volatility that are added to plastics to increase their flexibility: they work by lubricating the internal movement of the molecules. For example, unplasticized polyvinyl chloride (uPVC) is inherently a very hard, rigid substance which is used for making window frames and record discs. With a plasticizer added PVC becomes flexible and is used for the soles of shoes.

Di- and tri-hydric alcohols are used in the manufacture of **polymers** (large molecules made up of repeating smaller units). Ethane-1,2-diol reacts with benzene-1,4-dicarboxylic acid to give a **polyester** known as 'Terylene' or 'Dacron' (see Figure 29.7). The long-chain molecules line up in parallel rows, and the polyester can be spun into fibres. If some trihydric alcohol, such as propane-1,2,3-triol, is added then the chains can be **cross-linked** (see Figure 29.8), producing a rigid three-dimensional rather than a two-dimensional structure. Polyesters of this type are called **alkyd resins**, and are used in paints.

Ethane-1,2-diol ($HOCH_2CH_2OH$) is used as antifreeze. It mixes with water in all proportions, so that concentrated solutions can be used to lower the freezing point of water substantially below 0 °C (see Section 7.2). It also has a high boiling point (198 °C) so that it does not vaporize out of solution.

BOX 29.2

Toxicity of alcohols

As with many drugs, there is a fairly narrow range between the quantity of ethanol that produces a significant physiological response and the amount that is dangerous. A blood concentration of $0.017\,mol\,dm^{-3}$ is the legal limit for a car driver in the UK; with twice that concentration most people appear obviously drunk; at four times they are probably unconscious, and six times that concentration may well be fatal.

Higher alcohols, especially those with three to six carbon atoms, are called **fusel oils** and are produced in small concentrations during fermentation. They are more toxic than ethanol, but seldom cause fatalities because they rapidly induce vomiting. They are one of the main causes of the hangover that follows excessive drinking.

Methanol is converted in the liver into methanal, $CH_2{=}O$. It has a particularly serious effect on the optic nerves, causing blindness. Methanol is one of the compounds added to ethanol in order to make 'methylated spirits' unpalatable.

Laboratory ethanol often contains traces of the highly toxic benzene, used to remove the last traces of water from the azeotrope.

Summary

- ☐ Alcohols contain the hydroxyl functional group: their chemistry depends on the combined behaviour of the carbon–oxygen single bond and the oxygen–hydrogen bond.
- ☐ Ethers contain the functional group C—O—C; their chemistry depends on the behaviour of the carbon–oxygen single bond.
- ☐ The general formula for both ethers and alcohols is $C_nH_{2n+2}O$. Compounds with the same number of carbon atoms, *n*, are structural isomers.
- ☐ Ethers are named by the same rules as halogenoalkanes; RO— groups are called alkoxy groups.
- ☐ The names of alcohols are based on those of the alkanes with the same carbon skeleton, with the ending *-ane* changed to *-anol*.
- ☐ The physical properties of ethers resemble those of halogenoalkanes with similar molar mass.
- ☐ The physical properties of alcohols are dominated by hydrogen-bonded interactions of the hydroxyl group. The melting points, and especially the boiling points, are higher than those of isomeric ethers. The lower alcohols are good solvents for polar and ionic solutes.
- ☐ The hydroxyl group (—OH) in alcohols is chemically similar to that in water. Alcohols are weak bases ($\rightarrow ROH_2^+$) and weak acids ($\rightarrow RO^-$, alkoxides). Their reduction by sodium metal with the evolution of hydrogen is a test for the —OH group.
- ☐ The carbon–oxygen bond in unprotonated ethers and alcohols is unreactive.
- ☐ Alcohols react with excess concentrated sulphuric acid at 170 °C to give alkenes (by dehydration).
- ☐ An excess of alcohol with concentrated sulphuric acid at 140 °C gives the symmetrical ether: $2\,ROH \rightarrow R_2O + H_2O$.
- ☐ Alcohols and ethers react with concentrated aqueous HI and HBr to give halogenoalkanes.
- ☐ Alcohols react with PCl_5 and with SCl_2O to give chloroalkanes. The evolution of HCl with these reagents is a test for the presence of the —OH group.
- ☐ Alcohols react with carboxylic acids, in the presence of concentrated sulphuric acid as a catalyst, to give esters.
- ☐ Primary alcohols may be oxidized to aldehydes or carboxylic acids, secondary alcohols to ketones. Tertiary alcohols resist oxidation.
- ☐ Alcohols containing the group $CH_3CH(OH)$— undergo the haloform reaction with NaOHal to give $CHHal_3$. The formation of iodoform (triiodomethane, CHI_3), a yellow solid with a characteristic smell, may be used to test for the presence of $CH_3CH(OH)$—.
- ☐ Alcohols may be made by the hydrolysis of halogenoalkanes, by the reduction of C=O double bonds, or by the addition of H—OH to an alkene (catalysed by concentrated $(HO)_2SO_2$).
- ☐ Ethanol is important as the basis of intoxicating drinks, as an industrial solvent, and in the manufacture of ethanoic acid. It is potentially an important raw material that may supplement or replace petroleum as a fuel and as a source of organic chemicals.
- ☐ Higher alcohols are used in the manufacture of detergents and plasticizers.
- ☐ Dihydric and trihydric alcohols are used in the manufacture of polyester fibres and alkyd resins. Ethane-1,2-diol is used as an antifreeze.

PROBLEMS

1 Draw the structures of the following alcohols: ethanol, propan-1-ol, butan-2-ol, 2-methylpropan-1-ol, 2-methylpropan-2-ol, 2-methylbutan-2-ol, 3,3-dimethylbutan-2-ol. Classify them according to their reactions with acidic potassium dichromate(VI), and name the classes.

2 Draw the structures of all the isomers of $C_5H_{12}O$. Which of them would react with iodine in alkaline solution to give iodoform (triiodomethane)? What is the colour of iodoform?

3 Draw the structural formulae of the organic products that would result from treating $(CH_3)_2CHCH_2OH$ as follows:
(a) Warm with metallic sodium.
(b) Heat under reflux with potassium dichromate(VI) in dilute sulphuric acid.
(c) Heat under reflux with SCl_2O.
(d) Heat strongly with an excess of concentrated sulphuric acid.
(e) Heat with iodine in alkaline solution.

4 Show how the following compounds can be made from ethanol, using only inorganic reagents: (a) $CH_3COOC_2H_5$, (b) $(C_2H_5)_2O$, (c) CH_3CHO.

5 (a) For each of parts (i) to (ix) below *only one* of the alternatives **A**, **B**, **C**, **D**, **E** is correct. Answer each part by giving the appropriate letter.

A benzyl alcohol **B** butan-1-ol **C** ethanol
D methanol **E** propan-2-ol

(i) Which is most frequently used as a solvent in the laboratory?
(ii) Which has the highest vapour pressure at 20 °C?
(iii) Which is oxidized to a ketone?
(iv) Which is least soluble in water?
(v) Which burns with the smokiest flame?
(vi) Which is most readily oxidized by dilute acidified potassium manganate(VII) solution (permanganate) to carbon dioxide and water?
(vii) Which, when oxidized strongly, forms a solid product?
(viii) Which reacts with methanoic acid to form an ester which is isomeric with propanoic acid?
(ix) Which reacts additively with chlorine under suitable conditions?

(b) Indicate the reagents and conditions by which butan-1-ol may be converted to butan-2-ol.
(You are not expected to discuss the purification of either intermediates or the final product.) *(SUJB)*

6 Deduce the structures of compounds **A**, **B** and **C** in the following reaction sequence:

$$\underset{\mathbf{A}}{C_4H_8O_2} \xrightarrow[\text{(ii) } H_2O]{\text{(i) } LiAlH_4 \text{ in dry ethoxyethane}} \underset{\mathbf{B}}{C_4H_{10}O} \xrightarrow[170\,°C]{\text{excess conc. } (HO)_2SO_2} \underset{\mathbf{C}}{C_4H_8} \xrightarrow[\text{(ii) } H_2O]{\text{(i) cold conc. } (HO)_2SO_2} C_2H_5\text{—}\underset{\displaystyle OH}{\underset{|}{CH}}\text{—}CH_3$$

7 (a) Give the structural formula of each of the isomeric alcohols represented by C_4H_9OH. Indicate which of the isomers can display optical activity, pointing out the essential structural requirement.
(b) Discuss the extent to which it is possible to distinguish between the isomers of C_4H_9OH by examining the chemical properties of the products obtained by oxidation with acidified potassium dichromate(VI) solution.
(c) With the aid of equations, compare and contrast the reactions of propan-1-ol and phenol with (i) dilute, aqueous sodium hydroxide, (ii) ethanoic (acetic) anhydride. *(JMB)*

8 (a) Describe briefly an industrial method by which ethanol is manufactured.
(b) A chemist is considering ways of making samples of ethanol containing deuterium atoms in various positions. The following starting materials are available: D_2O (deuterium oxide), D_2SO_4, $LiAlD_4$, C_2H_4, C_2H_5OH, CH_3CHO, CH_3CO_2H and the usual laboratory inorganic chemicals. Describe using equations and essential experimental conditions (but not details of apparatus) how the chemist could obtain *four* different samples of ethanol, each of which contains one or more atoms of deuterium in the molecule. *(Oxford)*

9 Account for each of the following.
(a) Compounds **A**, **B** and **C** each have molecular formula C_5H_{10}. When catalytically hydrogenated, **A** and **B** give the same compound **D**, whereas **C** gives compound **E**.
(b) Of the aromatic compounds with formula $C_6H_4(COOH)_2$, only one readily forms an anhydride.
(c) Of the four alcohols of formula C_4H_9OH, only one gives a ketone on oxidation.
(d) When phenylethanone, $CH_3COC_6H_5$, reacts with hydroxylamine a derivative is produced which can exist in two forms.
(e) **X** and **Y** are compounds with molecular formula C_2H_6O. **X** is a liquid boiling at 78.5 °C and **Y** boils at −24.8 °C. *(AEB 1985)*

30 Carbonyl compounds: the C=O double bond

In this chapter we see that the chemistry of the carbon–oxygen double bond is governed largely by the attraction that the electronegative oxygen atom exerts on the bonding electrons, leading to nucleophilic attack on the carbon atom.

Compounds containing carbon–oxygen double bonds are widespread in nature; we consider some important examples. The related carbohydrates constitute the major energy reserves for animals and provide both the energy reserves and the structural skeletons of plants.

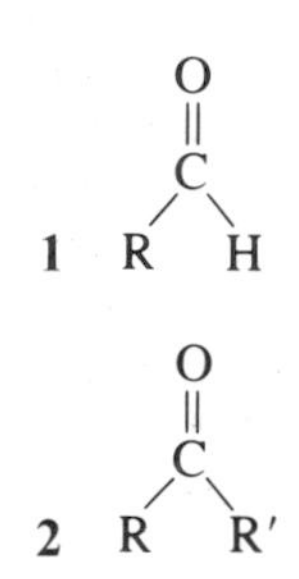

A carbon atom doubly bonded to an oxygen atom is known as a **carbonyl group**. In **aldehydes** (**1**, alkanals) the carbonyl carbon atom is attached to two hydrogen atoms, H—CO—H, or to an alkyl group and a hydrogen atom, R—CO—H, usually written R—CHO. Compounds with two alkyl groups attached to the carbonyl carbon atom (R—CO—R′, RR′CO) are **ketones** (**2**, alkanones). One can regard an aldehyde or ketone as having the structure of an alkane with one CH_2 group replaced by CO: their general formula is $C_nH_{2n}O$.

30.1 Nomenclature

Aldehydes

Aldehydes are named after the alkane with the same carbon skeleton: the ending *-ane* is changed to *-anal*. The carbonyl carbon atom is always given the positional number 1. Older names were based on non-systematic names for carboxylic acids, and the two examples given in parentheses below are still commonly used.

CH_2O	methanal (formaldehyde)
$\overset{2}{C}H_3$—$\overset{1}{C}HO$	ethanal (acetaldehyde)

Ketones

Ketones are also named after the corresponding alkane, using the ending *-anone*. The position of the carbonyl group in the chain is indicated, where necessary, by a number before the '-one'. The name 'acetone' is still very widely used in place of the systematic name 'propanone'.

CH_3—CO—CH_3	propanone (acetone)
CH_3—CO—CH_2—CH_3	butanone
$\overset{5}{C}H_3$—$\overset{4}{C}H_2$—$\overset{3}{C}O$—$\overset{2}{C}H_2$—$\overset{1}{C}H_3$	pentan-3-one
$\overset{5}{C}H_3$—$\overset{4}{C}H_2$—$\overset{3}{C}H_2$—$\overset{2}{C}O$—$\overset{1}{C}H_3$	pentan-2-one

30.2 Physical properties

Aldehydes and ketones have slightly higher melting points and markedly higher boiling points than ethers of similar molar mass:

	CH_3CHO	CH_3OCH_3	$CH_3CH_2CH_2CHO$	$CH_3CH_2COCH_3$	$CH_3CH_2OCH_2CH_3$
$t_m/°C$	−121	−138.5	−99	−86	−116
$t_b/°C$	+21	−23	+76	+80	+34.5

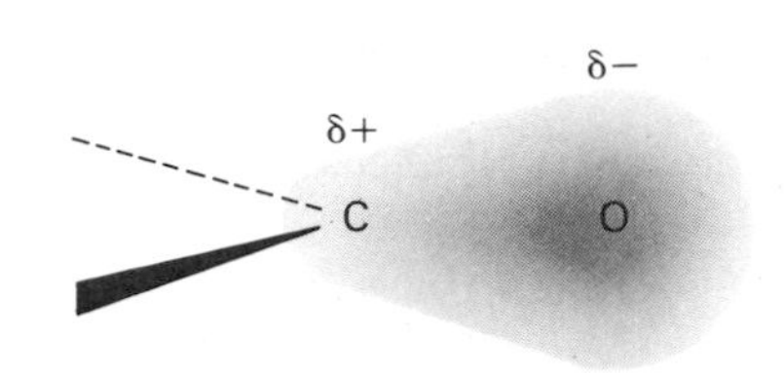

Figure 30.1 Electron distribution in the carbonyl group: electrons are attracted to the electronegative oxygen atom

The differences arise because the π-electrons of a carbon–oxygen double bond are more polarizable than the σ-electrons of a carbon–oxygen single bond, less tightly held by the nuclei, and more easily attracted to the electronegative oxygen atom (see Figure 30.1). This is also shown by the much larger electric dipole moment of the double bond: $\mu(CH_3CHO) = 2.7\,D$; $\mu(CH_3OCH_3) = 1.3\,D$. Thus both dispersion and dipole–dipole forces are larger than in ethers of comparable molar mass.

The polarity and polarizability of the carbonyl bond enable the lower ketones, particularly propanone and butanone, to dissolve a wide range of compounds, from inorganic salts to hydrocarbons. This makes them particularly important solvents in which to carry out many organic reactions. For example, the reaction of 1-bromooctane with potassium cyanide to give the nitrile (see Section 28.3) cannot be carried out in water, because the bromooctane is too insoluble, nor in an ether because that does not dissolve the potassium cyanide. Propanone dissolves both reactants. Furthermore, most nucleophiles, which in water or alcohol are surrounded by a shell of hydrogen-bonded solvent molecules, are much more reactive in ketones: some nucleophilic substitution reactions that require a long period of reflux in ethanol take place rapidly at room temperature in propanone. Such solvents, having high dipole moments but without hydrogen-bonding hydrogen atoms, are said to be **dipolar non-protolytic** or **dipolar aprotic**. Aldehydes are less useful than ketones as solvents because they are more reactive.

The high dipole moment of the carbonyl group leads to a very strong absorption at around $1700\,cm^{-1}$ in the i.r. spectrum (see Section 3.2) of aldehydes and ketones.

EXAMPLE

Show how i.r. and n.m.r. spectroscopy could be used to distinguish the following isomers, of formula C_3H_6O:

$CH_3{-}CH_2{-}CHO$ — propanal

$CH_3{-}CO{-}CH_3$ — propanone

$(CH_2)_2CH{-}OH$ (three-membered ring) — cyclopropanol

METHOD Decide which groups in the three molecules give i.r. absorption bands at characteristic frequencies. Then decide how many differently situated hydrogen atoms each molecule possesses, and hence the number of different n.m.r. signals. If necessary, consult Tables 26.2 and 26.3.

ANSWER Propanal and propanone show strong i.r. absorption around $1700\,cm^{-1}$, due to the stretching vibrations of the carbon–oxygen double bond. This is absent from the i.r. spectrum of cyclopropanol; however, the —OH group absorbs strongly at about $3500\,cm^{-1}$. Propanone has only one type of hydrogen atom (the two methyl groups are indistinguishable) and gives a single n.m.r. signal at $\delta \approx 2.2$. Propanal and cyclopropanol each contain three different types of hydrogen atom and give more complicated spectra.

COMMENT The exact stretching frequency of a carbonyl group is strongly influenced by the groups attached to it. Simple aldehydes absorb at slightly higher frequency than simple ketones ($1730\,cm^{-1}$ and $1715\,cm^{-1}$ respectively). Such distinctions are routinely used in research laboratories in preference to the test-tube reactions described in the next section.

30.3 Reactivity patterns and important reactions

Addition reactions

Many reactions of aldehydes and ketones are governed by nucleophilic attack on the carbonyl carbon atom. This is usually accompanied by a proton transfer from the solvent to the carbonyl oxygen atom, so that the overall reaction is **addition** of Nu, H across the carbon–oxygen double bond:

$$R(Y)C{=}\ddot{O}: \;+\; :Nu^- \longrightarrow R(Y)(Nu)C{-}\ddot{O}:^- \xrightarrow{H^+} R(Y)(Nu)C{-}\ddot{O}H$$

[Y = R′ or H]

(We use the symbol $:Nu^-$ as a general representation for a nucleophile, but remember that the nucleophile is not necessarily negatively charged.)

For example, propanone forms a crystalline addition product with excess concentrated aqueous sodium hydrogensulphite. The nucleophile is the sulphite ion, SO_3^{2-} (see Section 22.5).

$$(H_3C)(CH_3)C{=}\ddot{O}: \;+\; :SO_3^{2-} \longrightarrow H_3C(CH_3)(SO_3^-)C{-}\ddot{O}:^- \xrightarrow{H^+} H_3C(CH_3)(SO_3^-)C{-}\ddot{O}H$$

The three groups attached to the carbonyl carbon atom take up a roughly trigonal planar arrangement, whereas the addition product is roughly tetrahedral around the same carbon atom. The reaction is analogous to the nucleophilic substitution reactions of halogenoalkanes (see Section 28.3). The fact that the functional group is doubly bonded has two consequences. First, the oxygen atom is not expelled from the molecule, because only one of the bonds is broken. Secondly, carbonyl compounds react readily, in contrast to the unreactivity of the singly bonded ethers. The polarizability of the double bond π-electrons facilitates their displacement towards the electronegative oxygen atom.

The reactions are often catalysed by acids or bases. Acids generate small concentrations of the cation

$$>C{=}OH^+$$

This is more reactive than the neutral carbonyl compound because of the attraction of the bonding electrons towards the positive charge. Bases can convert a nucleophilic reagent, H—Nu, into the more reactive Nu^-.

Many of these reactions form an equilibrium mixture:

$$R(Y)C{=}O \;+\; H{-}Nu \rightleftharpoons R(Y)(Nu)C{-}OH$$

The position of such equilibria depends, among other things, on the groups attached to the carbonyl carbon atom. When both are hydrogen (i.e., in methanal) the equilibrium lies furthest to the right. When both are alkyl groups (i.e., in ketones) it lies furthest to the left. With aldehydes other than methanal its position is intermediate. The difference is partly steric and partly electronic. Four groups around the central carbon atom are more crowded together than three, so that addition is disfavoured by bulky groups. The polarization of the carbonyl group leaves a partial positive charge on the carbonyl carbon atom that is stabilized by electron donation from alkyl groups (just as alkyl groups stabilize carbocations, as described in Section 28.3).

Similar steric and electronic effects influence the *rates* of nucleophilic attack. Methanal is the most reactive carbonyl compound, other aldehydes react more slowly, and ketones are still less reactive.

Important examples of nucleophilic addition reactions are summarized below.

1 Reduction: addition of H, H Aldehydes are reduced to primary alcohols, ketones to secondary alcohols:

$$RCH{=}O \rightarrow RCH_2{-}OH$$

$$R_2C{=}O \rightarrow R_2CH{-}OH$$

The usual reagents are sodium tetrahydridoborate (sodium borohydride, $NaBH_4$) in water or lithium tetrahydridoaluminate (lithium aluminium hydride, $LiAlH_4$) in dry ethoxyethane, followed by the addition of water. The BH_4^- or AlH_4^- ions act as a source of nucleophilic H^-, and the water provides H^+.

2 Addition of CN, H Aldehydes and ketones react with potassium cyanide in water to give 2-hydroxynitriles (also called cyanohydrins).

$$R(Y)C{=}O \xrightarrow[H_2O]{CN^-} R(Y)C(OH)CN$$

The nitrile group may be converted into other functional groups (—COOH by hydrolysis, $—CH_2NH_2$ by reduction see Section 31.4) so that this reaction provides an important way of making molecules with two functional groups and with a longer carbon chain than the original reactant.

3 Addition of $NaHSO_3$ ($SO_3^-Na^+$, H) Aldehydes and ketones in which at least one of the two alkyl groups is methyl (others are too bulky) react as described above to give **hydroxysulphonates** that are easily purified by recrystallization. They can be converted back to the carbonyl compound by dilute acid or alkali and therefore provide a useful way of purifying liquid aldehydes or methyl ketones that cannot be recrystallized.

4 Hydration: addition of OH, H When methanal is dissolved in water it is almost totally hydrated to give dihydroxymethane, $HO{-}CH_2{-}OH$. It is not possible to obtain the pure diol out of solution, however: if attempts are made to remove the excess water it decomposes back to methanal. Other aldehydes are about half hydrated in aqueous solution, ketones hardly at all.

5 Hemiacetal formation: addition of OR, H When methanal is dissolved in methanol it forms the addition product methoxymethanol:

$$H_2C{=}O \xrightarrow{CH_3OH} H_2C(OH)OCH_3$$

Such compounds, with hydroxy and alkoxy groups attached to the same carbon atom, are called **hemiacetals**. Like the diols, they are normally obtained only in solution. Some cyclic hemiacetals may be isolated as pure compounds, however, including the very important **carbohydrates** (see Section 30.5).

BOX 30.1

Grignard reagents

Most organic compounds with functional groups contain bonds between carbon and a more electronegative element. **Organometallic** compounds, on the other hand, contain a carbon–metal bond, which is polarized in the direction $\overset{\delta-}{C}—\overset{\delta+}{M}$. When the bond is broken heterolytically the electron pair moves towards the carbon atom, which is therefore nucleophilic.

An important group of organometallic compounds is made by reducing halogenoalkanes with magnesium in dry ethoxyethane:

$$C_2H_5—Br + Mg \rightarrow \underset{\text{ethylmagnesium bromide}}{C_2H_5—Mg—Br}$$

The compounds are called **Grignard reagents** after the French chemist Victor Grignard, who studied their reactions. Their importance to synthetic chemists lies in their use to form carbon–carbon bonds, so building up carbon skeletons. They add to the carbonyl group of aldehydes and ketones in the **Grignard reaction**. The first part of the reaction is carried out in dry ethoxyethane; water is then added to complete the addition:

$$R(Y)C{=}O \xrightarrow[\text{(ii) } H_2O]{\text{(i) } C_2H_5MgBr} R(Y)C(OH)C_2H_5$$

A similar reaction occurs with carbon dioxide to give, after the addition of water, a carboxylic acid:

$$CO_2 + C_2H_5MgBr \xrightarrow[\text{(ii) } H_2O]{\text{(i) ethoxyethane}} C_2H_5COOH + HOMgBr$$

Addition–elimination reactions (condensations)

Compounds of the type $X—NH_2$ add to a carbonyl group (addition of X—NH, H), but the resulting adduct readily loses a molecule of water (elimination) to form a carbon–nitrogen double bond:

$$R(Y)C{=}O + :NH_2—X \longrightarrow R(Y)C(O^-)\overset{+}{N}H_2—X \xrightarrow{\pm H^+} R(Y)C(OH)NH—X \xrightarrow{-H_2O} R(Y)C{=}N—X$$

(The step marked $\pm H^+$ involves proton exchanges to and from the solvent.) Important examples of this reaction are tabulated below.

$NH_2—X$		Product	
$NH_2—OH$	hydroxylamine	$>C{=}N—OH$	oximes
$NH_2—NHC_6H_5$	phenylhydrazine	$>C{=}N—NHC_6H_5$	phenylhydrazones
$NH_2—NHCONH_2$	semicarbazide	$>C{=}N—NHCONH_2$	semicarbazones

3 (structure: $R_2C=N-NH-C_6H_3(NO_2)_2$)

Oximes, phenylhydrazones and semicarbazones are crystalline solids with characteristic melting points, and are therefore used for identification of the parent aldehyde or ketone. Particularly important for this purpose are the 2,4-dinitrophenylhydrazones (DNPs, **3**). 2,4-Dinitrophenylhydrazine dissolves in methanol and dilute sulphuric acid (forming the cation $X—NH_3^+$) to give what is known as **Brady's reagent**. When this is added to a solution of an aldehyde or ketone it gives, in high yield, a yellow-orange precipitate of the DNP. These are only very weakly basic because the NH electron pair is delocalized into the ring and the electron-attracting nitro groups (see Section 35.4) and they do not dissolve in dilute acid. They are easily recrystallized and have sharp melting points.

Reactions at the α-carbon atom

Some reactions of carbonyl compounds begin with the loss of a proton from the so-called α- (Greek: *alpha*) carbon atom, the one next to the carbonyl group, to form an **enolate** anion, which is nucleophilic. An electrophile can then attack the α-carbon atom, replacing the hydrogen ion removed in the first step:

$$\ddot{B}^- + H—C_\alpha RR'—C(=\ddot{O}:)R'' \longrightarrow RR'C=C(—\ddot{O}:^-)R'' \xrightarrow{E^+} E—CRR'—C(=\ddot{O}:)R''$$

enolate anion

The base need not be negatively charged: water can act as the base if the carbonyl oxygen is protonated by acid. Likewise the electrophile need not be positively charged.

The following are examples of this type of reaction.

1 Halogenation When an aldehyde or ketone is treated with a halogen in dilute acid or alkali an α-halogeno-compound is formed.

$$CH_3—CO—CH_3 + Cl_2 \xrightarrow{H^+ \text{ or } HO^-} CH_3—CO—CH_2—Cl + HCl$$

2 The haloform reaction In alkaline solution and with an excess of halogen the group $—COCH_3$ is converted to $—COCHal_3$. This is then hydrolysed in the same alkaline solution to give the trihalogenomethane (haloform) and the carboxylate anion, in a reaction that resembles ester hydrolysis, discussed in Section 31.4.

$$Y—COCH_3 \xrightarrow[NaOH]{I_2} Y—COCI_3 \xrightarrow[NaOH]{H_2O} Y—CO_2^- + \underset{\text{yellow precipitate}}{CHI_3} \quad [Y = R \text{ or } H]$$

The precipitate of triiodomethane (iodoform), with its characteristic pale yellow colour and 'hospital' smell, is used as a test for the presence of the $CH_3CO—$ group. The same reagents (halogen plus alkali), when heated with alcohols containing the group $CH_3CH(OH)—$, oxidize them to $CH_3CO—$, so that the haloform reaction provides a test for these alcohols too (see Section 29.4). Carbonyl compounds usually react in the cold, the alcohols only when heated.

3 The aldol reaction In this reaction two aldehyde molecules react together to form a hydroxyaldehyde or **aldol** (*ald-* for $—CHO$, *-ol* for $—OH$). The α-carbon atom of the enolate anion from one molecule is the nucleophile, attacking the electrophilic carbonyl carbon atom of the other. For example, in dilute alkaline solution ethanal reacts as follows:

$$H_3C—CH=\ddot{O} + CH_2=CH—\ddot{O}:^- \longrightarrow H_3C—CH(—\ddot{O}:^-)—CH_2—CH=\ddot{O}: \xrightarrow{H_2O} CH_3—CH(OH)—CH_2—CH=O$$

3-hydroxybutanal

One molecule undergoes addition across the carbon–oxygen double bond; the other is substituted at the α-position. (In concentrated alkali further reaction leads to a polymeric solid.)

When the reaction is carried out with an acid catalyst the aldol is dehydrated to give an unsaturated aldehyde:

$$CH_3{-}CH(OH){-}CH_2{-}CHO \xrightarrow{H^+} \underset{\text{but-2-enal}}{CH_3{-}CH{=}CH{-}CHO} + H_2O$$

Reactions that form a new carbon–carbon bond are important tools of the synthetic chemist, who can thereby build up large carbon skeletons from smaller molecules.

Ketones, for the steric and electronic reasons outlined above, have much less tendency to form aldols than aldehydes have, but they do react in acidic solution to give the unsaturated ketone products.

Oxidation

Ketones are not easily oxidized. Vigorous oxidizing agents, such as hot aqueous potassium manganate(VII), may give a mixture of carboxylic acids containing fewer carbon atoms than the original ketone. Aldehydes, on the other hand, are readily oxidized to the corresponding acid (see Section 29.4):

$$R{-}C(=O){-}H \xrightarrow[H^+(aq)]{Cr_2O_7^{2-}} R{-}C(=O){-}OH$$

Typical reagents used for the preparation of acids by this means are acidified potassium dichromate(VI) or acidified potassium manganate(VII). Milder reagents also perform the oxidation, however: for example, oxygen in the air can produce the acid as an impurity in stored aldehydes. Some mild oxidizing agents are used to distinguish aldehydes from ketones: two common tests are as follows.

1 **Fehling's solution** contains an alkaline solution of a copper(II) complex, made by mixing aqueous copper(II) sulphate with an alkaline solution of potassium sodium 2,3-dihydroxybutanedioate (tartrate). When heated with aldehydes it is reduced to a red precipitate of copper(I) oxide, Cu_2O. Methanal produces some metallic copper as well.

2 **Tollens' reagent** is a solution containing the complex ion $[Ag(NH_3)_2]^+$. It is prepared by adding a small quantity of aqueous sodium hydroxide to aqueous silver nitrate and dissolving the resulting brown precipitate of silver oxide in aqueous ammonia. When it is warmed with aldehydes the Ag(I) is reduced to metallic silver, forming a mirror on the walls of the test-tube.

Although these tests do not give positive results with simple ketones, they do with α-hydroxyketones, which contain the grouping —CH(OH)CO— found in some sugars (see Section 30.5).

EXAMPLE

What reactions could be used to prepare the following compounds from ethanal, CH_3CHO?

$$CH_3{-}\underset{\displaystyle OH}{\underset{|}{CH}}{-}\underset{\displaystyle OH}{\underset{|}{CH_2}} \qquad CH_3{-}\underset{\displaystyle OH}{\underset{|}{CH}}{-}CH_2{-}\underset{\displaystyle OH}{\underset{|}{CH_2}}$$

METHOD Select reactions that will extend the carbon skeleton from two atoms to three and four, respectively. Then decide how to introduce the hydroxyl functional groups.

ANSWER

$$CH_3{-}CHO \xrightarrow[H_2O]{KCN} CH_3{-}\overset{\displaystyle H}{\overset{|}{\underset{\displaystyle OH}{\underset{|}{C}}}}{-}CN \xrightarrow[H_2O\ \text{reflux}]{(HO)_2SO_2} CH_3{-}\overset{\displaystyle H}{\overset{|}{\underset{\displaystyle OH}{\underset{|}{C}}}}{-}COOH$$

$$\xrightarrow{\text{(i) } LiAlH_4 \text{ (ii) } H_2O} CH_3{-}\overset{\displaystyle H}{\overset{|}{\underset{\displaystyle OH}{\underset{|}{C}}}}{-}\underset{\displaystyle OH}{\underset{|}{CH_2}}$$

$$2\ CH_3{-}CHO \xrightarrow[\text{dil. NaOH(aq)}]{\text{aldol reaction}} CH_3{-}\overset{\displaystyle H}{\overset{|}{\underset{\displaystyle OH}{\underset{|}{C}}}}{-}CH_2{-}CHO \xrightarrow[\text{(ii) } H_2O]{\text{(i) } LiAlH_4} CH_3{-}\overset{\displaystyle H}{\overset{|}{\underset{\displaystyle OH}{\underset{|}{C}}}}{-}CH_2{-}\underset{\displaystyle OH}{\underset{|}{CH_2}}$$

COMMENT The first reaction sequence illustrates an important feature of **synthetic organic chemistry** (which is concerned with how to prepare organic compounds), namely that it is usually necessary to draw together aspects of the chemistry of several different classes of compound.

30.4 Preparative methods for carbonyl compounds

The methods of preparing carbonyl compounds are described in other chapters and are only summarized here.

Aldehydes

1 Controlled oxidation of primary alcohols (see Section 29.4), with the aldehyde being distilled from the reaction mixture:

$$RCH_2OH \xrightarrow[H^+(aq)\ \text{distil}]{Cr_2O_7^{2-}} RCHO$$

2 Rosenmund reduction of acid chlorides (see Section 31.4):

$$RCOCl \xrightarrow[Pd/BaSO_4]{H_2} RCHO$$

Ketones

Oxidation of secondary alcohols (see Section 29.4):

$$RR'CHOH \xrightarrow[H^+(aq)\ 30°C]{Cr_2O_7^{2-}} RR'CO$$

30.5 The importance of carbonyl compounds. Carbohydrates

Methanal

Methanal (formaldehyde, CH_2O) is used in the manufacture of **thermosetting plastics**, discussed in Section 35.6. It is difficult to handle because it is a gas that irritates the nose, eyes and throat. It is therefore usually used either as a 40 per cent solution in water and methanol known as **formalin** (used for disinfecting warehouses and ships and for preserving biological specimens) or as a white solid polymer, **poly(methanal)** or **paraformaldehyde**, that gives off free methanal when heated:

$$HO{-}(CH_2{-}O)_n{-}H \xrightarrow{\text{heat}} nCH_2O + H_2O$$

Carbohydrates

Cyclic hemiacetals (see Section 30.3), with five- or six-membered rings (**4**) are always in equilibrium with a small proportion of the corresponding open-chain hydroxyaldehyde (**5**) or hydroxyketone, and many of the reactions of the mixture actually occur via this minor component. **Carbohydrates** are molecules of this type, containing further hydroxy groups. They are so called because many of them have the general formula $C_nH_{2m}O_m$ which, if written $C_n(H_2O)_m$, looks like a hydrate of carbon: the name, of course, relates *only* to the formula and not in any way to the actual structure of the molecules.

1 Monosaccharides, such as the isomers **fructose** (**6**) (fruit sugar) and **glucose** (**7**), contain a single carbonyl or hemiacetal group. Glucose plays a central role in the biochemistry of almost all living things: its oxidation, through a series of reactions, to carbon dioxide and water is the main source of their energy:

$$C_6H_{12}O_6(s) + 6O_2(g) \rightarrow 6CO_2(g) + 6H_2O(l)$$
$$\Delta_cH^{\ominus}(298\ K) = -2802\ kJ\ mol^{-1}$$

Both glucose and fructose reduce Fehling's solution and Tollens' reagent and are therefore called **reducing sugars**. Glucose is, in fact, often used in the commercial manufacture of silver mirrors. A variant of Fehling's solution, known as **Benedict's solution**, has been used to test for the presence of excess glucose in the urine of people suspected of having diabetes.

2 Disaccharides (**8**) contain two monosaccharide units joined together via a C—O—C linkage. They can be hydrolysed to the monosaccharides by dilute acids and by enzymes. **Sucrose** (**9**), common cane or beet sugar, contains one glucose and one fructose unit, linked together via their hemiacetal carbon atoms. The loss of the hemiacetal hydroxyl groups prevents the ring from opening; sucrose does not reduce Fehling's solution or Tollens' reagent, and is therefore called a **non-reducing sugar**. About a hundred million tonnes of sucrose are produced each year throughout the world: white granulated sugar is almost 100 per cent pure sucrose.

3 Polysaccharides contain long chains of monosaccharide units. The commonest examples are **glycogen, starch** and **cellulose**, all of which are built up from glucose units.

4 hemiacetal

5

6 fructose

glucose

7 monosaccharides

8 disaccharide

9 sucrose

Glycogen is the main storage depot for glucose in animal cells; starch plays the same role in plants. In both compounds the chains are coiled, as shown diagrammatically in Figure 30.2, so that the molecules are globular in shape. Raw starch – as found in flour, for example – is granular and insoluble in water. When it is cooked the granules swell and burst, allowing water molecules to penetrate the coiled chains and to form hydrogen bonds to the many hydroxyl groups; the starch dissolves, thickening the solution as it does so. Starch solution gives a characteristic blue-black colour with iodine – a test that is used for iodine or for starch. The iodine molecules line up in the starch coils: the exact colour of the complex they form depends on the length of the coil.

Figure 30.2 Representation of the glycogen molecule: the circles stand for glucose units

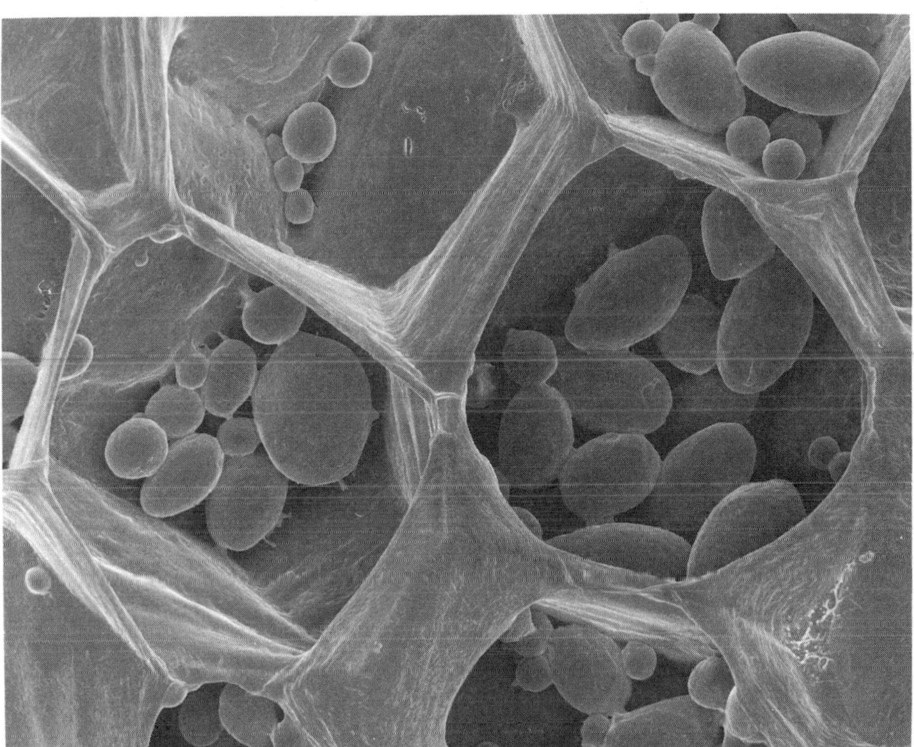

Figure 30.3 Photomicrograph showing starch grains within the cells of a raw potato

10 cellulose

Cellulose (**10**), the main constituent of plant cell walls, is said to be the most abundant organic chemical on Earth. Its molecule is a single unbranched chain of at least 3000 glucose units (shown as circles). The molecules are not coiled: the hydroxyl groups of one molecule form hydrogen bonds to those of another, producing the familiar fibrous structure of cotton, which is almost pure cellulose. One consequence of the different shapes of starch and cellulose molecules is that the digestive enzymes that break down starch molecules to glucose are unable to attack cellulose. We can write on it (paper), wear it (cotton) and build with it (wood); we eat it ('dietary fibre'), but we cannot digest it.

When the hydroxyl groups of cellulose are acylated to form ester groups (see Section 31.4), the molecular chains are no longer held together by hydrogen bonds and the resulting esters dissolve in propanone or dichloromethane to give viscous, syrupy solutions. These can be forced through fine nozzles (spinnerets) to give fibres that are less stiff than those of cotton. The ethanoate ester is commonly called cellulose acetate. It was first marketed as 'artificial silk' or rayon, and is now usually known as **viscose**.

Summary

- ☐ Carbonyl compounds contain the C═O double bond. Aldehydes contain the functional group —CHO; ketones contain two alkyl groups attached to the carbonyl carbon atom.
- ☐ The names of aldehydes end in *-anal*, and those of ketones in *-anone*.
- ☐ The physical properties of aldehydes and ketones reflect the high polarizability of the C═O double bond, in which the oxygen atom bears a substantial partial negative charge.
- ☐ Ketones, particularly propanone and butanone, are useful solvents.
- ☐ Base-catalysed addition reactions of carbonyl compounds are usually started by nucleophilic attack on the carbonyl carbon atom. The reaction normally continues with the transfer of a proton from the solvent to the carbonyl oxygen atom. In many cases the reactions can also be catalysed by acids, when the proton transfer to oxygen occurs first.
- ☐ Methanal reacts faster than other aldehydes, which in turn react faster than ketones.
- ☐ Addition reactions of aldehydes and ketones include the following:

  $$\rangle C{=}O \rightarrow \rangle C(OH)Nu$$

  where Nu = H, CN, SO_3Na, OH.
- ☐ Addition–elimination reactions (condensations) of aldehydes and ketones include the following:

  $$\rangle C{=}O + NH_2X \rightarrow \rangle C{=}NX$$

  where X = OH, NHC_6H_5, $NHCONH_2$, $NHC_6H_3(NO_2)_2$. The yellow-orange crystalline 2,4-dinitrophenylhydrazones (DNPs) are used to characterize aldehydes and ketones.
- ☐ A second type of reaction involves replacement of an α-hydrogen atom by an electrophile, as in halogenation: for example,

  $$CH_3COCH_3 + Cl_2 \rightarrow CH_3COCH_2Cl + HCl$$
- ☐ Ethanal and ketones containing the CH_3CO— group undergo the haloform reaction with a halogen in alkaline solution to give $CHHal_3$. The iodoform test is a useful method of recognizing this group.
- ☐ The aldol reaction of aldehydes involves two molecules. One undergoes addition across the C═O bond, the other substitution at the α-position:

  $$2CH_3CHO \rightarrow CH_3CH(OH)CH_2CHO$$
- ☐ Aldehydes, unlike ketones, are readily oxidized to carboxylic acids. This is the basis of tests for aldehydes using Fehling's solution or Tollens' reagent, which are mild oxidizing agents.
- ☐ Aldehydes may be made by the oxidation of primary alcohols (acidic potassium dichromate(VI)) or by the reduction of acid chlorides (hydrogen with a palladium catalyst supported on barium sulphate).
- ☐ Ketones may be made by the oxidation of secondary alcohols (acidic potassium dichromate(VI)).
- ☐ Methanal is used in the manufacture of thermosetting plastics.
- ☐ Carbohydrates are cyclic hemiacetals derived from polyhydroxy-aldehydes or -ketones. Monosaccharides such as glucose and fructose contain a single carbonyl group in the open-chain form or a single hemiacetal group in the ring form. Disaccharides, such as sucrose, contain two monosaccharide units joined by a C—O—C linkage. Polysaccharides, such as starch, glycogen and cellulose, contain many monosaccharide units.

PROBLEMS

1 Draw the structural formulae of: hexanal, hexan-3-one, propanone oxime, 3-hydroxybutanal, 2,4-dinitrophenylhydrazine, hydroxylamine.

2 (a) What is the organic product, **A**, obtained from the gentle oxidation of propan-1-ol? Giving brief but essential experimental details, explain how the oxidation may be carried out in practice.
(b) Name and draw the structure of one isomer of **A**. How would you use 2,4-dinitrophenylhydrazine to distinguish between the isomers? Give two other tests that could be used to distinguish between them, stating what would be observed in each case.

3 State which of the compounds **A** to **D** listed below react with:
(a) $I_2 + NaOH(aq)$,
(b) $[Ag(NH_3)_2]^+(aq)$,
(c) PCl_5, giving fumes of HCl.
If a reaction occurs, draw the structure of the organic product(s).
A propan-1-ol **B** propan-2-ol
C propanone **D** propanal

4 Name one compound, **X**, other than water, that will add to aldehydes. What general type of reagents are water and **X** in these reactions? Draw the structure of the product of the reaction of **X** with ethanal.

5 A compound **A**, C_4H_9Br, was boiled with NaOH(aq) to give **B**, $C_4H_{10}O$. **B** was oxidized to **C**, C_4H_8O. **C** gave a precipitate with 2,4-dinitrophenylhydrazine but did not react with Fehling's solution.
Give structural formulae for **A**, **B** and **C**. What reagent can convert **B** into **C**? Draw the structures of two isomers of **C** that give precipitates with both 2,4-dinitrophenylhydrazine and with Fehling's solution.

6 Explain the term 'α-carbon atom' as applied to ethanal. Give an example of a reaction in which ethanal undergoes a substitution at the α-carbon atom.

7 What is the formula of 'aldol'? What reactants are used to prepare aldol?

8 Naming the starting materials and giving appropriate equations, give *one* example each to illustrate the essential differences between addition and condensation polymerization. Further give *two* examples of condensation polymerization reactions in which other typical bonds are formed.
Giving all necessary practical details, describe how *one* of the condensation polymers may be formed in the laboratory.
Explain how melting temperatures of polymers can be determined experimentally and outline, giving *two* examples, how the temperatures vary with polymer structure.
Calculate the mass of polymer formed from 60 g of urea on condensation with the appropriate amount of methanal (formaldehyde) (i) when no cross-linking takes place and (ii) if the maximum amount of cross-linking occurs.

$[A_r(H) = 1; A_r(C) = 12; A_r(N) = 14; A_r(O) = 16]$ (*Welsh*)

9 (a) (i) Write down, giving equations and essential conditions, *one* reaction scheme by which a sample of ethanal (acetaldehyde) could be prepared from ethanol.
(ii) State how you would decide whether your sample of ethanal was pure and suggest *two* impurities which might be present.
(b) Describe what happens when ethanal is treated with an alkaline solution of iodine, and calculate the mass of iodine needed to react completely with 2.2 g of ethanal.
(c) Outline the mechanism of the reaction between ethanal and hydrogen cyanide, and explain (i) the function of any catalyst used and (ii) why no such reaction occurs between ethene and hydrogen cyanide. (*Oxford*)

10 Describe a procedure by which you would attempt to prepare in the laboratory a pure sample of propanal (CH_3CH_2CHO) from propan-1-ol, stating the reagents involved.
Give at least *four* of the typical reactions of aldehydes and ketones.
State briefly how you would attempt to identify an unknown aldehyde or ketone. (*Nuffield*)

11 (a) 'Aldehydes and ketones can be used to prepare a wide range of organic compounds.' Discuss this statement, giving *five different* types of reaction.
(b) For each of the following, describe and explain *one* simple experiment which would enable you to demonstrate:
(i) similar chemical behaviour of ethanal and glucose;
(ii) different chemical behaviour of ethanal and benzaldehyde. (*Oxford*)

12 Iodomethane in ethoxyethane solutions reacts with magnesium to form the compound CH_3MgI, **A**. **A** reacts with propanone to form $[(CH_3)_3CO]^-[MgBr]^+$. Predict the reaction of **A** with the following substances:
(a) CH_2O (b) CH_2—CH_2 bridged by O (epoxyethane)
(c) $CH_3CO_2C_2H_5$ (d) I_2
(e) CO_2 (f) CH_3CO_2H
(g) $C_6H_5CH_2Br$ (h) $(CH_3)_3CCl$
(*Oxford Entrance*)

13 The bromination of propanone, according to the equation

$$CH_3COCH_3 + Br_2 \rightarrow CH_3COCH_2Br + HBr$$

is accelerated by base. In the presence of hydroxide ion, the rate is given by

$$rate = k[CH_3COCH_3][OH^-]$$

Suggest a possible mechanism for the reaction. Predict what would happen if the propanone were treated with a solution of NaOD in D_2O. (*Cambridge Entrance part question*)

31 Carboxylic acids and their derivatives

When an electronegative atom or group is attached to a carbon–oxygen double bond it modifies the carbonyl group reactions. Most reactions still start with nucleophilic attack on the carbonyl carbon atom, but they continue with the expulsion of the electronegative atom or group so that the net reaction is substitution. Conversely, the carbonyl group modifies the properties of the electronegative group, so that, for example, hydroxyl groups are much more acidic in carboxylic acids than they are in alcohols.

Compounds containing a hydroxyl group attached to a carbonyl carbon atom are **carboxylic acids**. The C=O and O—H groups each modify the behaviour of the other, so that the chemistry of the carboxylic acids differs from that of aldehydes and ketones and from that of alcohols. The —COOH is therefore treated as a single functional group, the **carboxyl group**. The general formula of RCOOH is $C_nH_{2n}O_2$. Compounds of the type R—CO—Z, in which a carbonyl carbon atom is attached to electronegative groups other than hydroxyl, are closely related to the carboxylic acids, and are usually called **carboxylic acid derivatives**, or simply **carboxylic compounds**. **Nitriles** contain a carbon–nitrogen triple bond, R—C≡N: their chemistry, as we shall see, is related to that of the carboxylic compounds, and they too are usually classified as carboxylic acid derivatives.

31.1 Nomenclature

Carboxylic acids are named after the corresponding alkanes, using the ending *-anoic acid*. As with the aldehydes, the carbonyl carbon atom is numbered 1. Traditional names, usually descriptive of their natural origin, are especially common for the acids: in particular, 'acetic acid' is very widely used for ethanoic acid.

H—COOH	methanoic acid (formic acid – Latin: *formica*, ant)
$\overset{2}{C}H_3\text{—}\overset{1}{C}OOH$	ethanoic acid (acetic acid – Latin: *acetum*, vinegar)
$\overset{3}{C}H_3\overset{2}{C}H_2\text{—}\overset{1}{C}OOH$	propanoic acid (propionic acid – Greek: *protos*, first; *pion*, fat*)
$\overset{4}{C}H_3\overset{3}{C}H_2\overset{2}{C}H_2\text{—}\overset{1}{C}OOH$	butanoic acid (butyric acid – Latin: *butyrum*, butter)

* Because it is the 'first' acid (lowest M_r) that forms fats: see Section 31.6

Acyl groups The group R—CO— is called an *acyl* group. Specific acyl groups are named from the corresponding carboxylic acid: for example, CH_3CO— is called **ethanoyl** (acetyl).

Carboxylic acid derivatives, R—CO—Z, are named in different ways, depending on the nature of Z. The table below gives the general names of the classes of compounds and, as examples, the names of particular individual compounds. Nitriles are also included.

General formula	Class name	Specific compound	Name
R—CO—Cl	acid chloride or acyl chloride	CH_3COCl	ethanoyl chloride
R—CO—O—CO—R	acid anhydride	$(CH_3CO)_2O$	ethanoic anhydride
R—CO—OR′	ester	$CH_3COOC_2H_5$	ethyl ethanoate
R—CO—NH_2	amide	CH_3CONH_2	ethanamide
R—CN	nitrile	CH_3CN	ethanenitrile

1 methanoic acid
$t_m = 8°C$
$t_b = 101°C$

2 benzene
$t_m = 5°C$
$t_b = 80°C$

31.2 Physical properties

Except in aqueous solution, carboxylic acids exist largely as hydrogen-bonded dimers (a **dimer** consists of two molecules held tightly together – see Section 5.1). This is shown, for example, by the fact that if the molar mass of a carboxylic acid is determined from the elevation of the boiling point of a solution in benzene (see Section 7.2), the result is twice the calculated value for a single molecule. The dimeric structure has a symmetry that allows the paired molecules to pack efficiently in the solid so that the melting points of carboxylic acids are relatively high (see Figures 31.1 and 31.2). The hydroxyl hydrogen atoms are not free to form hydrogen bonds to neighbouring dimers, however, and the boiling points of the acids are lower than might have been expected. The comparison of the properties of methanoic acid (**1**) with those of benzene (**2**) (with a geometrically similar molecule) is interesting.

The lower carboxylic acids have characteristic pungent smells. Vinegar smells of the ethanoic acid it contains. Butanoic acid gives its smell to rancid butter. The old names for the C_6, C_8 and C_{10} acids were – confusingly – caproic, caprylic and capric acids, all derived from the Latin *caper*, a goat: again, their smells are characteristic!

Many esters have strongly fruity smells or tastes, and a large number of them occur naturally in plants. The characteristic taste and smell of most fruits derive from a complicated mixture of volatile compounds: over seventy esters have been detected in apples, together with alcohols, acids, aldehydes, ketones and other odorous compounds. Nowadays many esters are synthesized commercially and used in artificial food scents and flavourings. For example, ethyl methanoate can be used to give a rum-like flavour, 3-methylbutyl ethanoate smells of pears and octyl ethanoate of oranges, and the methyl, ethyl, pentyl and 3-methylbutyl esters of butanoic acid smell of apples, pineapples, apricots and plums respectively (see also Figure 31.3).

Ethanamide is a white, crystalline solid, in which the molecules are held together by hydrogen bonding.

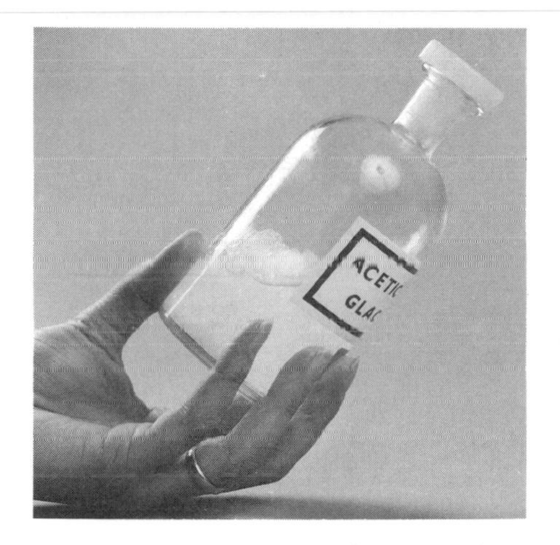

Figure 31.1 Ethanoic (acetic) acid melts at about room temperature, 16.6 °C: its transparent crystals give rise to the description *glacial*

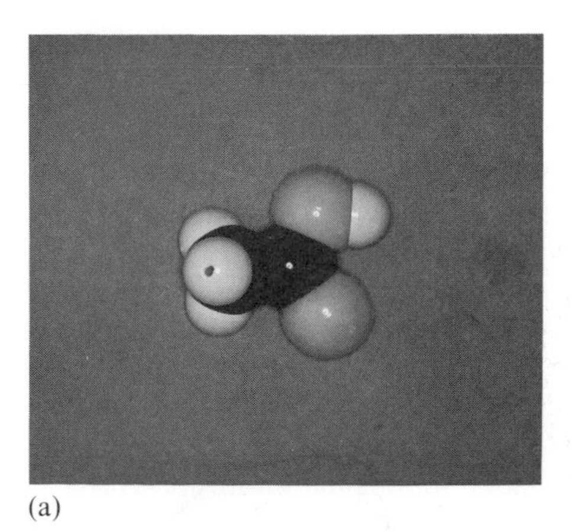

(a)

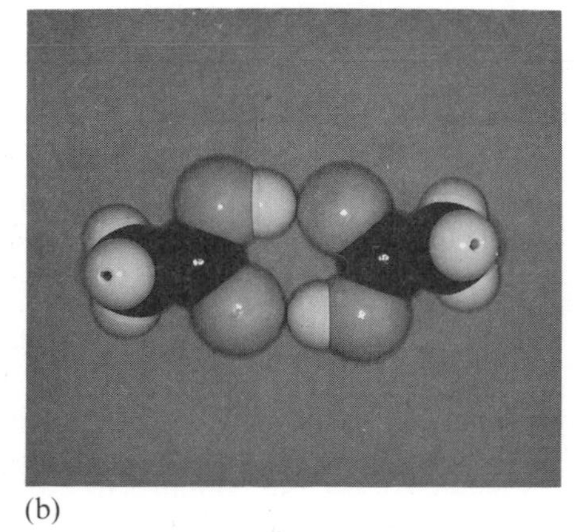

(b)

(c)

Figure 31.2 Space-filling models of ethanoic acid: (a) the monomer, (b) the dimer, and of (c) the ethanoate anion

Figure 31.3 Part of the vast collection of natural and synthetic fragrances in a perfumery laboratory

31.3 The acidity of carboxylic acids

Carboxylic acids are weak acids: they are partially ionized in water with the formation of the **carboxylate** anion:

$$\text{R—C}\begin{matrix}\text{=O}\\ \text{—OH}\end{matrix} + H_2O \rightleftharpoons H_3O^+ + \text{R—C}\begin{matrix}\text{O}\\ \text{O}\end{matrix}^-$$

In the carboxylate anion (see Figure 31.2(c)) the two oxygen atoms are equivalent: the C—O bond lengths are intermediate between those of C—O single and C═O double bonds, and each oxygen atom carries half the negative charge (see Figure 31.4 and Section 2.5). The carboxylate anion with its delocalized charge is more stable than an alkoxide anion, RO^-. The hydroxyl group of carboxylic acids therefore ionizes more readily than that of alcohols, and the carboxylic acids are much stronger acids. They neutralize alkalis and release carbon dioxide from carbonates:

$$CH_3COOH(aq) + HO^-(aq) \rightarrow CH_3CO_2^-(aq) + H_2O(l)$$

$$2CH_3COOH(aq) + CO_3^{2-}(aq) \rightarrow \underset{\text{ethanoate ion}}{2CH_3CO_2^-(aq)} + H_2O(l) + CO_2(g)$$

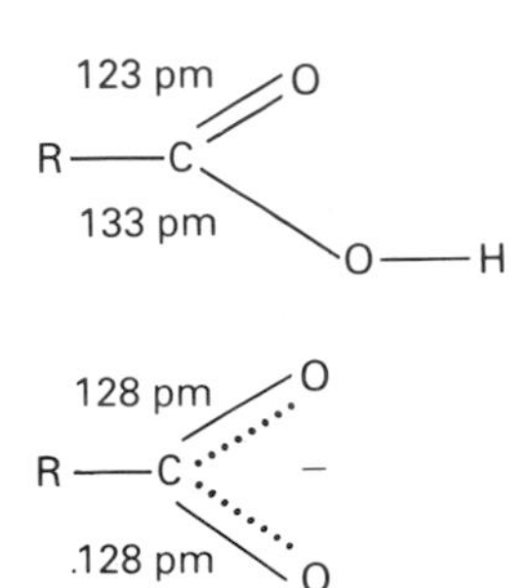

Figure 31.4 Carbon–oxygen bond lengths in carboxylic acids and carboxylate anions

Table 31.1 pK_a values of some carboxylic acids

H—COOH	3.75
CH_3—COOH	4.75
$HOCH_2$—COOH	3.83
FCH_2—COOH	2.66

Electron-attracting groups in the R group of RCO_2^- further stabilize the anion and increase the acid strength (see Table 31.1). Conversely, the difference between the strengths of methanoic and ethanoic acids is a consequence of the electron-donating properties of the methyl group (compare the stabilizing effects of methyl groups in carbocations, discussed in Section 28.3).

31.4 Reactivity patterns and important reactions

As with aldehydes and ketones, most of the reactions of carboxylic compounds start with nucleophilic attack on the carbonyl carbon atom:

$$\text{R(Z)C=}\ddot{\text{O}}\text{:} + \text{:Nu}^- \rightarrow \text{R(Z)(Nu)C—}\ddot{\text{O}}\text{:}^-$$

(As before we use the symbol $:Nu^-$ as a general representation for a nucleophile, but remember that nucleophiles are not necessarily negatively charged.)

Unlike the reactions of aldehydes and ketones, however, this reaction is not completed by attachment of a proton to the negatively charged oxygen atom, but by the expulsion of the electronegative group Z:

$$R-C(-O^-)(Nu)(Z) \longrightarrow R-C(=O)-Nu + :Z^-$$

The complete reaction is therefore a **nucleophilic substitution**:

$$R-C(=O)-Z + Nu^- \longrightarrow R-C(=O)-Nu + Z^-$$

Interconversion of carboxylic acid derivatives

If both Z and Nu in the above equation are electronegative groups, the reaction represented is the conversion of one carboxylic acid derivative into another. In general the reactivity of the compounds depends on the ease of expulsion of Z^-, so that the order of reactivity follows the order of stabilities of Z^-, as follows:

$$\underset{\text{acid chloride}}{RCOCl} > \underset{\text{acid anhydride}}{RCO{-}O{-}COR} > \underset{\text{acid/ester}}{\begin{Bmatrix} RCOOH \\ RCOOR' \end{Bmatrix}} > \underset{\text{amide}}{RCONH_2} > \underset{\text{carboxylate anion}}{RCO_2^-}$$

The position of equilibrium between any two derivatives also lies in favour of the compound to the right in the series. It is normally easy to make any carboxylic acid derivative from one shown to the left of it but, with three exceptions discussed below, not from one on its right.

Acid chlorides (acyl chlorides) The first exception to the general rule is that the hydroxyl group of carboxylic acids (like that of alcohols – see Section 29.3) can be chlorinated using phosphorus pentachloride or sulphur dichloride oxide; for example:

$$CH_3COOH + SCl_2O \rightarrow CH_3COCl + SO_2 + HCl$$

The resulting acid chlorides are much more reactive than acids. The chlorine atom is readily replaced by other nucleophiles, as in the following reactions, so that acid chlorides are very useful intermediates in organic synthesis.

1 $\underset{\text{ethanoyl chloride}}{CH_3COCl} + H_2O \rightarrow CH_3COOH + HCl$

Acid chlorides are readily hydrolysed in cold water: they fume in moist air as droplets of ethanoic and hydrochloric acids are formed. If the vapour of an acid chloride reaches the fluid round an eye the acids induce tears; compounds that have this effect are called **lachrymatory**.

2 $CH_3COCl + C_2H_5OH \rightarrow CH_3COOC_2H_5 + HCl$

$CH_3COCl + C_6H_5OH \rightarrow CH_3COOC_6H_5 + HCl$

These rapid reactions form esters (see below): they are similar to hydrolysis. The replacement of a hydrogen atom by RCO— is **acylation**: acid chlorides are **acylating agents**. Specifically, ethanoyl chloride, which replaces H— by CH_3CO—, is an **ethanoylating agent**.

3 $CH_3COCl + NH_3 \rightarrow CH_3CONH_2 + HCl$

$CH_3COCl + RNH_2 \rightarrow CH_3CONHR + HCl$

$CH_3COCl + RR'NH \rightarrow CH_3CONRR' + HCl$

These rapid reactions are ethanoylations of ammonia or amines (see Section 32.3). The amide products, discussed further below, are white solids.

4 $CH_3COCl + CH_3CO_2Na \rightarrow CH_3CO—O—COCH_3 + NaCl$

When an acid chloride and a salt of the acid are heated together they give an acid anhydride, discussed below.

Acid anhydrides The acid anhydrides are so called because they can be regarded as the result of the loss of one molecule of water from two molecules of carboxylic acid:

$$2CH_3COOH - H_2O \rightarrow CH_3CO—O—COCH_3$$

(This is not normally a method of preparing an anhydride, however.)

The chemistry of the acid anhydrides is very similar to that of the acid chlorides, except that the anhydrides are a little less reactive. Their main reactions are as follows:

$$(CH_3CO)_2O + H_2O \rightarrow 2CH_3COOH$$

$$(CH_3CO)_2O + C_2H_5OH \rightarrow CH_3COOC_2H_5 + CH_3COOH$$

$$(CH_3CO)_2O + NH_3 \rightarrow CH_3CONH_2 + CH_3COOH$$

ethanoic anhydride

Ester formation and hydrolysis When a carboxylic acid is refluxed with an alcohol with a trace of mineral acid catalyst (commonly concentrated sulphuric acid), an equilibrium is set up with an ester:

$$CH_3COOH + C_2H_5OH \underset{}{\overset{H^+}{\rightleftharpoons}} CH_3C(=O)OC_2H_5 + H_2O$$

ethanoic acid ethanol ethyl ethanoate

If equal amounts of acid and alcohol are used, then at equilibrium all four species – acid, alcohol, ester and water – are present in similar amounts (see Section 10.1). The ester can be obtained in good yield by increasing the amount of inorganic acid above that needed just for catalysis. This absorbs the water and so displaces the equilibrium to the right. Also the ester, which lacks a hydroxyl group and consequently cannot act as a hydrogen bond donor, often has a lower boiling point than the acid or alcohol, and can be distilled out of the reaction mixture.

Esters can be hydrolysed back to a mixture of acid and alcohol either by using dilute aqueous mineral acid (high concentration of water to displace the position of the equilibrium to the left) or better by using aqueous sodium hydroxide. This gives the carboxylate anion rather than the free acid and the reaction goes effectively to completion.

$$H_3C{-}C(=\ddot{O}){-}\ddot{O}C_2H_5 + :\ddot{O}H^- \rightleftharpoons H_3C{-}C(-\ddot{O}:^-)(OH)(\ddot{O}C_2H_5) \rightleftharpoons H_3C{-}C(=\ddot{O}){-}\ddot{O}H + C_2H_5\ddot{O}:^- \xrightarrow[\text{transfer}]{\text{proton}} CH_3CO_2^- + C_2H_5OH$$

The formation of soap, described in Section 31.6, is an example of this type of reaction, and the alkaline hydrolysis of esters is often called **saponification** (Latin: *sapo*, soap). The carboxylate salt may, of course, be converted into the acid by treating it with a stronger, mineral acid such as HCl(aq), the second exception to the general rule that one may move only to the right in the series of carboxylic compounds shown above.

BOX 31.1

Mechanisms of ester hydrolysis

$$R{-}CO{-}O{-}R' + HO{-}H \longrightarrow R{-}CO{-}OH + H{-}O{-}R'$$

3 acyl–oxy fission

$$R{-}CO{-}O{-}R' + H{-}OH \longrightarrow R{-}CO{-}O{-}H + R'{-}OH$$

4 alkyl–oxy fission

We have described the mechanism of saponification as involving the attack of hydroxide ion on the carboxyl carbon atom, followed by displacement of an alkoxide ion. This is described as **acyl–oxy fission (3)** because the bond between the acyl group and the alkoxy oxygen atom is broken. We could also imagine a mechanism involving **alkyl–oxy fission (4)**.

The evidence for acyl–oxy fission comes from **isotopic labelling**. If the hydrolysis of ethyl ethanoate is carried out in water enriched (labelled) with the isotope ^{18}O, all the label is found in the ethanoic acid and none in the ethanol.

$$H_2{}^{18}O + CH_3CO{-}OC_2H_5 \rightarrow CH_3CO{-}{}^{18}OH + C_2H_5OH$$

The earliest studies were done in the 1930s when isotopically enriched compounds were rare and expensive. They were carried out on a very small scale, using water enriched with only 0.35 per cent of ^{18}O, and analysing the extent of labelling from density measurements!

By contrast, sulphonate esters are hydrolysed via alkyl–oxy fission, that is, by an SN2 attack (see Section 28.3) on the alkyl group. The difference arises, at least in part, because the sulphonate anion is more stable than the carboxylate anion, and makes a better leaving group.

$$HO^- + H_3C{-}CH_2{-}OSO_2CH_3 \longrightarrow CH_3CH_2{-}OH + CH_3SO_3^-$$

Amides Amides are readily formed, as described above, by the reaction of ammonia or an amine with an acid chloride or anhydride. They are also formed, more slowly, from esters:

$$CH_3COOC_2H_5 + NH_3 \rightarrow \underset{\text{ethanamide}}{CH_3CONH_2} + C_2H_5OH$$

The reaction of ammonia (a base) with a carboxylic acid gives the corresponding ammonium salt. This can be dehydrated by heating it strongly, typically for several hours under reflux, to give the amide:

$$CH_3COOH + NH_3 \rightarrow \underset{\text{ammonium ethanoate}}{NH_4{}^+\,CH_3CO_2{}^-} \xrightarrow{\text{heat}} CH_3CONH_2 + H_2O$$

Amides, like esters, are hydrolysed when they are refluxed in strongly alkaline solution to give the carboxylate salts:

$$CH_3CONH_2(aq) + HO^-(aq) \rightarrow CH_3CO_2{}^-(aq) + NH_3(aq)$$

They are also hydrolysed in aqueous acid to the carboxylic acid. This reversal of the general rule, the third of the exceptions referred to above, occurs because the ammonia is protonated:

$$CH_3CONH_2(aq) + H_3O^+(aq) \rightarrow CH_3COOH(aq) + NH_4{}^+(aq)$$

The amide itself is not significantly protonated in aqueous acid: it is much less basic than ammonia (and amines, see Section 32.3) because the nitrogen lone electron pair is delocalized with the carbonyl group (see Figure 31.5). Amides are also converted into carboxylic acids by treatment with nitrous acid (produced in the reaction mixture from sodium nitrite and dilute sulphuric acid); molecular nitrogen is produced:

$$RCONH_2 + HONO \rightarrow RCOOH + N_2 + H_2O$$

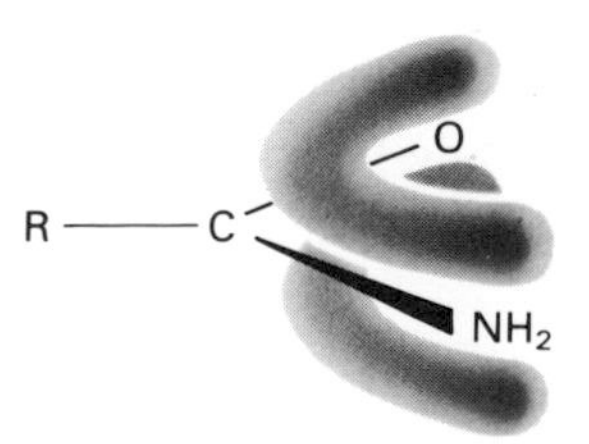

Figure 31.5 Delocalized electrons from the nitrogen lone pair and the carbonyl π-bond of amides

When amides are heated with bromine and alkali they are converted into amines with one carbon atom fewer (**Hofmann degradation**):

$$RCONH_2 \xrightarrow[NaOH]{Br_2} RNH_2(+CO_2 + HBr + NaBr)$$

Reduction of carboxylic acids and derivatives

All the carboxylic compounds except amides are reduced to primary alcohols by lithium tetrahydridoaluminate ($LiAlH_4$) in dry ethoxyethane, followed by water. The first step is effectively a substitution reaction by hydride ion, H^-, to give the aldehyde, which is then reduced further (see Section 30.3).

$$R{-}C(=O){-}Z \xrightarrow[\text{(ii) } H_2O]{\text{(i)} LiAlH_4} R{-}CH_2{-}OH$$

($Z = Cl, OCOR, OH, OR'$)

The aldehyde is reduced faster than the acid derivative, so that it is not usually possible to stop the reduction at the aldehyde. If the aldehyde is the desired product it can be made by reoxidizing the primary alcohol (see Section 29.4). Alternatively it can be made by the **Rosenmund reduction** of an acid chloride, using hydrogen gas over a catalyst of palladium supported on barium sulphate, intentionally poisoned to prevent reduction of the aldehyde:

$$RCOCl + H_2 \xrightarrow{Pd/BaSO_4} RCHO + HCl$$

Amides are reduced by $LiAlH_4$ without loss of the amino group, because the anion NH_2^- is a very poor leaving group. The products are amines, described in Chapter 32.

$$RCONH_2 \xrightarrow[\text{(ii)} H_2O]{\text{(i)} LiAlH_4} RCH_2NH_2$$

The more selective reducing agent sodium tetrahydridoborate ($NaBH_4$) reduces aldehydes and ketones (see Section 30.3) but not carboxylic acids or their derivatives.

Nitriles

When amides are heated with phosphorus(V) oxide, nitriles are formed by dehydration. These are compounds containing a carbon–nitrogen triple bond:

$$CH_3CONH_2 \xrightarrow[-H_2O]{P_4O_{10}} CH_3{-}C{\equiv}N$$

ethanenitrile

Conversely, nitriles can be hydrolysed by dilute mineral acids or alkalis, via the amide, to carboxylic acids or carboxylate salts:

$$RCN \xrightarrow{H_3O^+} RCONH_2 \xrightarrow{H_3O^+} RCOOH + NH_4^+$$

$$RCN \xrightarrow[HO^-]{H_2O} RCONH_2 \xrightarrow{HO^-} RCO_2^- + NH_3$$

Because of their ready interconversion with acids and amides, nitriles are usually regarded as carboxylic acid derivatives even though they do not contain a —CO—Z group.

Nitriles, like amides, are reduced to amines with lithium tetrahydridoaluminate:

$$RCN \xrightarrow[\text{(ii)} H_2O]{\text{(i)} LiAlH_4} RCH_2NH_2$$

Nitriles are valuable intermediates in the synthesis of organic compounds. In particular they provide a way of adding an extra carbon atom to a chain (**5**).

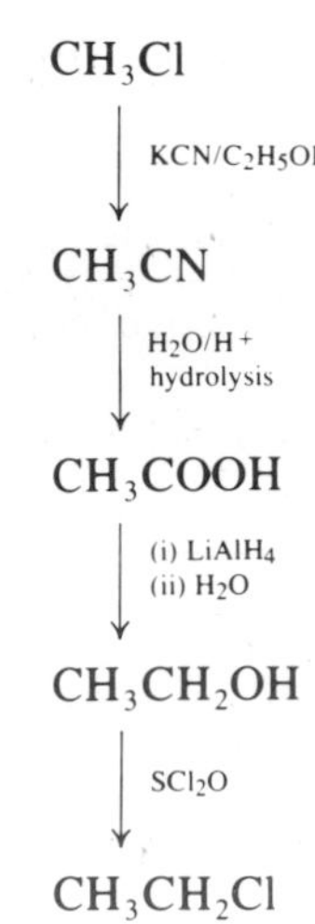

EXAMPLE

Show how propan-1-ol may be converted into (a) propanoic acid, (b) butanoic acid, (c) 2-hydroxybutanoic acid.

METHOD Note that the preparation of (a) does not involve increasing the length of the carbon chain whereas preparing (b) and (c) requires the addition of an extra carbon atom in both cases. Preparations of (a) and (b) involve the formation of carboxylic acids, but (c) has two functional groups: consider reactions that both extend the carbon skeleton and introduce a second functional group.

ANSWER (a) Direct oxidation:

$$CH_3CH_2CH_2OH \xrightarrow[\text{under reflux}]{Cr_2O_7^{2-}/H^+(aq)} CH_3CH_2COOH$$

(b) Extend the chain with CN^-:

$$CH_3CH_2CH_2OH \xrightarrow{SCl_2O} CH_3CH_2CH_2Cl$$

$$\xrightarrow[C_2H_5OH]{KCN} CH_3CH_2CH_2CN$$

$$\xrightarrow[H_2O/H^+]{\text{hydrolysis}} CH_3CH_2CH_2COOH$$

(c) Use the addition of hydrogen cyanide to an aldehyde (see Section 30.3):

$$CH_3CH_2CH_2OH \xrightarrow[\text{distil}]{Cr_2O_7^{2-}/H^+(aq)} CH_3CH_2CHO$$

$$\xrightarrow[H_2O]{KCN} CH_3CH_2CH(OH)CN \xrightarrow[H_2O/H^+]{\text{hydrolysis}} CH_3CH_2CH(OH)COOH$$

COMMENT Reactions (b) and (c) illustrate the versatility of the nitrile group in synthesis.

31.5 Preparative methods for carboxylic compounds

Most of the methods of preparing carboxylic compounds are described in Section 4 of this chapter; three preparations, described in other chapters, are summarized here.

Carboxylic acids can be made by the oxidation of primary alcohols (see Section 29.4) or of aldehydes (Section 30.3), using acidified dichromate(VI):

$$RCH_2OH \xrightarrow[H^+(aq)]{Cr_2O_7^{2-}} RCHO \xrightarrow[H^+(aq)]{Cr_2O_7^{2-}} RCOOH$$

Carboxylic acids can also be made via the reaction of Grignard reagents with carbon dioxide (see Box 30.1).

Nitriles can be made by the reaction of a halogenoalkane with potassium cyanide in ethanol (Section 28.3):

$$RHal + CN^- \rightarrow RCN + Hal^-$$

31.6 The importance of carboxylic compounds

Ethanoic acid, which is the acid component of vinegar, is used in the manufacture of the polymers PVA and viscose (see Sections 33.5 and 30.5).

$$\begin{array}{l} R\text{—CO—O—}CH_2 \\ \qquad\qquad\quad | \\ R'\text{—CO—O—}CH \\ \qquad\qquad\quad | \\ R''\text{—CO—O—}CH_2 \end{array} \quad (6)$$

Oils and fats Most animal and vegetable oils and fats (**lipids**) are esters of the trihydric alcohol propane-1,2,3-triol (glycerol). They are commonly called **triacylglycerols** (**6**). The acids RCOOH, R′COOH and R″COOH are called **fatty acids** because of their origin. They are nearly all straight-chain compounds with the alkyl groups R, R′ and R″ either fully saturated or containing one or more double bonds (see below and Box 31.2). The fatty acids are still usually known by traditional names derived from one of the fats or oils from which they are obtained. Examples of the most commonly occurring acids are given in Table 31.2.

Table 31.2 Some common fatty acids

	Traditional name	Origin of name
Saturated		
$CH_3(CH_2)_{10}COOH$	lauric acid	laurel seed
$CH_3(CH_2)_{12}COOH$	myristic acid	nutmeg (*Myristica fragrans*)
$CH_3(CH_2)_{14}COOH$	palmitic acid	palm oil
$CH_3(CH_2)_{16}COOH$	stearic acid	suet (Greek: *stear*)
Unsaturated		
$CH_3(CH_2)_7CH{=}CH(CH_2)_7COOH$	oleic acid	oil
$CH_3(CH_2)_4CH{=}CHCH_2CH{=}CH(CH_2)_7COOH$	linoleic acid	oil of flax (Latin: *linum*)

BOX 31.2

Saturated and unsaturated fats and oils

Only their melting points distinguish oils and fats: oils are liquid at normal temperatures, while fats are solid. In general oils contain a high proportion of unsaturated triacylglycerols, because the double bonds stiffen the hydrocarbon chains against bending and the molecules pack less easily. Mammals usually contain a higher proportion of saturated fat than do fish or plants. Fish are 'cold-blooded'; plants have to cope with low temperatures at night or in the winter: in both kinds of organism the oils need to have a fairly low melting point. Mammals, on the other hand, maintain a high body temperature at which even saturated fats remain semi-fluid. (Compare the properties of butter on a cold day with those of a packet held too long in the hand.)

Margarine was first produced in France in 1869, when Napoleon III held a competition to find a cheap edible fat to replace butter, 'suitable for the working class and the Navy'. For many years all margarine was produced by hydrogenating vegetable and fish oils, using a nickel catalyst, to saturate many of the carbon–carbon double bonds and give a product that resembled butter. Recently there have been moves to reduce the proportion of saturated fats in food and to increase the proportion of polyunsaturated fats (those in which the fatty acid groups contain two or more double bonds, such as linoleic). The theory is that saturated fats in the diet encourage build-up of **cholesterol** in the blood, and that polyunsaturated fats help to keep cholesterol levels down. Cholesterol is a fat-like substance, not a triacylglycerol but a steroid, related chemically to the sex hormones and bile acids. It tends to line the walls of blood vessels, hindering the flow of blood and possibly provoking heart disease. This theory has led to the development of margarines that, far from being more saturated than the oils from which they are made, are often less saturated. Such margarines are usually much softer than butter or the older-style 'block' margarines because of the lower melting points of the unsaturated triacylglycerols.

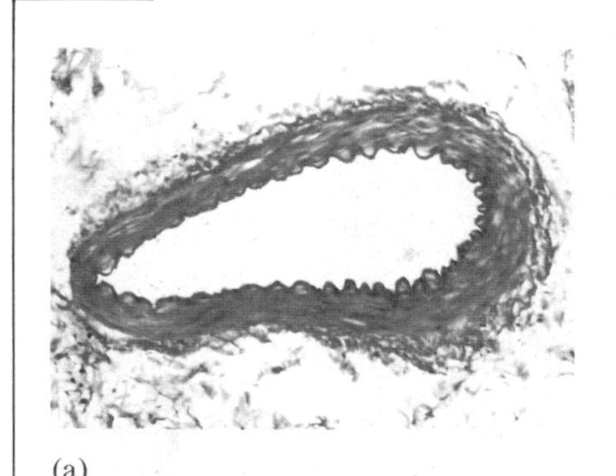

(a)

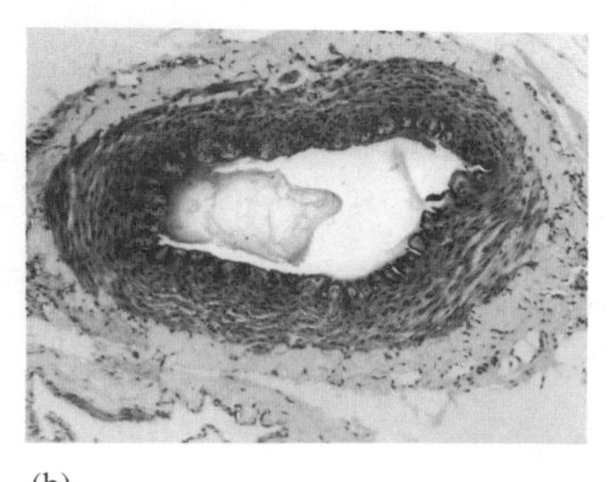

(b)

Figure 31.6 Cross-sections through (a) a healthy artery (the small blood vessels are veins), (b) an artery of which the walls have been thickened by deposits of cholesterol

Any natural oil contains a mixture of triacylglycerols, some with all the fatty acid groups the same, some with them different. However, the proportions in which the acid groups occur are more or less constant for any particular oil or fat. Human body fat, for example, contains mainly oleic (47 per cent), palmitic (24 per cent), linoleic (10 per cent) and stearic (8 per cent) groups. Linoleic acid is an **essential fatty acid** in the human diet: the body needs it for the manufacture of prostaglandins, but mammals lack the enzymes to synthesize acids with double bonds beyond carbon atom 9.

Hydrolysis of an oil or fat with aqueous sodium hydroxide (saponification) gives propane-1,2,3-triol and the carboxylate salts. These are **soaps**: the main constituents of ordinary soap are sodium stearate, sodium palmitate and sodium oleate. Their cleansing properties come from the presence of the long oil-soluble hydrocarbon chain and the ionic water-soluble carboxylate group (see Box 34.3).

Summary

- ☐ Carboxylic acids contain the functional group —COOH. Other compounds of the type R—CO—Z, where Z is an electronegative atom or group, are called carboxylic acid derivatives.
- ☐ Carboxylic acids are named as alkanoic acids. The group RCO— is an acyl group: specific acyl groups are named alkanoyl. Carboxylic acid derivatives are named as in the following examples:

RCOCl, acid chlorides	CH_3COCl, ethanoyl chloride
$(RCO)_2O$, acid anhydrides	$(CH_3CO)_2O$, ethanoic anhydride
RCOOR′, esters	$CH_3COOC_2H_5$, ethyl ethanoate
$RCONH_2$, amides	CH_3CONH_2, ethanamide
RCN, nitriles	CH_3CN, ethanenitrile

- ☐ Traditional names are common for many carboxylic acids and derivatives: in particular, ethanoic acid is known as acetic acid.
- ☐ Carboxylic acids exist largely as hydrogen-bonded dimers, except in aqueous solution.
- ☐ Carboxylic acids are acidic because of the relative stability of the negative charge in the carboxylate anion, shared between the two equivalent oxygen atoms.
- ☐ Most reactions of carboxylic acids and derivatives start with nucleophilic attack on the carbonyl carbon atom: displacement of the electronegative atom or group attached to the C═O carbon atom leads to substitution products.
- ☐ It is possible to prepare any carboxylic compound in the following sequence from a compound to its left by direct substitution:

RCOCl	RCO—O—COR	$\left\{\begin{matrix} RCOOH \\ RCOOR' \end{matrix}\right\}$	$RCONH_2$	RCO_2^-
acid chloride	acid anhydride	acid/ester	amide	carboxylate anion

- ☐ Moving to the left in the above sequence is possible in the following two reactions (and also by protonating the carboxylate salt).

$$RCOOH + PCl_5 \text{ (or } SCl_2O) \rightarrow RCOCl$$

$$RCONH_2 + H_3O^+ \rightarrow RCOOH + NH_4^+$$

- ☐ The Hofmann degradation of amides with bromine and sodium hydroxide converts $RCONH_2$ to RNH_2.
- ☐ Amides react with nitrous acid in the cold to give carboxylic acids and molecular nitrogen.
- ☐ Amides can be dehydrated by phosphorus(V) oxide to nitriles, RC≡N.
- ☐ Nitriles can be hydrolysed to amides and carboxylic acids.

- ☐ Carboxylic compounds except amides and nitriles are reduced by lithium tetrahydridoaluminate ($LiAlH_4$) to RCH_2OH.
- ☐ Amides and nitriles are reduced by $LiAlH_4$ to RCH_2NH_2.
- ☐ Carboxylic acids may be made by the oxidation of primary alcohols or aldehydes, or by the hydrolysis of esters or nitriles.
- ☐ Carboxylic acid derivatives are made from acids or other acid derivatives as outlined above.
- ☐ Nitriles are made by dehydrating amides or by the reaction of a halogenoalkane with cyanide ion.
- ☐ Ethanoic acid (acetic acid) is used for making the plastics materials PVA and viscose. It is the acid component of vinegar.
- ☐ Natural oils and fats are esters of the trihydric alcohol propane-1,2,3-triol (glycerol) with fatty acids. They are known as triacylglycerols.

PROBLEMS

1 Draw the structural formulae of: butanoic acid, propanoyl chloride, ethanoic anhydride, methyl pentanoate, methanamide, hexanenitrile.

2 Explain how the following conversions could be carried out, briefly describing the essential experimental conditions:
(a) ethanol → ethanoic acid
(b) ethanoic acid → ethanol
(c) ethanoic acid → ethanoyl chloride
(d) ethanamide → ethanoic acid

3 Ethanoyl chloride is said to be a good 'ethanoylating agent': what is the meaning of this term? Why is ethanoyl chloride a good ethanoylating agent? Why is it more susceptible to attack by nucleophiles than is chloroethane? Give three equations to illustrate these reactions of ethanoyl chloride.

4 How, and under what conditions, does NaOH(aq) react with: (a) ammonium ethanoate, (b) ethanamide, (c) ethanoyl chloride, (d) tristearoylglycerol (a typical fat)? What are the roles of the hydroxide ion in reactions (a) and (b)?

5 Give the organic products of the reactions between the following pairs of compounds, stating one important reaction condition for each reaction:
(a) ethanoic acid and lithium tetrahydridoaluminate,
(b) ethanamide and nitrous acid,
(c) ethyl ethanoate and water.

6 Describe tests that would distinguish between each of the pairs of compounds below, saying what is observed with each member of the pair. Where possible, write equations for the reactions.
(a) CH_3CH_2CHO and CH_3CH_2COOH
(b) $CH_3COOCOCH_3$ and $CH_3COOCH_2CH_3$
(c) $CH_3COCH_2CH_3$ and $CH_3COOCH_2CH_3$
(d) $CH_3CO_2NH_4$ and CH_3CONH_2

7 Deduce the structures of compounds **A** to **F** in the following reaction sequences:

$$\underset{\mathbf{A}}{C_4H_7N} \xrightarrow[\text{(ii) } H_2O]{\text{(i) } LiAlH_4 \text{ in dry ethoxyethane}} \underset{\mathbf{B}}{C_4H_{11}N} \xrightarrow{CH_3COCl} \underset{\mathbf{C}}{C_6H_{13}NO}$$

$$\mathbf{A} \xrightarrow{\text{hot } (HO)_2SO_2(aq)} \mathbf{D}$$

$$\underset{\mathbf{D}}{C_4H_8O_2} \xrightarrow[\text{(ii) } H_2O]{\text{(i) } LiAlH_4 \text{ in dry ethoxyethane}} \underset{\mathbf{E}}{C_4H_{10}O} \xrightarrow[300\,^\circ C]{Al_2O_3} \underset{\mathbf{F}}{C_4H_8} \xrightarrow{HBr} (CH_3)_3C\text{—}Br$$

8 Show how (a) CH_3CH_2COOH and (b) $CH_3CH_2CH_2COOH$ could be prepared from 1-bromopropane, using only inorganic reagents.

9 A compound containing an amide group was hydrolysed and gave two organic compounds as products.

One compound was sparingly soluble in water, burnt with a smoky flame and was a monocarboxylic acid; 0.21 g of this monobasic acid when titrated with 0.10 M sodium hydroxide gave an end-point of 17.2 cm^3. Calculate the relative molecular mass of the organic acid and suggest a structural formula.

The second compound was readily soluble in water and gave a purple coloration when warmed with ninhydrin; but the solution was not optically active when tested in a polarimeter. Analysis by mass of a sample produced the result C, 32.0%; H, 6.7%; O, 42.7%; N, 18.7%. Calculate the empirical formula of the second compound and suggest a possible structural formula.

Draw a graphical formula of the original compound (a structural formula showing all atoms and bonds).

Give *two* examples of chemical reactions by which simple amides can be formed. (*Nuffield*)

10 A substance containing carbon, hydrogen and oxygen only, when heated under reflux with aqueous acid for several hours, could be split into two components.

The lower-boiling component was a liquid at room temperature. 1.0 g, on complete combustion, produced 2.2 g of carbon dioxide and 1.2 g of water. On volatilization in a gas syringe at 100 °C and one atmosphere pressure, 0.10 g of the liquid produced 51 cm^3 of vapour. When the liquid was oxidized, the product gave a crystalline derivative with 2,4-dinitrophenylhydrazine, but would not react with Fehling's solution.

The higher-boiling component from the heating under reflux was neutralized with ammonia and the product heated with phosphorus(V) oxide, P_2O_5. This yielded a product of relative molecular mass 97, whose composition by mass was 74.2% carbon, 11.3% hydrogen and 14.4% nitrogen.

Using the information provided, suggest a possible identity for each of the substances involved, and explain the reaction of the original substance with aqueous acid when heated under reflux. *(Nuffield)*

11 (a) By means of general formulae (using R to denote alkyl groups) indicate the chemical structure of (i) proteins, (ii) fats and oils. Mark clearly, and name, the important linkages present in these food components.
(b) Explain the functions of proteins and fats and oils in the diet and give one important example of each.
(c) Give equations to show the products of hydrolysis of proteins and fats and oils. Indicate the conditions necessary to bring about the hydrolysis in the laboratory.
(d) Glucose, maltose and starch are carbohydrates. By means of simple diagrams indicate the relationship between them and state the product of hydrolysis of these carbohydrates.
(e) 'Cellulose is a carbohydrate similar in many respects to starch. It has no nutritional value but is essential in the human diet.' Explain this statement. *(Cambridge)*

12 Discuss the influence of structure on the relative acid strength of ethanoic acid, chloroethanoic acid and trichloroethanoic acid.
Outline a possible synthesis for *each* of the following compounds from the given starting materials:

(a) $CH_3CH_2CH_2CO_2H$ from $CH_3CH_2CH_2CH_2Br$,
(b) $CH_3CH_2CH_2CO_2H$ from $CH_3CH_2CH_2Br$,
(c)

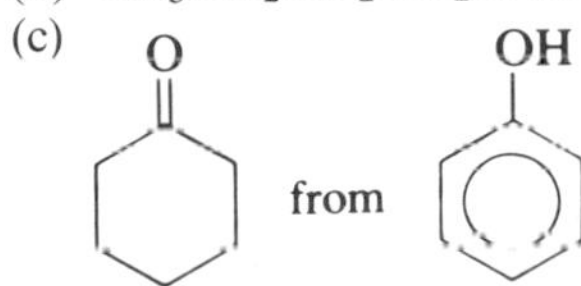

(Cambridge)

13 Match the boiling points (°C), 118, 21, 141, −88, 78 to the following compounds, giving explanations for your choices:

CH_3CH_3, CH_3CH_2OH, $CH_3C(\!=\!O)H$,

$CH_3C(\!=\!O)OH$, $CH_3CH_2C(\!=\!O)OH$

Starting with *each* of the compounds, show how it could be converted into *one* of the other compounds in the list. Essential reaction conditions and equations should be given.
Two of the compounds reacted together in the presence of mineral acid to give a compound with a sweet smell and containing C 58.8%, H 9.8%. Identify the compounds and explain the reaction which occurred.

$[A_r(C) = 12,\ A_r(H) = 1,\ A_r(O) = 16.]$ *(Welsh)*

14 (a) Give the names, and characteristic structural features, of *three* different types of organic compound containing a carbonyl group and no elements other than carbon, hydrogen and oxygen. Give a specific example of each type of compound.
(b) Given samples of pure carbonyl compounds, known to belong to the three types of compound which you have given in (a), suggest a series of chemical tests which would enable you to assign each to its proper type. Quote tests which will enable you to identify each of the three types.
(c) Ethanamide (acetamide) and ethanoyl (acetyl) chloride are carbonyl compounds containing carbon, hydrogen and oxygen as well as nitrogen or chlorine. Show, in outline, how these compounds may be (i) obtained from organic compounds containing only carbon, hydrogen and oxygen, and (ii) converted to compounds containing only these three elements. *(Oxford and Cambridge)*

15 Explain the following observations.
(a) Ethene (ethylene) does not react with water, but when it is passed into concentrated sulphuric acid and then the acid is diluted with water, ethanol is formed.
(b) Three isomers of formula $C_2H_2Br_2$ exist, but only two isomers of formula $C_2H_4Br_2$.
(c) Hydrogen cyanide adds to propanone (acetone) in the presence of bases but not in the presence of acids.
(d) Esterification of ethanoic (acetic) acid is catalysed by acids but not by bases; whereas hydrolysis of ethyl ethanoate (acetate) is catalysed by both acids and bases. *(Oxford and Cambridge)*

16 (a) Given a sample of ethanol as your only organic compound, describe briefly how you would obtain a sample of ethyl ethanoate.
(b) Write down the mechanism for the hydrolysis of an ester.
(c) Explain the following:
(i) The hydrolysis of simple esters is usually carried out in the presence of base rather than acid.
(ii) Ketones and esters both have carbonyl groups but only esters are hydrolysed easily. *(Oxford)*

17 Identify by name and structural formula the compounds **A**, **B**, **C**, **D**, **E**, **F**, **G** and **H**. Give reasons for your identifications and write equations for the reactions described.
(a) When heated together **A** and **B** give a volatile liquid **C**, $C_4H_6O_3$, and a residue of sodium chloride. If either **A** or **C** is treated with phenylamine (aniline) and the mixture poured into water, a solid **D**, C_8H_9NO, can be collected by filtration.
(b) A gas **E** has relative molecular mass 28. A mixture of **E** and hydrogen passed over a heated catalyst gives a liquid **F**. Treatment of **F** with a mixture of concentrated sulphuric acid and potassium bromide gives **G**. Reaction of potassium cyanide with an alcoholic solution of **G** gives **H**, which can be hydrolysed under acidic conditions to ethanoic (acetic) acid. *(Oxford)*

18 How may fats and oils be converted into soaps? Describe the main structural features of molecules in soaps and explain how soap action is related to these features.
What is the effect of hard and soft water on soaps? *(Nuffield)*

32 Amines: the nitrogen lone pair. Chiral molecules

In this chapter we see that the chemistry of the amines is dominated by the presence of the lone pair of electrons on the nitrogen atom. Reactions in which the carbon–nitrogen bond is broken are uncommon. We also consider the amino-acids, the building blocks of protein, and use them to introduce a special type of isomerism.

If one or more of the hydrogen atoms of ammonia, NH_3, is replaced by an alkyl group the resulting compounds are called **amines**. They are classified as **primary**, **secondary** and **tertiary** amines depending on the number of alkyl groups attached to the nitrogen (see Figure 32.1), *not* – as with alcohols and halogenoalkanes – on the nature of those alkyl groups. Adrenalin (**1**), for example, is both a secondary amine and a secondary alcohol. If the hydrogen atoms of the ammonium cation, $NH_4{}^+$, are replaced by alkyl groups the resulting cations are called primary, secondary, tertiary or **quaternary** ammonium cations.

The general formula for an alkyl group is C_nH_{2n+1}, so that the general formula for an amine is $C_nH_{2n+3}N$. Since the relative atomic mass of nitrogen is even (14), the relative molecular mass of an amine is odd – the same holds for other compounds with single nitrogen atoms, such as amides and nitriles. The appearance in the mass spectrum of a molecular ion (see Section 3.1) with an *odd* value of m/z is therefore often an indication that the compound contains nitrogen.

primary amine

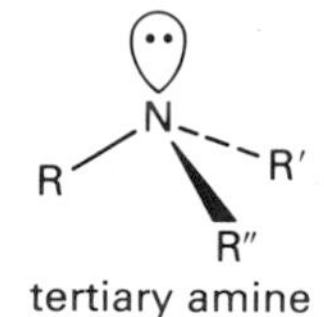

secondary amine

tertiary amine

quaternary ammonium cation

Figure 32.1 Classes of amines, and a quaternary ammonium cation

BOX 32.1

Phase-transfer catalysis

We saw in Section 30.2 that ketones are used as solvents in reactions in which one of the reactants is an ionic nucleophile that dissolves in water, and the other is a non-polar organic compound that does not. A more recent technique is the use of **phase-transfer catalysts**, of which quaternary ammonium salts are important examples, to carry the nucleophile from the water into the organic phase.

If a heated mixture of 1-bromooctane and aqueous potassium cyanide is stirred even for several days there is no significant reaction. If a little tetrabutylammonium bromide is added, the reaction takes only about an hour. Unlike K^+, the quaternary ammonium cation dissolves in the bromooctane, because the positive charge is buried deep inside a shell of alkyl groups so that dispersion forces dominate its interactions with solvent molecules. To maintain charge neutrality, the cyanide anion is dragged into the bromooctane phase, where the reaction takes place:

$$\text{aqueous phase} \quad CN^-(aq) + Br^-(aq) + K^+(aq)$$

$$\text{organic phase} \quad CN^- + Br^- + (C_4H_9)_4N^+$$

32.1 Nomenclature

Amines are named from the alkyl groups attached to the nitrogen followed by the ending *-amine*, as in the following examples.

CH_3NH_2	$(C_2H_5)_2NH$	$C_2H_5NHCH_3$	$(CH_3CH_2CH_2)_3N$
methylamine	diethylamine	ethylmethylamine	tripropylamine

Substituted ammonium cations are named as follows:

$(C_2H_5)_2NH_2^+$	$(CH_3)_4N^+$
diethylammonium	tetramethylammonium

32.2 Physical properties

Hydrogen bonding to nitrogen is weaker than that to the more electronegative oxygen. Primary and secondary amines therefore have lower boiling points than alcohols of similar molar mass, though higher than ethers and alkanes. In tertiary amines there are no hydrogen atoms attached directly to nitrogen and hydrogen bonding is absent: their boiling points are still lower. For example:

	n-C_3H_7OH	n-$C_3H_7NH_2$	$C_2H_5NHCH_3$	$(CH_3)_3N$	$(CH_3)_3CH$
t_b/°C	97	48	37	3	−12

Many amines have very powerful smells. The characteristic smell of less-than-fresh fish comes from lower amines such as ethylamine. Rotting animal flesh gives off the diamines $H_2N(CH_2)_4NH_2$ and $H_2N(CH_2)_5NH_2$, which have the trivial (and descriptive) names of putrescine and cadaverine.

32.3 Reactivity patterns and important reactions

Nitrogen (χ = 3.04 – see Table 2.1) is more electronegative than carbon (2.55) but much less electronegative than oxygen (3.44) or fluorine (3.98). This has two consequences for the chemistry of amines. First, there is little tendency for the carbon–nitrogen bond to cleave heterolytically: amines do not undergo nucleophilic substitution or elimination reactions. Secondly, the lone electron pair is less strongly held by the nucleus than are the lone pairs of oxygen, so that amines are stronger bases than alcohols are. The characteristic reaction of an amine is therefore with an electron pair acceptor, that is, a Lewis acid or an organic electrophile (compare the non-redox reactions of ammonia, described in Section 21.5):

$$R\ddot{N}H_2 \curvearrowright E^+ \rightarrow R\overset{+}{N}H_2E$$

(The symbol E^+ for an electrophile does not mean that all electrophiles are necessarily positively charged.)

The basicity of amines

Amines are weak bases, turning moist litmus blue. They are partially ionized in water:

$$RNH_2(aq) + H_2O(l) \rightleftharpoons RNH_3^+(aq) + OH^-(aq)$$

They react with Brønsted acids to give substituted ammonium salts:

$$RNH_2 + HCl \rightarrow RNH_3^+Cl^-$$

A test for an amine, as for ammonia, is the appearance of a white cloud of its salt when a drop of concentrated hydrochloric acid on a glass rod is held in its vapour.

Both the reactions shown above for primary amines also occur with secondary and tertiary amines. Primary amines are stronger bases than ammonia, and secondary amines are a little stronger still. Tertiary amines, however, are usually weaker than primary (see Table 32.1). This irregular

Table 32.1 Illustrative pK_b values of amines (see also Section 12.2)

	pK_b
NH_3	4.75
CH_3NH_2	3.35
$(CH_3)_2NH$	3.25
$(CH_3)_3N$	4.20

behaviour arises because there are two opposing trends at work. When the ionization equilibria are measured in the *gas phase* the base strength is observed to increase regularly from ammonia through to tertiary amines. Alkyl groups attached to the nitrogen atom stabilize the positive charge on the nitrogen in much the same way as they stabilize carbocations (see Section 28.3) and thereby favour the ionization. In water, however, solvation by hydrogen bonding also contributes to the stabilization of the cations. This is most effective for the NH_4^+ ion because it has four hydrogen atoms that can form hydrogen bonds, and least effective for the tertiary ammonium cation which has only one.

strongly solvated

weakly solvated

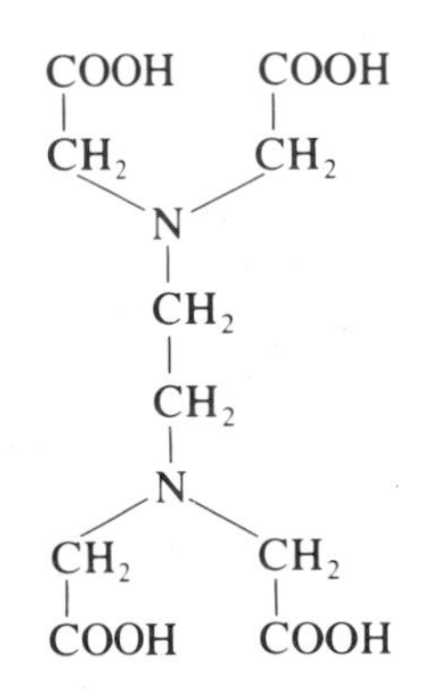

2

3

Amine complexes

Amines readily form complexes with Lewis acids, such as the ions of d-block elements (see Section 25.7) and other metals: like ammonia, amines form deep blue complex ions with copper(II) ions. 'Edta' (**2**) is named from '**e**thylene**d**iamine**t**etra**a**cetic acid'. It is a very powerful complexing agent, forming **chelates** – compounds in which an ion is enveloped by complexation with electron pairs on several different atoms of a Lewis base. In the calcium chelate (**3**), for example, the calcium ion is surrounded by four oxygen atoms and two nitrogen atoms. The chelate is very soluble in water because of the negative charges it bears and through hydrogen bonding to the oxygen atoms that stick out from the surface of the complex. It is used in the treatment of lead poisoning: the lead ions become similarly chelated and the soluble complex is excreted harmlessly. Edta is widely used in **complexometric titration** for the analysis of metal ions. The hardness of water can be determined by titrating with edta of known concentration in a slightly alkaline buffer solution, using the dye eriochrome black T as an indicator. The dye forms a red complex with calcium and magnesium ions: edta extracts the ions from the complex forming colourless chelates and leaving the free dye, which is blue at pH = 9.

Alkylation of ammonia and amines

The reaction of aqueous ammonia with a halogenoalkane was mentioned in Section 28.3. The nucleophilic nitrogen atom displaces the halide ion in a nucleophilic substitution reaction; for example:

$$NH_3 + C_2H_5Br \rightarrow \underset{\text{ethylammonium bromide}}{C_2H_5NH_3^+\,Br^-}$$

The ethylammonium cation is slightly dissociated in water (compare Section 12.2) so that the solution contains some free ethylamine:

$$C_2H_5NH_3^+ + H_2O \rightleftharpoons C_2H_5NH_2 + H_3O^+$$

This in turn can enter into a nucleophilic substitution reaction with some of the unreacted halogenoalkane:

$$C_2H_5NH_2 + C_2H_5Br \rightarrow (C_2H_5)_2NH_2^+\,Br^-$$

Similar reactions can then give the tertiary ammonium salt (in equilibrium with free tertiary amine) and finally the quaternary ammonium salt, which cannot react further:

$$(C_2H_5)_2NH + C_2H_5Br \rightarrow (C_2H_5)_3NH^+\,Br^-$$

$$(C_2H_5)_3N + C_2H_5Br \rightarrow (C_2H_5)_4N^+\,Br^-$$

The reactions may, of course, be started with an amine instead of with ammonia. They may also be carried out by heating the reagents together in a sealed tube, in the absence of water.

The reactions are not generally useful for making primary and secondary amines. With three moles of halogenoalkane to each mole of ammonia the main product is the tertiary amine, but smaller amounts of halogenoalkane give mixtures of products. When an excess of halogenoalkane is used, the main product is the quaternary salt (and the process is called **quaternization**).

Acylation of ammonia and amines

Ammonia and primary and secondary amines react with acid chlorides and anhydrides to give **amides** (see Section 31.4). Examples are:

$$CH_3COCl + (C_2H_5)_2NH \rightarrow CH_3CON(C_2H_5)_2 + HCl$$

$$(CH_3CO)_2O + C_2H_5NH_2 \rightarrow CH_3CONHC_2H_5 + CH_3COOH$$

Amides are much less nucleophilic than amines, because the nitrogen lone pair is delocalized with the carbonyl group; there is little tendency for further reaction. Acyl derivatives are useful both for **protecting** amine groups in complicated syntheses and – because their melting points are sufficiently sharp – for the identification of unknown amines.

Reaction of primary amines with nitrous acid

Primary amines react in the cold with nitrous acid (nitric(III) acid, formed in the reaction mixture from sodium nitrite and dilute sulphuric acid) to give molecular nitrogen, together with a mixture of organic products, usually with the alcohol predominating:

$$RNH_2 + HONO \rightarrow N_2 + ROH\,(+\text{other organic products}) + H_2O$$

The effervescence is a useful test for primary amines, and as the nitrogen is formed almost quantitatively it can be used to determine the amount of $—NH_2$ in an unknown sample. The reaction is not a good general route to alcohols, however: the yield is often very low.

32.4 Preparative methods for amines

Tertiary amines can be made from the reaction of halogenoalkanes with ammonia, as described above. Other preparations are discussed in detail in previous chapters and are only summarized here.

The reduction of nitriles or amides with lithium tetrahydridoaluminate in dry ethoxyethane, followed by water (see Section 31.4) leads to amines:

$$RCN \xrightarrow[\text{(ii) } H_2O]{\text{(i) } LiAlH_4} \underset{\text{primary amine}}{RCH_2NH_2}$$

$$RCONH_2 \xrightarrow[\text{(ii) } H_2O]{\text{(i) } LiAlH_4} \underset{\text{primary amine}}{RCH_2NH_2}$$

$$RCONHR' \xrightarrow[\text{(ii) } H_2O]{\text{(i) } LiAlH_4} \underset{\text{secondary amine}}{RCH_2NHR'}$$

$$RCONR'R'' \xrightarrow[\text{(ii) } H_2O]{\text{(i) } LiAlH_4} \underset{\text{tertiary amine}}{RCH_2NR'R''}$$

Amines may also be prepared from amides, via the Hofmann degradation (see Section 31.4):

$$RCONH_2 \xrightarrow[\text{NaOH(aq) heat}]{Br_2} \underset{\text{primary amine}}{RNH_2}$$

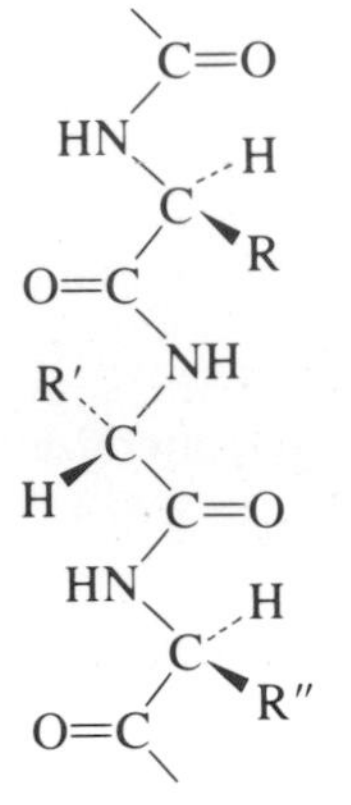

32.5 Proteins, peptides and amino-acids

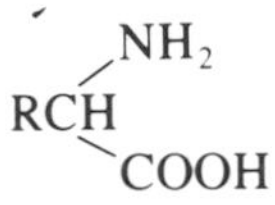

Protein, it could be said, is the stuff of life, of flesh and blood. Muscle and skin, haemoglobin, enzymes are all largely protein. So are some of the stuffs of death – bacterial toxins, which include some of the deadliest poisons known, and the surface membranes of viruses. Proteins are long-chain molecules of the general structure shown in **4**, with repeating —NH—CHR—CO— groups. When proteins are hydrolysed in acidic solution the amide groups are cleaved:

$$\text{—CO—NH—CHR—CO—NH—} + 2H_2O \rightarrow \text{—COOH} + NH_2\text{—CHR—COOH} + NH_2\text{—}$$

4

$RCH(NH_2)COOH$

5 α-amino-acid

The products of hydrolysis are **α-amino-acids** (**5**), the 'α' signifying that the amino group is attached to the α-carbon atom, the one next to the carboxyl group. The multitude of proteins occurring in nature are constructed from only about twenty different amino-acids, differing in the nature of the R group. R may be small and non-polar, H— or CH_3—, for example; it may contain a second —NH_2 or —COOH group; or it may contain a different functional group, such as the **thiol** group, —SH (see Table 32.2).

Table 32.2 Eight important α-amino-acids, R—CH(NH_2)COOH

R—	Name
H—	glycine
CH_3—	alanine
$C_6H_5CH_2$—	phenylalanine
$(CH_3)_2CH$—	valine
$HOCH_2$—	serine
$HOCOCH_2CH_2$—	glutamic acid
$NH_2(CH_2)_4$—	lysine
$HSCH_2$—	cysteine

$RCH(NH_3^+)CO_2^-$

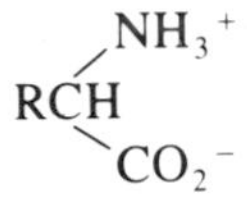

6 zwitterionic form of an amino-acid

Although amino-acids are normally represented and named as if they contained the basic —NH_2 and acidic —COOH groups, they exist largely as **inner salts** or **zwitterions** (**6**).

In acidic solution both functional groups are largely protonated (—NH_3^+ and —COOH) and the major species is positively charged (cationic). In alkali the —NH_2 group exists mainly as such, the carboxyl group as —CO_2^-, and the major species is negatively charged (anionic). At some intermediate pH value, the numbers of cationic and anionic groups will be exactly equal: this is called the **isoelectric point**. It is a characteristic of each amino-acid. Proteins, too, have characteristic isoelectric points at which they have no net electric charge.

If a solution of an amino-acid or protein is placed in an electric field, then the species will, on average, move towards the anode if there is a predominance of anionic groups and towards the cathode if there is a predominance of cationic groups. The rate of movement depends on the difference between the pH of the solution and the isoelectric point: the bigger the difference, the greater the proportion of molecules bearing a net charge. This is the basis of the analytical technique called **paper electrophoresis**, illustrated in Figure 32.2. A solution containing a mixture of amino-acids is prepared from the hydrolysis of a protein, and is spread in a line across a strip of thick filter paper soaked with a buffer solution of known pH. This is laid between two troughs filled with the buffer and containing electrodes. When a current is passed the amino-acids separate according to their isoelectric points. After a while the paper is removed and dried. The positions of the different amino-acids can be revealed by staining them with a dye, and they can be characterized by the distance they have moved.

Proteins, too, can be characterized by electrophoresis. It is used, for example, as a quick and simple test to screen people who may suffer from sickle cell trait, a congenital disorder common in black Africans.

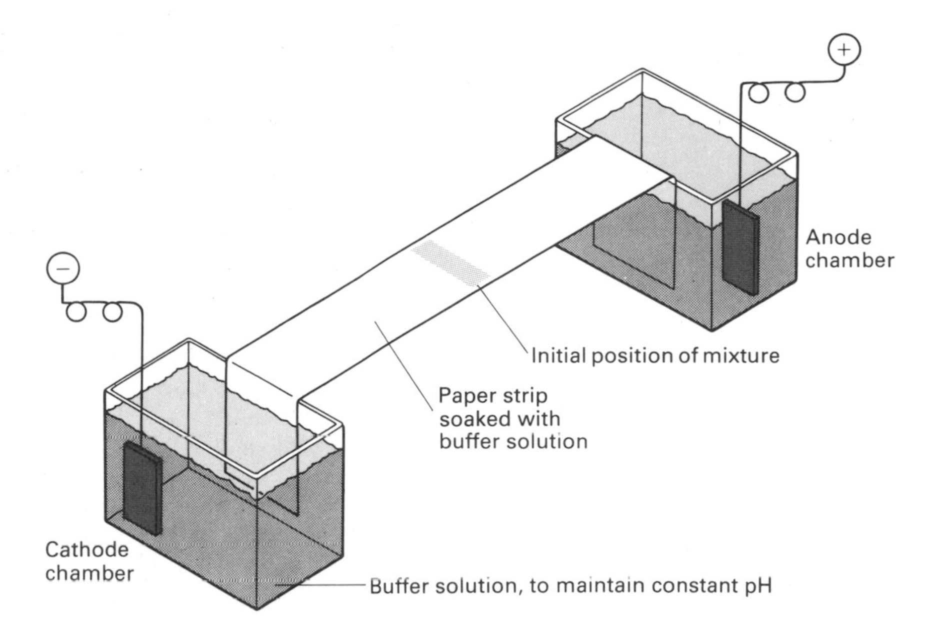

Figure 32.2 Paper electrophoresis

Electrophoresis can detect in the blood of carriers of the trait the presence of an abnormal haemoglobin, called HbS, which differs from ordinary haemoglobin, HbA, by just two amino-acid groups in the 574 that make up the haemoglobin molecule (two uncharged valines instead of anionic glutamates). That small difference is enough to change the structure of the cell (see Figure 32.3). It also changes the isoelectric point. The red protein needs no staining on the electrophoretogram, on which a pair of lines immediately shows that the patient's blood contains two types of haemoglobin.

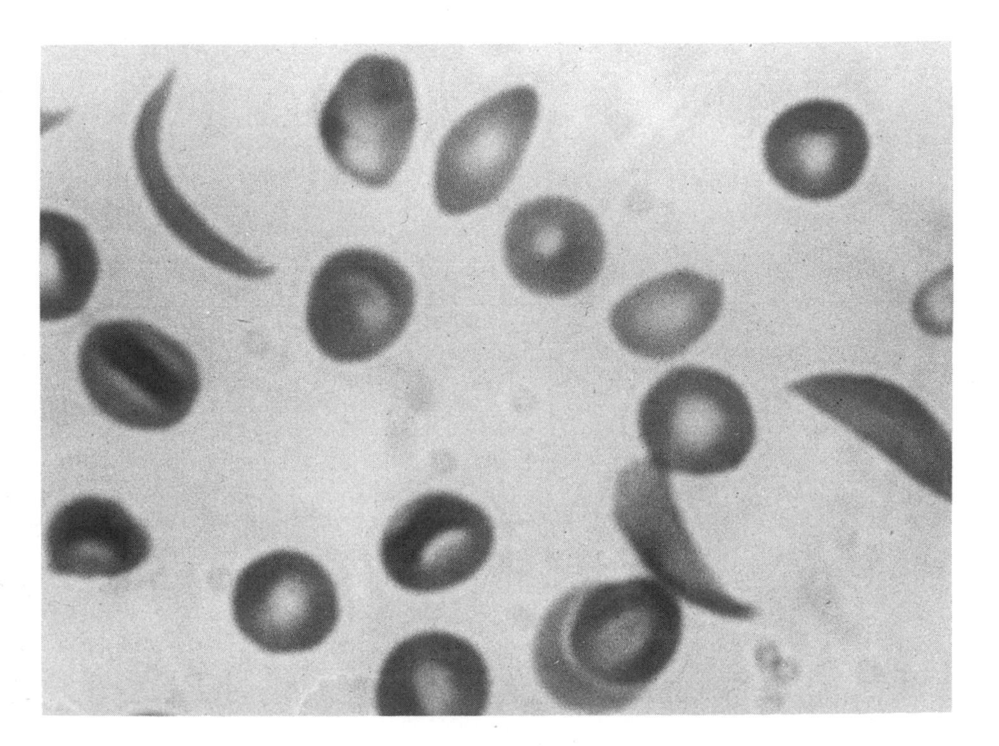

Figure 32.3 Photomicrograph of a sample of human blood, showing both normal and sickle red blood cells

EXAMPLE

The pK_a value for the carboxylic acid group of glycine (aminoethanoic acid), NH_2CH_2COOH, is 2.35. The pK_b value for the amine group is 4.22. In an aqueous solution of aminoethanoic acid with the pH adjusted to 7.00 (at 25 °C) by the addition of alkali, what is the proportion of molecules containing unionized COOH groups, and what is the proportion of molecules containing free (unprotonated) NH_2 groups?

METHOD Use equations analogous to 12.2.1 and 12.2.7, and the definition of pK_a in equation 12.2.2.

ANSWER From equation 12.2.2, $K_a = 10^{-2.35}$ mol dm^{-3} = 4.5×10^{-3} mol dm^{-3}. Now $[H_3O^+][RCO_2^-]/[RCOOH] = K_a$ (as in equation 12.2.1) and at pH = 7.00, $[H_3O^+] = 1.0 \times 10^{-7}$ mol dm^{-3}. Therefore the ratio $[RCO_2^-]/[RCOOH] = (4.5 \times 10^{-3})/(1.0 \times 10^{-7}) = 4.5 \times 10^4$. In other words, only one carboxyl group in 45 000 remains unionized.

Similarly, $[R'NH_3^+][OH^-]/[R'NH_2] = K_b = 10^{-4.22}$ mol dm^{-3} = 6.0×10^{-5} mol dm^{-3}. Setting $[OH^-] = 1.0 \times 10^{-7}$ mol dm^{-3} gives the ratio $[R'NH_3^+]/[R'NH_2] = 600$: one amino group in 600 remains unprotonated.

COMMENT This calculation shows that at pH = 7.00 nearly all the molecules are in the zwitterionic form $NH_3^+CH_2CO_2^-$, but that there are a few more molecules bearing CO_2^- groups than there are with NH_3^+ groups. If the solution is made slightly more acidic, the isoelectric point is reached when the concentrations of cationic and anionic groups are equal. For glycine, this is at pH = 6.06.

BOX 32.2

7 (structure: benzene ring fused with CO, NH, SO_2)

8 (structure: cyclohexyl–$NH{-}SO_3^-\ Na^+$)

9 (structure: $HOCOCH_2$, NH_2, H, C=O, HN, H, CH_3OCO, $CH_2C_6H_5$)

Artificial sweeteners

As Western society becomes increasingly weight-conscious many people seek to satisfy their desire for sweet foods without consuming too much sugar. Most artificial sweeteners are derived from amines. The earliest, **saccharin (7)**, was discovered accidentally in 1879 when a chemist who prepared it did not wash his hands before a meal! It is some 500 times sweeter than sucrose but has a bitter after-taste. **Sodium cyclamate (8)**, discovered in 1937, is about 30 times sweeter than sucrose and does not have an after-taste. More recently **aspartame (9)**, the methyl ester of a dipeptide (two amino-acids joined by a peptide link), has been found to be about 200 times sweeter than sucrose; it too has no after-taste, but unfortunately it cannot be cooked because on hydrolysis it loses its sweetness.

There has always been concern about the possibility of adverse side effects from food additives. Cyclamates have been banned in many countries, including the UK, following reports that rats fed on large doses developed bladder tumours, perhaps because the cyclamate is converted in the body to cyclohexylamine, $C_6H_{11}NH_2$, which is a known carcinogen. Saccharin is also suspect: rats fed on very high saccharin diets (equivalent to a human eating over 10 000 tablets a day!) developed bladder tumours. The relevance of such experiments to human welfare is dubious: the dangers of obesity appear far greater than the miniscule risks associated with cyclamates or saccharin, but many people still believe that sucrose, because it is 'natural', must be inherently better than anything 'artificial'.

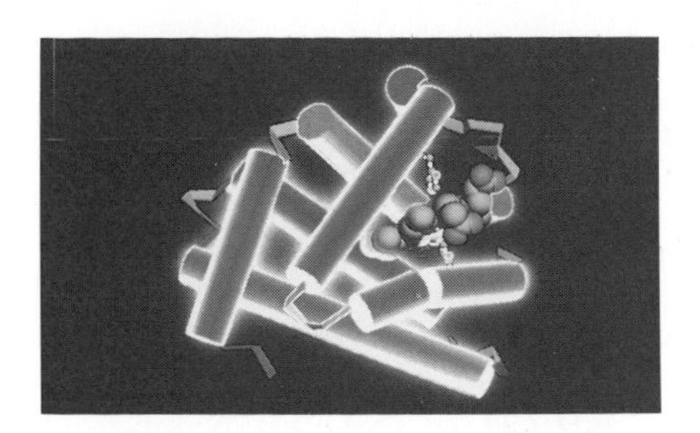

Figure 32.4 A computer-generated model of the tertiary structure of the protein myoglobin: the cylindrical sections represent regularly coiled parts of the chain

The amide group in proteins, —CONH—, is usually called the **peptide link**. Molecules with only a few amino-acid residues, such as the artificial sweetener aspartame (see Box 32.2), are called **polypeptides** (or just **peptides**). Proteins may have only about fifty amino-acid residues, as for insulin, or they may have thousands.

The characteristic properties of a protein are a consequence of its detailed three-dimensional structure. The **primary structure** of a protein is the sequence of amino-acids in the chain. The **secondary structure** describes the local shape of the chain – a zig-zag or a coil, for example (see Box 32.3). The chain is itself twisted into irregular convolutions, called the **tertiary structure** (see Figure 32.4), the shape of which is governed mainly by hydrogen bonding between groups in the protein chain and by occasional 'cross-links' where two thiol groups are oxidized to form a disulphide bridge:

$$>CH{-}SH + HS{-}CH< \xrightarrow{\text{oxidation}} >CH{-}S{-}S{-}CH<$$

BOX 32.3

Silk, wool and nylon

Silk and **wool** are protein fibres. In silk the molecules are held together in zig-zag chains by hydrogen bonding between the NH groups of one chain and the CO groups of the next. In wool the protein chain is coiled, like a telephone cord, with hydrogen bonding between groups in adjacent turns of the coil (see Figure 32.5). Wool fibres, unlike those of silk, can be stretched by pulling out the coil and lengthening the hydrogen bonds. If the structure is too severely disrupted, however – for example, by washing the wool at too high a temperature – the coils cannot revert to their original position and the fibres lose their springy texture.

(a)

```
     \                    /
      N   H  ........ O=C
     /                    \
  CHR                      CHR
     \                    /
      C=O.......H—N
     /                    \
---H—N                     C=O---
     \                    /
      CHR              CHR
     /                    \
---O=C                     N—H---
     \                    /
      N—H......O=C
     /                    \
  CHR                      CHR
     \                    /
```

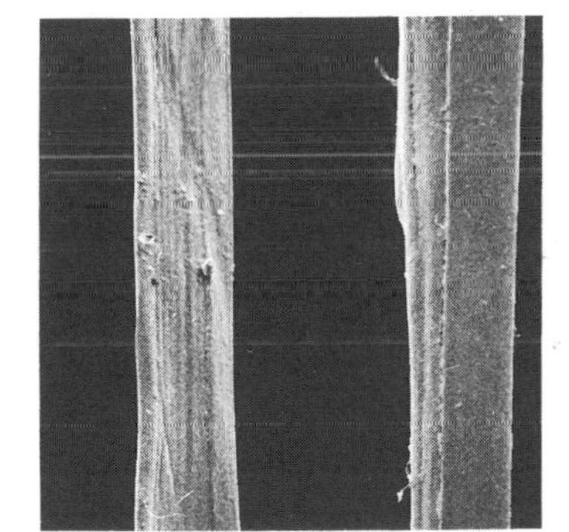

(b)

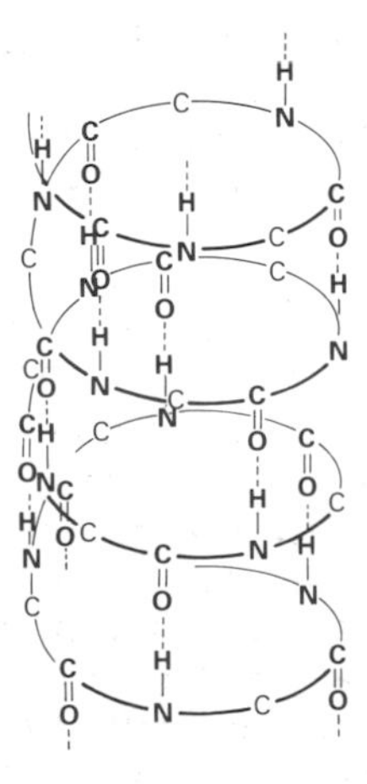

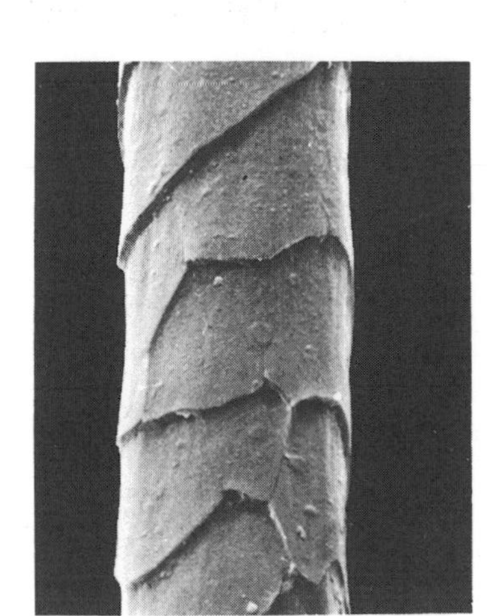

Figure 32.5 Representations of the secondary structures of (a) silk and (b) wool, and photomicrographs of their fibres: the dashed lines represent hydrogen bonds between CO and NH, and the faint C symbols represent the α-carbon atoms (the H and R groups attached to these are omitted, for clarity)

$$-NH-(CH_2)_5-CO-NH-(CH_2)_5-CO-$$

10 nylon-6

$$-NH-(CH_2)_6-NH-CO-(CH_2)_4-CO-NH-$$

11 nylon-6,6

Nylon was developed in the 1930s, originally as an alternative to silk for making parachute canopies (see Figure 32.6). Like protein fibres nylon is a **polyamide**, but unlike protein there is not just a single carbon atom between the —CONH— groups. In fact there is a range of nylons, with different carbon chains between the amide links. All share the same properties of strength, because in all of them the chains are held firmly together by hydrogen bonding, much as they are in silk. The two most common forms of nylon are called nylon-6 (**10**) and nylon-6,6 (**11**). Nylon-6, like proteins, is derived from an amino-acid: the 6 in the name refers to the number of carbon atoms in the repeating $—NH(CH_2)_5CO—$ unit. Nylon-6,6, on the other hand, is made from two different molecules, a diamine $H_2N(CH_2)_6NH_2$ and a diacid $HOCO(CH_2)_4COOH$: the two 6s in the name refer to the number of carbon atoms in the two fragments. Nylon-6,6 is an example of a **condensation polymer**. The long chains are built up from the original molecules by a condensation reaction, that is, one in which water (or more generally any small molecule) is expelled as the organic parts combine (compare Section 30.3):

$$—COOH + H_2N— \xrightarrow{-H_2O} —CO—NH—$$

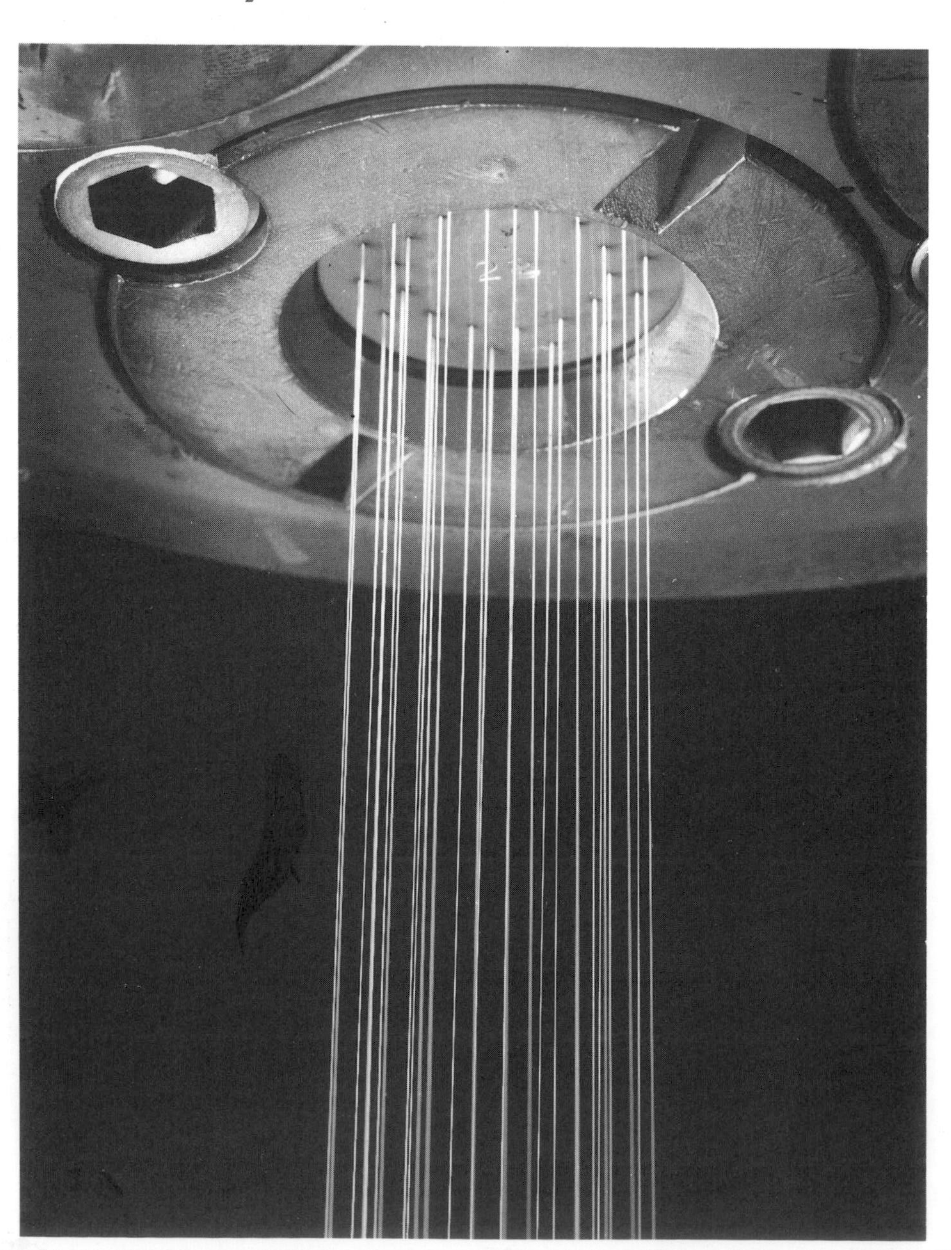

Figure 32.6 Spinning nylon fibres

Changes in the secondary or tertiary structure can totally alter the properties of the protein: it is said to be **denatured**. For example, the changes that happen when an egg is boiled arise mainly because the protein chains flail around with their increased kinetic energy; hydrogen bonds are broken and reformed in different places, and the original structure is lost.

Enzymes are protein molecules that have remarkable catalytic powers: reactions that in the test-tube need high temperature or extremes of acidity or alkalinity can take place rapidly with enzyme catalysis at moderate pH and at body temperature. The key to their activity lies in their precise shape, each specially shaped enzyme affecting just one reaction. The reacting molecules are bonded to functional groups in the amino-acid side chains in such a way that they are precisely aligned with each other for reaction to take place (see Section 14.5).

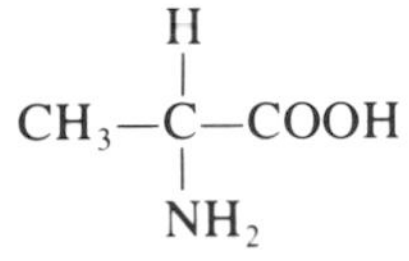

$$CH_3-\overset{\displaystyle H}{\overset{|}{C}}-COOH \quad (\underset{\displaystyle NH_2}{\text{ on C}})$$

12

32.6 Chiral molecules. Optical isomerism

Alanine (2-aminopropanoic acid, **12**) is one of the twenty or so amino-acids that go to make up protein. Models show that there are two different ways that the four groups H, CH_3, NH_2 and COOH can be attached to the central carbon atom. They are represented in the perspective diagrams of Figure 32.7. These structures are not the same. A model of one cannot be superimposed on a model of the other. In fact, they are related as the right hand is to the left. If we imagine a mirror placed along the dotted line between the structures in Figure 32.7 we can see that the bottom diagram is a reflection of the top one, and vice versa.

The property of 'left- and right-handedness' – characteristic of an entity that is not identical with its mirror image – is called **chirality**, and the entities are said to be **chiral** (Greek: *cheir*, hand). Any compound with four *different* atoms or groups attached to a single carbon atom is chiral. The carbon atom bearing the four different groups is called a **chiral centre**.

The two mirror image isomers are said to be **enantiomers** of each other (Greek: *enantios*, face-to-face). A pair of enantiomers are identical in all respects except for their orientation in space: for example, in alanine the —NH_2 and —COOH groups in the two enantiomers are the same distance apart and influence each other in exactly the same way. The enantiomers have the same physical and chemical properties as each other, with two exceptions.

First, if a beam of plane-polarized light is passed through two solutions containing equal concentrations of the different enantiomers, then the plane of polarization is rotated to the right (clockwise) by one enantiomer and by an equal amount to the left by the other (see Figure 32.8). This behaviour, which arises from the lack of symmetry within the molecules, gives rise to the term **optical isomerism** as an alternative to chirality. The enantiomer that rotates to the right (clockwise) is denoted by a (+) sign, as in (+)-alanine; the other enantiomer is distinguished by a (−) sign.

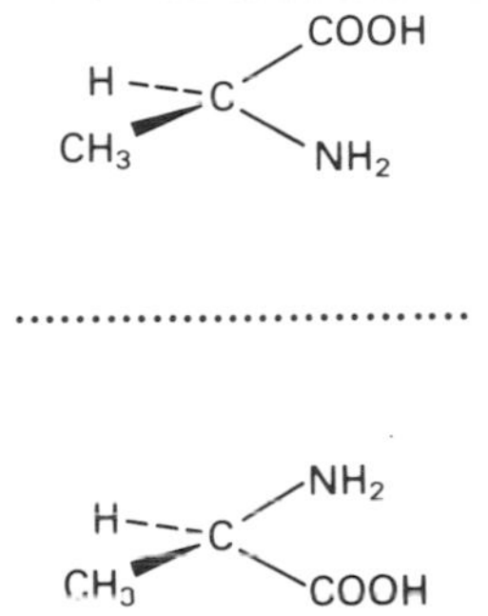

Figure 32.7 Enantiomers of alanine

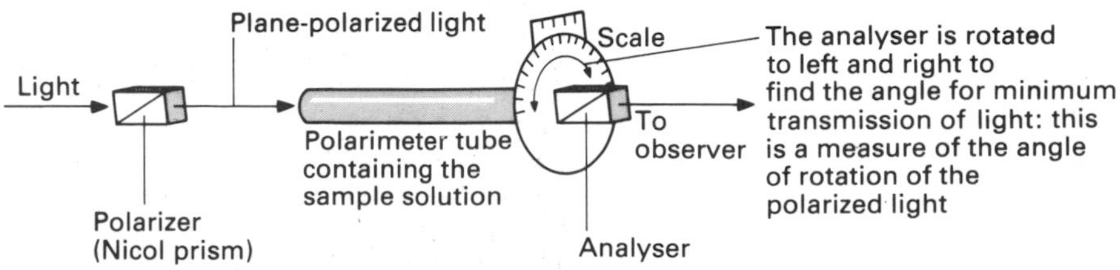

Figure 32.8 A polarimeter, for measuring the rotation of plane-polarized light by an optically active solution

The second difference is that the chemical reactions of the two enantiomers *with another chiral molecule* are different: it is like the difference in the interactions between left hand and left glove and those between right hand and left glove. This is very important in naturally occurring compounds. Nearly all of these are chiral and in nature normally only one of the enantiomers is formed. The other may have very different properties. Naturally occurring alanine is much less sweet-tasting than its unnatural enantiomer: the difference arises because the taste-receptor molecules on the tongue are also chiral and interact differently with the two enantiomers. It was discovered too late that the drug thalidomide, used to counteract nausea in pregnant women, is **teratogenic**, that is, it damages the foetus. Many badly deformed babies were born in the early 1960s after their mothers had taken the drug. In fact it is only the (−) enantiomer that has this dreadful side effect; the (+) enantiomer is just a mild sedative.

EXAMPLE

Which of the structurally isomeric alcohols C_4H_9OH is chiral and exists as a pair of enantiomers?

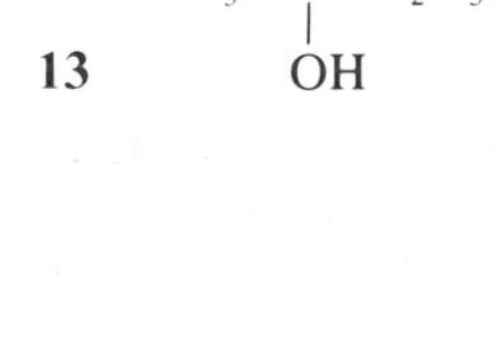

METHOD Consider the structural formulae of the different isomers (compare the introductory paragraphs of Chapter 29). Decide which of these possesses a chiral centre.

ANSWER Butan-2-ol (**13**) is chiral, because carbon atom 2 has four different groups attached to it: CH_3, H, OH, C_2H_5.

COMMENT The but-2-yl group is the simplest alkyl group whose derivatives are chiral.

When the organic chemist makes a chiral compound from non-chiral starting materials, the product is always a 1:1 mixture of the two enantiomers: this is known as a **racemic mixture**. For example, Figure 32.9 shows the addition of hydrogen cyanide to ethanal to give the chiral compound CH_3—CH(OH)—CN; this can be hydrolysed to 2-hydroxypropanoic acid (lactic acid), CH_3—CH(OH)—COOH, which is found in milk. Ethanal is a flat molecule; the cyanide ion can attack the carbonyl group from above or below with equal ease to give equal amounts of the two enantiomers. A racemic mixture has the same chemical properties as the single enantiomers, but physical properties that depend on the packing of molecules in the solid are different – solubilities and melting points, for example. To continue with an analogy we used before, the way a collection of pairs of gloves could be stacked together is different from the way a collection of left gloves could be stacked. The melting points of both (+) and (−) enantiomers of lactic acid are, of course, the same: 53 °C. That of the racemic mixture, (±)-lactic acid, is 18 °C, however. A racemic mixture does not rotate plane-polarized light: for every (+) molecule that rotates it to the right, there is a (−) molecule that rotates it an equal amount to the left.

It is often necessary to separate a racemic mixture into the two separate enantiomers, a process called **resolution** – in order to compare a synthetic compound (racemic) with a naturally occurring single enantiomer. This is done by using the different reactions of enantiomers with other chiral molecules. We could separate pairs of gloves by sticking a left hand into each glove: the fit would be different. In the same way, if we treat racemic lactic acid with the naturally occurring chiral base (−)-strychnine (from the poisonous seeds of the tree *Nux vomica*) we obtain a mixture of two different salts, (−)-acid/(−)-base and (+)-acid/(−)-base. These are not mirror images of each other and therefore have different physical properties. For example, the difference in their solubilities can be used to crystallize the less soluble one preferentially out of solution. Then hydrolysis of the separated salts with dilute sulphuric acid will release the free lactic acids as separate enantiomers.

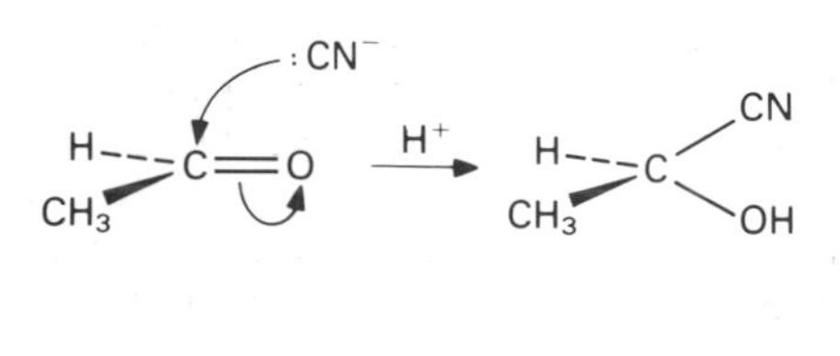

Figure 32.9 Formation of a racemic mixture by the attack of cyanide ion on either side of non-chiral ethanal molecules

Summary

- Primary amines have the formula RNH_2. Secondary amines are R_2NH. Tertiary amines are R_3N. Quaternary ammonium cations are R_4N^+. (The R groups attached to a nitrogen atom do not have to be the same.)
- The general formula of amines in $C_nH_{2n+3}N$. Their relative molecular masses are odd, giving mass spectra with molecular ions having characteristically odd values of m/z.
- Amines are named from the alkyl groups attached to the nitrogen atom, followed by *-amine*.
- Weak hydrogen bonding in primary and secondary amines leads to boiling points that are higher than those of alkanes or ethers of comparable molar mass but lower than those of alcohols. In tertiary amines hydrogen bonding is absent: their boiling points are lower than those of primary and secondary amines.
- Amines are basic and nucleophilic, reacting mainly via the nitrogen lone electron pair. They are partially ionized in water.
- The order of basicity of amines is $NH_3 < RNH_2 < R_2NH > R_3N$. This reflects a balance in stabilization of the ammonium cation between electron donation from the attached alkyl groups (favouring R_3NH^+) and hydrogen bonding from solvent water molecules (favouring $NH_4{}^+$).
- Amines readily form complexes with Lewis acids, such as d-block ions. The compound edta is a strong chelating agent.
- Ammonia and amines react with halogenoalkanes to give a mixture of primary, secondary and tertiary amines and quaternary ammonium salts. With a 3:1 ratio of halogenoalkane to ammonia, the tertiary amine is the major product. With an excess of halogenoalkane it is the quaternary salt.
- The conversion of an amine to its quaternary salt is called quaternization.
- Ammonia and primary or secondary amines react with acid chlorides and anhydrides to give amides.
- Primary amines react with nitrous acid to give molecular nitrogen and low yields of alcohols.
- Amines may be prepared by the reduction of nitriles or amides. Primary amines may be made by the Hofmann degradation of amides. The alkylation of ammonia may be satisfactory for preparing tertiary amines and quaternary ammonium salts.
- Protein molecules contain the repeating unit —NH—CHR—CO—.
- Hydrolysis of the peptide link in proteins gives α-amino-acids, $RCH(NH_2)COOH$. Only about twenty amino-acids, differing in the nature of R, occur in most proteins.
- The α-amino-acids exist largely as zwitterions, $RCH(NH_3{}^+)CO_2{}^-$. The isoelectric point is the pH at which an amino-acid or protein has no net charge. Electrophoresis characterizes amino-acids or proteins by allowing them to migrate under an electric field.
- The properties of enzymes depend not only on the sequence of amino-acids (primary structure) but also on the local shape of the chain (secondary structure) and the way the chain is folded (tertiary structure). Enzymes function as catalysts because reacting molecules can be held together in exactly the right orientation.
- Molecules which can exist as two non-identical mirror image enantiomers are said to be chiral and to show the property of chirality. A carbon atom with four different groups attached to it is a chiral centre.
- Enantiomers have identical physical properties except that one enantiomer rotates plane-polarized light to the right (+), and the other rotates it to the left (−). They have identical chemical reactions, except those with other chiral molecules.
- Many naturally occurring compounds are chiral: only one enantiomer is normally found in nature.
- Synthesis of a chiral compound from non-chiral reactants gives a racemic mixture of the two enantiomers. This does not rotate plane-polarized light.
- A racemic mixture can be resolved by reaction with a chiral molecule: the products from the two enantiomers are not mirror images and can be separated through differences in their physical properties.

PROBLEMS

1 Draw structures for all the isomers with molecular formula $C_4H_{11}N$. Classify each isomer as (a) a primary, secondary or tertiary amine, and (b) containing primary, secondary and/or tertiary alkyl groups.

2 Explain why ethanamide is a weaker base than ethylamine and a weaker acid than ethanoic acid.

3 A compound **A** contains 31.2% C, 9.1% H and 18.2% N, by mass. When heated strongly it loses water to give **B**, which contains 40.7% C, 8.5% H and 23.7% N. When **B** is distilled with phosphorus(V) oxide it gives **C** which contains 58.5% C, 7.3% H and 34.1% N, and which gives a peak in the mass spectrum at $m/z = 41$ for the molecular ion. Calculate the empirical and molecular formulae of **A**, **B** and **C**, and suggest structural formulae for each.

4 0.505 g of an amide reacts with an excess of nitrous acid with the evolution of 120 cm^3 of nitrogen gas (at 20 °C and 1 atm). The only organic product is optically active. What are the molecular and structural formulae of the amide?

5 Show how the following compounds may be prepared from 1-bromopropane, using only inorganic reagents:
(a) $(CH_3CH_2CH_2)_4N^+ Br^-$
(b) $CH_3CH_2CH_2CH_2NH_2$
(c) $CH_3CH_2CH_2CONH_2$
(d) $CH_3CH_2CH_2NH_2$ (remembering that the direct reaction of ammonia with a halogenoalkane is not usually a satisfactory way of making a primary amine).

6 In the context of amino-acid chemistry, explain what is meant by the terms 'zwitterion' and 'isoelectric point'. Outline how the technique of paper electrophoresis could be used to determine the isoelectric point of an amino-acid.

7 Discuss the isomerism of the following compounds:
(a) C_7H_{16}
(b) $C_3H_6Br_2$
(c) $CH_3CH(OH)COOH$
(d) C_4H_9OH
(e) $CH_3CH(OH)COOC_4H_9$

8 (a) What is the effect of each of the following reagents upon primary amines?
(i) hydrogen chloride
(ii) ethanoyl (acetyl) chloride
(iii) iodomethane (methyl iodide)
(iv) nitrous acid
(b) On the basis of their reactions with reagents (ii) and (iii), to what mechanistic class of reagent do amines belong?
(c) To what classes of compound do the products from (i), (ii) and (iii) belong?
(d) Indicate in each case in (a) how, if at all, the original amine may be regenerated from the product in a single reaction. *(Oxford and Cambridge)*

9 Write an account of some aspects of the chemistry of proteins. Your answer should consider at least *two* of the following:
(a) the chemical composition of proteins,
(b) the structure of protein molecules,
(c) the relationship between the structure and the physical and chemical properties of proteins,
but you need not restrict your answer to these aspects. *(Nuffield)*

10 Amino-acids contain two different functional groups in the same molecule. Describe some typical physical and chemical properties of this class of compound and, where possible, interpret these properties in terms of the structure possessed by amino-acids. *(Nuffield)*

11 When the neutral compound **A**, $C_{10}H_{13}NO$, was refluxed with dilute acid it formed two products **B**, C_2H_7N, and **C**.
On analysis, **C** was found to contain 70.59% carbon, 23.53% oxygen and 5.88% hydrogen by weight. The relative molecular mass of **C** was found to be 136.
On reaction with alkaline potassium manganate(VII) (permanganate) solution, **C** was oxidized to **D**, $C_8H_6O_4$.
D, which was acidic, was readily dehydrated to the neutral substance **E**, $C_8H_4O_3$.
B reacted with gaseous hydrogen chloride to form the ionic solid **F**, C_2H_8NCl. When **F** was dissolved in dilute hydrochloric acid and sodium nitrite solution added, a yellow oil, **G**, was formed and no effervescence occurred.
(a) What is the empirical formula of **C**?
(b) What is the molecular formula of **C**?
(c) Write the structural formulae for substances **A** to **G**.
(d) What are the names of substances **B**, **C** and **F**? *(SUJB)*

12 (a) Give the reagents and essential conditions needed to convert ethanoic acid into ethanamide and give equations for the reactions concerned.
(b) Explain why ethanamide and aminoethanoic acid are both weaker bases than aminoethane.
(c) Using the isomeric amino-acids $C_3H_7NO_2$ as your examples, explain the meaning of the terms *structural isomerism* and *optical isomerism*. *(Oxford)*

13 Answer *five* parts of this question. In each case draw the structure(s) of the initial compound(s) which satisfy the requirements and justify your answer.
(a) A compound C_4H_9NO which upon treatment with either boiling aqueous acid or NaOH gives CH_3CH_2COOH (or its sodium salt) as one of the products.
(b) Three isomers C_4H_8 which give the same product on treatment with hydrogen at atmospheric pressure in the presence of a palladium catalyst.
(c) Two isomeric alcohols $C_4H_{10}O$ which differ from each other only in the direction in which they rotate plane-polarized light.
(d) A compound $C_4H_4O_3$ which reacts with $NaOCH_3$ in CH_3OH to give the product $C_5H_7O_4Na$.
(e) A compound $C_7H_{12}O$ which forms a crystalline derivative upon treatment with 2,4-dinitrophenyl-hydrazine reagent but does not absorb hydrogen at atmospheric pressure in the presence of a palladium catalyst and cannot be resolved into optical isomers.
(f) A compound C_3H_9N which reacts with an excess of CH_3I to give a salt and with HNO_2 in water to give $CH_3CH(OH)CH_3$.
(g) A compound C_4H_9Br which reacts rapidly in boiling water to give $C_4H_{10}O$ and HBr. *(Cambridge Entrance)*

33

Alkenes: the C═C double bond

Alkenes contain a carbon–carbon double bond. This is more reactive than carbon–carbon or carbon–hydrogen single bonds and consequently is the site of most of the reactions of the alkene molecule. Many of the reactions begin with electrophilic attack on the electron-rich double bond. A few of the lower alkenes, derived from cracking petroleum fractions, are the raw materials for most of the organic chemical industry.

1 R—CH═CH—R′

2 R—CH_2—CH_2—R′

An **alkene** with one double bond in its molecule (**1**) contains two hydrogen atoms fewer than the alkane with the same carbon skeleton (**2**): the general formula is therefore C_nH_{2n} and the empirical formula of all such alkenes is CH_2.

The distribution of the four electrons of the carbon–carbon double bond may be described in terms of σ- and π-bonds (see Figure 33.1 and Section 2.5). The double bond is weaker than two carbon–carbon single bonds (see Table 9.4), so that it takes relatively little energy to half-break a double bond and leave the carbon atoms bonded by a single bond. The carbon–carbon double bond is therefore a functional group in that it is the site of most alkene reactions; the rest of the carbon skeleton remains, as usual, an unreactive support.

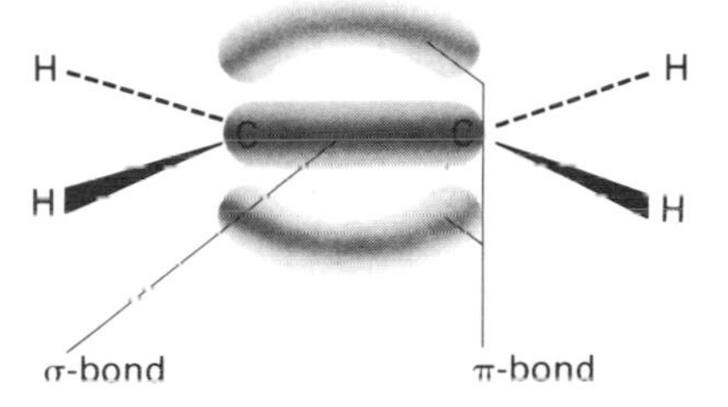

Figure 33.1 Electron distribution in the carbon–carbon double bond

EXAMPLE

Estimate the values of $\Delta_r H^\ominus$ for the following reactions:

(a) CH_3—$CH_3(g) + H_2(g) \rightarrow 2CH_4(g)$

(b) CH_2═$CH_2(g) + H_2(g) \rightarrow CH_3\quad CH_3(g)$

METHOD The standard atomization enthalpy of hydrogen is 218 kJ mol^{-1}. Use the mean bond enthalpies of Table 9.4 for carbon–carbon single and double bonds and for carbon–hydrogen bonds.

ANSWER Figure 33.2 illustrates the cycle of enthalpy changes. In reaction (a) the bonds broken are C—C and H—H, and two C—H bonds are made: $\Delta_r H^\ominus = 348 + (2 \times 218) - (2 \times 413)\,\text{kJ mol}^{-1} = -42\,\text{kJ mol}^{-1}$. In reaction (b) the bonds broken are C═C and H—H, and two C—H bonds and a C—C bond are made. (Of course, the carbon–carbon bond is not really completely broken, but the enthalpy change is calculated *as if* C═C were broken and C—C reformed.)

$$\Delta_r H^\ominus = 612 + (2 \times 218) - (2 \times 413) - 348\ \text{kJ mol}^{-1} = -126\ \text{kJ mol}^{-1}$$

COMMENT The reduction of ethane into two molecules of methane occurs only under fairly drastic conditions: the so-called **hydrocracking** of alkanes is carried out catalytically at around 400 °C and pressures of around 150 atmospheres. Alkenes, on the other hand, may be reduced at room temperature and pressure (see Section 33.3). Alkenes are said to be **unsaturated** because they can readily take up more hydrogen: alkanes are **saturated** because they cannot.

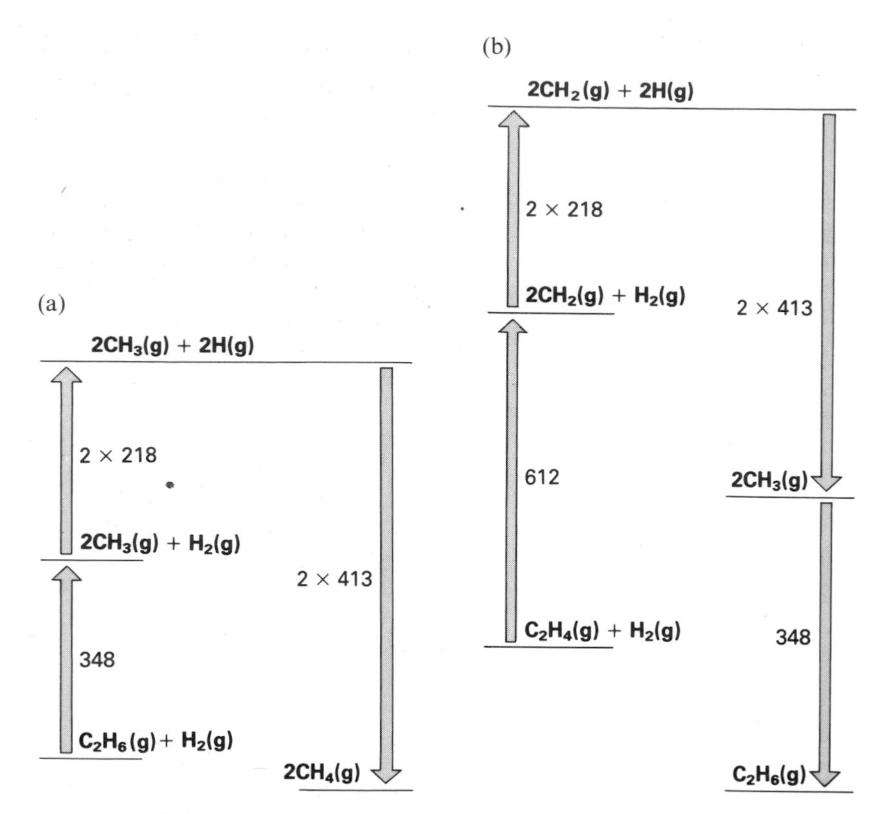

Figure 33.2 Estimating the enthalpy changes for (a) hydrocracking ethane and (b) addition of hydrogen to ethene

The physical properties of alkenes, such as melting and boiling points, are not greatly different from those of alkanes with the same carbon skeleton (see Section 27.2):

	C_2H_4	C_2H_6	C_3H_6	C_3H_8
$t_m/°C$	−169	−183	−185	−190
$t_b/°C$	−104	−89	−47	−42

33.1 Nomenclature

Alkenes are named using the same principles as are alkanes, with the following modifications:

1 The names end in *-ene* instead of *-ane*.

$CH_2{=}CH_2$ ethene $\quad$ $CH_3{-}CH{=}CH_2$ propene

(The older names ethylene and propylene are still used.)

2 The 'spine' is the longest chain that contains the double bond. For chains more than three carbon atoms long the position of the carbon atom at which the double bond starts is given by a number before '-ene'. Where there is a choice, the direction of numbering is chosen so that the double bond is nearer the end of the chain numbered 1.

$CH_3{-}CH_2{-}CH{=}\overset{1}{C}H_2$ but-1-ene $\quad$ $CH_3{-}CH{=}\overset{2}{C}H{-}\overset{1}{C}H_3$ but-2-ene

$CH_3{-}\overset{3}{C}H(CH_3){-}\overset{2}{C}H{=}\overset{1}{C}H_2$ 3-methylbut-1-ene (not 2-methylbut-3-ene)

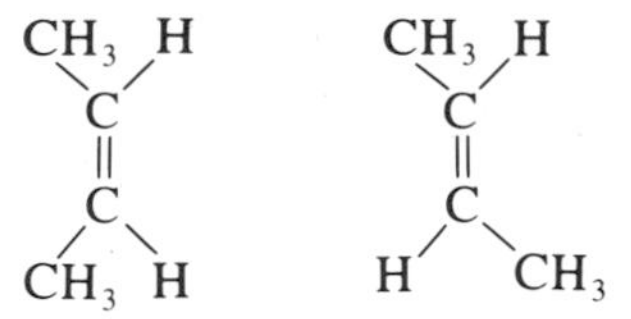

$$\begin{array}{cc} CH_3\ \ H & CH_3\ \ H \\ \diagdown\ \diagup & \diagdown\ \diagup \\ C & C \\ \| & \| \\ C & C \\ \diagup\ \diagdown & \diagup\ \diagdown \\ CH_3\ \ H & H\ \ CH_3 \end{array}$$

3 *cis*-but-2-ene *trans*-but-2-ene

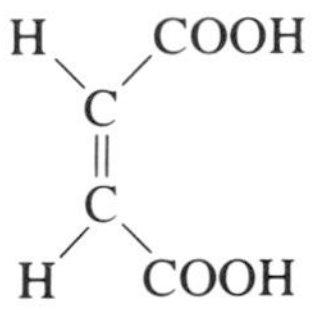

$$\begin{array}{c} H\ \ \ COOH \\ \diagdown\ \diagup \\ C \\ \| \\ C \\ \diagup\ \diagdown \\ H\ \ \ COOH \end{array}$$

4 *cis*-butenedioic acid

33.2 Geometric isomerism

Because the carbon–carbon double bond resists rotation, as described in Section 2.4 (in contrast to the easy rotation about carbon–carbon single bonds – see Section 26.3), the two structures shown in **3** do not represent the same molecule. Unlike structural isomers, discussed in Section 26.4 (of which but-1-ene and but-2-ene are examples), they contain the *same groups* bonded together in the *same order*: the only difference is in the geometry of the bonds. This type of isomerism is therefore called **geometric isomerism**. The terms *cis* and *trans* (Latin: 'on this side of' and 'across') are used as shown in **3** to name simple compounds in which the groups at each end of the double bond are the same (or similar): hence the alternative name of ***cis*/*trans*-isomerism**.

Geometric isomers may differ significantly in their chemical reactions. For example, *cis*-butenedioic acid (**4**), in which the two carboxylic acid groups are close together, is readily dehydrated by heating it to near its melting point, 140 °C. The *trans* acid is stable to a much higher temperature. Geometric isomerism and optical isomerism (discussed in Section 32.6) are together called **stereoisomerism**, because they are concerned with the shapes of the isomers (Greek: *stereos*, solid).

33.3 Reactivity patterns and important reactions

Electrophilic addition

Because of the high electron density in the carbon–carbon double bond, alkenes are **nucleophilic**. Most nucleophiles – amines, for example – use a lone pair of electrons to form a new bond: alkenes use one of the electron pairs from the double bond. For example, ethene reacts with concentrated hydrobromic acid. A new bond is formed from one carbon atom to the electrophilic proton, leaving a single carbon–carbon bond and forming a carbocation (see Section 28.3):

$$CH_2{=}CH_2 + H_3O^+ \rightarrow CH_3\text{—}CH_2{}^+ + H_2O$$

The reaction is completed by attack of the bromide anion on the carbocation:

$$CH_3\text{—}CH_2{}^+ + Br^- \rightarrow \underset{\text{bromoethane}}{CH_3\text{—}CH_2\text{—}Br}$$

The complete reaction is thus **addition** to the carbon–carbon double bond, and because it is triggered by electrophilic attack on the alkene it is called **electrophilic addition**.

A similar reaction occurs with non-ionized hydrogen bromide in the gas phase or in an organic solvent. The alkene attacks the positive end of the $H^{\delta+}\text{—}Br^{\delta-}$ dipole, and the H—Br bond is broken during the first step of the reaction:

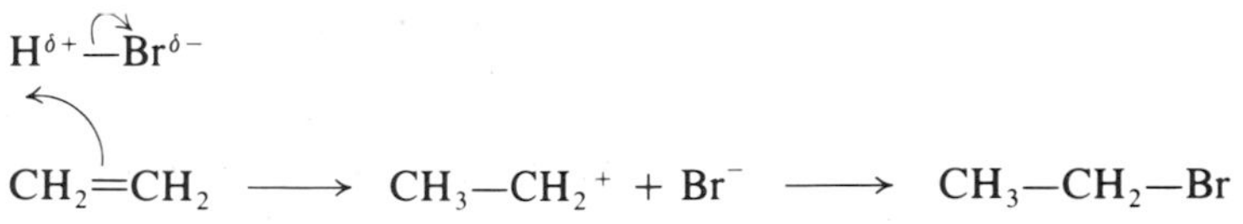

$$H^{\delta+}\text{—}Br^{\delta-}$$

$$CH_2{=}CH_2 \longrightarrow CH_3\text{—}CH_2{}^+ + Br^- \longrightarrow CH_3\text{—}CH_2\text{—}Br$$

Alkenes also react with bromine in the gas phase or in organic solvents such as tetrachloromethane. The bromine molecule, of course, is not polar. However, traces of water or ions in the surface of the reaction vessel can polarize it and thereby start the reaction: evidence for this comes from the observation that if gaseous ethene and bromine are meticulously dried, and mixed in a vessel whose walls are coated with paraffin wax they do not react.

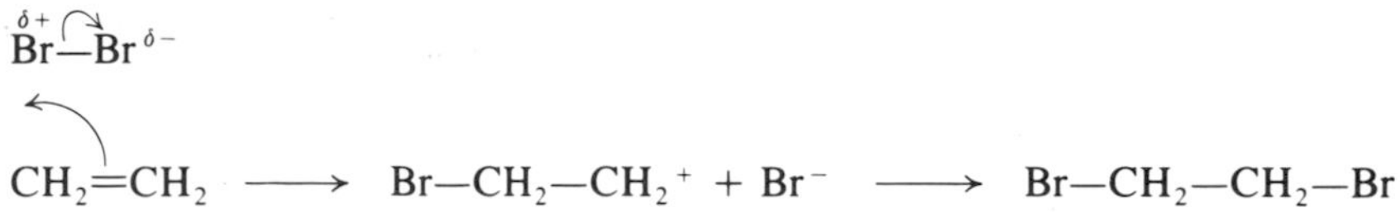

$$\overset{\delta+}{Br}\text{—}Br^{\delta-}$$

$$CH_2{=}CH_2 \longrightarrow Br\text{—}CH_2\text{—}CH_2{}^+ + Br^- \longrightarrow \underset{\text{1,2-dibromoethane}}{Br\text{—}CH_2\text{—}CH_2\text{—}Br}$$

If lithium chloride is present in the reaction mixture then some Br—CH_2—CH_2—Cl is also formed, by the attack of Cl^- on the intermediate cation.

Aqueous solutions of bromine contain bromic(I) acid (hypobromous acid, HOBr, Section 23.3). The weak $Br^{\delta+}$—$OH^{\delta-}$ bond is easily broken by attack of the nucleophilic carbon–carbon double bond on the electrophilic bromine:

$$\overset{\delta+}{Br}—\overset{\delta-}{OH}$$

$$CH_2{=}CH_2 \longrightarrow Br—CH_2—CH_2^+ + HO^- \longrightarrow Br—CH_2—CH_2—OH$$

2-bromoethanol

The consequent *decolorization of bromine water* is a test for the presence of a carbon–carbon double bond.

Other important electrophilic addition reactions

1 Addition of hydrogen chloride is closely similar to the addition of hydrogen bromide, and takes place with concentrated hydrochloric acid, or with hydrogen chloride in the gas phase or in an organic solvent:

$$CH_2{=}CH_2 + HCl \rightarrow CH_3—CH_2—Cl$$

chloroethane

2 Addition of chlorine is analogous to that of bromine. In the gas phase or in an organic solvent such as tetrachloromethane it gives a dichloroalkane:

$$CH_2{=}CH_2 + Cl_2 \rightarrow Cl—CH_2—CH_2—Cl$$

1,2-dichloroethane

Chlorine water gives the chloroalcohol:

$$CH_2{=}CH_2 + Cl—OH \rightarrow Cl—CH_2—CH_2—OH$$

2-chloroethanol

3 Hydration of alkenes (addition of water) is a two-stage process (see Section 29.3). If an alkene is treated with concentrated sulphuric acid it reacts much as it does with concentrated hydrobromic acid to give an **alkyl hydrogensulphate**:

$$CH_2{=}CH_2 + (HO)_2SO_2 \rightarrow CH_3—CH_2—OSO_2OH$$

ethyl hydrogensulphate

When the reaction mixture is diluted and warmed the alkyl hydrogensulphate is rapidly hydrolysed to the alcohol:

$$CH_3—CH_2—OSO_2OH + H_2O \rightarrow CH_3—CH_2—OH + (HO)_2SO_2$$

ethanol

This process (or the similar reaction with phosphoric acid) is used industrially to manufacture alcohols from petroleum-derived alkenes (see Section 29.5).

Direction of electrophilic addition

The addition of hydrogen bromide to propene, $CH_3CH{=}CH_2$, could apparently be expected to give two products, CH_3CH_2—CH_2Br and CH_3CHBr—CH_3. In fact only the second of these is formed. Similarly the additions of hydrogen chloride and of water (in the presence of sulphuric acid) give only CH_3CHCl—CH_3 and $CH_3CH(OH)$—CH_3 respectively. These observations were first made by a Russian chemist, Vladimir Markovnikov, and **Markovnikov's rule** states that in the addition of HZ to an unsymmetrical alkene, the hydrogen goes to the end of the double bond that already bears the greater number of hydrogen atoms.

The reason for such behaviour can be found in the mechanism. The initial attack of the electrophile (H^+ in these cases) generates a carbocation. Attack on the carbon atom at position 2 of propene gives a primary carbocation, whereas attack at position 1 gives a more stable secondary carbocation (see Section 28.3).

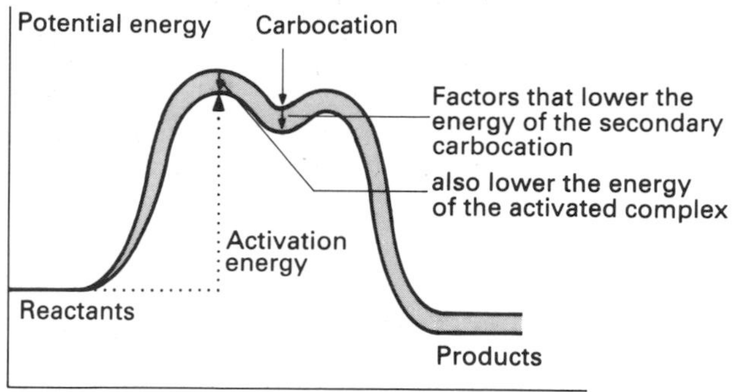

Figure 33.3 Changes in potential energy during electrophilic addition: the origin of Markovnikov's rule

$$CH_3CH{=}CH_2 \xrightarrow{H^+} CH_3CH^+{-}CH_3 \text{ (secondary)} \quad [\text{not } CH_3CH_2{-}CH_2^+ \text{ (primary)}]$$

Electron donation from the methyl groups (see Section 28.3) makes the secondary ion more stable than the primary one: it has a similar effect on the partial positive charge of the activated complex on the path from alkene to carbocation. Therefore the activation energy for forming the secondary ion is lower (see Figure 33.3). The secondary ion, and the products from it, are produced so much faster that the reaction is complete before any significant amount of the 'anti-Markovnikov' product is formed.

Oxidation of alkenes

Alkenes, like alkanes, burn in air to give carbon dioxide and water.

With alkaline potassium manganate(VII) alkenes are oxidized in the cold to diols. The purple manganate(VII) is reduced to green manganate(VI) and then to a brown precipitate of manganese(IV) oxide. The reaction provides a useful colour test for the detection of alkenes.

$$CH_2{=}CH_2 \xrightarrow[NaOH]{MnO_4^-} CH_2(OH){-}CH_2(OH)$$

ethane-1,2-diol

Alkenes are oxidized by trioxygen (ozone, O_3) in a complicated reaction that completely breaks the carbon–carbon double bond. The initial **ozonide** is readily hydrolysed to give carbonyl compounds:

$$>C{=}C< + O_3 \longrightarrow \text{ozonide} \xrightarrow{H_2O} >C{=}O + O{=}C< + H_2O_2$$

The hydrolysis is often carried out under mildly reducing conditions (for example, with zinc and dilute acid) so that the hydrogen peroxide does not oxidize any aldehydes that are formed to carboxylic acids. The complete reaction sequence is called **ozonolysis** and is useful for determining the structure of unknown alkenes, as in the following Example.

EXAMPLE

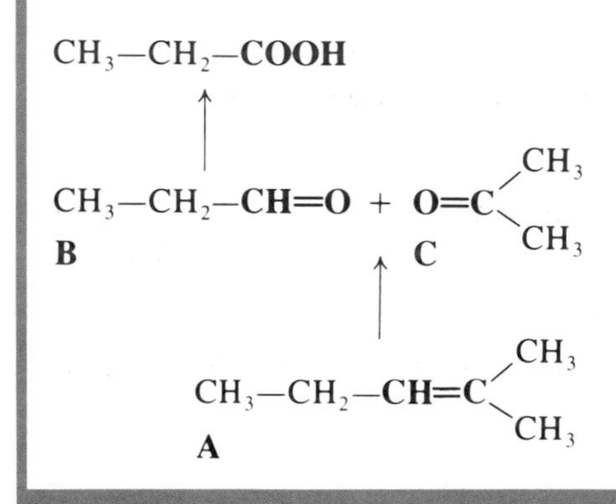

An alkene, **A**, has the molecular formula C_6H_{12}. Ozonolysis of **A** gives two organic products **B** and **C**, both with the molecular formula C_3H_6O. **B** can be oxidized by acidic dichromate(VI) to a carboxylic acid, $C_3H_6O_2$. **C** resists further oxidation. Suggest a structure for **A**.

METHOD Deduce the structures of compounds **B** and **C**, given that they contain C=O double bonds and that only **B** can be further oxidized. Then decide how these two compounds could have been formed by cleavage of a carbon–carbon double bond.

ANSWER There are only two isomers with the molecular formula C_3H_6O that contain carbon–oxygen double bonds: these are CH_3CH_2CHO, propanal, and CH_3COCH_3, propanone. **B** must be propanal because it can be oxidized to propanoic acid. Ozonolysis of **A** breaks the carbon–carbon double bond and leaves each of the doubly bonded carbon atoms as carbonyl groups. **A** was therefore $CH_3CH_2CH{=}C(CH_3)_2$.

COMMENT Ozonolysis followed by the iodoform reaction (see Section 30.3) provides a good way of detecting a methyl group attached to a carbon–carbon double bond, using the following reaction sequence:

$$(CH_3)(R)C{=}C(R')(R'') \xrightarrow{\text{ozonolysis}} (CH_3)(R)C{=}O \xrightarrow[\text{NaOH}]{I_2} CHI_3 + RCO_2^-$$

Hydrogenation

The carbon–carbon double bond can usually be reduced at room temperature and pressure by the addition of hydrogen using a catalyst of finely divided metal:

$$\underset{\text{alkene}}{R_2C{=}CR_2} + H_2 \xrightarrow{\text{catalyst}} \underset{\text{alkane}}{R_2CH{-}CHR_2}$$

Commonly used catalysts are platinum, palladium and **Raney nickel** (a highly porous nickel made by treating a nickel/aluminium alloy with aqueous sodium hydroxide to remove the aluminium). When an ordinary nickel catalyst is used, the reaction requires a higher temperature, around 150 °C. The reaction takes place on the metal surface where the alkene and hydrogen molecules are both adsorbed (see Section 14.5). Hydrogenation is useful for determining the number of double bonds in an unknown compound, by measuring the volume of hydrogen consumed.

BOX 33.1

The discovery of poly(ethene)

In the early 1930s some ICI chemists were carrying out basic research into the kinetics of reactions under very high pressures. One reaction they studied was that between benzaldehyde (see Section 35.1) and ethene. When they opened the pressure vessel they found a fraction of a gram of a white waxy solid. Its empirical formula, CH_2, showed it to be a polymer of ethene: the benzaldehyde had not reacted at all. This was a great surprise, because it had been thought that ethene was too unreactive to polymerize: indeed, when the experiment was described at a scientific meeting, the leading polymer chemist of the day dismissed it casually. Attempts to repeat the preparation failed, until it was realized that the ethene used in the original experiment had been accidentally contaminated with a trace of oxygen, in just the right concentration to catalyse the polymerization. It took many experiments, and several disastrous explosions, to find the right conditions, but eventually the first deliberate preparation of poly(ethene) was carried out, producing 8 g of the polymer.

At first it was thought that the new polymer might have a small-scale commercial use as an insulator for submarine cables, an improvement on gutta-percha (Box 33.3), which had been used till then. No one foresaw the enormous growth in its use, from the first commercial batch of 100 tonnes in 1939 to well over a million tonnes a year today.

Polymerization

When ethene is compressed to about 1500 atm at 190 °C with a trace of oxygen, it polymerizes to **poly(ethene)**, commonly called polythene:

$$n\mathrm{CH_2{=}CH_2} \rightarrow -(\mathrm{CH_2{-}CH_2})_n-$$

The reaction is an example of **radical polymerization**. An unpaired electron of the oxygen molecule initiates the reaction by attacking an ethene molecule:

$$\mathrm{O_2 + CH_2{=}CH_2} \rightarrow \mathrm{{}^{\bullet}O{-}O{-}CH_2{-}CH_2{}^{\bullet}}$$

The resulting radical attacks a second ethene molecule, and so on in a chain reaction (compare Section 27.3):

$$\mathrm{{}^{\bullet}O{-}O{-}CH_2{-}CH_2{}^{\bullet} + CH_2{=}CH_2} \rightarrow \mathrm{{}^{\bullet}O{-}O{-}CH_2{-}CH_2{-}CH_2{-}CH_2{}^{\bullet}}\ \text{etc}$$

Some rearrangement of the carbon skeleton also occurs during the reaction, giving rise to branched-chain polymers. The chain reaction is eventually terminated by one of a variety of reactions, but only after about one to ten thousand ethene molecules have been joined together so that the groups at the end of the chains are an insignificant part of the whole molecule. Poly(ethene) is effectively just a very large alkane.

A more recent method of polymerizing ethene uses a **Ziegler–Natta catalyst**, consisting of titanium(IV) chloride with triethylaluminium ($Al(C_2H_5)_3$), in an inert solvent such as heptane and at lower temperatures and pressures, typically 60 °C and 2 atmospheres. The detailed mechanism of the polymerization reactions is not fully understood. The product has fewer branched chains than that from the high-pressure process. The chains can pack together more tightly giving a denser and tougher material.

$$n\mathrm{CH_3{-}CH{=}CH_2} \downarrow -(\underset{|\atop \mathrm{CH_3}}{\mathrm{CH}}{-}\mathrm{CH_2})_n-$$

5

Using the Ziegler–Natta catalyst, propene can be polymerized to **poly(propene)** (**5**), commonly called polypropylene. This is even more rigid than high-density poly(ethene) and is used for moulded furniture. It also has a high mechanical strength and good resistance to abrasion. It can be hinged, and is therefore used for cases – for darts or make-up, for example. It is also spun into fibres for making ropes and hard-wearing carpets.

CH$_2$
H–C–CH$_3$
CH$_2$
H–C–CH$_3$

6

These properties arise from the regular structure of the polymer, with the methyl groups all arranged on one side of the carbon chain (**6**) enabling the molecules to pack together closely. Such a structure is called **isotactic** (Greek: arranged alike). Radical polymerization produces a sticky material, unsuitable for manufacture, that has a random (**atactic**) arrangement of methyl groups. Some polymers have alternating arrangements of side groups, as in one form of PVC: they are called **syndiotactic** (Greek: arranged in pairs).

Other compounds containing carbon–carbon double bonds can be polymerized. Important examples are the following:

$$-(\underset{|\atop \mathrm{C_6H_5}}{\mathrm{CH}}{-}\mathrm{CH_2})_n-$$

7

1 **Poly(phenylethene)**, or polystyrene (**7**), forms a stiffer polymer chain than that of poly(ethene), because of dispersion forces between the benzene rings. Polystyrene is therefore a rigid material, used for injection moulding and vacuum forming. When polystyrene is impregnated with pentane and then

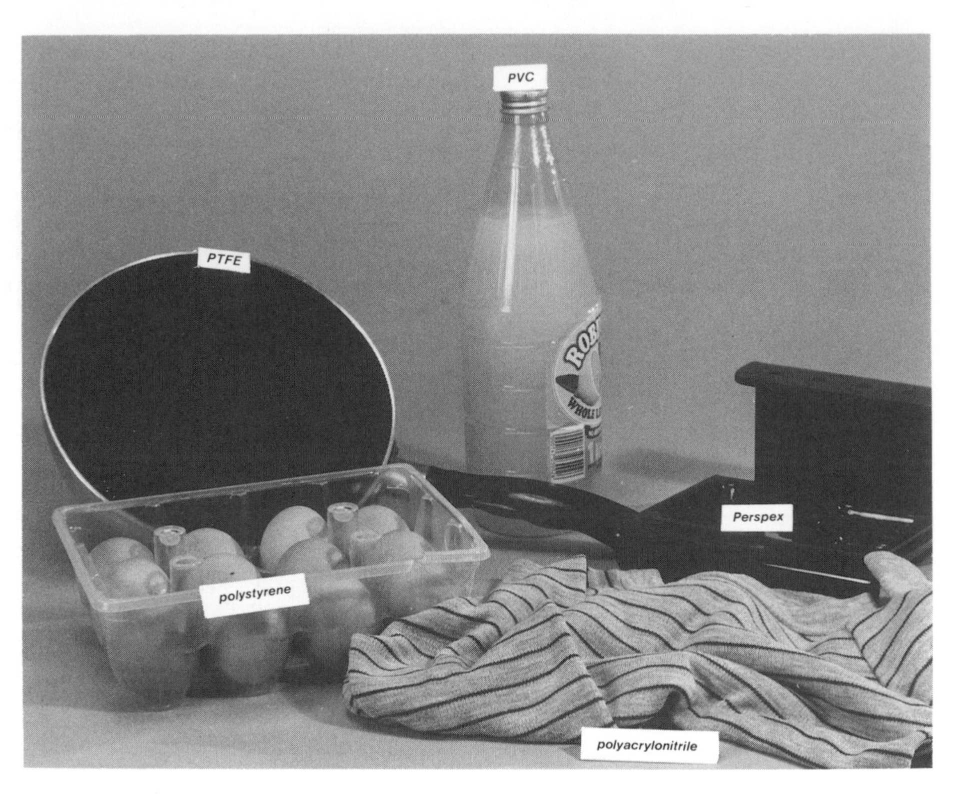

Figure 33.4 Products made from common plastics materials

heated in steam, the vapour blows it into a solid foam that is used for packaging and for heat and sound insulation.

8 $-(CH(Cl)-CH_2)_n-$

2 Poly(chloroethene), or polyvinyl chloride (PVC) (**8**), is also more rigid than poly(ethene), because of increased dispersion and dipole–dipole forces between the chains. Unplasticized PVC is rigid and tough: when plasticizers are added (see Section 29.6) the polymer becomes flexible.

9 $-(C(CH_3)(CO-OCH_3)-CH_2)_n-$

3 Poly(methyl 2-methylpropenoate), or poly(methyl methacrylate) ('Perspex', **9**), is even more rigid than polystyrene or PVC, partly because there are two substituent groups to hinder the movement of one chain against another, and partly because of the strong dipole–dipole interactions between the ester groups. It forms hard transparent sheets that were first used in the Second World War for the canopies of aeroplanes. It is now used in windows and shop signs.

10 $-(CH(CN)-CH_2)_n-$

4 Poly(propenenitrile), or polyacrylonitrile (**10**), is usually made as a **copolymer**; that is, a small amount of some other ethene derivative such as ethenyl ethanoate ($CH_2{=}CHOCOCH_3$, vinyl acetate) is mixed with the propenenitrile and the two are polymerized together. Fibres are made containing two slightly different copolymers. They are stretched while being heated, and when they cool the two polymers shrink to slightly different extents, giving a curly yarn that can be used for wool-like fabrics (**acrylics**).

11 $-(CF_2-CF_2)_n-$

5 Poly(tetrafluoroethene) (PTFE, **11**) is chemically inert (compare the stability of the chlorofluoromethanes, discussed in Section 28.5). It has high electrical resistance and a very low coefficient of friction and is used for protective and non-stick surfaces.

All these are examples of **addition polymers** (in contrast to the condensation polymers, such as nylon, described in Box 32.3). The polymer chain is built up by adding the monomer units one to another.

33.4 Preparative methods for alkenes

Industrially many alkenes are obtained by cracking petroleum (see Section 27.4). In the laboratory the two main preparative reactions are eliminations:

1 Dehydration of alcohols (see Section 29.3):

$$CH_3\text{—}CH_2OH \xrightarrow[170\,^\circ C]{\text{excess conc. }(HO)_2SO_2} CH_2\text{=}CH_2$$

2 Dehydrohalogenation of halogenoalkanes (see Section 28.3):

$$CH_3\text{—}CHBr\text{—}CH_3 \xrightarrow[55\,^\circ C]{KOH/C_2H_5OH} CH_3\text{—}CH\text{=}CH_2$$

33.5 The importance of alkenes

The lower alkenes, obtained from petroleum by cracking (see Box 27.2) are the key building blocks of the petrochemical industry. Nearly all the vast range of organochemical products – plastics, synthetic fibres, detergents, paints, pharmaceuticals and many others – are derived from ethene, propene or butadiene (but see Box 33.2). Figures 33.6 and 33.7 illustrate some of the major petrochemical processes. The last reaction in Figure 33.7 is an interesting example of industrial economy. It is cheaper to make the two products, which are required in roughly equal quantities, by this single **cumene process** than to carry out the separate manufacture of propanone from propene and of phenol from benzene.

BOX 33.2

Figure 33.5 A charge of hot coke being pushed out of a coke oven

What happens when the oil runs out?

The world has become almost totally dependent on petroleum for its energy and for its organic chemicals (see Section 27.4). The oil will not last for ever. As the supply dwindles we shall look to three main sources of energy: nuclear reactions, coal and 'biomass' (including vegetable matter), together with contributions from solar, wave, wind, geothermal, hydroelectric and other sources. Coal and biomass will also have to supply our organic chemicals – as they did in the past, though on a much smaller scale.

When coal is heated to about 1000 °C in the absence of air it is broken down into volatile products, **coal gas** (which was an important domestic and industrial fuel) and **coal tar**, leaving a residue of **coke**, which is largely carbon mixed with inorganic material. 100 kg of coal produces only about 5 kg of tar, a mixture of benzene and related compounds, so that this can never be a large-scale source of organic chemicals. Coke, however, can lead to many compounds that we now obtain from petroleum, either through calcium dicarbide (see Section 18.6) or through **synthesis gas**, a 1:2 mixture of carbon monoxide and hydrogen, which was named because it can be used to synthesize methanol, and thence other organic chemicals.

The main means for converting biomass to organic chemicals, both for fuel and for chemical manufacturing, is fermentation, discussed in Section 29.6, by which sugar, starch and, after initial breakdown with acid, cellulose are converted to ethanol. At present industrial ethanol is made largely from ethene. In the future, as in the past, the reverse is likely to be more important.

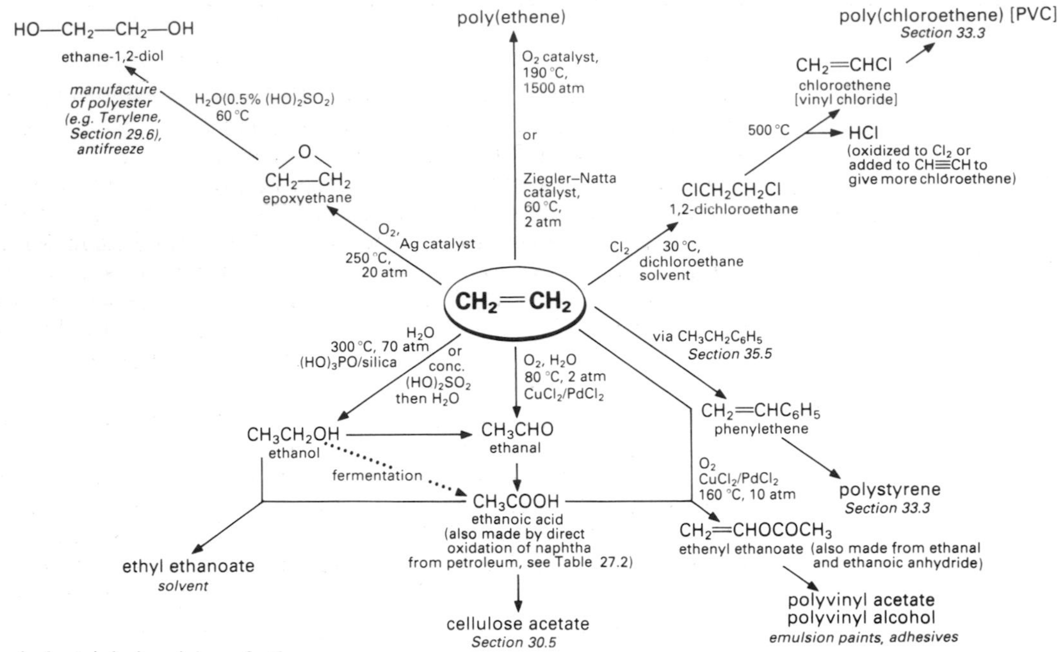

Figure 33.6 The industrial chemistry of ethene

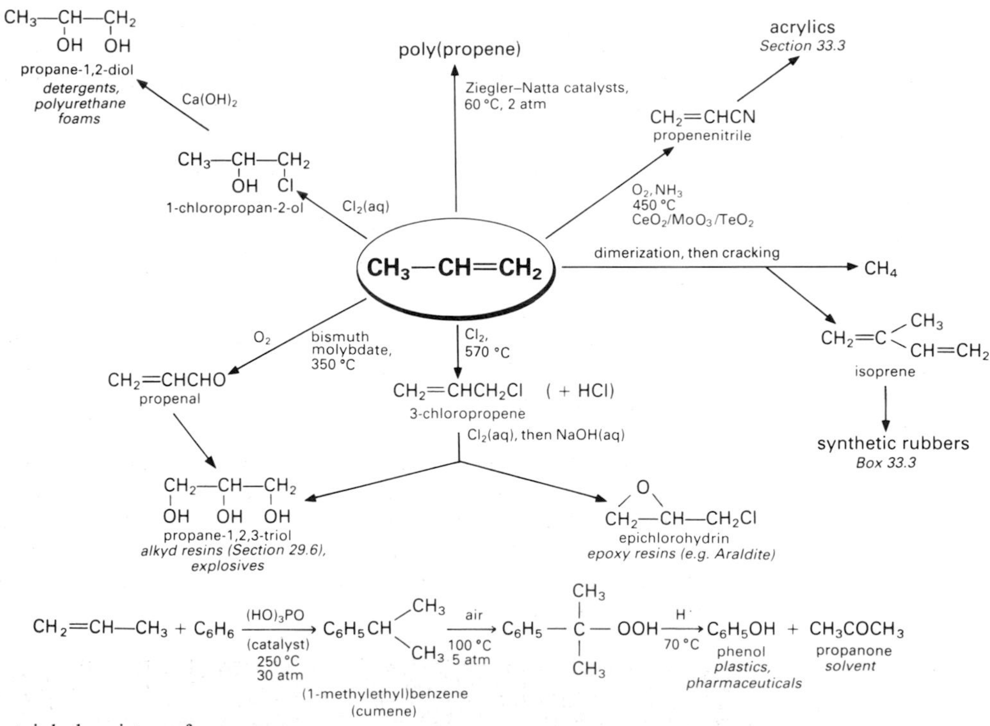

Figure 33.7 The industrial chemistry of propene

Butadiene

$$\mathrm{CH_2{=}CH{-}CH{=}CH_2}$$

12

Butadiene (**12**) is an example of a molecule containing two double bonds. It is a product of petroleum cracking, and is most commonly used to give copolymers with phenylethene (styrene) to make **SBR** (styrene–butadiene rubber) or with phenylethene and propenenitrile (acrylonitrile) to make **ABS** (acrylonitrile–butadiene–styrene). SBR has rubbery properties because the butadiene fragments contain double bonds that can be used to cross-link the polymer chains, just as in the vulcanization of natural rubber (see Box 33.3). ABS is rigid and tough; it is used for telephone receivers, suitcases and body panels of motor cars.

BOX 33.3

Rubber

Rubber originated in South America. The Conquistadores found the South American Indians playing with balls that bounced far better than the straw-packed leather or inflated pigs' bladders used in Europe.

Early rubber was soft and sticky when hot, hard and brittle when cold. In 1839 Charles Goodyear, an American, found by chance that when it was heated with a little sulphur, a process he called **vulcanization**, these undesirable properties were greatly reduced. This led to such an enormous demand that by the end of the nineteenth century there was a great shortage of rubber. Brazil, the home of the rubber tree, closely guarded its assets, but seeds were smuggled out and the great rubber plantations of Malaysia and Indonesia were started.

Even the new supplies of plantation rubber could not fully satisfy the demand, especially for motor car tyres. Attempts to develop a synthetic rubber-like material were frustrated by ignorance of the structure of natural rubber. Michael Faraday, as far back as 1826, had shown that its empirical formula is C_5H_8, and by 1860 it was known that when rubber is heated in the absence of air it gives off methylbutadiene (**isoprene, 13**), which has the molecular formula C_5H_8. The first synthetic rubber was a polymer of dimethylbutadiene, made in Germany during the First World War. All attempts to imitate natural rubber by polymerizing isoprene failed, however.

13

14

Rubber is indeed a polymer of isoprene, but the key to its structure is that all the double bonds have the same *cis* geometry (**14**): a random mixture of *cis* and *trans* arrangements gives a sticky material that has little strength or elasticity. This was what the early attempts had failed to achieve, and what had occurred by chance in the polymerization of dimethylbutadiene. In 1955, Ziegler–Natta catalysts were successfully used to produce synthetic poly(*cis*-isoprene). The geometric isomer of natural rubber with all the double bonds *trans* is **gutta-percha**. Much harder than rubber, it is used for insulating electrical cables and to cover golf balls. Like rubber latex, it comes from the sap of trees, in this case native to Malaysia.

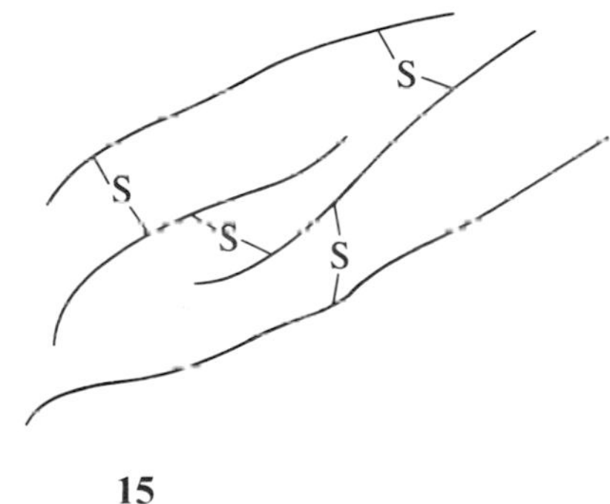

15

Once the structure of rubber was known, it could be seen that vulcanization works by cross-linking the polymer chains with bridges of sulphur atoms that join carbon atoms next to the double bond (**15**). This prevents them from being pulled apart when the rubber is stretched, giving it its great elasticity.

During the Second World War all the main sources of natural rubber fell into Japanese hands, encouraging the rapid development of new synthetic rubbers in the United States, Britain, the Soviet Union and Germany. Nowadays the total production of synthetic rubbers exceeds that of natural rubber, though the natural material and SBR are the two most important rubber substances.

Figure 33.8 Tapping natural latex from a rubber tree in Malaysia

Summary

- ☐ Alkenes contain a carbon–carbon double bond.
- ☐ The general formula of alkenes is C_nH_{2n}.
- ☐ Their names are derived from those of the alkanes, using the ending *-ene*.
- ☐ The torsional rigidity of the double bond leads to geometric isomerism (*cis/trans* isomerism). Geometric isomers have the same groups bonded in the same order; they differ only in the geometry of the bonds.
- ☐ Alkenes undergo electrophilic addition across the carbon–carbon double bond. Important additions occur with HBr and HCl, Br_2 and Cl_2, HOBr and HOCl, and (catalysed by $(HO)_2SO_2$) with H_2O. Decolorization of bromine water is used to test for the presence of carbon–carbon double bonds.
- ☐ The addition of HZ to unsymmetrical alkenes follows Markovnikov's rule: the hydrogen goes to the end of the double bond that already bears the greater number of hydrogen atoms.
- ☐ Alkenes are oxidized with alkaline potassium manganate(VII) to diols. The brown precipitate of MnO_2 is used as a test for carbon–carbon double bonds.
- ☐ The carbon–carbon double bond is broken in ozonolysis to give carbonyl compounds.
- ☐ Alkenes are reduced to alkanes by hydrogenation on the surface of finely divided metal catalysts such as platinum, palladium and nickel.
- ☐ Ethene, propene and other compounds containing carbon–carbon double bonds undergo addition polymerization to give a wide range of commercially important polymers.
- ☐ Ethene, propene and butadiene, obtained from the cracking of petroleum, are the building blocks for most of the organochemical industry.

PROBLEMS

1 Draw the structures of all the alkenes with molecular formula C_5H_{10}. Give one example of a compound C_5H_{10} that does *not* contain a double bond. Name all the isomers.

2 Give the names and structures of the following compounds: (a) the major product of the reaction between ethene and bromine water, (b) the products of ozonolysis of 2-methylpent-2-ene.

3 Give the names and structures of the possible products of reaction of ethene with concentrated sulphuric acid, and state the conditions that favour the formation of each.

4 Explain why the addition of bromine to a solution of $CH_3CH{=}CHCH_3$ in methanol gives a mixture of $CH_3CHBrCHBrCH_3$ and $CH_3CHBrCH(OCH_3)CH_3$.

5 State Markovnikov's rule. Show the mechanism of the reaction between propene and hydrogen chloride, and explain how it accounts for Markovnikov's observations.

6 Describe how propene may be converted into the following compounds:
(a) $CH_3CH_2CH_3$,
(b) $CH_3CHBrCH_3$,
(c) CH_3COCH_3,
(d) CH_3COOH.

7 An alcohol, **A**, $C_4H_{10}O$, can be oxidized to a carboxylic acid by refluxing with acidic aqueous potassium dichromate(VI). When **A** is heated to 170 °C with concentrated sulphuric acid it gives an alkene, **B**, C_4H_8. **B** is converted into **C**, an isomer of **A**, by treatment with cold concentrated sulphuric acid, followed by dilution with water. **C** is resistant to oxidation. Account for the above observations and draw the structures of **A**, **B** and **C**.

8 Outline the conditions under which ethene can be polymerized. Name two compounds other than ethene that can be polymerized. Why is poly(ethene) suitable for lining vessels containing alkaline solutions whereas Terylene and nylon are not?

9 Discuss the isomerism of the compounds $CH_3CH{=}CHR$, where $R = C_2H_5$, C_3H_7 and C_4H_9.

10 An alcohol, **A**, $C_5H_{12}O$, is optically active. It is dehydrated by passing its vapour over Al_2O_3 at 300 °C to give **B**, C_5H_{10}. **B** is treated with trioxygen (ozone) followed by zinc and dilute hydrochloric acid to give two products, **C**, C_3H_6O, and **D**, C_2H_4O. All four compounds, **A**, **B**, **C** and **D**, give iodoform when treated with iodine and aqueous sodium hydroxide. Account for these observations, and draw the structures of compounds **A** to **D**.

11 Comment on the following observations.
(a) When ethene (ethylene) is passed into bromine water containing sodium chloride, CH_2ClCH_2Br is one of the products.
(b) When ethene is kept at high temperature and pressure in a steel container a solid material is formed.
(c) The rate of addition of hydrogen cyanide to propanone (acetone) is greatly increased by the addition of a trace of potassium cyanide.
(d) Monochloroethanoic acid is a stronger acid than ethanoic acid in aqueous solution.

(e) The addition of pure hydrogen bromide to but-1-ene gives a mixture of two isomers of formula C_4H_9Br in equal proportions. *(Oxford and Cambridge)*

12 (a) Petroleum fractions such as naphtha are catalytically cracked in the industrial production of monomers such as ethene and propene which may then be polymerized to obtain plastics.
(i) Explain what is meant by catalytic cracking of naphtha and give an equation to illustrate the types of compound formed.
(ii) Describe the polymerization of propene, commenting on the importance of Ziegler–Natta catalysts in the production of poly(propene).
(iii) Discuss the economics of the industrial catalytic cracking of naphtha to provide monomers for the production of plastics.
(b) Give an equation for the reaction of propene with hydrogen bromide and explain, by reference to the detailed mechanism of this reaction, the nature of the product(s). *(AEB 1986)*

13 (a) Discuss what is meant by the terms *isotopes* and *isotopic abundance*.
(b) Describe in outline how in the mass spectrometer:
(i) positive ions are formed from organic molecules,
(ii) the ions are separated according to mass and charge.
(c) Separate samples of ethene gas were reacted with bromine vapour, aqueous bromine (bromine water) and aqueous bromine containing some dissolved sodium chloride; the organic products were examined using a mass spectrometer, peaks being observed as follows:

Experiment	Reagent	Mass/charge ratio	Information on peaks
1	Bromine vapour	186, 188, 190	Peak heights in ratio 1:2:1
2	Aqueous bromine	124, 126 186, 188, 190	All five are principal peaks
3	Aqueous bromine with dissolved sodium chloride	124, 126 142, 144, 146 186, 188, 190	All eight are principal peaks

Assuming that natural samples of carbon, hydrogen and oxygen contain only one isotope whilst chlorine contains ^{35}Cl and ^{37}Cl:
(i) use the results of experiment 1 to determine the isotopic composition of bromine,
(ii) give the names and structural formulae of the compounds responsible for the additional peaks in experiments 2 and 3,
(iii) show how the mechanism of the reaction accounts for the formation of these compounds. *(AEB 1985)*

14 (a) Copolymerization of monomers A and B can produce a variety of polymers with a range of properties which are related to the proportions of A and B used.
Condensation polymerization of monomers C and D produces a single polymer with a fixed property.
Give a specific example, with equations, of the formation of (i) a copolymer, (ii) a condensation polymer, and account for the differences indicated above.
(b) Explain how
(i) ethene can produce both low-density and high-density polymers,
(ii) propene can produce polymers with three different types of structure.
(c) Explain the differences between thermosetting and thermoplastic materials and give an example of each. *(Cambridge)*

15 (a) Show how the molecular shapes adopted by but-2-yne and but-2-ene can be interpreted in terms of their bonding. Include in your answer an explanation of the term *geometrical isomerism*.
(b) Write down the name of compound J. Briefly describe and explain an important industrial use of J.

H, H / C=C / H, Cl

Compound **J**

(c) Describe simple chemical experiments which would enable you to differentiate between
(i) but-2-ene and but-1-ene;
(ii) but-2-yne and but-1-yne. *(Oxford)*

16 A hydrocarbon **D** reacted with bromine to give **E**. On analysis **E** gave C 26.20%, H 3.93% and Br 69.87%. Ozonolysis of **D** followed by mild oxidation gave two compounds **F** and **G**.

F, C_3H_6O, is a neutral liquid which gives a pale yellow crystalline precipitate when treated with an alkaline solution of iodine.

G is a crystalline solid; 2.00 g of **G** are exactly neutralized by 33.9 cm^3 of aqueous NaOH solution (1.00 M). Dry distillation of **G** gave **H**, $C_4H_4O_3$, which on addition of water re-formed **G**. Compound **G** can be made in two stages from 1,2-dibromoethane by treatment with KCN followed by hydrolysis of the product.

Deduce the structures of compounds **D**, **E**, **F**, **G** and **H** and explain your reasoning carefully. *(Oxford)*

17 (a) Discuss the relative importance and availability of coal, oil and natural gas as raw materials in the world today and in the future.
(b) 'Organic chemists, working in industry, consume scarce natural resources at an alarming rate and produce materials which pollute our environment. Living would be cleaner and more enjoyable without their activities.' Discuss this statement. *(SUJB)*

34 Arenes: the benzene ring

In this chapter we examine the chemistry of the benzene ring. Its special aromatic stability gives it properties that are different from those of other unsaturated compounds. In particular it normally reacts by substitution, which allows the electron delocalization to be retained, rather than by addition.

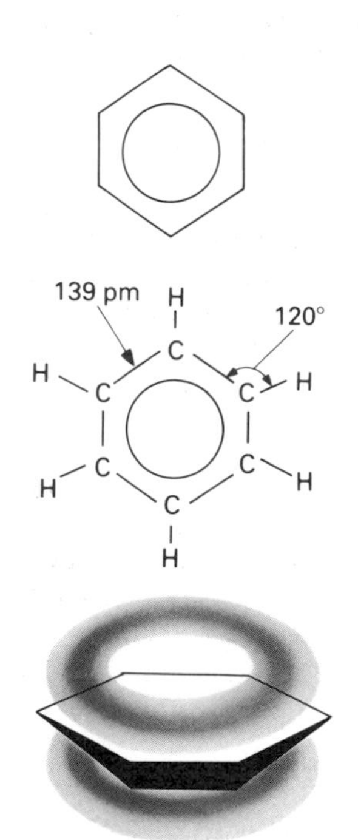

Figure 34.1 Representations of the benzene ring

In the mid-nineteenth century chemists were puzzled by the properties of a group of fragrant compounds that they obtained from various plant oils such as oil of bitter almonds or oil of wintergreen. They were very different from the triacylglycerol oils (discussed in Section 31.6), and they seemed to contain too few hydrogen atoms for the number of carbon atoms in their molecules. The unsaturated alkenes were known to be reactive, readily undergoing addition reactions. These compounds seemed to be even more unsaturated, but were nothing like as reactive and gave substitution products rather than addition. August Kekulé, the man who had first recognized the tetracovalency of carbon and who realized that organic compounds were built around extended carbon chains, showed that these strange oils were chemically related to benzene. It was he who proposed that benzene had a cyclic structure, which we now recognize as the origin of its stability through the delocalization of π-electrons (see Section 2.5). The benzene ring is a regular hexagon (see Figure 34.1). All the bond angles are 120°. The carbon–carbon bonds are all of equal length (139 pm), intermediate between the lengths of carbon–carbon single bonds (154 pm) and double bonds (134 pm). Previously in this book we have used dotted lines to represent delocalized electrons: for benzene, the usual representation is the solid circle shown in Figure 34.1.

Because of their fragrance, Kekulé called the benzene oils **aromatic compounds**. The term is now used for all compounds containing benzene rings, but in this context the word 'aromatic' has lost its original meaning: some benzene derivatives smell just as offensive as any other type of compound. Hydrocarbons containing a benzene ring are called **arenes**, *ar-* from 'aromatic' and *-ene* meaning an unsaturated hydrocarbon, as in 'alkene'. (Compounds that do not contain a benzene ring are called **aliphatic,** a word derived from the Greek for 'fat': that is, they are compounds that, like the fats, contain chains of carbon atoms.)

34.1 Nomenclature

The parent hydrocarbon, C_6H_6, is called benzene (see Box 34.1). Hydrocarbons with alkyl groups attached to a benzene ring are named as in the following examples:

CH_3

methylbenzene
(older name: toluene)

CH_3 CH_3

1,2-dimethylbenzene
(numbers chosen to be as low as possible)

CH_2CH_3 CH_3

1-ethyl-3-methylbenzene

The terms *ortho*, *meta* and *para* are commonly used to describe 1,2-, 1,3- and 1,4-disubstituted benzenes respectively, so that, for example, 1,2-dimethylbenzene is also called *ortho*-dimethylbenzene, abbreviated to *o*-dimethylbenzene.

The group C_6H_5—, derived from benzene by removing a hydrogen atom, is named unsystematically as *phenyl* (see Box 34.1).

BOX 34.1

Figure 34.2 Gas street lighting used in London's Charing Cross in the 1840s

Benzene and 'phene'

In the Middle Ages Arab traders brought to Europe a fragrant resin they obtained from the bark of a tree that grows on the island of Sumatra. With a fine disregard for geography they called it *laban jawi*, incense of Java. The French merchants with whom they dealt, assuming that 'la' must mean 'the', referred to it as *la banjawi*. In English the word was gradually changed to benjoin and later to gum benzoin, which is still the term in use today.

The main constituent of the resin is C_6H_5—CO—CH(OH)—C_6H_5, known chemically by the non-systematic name of **benzoin**, and a second component, C_6H_5—COOH, is called **benzoic acid** (see Section 35.1). In the 1830s it was found that when the acid was heated with quicklime (CaO) it gave the hydrocarbon C_6H_6, and this was called benzine, benzol (a name that is still used commercially for a mixture of benzene with other arenes) or **benzene**.

The French chemist Auguste Laurent, who first used the name benzene, also suggested the alternative *phene* (Greek, *phaino*, I give light) because the hydrocarbon was found in coal tar, formed in the manufacture of coal gas ('illuminating gas'). The name phene was never adopted, but its derivatives **phenyl**, for the group C_6H_5—, and **phenol** for C_6H_5—OH (see Section 35.1) are a part of standard nomenclature.

34.2 Physical properties

The melting points of some simple arenes clearly illustrate the importance of the shape of molecules in determining how tightly they pack together in crystals. This in turn has a major effect on the melting enthalpies and melting points (see Section 9.2). Table 34.1 shows that disrupting the symmetrical benzene ring by adding a single methyl group leads to a large decrease in melting point, but that the difference between the effects of a methyl group and an ethyl group is insignificant. Boiling points depend more on the size of molecules – in general, the bigger the molecules the larger the dispersion forces between them – and less on their shape.

Table 34.1 Melting and boiling points of three arenes

	t_m/°C	t_b/°C
benzene	5	80
$C_6H_5CH_3$	−95	111
$C_6H_5CH_2CH_3$	−95	136

Benzene was for many years used widely as a laboratory solvent. It is now known to be highly dangerous: in high concentrations its vapour can cause respiratory failure; prolonged exposure to lower amounts can induce anaemia or cancer. The less poisonous methylbenzene is now used instead.

34.3 Reactivity patterns and important reactions

Arenes, like alkenes, are unsaturated compounds, and like the alkenes they undergo addition reactions, forming cyclohexane derivatives. As we shall see, however, these are not their most important reactions.

Arenes can be reduced using hydrogen and a catalyst of finely divided platinum, palladium or nickel:

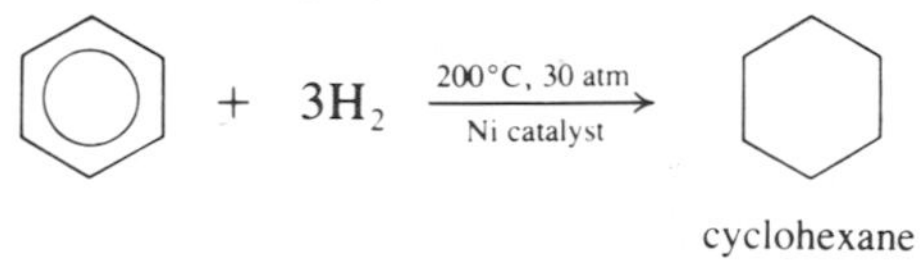

$$C_6H_6 + 3H_2 \xrightarrow[\text{Ni catalyst}]{200^\circ C,\ 30\ atm} C_6H_{12}$$

cyclohexane

Because of the stability of the aromatic ring, discussed in Section 2.5, the reaction conditions must be considerably more vigorous than those needed to reduce an alkene. It is not usually possible to reduce benzene stepwise, leaving one or two double bonds in the ring, because the reduction of these intermediate compounds is faster than that of benzene itself.

Benzene reacts with chlorine in bright sunlight or under ultraviolet irradiation (both of which produce chlorine atoms – compare Section 27.3) to give 1,2,3,4,5,6-hexachlorocyclohexane, commonly known as benzene hexachloride or BHC:

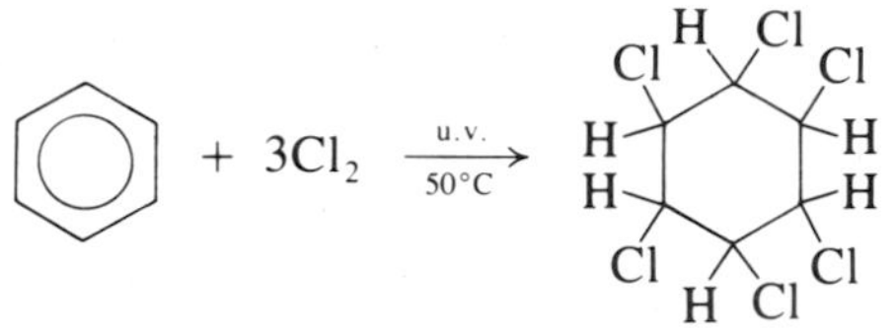

$$C_6H_6 + 3Cl_2 \xrightarrow[50^\circ C]{\text{u.v.}} C_6H_6Cl_6$$

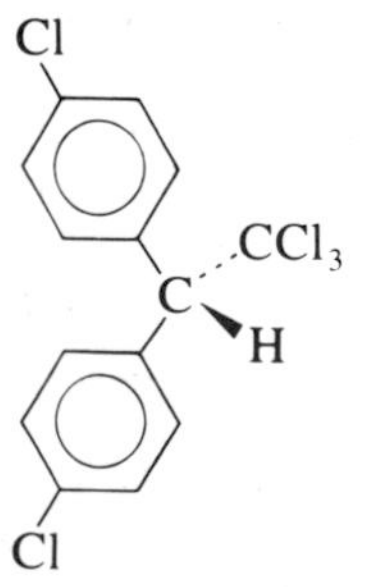

BHC

BOX 34.2

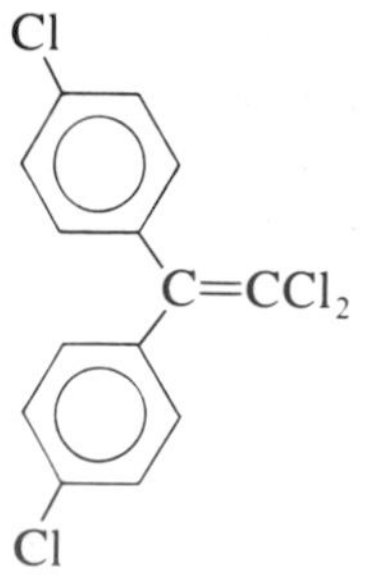

Organochlorine insecticides

BHC was discovered by Michael Faraday in 1825, the same year in which he discovered benzene. Over a hundred years later, during the Second World War, it was found to be a powerful insecticide. BHC is not a single compound: there are nine isomers with different relative arrangements of the hydrogen and chlorine atoms above and below the ring. Only one isomer, comprising about 15 per cent of the mixture, is actively insecticidal: it is known as **gamma-BHC** – whence the commercial name, Gammexane.

BHC is one of a group of organochlorine insecticides that includes DDT (**1**). (The abbreviation comes from **d**ichloro**d**iphenyl**t**richloroethane but the systematic name is 1,1,1-trichloro-2,2-di(4-chlorophenyl)ethane.) They are all extremely potent: they modify the structure of nerve cell walls, letting sodium and potassium ions escape and paralysing the insect by stopping nerve action. DDT has been the major weapon in the continuing war against malarial and yellow-fever mosquitoes and against the tsetse fly that carries sleeping sickness. The World Health Organization estimated that 5 million lives were saved in the first eight years of its use.

The organochlorine insecticides are chemically very stable, and this is now seen as a disadvantage. They dissolve in the body fat of animals, and levels build up along the food chain (**biological magnification**). When DDT was used to eradicate mosquitoes in Long Island (New York) it was found

1 $(4\text{-}ClC_6H_4)_2CH\text{–}CCl_3$

2 $(4\text{-}ClC_6H_4)_2C{=}CCl_2$

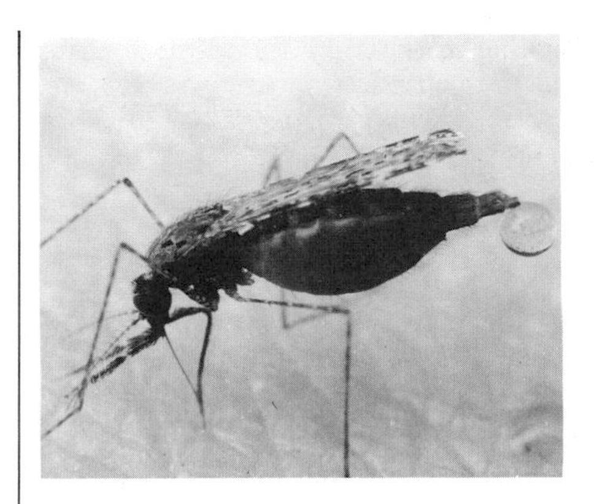

Figure 34.3 A malarial mosquito, drawing blood

in the fat of plankton (minute aquatic plants and animals) at only 0.04 parts per million, in small fish that fed on the plankton at 0.5 p.p.m., in larger fish that ate the small ones at 2 p.p.m., and in fish-eating birds such as ospreys at 25 p.p.m. In birds the DDT leads to eggs with very thin shells that break in the nest, possibly by causing calcium deficiency.

It has not been shown that DDT has any harmful effects in mammals. They have enzymes that convert DDT into the less poisonous DDE (**d**ichlorodiphenyl**d**ichloro**e**thene – **2**); the different shape of its molecule reduces the interference with cell walls. However, the observation of high levels of DDT in, for example, human milk has led to a reduction in its use and in that of other, more poisonous, organochlorine insecticides and to a search for similar compounds that are less soluble in body fat and more rapidly destroyed by bacteria. To date such biodegradable alternatives are much more expensive.

Meanwhile the dilemma faced by the 'third world' countries where insect-borne disease is rife is this: should they continue to make massive use of DDT on a scale that would no longer be acceptable in the West and with uncertain long-term risks, or should they scale down the fight against the insects?

Electrophilic aromatic substitution

The benzene ring, like the double bond of alkenes, is a region of high electron density and is attacked by electrophilic reagents. However, whereas alkenes react rapidly with reagents such as chlorine or bromine water, the conditions needed for the more stable benzene ring to react are more vigorous. A catalyst (sometimes called a **halogen carrier**) is needed to convert the halogen into a more reactive electrophile. Typical catalysts are the Lewis acids iron(III) chloride (which can be generated in the reaction mixture from metallic iron) and aluminium chloride. They react with the chlorine or bromine by accepting a lone pair of electrons into a vacant orbital:

$$Cl{-}Cl + AlCl_3 \rightarrow Cl^+ \, AlCl_4^-$$

The benzene ring then reacts with the electrophilic Cl^+ as follows:

$$C_6H_6 + Cl^+ \longrightarrow C_6H_6Cl^+$$

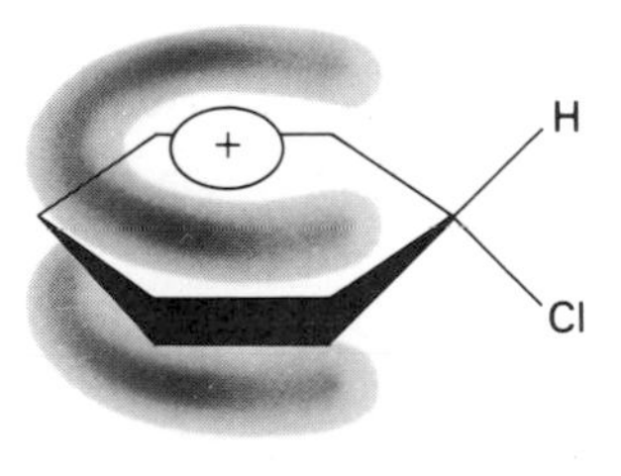

Figure 34.4 π-electron distribution in the Wheland intermediate: four π-electrons are shared by five carbon atoms

The 'horseshoe-in-a-hexagon' representation of the cation shows that the electrons are delocalized, as in benzene itself. Two of the six π-electrons are used for the new bond between one of the carbon atoms and the chlorine atom. The other four are spread round the remaining five carbon atoms. These originally contributed one electron each to the π-system, so that with only four electrons they are left sharing a positive charge (see Figure 34.4). The cation is sometimes called a **Wheland intermediate**, after an American chemist who studied the mechanism of the reaction.

Thus far the reaction resembles that of an alkene with a halogen (discussed in Section 33.3), electrophilic attack on the π-electrons giving a carbocation. The energy required is greater for benzene because the delocalization stabilization of the Wheland intermediate is much less than that of the benzene ring: hence the need for the catalyst. Whereas the reaction of an alkene is completed by nucleophilic attack to give an addition product, however, the Wheland intermediate reacts by shedding a proton, H^+, to restore the stable aromatic π-ring:

$$C_6H_6Cl^+ + AlCl_4^- \longrightarrow C_6H_5Cl + AlCl_3 + HCl$$

The complete reaction is therefore **electrophilic substitution**: the chlorine atom has replaced one of the hydrogen atoms of the benzene ring to give chlorobenzene (used in making the insecticide DDT):

$$C_6H_5H + Cl_2 \xrightarrow{AlCl_3} C_6H_5Cl + HCl$$

chlorobenzene

Benzene reacts in a similar way with other electrophiles. With bromine and a halogen carrier the reaction is similar to that with chlorine:

$$C_6H_5H + Br_2 \xrightarrow{AlCl_3} C_6H_5Br + HBr$$

bromobenzene

Acid chlorides react with benzene to give ketones. Again, aluminium chloride is used as a catalyst: it reacts with the acid chloride to give the strongly electrophilic acyl cation, RCO^+:

$$CH_3COCl + AlCl_3 \rightarrow CH_3\overset{+}{C}{=}O + AlCl_4^-$$

The complete reaction is known as the **Friedel–Crafts acylation**:

$$C_6H_5H + CH_3COCl \xrightarrow{AlCl_3} C_6H_5COCH_3 + HCl$$

phenylethanone

A similar reaction, **Friedel–Crafts alkylation**, occurs with halogenoalkanes, but it is less useful than the acylation because it is hard to stop several alkyl groups joining the benzene ring (for reasons discussed in the note on the effects of substituents on electrophilic substitution, below).

$$C_6H_5H + CH_3CH_2Cl \xrightarrow{AlCl_3} C_6H_5CH_2CH_3 + HCl$$

ethylbenzene

$O{=}\overset{+}{N}{=}O$

nitronium ion, isoelectronic with

3 $O{=}C{=}O$

One of the most important reactions of arenes is **nitration**. This is carried out with a mixture of concentrated nitric and concentrated sulphuric acids which react together to give the strongly electrophilic **nitronium ion**, NO_2^+ (**3**) (as shown by cryoscopy – see Section 8.2):

$$HO{-}NO_2 + 2(HO)_2SO_2 \rightarrow NO_2^+ + H_3O^+ + 2HOSO_3^-$$

The nitronium ion reacts with benzene at about 50 °C to give the oily nitrobenzene (**4**). Nitration is an important first step on the way to a wide range of compounds.

$C_6H_5NO_2$

4 nitrobenzene

$$C_6H_6 + NO_2^+ \longrightarrow C_6H_6NO_2^+$$

$$C_6H_6NO_2^+ + HOSO_3^- \longrightarrow C_6H_5NO_2 + (HO)_2SO_2$$

nitrobenzene

Fuming sulphuric acid (oleum, Section 22.6) reacts with benzene near room temperature: concentrated sulphuric acid reacts under reflux. In both cases the electrophile is SO_3 and this **sulphonation** gives benzenesulphonic acid:

$$C_6H_5\text{–H} + SO_3 \longrightarrow [C_6H_5(H)SO_3^-]^{+} \xrightarrow{\pm H^+} C_6H_5S(=O)_2OH$$

benzenesulphonic acid

(The step marked '$\pm H^+$' involves the transfer of protons to and from sulphuric acid.)

The reaction is used in making detergents and dyes: in the latter, the sulphonic acid group binds the dye molecule to basic groups in the fabric, for example to amine groups in proteins such as wool.

BOX 34.3

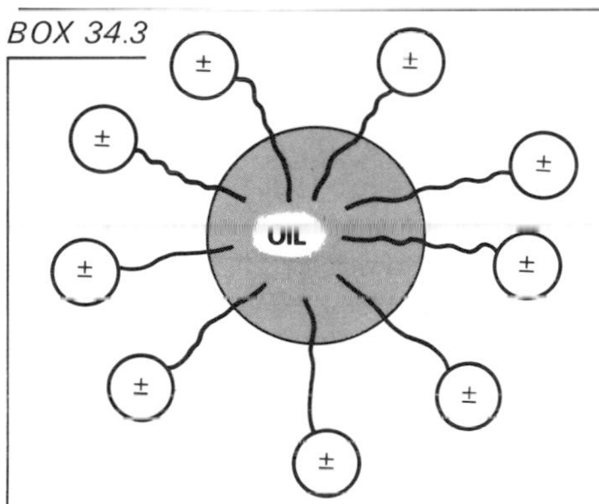

Figure 34.5 Detergent action: an oil droplet surrounded by detergent molecules that give it an ionic surface and render it water-soluble

5 $SO_3^-Na^+$

Detergents

Detergent molecules have long hydrocarbon chains with an ionic group (usually negatively charged) at one end: the hydrocarbon part becomes attached to globules of grease, which thereby become surrounded by the ionic groups and can dissolve in water (see Figure 34.5).

In soap (see Section 31.6), the hydrocarbon part is a straight chain alkyl group of about seventeen carbon atoms and the ionic group is —$CO_2^-Na^+$. In hard water, insoluble calcium and magnesium salts of these compounds precipitate out as scum: this is unpleasant, wasteful of soap and harmful to fabrics. Synthetic detergents with soluble calcium and magnesium salts were therefore developed.

Early synthetic detergents were derived from a tetramer of propene linked to a benzene ring carrying a sulphonate (—SO_3^-) ionic group (**5**). These detergents were unaffected by the bacteria in sewage works that break down soap molecules (that is, they were not **biodegradable**) and led to the pollution of rivers by detergent foam.

The problem was solved by a combination of chemistry, biochemistry and economics. The main reason the detergents were non-biodegradable was that the bacterial enzymes could not break down a branched chain. The propene tetramer had been used mainly because there was a surplus of propene being produced in naphtha crackers. New polymerization processes led to the development of poly(propene), and the branched-chain hydrocarbons also came into demand as anti-knock additives for petrol (see Section 27.4): straight-chain hydrocarbons became cheaper than the branched-chain ones, and unbranched sulphonate detergents came into use. These are broken down by bacteria much faster than the branched-chain compounds are, albeit still more slowly than soap.

(a)

(b)

Figure 34.6 The river Avon, showing (a) detergent foam in 1967, (b) cleaner water in the 1980s

The effects of substituents on electrophilic substitution The reactions considered so far have all used benzene itself as the arene. If there is a substituent already in the ring then there are three possible products:

1,2- (*ortho*) 1,3- (*meta*) 1,4- (*para*)

In practice all three isomers are formed, but their proportions depend very much on the nature of the group X.

Consider first the nitration of methylbenzene. Attack by NO_2^+ can give three different Wheland intermediates:

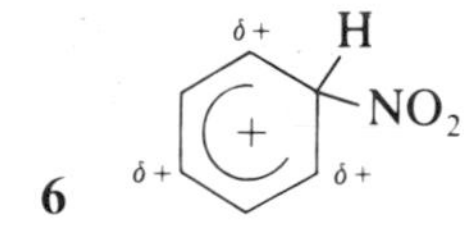

6

The 'horseshoe' representation hides the fact that the positive charge is not shared equally by all five carbon atoms. It is concentrated mainly on the alternate carbon atoms *ortho* and *para* to the position of electrophilic attack (**6**). If one of these charged positions bears an electron-donating methyl group the cation is stabilized, just as in the SN1 reactions of halogenoalkanes, described in Section 28.3, and in additions to alkenes (Markovnikov's rule – see Section 33.3). The *ortho* and *para* Wheland intermediates are thus of lower energy than the *meta*:

In turn, the activated complexes formed *en route* to the *ortho* and *para* intermediates are of lower energy than that for the *meta* reaction (see Figure 34.7), because the charge distribution in the activated complexes is similar to that in the intermediates. Therefore, the *ortho* and *para* products are formed faster than the *meta* compound, and predominate over it.

63% 3% 34%

The temperature at which the reaction takes place (about 25 °C) is lower than that needed to nitrate benzene at a comparable rate (about 50 °C). This is a second consequence of electron donation from the methyl group. The stabilization of the activated complex lowers the activation energy (see Figure 34.7, and compare Section 14.4) and accelerates the reaction.

This example can be generalized. Any *electron-donating* substituent attached to the benzene ring will *accelerate* electrophilic substitution relative to the reaction of unsubstituted benzene and will give mainly *2- and 4-substituted products* (*ortho* and *para*). Alkyl groups are the commonest electron donors. This is why the Friedel–Crafts alkylation reaction is difficult to control. An alkyl group entering the benzene ring makes it more reactive, so that further

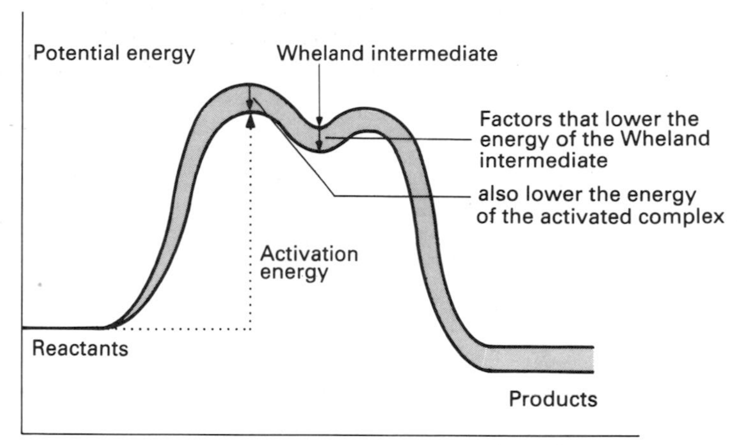

Figure 34.7 Changes in potential energy during electrophilic aromatic substitution

reaction tends to occur by a second attack on the alkylated ring rather than by alkylation of unreacted benzene molecules.

The effect of an *electron-attracting* substituent is precisely the opposite. The withdrawal of electrons from the positively charged ring of the Wheland intermediate, or from the partly charged ring of the activated complex, raises its energy and slows the reaction. The *ortho* and *para* positions, bearing the partial positive charges, are most susceptible, so that reaction occurs mainly at the *meta* position.

$$-\overset{|}{\underset{|}{C}}-\overset{+}{N}\genfrac{}{}{0pt}{}{\cdots O^{1/2-}}{\cdots O^{1/2-}}$$

nitro group, isoelectronic with

$$-\overset{|}{\underset{|}{C}}-C\genfrac{}{}{0pt}{}{\cdots O^{1/2-}}{\cdots O^{1/2-}} \qquad \mathbf{7}$$

Several examples of electron-attracting groups are discussed in the next chapter, but we may illustrate this behaviour by considering the effect of the nitro group (7). This attracts electrons because its nitrogen atom bears a positive charge (with a counterbalancing negative charge shared between the oxygen atoms – it is isoelectronic with the carboxylate group). If benzene is nitrated at 100 °C the main product is 1,3-dinitrobenzene: the second attack is *meta* to the first and the temperature must be higher than the 50 °C needed for mononitration.

$$C_6H_6 \xrightarrow[\text{conc. }(HO)_2SO_2,\ 100\,^\circ C]{\text{conc. }HONO_2} C_6H_4(NO_2)_2 \ (1,3\text{-dinitrobenzene})$$

The following table summarizes the general effects of substituents: further specific examples are discussed in the next chapter.

Substituents	Major products	Rate (relative to that of benzene)
electron-donating	1,2- (*ortho*) + 1,4- (*para*)	fast
electron-attracting	1,3- (*meta*)	slow

Reactions of alkyl side chains

Alkyl groups attached to a benzene ring react in general in the same way as alkanes do. When ethylbenzene is treated with chlorine or bromine under ultraviolet irradiation, *without* a Lewis acid catalyst, substitution occurs in the CH_2 group. The reaction is usually carried out by passing the halogen into the refluxing arene:

$$C_6H_5{-}CH_2{-}CH_3 + Cl_2 \xrightarrow[\text{reflux}]{\text{u.v.}} C_6H_5{-}\underset{\underset{\displaystyle Cl}{|}}{C}H{-}CH_3 + HCl$$

(1-chloroethyl)benzene

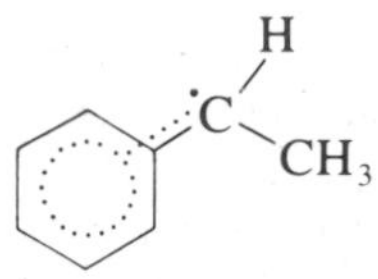

The mechanism, like that of the chlorination of alkanes discussed in Section 27.3, is a radical chain reaction. Alkanes give a mixture of products, with reaction occurring indiscriminately at different sites; alkylbenzenes react specifically at the site next to the benzene ring, however, because the radical intermediate is stabilized by delocalization of the unpaired electron with the π-orbitals of the ring (**8**).

8

Further reaction replaces the second hydrogen atom of the CH_2 group:

$$C_6H_5{-}CH(Cl){-}CH_3 + Cl_2 \xrightarrow[\text{reflux}]{\text{u.v.}} C_6H_5{-}CCl_2{-}CH_3 + HCl$$

(1,1-dichloroethyl)benzene

Under the same conditions methylbenzene reacts with chlorine or bromine to substitute one, two or three hydrogen atoms (the extent of reaction is determined from the increase in weight of the reaction mixture):

$$C_6H_5{-}CH_3 \xrightarrow[\text{u.v.}]{Cl_2} C_6H_5{-}CH_2Cl + HCl \xrightarrow[\text{u.v.}]{Cl_2} C_6H_5{-}CHCl_2 + HCl \xrightarrow[\text{u.v.}]{Cl_2} C_6H_5{-}CCl_3 + HCl$$

The benzene ring strongly resists oxidation, but the side chain is readily oxidized by refluxing with potassium manganate(VII) in either acidic or alkaline solution, or by acidic potassium dichromate(VI). *Regardless of the nature of the alkyl side chain*, the product of reaction is benzoic acid (see Section 35.1):

$$C_6H_5{-}R \xrightarrow[\text{reflux}]{MnO_4^-(aq)} C_6H_5{-}COOH$$

EXAMPLE

Two isomeric hydrocarbons, **A** and **B**, C_8H_{10}, are oxidized with potassium manganate(VII) under reflux. The product from **A** has the molecular formula $C_7H_6O_2$. **B** gives **C**, $C_8H_6O_4$. When **C** is strongly heated it is dehydrated to give an acid anhydride, $C_8H_4O_3$. Deduce the structures of **A**, **B** and **C**.

METHOD Decide what structural difference causes **A**, but not **B**, to lose a carbon atom on oxidation. Then consider the structural feature that allows **C** to form an acid anhydride.

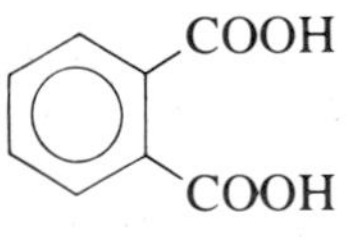

ANSWER **A** must be ethylbenzene, $C_6H_5{-}CH_2{-}CH_3$, which loses the CH_3 group on oxidation, giving $C_6H_5{-}COOH$. In **B** the two carbon atoms outside the benzene ring must both be methyl groups: oxidation gives a dicarboxylic acid. The formation of the anhydride shows that the two carboxylic groups must be adjacent (*ortho*), and that **B** is therefore 1,2-dimethylbenzene.

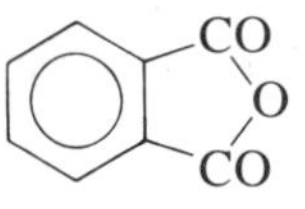

9 (1,2-$C_6H_4(COOH)_2$)

10 (1,2-$C_6H_4(CO)_2O$)

COMMENT Acid **C** is commonly called phthalic acid (**9**). Phthalic anhydride (**10**) is made industrially by catalytic air-oxidation of 1,2-dimethylbenzene and is used for cross-linking alkyd resins and to make phthalate ester plasticizers for PVC (see Section 29.6).

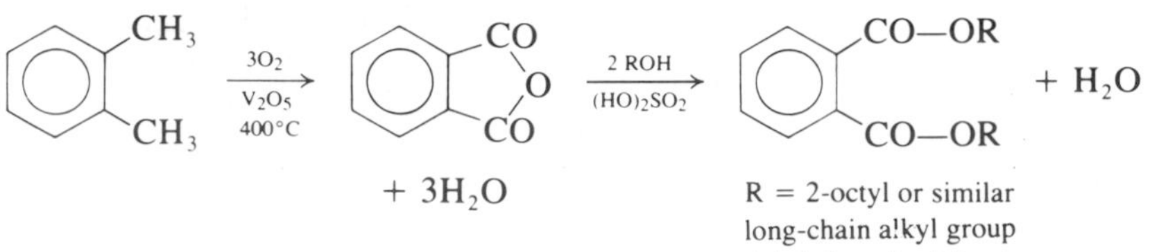

$$C_6H_4(CH_3)_2 \xrightarrow[V_2O_5,\ 400°C]{3O_2} C_6H_4(CO)_2O + 3H_2O \xrightarrow[(HO)_2SO_2]{2\,ROH} C_6H_4(CO{-}OR)_2 + H_2O$$

R = 2-octyl or similar long-chain alkyl group (phthalate plasticizer)

34.4 Preparative methods for arenes

The simple arenes are not normally obtained by laboratory preparations. For a long time they were obtained from coal tar, of which benzene and methylbenzene are major components. With the change from coal gas to natural gas as a domestic and industrial fuel and with the decline in the amount of coke used for steelmaking, this source has for the time being become much less important. Today benzene and the simple alkylbenzenes come mainly from the catalytic reforming of petroleum fractions, discussed in Section 27.4.

34.5 The importance of arenes

The main uses of arenes are for the preparation of other aromatic compounds, as discussed in the next chapter.

Summary

- ☐ Compounds containing the benzene ring are called aromatic.
- ☐ Aromatic hydrocarbons are called arenes.
- ☐ Disubstituted benzenes with groups at positions 1,2-, 1,3- and 1,4- are called *ortho*, *meta* and *para* compounds respectively.
- ☐ The group C_6H_5— is the phenyl group.
- ☐ The benzene ring is stabilized by electron delocalization.
- ☐ Arenes undergo a few addition reactions, for example, with hydrogen and a nickel catalyst or with chlorine in bright sunlight or under ultraviolet irradiation, giving cyclohexane derivatives.
- ☐ The characteristic reaction of arenes is electrophilic substitution, in which the stability of the benzene ring is retained in the products. Common examples are as follows:

	—H replaced by:
chlorination and bromination	—Cl, —Br
Friedel–Crafts acylation	—COR
Friedel–Crafts alkylation	—R
nitration	$—NO_2$
sulphonation	$—SO_2OH$

- ☐ Electrophilic substitution into a ring that bears an electron-donating substituent, such as an alkyl group, gives mainly the *ortho* and *para* products. The reaction is faster than that of benzene.
- ☐ Electrophilic substitution into a ring that bears an electron-attracting substituent, such as a nitro group, gives mainly the *meta* product. The reaction is slower than that of benzene.
- ☐ Alkyl side chains undergo chlorination and bromination, via a radical chain mechanism, at the position next to the benzene ring.
- ☐ Arenes with alkyl side chains are oxidized by potassium manganate(VII) in acid or alkali, or by acidic potassium dichromate(VI), to benzoic acid.

PROBLEMS

1 Describe two characteristic features of the shape of the benzene molecule. Give two pieces of experimental evidence, other than the shape of the molecule, which indicate that benzene does not have the structure of 'cyclohexatriene', with alternating single and double bonds.

2 Give the names and structures of all the arenes of molecular formula C_8H_{10}.

3 Describe the preparation of chlorobenzene from benzene, stating the reaction conditions. Describe the mechanism of the reaction. Under what conditions does benzene react with chlorine to give 1,2,3,4,5,6-hexachlorocyclohexane (benzene hexachloride)?

4 Describe the conditions, mechanism and products of the reaction of ethanoyl chloride with methylbenzene. What is this reaction called?

5 Explain why the mononitration of methylbenzene gives a mixture of two main products, whereas the mononitration of nitrobenzene gives only one. Describe the conditions under which the reactions are carried out.

6 Under what conditions does chlorine react with the ethyl group of ethylbenzene, and under what conditions does it react with the benzene ring? Account for the difference. State the product(s) of the two types of reaction.

7 (a) How and under what conditions does chlorine react with (i) benzene, (ii) methylbenzene? Your answer should include some explanation of the mechanisms by which the reactions might take place.
(b) Give a simple test-tube reaction by which propan-1-ol may be distinguished from propan-2-ol.
(c) Give one way in which 2-chloropropane differs *chemically* from chlorobenzene, suggesting a reason for this difference in behaviour. (*AEB 1986*)

8 Give an account of the nitration of benzene. Your account should include the essential conditions for the reaction as well as an indication of the mechanism of the reaction.

In which *two* ways does the nitration of methylbenzene (toluene) differ from the nitration of benzene? (Mechanisms are not required here.)

Giving the reaction sequences required in the form of equations and the essential conditions and reagents for each, indicate how benzene may converted into
(a) phenylamine (aniline),
(b) benzonitrile,
(c) (phenylmethyl)amine, $C_6H_5CH_2NH_2$.
(*Oxford and Cambridge*)

9 (a) In what respects does the mononitration of nitrobenzene differ from the mononitration of methylbenzene (toluene)?
(b) The mononitration of nitrobenzene yields a product, $C_6H_4N_2O_4$, which is reduced by ammonium sulphide to compound **A**, $C_6H_6N_2O_2$. On treatment with aqueous nitrous acid, in aqueous hydrochloric acid at 0 °C, **A** gives a solution, **B**, which, with KCN/CuCN, yields compound **C**, $C_7H_4N_2O_2$. Solution **B**, when boiled, gives compound **D**, $C_6H_5NO_3$. Solution **B** also gives strongly coloured precipitates when poured into alkaline solutions of phenols.
Giving your reasons, suggest structures for compounds **A**, **C** and **D**.
(c) Give the name and the structural formula of the organic species in solution **B**. Suggest *two* reasons why such solutions are important. (*Oxford and Cambridge*)

10 Compound **A** ($C_{10}H_{10}$) gives a precipitate with ammoniacal silver nitrate solution. **A** reacts with CH_3MgBr in ether to give off a gas. Addition of propanone to the ethereal mixture followed by water gives **B**. **B** with one mole of H_2 in the presence of Pt gives **C**, which with another mole of H_2 gives **D**. **D** heated with hot concentrated acid gives **E**. **E** with one mole of H_2 in the presence of Pt gives **F** ($C_{13}H_{20}$). **C** with hot concentrated acid gives **G**. **G** will take up two moles of H_2 to give **F**. **G** with one mole of dry HCl gives **H** as the major product and a small amount of an isomer **I**.

A reacts with two moles of H_2 in the presence of Pt to give **J**. Nitration of **J** gives a single mononitro-derivative. Write structural formulae for compounds **A** to **I** and comment on any reactions you find interesting.
(*Oxford Entrance*)

35 Aromatic compounds with functional groups

When a functional group is attached to a benzene ring there is an interaction between the π-electrons of the ring and lone pair or π-electrons of the functional group. This modifies the reactions both of the group and of the ring. For example, when a hydroxyl group is attached to a benzene ring the resulting phenol is much more acidic than an aliphatic alcohol, and the benzene ring is much more reactive in electrophilic substitutions than is benzene itself.

35.1 Nomenclature

Aromatic compounds containing alkyl, halogeno and nitro groups are named in a similar manner to the related aliphatic compounds, as in the following examples:

CH_3 Cl NO_2 Cl Cl

methylbenzene (toluene) chlorobenzene nitrobenzene 1,3-dichlorobenzene (*meta*-dichlorobenzene)

Different substituents in the same ring are named in alphabetical order: for example, 'methyl' comes before 'nitro-' in

CH_3 O_2N NO_2 NO_2

2-methyl-1,3,5-trinitrobenzene (trinitrotoluene: 'TNT')

Aromatic amines are also named in the same way as aliphatic ones:

NH_2

phenylamine

(Phenylamine is still often called by its traditional name **aniline**, however.)

The compound C_6H_5OH is not 'benzenol': it is called **phenol** (see Box 34.1). Molecules with more than one hydroxyl group are named systematically, however:

OH OH NO_2 OH OH

phenol 2-nitrophenol (*ortho*-nitrophenol) benzene-1,2-diol

Aromatic compounds containing the functional groups —CHO, —CN and —COOH (and the carboxylic acid derivatives) are named differently from aliphatic ones. Whereas ethanoic acid has the same carbon skeleton as the parent hydrocarbon ethane, the compound C_6H_5COOH has one carbon atom more than benzene has and the name must indicate this, as shown below. The older, non-systematic names shown in parentheses are still very commonly used: we shall use them in this chapter.

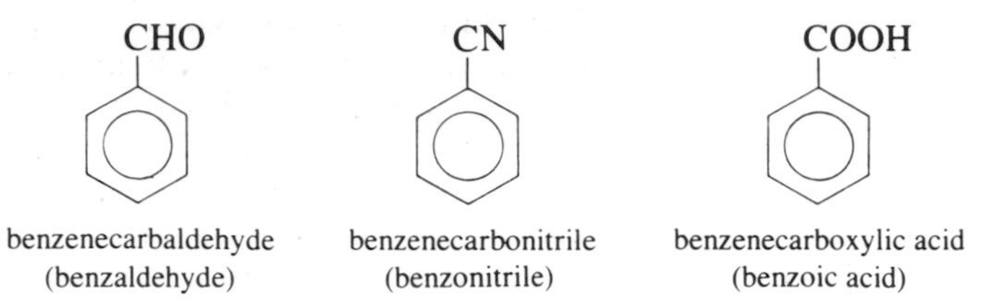

benzenecarbaldehyde (benzaldehyde) | benzenecarbonitrile (benzonitrile) | benzenecarboxylic acid (benzoic acid)

Aromatic compounds with carbon side chains can be named either as substituted benzenes or as phenyl-substituted aliphatic compounds. The name usually used is the one that emphasizes the common chemical behaviour of the compound. Thus the name **methylbenzene** – rather than phenylmethane – emphasizes that most of the common reactions of the compound involve the benzene ring.

CH_3

methylbenzene
(rather than phenylmethane)

On the other hand when a benzene ring is attached to a carbon–carbon double bond the name used indicates that the common reactions are those of the alkene part of the molecule:

$CH{=}CH_2$

phenylethene
(rather than ethenylbenzene)

(The traditional name of phenylethene is **styrene**.)

Similarly, ketones with a phenyl group attached to the carbonyl group are named as in the following example:

$CO{-}CH_3$

phenylethanone
(rather than ethanoylbenzene)

'Phenylethanone' is correct, even though CH_3CHO is ethanal, not ethanone: the compound above is a ketone, so the appropriate ending is *-one*.

This convention seems to imply that one must know about the reactivity of a compound before it can be named. In practice, either name is acceptable.

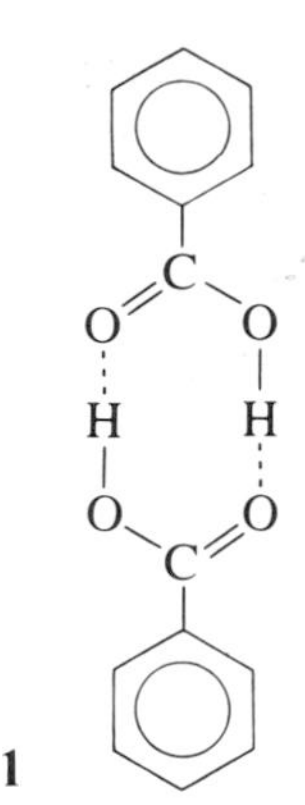

35.2 Physical properties

Most aromatic compounds that have a single small group attached to the benzene ring are liquids at room temperature. Two important exceptions are the solids phenol, C_6H_5OH ($t_m = 43\,^\circ C$) and benzoic acid, C_6H_5COOH ($t_m = 122\,^\circ C$), in both of which the molecules are held together by hydrogen bonding.

Above its melting point, phenol is partially miscible with water (see Section 7.1), because of hydrogen bonding. Similarly phenylamine is sparingly soluble in water. Compounds without hydrogen bonding groups are insoluble in water. Benzoic acid dissolves in hot water but is much less soluble in the cold, partly because the acidic hydrogen atoms are occupied in hydrogen bonding that holds two molecules together (**1** – see also Section 31.2) and are unavailable for hydrogen bonding to solvent molecules. Most aromatic compounds are soluble in organic solvents such as methylbenzene because the mutual dispersion forces are favourable.

All aromatic compounds absorb ultraviolet light (see Section 3.2), which can excite an electron in the delocalized π-system to a higher level. When several chromophores are attached to a benzene ring, the excitation energy is lower and can be supplied by photons of visible light: as a result the compounds are coloured. Most dyestuffs are aromatic compounds (see Box 35.3). Some simple aromatic compounds, colourless when pure, are readily oxidized to coloured products; the presence of these as trace impurities colours the material in 'off the shelf' bottles. Phenol, for example, is often pink and phenylamine may be quite dark brown.

35.3 Reactions of the benzene ring

The chemistry of aromatic compounds is largely controlled by the interaction between the groups attached to the benzene ring and the delocalized π-electrons. These act like a reservoir, accepting electrons from electron-donating groups and providing a source for electron attracting groups. We need to consider the consequences of this interaction under two headings: the effect of the functional group on the reactions of the benzene ring, and (in the following section) the effect of the ring on the reactions of the functional group.

In Section 34.3 we saw that electron-donating groups accelerate electrophilic substitution reactions of the benzene ring and lead mainly to *ortho* and *para* substitution and that, conversely, electron-attracting groups slow down substitution and lead mainly to *meta* substitution. Alkyl groups attached to the benzene ring are, as we have seen, electron-donating. So are carbon–carbon double bonds, but phenylethene reacts with electrophiles at the double bond rather than undergoing substitution in the ring.

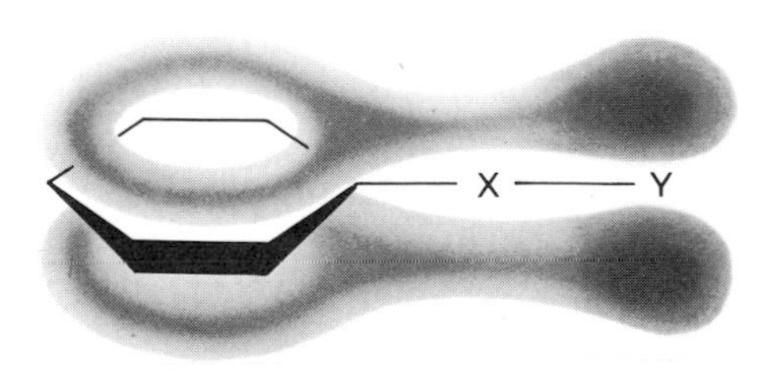

Figure 35.1 π-electron distribution in aromatic compounds with double-bonded functional groups

In general all other common multiply bonded functional groups, —X=Y, are electron-attracting, as illustrated in Figure 35.1. Atom Y is more electronegative than X, as in the carbonyl group, so that the bonding electrons are attracted away from the ring. Compounds with these groups attached to the benzene ring follow the pattern described above: they undergo electrophilic substitution more slowly than does benzene (or require higher reaction temperatures to go at the same rate) and they give the *meta* product, as in the following examples.

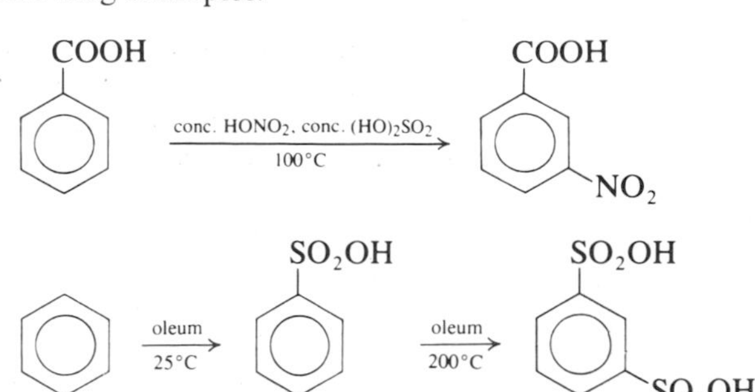

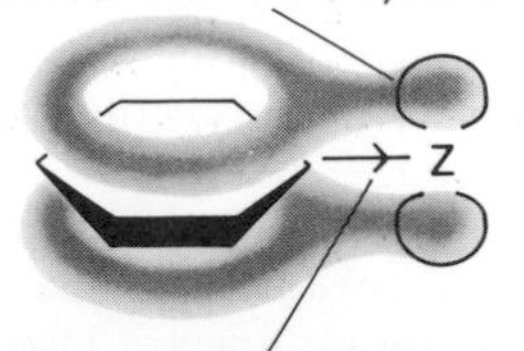
Lone pair electrons feed into π-system
Z
Electronegative Z atom withdraws electrons from the benzene ring

Figure 35.2 π-electron distribution in aromatic amines, phenols and halogen compounds

The deactivating effect of these groups is enough to prevent some substitution reactions (for example, the Friedel–Crafts alkylations and acylations discussed in Section 34.3) from occurring at all. Indeed, nitrobenzene is often used as a solvent for Friedel–Crafts reactions.

Hydroxy and amino groups are electron-donating towards the benzene ring. Although they are more electronegative than carbon, and therefore attract electrons more strongly, they also carry lone pair electrons which can be fed back into the delocalized π-ring (see Figure 35.2). A total of eight electrons, six from the benzene ring and the two lone pair electrons, are delocalized over seven atoms. The benzene ring has more than its original six and the donor atom is left with less than its original two. Consequently, both phenol and phenylamine undergo electrophilic substitution reactions very rapidly (much more rapidly than methylbenzene does) to give *ortho* and *para* substituted products. In fact, they are so reactive that bromine water is rapidly decolorized (no catalyst is needed) and it is hard to stop the reaction short of replacing all three *ortho* and *para* hydrogen atoms.

$$C_6H_5OH \xrightarrow{3Br_2(aq)} C_6H_2Br_3OH + 3HBr$$

2,4,6-tribromophenol

The white precipitate formed with phenol has a characteristic 'antiseptic' smell: the chlorine analogue, which smells similar, is TCP.

The patterns of electrophilic substitution are summarized in the following table.

Group attached to ring	Main products	Rate of reaction relative to benzene
NH_2, OH	*ortho* and *para*	much faster
R	*ortho* and *para*	faster
F, Cl, Br, I	*ortho* and *para*	slower (see Box 35.1)
NO_2 $COCH_3$, CHO COOH, COOR CN SO_2OH	*meta*	much slower

BOX 35.1

Electrophilic substitution of halogenobenzenes

In the interaction between ring and substituent (see Figure 35.2), the balance between electron donation and electron attraction depends on the electronegativity of the atom bonded to the ring and on its size. In phenylamine the weakly electronegative nitrogen atom neither attracts electrons strongly from the ring, nor holds tightly to its own lone pair: electron donation is therefore dominant. In phenol electron attraction by the oxygen atom is greater, but electron donation still outweighs it.

The halogenobenzenes behave differently (see Figure 35.3). Consider first fluorobenzene: the electronegative fluorine atom attracts electrons strongly and also holds tightly on to its lone pairs. Electron attraction and donation are finely balanced. The outcome is that fluorobenzene is less reactive than benzene in electrophilic substitution reactions, but nevertheless gives mainly *ortho* and *para* products. (Electron attraction by the fluorine atom raises the energy of all three activated complexes – *ortho*, *meta* and *para*, see Section 34.3 – but the feedback of lone pair electrons lowers the energy of the *ortho* and *para* activated complexes relative to the *meta*.) In the halogen series the electronegativity falls in the order fluorine > chlorine > bromine > iodine, but because of the increasing size of the atoms the overlap between the lone pairs and the π-system also falls off. Coincidentally, the balance between electron donation and attraction remains about the same: all the halogenobenzenes react in the same unusual manner, more slowly than benzene but giving *ortho* and *para* products.

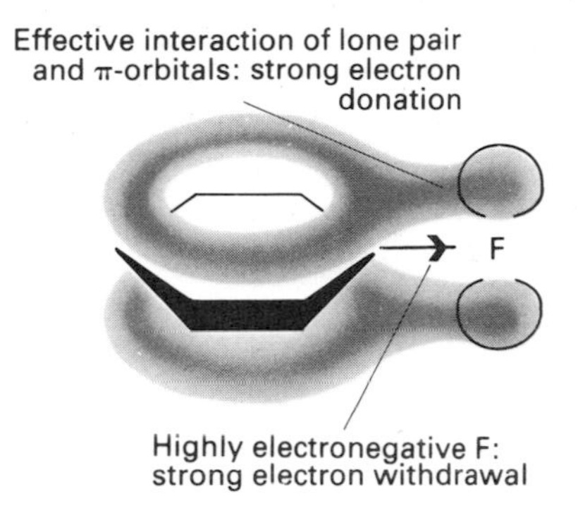
Effective interaction of lone pair and π-orbitals: strong electron donation
F
Highly electronegative F: strong electron withdrawal

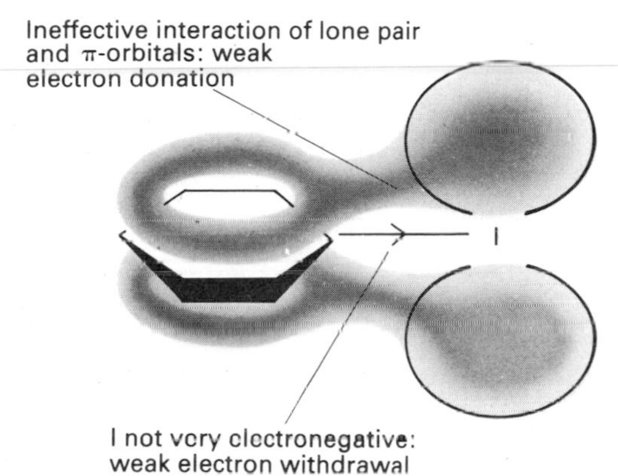
Ineffective interaction of lone pair and π-orbitals: weak electron donation
I
I not very electronegative: weak electron withdrawal

Figure 35.3 Balance of strong electron donation and attraction in fluorobenzene, and of weak electron donation and attraction in iodobenzene

EXAMPLE

Suggest reaction sequences for the preparation of the three isomers of chloronitrobenzene, starting from benzene.

METHOD Consider ways of introducing the chloro and nitro groups into the benzene ring, and decide in which order to carry out these reactions to obtain the required isomers.

ANSWER The chloro group can be introduced using chlorine with aluminium chloride as a catalyst. The nitro group can be introduced using concentrated nitric and sulphuric acids. If the chlorination is done first, the resulting chlorobenzene undergoes nitration in the *ortho* and *para* positions, and the two isomers may then be separated by, for example, chromatography. If the nitration is done first, subsequent chlorination gives the *meta* isomer.

COMMENT Where a mixture of *ortho* and *para* products is formed there is no simple way of carrying out the reaction so as to produce only one of the isomers. The mixture has to be separated by chromatography, fractional distillation or some other appropriate method.

BOX 35.2

OH

$NHCOCH_3$

2

OCH_2CH_3

$NHCOCH_3$

3

GO

$CH_2—CH_2$

$N—CH_3$

O

G′O

morphine: G=H; G′=H
codeine: G=H; G′=CH_3
4 heroin: G=CH_3CO; G′=CH_3CO

Analgesics

Aspirin is probably the best known of the pain-killing drugs (**analgesics**). It is also antipyretic (reduces fever) and anti-inflammatory (eases rheumatic and arthritic inflammations). Other drugs with similar properties include paracetamol (**2**) and phenacetin (**3**). They are examples of **proprietary** drugs that can be bought over the counter without a doctor's prescription, though some of them would probably not be allowed on the market if they were only newly discovered: indeed phenacetin, which is a suspected carcinogen and can cause serious kidney damage, is no longer sold in the UK. Most proprietary drugs can be bought either as a brand-named product (such as 'Aspro') or, more cheaply, as an unbranded drug; these are usually designated in the UK by the letters BP, which stand for the reference book called *The British Pharmacopeia.*

The most effective pain-killers so far discovered are the **narcotic** (sleep-inducing) drugs related to morphine (**4**), a constituent of opium. These have major side effects on brain function, however, notably their dangerous addictiveness. Such drugs are therefore kept under strict control and administered only under medical supervision.

Figure 35.4 Drug manufacture in ultra-sterile conditions

35.4 Reactions of the functional groups

Halogenobenzenes

The interaction between a lone electron pair of the halogen and the π-electrons of the benzene ring (see Figure 35.2) strengthens the carbon–halogen bond relative to that in halogenoalkanes (bond dissociation enthalpies: CH_3—Cl, 350 kJ mol^{-1}; C_6H_5—Cl, 400 kJ mol^{-1}). The result is that halogenobenzenes are much less reactive compounds; in marked contrast to the halogenoalkanes, they do *not* normally undergo nucleophilic substitution reactions – for example, with aqueous ammonia or sodium hydroxide. This is one reason why DDT is so stable (see Box 34.2).

Phenol

Because of its hydroxyl group, phenol has many reactions in common with alcohols. For example, it forms esters with acid chlorides or anhydrides (though *not* usually with carboxylic acids under sulphuric acid catalysis). For example, the phenol 2-hydroxybenzoic acid, commonly known as salicylic acid, is ethanoylated to give **aspirin** (see Box 35.2).

$$(CH_3CO)_2O + \text{HO}-C_6H_4-\text{COOH} \longrightarrow \text{CH}_3\text{CO}-\text{O}-C_6H_4-\text{COOH} + CH_3COOH$$

ethanoic anhydride

2-ethanoyloxybenzoic acid (acetylsalicylic acid, aspirin)

In phenol, as in the halogenobenzenes, the interaction between a lone electron pair and the ring π-system strengthens the carbon–oxygen bond. Consequently phenols *cannot be converted into halogenobenzenes* by reaction with phosphorus halides or similar reagents.

Phenols are *much more acidic than alcohols*, though less acidic than carboxylic acids (compare Table 35.1). Phenol itself is caustic, and contact with the skin should be avoided. It is insoluble in water at room temperature, but gives a clear solution in alkali by forming the **phenoxide** ion:

Table 35.1 pK_a values of three hydroxy compounds

CH_3—OH	15.2
C_6H_5—OH	9.89
CH_3—COOH	4.75

$$C_6H_5OH + NaOH \longrightarrow Na^+\,{}^-O-C_6H_5 + H_2O$$

sodium phenoxide

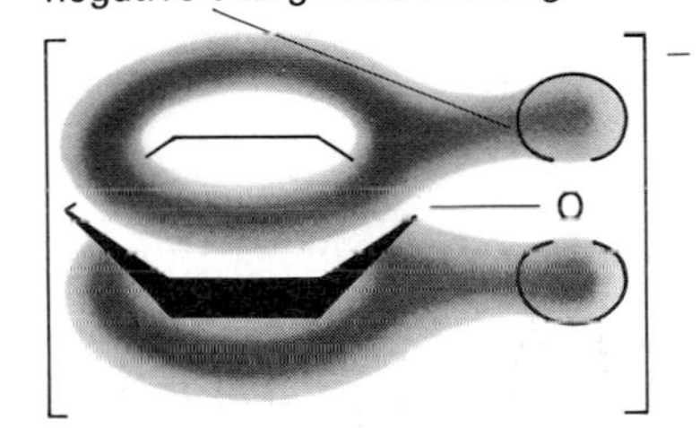
Electron delocalization disperses negative charge into the ring
O
−

Figure 35.5 Delocalization stabilization of the phenoxide anion

The difference in acid strength between phenols and carboxylic acids can be used in a quick diagnostic test: carboxylic acids are acidic enough to release carbon dioxide from a solution of sodium carbonate or sodium hydrogencarbonate, phenols are not.

The acidity of phenols has a similar origin to that of carboxylic acids, discussed in Section 31.3. The negative charge in the anion can be dispersed because of the interaction with the π-electron ring – see Figure 35.5. This stabilizes the anion, making the ionization relatively more favourable. Compare the following:

$$CH_3OH + H_2O \rightleftharpoons H_3O^+ + CH_3O^-$$

For alcohols, this equilibrium lies far to the left.

$$C_6H_5OH + H_2O \rightleftharpoons H_3O^+ + C_6H_5O^-$$

delocalization

Stabilization of the phenoxide ion means that the equilibrium, while still well to the left, is less so than for alcohols.

$$CH_3-C(=O)OH + H_2O \rightleftharpoons H_3O^+ + CH_3-CO_2^-$$

The carboxylate ion is still more stabilized, with the charge shared between the two electronegative oxygen atoms.

The dispersal of the charge into the benzene ring means that the acid strength of a phenol is very sensitive to the nature of other groups in the ring. Electron-attracting groups further disperse the charge and increase the acid strength: electron-donating groups have the opposite effect.

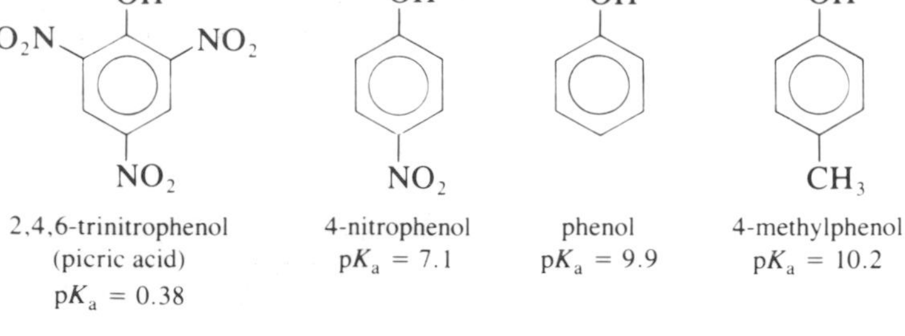

4-Nitrophenol is used as an acid–base indicator. The phenol itself is almost colourless; its anion is intensely yellow.

Many phenols form coloured (often violet) complexes with neutral iron(III) chloride. This provides a simple diagnostic test for phenols.

Amines

Phenylamine is a base: it forms a clear solution in dilute hydrochloric acid, though it is only sparingly soluble in water. However, just as phenol is a stronger acid than aliphatic alcohols, so phenylamine ($pK_b = 9.4$) is a *weaker base* than aliphatic amines ($pK_b \approx 3$). The extended delocalization of the nitrogen lone pair and the ring π-electrons stabilizes the free amine (see Figure 35.6), so that in the reaction

$$C_6H_5NH_2 \text{ (delocalization)} + H_3O^+ \rightleftharpoons C_6H_5NH_3^+ + H_2O$$

stabilization shifts the equilibrium further to the left of its position for aliphatic amines. In the ammonium salt there is no lone pair, so no such delocalization is possible.

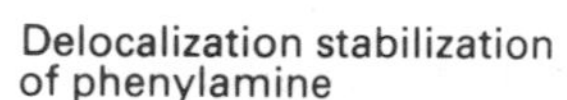

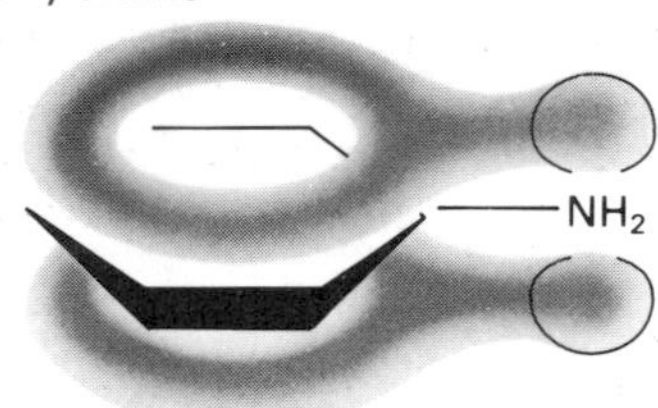

No lone pair electrons:
no delocalization stabilization

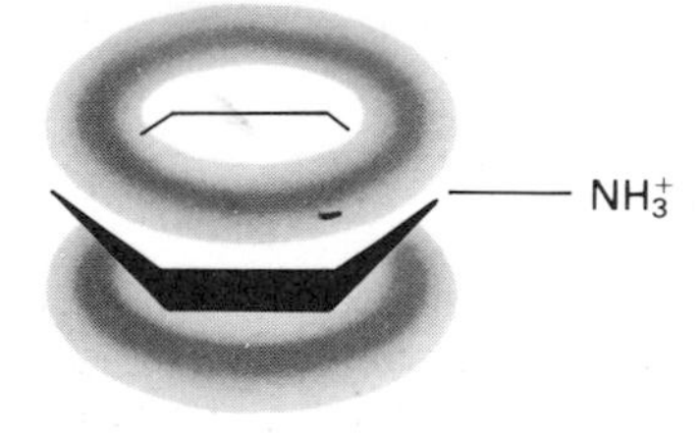

Figure 35.6 Delocalization stabilization of phenylamine is absent from the phenylammonium cation

Phenylamine reacts like aliphatic amines, though more slowly, with halogenoalkanes to give secondary and tertiary amines and quaternary ammonium salts (compare Section 32.3). Similarly, acid chlorides give solid amides, which can be used to characterize the amines:

$$C_6H_5{-}NH_2 + CH_3COCl \rightarrow C_6H_5{-}NHCOCH_3 + HCl$$

An important difference between aromatic and aliphatic amines is their reaction with nitrous acid (HONO, nitric(III) acid), made in the reaction mixture from aqueous sodium nitrite and dilute sulphuric acid, which is also used to dissolve the amine. Under these conditions, primary aliphatic amines give molecular nitrogen together with a variety of organic products (see Section 32.3). Primary aromatic amines react at low temperature (0 °C) to give **diazonium salts**, in a procedure called **diazotization**:

$$C_6H_5NH_2 + HONO + (HO)_2SO_2 \xrightarrow{0\,°C} C_6H_5N^+{\equiv}N\ HOSO_3^- + 2H_2O$$

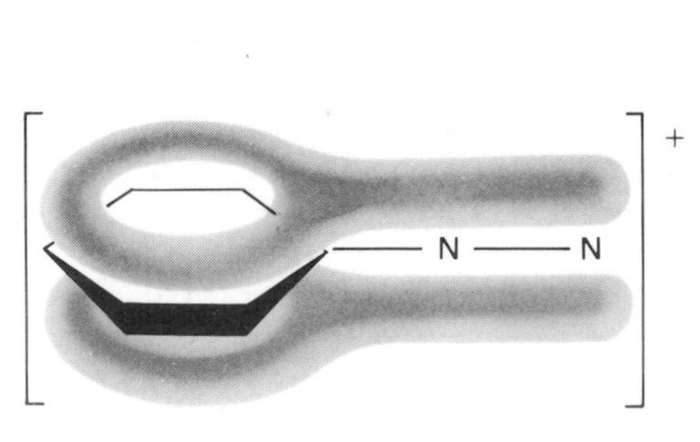

Figure 35.7 Delocalization stabilization of the aromatic diazonium ion

The difference between aromatic and aliphatic amines in their reactions with nitrous acid is again due to electron delocalization. Aliphatic diazonium salts are very unstable and decompose rapidly. In the aromatic diazonium ion the π-electrons of the nitrogen–nitrogen bond are delocalized with those of the ring (see Figure 35.7) and the salts are less unstable.

Diazonium ions are key intermediates in making a wide range of other aromatic compounds. They are easily made via the following reaction sequence:

$$C_6H_6 \xrightarrow[\text{conc. }HONO_2,\ \text{conc. }(HO)_2SO_2,\ 50°C]{\text{nitration}} C_6H_5NO_2 \xrightarrow[\text{Sn/conc. HCl (Section 35.5)}]{\text{reduction}} C_6H_5NH_2 \xrightarrow[NaNO_2,\ \text{dil. }(HO)_2SO_2,\ 0°C]{\text{diazotization}} C_6H_5N^+{\equiv}N$$

The $—N^+{\equiv}N$ group can be replaced by other groups in a reaction that is effectively a nucleophilic substitution (though the actual mechanism varies from one reaction to another and is often complicated):

$$C_6H_5N^+{\equiv}N + Nu^- \longrightarrow C_6H_5Nu + N_2$$

(The symbol Nu^- for the nucleophile does not imply that it necessarily bears a negative charge.) As we saw above, halogenoarenes do not normally undergo nucleophilic substitutions. The diazonium salts therefore play the role in aromatic chemistry that the halogenoalkanes do in aliphatic chemistry. The following replacement reactions of diazonium salts are important examples of their use.

Replacement by hydroxyl If a solution of a diazonium salt in water is heated, a phenol is formed, and can readily be detected by its smell.

$$C_6H_5N^+{\equiv}N\ HOSO_3^- + H_2O \xrightarrow{50°C} C_6H_5OH + N_2 + (HO)_2SO_2$$

Replacement by chlorine or bromine This reaction requires catalysis by copper(I). It is carried out by heating the diazonium salt solution with copper(I) chloride or bromide (when it is called the **Sandmeyer reaction**) or by using the hydrohalic acid in place of sulphuric acid to make the diazonium salt and heating the resulting solution over finely powdered copper or bronze.

$$C_6H_5N^+{\equiv}N + Cl^- \xrightarrow[\text{CuCl, HCl}]{50°C} C_6H_5Cl + N_2$$

Replacement by iodine does not require a catalyst: aqueous potassium iodide is used.

$$C_6H_5N^+{\equiv}N + I^- \xrightarrow{50°C} C_6H_5I + N_2$$

Replacement by cyanide This again needs copper(I) catalysis, using potassium tetracyanocuprate(I).

$$C_6H_5N^+{\equiv}N + CN^- \xrightarrow[K_3[Cu(CN)_4]]{50°C} C_6H_5CN + N_2$$

benzonitrile

Diazo coupling reactions In these reactions the $—N^+{\equiv}N$ group is not lost but instead acts as the electrophile attacking a second benzene ring; reactions like these occur only with the most reactive aromatic compounds, amines and phenols (see above). The phenols react in alkaline solution, in which they are converted to the strongly nucleophilic phenoxide:

$$C_6H_5N^+{\equiv}N + C_6H_5O^- \xrightarrow[H_2O]{0°C} HOC_6H_4N{=}NC_6H_5$$

4-(phenylazo)phenol
(orange-yellow precipitate)

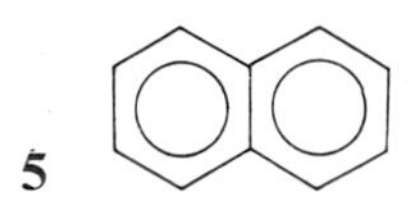
5

The corresponding reaction with naphthalen-2-ol (naphthalene (**5**) is a hydrocarbon with two benzene rings joined together) gives a characteristic scarlet precipitate, insoluble in alkali because the hydroxyl hydrogen atom is tied up in internal hydrogen bonding: the reaction is used as a test for primary aromatic amines.

$$C_6H_5N^+{\equiv}N \xrightarrow[\text{dil. NaOH, } 0°C]{\text{naphthalen-2-ol}} \text{scarlet precipitate}$$

The reaction with amines is exemplified by the preparation of the acid–base indicator methyl orange (see Section 12.3).

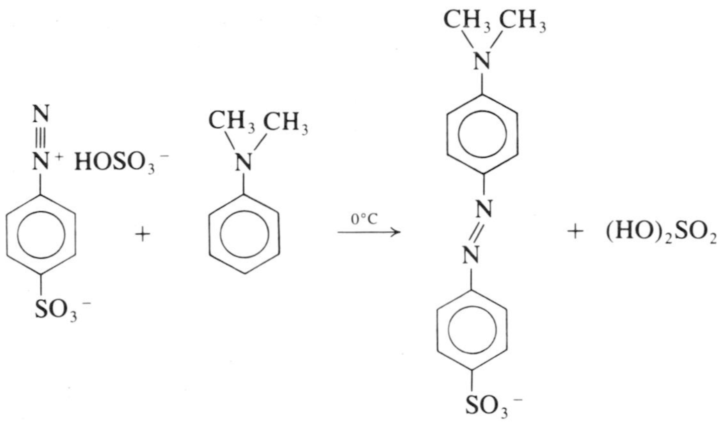

'methyl orange'

All these diazo-coupled products are brightly coloured, and the reactions are the basis of the so-called azo or aniline dyes, discussed in Box 35.3.

BOX 35.3

Dyes

Coloured compounds absorb photons in the visible region of the electromagnetic spectrum: the energy raises an electron to a higher energy level (see Section 3.2). Molecules with extended systems of π-electrons have orbitals of appropriate energies. For example, the orange colour of carrots comes from a group of hydrocarbons known as **carotenes** which have a chain of eleven carbon–carbon double bonds, as in β-carotene:

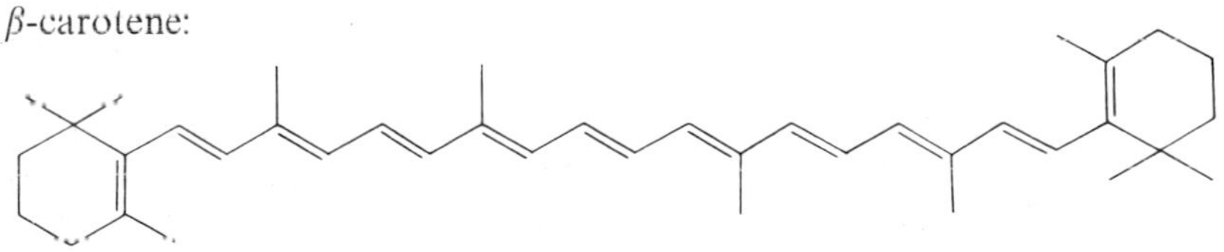

line diagram (see Section 26.3) of β-carotene

Dyes, besides having the desired colour, must be able to be attached to the substance to be dyed. **Vat dyes** are highly insoluble compounds that are formed by a precipitation reaction in the dyeing vat: the particles are physically enmeshed within the fibres. Many dyes, however, are chemically bonded to the dyed material: in **acid dyes**, for example, an acidic group in the dye is used to attach it to basic amino groups in wool or silk, or to esterify hydroxyl groups in the cellulose fibres of cotton.

Azo compounds, containing two benzene rings joined by the —N=N— group to form an extended π-system, are the commonest dyes. They are easily made via the diazo coupling reaction, the starting compounds are cheap, and it is easy to introduce groups to modify the colour and to bond to the material. Because the simplest ones derive from phenylamine (aniline) they are also commonly referred to as 'aniline dyes'.

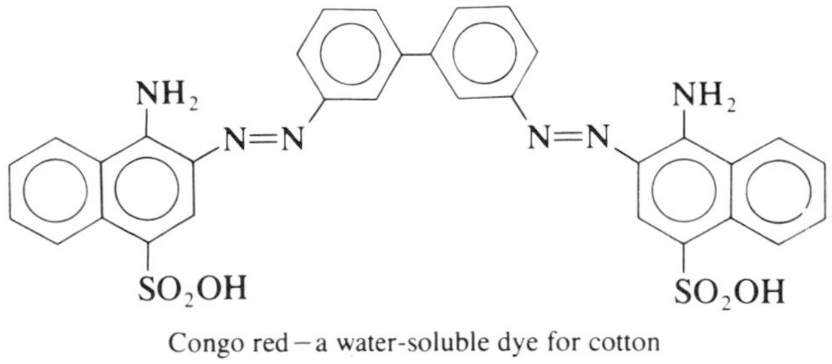

Congo red – a water-soluble dye for cotton

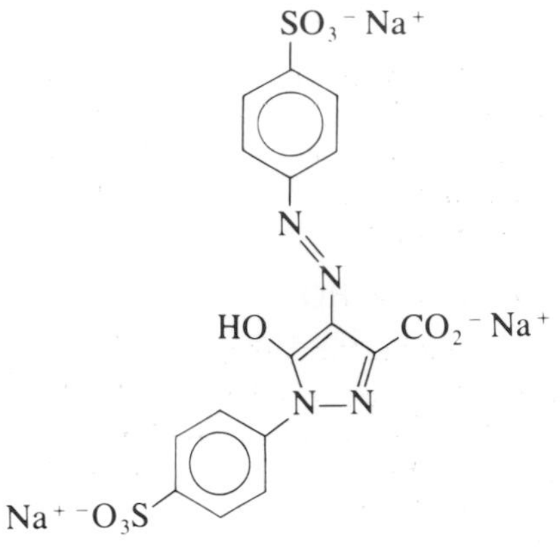

$SO_3^- Na^+$ — N=N — HO — $CO_2^- Na^+$ — N–N — $Na^+ {}^-O_3S$

6

Certain azo dyes are permitted food additives, but there is considerable anxiety that some of them may be dangerous to asthmatics and that they may be a cause of hyperactivity in young children. Tartrazine (**6**), a very widely used orange dye that has the food code E102, is especially suspect.

EXAMPLE

Show how phenylamine may be converted into methyl orange.

METHOD The two compounds needed to make methyl orange are dimethylphenylamine and 4-aminobenzenesulphonic acid (which must then be diazotized). Decide how these can be made from phenylamine.

ANSWER

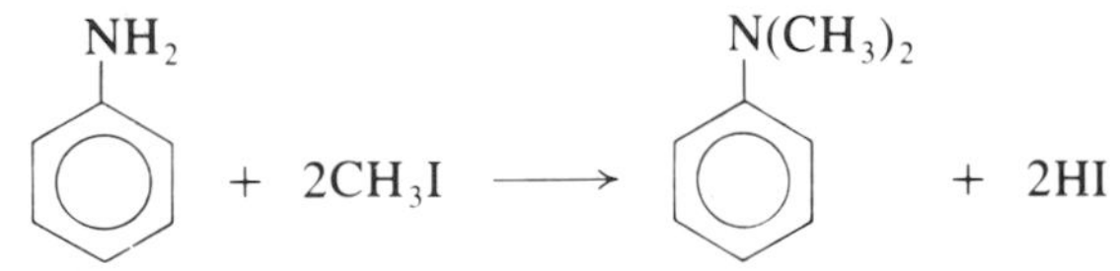

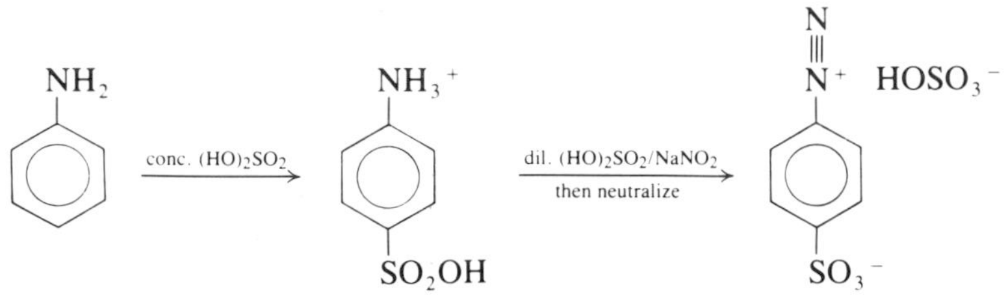

$$C_6H_5NH_2 + 2CH_3I \longrightarrow C_6H_5N(CH_3)_2 + 2HI$$

$$C_6H_5NH_2 \xrightarrow{\text{conc. }(HO)_2SO_2} {}^+NH_3C_6H_4SO_2OH \xrightarrow[\text{then neutralize}]{\text{dil. }(HO)_2SO_2/NaNO_2} N{\equiv}N^+C_6H_4SO_3^- \quad HOSO_3^-$$

The coupling reaction is then performed as shown in the text.

COMMENT Aromatic amines are less basic and less nucleophilic than aliphatic amines: it is easy to stop the reaction with a halogenoalkane at the tertiary amine stage. The yield in the sulphonation can be increased if the amine is first converted to the amide, using ethanoyl chloride. This is less reactive than the amine and undergoes fewer side reactions. The sulphonated amide can be hydrolysed back to the amine by refluxing it with dilute sulphuric acid.

Carbonyl and carboxyl compounds

Aromatic aldehydes, ketones, carboxylic acids and acid derivatives generally behave in a similar manner to their aliphatic counterparts, described in Sections 30.3 and 31.4. We have seen how the properties of halogeno, hydroxyl and amino groups are considerably modified when their lone pair electrons are incorporated into the delocalized π-system, because that opposes the normal polarization of the bond between the carbon atom and the more electronegative element. The effects are less dramatic when the π-electrons from the ring and the carbon–oxygen double bond interact. There are eight electrons to cover eight atoms. The electronegative oxygen atom takes more than its share and gains a partial negative charge, leaving a partial positive charge on the carbon atoms (**7**; compare Figure 35.1). The resulting polarization is not very different from that in aliphatic carbonyl compounds (**8**, Figure 30.1) and the chemistry of the two groups of compounds is much the same. For example, the pK_a of benzoic acid (4.19) is similar to that of ethanoic acid (4.75).

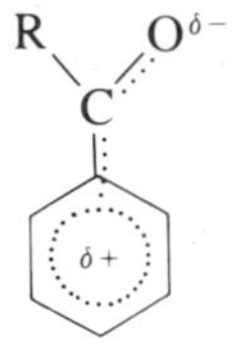

R — C — $O^{\delta-}$ — $\delta+$

7

R — $\delta+$ — $O^{\delta-}$ — C — CH_3

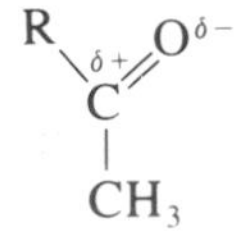

8

Aralkenes

Compounds with a carbon–carbon double bond attached to the benzene ring are called **aralkenes**. Because alkenes are more reactive towards electrophiles than arenes are, the reactions of aralkenes occur mainly at the double bond, and the compounds behave much like aliphatic alkenes. For example, they react readily with bromine to give an addition product:

$$C_6H_5—CH{=}CH_2 + Br_2 \rightarrow C_6H_5—CHBr—CH_2Br$$

Additions of hydrogen chloride and other unsymmetrical reagents give the product in which the electronegative group becomes bonded to the carbon atom next to the benzene ring. This is in accord with the principle underlying Markovnikov's rule (given in Section 33.3): initial attack of the electrophile (H^+) at the carbon atom further from the benzene ring gives an intermediate that can be stabilized by electron delocalization.

$$C_6H_5—CH{=}CH_2 \xrightarrow{HCl} C_6H_5—\overset{+}{C}H—CH_3 + Cl^- \longrightarrow C_6H_5—CHCl—CH_3$$

The important polymerization of phenylethene to poly(phenylethene), polystyrene, was discussed in Section 33.3.

35.5 Preparative methods for aromatic compounds

Many of the methods of preparing aromatic compounds are described in earlier sections of this chapter. Those that are discussed in other chapters are summarized here.

Halogenobenzenes

Chlorobenzenes and bromobenzenes may be made by direct halogenation of benzene using a Lewis acid catalyst as described in Section 34.3:

$$C_6H_6 + Cl_2 \xrightarrow{AlCl_3} C_6H_5Cl + HCl$$

chlorobenzene

Alternatively they may be made from diazonium salts (see Section 35.4).

Phenols

Phenols are made by warming aqueous solutions of diazonium salts (see Section 35.4), *not* by the reaction of hydroxide ion with chlorobenzene: halogenobenzenes do not readily undergo nucleophilic substitution reactions, as explained in Section 35.4.

Phenol itself is made industrially, together with propanone, from the oxidation of cumene, (1-methylethyl)benzene (see Section 33.5).

Phenylamine

Phenylamine is made by the reduction of nitrobenzene, which is in turn made by the nitration of benzene (see Section 34.3). In the laboratory the usual reagent for the reduction is tin and concentrated hydrochloric acid. The product of reaction is phenylammonium chloride. The solution is made alkaline and the free amine is steam distilled (see Section 7.1).

$$C_6H_6 \xrightarrow[\text{conc. } HONO_2,\ \text{conc. } (HO)_2SO_2,\ 50°C]{\text{nitration}} C_6H_5NO_2 \xrightarrow[\text{Sn/conc. HCl}]{\text{reduction}} C_6H_5NH_3^+\,Cl^- \xrightarrow{\text{NaOH(aq)}} C_6H_5NH_2 + H_2O + NaCl$$

phenylamine

Phenylamine is *not* made from the reaction of ammonia with chlorobenzene: halogenobenzenes do not readily undergo nucleophilic substitution reactions, as explained in Section 35.4.

Aromatic aldehydes and ketones

Aromatic carbonyl compounds – like aliphatic ones (see Section 30.4) – may be made by the oxidation of alcohols. The usual reagent is acidified potassium dichromate(VI).

$$C_6H_5\text{—}CH_2OH \xrightarrow[\text{distil}]{Cr_2O_7^{2-}/H^+(aq)} C_6H_5\text{—}CHO$$

benzaldehyde

$$C_6H_5\text{—}CH(OH)\text{—}CH_3 \xrightarrow[\text{reflux}]{Cr_2O_7^{2-}/H^+(aq)} C_6H_5\text{—}CO\text{—}CH_3$$

phenylethanone

They can also be made by the hydrolysis of (dichloroalkyl)benzenes, which are in turn made by the side-chain chlorination of alkylbenzenes (see Section 34.3); for example:

$$C_6H_5\text{—}CH_3 \xrightarrow[\text{u.v., reflux}]{2Cl_2} C_6H_5\text{—}CHCl_2 + 2HCl \xrightarrow{Ca(OH)_2(aq)} C_6H_5\text{—}CHO + H_2O + CaCl_2$$

Aromatic carboxylic acids and derivatives

Aromatic carboxylic acids can be made, like aliphatic ones, by the oxidation of aldehydes or primary alcohols as described in Section 31.5. Because the stable benzene ring resists oxidation, however, *any* aliphatic side chain can be oxidized by powerful oxidizing agents such as concentrated nitric acid, potassium manganate(VII) in acid or alkali, or acidic potassium dichromate(VI) to give the carboxylic acid group (see Section 34.3). No further oxidation is possible without disintegrating the benzene ring. When the reaction mixture is cooled the acids precipitate, since they are only slightly soluble in the cold.

$$C_6H_5\text{—}CH_2OH \xrightarrow[\text{reflux}]{Cr_2O_7^{2-}/H^+(aq)} C_6H_5\text{—}COOH$$

$$C_6H_5\text{—}R \xrightarrow[\text{reflux}]{\text{conc. } HONO_2} C_6H_5\text{—}COOH$$

Acids may also be made by the hydrolysis of nitriles, which can in turn be made from diazonium salts, as described in Section 35.4.

Carboxylic acid derivatives – acid chlorides, anhydrides, esters and amides – are made by the standard methods used for aliphatic derivatives (see Section 31.4).

Aralkenes

Aralkenes may be made by the normal elimination reactions used for aliphatic alkenes (see Section 33.4). Phenylethene is made industrially by the catalytic dehydrogenation of ethylbenzene, which in turn is made from the addition of ethene to benzene using aluminium chloride as a catalyst.

$$C_6H_6 \xrightarrow[AlCl_3,\ 100^\circ C]{CH_2{=}CH_2} C_6H_5CH_2CH_3 \xrightarrow[Fe_2O_3,\ Cr_2O_3]{630^\circ C} C_6H_5CH{=}CH_2 + H_2$$

EXAMPLE

How can 4-aminobenzoic acid (a component of the vitamin folic acid and a constituent of sun-tan lotions) be made from methylbenzene?

METHOD Consider methods of preparing aromatic amines and aromatic carboxylic acids. Decide which are appropriate for the given starting material.

ANSWER The amino group can be introduced by nitration followed by reduction. The carboxylic acid can be made by oxidizing the methyl group. The following sequence will therefore produce 4-aminobenzoic acid:

$$C_6H_5CH_3 \xrightarrow[\text{conc. } HONO_2,\ \text{conc. } H_2SO_4,\ 25^\circ C]{\text{nitrate}} 4\text{-}NO_2C_6H_4CH_3 \xrightarrow[KMnO_4,\ H^+(aq)]{\text{oxidize}} 4\text{-}NO_2C_6H_4COOH \xrightarrow[\text{neutralize } NaOH(aq)]{\text{reduce Sn/conc. HCl}} 4\text{-}NH_2C_6H_4COOH$$

COMMENT The nitration must be done before the methyl group is oxidized: nitration of benzoic acid occurs in the 3 (*meta*) position. The oxidation is best done before the nitro group is reduced, since aromatic amines are easily oxidized to polymeric tarry materials.

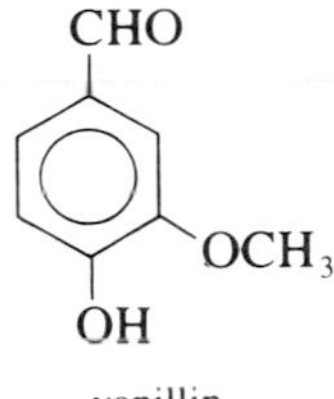

9 vanillin (natural and synthetic flavouring)

10 musk xylene (a synthetic perfume that smells of musk; structurally it is not related to natural musk, which is a ketone with a 15-membered carbon ring)

11 oestrone (a female sex hormone)

12 pyridoxine (vitamin B_6)

35.6 The importance of aromatic compounds

Aromatic compounds are of widespread importance. Many of those that occur in nature or that are manufactured commercially are very complicated molecules: they may have several groups attached to the benzene ring, or more than one ring, or some of the carbon atoms of the ring may be replaced by other elements. A few examples are illustrated (**9**–**12**). The following paragraphs mention some of the uses of the simpler aromatic compounds.

Halogenobenzenes

Chlorobenzene is used in the manufacture of DDT (see Box 34.2).

Phenol

Phenol finds a major use in the manufacture of **thermosetting plastics**. These can be moulded into shape and then when heated undergo extensive cross-linking (compare the cross-linking of alkyd resins and of vulcanized rubber, discussed in Section 29.6 and Box 33.3 respectively) to set them permanently rigid. Common examples are the casings of electric plugs and sockets. (By contrast, polymers without cross-links, such as poly(ethene), are **thermoplastic**: when heated they soften and when cooled they harden again in a repeatable manner.) The most important phenolic plastics are the **phenol–formaldehyde resins** (formaldehyde being an old name for methanal, $CH_2{=}O$). They are formed by the acid- or base-catalysed reaction of phenol with methanal. The ratio of the two is chosen to give a polymer with little cross-linking to start with. Then, during the heat treatment, further methanal is added to link the chains together.

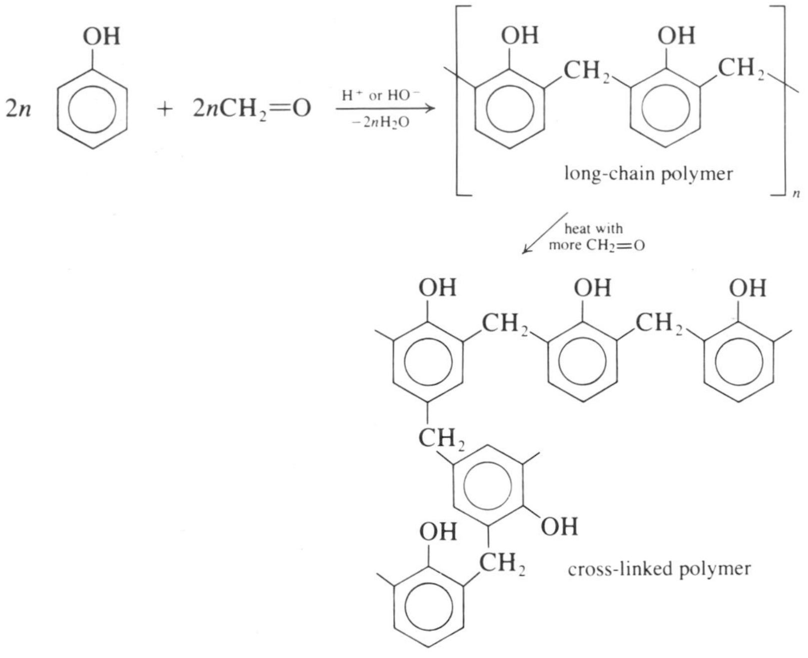

$$2n\,C_6H_5OH + 2nCH_2{=}O \xrightarrow[-2nH_2O]{H^+ \text{ or } HO^-}$$

long-chain polymer

heat with more $CH_2{=}O$

cross-linked polymer

Phenylamine

Phenylamine is an important starting material in the manufacture of dyes (see Box 35.3), pharmaceuticals and photographic chemicals. These, however, require relatively small amounts of the compound. The major use is in the manufacture of antioxidants in the rubber industry.

Carboxylic acids and derivatives

Benzoic acid, sodium benzoate and several esters of 4-hydroxybenzoic acid are widely used as food preservatives: they inhibit the growth of bacteria and yeasts, though the way they work is not understood. There are suggestions that asthmatic people and hyperactive children may be sensitive to them, but these have not been confirmed.

Benzene-1,4-dicarboxylic acid (terephthalic acid) is used in the manufacture of 'Terylene' and other polyesters (see Section 29.6).

Summary

- ☐ The names of aromatic compounds are illustrated by the following examples: methylbenzene, chlorobenzene, nitrobenzene, phenol, phenylamine, benzaldehyde, benzoic acid, phenylethene, phenylethanone.
- ☐ Most aromatic compounds with a single small group attached to the ring are liquids at room temperature, except for phenol and benzoic acid.
- ☐ The behaviour of aromatic compounds in electrophilic substitution reactions is summarized below.

Group attached to ring	Main products	Rate of reaction relative to benzene
NH_2, OH	*ortho* and *para*	much faster
R	*ortho* and *para*	faster
F, Cl, Br, I	*ortho* and *para*	slower
NO_2 $COCH_3$, CHO COOH, COOR CN SO_2OH	*meta*	much slower

- ☐ The halogen atom of halogenobenzenes is not normally displaced by nucleophilic substitution.
- ☐ The hydroxyl group of phenols is more acidic than that of alcohols.
- ☐ Phenols may be acylated with acid chlorides or acid anhydrides, but not with a carboxylic acid and concentrated sulphuric acid.
- ☐ Phenols do not give halogenoarenes with PCl_5 and similar reagents.
- ☐ Phenylamine is less basic than aliphatic amines.
- ☐ Phenylamine reacts with nitrous acid at 0 °C to give benzenediazonium salts. These react with nucleophiles, usually at about 50 °C, to give substituted benzenes, as summarized below.

Reagent	Catalyst	Product
H_2O		phenol
Cl^-, Br^-	Cu(I)	halogenobenzene
I^-		iodobenzene
CN^-	Cu(I)	benzonitrile

- ☐ Benzenediazonium salts also couple with phenols and aromatic amines to give coloured azo compounds.
- ☐ The chemistry of carbonyl and carboxyl groups attached to benzene rings is closely similar to that of their aliphatic counterparts.
- ☐ Electrophilic addition occurs readily across the carbon–carbon double bond of aralkenes.
- ☐ Phenylethene is polymerized to poly(phenylethene) (polystyrene).
- ☐ Halogenobenzenes are made by direct halogenation or from diazonium salts.
- ☐ Phenols are made in the laboratory by warming aqueous diazonium salts.
- ☐ Phenylamine is made by the reduction of nitrobenzene.
- ☐ Aromatic aldehydes and ketones are made by the oxidation of primary and secondary alcohols, respectively, or by the hydrolysis of (dichloroalkyl)-benzenes.
- ☐ Aromatic carboxylic acids are made by the oxidation of primary alcohols or aldehydes or the vigorous oxidation of any aromatic compounds with alkyl side chains.
- ☐ Aralkenes are made in the laboratory by elimination reactions, as for aliphatic alkenes.
- ☐ Phenols are widely used in the manufacture of thermosetting plastics.
- ☐ The diazotization of aromatic amines followed by diazo coupling is used to manufacture azo dyes.

PROBLEMS

1 Give two examples of reactions in which phenols and alcohols behave similarly and two in which they differ. Give equations and reaction conditions in each case.

2 Explain clearly how you could distinguish between the following pairs of compounds:
(a) phenol and benzoic acid;
(b) $C_6H_5CH_2NH_2$ and $C_6H_5NH_2$;
(c) $C_6H_5CH_2Cl$ and C_6H_5Cl.

3 Show how the following conversions could be carried out, giving reaction conditions in each case. (Some of the conversions require more than one reaction step.)
(a) benzene to phenylamine;
(b) phenylamine to benzonitrile (benzenecarbonitrile);
(c) benzonitrile to benzoic acid;
(d) benzene to phenylethanone;
(e) phenylethanone to benzoic acid.

4 Explain the following observations.
(a) Phenylamine is a weaker base than ethylamine.
(b) Phenylamine reacts with bromine water to give 2,4,6-tribromophenylamine.
(c) Phenol is formed when phenylamine is treated with $NaNO_2$ in cold dilute sulphuric acid and the solution is then warmed to 50 °C.

5 Deduce the structures of compounds **A** to **G** in the following reaction sequence.

$$\underset{\textbf{A}}{C_8H_7NO_2} \xrightarrow{H_2/Ni} \underset{\textbf{B}}{C_8H_9NO_2} \xrightarrow[\text{(ii) NaOH(aq)}]{\text{(i) Sn/conc. HCl}}$$

$$\underset{\textbf{C}}{C_8H_{11}N} \xrightarrow[\text{(ii) } K_3[Cu(CN)_4]]{\text{(i) } NaNO_2\text{/dil. }(HO)_2SO_2} \underset{\textbf{D}}{C_9H_9N} \xrightarrow[100\,^\circ C]{H_2O/(HO)_2SO_2}$$

$$\underset{\textbf{E}}{C_9H_{10}O_2} \xrightarrow[100\,^\circ C]{MnO_4^-/H^+(aq)} \underset{\textbf{F}}{C_8H_6O_4} \xrightarrow{200\,^\circ C} \underset{\textbf{G}}{C_8H_4O_3}$$

6 (a) Describe, with essential practical details, how a sample of phenylamine (aniline) can be prepared from benzene.
(b) Give reaction schemes for the formation of ethanamide (acetamide) and 1-aminobutane (1-butylamine) from butyl ethanoate (n-butyl acetate), $CH_3COOCH_2CH_2CH_2CH_3$, as the only source of organic compounds.
(c) Compare the base strengths of phenylamine, ethanamide and 1-aminobutane.
(d) Give an example of the acylation of *one* of the above amines. (*Welsh*)

7 (a) For *each* of the species listed below give *one* example of a typical organic reaction. For *each* example include an equation, the essential reaction conditions and an explanation as to whether the species is acting as a nucleophile or an electrophile.

NO_2^+, CN^-, OH^-, $C_2H_5O^-$, H^+

Either by choosing from the examples you have given above or by using other examples, give *one* reaction to illustrate the meaning of *each* of the terms: *addition*, *substitution* and *elimination*.
(b) State how, and under what conditions, bromine reacts with (i) phenol, (ii) benzene and (iii) ethene. Give an equation for *each* process.
(c) Suggest how you could show that a compound was 1-chloropropane and not 2-chloropropane. (*Welsh*)

8 (a) Describe the manufacture of phenol from benzene by the cumene process and outline the industrial importance of phenol.
(b) How, and under what conditions, would you expect a compound with the formula

OH

$CH{=}CHCH_2OH$

to react, if at all, with
(i) sodium,
(ii) sodium hydroxide,
(iii) phosphorus pentachloride,
(iv) ethanoic acid,
(v) potassium manganate(VII)? (*Cambridge*)

9 (a) Give an account of the industrial importance of benzene, referring to at least two organic chemicals which are manufactured from benzene by different processes.
(b) What is the order of increasing acid strength for the compounds ethanoic acid, ethanol and phenol? Give what explanation you can in terms of the structures of these compounds.
(c) What do you understand by the term *acylating agent*? Illustrate your answer by reference to the use of ethanoic anhydride in the manufacture of aspirin. (*Cambridge*)

10 'The chemical properties of a functional group depend on whether it is attached to an alkyl group, an acyl group or an aryl group'. Discuss this statement with reference to the —OH, —Cl and —NH_2 groups.
Explain any differences in properties as far as you can in terms of the electron distribution in the molecules concerned. (*Cambridge*)

11 Explain carefully what is meant by *each* of the following types of organic reaction:
(a) electrophilic addition,
(b) electrophilic substitution,
(c) nucleophilic substitution.
Illustrate your answer by reference to the mechanism of *one* example of *each* type of reaction. (*Cambridge*)

12 Describe and briefly explain what happens in *each* of the following experiments and write balanced equations for the reactions that occur.
(a) Propan-2-ol is warmed with acidified aqueous potassium dichromate(VI).
(b) Ethanedioic acid is heated with concentrated sulphuric acid.
(c) Aqueous bromine is added to aqueous phenol.
(d) Phenylethene (a liquid) is allowed to stand in the air for some time.
(e) Cold, aqueous sodium nitrite is slowly added to a solution of phenylamine in hydrochloric acid at 5 °C and the resulting mixture is poured into an alkaline solution of phenol. (*Cambridge*)

13 Write down, giving equations and essential reaction conditions, a reaction scheme by which phenylamine (aniline) can be made into an azo dye.

Give *evidence* in support of the following statements:

(a) Phenylamine gives reactions characteristic of a primary amine.

(b) The benzene nucleus of phenylamine is more reactive towards electrophiles than that of benzene itself.

(c) There is delocalization of electrons in benzene.

(Oxford)

14 (a) Describe how you would prepare a pure sample of phenylamine (aniline) from nitrobenzene in the laboratory. Your account should include

(i) the reagents used;
(ii) the conditions of the reactions;
(iii) the techniques used;
(iv) the equation(s) for the reaction(s);
(v) the method of purifying your sample of phenylamine.

(b) Write the full *structural* formula of the principal organic substance produced when phenylamine reacts with

(i) bromine water;
(ii) ethanoyl chloride (acetyl chloride). *(SUJB)*

15 Outline clearly how you would distinguish between the members in each of the following pairs of compounds using *one* simple chemical test for each pair.

In each case, state the reagents and conditions used, describe what happens for each compound during the test and write an equation (or equations) for the reactions involved.

(a) C_6H_6 (benzene) and C_6H_{12} (hexene);
(b) $CH_3CH_2CH_2Cl$ and $CH_3CH_2CH_2I$;
(c) CH_3CH_2CHO and $CH_3CH_2COCH_3$;
(d) $C_6H_5NO_2$ and $C_6H_5NH_2$. *(SUJB)*

16 The organic compound **P**, C_8H_8O, gave the triiodomethane (iodoform) reaction and on vigorous oxidation yielded **Q**, $C_7H_6O_2$, which was sparingly soluble in cold water.

After neutralization, **Q** gave a buff precipitate with aqueous iron(III) chloride.

Treatment of **Q** with phosphorus pentachloride gave **R** which reacted with concentrated aqueous ammonia to give a precipitate of **S**, C_7H_7ON.

Prolonged treatment of **S** with phosphorus(V) oxide gave **T** which could be reduced to **U**, C_7H_9N.

On reduction with sodium amalgam, **P** gave **V**, $C_8H_{10}O$. **V** gave the triiodomethane (iodoform) reaction, but did not give a coloration with aqueous iron(III) chloride.

(a) Write the structural formulae for substances **P** to **V**.

(b) Write equations for

(i) the reaction of **P** with iodine and alkali (the iodoform reaction);
(ii) the reaction of **Q** with phosphorus pentachloride;
(iii) the reaction of **R** with water. *(SUJB)*

17 (a) Describe briefly the laboratory preparation of a solution of benzenediazonium chloride, starting from phenylamine (aniline).

(b) Describe and explain what would be seen when a solution of benzenediazonium chloride is treated with:

(i) an alkaline solution of phenol;
(ii) warm dilute aqueous sulphuric acid.

(c) The compound $C_6H_5NNC_6H_5$ exists as two stereoisomeric forms. Draw clear diagrams to show the molecular shapes of these two isomers, and explain the type of isomerism involved. *(Oxford)*

18 Suggest, giving your reasons, the structures of the products from the following reactions.

(a)
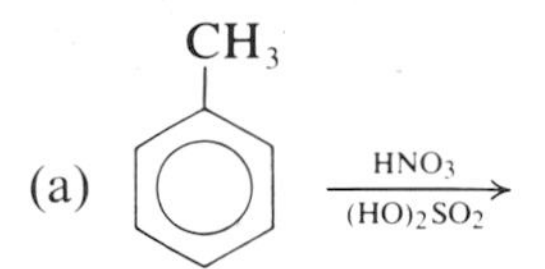

(b)
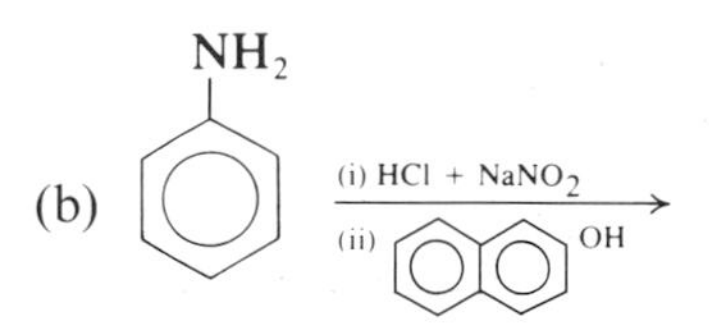

(c) $CH_3CH_2CH_2Cl \xrightarrow[\text{(ii) } H_2O,\ H^+(aq)]{\text{(i) KCN}}$

(Cambridge Entrance part question)

19 Answer *four* parts of this question.

Draw the structures corresponding to the following pairs of compounds and give a brief explanation of your reasoning:

(a) Molecular formula C_3H_6 – compound **A** rapidly decolorizes aqueous bromine, but compound **B** is unchanged under these conditions.

(b) Molecular formula C_2H_4O – compound **C** reacts with $NaBH_4$ to give a polar product C_2H_6O, but compound **D** is unchanged under these conditions.

(c) Molecular formula $C_2H_2Cl_2$ – compound **E** has a dipole whereas compound **F** has no dipole.

(d) Molecular formula C_7H_7Cl – compound **G** gives a white precipitate with cold ethanolic silver nitrate, whereas compound **H** does not react under these conditions.

(e) Molecular formula $C_3H_5ClO_2$ – compound **I** has pK_a 2.83 (K_a of 1.47×10^{-3} mol dm^{-3}) while compound **J** has pK_a 3.98 (K_a of 1.04×10^{-4} mol dm^{-3}).

(f) Molecular formula C_4H_9Br – compound **K** reacts rapidly with $NaOC_2H_5$ in ethanol to give C_4H_8, while compound **L** gives mainly $C_6H_{14}O$ under the same conditions. *(Cambridge Entrance)*

20 Describe briefly the methods which are available for the determination of the structure of an unknown organic compound.

A natural product, **A**, $C_9H_8O_4$, is optically inactive and reacts with aqueous sodium carbonate with evolution of a gas. Treatment of **A** with aqueous sodium hydroxide followed by hydrochloric acid results in the formation of acid **B**, $C_7H_6O_3$. **B** gives a strong purple coloration with iron(III) salts. On heating **B** gives **C**, C_6H_6O, which gives a weaker coloration with iron(III) salts.

Suggest possible structures for **A**, **B** and **C** and outline further experiments you might perform to confirm the structure of **A**. *(Cambridge Entrance)*

21 (a) Give the essential chemical nature of (i) a soap, (ii) a synthetic detergent , (iii) a synthetic fibre.

(b) Paracetamol (an analgesic used as an alternative to aspirin) has the structure

$$H-\underset{H}{\overset{H}{C}}-\underset{O}{\overset{}{C}}-\underset{H}{N}-C_6H_4-OH$$

Predict the reaction(s) which paracetamol would undergo with each of the following reagents, drawing structures wherever possible: (i) boiling aqueous sodium hydroxide, (ii) ethanoyl chloride, (iii) aqueous iron(III) chloride.

(c) Outline the principles of gas–liquid chromatography (g.l.c.) and explain why it is a useful analytical technique.

(AEB 1984)

22 For the following reaction scheme give:

(a) the names and structural formulae for the compounds **A** to **F**;

(b) the reagents and reaction conditions in reactions (i) to (vi):

benzene $\xrightarrow{(i)}$ $C_6H_5NO_2$ (**A**) $\xrightarrow{(ii)}$ C_6H_7N (**B**)

C_6H_7N (**B**) $\xrightarrow{(iii)}$ $C_6H_5N_2Cl$ (**C**)

$C_6H_5N_2Cl$ (**C**) $\xrightarrow{(iv)}$ C_6H_6O (**D**) $\xrightarrow{(v)}$ $C_{13}H_{10}O_2$ (**E**)

$C_6H_5N_2Cl$ (**C**) $\xrightarrow{(vi)}$ C_6H_5I (**F**)

(AEB 1985)

23 (a) Compare the reaction of nitrous acid with an aliphatic amine (e.g. 1-aminobutane) and with phenylamine by giving (i) reaction conditions, (ii) equations, (iii) reasons for differences in behaviour.

(b) The following reaction scheme shows a conversion of nitrobenzene to 4-hydroxyphenylazobenzene:

nitrobenzene $\xrightarrow{A}$ phenylamine $\xrightarrow{B}$ **X**

X $\xrightarrow{\text{hot water}}$ phenol

X + phenol $\xrightarrow{C}$ 4-hydroxyphenylazobenzene

Give (i) the experimental conditions required for stages **A** and **C**, (ii) the identity of the organic molecule/ion present in **X** and an equation for reaction **C**, (iii) the reason why this synthesis has industrial importance.

(c) Compare the relative base strengths of phenylamine, ammonia and 1-aminobutane, explaining the differences.

(d) Give formulae of suitable monomers for the formation of (i) a synthetic polyamide (e.g. nylon) and (ii) a natural polyamide, i.e. a protein. Give the name and structural formula of the linkage present in proteins, and comment on the nature of the intermolecular and intramolecular bonding in proteins.

(AEB 1986)

REFERENCE INFORMATION

Quantities and equations

Symbols

We have used internationally accepted symbols and subscripts. The states of substances are denoted g (gas), l (liquid), s (solid), aq (aqueous solution). Standard states are indicated by the superscript $^{\ominus}$. The sign $\approx$ means 'approximately equal to'. $x > y$ means x greater than y; $x < y$ means x less than y. $\ln x$ is the natural logarithm (base e) of x, and $\lg x$ is the logarithm to the base 10 of x.

Units

We have used the international system of units ('SI') based on the kilogram (kg), metre (m), second (s), ampere (A), kelvin (K) and mole (mol). The prefixes used are as follows:

p (pico)	n (nano)	μ (micro)	m (milli)
10^{-12}	10^{-9}	10^{-6}	10^{-3}
c (centi)	d (deci)	k (kilo)	M (mega)
10^{-2}	10^{-1}	10^{3}	10^{6}

The units of quantities include the following:

energy: joule, $1\,\text{J} = 1\,\text{kg}\,\text{m}^2\,\text{s}^{-2}$
power: watt, $1\,\text{W} = 1\,\text{J}\,\text{s}^{-1}$
force: newton, $1\,\text{N} = 1\,\text{kg}\,\text{m}\,\text{s}^{-2}$
pressure: pascal, $1\,\text{Pa} = 1\,\text{N}\,\text{m}^{-2}$
charge: coulomb, $1\,\text{C} = 1\,\text{A}\,\text{s}$
potential difference: volt, $1\,\text{V} = 1\,\text{J}\,\text{A}^{-1}\,\text{s}^{-1} = 1\,\text{J}\,\text{C}^{-1}$

In numerical calculations we include the units and cancel and multiply them just like ordinary algebraic symbols. This is a good way of avoiding mistakes because if the units turn out to be wrong the calculation is certainly wrong (but the opposite is not necessarily true!) An example of this **quantity calculus** is the calculation of the value of R in Section 4.1.

Concentration, molarity and molality

Concentration is the **amount of substance per unit volume**. If the amount of substance is expressed in moles and the volume of the solution expressed in cubic decimetres (litres), the concentration is in $\text{mol}\,\text{dm}^{-3}$. **Molarity** is a synonym for concentration, but its use is now frowned upon; the statement 'a 2.0 M solution' is still sometimes encountered, but the modern expression would be 'a solution of concentration $2.0\,\text{mol}\,\text{dm}^{-3}$'. **Molality** is the amount of substance of solute per unit mass of solvent. It is normally expressed as so many $\text{mol}\,\text{kg}^{-1}$, meaning so many moles of solute per kilogram of solvent.

Molar quantities

The molar mass, M, is the mass per unit amount of substance. It is normally expressed as so many grams per mole ($\text{g}\,\text{mol}^{-1}$). It has the same numerical value as M_r; that is, $M = M_r\,\text{g}\,\text{mol}^{-1}$. If the mass of a sample is m, the amount of substance present is $n = m/M$. In general a molar quantity (molar entropy, for example) is the quantity per unit amount of substance. Therefore if a sample contains an amount of substance n and its entropy is S, the molar entropy is $S_m = S/n$. In the text we encounter molar enthalpy, entropy, Gibbs energy and heat capacity. The molar conductivity is the odd man out; it is conductivity divided by concentration.

Chemical equations

The arrow $\rightarrow$ is used to denote a definite direction of reaction; the double half-arrow $\rightleftharpoons$ is used to denote an equilibrium. We refer to the substances on the left of a chemical equation as the 'reactants' and those on the right as the 'products', even though at equilibrium the terms lose their distinction. We have normally specified the phase of the reactants and the products, as in $2H_2(g) + O_2(g) \rightarrow 2H_2O(l)$. The interpretation of the reaction is as follows. *Either* we refer to individual molecules, when it states that two molecules of hydrogen gas react with one molecule of oxygen gas to produce two molecules of liquid water. *Or* we refer to amounts of substance. Then we say that 2 mol H_2 molecules (in the gas phase) react with 1 mol O_2 molecules (in the gas phase) to produce 2 mol H_2O molecules (as liquid). It is an elementary but necessary requirement that the equation should *balance*, that is, the number of atoms of each element must be the same on both sides of the equation. Furthermore, if charged species are involved, the net charge must be the same on each side of the equation.

Tables and graphs

All tables have pure numbers for their entries. This is achieved by dividing the quantity at the head of the column (for example, the first ionization energy I_1 in Table 1.2) by its units. Thus $I_1/\text{kJ mol}^{-1}$ is a dimensionless number because the units of I_1 are cancelled. Since Table 1.2 shows that $I_1/\text{kJ mol}^{-1} = 1312$ for hydrogen, the first ionization energy itself is $I_1 = 1312\,\text{kJ mol}^{-1}$. Graphs are drawn with their axes labelled with pure numbers (obtained in the same way). The slope of a graph is therefore also a dimensionless number.

Physical chemistry

Amount of substance (n)

$n = m/M$, where m = mass of sample, and M = molar mass

$n = V/V_m$, where V = volume of gaseous sample, and V_m = molar volume

$n = N/L$, where N = number of entities, and L = Avogadro's constant

Structure and bonding

A nuclide $^A_Z\mathrm{E}$ has Z protons and $A-Z$ neutrons in a central nucleus, and Z electrons around the nucleus.

The numbers of each type of orbital are:
s: 1, p: 3, d: 5

First ionization energy: $\mathrm{E(g)} \rightarrow \mathrm{E^+(g)} + \mathrm{e^-(g)}$

Electron-gain energy: $\mathrm{E(g)} + \mathrm{e^-(g)} \rightarrow \mathrm{E^-(g)}$

Forces between molecules include:
dipole–dipole forces, dispersion forces, hydrogen bonding

Instrumental techniques

Spectroscopy, for the determination of identity, structure, shape and energy levels:
infrared, ultraviolet, visible, nuclear magnetic resonance

Mass spectrometry, for the determination of identity, isotopic abundance and molecular formula.

X-ray diffraction, for the determination of crystal structures and the arrangement of atoms in molecules.

Chromatography (h.p.l.c., t.l.c. and g.l.c.), for the separation and identification of mixtures.

Thermodynamics

Standard state (denoted $\ominus$): The pure substance at 1 bar pressure and at the temperature specified.

First law: Energy can be neither created nor destroyed.

Enthalpy change: ΔH = energy absorbed by heating, at constant pressure.

Hess's law: The overall reaction enthalpy is independent of the route between reactants and products.

Second law: The entropy of the universe increases during a natural change.

Gibbs energy: $\Delta G = \Delta H - T\Delta S$

At constant pressure and temperature, a reaction tends towards lower Gibbs energy.

Equilibrium: $\Delta_r G^\ominus = -RT \ln K$
$= -zFE^\ominus$

Nernst equation: $E = E^\ominus - (RT/zF) \ln Q$

Kinetics

If $rate = k[\mathrm{A}]^x[\mathrm{B}]^y$
the reaction is of order x in A, order y in B and of overall order $x + y$.

Zeroth-order reaction: $rate = k$

First-order reaction: $rate = k[\mathrm{A}]$, half-life $= (\ln 2)/k$

Second-order reaction: $rate = k[\mathrm{A}]^2$ or $rate = k[\mathrm{A}][\mathrm{B}]$

Arrhenius equation: $k = Ae^{-E_a/RT}$

Arrhenius plot: $\ln k$ against $1/T$: slope $= -E_a/R$

Equilibrium constants

For the reaction $a\mathrm{A} + b\mathrm{B} \rightleftharpoons c\mathrm{C} + d\mathrm{D}$,

$$K_c = \left\{\frac{[\mathrm{C}]^c[\mathrm{D}]^d}{[\mathrm{A}]^a[\mathrm{B}]^b}\right\}_{\mathrm{eq}}$$

Acid and base definitions:

	Acid	Base
Brønsted	Proton donor	Proton acceptor
Lewis	Electron pair acceptor	Electron pair donor

Acid ionization constant: $K_a = \left\{\frac{[\mathrm{H_3O^+}][\mathrm{A^-}]}{[\mathrm{HA}]}\right\}_{\mathrm{eq}}$, $\mathrm{p}K_a = -\lg K_a$

Base ionization constant: $K_b = \left\{\frac{[\mathrm{BH^+}][\mathrm{OH^-}]}{[\mathrm{B}]}\right\}_{\mathrm{eq}}$, $\mathrm{p}K_b = -\lg K_b$

Ionic product for water: $K_w = \{[\mathrm{H_3O^+}][\mathrm{OH^-}]\}_{\mathrm{eq}}$, $\mathrm{p}K_w = -\lg K_w$
$\mathrm{pH} = -\lg[\mathrm{H_3O^+}]$

Solubility product: $K_{sp} = \{[\mathrm{M}^{m+}]^x[\mathrm{A}^{a-}]^y\}_{\mathrm{eq}}$

Miscellaneous useful equations

Bohr frequency condition: $\Delta E = h\nu$

Perfect gas equation: $pV = nRT$
(Boyle: $V \propto 1/p$, Charles: $V \propto T$, Avogadro: $V \propto n$)

Root mean square speed of gas molecules: $c_{\mathrm{rms}} \propto \sqrt{(T/M)}$

Dalton's law: $p = p_{\mathrm{A}} + p_{\mathrm{B}}$, with $p_{\mathrm{A}} = x_{\mathrm{A}}p$ and $p_{\mathrm{B}} = x_{\mathrm{B}}p$

Mole fraction: $x_{\mathrm{A}} = n_{\mathrm{A}}/n$, with $n = n_{\mathrm{A}} + n_{\mathrm{B}}$

Raoult's law: $p_{\mathrm{A}} = x_{\mathrm{A}}p_{\mathrm{A}}^*$, *where* x_{A} is the mole fraction of A in the liquid mixture

Osmotic pressure: $\Pi V = nRT$

Colligative properties: $\delta T = km_{\mathrm{A}}$ ($k = k_f$ or k_b)

Molar conductivity: $\Lambda = \kappa/c$

Kohlrausch's law: $\Lambda^{\infty}(\mathrm{AB}) = \lambda^{\infty}(\mathrm{A^+}) + \lambda^{\infty}(\mathrm{B^-})$

Bragg equation: $\lambda = 2d \sin\theta$

Inorganic chemistry

Oxidation number ranges of some elements

Period 2

	Li	Be	B	C	N	O	F
Highest oxidation number	+1	+2	+3	+4	+5	+2	−1
Lowest oxidation number	0	0	0	−4	−3	−2	−1

Period 3

	Na	Mg	Al	Si	P	S	Cl
Highest oxidation number	+1	+2	+3	+4	+5	+6	+7
Lowest oxidation number	0	0	0	−4	−3	−2	−1

Oxidation numbers of the d-block elements are presented in Table 25.1.

Covalent and ionic structures

Covalent structures

Number of neighbouring atoms in molecule or ion	Shape	Examples
2	linear	$BeCl_2$, CO_2, NO_2^+, I_3^-
	angular	$SnCl_2$, NH_2^-, O_3, SO_2
3	trigonal planar	BF_3
	pyramidal	NF_3, NH_3, H_3O^+, ClO_3^-
4	tetrahedral	BF_4^-, CCl_4, PCl_4^+, ClO_4^-, $CoCl_4^{2-}$
	square planar	ICl_4^-, XeF_4
5	trigonal bipyramidal	PCl_5
6	octahedral	SiF_6^{2-}, PF_6^-, SF_6, $[Al(H_2O)_6]^{3+}$, $[Fe(CN)_6]^{4-}$, $[Co(NH_3)_6]^{3+}$

Ionic structures

Structure	Coordination	Examples
rock-salt structure (Figure 17.6(a))	6:6	NaCl and most other alkali metal halides ($r_+/r_- < 0.732$) CaO and other MO oxides
caesium chloride structure (Figure 17.6(b))	8:8	CsCl and some other caesium halides
rutile structure (Figure 18.7(a))	6:3	MgF_2 and some other MF_2 fluorides TiO_2 and some other MO_2 oxides
fluorite structure (Figure 18.7(b))	8:4	CaF_2 and some other MF_2 fluorides

Trends in the characteristics of oxides and fluorides

Oxides

Empirical formula	Na_2O	MgO	Al_2O_3	SiO_2	P_2O_5	SO_3	Cl_2O_7
Structure and bonding*	i	i	i	ce	ce/cm	ce/cm	cm
Acid–base properties†	b	b	am	a	a	a	a

Fluorides

Empirical formula	NaF	CaF_2	AlF_3	SiF_4	PF_5	SF_6	ClF_5
Structure and bonding*	i	i	i	cm	cm	cm	cm

* i = ionic; ce = covalent extended network; cm = covalent molecular
† b = basic; am = amphoteric; a = acidic

Industrial processes

Product(s)	Name of process	Equations	Conditions
Electrolytic processes			
1 Sodium hydroxide and chlorine	Diaphragm cell	Cathode: $2H_2O(l) + 2e^- \rightarrow 2OH^-(aq) + H_2(g)$ Anode: $2Cl^-(aq) \rightarrow Cl_2(g) + 2e^-$	Electrolyte: brine, NaCl Cathode: steel Anode: titanium
2 Aluminium	Hall–Héroult	Cathode: $Al^{3+}(l) + 3e^- \rightarrow Al(l)$	Electrolyte: Al_2O_3/Na_3AlF_6 Electrodes: graphite Temperature: 950 °C
Catalytic processes			
3 Ammonia	Haber–Bosch	$N_2(g) + 3H_2(g) \rightarrow 2NH_3(g)$	Temperature: 450 °C Pressure: 250 atm Catalyst: iron
4 Nitric acid	Ostwald	$4NH_3(g) + 5O_2(g) \rightarrow 4NO(g) + 6H_2O(g)$ $4NO(g) + 3O_2(g) + 2H_2O(l) \rightarrow 4HONO_2(aq)$	Temperature: 850 °C Pressure: 5 atm Catalyst: platinum/rhodium
5 Sulphuric acid	Contact	$2SO_2(g) + O_2(g) \rightarrow 2SO_3(g)$ Dissolve in 98% concentrated $(HO)_2SO_2$ and dilute with water	Temperature: 450 °C Catalyst: vanadium(V) oxide
Other processes			
6 Sodium carbonate	Solvay	(i) $CO_2(g) + NH_3(aq) + NaCl(aq) + H_2O(l) \rightarrow NaHOCO_2(s) + NH_4Cl(aq)$ (ii) $2NaHOCO_2(s) \xrightarrow{heat} Na_2CO_3(s) + H_2O(g) + CO_2(g)$ (iii) $Ca(OH)_2(aq) + 2NH_4Cl(aq) \rightarrow 2NH_3(g) + CaCl_2(aq) + 2H_2O(l)$	
7 Iron	Blast furnace, followed by basic oxygen process for steel	$Fe_2O_3(s) + 3CO(g) \rightarrow 2Fe(l) + 3CO_2(g)$ CO from $CO_2(g) + C(s) \rightarrow 2CO(g)$ CO_2 from $CaCO_3(s) \rightarrow CO_2(g) + CaO(s)$ $CaO(s) + SiO_2(s) \rightarrow CaSiO_3(l)$	

Organic chemistry

Methods for introducing functional groups

Photochemical halogenation

$R—H \xrightarrow[\text{light}]{Cl_2} R—Cl$ (Br_2 reacts similarly)

Alkanes: substitution takes place indiscriminately with respect to position.

Arene side chains: substitution takes place preferentially at the position next to the benzene ring.

Halogenation at α-carbon atom

$—CO—\overset{|}{\underset{|}{C}}—H \xrightarrow[HO^-(aq) \text{ or } H^+(aq)]{Cl_2} —CO—\overset{|}{\underset{|}{C}}—Cl$ (Br_2 reacts similarly)

Substitution in benzene ring

$C_6H_5—H \rightarrow C_6H_5—X$

	Reaction conditions required:
X = Cl, Br	$Cl_2, Br_2/AlCl_3$
NO_2	conc. $HONO_2$/conc. $(HO)_2SO_2$/50 °C
SO_2OH	SO_3/conc. $(HO)_2SO_2$/30 °C

Methods for extending the carbon skeleton

Using KCN

$CN^- + R—Br \xrightarrow[\text{reflux}]{C_2H_5OH} R—CN + Br^-$

$RCHO + CN^-(aq) \xrightarrow{H_2O} RCH(OH)CN$ (ketones react similarly)

$C_6H_5—N^+{\equiv}N + CN^-(aq) \xrightarrow[K_3[Cu(CN)_4]]{50\ °C} C_6H_5—CN + N_2$

Using Grignard reagents

$R—CHO \xrightarrow[\text{(ii) } H_2O]{\text{(i) } R'MgBr} R—CH(OH)—R'$ (ketones react similarly)

$CO_2 \xrightarrow[\text{(ii) } H_2O]{\text{(i) } RMgBr} R—COOH$

The aldol reaction

$$2R—CH_2—CHO \xrightarrow{HO^-(aq)} \begin{array}{l} R—CH_2—CH—OH \\ \qquad\quad\ \ | \\ \qquad\ \ R—CH—CHO \end{array}$$

$2R—CH_2—CHO \xrightarrow{H^+(aq)} R—CH_2—CH{=}CR—CHO$

Friedel–Crafts reactions

$C_6H_6 + RCl \xrightarrow{AlCl_3} C_6H_5—R + HCl$

$C_6H_6 + RCOCl \xrightarrow{AlCl_3} C_6H_5—CO—R + HCl$

Methods for contracting the carbon skeleton

Hofmann degradation

$$R{-}CO{-}NH_2 \xrightarrow{Br_2/NaOH(aq)} R{-}NH_2$$

Haloform reaction

$$\left.\begin{array}{r} Y{-}CH(OH){-}CH_3 \\ \textit{or}\ Y{-}CO{-}CH_3 \end{array}\right\} \xrightarrow{Hal_2/NaOH(aq)} YCO_2^- + CHHal_3 \qquad [Y = R \text{ or } H]$$

Ozonolysis

$$R{-}CH{=}CH{-}R \xrightarrow[(ii)\ Zn/H^+(aq)]{(i)\ O_3} 2R{-}CHO$$

Other alkenes react similarly.

Manganate(VII) oxidation

$$C_6H_5{-}R \xrightarrow[reflux]{MnO_4^-(aq)} C_6H_5{-}COOH$$

Other vigorous oxidizing agents can be used in place of manganate(VII) ion.

Interconversion of carboxylic acid derivatives

Reactant \ Product	RCOCl	$(RCO)_2O$	RCOOH	RCOOR′	$RCONH_2$	RCO_2^-	RCN
RCOCl	–	S	S	S	S	S	X
$(RCO)_2O$	X	–	S	S	S	S	X
RCOOH	SCl_2O		–	$R'OH/H^+$	$NH_3(aq)$ then heat	alkali	X
RCOOR′	X	X	acid	–	S	alkali	X
$RCONH_2$	X	X	acid or HONO(aq)		–	alkali	P_4O_{10}
RCN	X	X	acid		acid or alkali	alkali	–

Key: S: direct substitution of the type $RCOX + Y^- \rightarrow RCOY + X^-$
acid: acidic hydrolysis
alkali: alkaline hydrolysis
X: not a synthetic route
A blank indicates that the conversion is possible but unimportant

Preparation of substituted benzenes from nitrobenzene

$$C_6H_5{-}NO_2 \xrightarrow{Sn/conc.\ HCl} C_6H_5{-}NH_2 \xrightarrow[dil.\ (HO)_2SO_2,\ 0°C]{NaNO_2} C_6H_5{-}N^+{\equiv}N \xrightarrow[50°C]{CuCl\ or\ CuBr\ or\ KI} C_6H_5{-}Hal$$

$$C_6H_5{-}N^+{\equiv}N \xrightarrow{H_2O/50°C} C_6H_5{-}OH$$

$$C_6H_5{-}N^+{\equiv}N \xrightarrow{K_3[Cu(CN)_4]/50°C} C_6H_5{-}CN$$

Major routes for interconverting functional groups

Halogenoalkanes | **Alcohols** | **Aldehydes and ketones** | **Carboxylic acids** | **Alkenes** | **Amines**

$$\mathrm{R{-}Hal} \underset{\mathrm{SCl_2O,\ PBr_3,\ PI_3}}{\overset{\mathrm{HO^-(aq)/reflux}}{\rightleftarrows}} \mathrm{R{-}OH}$$

$$\mathrm{RCOOH} \xrightarrow[\text{then: (i) LiAlH}_4\text{ (ii) H}_2\text{O}]{\text{via RCONH}_2\text{ or RCN (see chart on page 525)}} \mathrm{RCH_2NH_2}$$

$$\left.\begin{matrix}\mathrm{RR'CHOH}\\ \mathrm{RCH_2OH}\end{matrix}\right\} \underset{\text{(i) LiAlH}_4\text{ (ii) H}_2\text{O}}{\overset{\mathrm{Cr_2O_7^{2-}/H^+(aq)}\ \text{(see note to right)}}{\rightleftarrows}} \left\{\begin{matrix}\mathrm{RR'CO}\\ \mathrm{RCHO}\end{matrix}\right.$$

$$\mathrm{RCHO} \xrightarrow[\text{reflux}]{\mathrm{Cr_2O_7^{2-}/H^+\ (aq)}} \mathrm{RCOOH}$$

(Note: oxidation of secondary alcohols to ketones is under reflux; that of primary alcohols to aldehydes is with distillation.)

$$\mathrm{RCH_2OH} \underset{\text{(i) LiAlH}_4\text{ (ii) H}_2\text{O}}{\overset{\mathrm{Cr_2O_7^{2-}/H^+(aq)/reflux}}{\rightleftarrows}} \mathrm{RCOOH}$$

$$\mathrm{RCHO} \xleftarrow[\text{(ii) H}_2\text{/Pd/BaSO}_4]{\text{(i) SCl}_2\text{O}} \mathrm{RCOOH}$$

$$\mathrm{R_2CH{-}CHalR_2} \underset{\text{HCl or HBr (conc., gas phase or organic solvent)}}{\overset{\mathrm{KOH/C_2H_5OH/55^\circ C}}{\rightleftarrows}} \mathrm{R_2C{=}CR_2}$$

$$\mathrm{R_2CH{-}C(OH)R_2} \underset{\text{(i) cold conc. (HO)}_2\text{SO}_2\text{ (ii) H}_2\text{O (warm)}}{\overset{\text{excess conc. (HO)}_2\text{SO}_2\text{/170°C}}{\rightleftarrows}} \mathrm{R_2C{=}CR_2}$$

$$\mathrm{R{-}Hal} \xrightarrow{\mathrm{NH_3(aq)}\text{ or }\mathrm{NH_3(g)}\text{(sealed tube) (poor yields: also gives secondary and tertiary amines and quaternary ammonium salts)}} \mathrm{R{-}NH_2}$$

$$\mathrm{R{-}OH} \xleftarrow[\text{NaNO}_2\text{/dil. (HO)}_2\text{SO}_2\text{ (poor yields)}]{} \mathrm{R{-}NH_2}$$

Answers to numerical problems

Chapter 1
4 32.06
7 3.6×10^{-19} J, 0.0072
8 1312 kJ mol^{-1}
9 496 kJ mol^{-1}
11 (a) 0.27 (b) 1.8×10^{-6}

Chapter 4
3 1.95 atm, 197 kPa
4 1.2×10^{17}
5 0.118 Pa
6 5.0×10^{5} mol, 8.5×10^{3} kg
7 27 kg
8 (a) 990 kPa, 990 kPa (H_2) (b) 1490 kPa, 990 kPa (H_2), 500 kPa (Cl_2)
(c) 1490 kPa, 990 kPa (HCl), 500 kPa (H_2)
9 (a) 324 m s^{-1} (b) 681 m s^{-1}; 0.973
10 5850 K
11 16 g mol^{-1}
13 (a) 20.0 atm (b) 18.6 atm

Chapter 5
10 LiBr 0.379, NaBr 0.523, KBr 0.708, RbBr 0.764, CsBr 0.872

Chapter 6
7 15.5 mm
8 346 g
9 1.1×10^{23}

Chapter 7
1 0.174, 0.826
4 (a) 2.619 kPa (methanol), 4.556 kPa (ethanol) (b) 7.175 kPa
(c) 0.365 (methanol), 0.635 (ethanol)
5 63 g
8 60 g mol^{-1}
9 100.05 °C, −0.20 °C
10 261 kPa

Chapter 8
3 60 C, 3.75×10^{20}, 6.22×10^{-4} mol
4 (a) 32.2 min (b) 96.5 min
8 (a) 267 S cm^2 mol^{-1} (b) 145 S cm^2 mol^{-1}; 6.9 kΩ
10 0.0188

Chapter 9
3 6.6 min
4 1.2 MJ
5 60 g
6 2.4 km
9 (a) −890 kJ mol^{-1} (b) −3351 kJ mol^{-1}
10 +4 kJ mol^{-1}

Chapter 10
10 2.38×10^{-3} dm^6 mol^{-2}
11 3.53×10^{-7} atm^{-2}
12 1.89×10^{-5} mol
13 8.36×10^{-7} atm

Chapter 11

8 (a) $-326.4\,J\,K^{-1}\,mol^{-1}$ (b) $-199.3\,J\,K^{-1}\,mol^{-1}$
(c) $-477.1\,J\,K^{-1}\,mol^{-1}$

9 (a) $-571.6\,kJ\,mol^{-1}$, $-474.3\,kJ\,mol^{-1}$ (b) $-92.2\,kJ\,mol^{-1}$, $-32.8\,kJ\,mol^{-1}$ (c) $+922\,kJ\,mol^{-1}$, $+1064\,kJ\,mol^{-1}$

10 (a) $-474.4\,kJ\,mol^{-1}$ (b) $-33.0\,kJ\,mol^{-1}$

11 (a) $8.0\,J\,K^{-1}\,mol^{-1}$ (b) $33.1\,J\,K^{-1}\,mol^{-1}$

Chapter 12

4 $1.7 \times 10^{-6}\,mol\,dm^{-3}$, $2.0 \times 10^{-9}\,mol\,dm^{-3}$

5 (a) 2.53 (b) $1.42 \times 10^{-4}\,mol\,dm^{-3}$

6 1.46

Chapter 13

5 (a) $+1.10\,V$ (b) $+0.22\,V$ (c) $-1.23\,V$ (d) $-0.15\,V$

6 (a) 2×10^{37} (b) 3×10^{7} (c) 7×10^{-84} (d) 3×10^{-3}

Chapter 14

4 624 s (a) 31.2 min (b) 259 s

5 (a) 0.983 (b) 0.018

6 13.7 min, $5.1 \times 10^{-2}\,min^{-1}$

7 $8.05 \times 10^{-3}\,min^{-1}$, 86.1 min

Chapter 16

1 (d) $1.7 \times 10^{8}\,kg$

Chapter 17

5 (b) $0.120\,dm^{3}$

7 (a) 275 tonnes (b) $7.48 \times 10^{11}\,C$

Chapter 18

6 (b) $7.7 \times 10^{2}\,m^{3}$ (c) 2.92 tonnes

Chapter 19

3 Al 14.0%, Cl 73.9%

8 $1.07 \times 10^{10}\,C$

Chapter 20

2 $-890\,kJ\,mol^{-1}$, $-1517\,kJ\,mol^{-1}$

3 $6 \times 10^{9}\,mol\,dm^{-3}$

Chapter 21

6 (c) 1 kg; $10^{8}\,kg$

7 (b) 2.86 tonnes

Chapter 23

9 3.1%

Index

Periodic Table

	I	II											III	IV	V	VI	VII	VIII
								$_{1}$H 1.008										$_{2}$He 4.003
2	$_{3}$Li 6.941	$_{4}$Be 9.012											$_{5}$B 10.81	$_{6}$C 12.01	$_{7}$N 14.01	$_{8}$O 16.00	$_{9}$F 19.00	$_{10}$Ne 20.18
3	$_{11}$Na 22.99	$_{12}$Mg 24.31											$_{13}$Al 26.98	$_{14}$Si 28.09	$_{15}$P 30.97	$_{16}$S 32.06	$_{17}$Cl 35.45	$_{18}$Ar 39.95
4	$_{19}$K 39.10	$_{20}$Ca 40.08	$_{21}$Sc 44.96	$_{22}$Ti 47.90	$_{23}$V 50.94	$_{24}$Cr 52.00	$_{25}$Mn 54.94	$_{26}$Fe 55.85	$_{27}$Co 58.93	$_{28}$Ni 58.71	$_{29}$Cu 63.55	$_{30}$Zn 65.37	$_{31}$Ga 69.72	$_{32}$Ge 72.60	$_{33}$As 74.92	$_{34}$Se 78.96	$_{35}$Br 79.90	$_{36}$Kr 83.80
5	$_{37}$Rb 85.47	$_{38}$Sr 87.62	$_{39}$Y 88.91	$_{40}$Zr 91.22	$_{41}$Nb 92.91	$_{42}$Mo 95.94	$_{43}$Tc (99)	$_{44}$Ru 101.1	$_{45}$Rh 102.9	$_{46}$Pd 106.4	$_{47}$Ag 107.9	$_{48}$Cd 112.4	$_{49}$In 114.8	$_{50}$Sn 118.7	$_{51}$Sb 121.8	$_{52}$Te 127.6	$_{53}$I 126.9	$_{54}$Xe 131.3
6	$_{55}$Cs 132.9	$_{56}$Ba 137.3	$_{57}$La 138.9	$_{72}$Hf 178.5	$_{73}$Ta 180.9	$_{74}$W 183.9	$_{75}$Re 186.2	$_{76}$Os 190.2	$_{77}$Ir 192.2	$_{78}$Pt 195.1	$_{79}$Au 197.0	$_{80}$Hg 200.6	$_{81}$Tl 204.4	$_{82}$Pb 207.2	$_{83}$Bi 209.0	$_{84}$Po (210)	$_{85}$At (210)	$_{86}$Rn (222)
7	$_{87}$Fr (223)	$_{88}$Ra (226)	$_{89}$Ac (227)															

Lanthanoids	$_{58}$Ce 140.1	$_{59}$Pr 140.9	$_{60}$Nd 144.2	$_{61}$Pm (147)	$_{62}$Sm 150.4	$_{63}$Eu 152.0	$_{64}$Gd 157.3	$_{65}$Tb 158.9	$_{66}$Dy 162.5	$_{67}$Ho 164.9	$_{68}$Er 167.3	$_{69}$Tm 168.9	$_{70}$Yb 173.0	$_{71}$Lu 175.0
Actinoids	$_{90}$Th 232.0	$_{91}$Pa (231)	$_{92}$U 238.0	$_{93}$Np (237)	$_{94}$Pu (242)	$_{95}$Am (247)	$_{96}$Cm (248)	$_{97}$Bk (247)	$_{98}$Cf (251)	$_{99}$Es (254)	$_{100}$Fm (253)	$_{101}$Md (256)	$_{102}$No (254)	$_{103}$Lr (257)

6(a)

(b)

(c)

7

8(a)

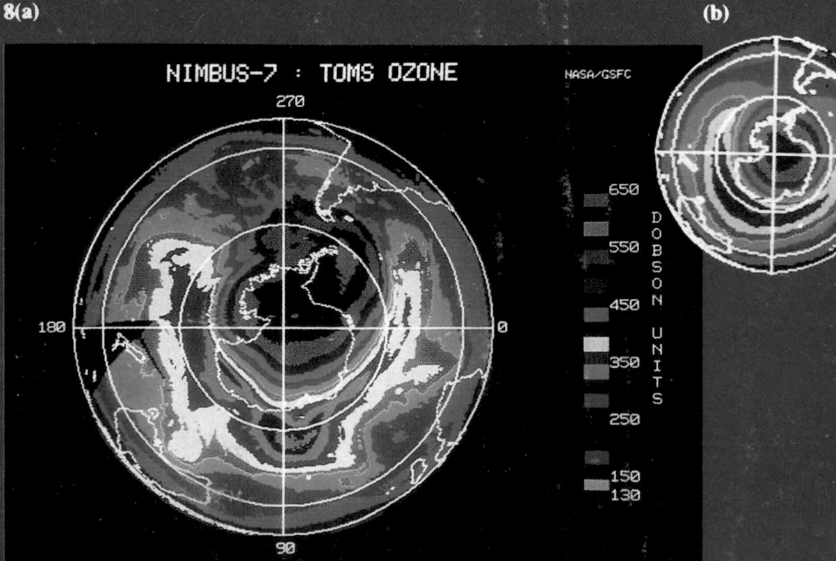

(b)

Colour 6 Group VII elements: (a) chlorine, which is a gas at room temperature, forms a liquid on a cold finger condenser, (b) bromine is a liquid with a dense brown vapour, (c) the purple vapour of iodine condenses to form purple–black crystals

Colour 7 1,2-Diaminoethane (ethylenediamine, en) complexes of nickel: when diaminoethane is added to a green aqueous solution of Ni^{2+}, the solution changes first to green–blue, then to purple–blue, and finally to magenta as the chelate complexes $[Ni(H_2O)_4]^{2+}$, $[Ni(H_2O)_2(en)_2]^{2+}$ and $[Ni(en)_3]^{2+}$ are successively formed

Colour 8 Satellite maps of the Earth produced by TOMS (Total Ozone Measurement Spectrometer), showing the colours used to represent units of atmospheric ozone in the scale on the right: (a) measurements made on 30 October 1986, and (b) the monthly mean for October 1980, the year in which a 'hole' in the ozone layer was first observed. The 'hole', visible here as a purple, black and grey area covering most of Antarctica, develops each year in the Antarctic spring and appears to be growing year by year

Colour 9 Steel manufacture: charging hot metal into a basic oxygen converter